Applied Mechanics
Progress and Applications

Proceedings of the

Third Australasian Congress on

Applied Mechanics

Applied Mechanics
Progress and Applications

Sydney, Australia

20 – 22 February 2002

edited by

Liangchi Zhang
The University of Sydney

Liyong Tong
The University of Sydney

John Gal
The University of Western Sydney

World Scientific
New Jersey • London • Singapore • Hong Kong

Published by

World Scientific Publishing Co. Pte. Ltd.
P O Box 128, Farrer Road, Singapore 912805
USA office: Suite 1B, 1060 Main Street, River Edge, NJ 07661
UK office: 57 Shelton Street, Covent Garden, London WC2H 9HE

British Library Cataloguing-in-Publication Data
A catalogue record for this book is available from the British Library.

APPLIED MECHANICS
Progress and Appplications

ISBN 981-02-4867-9

Printed in Singapore by Mainland Press

The Third Australasian Congress on Applied Mechanics (ACAM2002)

20th – 22nd February 2002, Sydney, Australia

Jointly organized by
The National Committee of Applied Mechanics, IEAust
The University of Sydney

Sponsored by
The University of Sydney
Asian Office of Aerospace Research and Development, Air Force Office of Scientific Research, USA
MSC SOFTWARE SIMULATING REALITY Australia

Preface

Applied mechanics is the study of the deformation of solids and the flow of fluids subjected to forces and motion and, in addition to the underlying science, with a particular emphasis on engineering applications. It not only contributes to the development of basic discipline-oriented areas but also plays a key role in the technological advancement of many cross-disciplinary fields. In recent decades, applied mechanics has also become increasingly important to the growth of some emerging fields such as nanoscience and nanotechnology.

Following the success of the previous two congresses in Melbourne in 1996 and in Canberra in 1999, the National Committee of Applied Mechanics is presenting the Third Australasian Congress on Applied Mechanics (ACAM2002) to provide a further international forum for researchers, industry practitioners, engineers and postgraduate scholars to promote, exchange and disseminate knowledge and experience of the most recent advances with a focus on the behaviour of solids.

The proceedings of the Congress contains over 110 technical papers selected from contributors from Oceania, Asia, America and Europe, covering a wide range of topics including biomechanics, constitutive modelling, damage, fracture, fatigue, dynamics, impact, vibration, geo-mechanics, tribology, machining, metal forming, particle mechanics, composites, structural mechanics, smart materials and nano and micromechanics. The volume thus presents much valuable results for the reader who is interested in the further study and application of applied mechanics.

Each of the papers in the proceedings was reviewed by experts in the relevant technical field and was published from the camera-ready manuscripts prepared by the authors. Despite the great effort of the referees and Editors, some deviation in the uniformity of the presentation quality could not be avoided. Tight publication schedules further precluded prolonged exchanges with authors of the papers that were deemed to require multiple revisions.

The Organizing and Technical Committee of ACAM2002 would like to sincerely thank all the keynote speakers, authors, participants and referees for their kind support, without which the Congress could never be successful. Thanks are also due to the National Committee of Applied Mechanics, IEAust, and the University of Sydney for their dedication in bringing the Congress to the professional community. Appreciation is expressed to the US Air Force Office of Scientific Research, Asian Office of Aerospace Research and Development, and to MSC Software Australia for their kind financial sponsorship of the Congress.

Lastly, on behalf of the Organizing and Technical Committee of ACAM2002, we heartily welcome all participants to Sydney and trust that they find the Congress and the trip rewarding and enjoyable.

Liangchi Zhang
Liyong Tong
John Gal
Alan Hellier
Khosrow Zarrabi

Sydney
18 September 2001

Contents

Constitutive Modelling

Fracture, Fatigue and Damage

Measurement and Case Studies

Machining and Surfacing

Metal Forming

SCALE EFFECTS IN MACRO-MECHANICS AND MICRO-MECHANICS IN SOLIDS

TADASHI SHIOYA
Department of Aeronautics and Astronautics, University of Tokyo, 7-3-1 Hongo, Bunkyo-ku, Tokyo 113-8656, Japan
E-mail: tshioya@mail.ecc.u-tokyo.ac.jp

Scale effects in mechanics are discussed focusing the concept of macroscopic aspects and microscopic aspects. Problems of scale effects in various fields are introduced. The mechanism of scale effects is analyzed and explained in each case. Non-dimensional method is introduced to meet the scale effect. Examples of scale effects are shown in the deformation and fracture of solids.

1 Introduction

There are many scale effect problems in mechanics. In solid mechanics, the behavior of large structure is often different from that of a smaller structure even if the shape and apparent conditions are similar. . In fluid dynamics, the flow pattern depends on the scale of the object and the flow velocity. The scale effects are found not only in purely mechanical problems, but also in electric and magnetic system (MEMS), thermal problems and also in chemical reaction problems. Experiments of smaller scale models in advance of the construction of a large-scale structure, often give useful information to the designers. In some cases, actual object is very small and it is more convenient to use a larger experimental model practically. For example, micro-machine system, sensor-actuator system, or, smart materials such as MEMS are often so small that a larger model is used in the simulation experiment.

In the process of developing a new engine, sometimes a small model engine is made as a simulation of actual scale engine. The mechanical parts can be downsized by examining the scale effect, however, the combustion part has its own characteristic length that cannot be downsized. For example, turbo machinery system is generally characterized by the peripheral velocity of rotation so that a small gas turbine and a large gas turbine with same geometry have similar characteristics if the peripheral velocities are the same, with exception of the combustion chamber.

In this presentation, the scale effects in mechanics are discussed with the concept of macroscopic and microscopic aspects and several examples are shown for understanding of the concept..

2 Concept of Scale Effects in Macro-Mechanics and Micro-Mechanics

2.1 Macro-Mechanics and Micro-Mechanics

The scale effects in mechanics can be approached by macroscopic analysis and/or microscopic analysis. Before proceeding with the discussion on the scale effects, it is better to clarify the definition of the terms "macroscopic" and "microscopic". In this text, "macroscopic" means "in continuum mechanics". In macroscopic aspects, various mathematical techniques in continuum theory can be utilized, such as differentiation, partial derivatives, interpolation, etc. Definition of the "strain" itself is substantially macroscopic. On the other hand, "microscopic means that this aspects concerns some small discrete objects, such as atoms, molecules, grains, inclusions, etc. In many cases, it is difficult to make use of the techniques in continuum mechanics.

2.2 Scale effects in macro-mechanics

In order to investigate the scale effect in mechanics, it is important to examine the constitutive (internal) relations of the material and external conditions that involve the phenomenon. The constitutive relations here include not only the stress-strain relation and the fracture criterion, but also include the mass density, thermal conductivity, electric properties etc if they have influences to the phenomenon. The external conditions also include not only mechanical conditions such as externally applied force, friction force, gravity etc, but also the thermal, electrical and other conditions if they have influences to the phenomenon. In the mechanical problems, forces acting on a body are generally considered. Some forces such as friction forces solid contact forces and fluid dynamic pressure act on the surface of the body and are in many cases regarded as proportional to the surface area, whereas some other forces acting as the body forces such as gravity and electro-magnetic forces are considered to be proportional to the magnitude of mass, i.e., proportional to the volume provided that the density is uniform. The mechanical characteristic nature of solid materials is generally expressed in terms of stress, i.e., elastic moduli, fracture stress, yield stress, etc, with some exceptions such as the stress intensity factor. Since the stress is a force divided by an area, if the system only contains forces proportional to the area, there may be no apparent scale effect at least as far as the deformation concerns. If a system has a mixed type, i.e., if it contains body forces proportional to the volume, the deformation itself has a scale effect. For example, in a large-scale structure supporting the weight of the structure itself such as buildings, bridges and etc, the larger the scale of the structure is, the larger the stress in the structure becomes. Another example may be found in a micro machine system. If the machine force is proportional to the volume of the micro engine, then, the smaller the system becomes, the more difficult to overcome the friction forces. There may be similar problems in smart materials that have mixed type conditions.

There will be another type of scale effect, which comes from the time effect. The velocities of elastic wave and plastic wave that are determined from the inertia and the elastic or plastic moduli relation influence the scale effect of the phenomenon [8,12]. Careful attention should be paid when an impact-shock absorber test is executed for different scale models. The velocity of heat conduction and transfer also influence the scale effect of the phenomenon. In some experiment, the temperature increase may be neglected in a small specimen, whereas, it may have substantial influence in a large-scale specimen.

In case, the above type of scale effect is neglected, still there may be scale effect induced by other type of influences. The brittle fracture in solids is one of the cases. Even if the internal stress is similar, the possibility of fracture becomes large in a very brittle solid if the scale of the element becomes large. The scale effect can be evaluated using the Weibull modulus of the material.

Using a different size model as a simulation of the actual phenomenon requires a careful check that the behavior of different scale model really represents that of the actual one. Generally non-dimensional analysis is employed to meet the actual phenomenon and the different scale model [4]. Namely the parameters that may concern the phenomenon are examined, from these parameters, smaller number of non-dimensional parameters are deduced, and the same non-dimensional parameters are used in the actual phenomenon and in the model. More concretely, governing equations, normally written in partial differential equation, are rewritten in non-dimensional form (although not in unique way), and the boundary and initial conditions are also expressed in non-dimensional form. As far as the phenomenon is governed only by these non-dimensional relations and the common values are used in the parameters, the scale effect is automatically cleared. Sometimes it is difficult to employ common values in the different size models to all the non-dimensional parameters at the same time. For example, in a fluid dynamics, Mach number and Reynolds number in different size models cannot be satisfied at the same time unless different fluids are used. In these cases, the parameters that have stronger effect should be employed

2.3 Scale effect in micro-mechanics

The above discussions on scale effect are in the continuum mechanics. There are also scale effect problems that may be explained in microscopic mechanism rather than in the macroscopic continuum mechanics. The strength of fiber reinforced composite material is influenced by the fiber thickness under the same fiber/matrix volume ratio [9]. Also the behavior of a material that contains dispersed particles has a scale effect of particle size or particle spacing besides the average concentration of the particles. The effect of grain size in a metal or in other materials is also fundamentally microscopic one. The strength of materials is sometimes influenced by the microscopic scale effects such as micro-crack size and spacing, each void growth, the atomic spacing in crystal lattice, dislocation width, precipitation hardening mechanism, work hardening mechanism due to dislocation multiplication. The scale effect explained by molecular dynamics and distance of chemical reaction etc are also considered to be microscopic.

In some cases, microscopic scale effect may be transformed into macroscopic scale effect and analyzed by the use of techniques in the continuum mechanics..

3 Examples of Scale Effect in Solid Mechanics

3.1 Yielding and Plastic Flow in Mild Steel

The theory of plasticity is based on the constitutive relation expressed in terms of stress-strain equation. Since the stress-strain relation can be expressed in the non-dimensional form, the stress being divided by the elastic constant, there should be no scale effect as far as the inertia effect can be neglected. The yielding patterns (plastic zone patterns) in a large element and a smaller element should be similar. However, the yield behavior of annealed mild steel is different from that of other metals. The cross-section of twisted round bar of mild steel has plastic zone wedges pattern called "strain figures", which is quite different from that of other metals. This phenomenon is occurred due to the particular constitutive relation of mild steel. The stress-strain curve of mild steel has a negative slope part after the upper yield point and this negative characteristic causes the instability of uniform yield pattern that appears in most other metals. This instability problem was analyzed by the authors [7,10], who showed the development of wedge type pattern by numerical analysis based on constitutive relations. As the constitutive relations, the stress-strain relation of Johnston-Gilman-Hahn type equation and the newly proposed yield condition were used. The stress-strain relation is expressed in the form below:

$$\frac{\mathrm{d}\varepsilon_{\mathrm{p}}}{\mathrm{d}t} = f(\sigma, \varepsilon_{\mathrm{p}}) = C_1 (C_2 + \varepsilon_{\mathrm{p}}{}^a)(\sigma - C\varepsilon_{\mathrm{p}})^n, \tag{1}$$

where σ and ε_{p} are the stress and the plastic strain, respectively, and all other variables are the material constants. Besides the stress-strain relation, the authors proposed a condition of plastic zone spreading and obtained the elastic-plastic boundary condition with material constants by the experiment of Lüders band expansion.

$$\sigma = \sigma_{\mathrm{u}} - Km, \tag{2}$$

where m is the plastic strain gradient across the elastic plastic boundary and σ_{u} and K are the material constants. This condition includes a characteristic length in K so that the plastic zone pattern should have a scale effect. The experiment and the calculation were performed changing the diameters of the twisted round bar which are shown in Figure 1 and Figure 2 [6,7]. The number of plastic zone wedges increases with the diameter in both cases.

The above two constitutive relations (1) and (2) are written in the continuum mechanics; however, their mechanism come from a consideration in microscopic aspects. The stress-strain relation of Johnston-Gilman-Hahn type equation (1) is based on the dislocation mechanics, i.e., rapid multiplication of mobile dislocations and the stress dependence of dislocation velocity together with a work hardening mechanism. In the case of yield initiation condition (2), the situation in the each grain at the elastic plastic boundary is considered. The multiplication of mobile dislocations in already yielded region piles up to the boundary to the adjacent non-yielded grain causing a kind of micro-stress concentration in the non-yielded grain. This micro- stress concentration together with the macroscopic applied stress causes the nucleation of mobile dislocations in the grain meeting the condition of yield initiation in the macroscopic aspects.

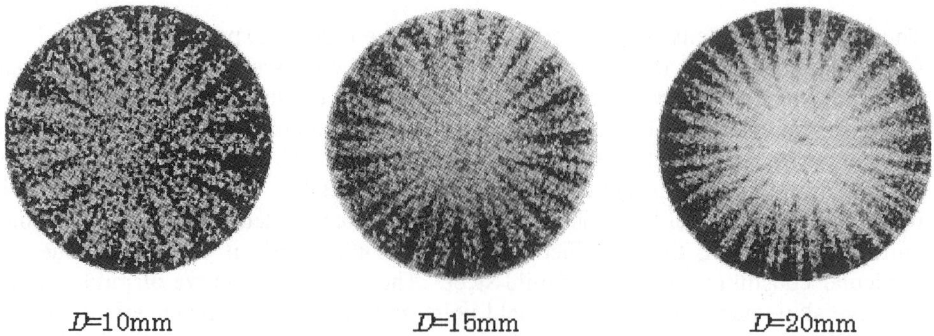

Figure 1. Observed strain figures of mild steel. The cross-section of twisted round bars with different diameters

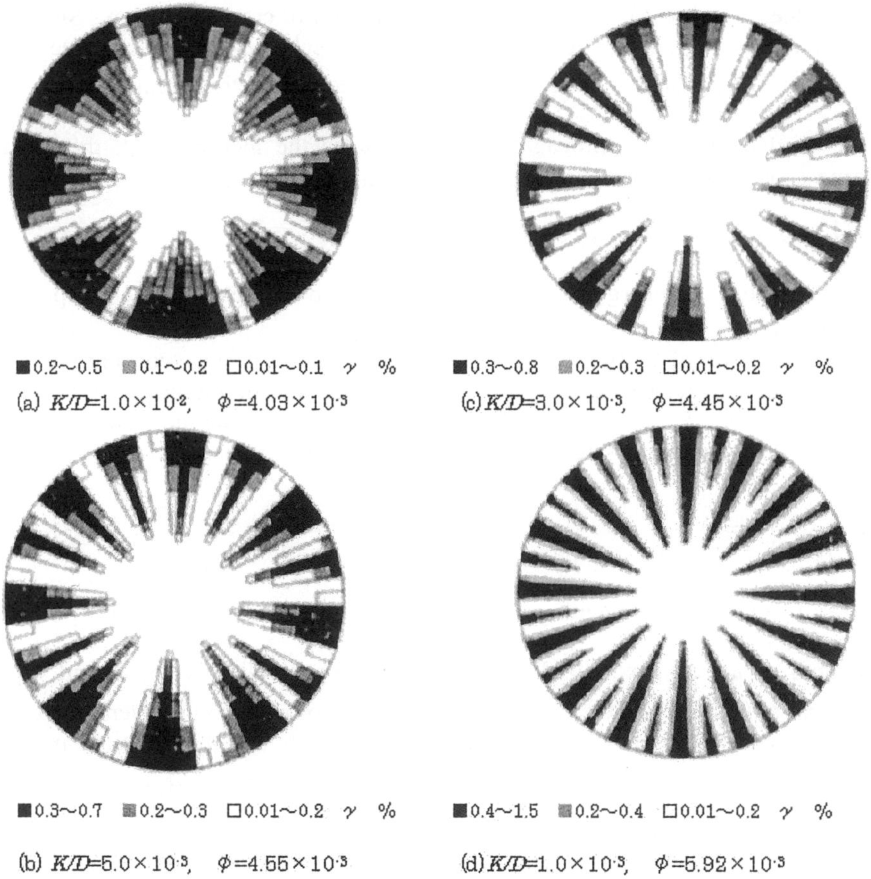

Figure 2. Calculated plastic zone patterns of mild steel. The cross-section of twisted round bars with different diameters

3.2 Dynamic Crack Propagations

Crack propagation in brittle material is an important part of fracture mechanics [2,3]. Macroscopic approach to the crack propagation problem is represented by the "Fracture Mechanics". The concept of the fracture mechanics was first introduced by Griffith who analyzed the crack extension problem in terms of energy balance of the system, which contains the crack.

$$\frac{\partial}{\partial c}(W - \Gamma) = 0, \qquad (3)$$

where c is the crack length, W is the potential energy decrease of the system (the release of stored elastic energy and work done by movement of the external loads), and $\Gamma = 2\gamma c$ is the energy of the surface of the crack, in which γ is the surface tension of the material or the surface formation energy. The equation implies that at the critical crack length, the total diminution of the potential energy of the system, $W - \Gamma$ is a balanced minimum value, and its slope becomes positive when the crack extends beyond the critical value. Equation (3) gives a necessary condition of crack extension. Since the constant γ includes a characteristic length, the fracture mechanics itself has a scale effect. This scale effect is examined in detail comparing with the ductile fracture in the next sub-section.

In case of dynamic crack propagation that includes the inertia effect, the kinetic energy term T is further added to the energy equation by Mott, as,

$$\frac{\partial}{\partial c}\left(U+\Gamma+T\right)=0\ , \tag{4}$$

where U is the potential energy of the system, which can be regarded to correspond to $-W$ in (3).

As a generalization of Griffith energy criterion, the energy balance concept has been applied to two-dimensional dynamic problems for prediction of the crack path by the author and co-workers [5]. The generalized equation of motion using the total kinetic and potential energies becomes the Lagrange's equation which is written as,

$$\frac{\mathrm{d}}{\mathrm{d}t}\left(\frac{\partial L}{\partial \dot{q}_i}\right)-\frac{\partial L}{\partial q_i}=N_i\ , \tag{5}$$

where L (= $T-U$) is the Lagrangian, T is the kinetic energy, U is the potential energy, q_i and $\dot{q}_i$ are the generalized coordinate and the generalized velocity, and N_i is the non-conservative force. In the two-dimensional crack propagation problem, the crack tip position x and y are taken as the generalized coordinate and the surface formation energy γ describes the non-conservative forces N_x and N_y. The differential equation (5) is integrated to obtain the crack path assuming that the crack surface formation energy γ is a function of the crack propagation velocity V (Fig. 3). In case of fixed sided plate, the crack tends to take a path towards the center line of the plate in case of low propagation velocity, however, it turns to go away from center line (to the fixed side) in case of high velocity under the assumption of no branching.

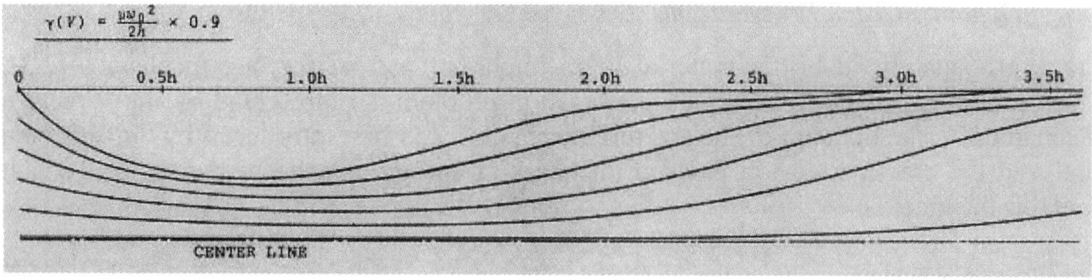

Figure 3. Calculated Crack Path

The oscillation phenomenon is also analyzed by the energy relations (Fig. 4) [11,13]. Since all the governing equations are written in a non-dimensional form, i.e., the length is normalized by the width of the plate and the velocity is normalized by the wave velocity in the material, the scale effect problem is automatically solved.

The microscopic process of crack surface formation in a homogeneous material such as PMMA can be realized by the observation of the fracture surface. The crack surface has different patterns depending on the crack propagation velocity V (Fig. 5). The parabolic line patterns appear at relatively low velocity. These parabolic lines are the intersection lines of the crack fronts of the main crack and the micro-cracks that nucleate, grow and coalesce into the main crack.

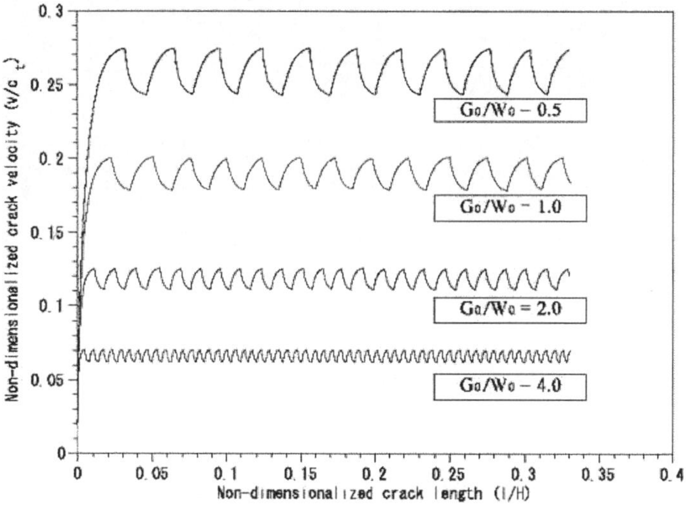

Figure 4. Simulation of Crack Propagation Oscillation (Mode III, Fixed sided)

The pattern changes in shape and density with the crack velocity V. Typical fracture surfaces in the parabolic region observed by SEM are shown for different crack velocities in Fig. 5 (a) and (b). These fracture surface patterns produced by the dynamic micro-crack coalescence can be simulated by the use of micro-crack initiation condition and propagation velocity condition (Fig. 6). Parabolic shapes have mathematically self-similarity so that the scale effect might correspond to its sharpness. Since the parabolas on the fracture surface become less sharp and more dense as the crack velocity increases, the fracture surface of higher velocity corresponds to that of a larger scale fracture surface with denser micro-crack initiation sites.

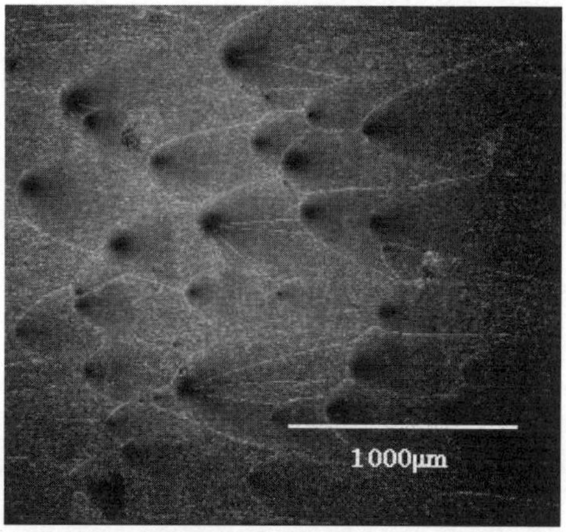

(a) $V = 250$ m/s

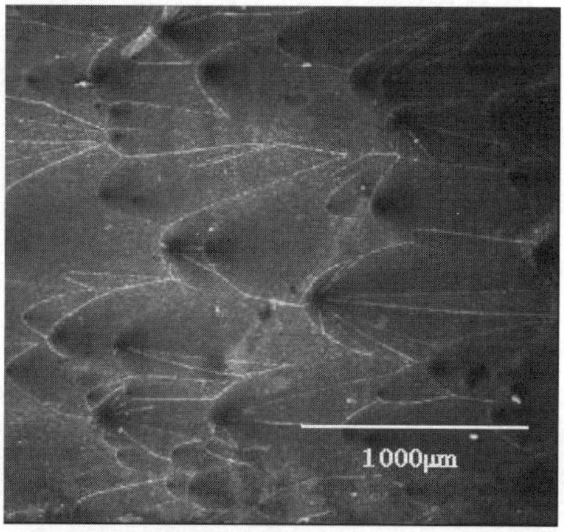

(b) $V = 340$ m/s

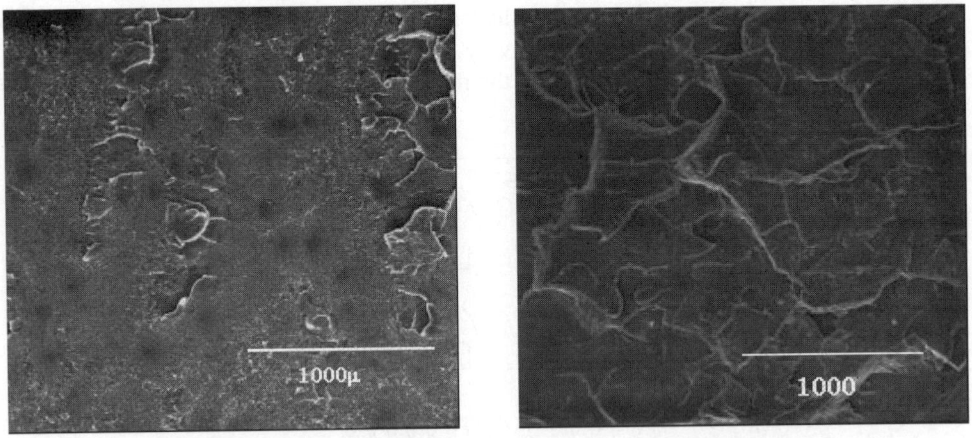

(c) $V = 420$ m/s (d) $V = 650$ m/s

Figure 5. Fracture surface of PMMA observed by SEM. V: Crack Velocity

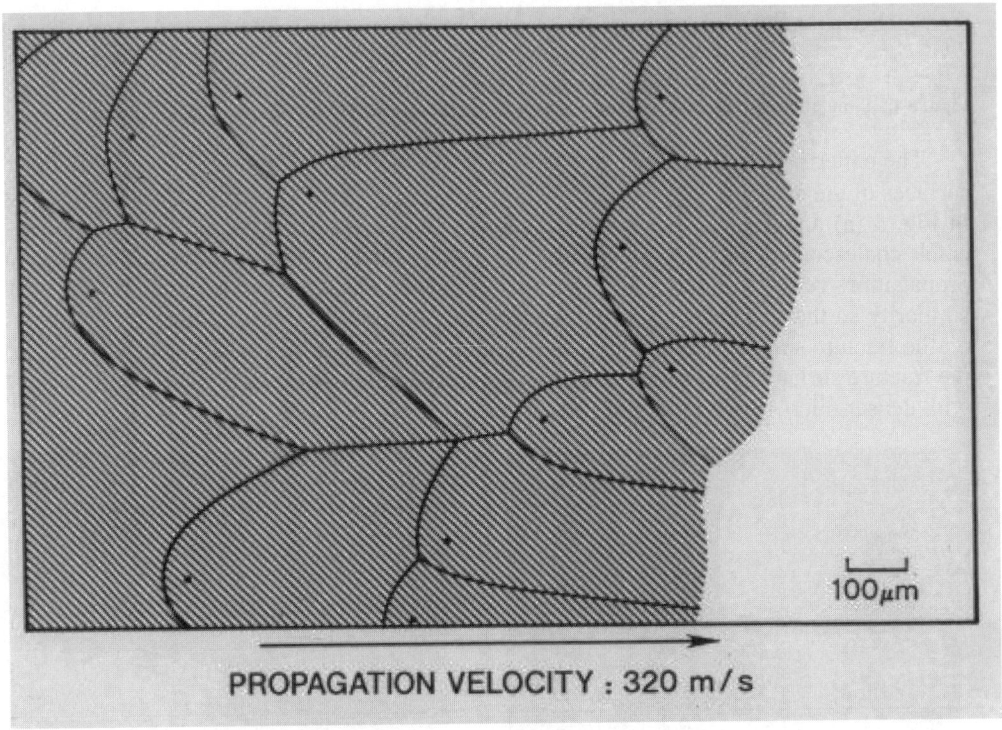

Figure 6. Simulated parabolic pattern on the fracture surface

3.3 Ductile Brittle Transition in Fracture

The fracture modes of solids are generally classified into two categories, i.e., ductile fracture and brittle fracture. The majority of materials are classified into ductile materials and brittle materials, however, there are also many materials that show ductile fracture in some conditions and brittle fracture in other conditions. The ductile-brittle transition parameters are the temperature, loading rate, chemical contents of the materials, geometrical shape including the size and etc. It is sometimes not easy to judge whether the fracture mode is brittle or ductile in the transient modes. The definition of brittle fracture is theoretically macroscopic one. It is that the fracture mechanics is applicable to the mode, i.e., the fracture criterion, such as the critical stress intensity factor $K_{\mathrm{I\,C}}$ or the surface formation energy γ is a material constant independent of the crack size. However, practically the brittle fracture is identified by the observation of fracture surface, i.e., the classification of brittle fracture and the ductile fracture is in the microscopic aspect and is theoretically different from the macroscopic one. These two identifications are empirically thought to be equivalent. However whether the macroscopic criterion and microscopic one are theoretically equivalent has not been proved yet.

Our study on brittle-ductile transition is focused to the relation between the macroscopic criterion and the microscopic criterion. Theoretically there is no fundamental difficulty to examine the macroscopic criterion, since the definition is clear, i.e., it is to check whether $K_{\mathrm{I\,C}}$ is a material constant independent of the crack size. Only the problem in macroscopic check is its cost. It would be more difficult to check the microscopic one, since the microscopic criterion, that is the fracture surface analysis depends on somewhat subjective empirical pattern recognition by the experts. This traditional pattern recognition has to be interpreted to more objective method without ambiguity. Our idea of distinguishing the fracture surface is a pattern recognition using a concept of "fractal" [1]. Fractal pattern means that the enlargement of the pattern is the same as the original pattern; in other words, there is no scale effect in the real fractal pattern. If the fracture surface has a fractal pattern, then there is no scale effect, meaning that the fracture mode has no characteristic length, i.e., the fracture is ductile. The fractal pattern is characterized by a constant fractal dimension. If the pattern is not fractal, then, the defined fractal dimension is not constant. Therefore whether the fracture mode is ductile or brittle corresponds to whether the fractal dimension is constant or not. The ductile-brittle transition is also understood in this aspect. The ductile-brittle transition thus obtained in microscopic aspect should be compared with the transition in macroscopic aspect.

In order to obtain the fractal dimension of the fracture surface profile, the height-height correlation to the scale t is taken in the second order. For the profile $z(x)$, the correlation $C(t)$ is defined as,

$$C(t) = \overline{\{z(x+t) - z(x)\}^2}\,, \tag{6}$$

where the upper bar means the average. If the profile has a fractal structure, there exists a power law between the scale t and the correlation $C(t)$, namely,

$$C(t) = Gt^{2-D}\,, \tag{7}$$

where D is the fractal dimension. In other words, if the slope in the logarithmic diagram of the above relation is linear, the profile can be regarded as a fractal pattern and the fracture mode should be ductile (Fig. 8).

In some cases, the linearity may be restricted to a certain interval of the scale t. For such cases, we can say that the fracture mode is ductile for the scale of some length and brittle for the larger scale. This is a newly proposed concept of ductile and brittle definition. The ductile-brittle transition occurs also in the scale of observation. In fact, the fracture of relatively ductile material but with a very large crack at very large-scale structure may have ductile fracture surface in a small scale observation, however, should have a brittle fracture mode at a larger scale.

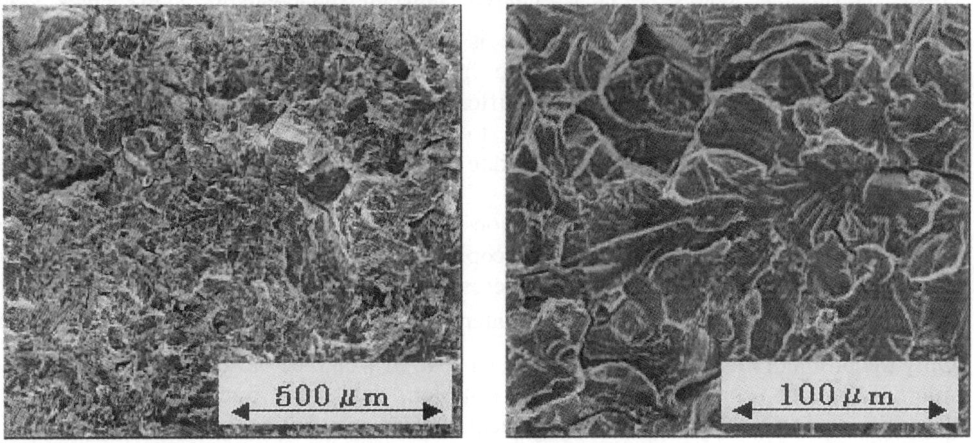

Stress tri-axiality large (relatively brittle)

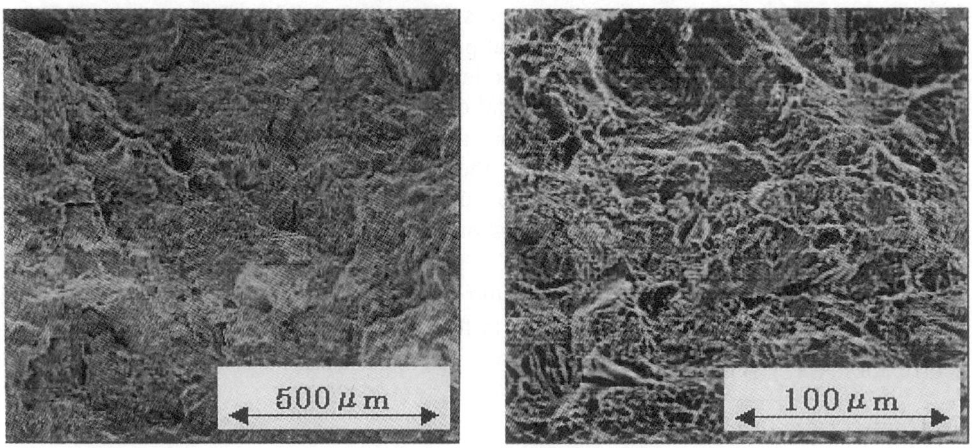

Stress tri-axiality small (relatively ductile)

Figure 7. Ductile-brittle transition observed by SEM photographs. Fracture surface of mild steel. Stress tri-axiality and observation scale changed

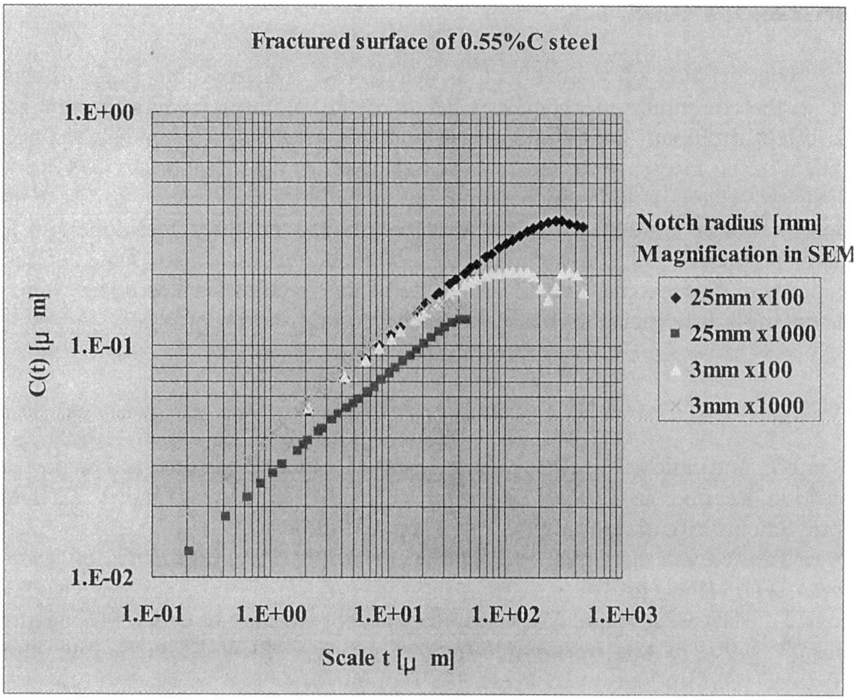

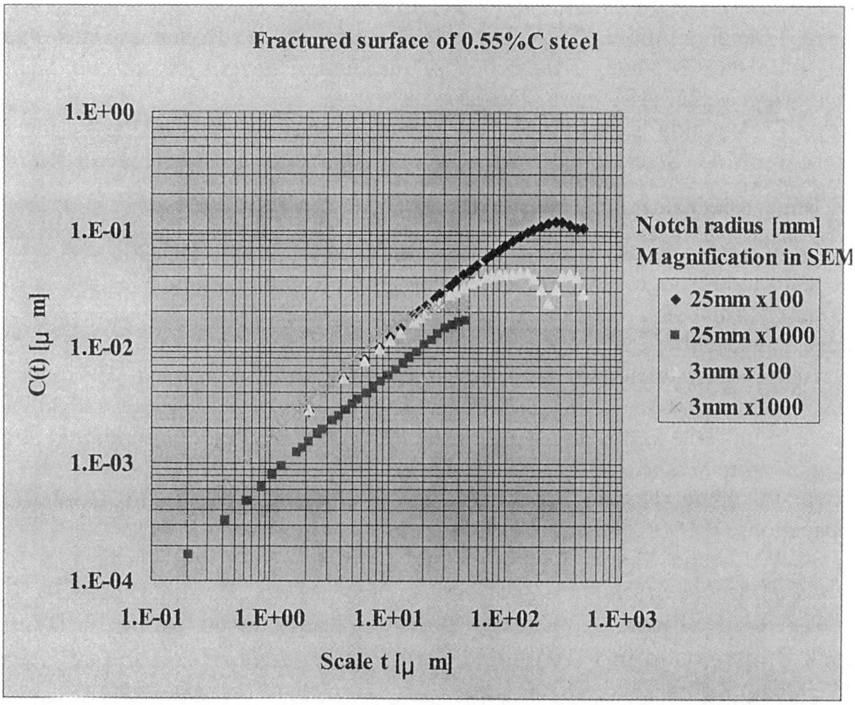

Figure 8. Height-height correlation $C(t)$ vs. observation scale t (in logarithmic scale)

4 Discussion and Conclusion

There are several kinds of scale effect in mechanics. Macroscopic scale effects are analyzed in the continuum mechanics. Characteristic length plays an important role in the scale effect problem. Non-dimensional analysis may help the understanding, and careful attention is necessary to check what parameter is most important. Microscopic scale effects are related with the microscopic process with discrete characteristic length.

The deformation of mild steel has a scale effect due to the characteristic length in the constitutive relation. The dynamic crack propagation has also scale effects in macroscopic and microscopic levels. The scale of the fracture surface observation may affect the ductile-brittle transition in the fracture of solid.

References

1. Ikeshoji T. and Shioya T., Fractal Dimension of Fracture Surface in Ductile-Brittle Transition Regime. in *Fractal Frontier*, ed. by M. M. Novak and T. G. Dewey, (World Scientific, Colorado, USA, 1997) pp. 255-263.
2. Shioya T., Analysis of Crack Propagation in Brittle Material, *JSME International Journal* **37A** (1994) pp.1-6.
3. Shioya T., Macroscopic and Microscopic Aspects of Dynamic Crack Propagation. in *Impact Response of Materials and Structures*, ed. by V. P. W. Shim, S. Tanimura and C. T. Lim, (Oxford University Press, 1999) pp.60-74,
4. Shioya T., Non-Dimensional Design Parameters for FOD Tolerant Fan Blade, *Chinese Journal of Aeronautics* **4**, (1991) pp.113-120.
5. Shioya T. and Fujimoto K., A Theoretical Study on the Running Crack Path by Energy Balance Method, *Transaction of the Japan Society for Aeronautical and Space Sciences*,.**25**, (1983) pp. 246-257
6. Shioya T., Machida T., Ishida R. and Fujiura Y., Scale Effect of the Plastic Zone Pattern in Mild Steel, *Proceedings of the 22nd Japan Congress on Materials Research* (1979) pp. 47-50.
7. Shioya T. and Shioiri J., Elastic-Plastic Analysis of the Yield Process in Mild Steel, *Journal of the Mechanics and Physics of Solids* **24** (1976) pp. 187-204.
8. Shioya T. and Stronge W. J., Impact Bending of a Rotating, Rigid-Plastic Fan Blade, *Journal of Propulsion and Power, AIAA* **1** (1985) pp. 375-380.
9. Shioya T. and Toh T., A Theoretical Study of Micro- Structural Size Effect on Crack Extension in Composite Materials, *Advanced Composite Materials* **1** (1991) pp. 123-135.
10. Shioya T. and Toh T., Instability of Plastic Flow in Mild Steel, *International Journal of Engineering Science*, **23** (1985) pp. 101-112.
11. Shioya T., Zhou F. and Ishida R., Micro-cracking Process in Brittle Crack Propagation, *DYMAT Journal* **2** (1995) pp.105-118.
12. Stronge W. J. and Shioya T., Impact and Bending of a Rigid Plastic Fan Blade, *Journal of Applied Mechanics, Transactions of the ASME* **51** (1984) pp. 501-504.
13. Zhou F. and Shioya T., Energy Balance Analysis on Mode-III Dynamic Crack Propagation in Fixed-sided Strip, *International Journal of Fracture* **80** (1996) pp.33-44.

THE INFLUENCE OF SECTIONAL STABILITY IN ENGINEERING DESIGN OF THIN-WALLED MEMBERS

G.J. HANCOCK

Department of Civil Engineering, University of Sydney, NSW Australia, 2006
Email: hancock@civil.usyd.edu.au

Sectional stability of thin-walled sections includes short and intermediate half-wavelength buckling modes commonly called **local** and **distortional.** Local modes involve only flexural deformations whereas distortional modes involve flexural and membrane deformations of the plates forming the cross-section and occur at longer half-wavelength. These modes may interact with each other or with long half-wavelength buckling modes such as flexural and flexural-torsional. Design methods to account for these modes in engineering design have been proposed and partially incorporated into current and pending design standards. The paper discusses these modes, their interactions and current and proposed design methods in modern design standards.

1 Introduction

The three basic modes for design of thin-walled structural members in compression and/or bending are the local, distortional and overall modes. The three modes are shown in Fig. 1 for a lipped channel bent about a horizontal axis where the top flange is in compression and the bottom flange is in tension. Fig. 1 was computed using a finite strip buckling analysis and is a graph of the buckling stress versus half-wavelength. The *local buckling mode* involves only plate flexure with the line junctions remaining straight and occurs at a short half-wavelength. For the section shown, the local mode is mainly in the top of the web and the flange element which are both in compression. The *overall buckling mode* in this case is flexural-torsional with the cross-section remaining undistorted. It occurs at long half-wavelengths typically beyond 3000 mm for this section. The *distortional buckling mode* consists of both membrane and flexural deformations of the plate elements and occurs at half-wavelengths intermediate between the local and overall modes. In this case, it involves rotation of the flange and lip about the flange-web junction with transverse flexure of the web. These three modes are characteristic of thin-walled open-section members in compression and/or bending.

Methods to account for local buckling of thin-walled members of metal structures have been thoroughly researched and incorporated in a range of Australian structural design standards for hot-rolled steel [1], cold-formed steel [2] and aluminium [3]. Equivalent USA Specifications are [4], [5] and [6] respectively. Methods to account for overall buckling including flexural (Euler) and flexural-torsional buckling have also been well researched [7] and incorporated in the structural design standards cited above. Methods to account for distortional buckling have only begun to be researched in the last 20 years. For flexural members, several useful references which summarise the work on distortional buckling to date are Hancock [8] and Hancock, Rogers and Schuster [9] and, Schafer and Pekoz [10]. For compression members, early work is given in Hancock [11], Lau and Hancock [12], Kwon and Hancock [13], and Hancock, Kwon and Bernard [14]. More recent work on distortional buckling of compression members including a thorough summary can be found in Schafer [15, 16].

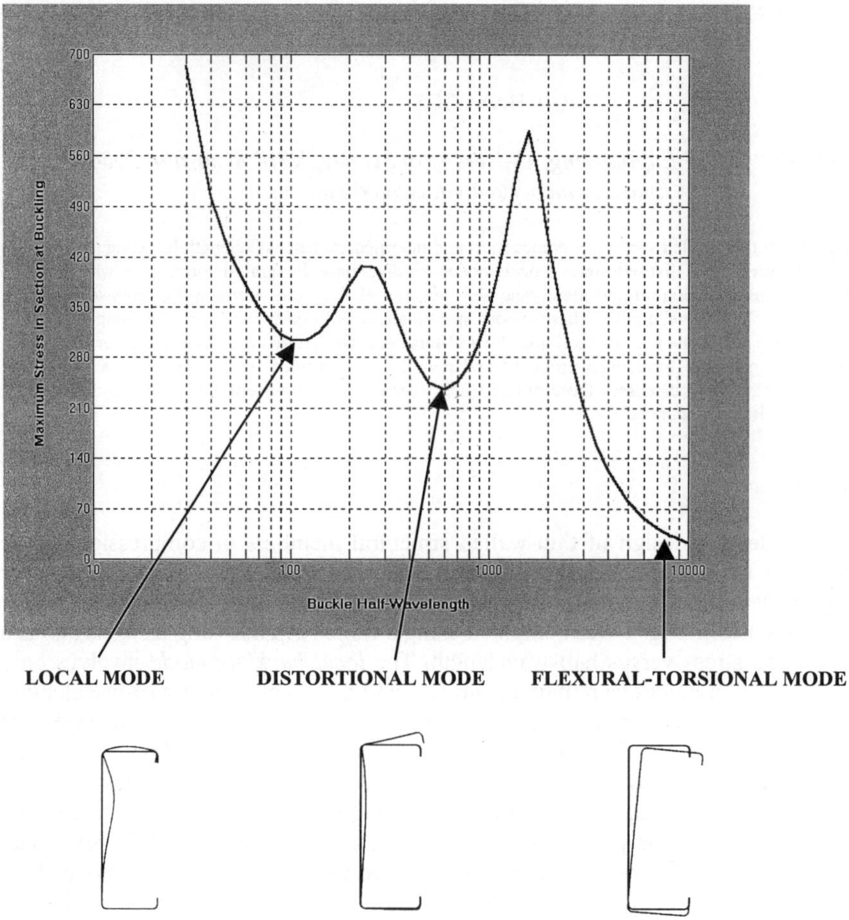

Figure 1. Buckling stress versus half-wavelength for a C-Section in bending about a horizontal axis

Design rules for distortional buckling of simple lipped channel and Z-sections have been incorporated in Clause 4.2 of the AISI Specification for Cold-Formed Steel Structural Members [5] using an effective width approach extended by Desmond, Pekoz and Winter [17] to account for distortional buckling. This clause also forms the basis of Clause 2.4.3 of the Australian/New Zealand Standard for Cold-Formed Steel Structural Members [2]. For more complex sections such as storage rack columns with additional rear flanges, design rules have been incorporated in Clause 3.3.3.3 for flexural members, and Clause 3.4.6 for compression members of the Australian/New Zealand Cold-Formed Steel Structures Standard [2]. An Appendix D of [2] gives explicit formulae for computing elastic distortional buckling stresses and is based on [8], [12]. The conversion of elastic distortional buckling stresses to design strengths is based on the methods in Hancock, Kwon and Bernard [14].

A new method, called the "Direct Strength Method", has been developed by Schafer and Pekoz [18], [19]. The method had its genesis in the design method for distortional buckling of thin-walled sections developed by Hancock, Kwon and Bernard [14]. This is the method that was incorporated in the Australian/New Zealand Standard for Cold-Formed Steel Structures [2]. However, the Direct Strength Method goes one step further and assumes that local buckling behavior can also be predicted using the elastic local buckling stress of the whole section with an appropriate strength design curve for local instability. The method has the advantage that calculations for complex sections are very simple provided elastic buckling solutions are available. The method is currently under consideration by the AISI Specification Committee. More detail of the method is given in Section 3 of this paper.

One major area requiring careful consideration is the interaction between buckling modes. Short half-wavelength buckling modes, such as local (L) and distortional (D), may interact with long half-wavelength buckling modes such as flexural or flexural-torsional (called E modes in this paper which is short for Euler). In this paper, local plus overall is designated L+E, distortional plus overall is designated D+E, local and distortional is designated L+D, and the interaction of all three modes is designated L+D+E. Research on these interaction modes is much more limited and detailed discussion forms the basis of much of this paper.

2 Mode Interaction

2.1 General

The interaction between buckling modes can take one of two forms. These are:

(a) Linear interaction where the interaction of the two modes occurs at the same half-wavelength.

(b) Non-linear interaction where the short half-wavelength mode (usually local) occurs over multiple half-wavelengths and interacts with a long half-wavelength mode by reducing its effective stiffness against buckling.

2.2 Linear Interaction Buckling

To explain the linear interaction mode, consider Fig. 2. The section is a 150 mm deep, 50 mm wide, 6 mm lip, 1.5 mm thick section and is subjected to uniform compression.

The local mode in the web and the distortional mode of the flanges and lips occur at the same half-wavelength. This is classified as a *linear interaction mode* since both modes occur in the linear eigenmode. The linear interaction is between the local and distortional modes in this case.

Linear interaction modes also occur between distortional and overall buckling, particularly flexural-torsional buckling. To demonstrate this mode, consider the section in Fig. 1 subjected to bending but with the bottom (tension) flange restrained against twisting as would occur if the flange were restrained by sheeting screw-fastened to it. The graph of buckling stress versus half-wavelength is shown in Fig. 3. At longer half-wavelengths, the flexural torsional mode (often called the lateral buckling mode for beams) interacts with the distortional mode of the flange to give a linear interaction mode called a lateral-distortional mode.

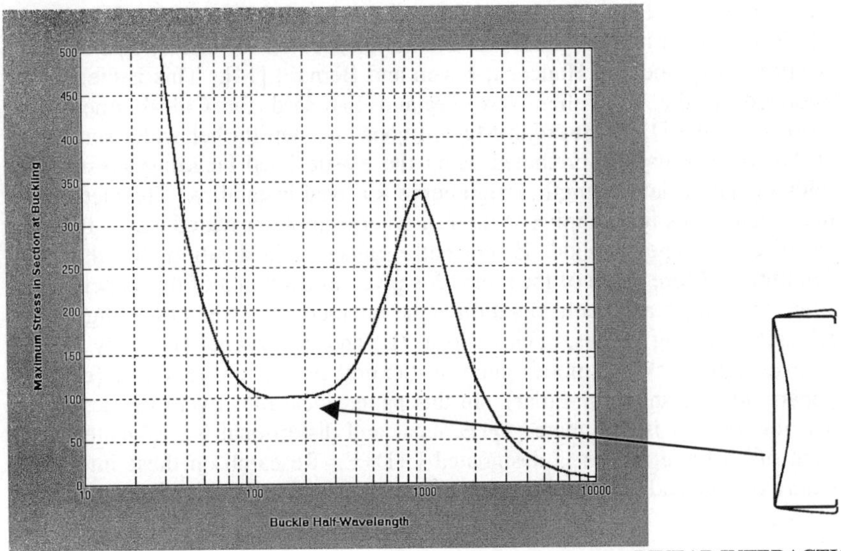

LINEAR INTERACTION MODE

Figure 2. Linear interaction buckling at short half-wavelength - Local and distortional buckling occurring at the same half-wavelength

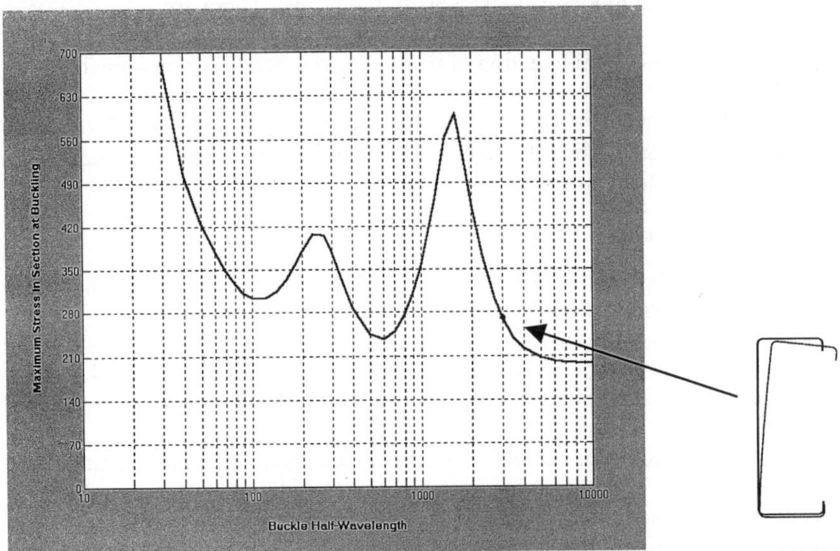

LATERAL-DISTORTIONAL MODE

Figure 3. Linear interaction at long half-wavelength – Lateral and distortional buckling occurring at the same half-wavelength

Linear interaction modes are generally easier to compute because they are simply based on a linear eigenvalue analysis. However, a method such as finite strip analysis [20] or generalised beam theory [21] is required to compute the buckling modes and corresponding half-wavelengths.

2.3 Non-linear Interaction Buckling

A *non-linear interaction mode* normally involves a study of the post-buckling behaviour of the local mode with change of stress distribution and elemental stiffnesses in the post-buckling range of the local mode. These changes of elemental stiffnesses reduce the effective rigidities of the section for distortional or overall buckling. The most fundamental work on the problem of local and overall buckling was performed by Koiter [22] and Van der Neut [23]. A large amount of literature exists on this subject and many papers can be found in the Proceeding of the 2nd and 3rd International Conferences on Coupled Instabilities [24], [25] including reference to earlier work. In [26], Pignataro discusses buckling and postbuckling analysis of continuous mechanical systems due to Koiter theory and analyses the influence of multiple interaction modes on the post buckling behaviour of thin-walled members. A theoretical and experimental study of a range of thin-walled I-, box and channel sections in compression is given in Hancock [27] where the modes of interaction are primarily local and flexural (L+E). Details of a non-linear analysis method based on a finite strip post-buckling analysis are also given in [27]. A method for analyzing the interaction of local and flexural-torsional buckling (L+E) is given in Rasmussen [28], [29], [30]. This method also uses a finite strip post-local buckling analysis but applies it to a general bifurcation analysis of thin-walled sections undergoing flexural, torsional or flexural-torsional buckling.

The interaction of local and distortional buckling is also possible as shown in Fig. 5 where the short half-wavelength local buckle of the lips as shown in Fig. 4 interacts with the longer half-wavelength distortional buckle of the flanges also shown in Fig. 4.

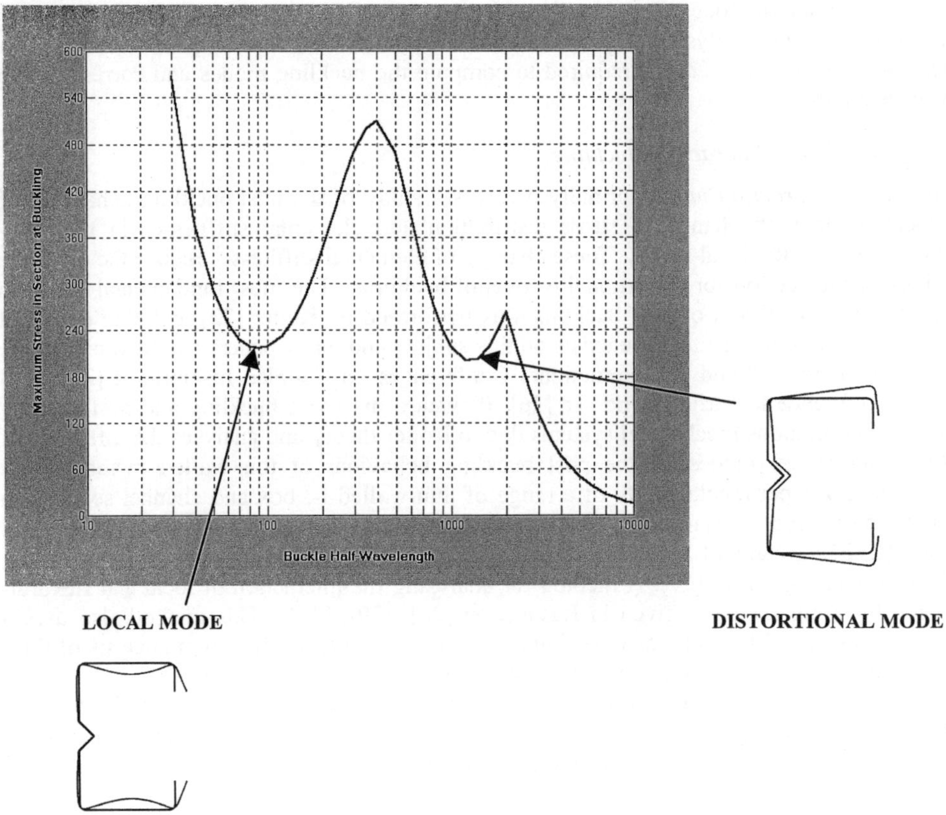

Figure 4 C-Section with web stiffener which causes local and distortional buckling to occur at approximately the same buckling stress in compression

Very little research exists on L+D interaction. At this time, different test programs indicate different results. The tests of Kwon and Hancock [31] for sections similar to those in Fig. 4 indicate little interaction whereas the tests of Baldassino and Hancock [32] for perforated rack sections indicate significant interaction. Schafer [15] has analysed the available data and concludes that L+D interaction does not appear significant except for that in [32].

Similarly, the interaction between distortional and overall modes (D+E) has not been widely researched. Schafer [15] states: "interaction of distortional with Euler buckling cannot be definitively recognised nor rejected from the available data". However, recent tests by Yan and Young [33] show significant interaction for distortional and overall buckling of lipped channels with return lips. As a result of the closer proximity of the buckling half-wavelengths for distortional and overall modes, the interaction is more likely to be a linear rather than a non-linear interaction mode.

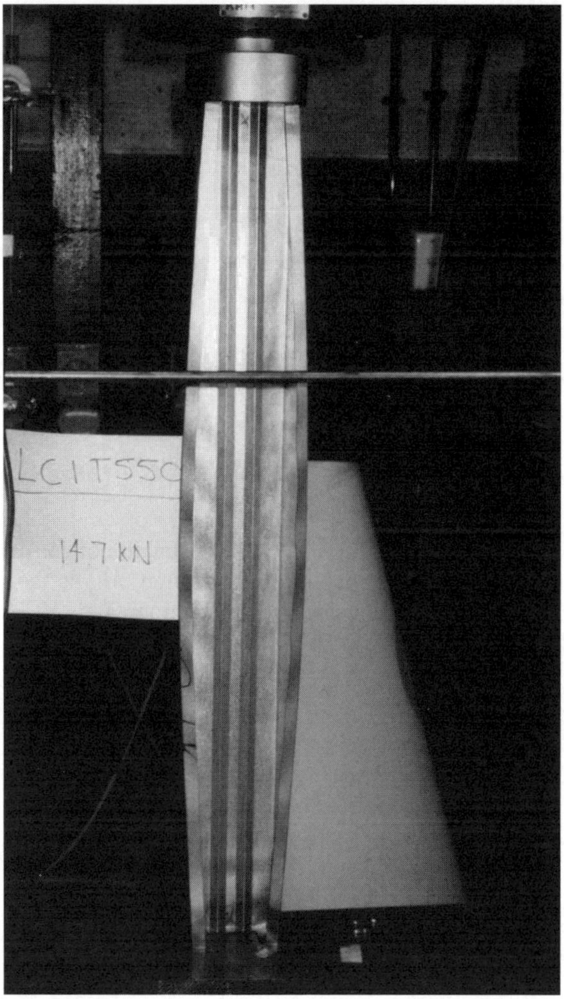

Figure 5. Test specimen with non-linear interaction of local and distortional buckling modes

3 Direct Strength Method

Based on the observations above, the Direct Strength Method as currently proposed by Schafer and Pekӡ [18], [19] only considers L+E and D+E interaction. A detailed summary of the method including its application is given in [34].

The Direct Strength Method determines the strength for L+E interaction and D+E interaction independently and takes the lesser of the two as the strength. For columns, these are the column axial strengths $\boldsymbol{P_{nl}}$ and $\boldsymbol{P_{nd}}$, and for beams, they are the beam bending strengths $\boldsymbol{M_{nl}}$ and $\boldsymbol{M_{nd}}$. The equations for $\boldsymbol{P_{nl}}$ and $\boldsymbol{M_{nl}}$ considering the L+E interaction are given by:

$$P_{nl} = P_{ne} \quad for \ \lambda_l \le 0.776 \tag{1}$$

$$P_{nl} = \left(1 - 0.15\left(\frac{P_{crl}}{P_{ne}}\right)^{0.4}\right)\left(\frac{P_{crl}}{P_{ne}}\right)^{0.4} P_{ne} \quad for \ \lambda_l > 0.776 \tag{2}$$

where $$\lambda_l = \sqrt{\frac{P_{ne}}{P_{crl}}} \tag{3}$$

$$M_{nl} = M_{ne} \quad for \ \lambda_l \le 0.766 \tag{4}$$

$$M_{nl} = \left(1 - 0.15\left(\frac{M_{crl}}{M_{ne}}\right)^{0.4}\right)\left(\frac{M_{crl}}{M_{ne}}\right)^{0.4} M_{ne} \quad for \ \lambda_l > 0.776 \tag{5}$$

where $$\lambda_l = \sqrt{\frac{M_{ne}}{M_{crl}}} \tag{6}$$

where $\boldsymbol{P_{ne}}$ is the inelastic overall column buckling strength not considering interaction, $\boldsymbol{P_{crl}}$ is the elastic buckling load for local buckling of a column, $\boldsymbol{M_{ne}}$ is the inelastic beam flexural-torsional buckling strength not considering interaction, and $\boldsymbol{M_{crl}}$ is the elastic buckling moment for local buckling of the beam.

The equations for the column axial strength $\boldsymbol{P_{nd}}$ and beam bending strength $\boldsymbol{M_{nd}}$ considering the D+E interaction are:

$$P_{nd} = P_{ne} \quad for \ \lambda_d \le 0.561 \tag{7}$$

$$P_{nd} = \left(1 - 0.25\left(\frac{P_{crd}}{P_{ne}}\right)^{0.6}\right)\left(\frac{P_{crd}}{P_{ne}}\right)^{0.6} P_{ne} \quad for \ \lambda_d > 0.561 \tag{8}$$

where $$\lambda_d = \sqrt{\frac{P_{ne}}{P_{crd}}} \tag{9}$$

$$M_{nd} = M_{ne} \quad for \ \lambda_d \le 0.561 \tag{10}$$

$$M_{nd} = \left(1 - 0.25\left(\frac{M_{crd}}{M_{ne}}\right)^{0.6}\right)\left(\frac{M_{crd}}{M_{ne}}\right)^{0.6} M_{ne} \quad for \ \lambda_d > 0.561 \tag{11}$$

where $$\lambda_d = \sqrt{\frac{M_{ne}}{M_{crd}}} \tag{12}$$

where P_{ne} is the inelastic overall column buckling strength not considering interaction, P_{crd} is the elastic buckling load for distortional buckling of a column, M_{ne} is the inelastic beam flexural-torsional buckling strength not considering interaction, and M_{crd} is the elastic buckling moment for distortional buckling of the beam.

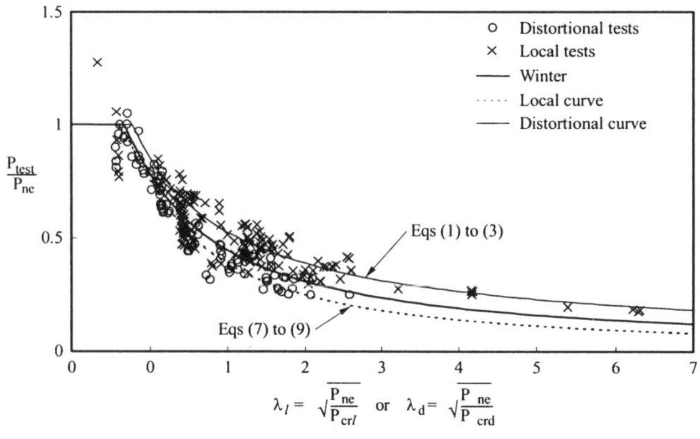

Figure 6. Column strengths considering L+E interaction and D+E interaction

A comparison of the test strengths for failure of columns in the L+E mode is shown by the crosses (x) in Fig. 6, and the failure of columns in the D+E mode is shown by the circles (o) in Fig. 6. Although there is a scatter in the data, the two different equations for columns provide a reasonable mean fit to the test data. Of importance is the fact that the L+E graph is higher than the D+E graph. This is a simple consequence of the fact that local buckling has a higher postbuckling strength reserve than distortional buckling.

4 Engineering Design

Structural design standards have been developed to assist design engineers to predict the strength of thin-walled sections accounting for sectional stability. As the most common form of sectional instability is local buckling, this was addressed first. Local buckling is known to have a strong postbuckling reserve and so the effective width method proposed by Von Karman [35] and calibrated for imperfect cold-formed steel elements by Winter [36] has been used in design standards such as [4] for over 50 years. When the distortional buckling mode was discovered to have an influence about 25 years ago, it was natural to extend the effective width method to account for distortional buckling of simple lip-stiffened elements. Although the method is accurate for local and distortional instability of a simple lipped channel of the type shown in Fig. 2, it presents considerable problems for more complex shapes such as those shown in Fig. 4 where intermediate stiffeners are added to the web. The Direct Strength Method therefore provides a more general method to predict the strength of complex shapes accurately accounting for the local and distortional modes and their interaction with overall modes. The Direct Strength Method does not require computation of effective sections but requires computer software such as THIN-WALL [20] to compute elastic buckling stresses of thin walled sections. Although this may seem straightforward, the challenge is to get structural design engineers to accept the use of computer software in place of their traditional effective width methods.

5 Conclusions

Sectional instability in the form of local and distortional buckling is often present in thin-walled sections undergoing compression and/or bending. Overall buckling modes such as flexural and flexural-torsional also occur if the member is unbraced and can interact with the sectional instability modes. The interaction of modes can take the form of linear or non-linear interaction where linear interaction implies buckling of the different modes in a linear eigenmode at the same half-wavelength. At this time, only some of the interaction modes have been thoroughly investigated, and there is much work to be done to understand the interaction of other modes, especially those involving distortional buckling. In the meantime, empirical methods based mainly on elastic bucking solutions and test data have been used to provide design engineers with reliable methods. At present, there is a move to replace the traditional Effective Width Method by the newly developed Direct Strength Method. This latter method is more general and applicable to complex shapes. However, its general application may require more research before designers have faith in its applicability.

References

1. Standards Association of Australia, Steel structures, AS 4100-1998.
2. Standards Australia/Standards New Zealand, Cold-formed steel structures, 1996.
3. Standards Australia/Standards New Zealand, Aluminium structures, Part 1: Limit state design, AS/NZS 1664.1:1997.
4. American Institute of Steel Construction, Load and resistance factor design specification for structural steel buildings, December 1998.
5. American Iron and Steel Institute, Specification for the design of cold-formed steel structural members, 1996.
6. The Aluminium Association, Specification for aluminium structures, 5th ed. December 1986.
7. Trahair N. S., Flexural-torsional buckling, (E. & F. Spon, 1993).
8. Hancock G. J., Design for distortional buckling of flexural members, *Thin-walled Structures*, **20**, (1997) pp. 3-12.
9. Hancock G. J., Rogers C. A. and Schuster R. M., Comparison of the distortional buckling method for flexural members with tests, *13th International Specialty Conference on Cold-Formed Steel Structures*, (University of Missouri-Rolla, St. Louis 1996) pp. 125-139.

10. Schafer B. and Pek $\bar{z}$ T., Laterally braced cold-formed steel flexural members with edge stiffened elements, *Journal of Structural Engineering*, ASCE **125(2)** (1999) pp.118-127.
11. Hancock G. J., Distortional buckling of steel storage rack columns, *Journal of Structural Engineering*, ASCE **111 (12)** (1985) pp. 2770-2783.
12. Lau S. C. W. and Hancock G. J., Distortional buckling formulas for channel columns, *Journal of Structural Engineering*, ASCE **113(5)** (1987) pp. 1063-1078.
13. Kwon Y. B. and Hancock G. J., Strength tests of cold-formed channel sections undergoing local and distortional buckling, *Journal of Structural Engineering* **118(7)** (1992) pp. 1786-1803.
14. Hancock G. J., Kwon Y. B. and Bernard E. S., Strength design curves for thin-walled sections undergoing distortional buckling, *Journal of Constructional Steel Research* **31(2/3)** (1994) pp. 169-186.
15. Schafer B. W., Distortional buckling of cold-formed steel columns, *Final Report* to the American Iron and Steel Institute, August 2000.
16. Schafer B. W., Thin-walled column design considering local, distortional and Euler buckling, *Proceedings* of the Annual Technical Session Structural Stability Research Council May (2001) Fort Lauderdale, Florida.
17. Desmond T. P., Pek $\bar{z}$ T. and Winter G., Edge stiffeners for thin-walled steel members, *Journal of Structural Engineering* **107(2)** (1981) pp. 329-353.
18. Schafer B. W. and Pek $\bar{z}$ T., Direct strength prediction of cold-formed steel members using numerical elastic buckling solutions, *Thin-Walled Structures Research and Development* ed. by N. E. Shanmugan, J. Y. R Liew and V. Thevendran (Elsevier 1998) pp. 137-144.
19. Schafer B. W. and Pek $\bar{z}$ T., Members using numerical elastic buckling solutions, *Proceedings, 14th International Specialty Conference on Cold-Formed Steel Structures*, (University of Missouri-Rolla, St. Louis 1998) pp. 69-76.
20. Papangelis J. P. and Hancock G. J., Computer analysis of thin-walled structural members, *Computers and Structures* **56, No 1** (1995) pp. 157-176.
21. Davies J. M., Generalised beam theory (GBT) for coupled instability problems, Part IV of *Coupled Instabilities in Metal Structures,* ed. by J. Rondal (1998) pp. 151-223.
22. Koiter W. T., A translation of the stability of elastic equilibrium, *Technical Report* **AFFDL-TR-70-25** (Air Force Flight dynamics Laboratory, Wright-Patterson Air Force Base, Ohio 1970).
23. Van de Neut A., The interaction of local buckling and column failure of thin-walled compression members, *Proceedings, 12th International Congress of Applied Mechanics* (1969) pp. Springerverlag, Germany.
24. *Proceedings, 2nd International Conference on Coupled Instabilities in Metal Structures*, ed. by J. Rondal, D. Dubina and V. Gioncu (Imperial College Press 1996).

25. *Proceedings, 3rd International Conference on Coupled Instabilities in Metal Structures*, ed. by D. Camotin, D. Dubina and J. Rondal (Imperial College Press 2000).
26. Pignatoro M., Stability, bifurcation and postbuckling analysis, Part II of *Coupled Instabilities in Metal Structures – Theoretical and Design Aspects*, ed. by J. Rondal (Springer Wien, New York 1998) pp. 29-83.
27. Hancock G. J., Finite strip buckling and nonlinear analyses and distortional buckling analysis of thin-walled structural members, Part V of *Coupled Instabilities in Metal Structures*, ed. by J. Rondal (1998) pp. 225-289.
28. Rasmussen K. J. R., Bifurcation of locally buckled members, *Thin-Walled Structures*, **28(2)** (1997) pp. 117-154.
29. Young B. and Rasmussen K. J. R., Bifurcation of singly symmetric columns, *Thin-Walled Structures*, **28(2)** (1997) pp. 155-177.
30. Rasmussen K.J.R. and Hasham A.S., Flexural and flexural-torsional buckling of locally buckled beam-columns, *Thin-Walled Structures*, **29(1-4)** (1997) pp. 203-233.
31. Kwon, Y. B. and Hancock G. J., Tests of channels with local and distortional buckling, *Journal of Structural Engineering*, ASCE **118(7)** (1992) pp. 1786-1803.
32. Baldassino N. and Hancock G. J., Distortional buckling of cold-formed steel storage rack sections including perforations, *Proceedings, Fourth International Conference on Steel and Aluminium Structures – Light-Weight Steel and Aluminium Structures*, ed. by P. M≅kel≅inen and P. Hassinen, Espoo, Finland (1999) pp. 131-154.
33. Yan, J. and Young, B., Compression tests of thin-walled lipped channels with return lips, *Proceedings, Third International Conference on Thin-Walled Structures – Advances and Developments,* ed. by J. Zara∇, K. Kowal-Michalska and J. Rhodes, Cracow, Poland (2001) pp.249-265.
34. Hancock G. J., Murray T. M. and Ellifritt D. S., Cold-formed steel structures to the AISI Specification, (New York, Marcel Dekker, Inc 2001).
35. Von Karman T., Sechler E. E. and Donnell L. H., The strength of thin plates in compression, *Transactions, ASME,* **54** MP54-5 (1932).
36. Winter G., Thin-walled structures – theoretical solutions and test results, *Preliminary Publications of the 8th Congress, IABSE* (1968) pp. 101-112.

VIRTUAL BIOMECHANICS: NON INVASIVE DYNAMIC INSIGHTS INTO THE MASTICATORY SYSTEM

L. M. GALLO

Clinic for Masticatory Disorders and Complete Dentures
Center for Dental and Oral Medicine and Maxillofacial Surgery
Plattenstrasse 11
CH-8028 Zurich, Switzerland
E-mail: luigi@zui.unizh.ch

The realistic replication of mandibular motion for the construction of dental prostheses was the main purpose for studying mandibular kinematics in the past. Meanwhile, the research interest has shifted to the analysis of the function of the masticatory system, especially of the temporomandibular joint (TMJ). With the development of sophisticated microelectronics, mandibular tracking has been made possible in a precise way by recording all six degrees-of-freedom relative to the skull. The application of motion to craniomandibular anatomical models gives a non invasive dynamic insight into the TMJ. This virtual biomechanical technique (dynamic stereometry) facilitates dynamic, three-dimensional measurements within the joint. The present work deals with new methods developed for an integrated in vivo analysis of mandibular biomechanics. A series of studies were performed using these new methods. The aims of these studies were the animation of real TMJ morphologies with real jaw tracking data (which permits the motion characterization by means of time-dependent variables describing the relationship between the articulating surfaces), the quantitative analysis of the intra-articular distances in the TMJ related to functional mandibular movements and the use of the dynamic stereometry of the TMJ for the study of the joint load by means of finite element analysis.

1 Introduction

The development of new diagnostic methods for the study of disorders of the masticatory system is of great importance, since signs and symptoms of these disturbances affect large portions of the general population [1-3]. One of the components of the masticatory system often diseased is the temporomandibular joint (TMJ), a compound joint consisting of two incongruent bony surfaces, the *fossa* (in the temporal bone of the skull) and the *condyle* (in the lower jaw, *i.e.* the mandible) divided by an *articular disc* (Fig. 1). The disc tissue consists of fibrocartilage and mainly functions as a stress distribution and lubrication mechanism. Due to the peculiar anatomy of the TMJ and to the complex arrangement of the masticatory muscles, mandibular movements are generally performed by combined rotations and translations. In a normal TMJ, during mouth opening/closing for instance, the movements between the condyle and the disk are mainly hinge-like, whereas those between the condyle-disk complex and the fossa are translatory [4-6]. During mandibular movements the disc is deformed by the variations of the intraarticular space, thus possibly experiencing local stresses.

Temporomandibular biomechanics is only partially understood. On one hand, TMJ kinematics has been extensively investigated at first mainly for the proper fabrication of tooth restorations, and, more recently, also to study the function of the masticatory system. Thus, it is now possible to assess the regularity or irregularity of mandibular movements in highly sophisticated ways [7-14]. However, the mechanics of the articulating surfaces of the TMJ, which ultimately influences the regularity of mandibular movements, has still many untapped aspects, especially those concerning the loading of soft tissues and in particular of the disc [15,16]. Degenerative diseases of the TMJ as well as of other human joints are believed to be due to a significant extent to traumas or

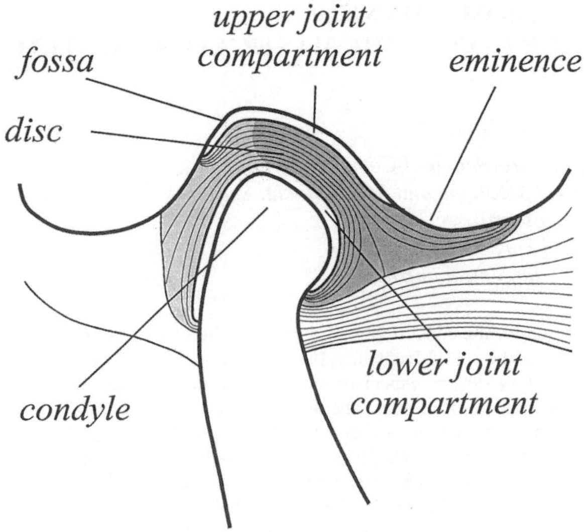

Figure 1. Temporomandibular joint.

overload with a consequent mechanical failure of the soft tissues [17-21]. In the TMJ, signs of osteoarthrosis of the bony surfaces as well as disc perforations have often been observed in cadavers [22,23]. In contrast to isolated animal experiments [24-27], the direct measurement *in vivo* of TMJ loading in humans still remains ethically and technically impossible, even with modern technology. TMJ loading can however be calculated indirectly by means of laboratory measurements of the material properties *in vitro* and computer simulations on the basis of models reconstructed from real anatomical data and corresponding real kinematic measurements [28-32]. If these models of the real anatomy can be animated and manipulated realistically on a workstation, this leads to the concept of *virtual biomechanical measurements*. Thus, in virtual biomechanics, a previous three-dimensional recording of anatomical structures and the acquisition of the joint(s) motion in all degrees-of-freedom is necessary.

Most medical imaging systems can deliver static information about the anatomy of hard and soft tissues of joint structures, allowing the analysis of their shapes, size and relative arrangement. However, conventional radiography, arthrography and fluoroscopy deliver only bidimensional, mostly unclear pictures of the TMJ, because they superimpose the images of all radio-opaque structures between the X-rays source and the sensitive medium. In contrast, computer tomography (CT) usually delivers a full set of three-dimensional information of the recorded structure, in which the volume elements are proportional to the X-ray absorption coefficients of the corresponding tissues. As CT scanners are also based on X-ray exposure, they are particularly suited for hard tissue imaging, which they do with high contrast and spatial resolution. Magnetic resonance imaging (MRI) records the free induction decay signal emitted by protons in the tissues, excited by means of radio frequency pulses superimposed to a strong magnetic field [33]. By means of appropriate small local variations of the strong magnetic field (gradients), the signal emitted by the protons can be modulated according to their spatial location. Thus, MRI is also capable of three-dimensional imaging without the need of any moving parts. In mandibular and TMJ biomechanical research, MRI has the advantage of not exposing the subject to X-ray hazard, so that asymptomatic volunteers can also be ethically recruited for studies. Furthermore, the images delivered offer a good representation of the soft tissues, such as the TMJ disc and the muscles, because they reflect the proton content (essentially water) of the tissues. Nowadays, MRI is routinely used for the analysis of TMJ disc shape and position [34,35] and is the preferred method for the acquisition of the anatomical data in virtual biomechanical studies [36,37].

Several electronic systems for jaw motion recording have been developed in the past, the most well-known being: the Case Gnathic Replicator [38], the Kinesiograph™ [39], the Sirognathograph™ [40], the Selspot™ system [41], the Axiograph [42] and the

String-LR-Recorder [43]. Several other systems have also been described [44,45,46,47,48,49,50]. The difficulty in constructing a system that meets all the requirements necessary to record jaw function may well explain why so many jaw tracking devices have been developed. Such a device should not be invasive, and it should not interfere with jaw and soft tissue function or restrict head movements. It should be simple to use, and, finally, it should record the movement of the whole mandible, not only of a single point. With this in mind, a noninvasive, optoelectronic system, called Jaws-3D, has been developed [51,52]. This system acquires three-dimensionally with six degrees-of-freedom and in real time the movement of the whole mandible relative to the head. Head movements are not restricted and lip function is minimally disturbed.

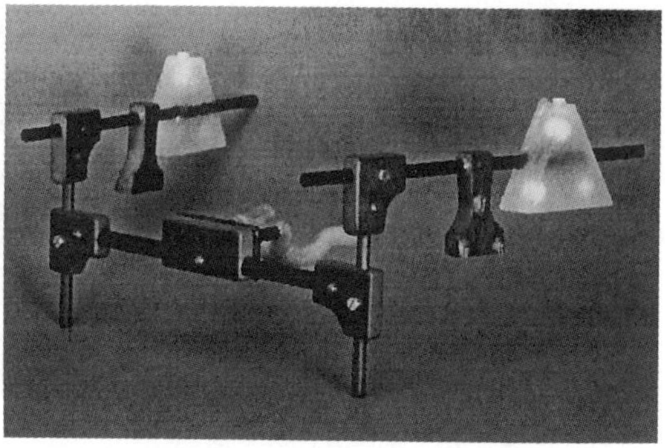

Figure 2. Reference system for MRI.

In this paper we present a non-invasive, virtual biomechanical method to ultimately analyze the mechanical loading patterns within the TMJ (*dynamic stereometry*). The method combines the data from a tomographic system with the kinematic information from a jaw tracking system. This technique yields a dynamic non invasive in vivo insight into the TMJ and gives the basis for subsequent experimental and numerical analyses.

2 Material and methods

A requirement for the development of the dynamic stereometry of the TMJ is the implementation of a reference, common to the imaging system as well as to the jaw tracking system, in order to combine the anatomical data with the motion recordings. This common reference consists of three extra-oral contrast spheres (∅ 10 mm) 30 mm distant from each other and fixed to the dental arches by means of a plastic face bow and an individualized monobloc (Fig. 2). The contrast spheres for MRI are made of PVC and are embedded in a prism filled with a Gadolinium solution (Magnevist®, Schering AG, D-13342 Berlin; 0.5 mmol/l). The contrast spheres for CT scans are made of lead. The reference spheres define a coordinate system, whose position is known in the imaging system as well as in the jaw tracking system.. The reference device is worn by the subject during the tomographic recording and during a static reference recording with the jaw tracking system. Subsequently, it is possible to record any movement without any further constraints.

2.1 Imaging systems

In order to obtain the anatomical information of the TMJ, a tomographic procedure is used, which can be either MRI or CT scanning. For ethical reasons, MRI is mostly used in order to assess the

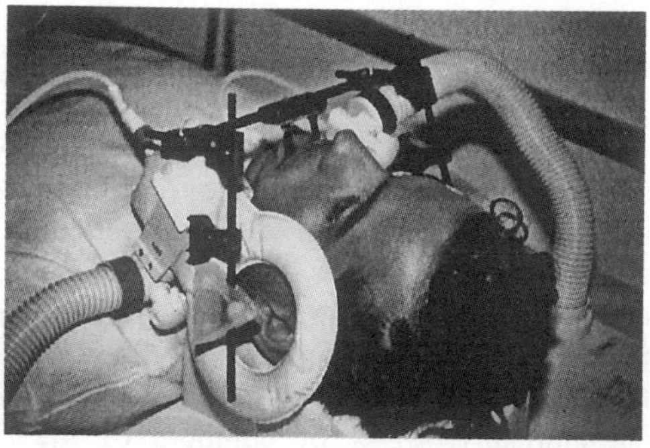

Figure 3. MRI recording.

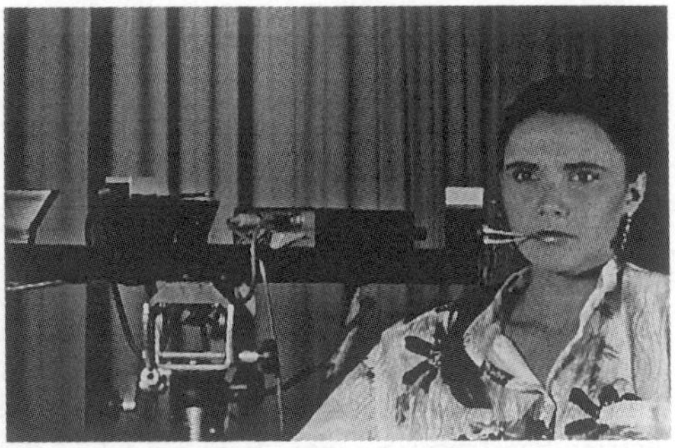

Figure 4. Jaw tracking.

normal functionality of asymptomatic subjects, since it produces no ionizing radiation. For an optimal representation of the TMJ we use an established protocol implemented on a 1.5 T machine (Gyroscan® NT, Philips Medical System, NL-5680 DA Best) with a double TMJ coil (∅ 12 cm) (Fig. 3). For the localization of the TMJ, we first take scout views, *i.e.* transversal slices 6 mm thick with 1 mm spacing. On these images we then set parasagittal slices perpendicular to the main condylar axis (*i.e.* with a 10°-15° angulation with respect to the sagittal plane). Each slice has a dimension of 130×130 mm with a resolution of 256 Pixel, each of size 0.51×0.51 mm. The recording volume is divided into two stacks, one stack with 14 contiguous slices each of 2 mm thickness in the TMJ area and one with the same slice thickness in the area of the references spheres (also 14 slices). The best contrast is obtained with a Fast-Field-Echo sequence, with a repetition time T_R of 300 ms and an echo time T_E of 9 ms for a total recording time of round 4 minutes.

2.2 *Kinematic recording*

The motion of the lower jaw is recorded with the jaw tracking device Jaws3D [51,52], which allows data acquisition with 6 degrees-of-freedom (Fig. 4). For this purpose, the lower jaw is assumed to be a rigid body. The system consists of 3 fixed linear CCD-cameras with 2048 elements each, that determine the spatial position of 2 target frames, each carrying 3 light-emitting diodes (LEDs). The target frames are rigidly connected to the upper and lower dental arches and determine a head related and a mandible related coordinate system. The target frames are connected to the teeth of both dental arches by means of special splints that do not interfere with occlusion. By calculating the motion of the lower jaw relative to the head, the head motion is eliminated, thus obtaining mandibular motion. The jaw tracking system scans a field-of-view of 140×140×140 mm and its geometrical resolution is better than 70 μm. The sampling frequency is 70 Hz.

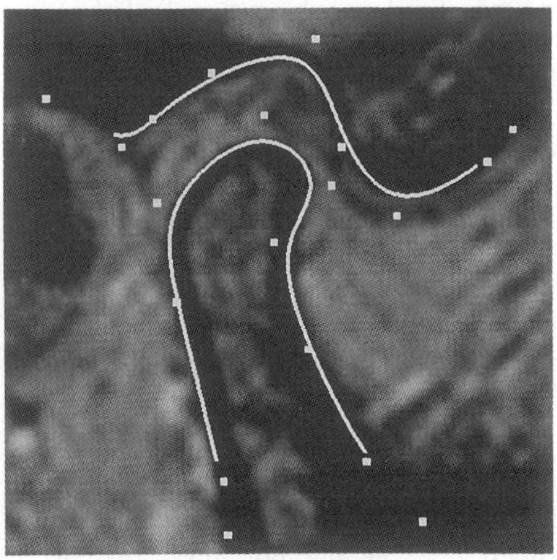

Figure 5. Contour definition.

2.3 *Reconstruction and animation*

The reconstruction and animation of the TMJ are performed on a graphic workstation by means of self-developed software [36,37].

The tomographic slices first undergo a segmentation process for the extraction and vectorial description of the surfaces of condyle and fossa as well as for the determination of the centers of the reference spheres. Whereas the centers of the reference spheres are determined automatically, a set of contours of the articulating surfaces is defined by manual

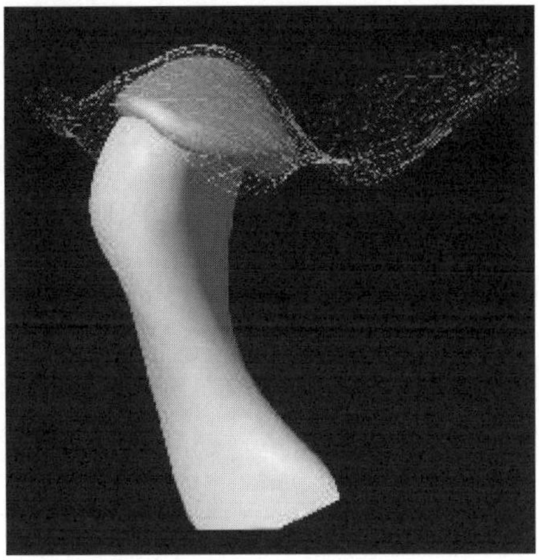

Figure 6. Reconstructed joint.

input of driving points of spline functions (Fig. 5). The contour sets are then triangulated and the resulting surfaces are represented realistically by means of shading algorithms (Fig. 6).

The reconstructed joints are then animated by the real jaw tracking data, either in full time resolution or after resampling in real time with around 24 images/sec. The animated images can be rotated and manipulated interactively and observed from any viewpoint. Beside visual observation of the animated images it is possible also to perform quantitative measurements, that allow a three-dimensional dynamic analysis of the relationship between the complete articulating surfaces.

Error analyses of the system show that the greatest geometric distortion occurs along the craniocaudal axis of the subject and amount to 0.9%. Noise effects cause errors in the surface geometry of ±0.17 mm for velocities between 10 and 230 mm/s. Altogether, the total measurement error lies in the order of magnitude of 1 mm.

3 Results

Fig. 7 shows a selection of images of a TMJ during a jaw opening/closing movement. The sequence begins at maximum intercuspation, reaches maximum opening and goes back to maximum intercuspation. It is also possible to reconstruct both TMJs of a subject and to animate them with the motion information recorded unilaterally. Because of the higher noise due to parallax errors, in the present version of the jaw tracking system it is not possible to perform precise intraarticular measurements in the TMJ contralateral to the target frames of the jaw tracking system.

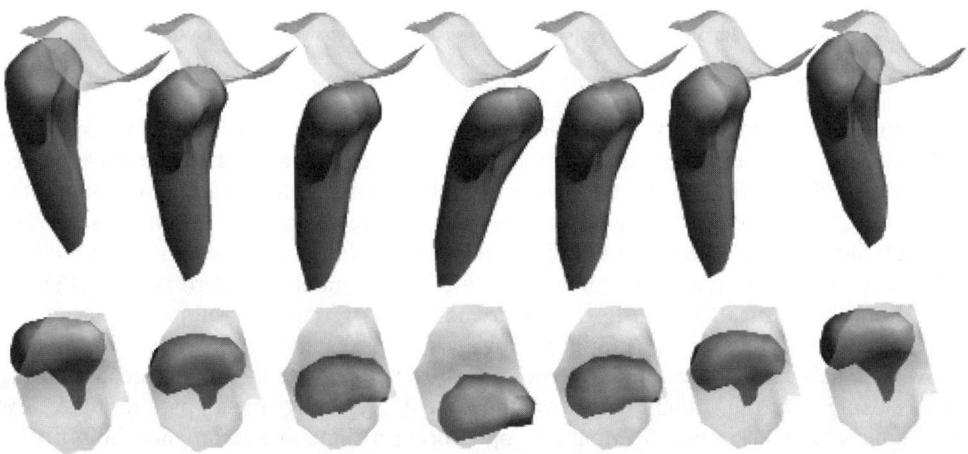

Figure 7. Jaw opening/closing sequence.

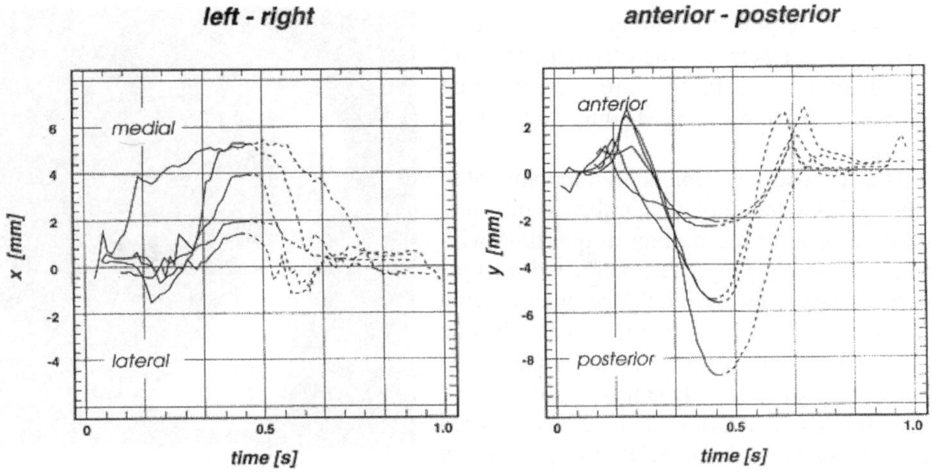

Figure 8. Path of minimum intraarticular distance during jaw opening/closing.

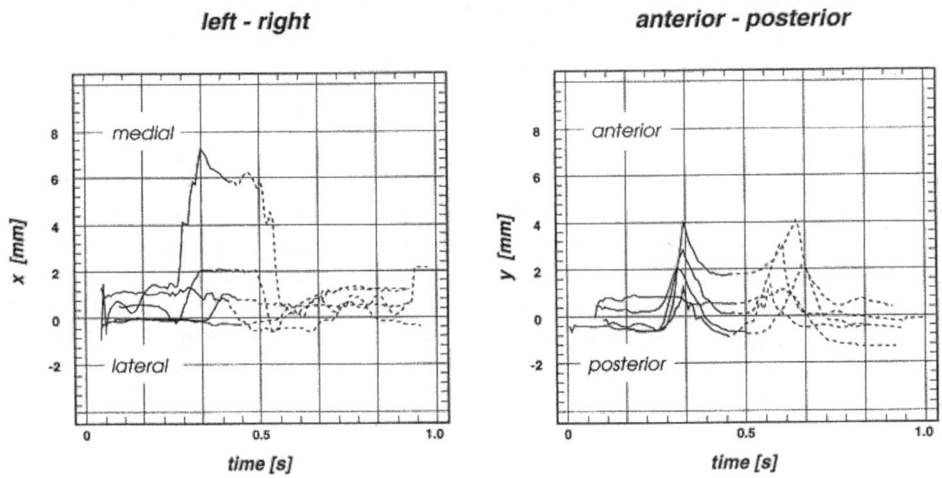

Figure 9. Path of minimum intraarticular distance during contralateral laterotrusion.

3.1 General relationship between the articulating surfaces

A simple way to quantitatively evaluate the motion of the animated joint models is to calculate the minimum intra-articular distance, *i.e.* the minimum distance between the surfaces of the fossa and the condyle. Generally, during mandibular movements, the minimum intra-articular distance varies and its location moves in space. The characteristic paths of the minimum distance location *relative to the condylar surface* are illustrated in Fig. 8 (jaw opening/closing), Fig. 9 (contralateral laterotrusion), Fig. 10 (ipsilateral laterotrusion) and Fig. 11 (protrusion) for a group of 5 subjects. During jaw opening, the location of the minimum intra-articular distance *relative to the condyle* moves medially and at the same time posteriorly. The opposite movement occurs during jaw closing. During other types of movement, typically clear patterns are less distinguishable than in jaw opening/closing. During ipsilateral laterotrusion only a very small condylar excursion of the minimum distance location can be observed.

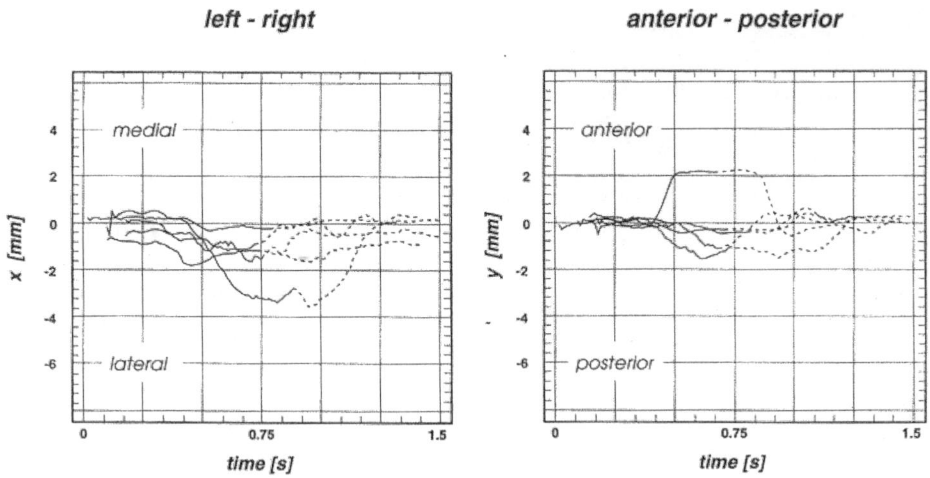

Figure 10. Path of minimum intraarticular distance during ipsilateral laterotrusion.

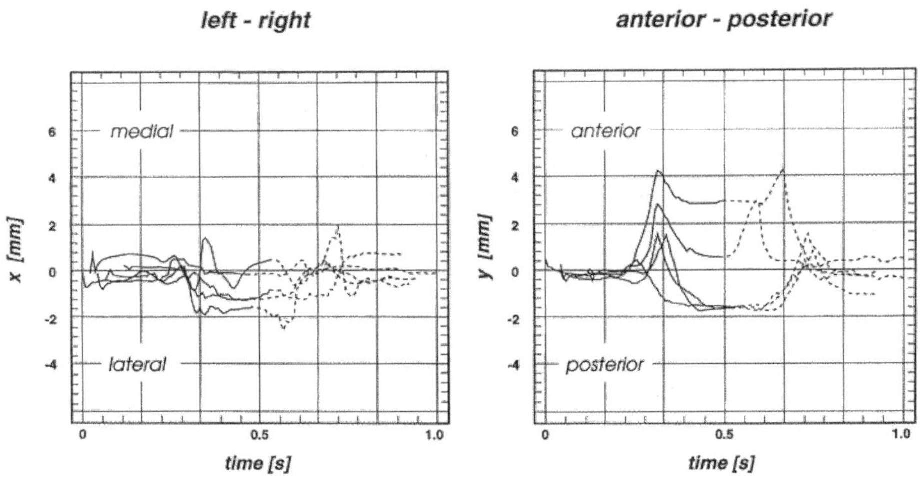

Figure 11. Path of minimum intraarticular distance during protrusion.

3.2 *Stress-fields in the healthy TMJ*

The fibers of the articular disc run dorsoventrally. Its mechanical strength is therefore anisotropic, the disc being weaker in mediolateral direction [53-55]. Therefore the stress-fields produced in the articular disc by the incongruency of the articulating surfaces represent a potential source of mechanical failure, especially if they have a relevant mediolateral component [56]. Under the assumption that the area with the minimum distance between the articulating surfaces corresponds to a stress-field, it is possible to show that the stress-fields translate mediolaterally during functional movements of the TMJ. Furthermore, one can estimate the mechanical energy dissipated in the disc tissue by the mediolateral stress-field movement. For this type of analysis we determine the position of the minimum distance between the articulating surfaces *relative to the fossa*.

Fig. 12 shows the path of the stress-field centroid in a TMJ during jaw opening/closing. Fig 13 shows the velocity profile of the stress-field centroid of another TMJ during five jaw opening/closing cycles. In fifteen asymptomatic TMJs an average

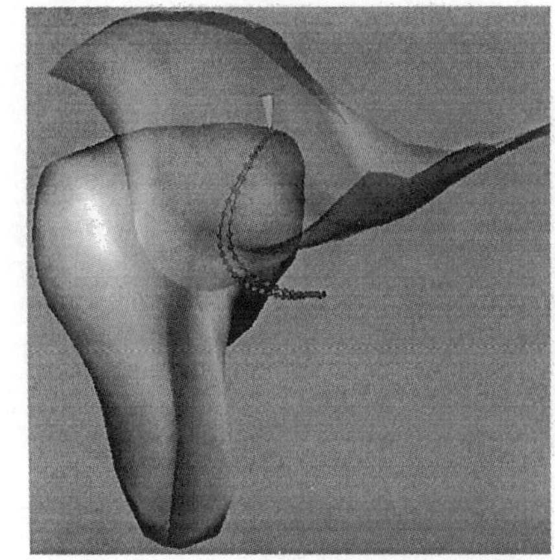

Figure 12. Stress-field path during jaw opening/closing.

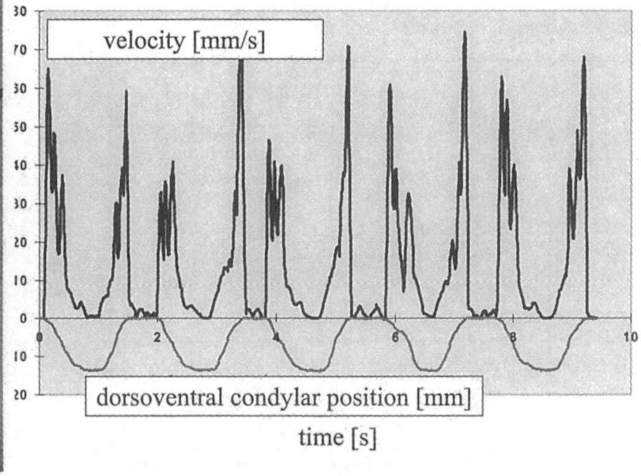

Figure 13. Velocity profile of the stress-field centroid in a TMJ during ten jaw opening/closing cycles.

mediolateral translation component of 6±3 mm has been found. The average radius of the stress-fields amounts to 2.4±1.7 mm and the maximum mediolateral velocity component of the stress-field centroid to 35±17 and 40±19 mm/s for a jaw opening/closing rate of 0.5 resp. 1.0 Hz. In two thirds of the cases the velocity reaches its maximum in the lateral portion of the articular disc. From experimental measurements it was possible to determine a relationship between the stress-field translational velocity and the resulting shear stress in the tissue, which shows high non-linearity. We could then calculate an average stress value of 1.4±0.5 and 2.2±0.4 MPa for the jaw opening/closing rates of 0.5 and 1.0 Hz. From all measured values we could estimate the energy dissipated in the in the disc tissue by the mediolateral stress-field movement. It shows a high variability, but with generally low values during jaw opening/closing movements, since it ranges between 10 and 700 mJ.

4 Future perspectives

The data reported above show that we can successfully couple anatomical structures with kinematic information, thus obtaining non-invasively a first set of quantitative data on the mechanical environment *in vivo* in the TMJ. Our virtual biomechanical modeling is however only the basis for further research. In the following we describe briefly two collaborative projects for further analysis of TMJ loading.

4.1 In vitro analysis of material properties of the articular disc

For the analysis of the mechanical properties of the articular disc tissue, a device reproducing the stress-field translation in tissue probes is under development at the University of Nebraska Medical Center (Lincoln NE, USA). The device moves cyclically a pseudocondyle on a tissue sample with velocities similar to those measured in our virtual biomechanical models. Underneath the tissue samples, a series of sensors detect the mechanical stress distribution depending on the plowing depth and velocity of the pseudocondyle. The ultimate goal of these measurements is the experimental determination of the effects of stress-field translation on the integrity of the disc surface. First results show an increase of the mechanical shear stresses with increasing translational velocity of the stress-field. However, the initial loading decreases with the number of loading cycles, apparently because of the reduced tissue hydration due to the free liquid flow.

4.2 Biphasic FE modeling of the articular disc

Since the articular disc of the TMJ consists of a cartilage-like structure, *i.e.* a soft hydrated tissue, it appears reasonable to choose a biphasic model for its numerical simulation. Biphasic models treat soft hydrated tissue as a continuous mixture of solid and liquid components in which time dependent viscous effects are modeled by means of the interaction between the phases [57]. In a collaboration with the Scientific Computation Research Center of the Renssalaer Polytechnic Institute (Troy NY, USA) first attempts have been performed in order to apply a three-dimensional velocity-pressure formulation of the biphasic theory to the articular disc for the first steps of jaw opening. In addition to the articular surfaces of the condyle and fossa of a subject, we also reconstructed the articular disc in maximum intercuspation (as in Fig. 6). For this preliminary study we assumed that a part of the cranial disc surface is attached to the fossa. Similarly, also a part of the caudal disc surface was considered attached to the condyle. A smaller motion range was chosen in order to keep the deformation in its linear formulation range. For the tissue, we chose an elasticity modulus of 0.675 MPa, a Poisson coefficient of 0.125 and a permeability constant of 4×10^{-2} mm^4/Ns as well as a solid volume fraction of 0.2. Within the first 90 ms of motion we found mainly tensile stress, although also the shear stress component is relevant. Furthermore, the fluid pressure accounts for a significant part of the total load.

4.3 Conclusion

These preliminary studies show that the data obtained from virtual biomechanical models can deliver useful information for *in vitro* experiments on TMJ disc tissues as well as be successfully used for FE analysis of TMJ loading. In addition to an accurate experimental determination of the material properties of these tissues, several important aspects have to be considered in the future. First, the modeling of the articular disc needs a true contact analysis in order to also simulate larger movements. Furthermore, the anisotropic nature of the disc must be implemented. Finally, the numerical modeling will still need an experimental validation.

5 Acknowledgments

The author thanks Jeff Nickel and Laura Iwasaki from the University of Nebraska Medical Center and Robert Spilker from the Scientific Computation Research Center of the Renssalaer Polytechnic Institute for the collaboration.

References

1. Nilner M., Epidemiologic studies in TMD. In: Current controversies in temporomandibular disorders, ed. C. McNeill, Chicago: *Quintessence* (1992) pp. 21-26.
2. Rugh J.D. and Solberg W.K., Oral Health Status in the United States: Temporomandibular Disorders *J Dent Educ*, **49** (1985) pp. 398-405.

3. Schiffman E.L., Fricton J.R., Haley D.P. and Shapiro B.L., The prevalence and treatment needs of subjects with temporomandibular disorders, *J Am Dent Assoc*, **120** (1990) pp. 295-303.
4. Freesmeyer W.B. and Stehle C.M., Biomechanics of temporomandibular joint movements (German), *Dtsch Zahnarztl Z* **43** (1988) pp. 199-208.
5. Pertes R.A., Attanasio R., Cinotti W.R. and Balbo M, The temporomandibular joint in function and dysfunction, *Clin Prev Dent* **10** (1988) pp. 23-29.
6. McKay G.S., Yemm R., Cadden S.W., The structure and function of the temporomandibular joint, *Br Dent J* **173** (1992) pp. 127-132.
7. Gibbs C.H., Lundeen H.C., Mahan P.E. and Fujimoto J., Chewing movements in relation to border movements at the first molar, *J Prosthet Dent* **46** (1981) pp. 308-322.
8. Gibbs C.H., Messerman T., Reswick J.B. and Derda H.J., Functional movements of the mandible *J Prosthet Dent* **26** (1971) pp. 604-620.
9. Hannam A.G., Musculoskeletal biomechanics in the human jaw. In: *Temporomandibular joint and masticatory muscle disorders*, eds. Zarb G.A., Carlsson G.E., Sessle B.J. and Mohl N.D., Copenhagen: Munksgaard (1994) pp. 101-127.
10. Proschel P., Feng H., Ohkava S., Ott R. and Hofmann M., Untersuchung zur Interpretation des Bewegungsverhaltens kondylärer Punkte, *Dtsch Zahnarztl Z* **48** (1993) pp. 323-326.
11. Proschel P. and Hofmann M., Frontal chewing patterns of the incisor point and their dependence on resistance of food and type of occlusion, *J Prosthet Dent* **59** (1988) pp. 617-624.
12. Gallo L.M., Airoldi G.B., Airoldi R.L. and Palla S., Description of mandibular finite helical axis pathways in asymptomatic subjects, *J Dent Res* **76** (1997) pp. 704-713.
13. Merlini L. and Palla S., The relationship between condylar rotation and anterior translation in healthy and clicking temporomandibular joints, *Schweiz Monatsschr Zahnmed* **98** (1988) pp. 1191-1199.
14. Salaorni C. and Palla S., Condylar rotation and anterior translation in healthy human temporomandibular joints, *Schweiz Monatsschr Zahnmed* **104** (1994) pp. 415-422.
15. Osborn J.W., The disc of the human temporomandibular joint: design, function and failure, *J Oral Rehabil* **12** (1985) pp. 279-293.
16. DeVocht J.W., Goel V.K., Zeitler D.L. and Lew D., A study of the control of disc movement within the temporomandibular joint using the finite element technique, *J Oral Maxillofac Surg* **54** (1996) pp. 1431-1437.
17. Chen C.T., Burton-Wurster N., Lust G., Bank R.A. and Tekoppele J.M., Compositional and metabolic changes in damaged cartilage are peak-stress, stress-rate, and loading-duration dependent, *J Orthop Res* **17** (1999) pp. 870-879.
18. Nuki G., Osteoarthritis: a problem of joint failure, *Z Rheumatol* **58** (1999) pp. 142-147.
19. Newberry W.N., Garcia J.J., Mackenzie C.D., Decamp C.E. and Haut R.C., Analysis of acute mechanical insult in an animal model of post-traumatic osteoarthrosis, *J Biomech Eng* **120** (1998) pp. 704-709.
20. Goodman S.B., Lee J., Smith R.L., Csongradi J.C. and Fornasier V.L., Mechanical overload of a single compartment induces early degenerative changes in the rabbit knee: a preliminary study, *J Invest Surg* **4** (1991) pp. 161-170.
21. Radin E.L., Burr D.B., Caterson B., Fyhrie D., Brown T.D. and Boyd R.D., Mechanical determinants of osteoarthrosis, *Semin Arthritis Rheum* **21** (1991) pp. 12-21.

22. Oberg T., Carlsson G.E. and Fajers C.M., The temporomandibular joint. A morphologic study on a human autopsy material, *Acta Odontol Scand* **29** (1971) pp. 349-384.
23. Jergenson M.A. and Barton J.M., The occurrence of TMJ Disc Perforations in an Aging Population, *J Dent Res* **77** (1998) SI.
24. Hylander W.L., Experimental analysis of temporomandibular joint reaction force in macaques, *Am J Phys Anthropol* **51** (1979) pp. 433-456.
25. Brehnan K., Boyd R.L., Laskin J., Gibbs C.H. and Mahan P., Direct measurement of loads at the temporomandibular joint in Macaca arctoides, *J Dent Res* **60** (1981) pp. 1820-1824.
26. Hohl T.H. and Tucek W.H., Measurement of condylar loading forces by instrumented prosthesis in the baboon, *J Maxillofac Surg* **10** (1982) pp. 1-7.
27. Boyd R.L., Gibbs C.H., Mahan P.E., Richmond A.F. and Laskin J.L., Temporomandibular joint forces measured at the condyle of Macaca arctoides, *Am J Orthod Dentofacial Orthop* **97** (1990) pp. 472-479.
28. Chen J., Akyuz U., Xu L. and Pidaparti R.M., Stress analysis of the human temporomandibular joint, *Med Eng Phys* **20** (1988) pp. 565-572.
29. Chin L.P., Aker F.D. and Zarrinnia K., The viscoelastic properties of the human temporomandibular joint disc, *J Oral Maxillofac Surg* **54** (1996) pp. 315-318.
30. Nickel J.C. and McLachlan K.R., In vitro measurement of the stress-distribution properties of the pig temporomandibular joint disc, *Arch Oral Biol* **39** (1994) pp. 439-448.
31. Tanne K., Tanaka E. and Sakuda M., The elastic modulus of the temporomandibular joint disc from adult dogs, *J Dent Res* **70** (1991) pp. 1545-1548.
32. Beek M., Koolstra J.H., van Ruijven L.J. and van Eijden T.M., Three-dimensional finite element analysis of the human temporomandibular joint disc, *J Biomech* **33** (2000) pp. 307-316.
33. Mansfield P. and Maudsley A.A., Planar spin imaging by NMR, *J Magn Resonance* **27** (1977) pp. 101-119.
34. Chen J. and Buckwalter K., Displacement analysis of the temporomandibular condyle from magnetic resonance images, *J Biomech* **26** (1993) pp. 1455-1462.
35. Price C., Connell D.G., MacKay A. and Tobias D.L., Three-dimensional reconstruction of magnetic resonance images of the temporomandibular joint by I-DEAS, *Dentomaxillofac Radiol* **21** (1992) pp. 148-153.
36. Krebs M., Gallo L.M., Airoldi R.L., Meier D., Bösiger P. and Palla S., Three-dimensional animation of the temporomandibular joint, *Technology Health Care* **2** (1994) pp. 193-207.
37. Krebs M., Gallo L.M., Airoldi R.L. and Palla S., A new method for the three-dimensional reconstruction and animation of the temporomandibular joint, *Ann Acad Med Singapore* **24** (1995) pp. 11-16.
38. Cannon D.C., Reswick J.B. and Messerman T. Instrumentation for the investigation of mandibular movements, *Eng Design Centre Rep* EDC 4-1964-8 (1964).
39. Jankelson B., Swain C.W., Crane P.F. and Radke J.C. Kinesiometric instrumentation: A new technology, *J Am Dent Assoc* **90** (1975) pp. 834-840.
40. Lewin A., Van Rensburg L.B. and Lemmer J., A method of recording the movement of a point on the jaws, *J Dent Assoc S Afr* **29** (1974) pp. 395-397.
41. Karlsson S., Recording of mandibular movements by intraorally placed light emitting diodes, *Acta Odont Scand* **35** (1977) pp. 111-117.
42. Slavicek R. Axiographie. In: *Kiefergelenk und Okklusion.* Druecke W., Klemt B. (editors). Berlin: Quintessenz, (1980) pp. 223-245.

43. Klett R. Electronic recording procedure for temporomandibular joint diagnosis (German), *Dtsch Zahnarztl Z* **37** (1982) pp. 991-998.
44. Hobo S. and Mochizuki S., A kinematic investigation of mandibular border movements by means of an electronic measuring system. Part I: Development of the measuring system, *J Prosthet Dent* **50** (1983) pp. 368-373.
45. Fischer-Brandies H., Burckhardt R. and Kluge G., Reproducibility of the results of optoelectronic axiography using the V-type stereognathograph (German), *Z Stomatol* **87** (1990) pp. 359-367.
46. Maurice K., Leberl F., Curry S. and Kober W., Real-time close-range 3-D motion measurements for dental medicine, *Proc SPIE* **1395** (1990) pp. 366-373.
47. Siegler S., Hayes R., Nicolella D. and Fielding A., A technique to investigate the three-dimensional kinesiology of the human temporomandibular joint, *J Prosthet Dent* **65** (1991) pp. 833-839.
48. Hayashi T., Ishioka K., A three-dimensional measurement system for human jaw movements using a photoelectric camera, *Front Med Biol Eng* **3** (1991) pp. 205-209.
49. Naeije M., van der Weijden J.J. and Megens C.C., OKAS-3D: optoelectronic jaw movement recording system with six degrees of freedom, *Med Biol Eng Comput* **33** (1995) pp. 683-688.
50. Guiard-Marigny T. and Ostry D.J., A system for three-dimensional visualization of human jaw motion in speech, *J Speech Lang Hear Res* **40** (1997) p. 1118-1121.
51. Mesqui F., Kaeser F. and Fischer P., On-line three-dimensional light spot tracker and its application to clinical dentistry, *Int Arch Photogrammetry Sens* **26** (1986) pp. 310-317.
52. Airoldi R.L., Gallo L.M. and Palla S., Precision of the jaw tracking system JAWS-3D, *J Orofac Pain* **8** (1994) pp. 155-164.
53. Tait D.C., Impulsive and static loading of the porcine temporomandibular joint disc, MS thesis, Winnipeg, Canada: University of Manitoba (1993).
54. Stormberg K.D., The effects of impulsive loading on the porcine temporomandibular joint disc, MS thesis, Lincoln NE: University of Nebraska Medical Center (1998).
55. Bruno M.J., Strain rate dependent anisotropic properties of the porcine temporomandibular joint disc, MS thesis, University of Nebraska Medical Center (1999).
56. Waldman S.D. and Bryant J.T., Dynamic contact stress and rolling resistance model for total knee arthroplasties, *J Biomech Eng*, **119** (1997) pp. 254-260.
57. Almeida E.S. and Spilker R.L., Mixed and penalty finite element models for the nonlinear behavior of biphasic soft tissues in finite deformation: Part I - Alternate formulations, *Comput Meth Biomech Biomed Eng*, **1** (1997) pp. 25-46.

REALIZATION OF COMPOSITES IN OFFSHORE APPLICATIONS PERFORMANCE, LIFE AND COST

OZDEN O. OCHOA

Department of Mechanical Engineering, Texas A&M University, College Station, Texas, USA

In the last two decades, the steady interest in utilizing composite structures in offshore oil exploration and production has led to the conceptual birth of various designs with limited opportunities for actual implementation in service. Nevertheless, they are at the edge of full implementation as the certification, design, and manufacturing triad is optimized for economic payoff. To date, various research entities both in Europe and US have demonstrated the technically enabling features of composites in rigid and flexible tubulars, pipelines, pressure canisters, and topside grating. The present technical challenge lies in the integration of manufacturing process and material selection that will enable an accurate long-term response prediction of these heterogeneous and anisotropic structures. The logical step forward for offshore is to form alliances to define the service conditions and their impact on life of structures in deep-water applications without compromising the competitiveness in design and manufacturing.

1 Introduction

Fiber reinforced polymeric composites are strong alternatives to metals for ultra-deep water drilling and production applications where lightweight, strength and durability are essential. The tailorability inherent in the reinforcement fibers and resin systems can readily be incorporated into components of high strength, high stiffness, low weight, high fatigue resistance, excellent insulation, high damping, and high corrosion resistance. These attributes are anticipated to directly translate to savings in capital and operational expenditures. The latest economic predictions propose that for a Gulf of Mexico TLP composite riser/composite tendon combination offered 37% reduction in total installed cost at 3000 meters.

Since the fundamental mechanical and material behavior characterization strictly depends on the development of constitutive models at multiple scales, the integration of rigorous representative tests to identify design allowables and their subsequent implementation in computational algorithms are the principle pursuits. Yet this must be coupled with cost effective material systems and manufacturing techniques throughout the production, installation and maintenance cycle. Assessing the role of processing, manufacturing and the long-term use in marine environment coupled with well fluids is of critical importance for successful application of advanced composites such as rigid and flexible(spoolable) tubulars, pipes, tanks, sandwich structures, panels, down hole tubing. In addition to structural load carrying ability, fire performance, impact resistance, wear, erosion and cavitation criteria must be built in the design phase.

The latest interest in ultra deep-water (10,000ft) operations, has expanded the studies from out fitting tension leg platforms with composite tubulars to other floating systems and subsea applications where composites may have the primary enabling role in the innovative solutions for ultra deep water. Federally assisted programs (NIST-ATP) in mid 1990s were able to bring teams together to focus on production risers, drilling risers and spoolables[1-3]. The Offshore Technology Research Center team contributed to testing and modeling efforts as a member of the NIST Drilling Riser Program where the principal industrial partners, ABB and Northrop Grumman manufactured risers & choke/kill lines. Extensive NDE characterization accompanied the tests conducted for tension fatigue, bending, collapse and burst composite risers. The numerical models for aging, and terminations were integrated with hygrothermal material characterization results. These efforts still need extensive field-testing and a better database for environmental (physical and chemical aging) characterization. Flexible risers, tethers and mooring lines, high-pressure flowlines and thick walled hybrid insulated subsea pipeline are excellent candidates for innovative design with composites. In the presentation, specific examples will be presented to reflect the relevance and competitiveness in the offshore vision shared by many innovative individuals.

Herein, the fundamental issues in relation to design and analysis will be presented from three different perspectives; service environment, testing rationale and computational simulations. The coupling of the design of these novel materials simultaneously with the design of the structures that offer unique mechanical and physical properties lead to challenging scenarios which can be studied through optimization and similitude principles. Multidisciplinary and concurrent methodology has the potential of empowering the designer, the materials and manufacturing engineer by enabling the flow of continuous and interactive feedback throughout the development; from the initial to the final stages of testing and evaluation. The ultimate expectation is to close the gap in generating database to enable a realistic design allowables and a design code.

2 Approach

The challenges to overcome reside in computational life prediction, materials system design, accelerated aging, full-scale testing, nondestructive evaluation, intelligent manufacturing, and certification and zero environmental impact. The focus of the brief examples included in the paper will enable the reader to be familiar with the development of experimental and computational foundation necessary for the structural assessment of different material systems, lay-ups and geometry when subjected to axial tension, bending, internal, and external pressure. The interaction between lamina-laminate-structure will be illustrated with research data generated at the Offshore Technology Research Center at Texas A&M University by the author on hybrid composites. The knowledge base from the lamina, laminate and rigid tube efforts are then extended to the design of flexible (spoolable) composite tubes capable of being coiled onto large spools for ease of storage, handling, and installation without inducing damage in the tube during storage [4-5].

A hybrid composite consists of multiple fibers, which may differ, in both material type and orientation embedded in a single resin system and offer greater choices for an economical design than traditional single fiber reinforced composites. Material and mechanical characterization through integrated testing and computational modeling is employed to address the major issues at laminate, lamina and constituent scales; (i) resin and fiber(s) selection, to ensure chemical compatibility and resistance to fluids, (ii) the lay-up and fiber orientation to develop a robust and economical reinforcement architecture, (iii) response to combined loads, and (iv) hygrothermal effects, and (v) residual stresses due to processing. Detailed structural response including hygrothermal loads are best simulated with finite element techniques utilizing special user elements and user material models with progressive damage criteria. At the lamina and micro scales, as needed, the damage initiation and growth must be modeled with nonlinear material and geometry considerations.

2.1 Environmental Response Characterization

Hygrothermal behavior of hybrid laminates coupled with mechanical loads and processing induced residual stresses are studied for a class of laminates which were aged in seawater at 140 F. First, the lamina properties are developed and then incorporated into the representative element volume RVE formulation to determine the response of flat coupons subjected to tensile and bending loads and tubes subjected to internal pressure. For small deformations, an incremental classical laminated plate theory formulation is utilized. The loading conditions consist of applying a force resultant to simulate tension, a moment resultant to simulate bending, and biaxial force resultants to simulate a tube subjected to internal pressure. The results showed that the correlation with simple models and experiments are quite realistic as observed in Figure 1.

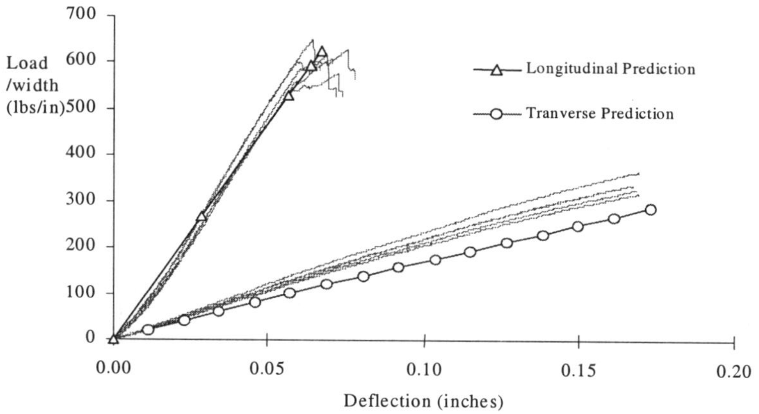

Figure1a Bending Response - $[0_2/90_2]_S$ Flat

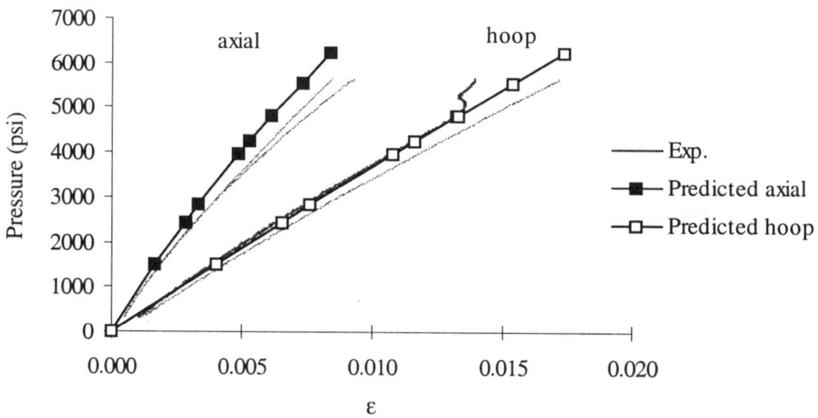

Figure 1b Pressure Response- $[90_3/\pm25/90_3]_T$ Tube

2.2 *Similitude Considerations- Testing Rationale*

The experimental data obtained from the internally pressurized tubes are used in a similitude study in an attempt to predict the burst pressure of a much larger diameter tube (e.g., a riser)with a significantly thicker wall. In this approach, the coefficients of the governing differential equation for the prototype (large tube) and the model (small tube) are compared to develop the scaling laws, which depend on the geometry, the constituent properties and the lay-up [4]. The stacking sequence for the prototype is $[\pm 90_3/\pm 15]_{NP}$ and for the model $[\pm 90_3/\pm 15]_{NM}$. In this scenario, the scaling ratios (model to prototype) of the radius, wall thickness and length are 0.131, 0.086, and 0.014 respectively. The effect of change in pressure ratio and change in the axial load ratio on the state of stress is presented clearly in Figure 2.

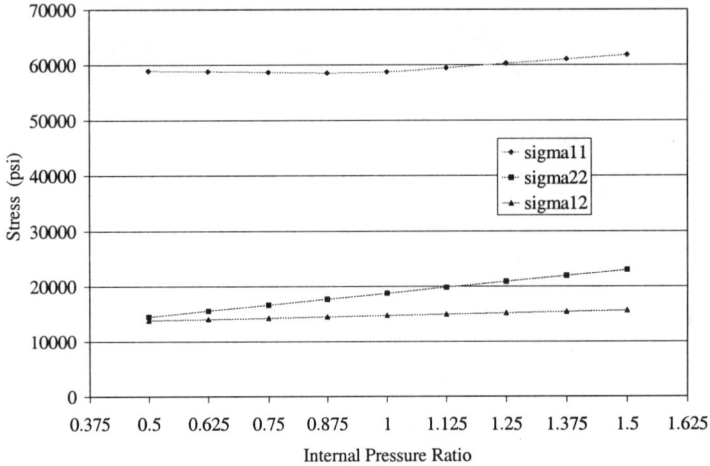

Figure 2a Stress vs Pressure Ratio

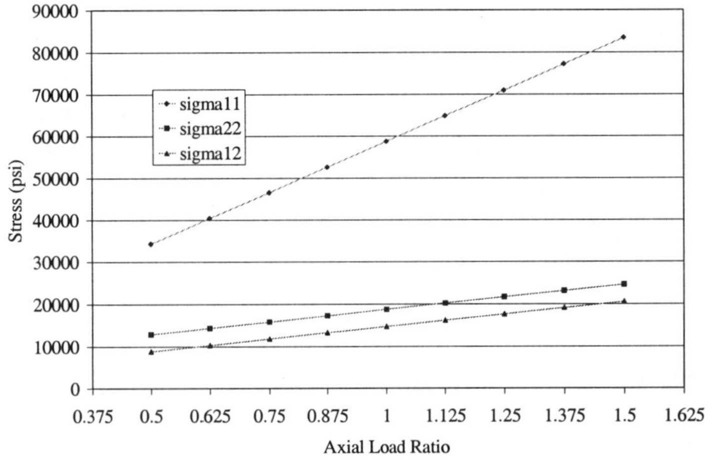

Figure 2b Stress vs Axial Load Ratio

2.3 Computational Assessment Example - Spoolable Tubular

The objective is to optimize the ratio of tube to spool diameter and thus select a more compliant wall configuration to prevent failure [6]. In general, these tubes have inner and outer liners and must sustain internal pressures of 5000-7500 psi, tensile loads up to 100 kips and bending strains up to 2%. The finite element method is utilized in simulating the laminated tubes under the four-point loading scenario. Once again, the layers are assumed to be perfectly bonded without manufacturing defects and residual stresses and the liner are neglected. The models incorporate non-linear geometry effects as well as a material subroutine to detect progressive damage accumulation, i.e., to predict damage initiation and its progression. Two-dimensional shell elements, are used to model the full tube. The outer supports in a four-point-bend configuration are simulated by constraining the translation in y and z axes of the nodes on the bottom half of the tube. The inner points where loads are introduced, is modeled with pressure distributed over a small number of elements. Angle-ply asymmetric glass, carbon and hybrid PMC tubes with various lay-up, and radius to thickness ratios are loaded until failure [7]. With the measured applied force, F, and the measured displacement, ω, the moment and curvature is calculated throughout the loading range until catastrophic failure occurs during testing. This is presented in Figure 3a for glass epoxy tubes. Typical shear stress distribution from analysis as shown in Figure 3b confirms the anticipated presence of high stress within the two internal load points. It is noted that for the same curvature, the carbon specimens require approximately 33% greater loading moment than the glass specimens for the $[\pm45]_4$ lay-up, and approximately 37% greater loading moment for the $[\pm55]_4$ lay-up.

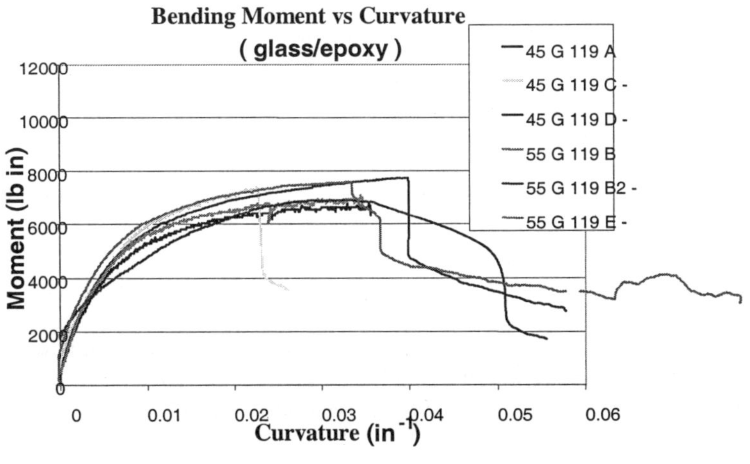

Figure 3(a) Moment vs curvature for four point flexure models

3 Conclusions

The multidisciplinary integration of optimization of design and performance in nonlinear and time dependent regimes, manufacturing, cost, environmentally acceptable material systems is complex yet essential in the low cost - high performance trade-off necessary in offshore applications. Understanding manufacturing options, simulating service loads and developing testing accordingly to support the computational models hold great promise for the confident acceptance of composite structures.

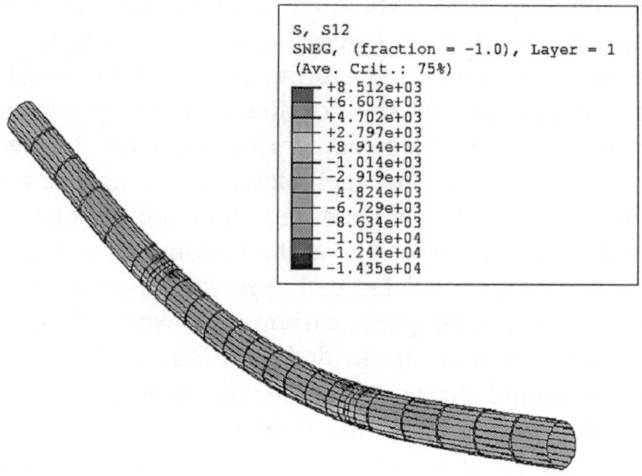

Figure 3(b) Shear stress distribution for four point flexure models

REFERENCES

1. Do, A.N, Beaudoin O., Odru, P., and Grosjean, F., "New Developments in High Strength Composite Materials for Lightweight Offshore Flexible Risers, CMOO-3, 2000.
2. Salama, M. M., Murali, J. and et al, "Design Considerations for Composite Drilling Riser", OTC 10565, Offshore Technology Conference, 1999.
3. Echtermeyer, A.T., "New Design Guidelines and Acceptance Criteria for Composites Offshore", CMOO-3, 2000.
4. Ochoa, O. O., and Ross, G. R., "Hybrid Composites: Models and Tests for Environmental Aging", *J. of Reinforced Plastics and Composites*, **17**, no. 9, 1998, pp. 787-799.
5. Chouchaoui, C. S., and Ochoa, O. O., "Similitude Study for Laminated Cylindrical Tube under Tension, Torsion, Bending, Internal & External Pressure; Part I: Governing Equations", and Part II: Scale Models", *J. of Composite Structures*, **44/4,** 1999, pp.221-229 and pp.231-236.
6. Fowler, S. Hampton, et al., "Development Update and Applications of an Advanced Composite Spoolable Tubing," OTC 8621, Offshore Technology Conference, 1998.
7. Rodriguez, D. and Ochoa, O. O., "Influence of Geometry, Lay-up, and Material on the Radius of Curvature of Spoolable Composite Tubulars", 16th ASC Annual Technical Meeting, 2001.

FRACTURE MECHANICS AND STRUCTURAL HEALTH MONITORING

FRANCIS ROSE

Aeronautical and Maritime Research Laboratory, Defence Science and Technology Organisation, 506 Lorimer Street, Fishermen's Bend VIC 3207, Australia.
E-mail: Francis.Rose@dsto.defence.gov.au

Structural health monitoring (SHM) offers the potential for a quantum gain in efficiency for asset maintenance and structural integrity management. A successful implementation of SHM will need to integrate two distinct strands that have traditionally been pursued separately, *viz.* damage detection and damage assessment. Both strands require a sound understanding of damage mechanisms, failure modes, and of the possible contrast mechanisms that can be exploited for damage detection. The present talk presents recently developed theoretical concepts and results for detecting delamination damage in fibre-composite, plate-like structures. The modelling is based on the premise that the primary effect of a delamination is to reduce locally the plate's flexural stiffness. Accordingly, for the purposes of detection, the delaminated region can be treated as an inhomogeneity, having a lower (and generally complex) flexural stiffness. The interaction of plate waves with such inhomogeneities is discussed within the framework of Mindlin plate theory. This theory provides an accurate representation of the exact dispersion relation for the lowest antisymmetric mode of the Rayleigh-Lamb spectrum up to quite high frequencies, corresponding to wavelengths of three times the plate thickness, thereby providing an adequate and practical basis for resolution of the in-plane dimensions of a delaminated region. The wave interaction problem is solved explicitly in the Born approximation, which is appropriate for detecting relatively low levels of damage (barely visible impact damage). The issue of deriving a micro-mechanical estimate of the (complex) flexural stiffness corresponding to a given geometry of delamination is addressed separately.

The influence of delamination damage on the residual strength after impact, particularly in the presence of secondary bending, is modelled by a circular inhomogeneity with a failure process-zone similar to the well-known Dugdale model. Analytical solutions are presented for the complex potentials corresponding to a dislocation interacting with a circular inhomogeneity under compression and flexure. These solutions can be used to reduce the problem of determining the residual strength to that of solving a one-dimensional singular integral equation for the dislocation density. Again, the issue of micro-mechanical modelling of the appropriate traction law for the process zone is addressed separately from the phenomenological model for predicting the residual strength. It is noted that the primary input for damage assessment (estimating the residual strength) is the local reduction in flexural stiffness due to delamination, which is provided by the proposed damage detection approach, thereby providing a plausible and consistent conceptual framework for SHM for this type of damage. This approach can be adapted to other types of laminar defects. Finally, the issues and challenges in detecting cracks in metallic plates, due for example to acoustic fatigue, are briefly discussed.

MECHANICS OF HUMAN JAW FUNCTION USING SCREW THEORY

J.A. GAL
School of Engineering & Industrial Design, University of Western Sydney, Australia
Email: ja.gal@uws.edu.au

L.M. GALLO, S. PALLA
Clinic for Masticatory Disorders & Complete Dentures, University of Zurich, Switzerland

G. MURRAY, I.J. KLINEBERG
Dental Clinical School, Westmead Hospital, Faculty of Dentistry, University of Sydney, Australia

C.W. JOHNSON
School of Computing Sciences, University of Technology, Sydney, Australia

In this paper the characterisation of jaw-mechanics by means of the finite screw and the wrench, is presented demonstrating its ability to provide a simplified, yet complete description of jaw function. With motion data and anatomical data from a single subject, the screw theory approach is implemented where the forces of the masticatory muscles are uniquely described at each instant by the wrench axis and its parameters. Incorporating the wrench axis description for muscle forces and moments during the motion with the finite screw representation of the motion, an integrated computer model of jaw mechanics is formulated. The results of an experimental program, involving Magnetic Resonance (MR) imaging for obtaining accurate anatomical representations of the masticatory muscles and the bony structures, as well as jaw-motion data collected via an optoelectronic recording instrument, are presented. The results show that the experimentally derived, screw-based model conforms to the clinically observed behaviour of jaw function and can provide a quantitatively accurate means of analyzing mandibular biomechanics.

1 Introduction

The complex spatial motion of the human mandible during common masticatory functions, such as chewing, is a result of the structure of the temporomandibular joint (TMJ) that permits movement in all six degrees-of-freedom. Previous research has shown that the kinematic concept of the screw axis, uniquely describes this movement at each instant [1, 2] where the motion is a combination of a rotation about and a translation along this axis. Other representations that identify a planar rotation axis [3] or the track of a single point in the mandible [4] do not contain all the features of the motion. Although the screw axis has been used for describing motion in biomechanical joints [5] it has not been used for jaw-motion by other dental researchers.

In considering the mechanics of the jaw the muscles that actuate the motion must also be considered. A number of mathematical models, based on anatomical data obtained from cadavers and animals, have been reported in the literature for estimating the magnitudes and lines-of-action of the muscle forces in humans [6, 7]. In this work a computer model based on real anatomical data from MRI (Magnetic Resonance Imaging) scans of muscle and of the bone structures has been used to putatively determine lines-of-action of forces by visually locating the origin and insertion points of each closing muscle group.

By combining jaw-motion recordings with the muscle forces derived from the MRI data, an integrated model of jaw mechanics is formulated where the changing magnitude, orientation and location of the resultant force throughout the movement is determined. The model is a compact and unique representation of the rigid-body characteristics of the jaw and implements the concept of finite screws, associated with the motion, and wrenches, associated with the forces acting on the jaw [8].

2 Materials and Methods

In order to demonstrate how the finite screw and the wrench representations may be implemented in practice, a single subject, identified as being asymptomatic, was used for recording data on open-close movement as well as MRI scans. The motion recordings were obtained via an optoelectronic instrument, Jaws-3D, consisting two lightweight triangular frames with an LED at each vertex and where one frame is cemented to the upper dental arch while the other frame fixed to the lower arch (see Fig. 1). Relative movement between the skull and the mandible is thus tracked at a sampling rate of 70Hz by three CCD (charge-coupled device) arrays providing position information in three-dimensional space for each sampled time interval.

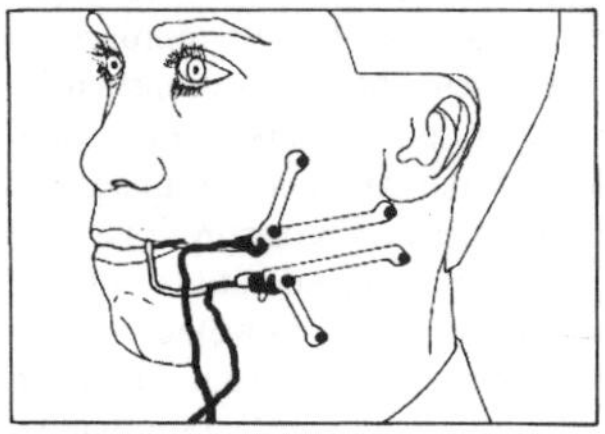

Figure 1. LED frames for motion recording

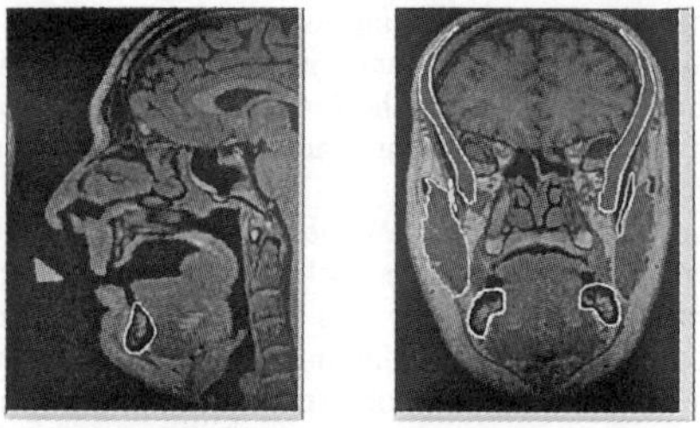

Figure 2. Sagittal and frontal MR images with reference block and segmented anatomical features

Using the same subject, a full MRI scan of the head was performed (Siemens Magnetom Vision, 1.5 Tesla) with a magnetically opaque prism-block held by clenched teeth providing the means by which a common co-ordinate system was established between the MRI and the Jaws-3D systems. A manual digitizing procedure was implemented to segment out the muscle and mandible cross-sections from which a three-dimensional computer model was constructed. A sagittal and frontal image including the reference block is shown in Figure 2. The model was used to estimate the origin and insertion points for four groups of muscles that contribute to the closing phase of the open-close jaw movement.

Based on the reconstructed computer model, shown in Figure 3, sixteen force lines-of-action, eight on each side, were putatively identified with an equal, arbitrary force magnitude of 20N assigned to each vector (see Figure 4). By incorporating the recorded mandible positions for a complete open-close cycle and tracking the changing locations of the origin and insertion points of each muscle-force line-of-action, the force vector positions and orientations were tracked throughout the closing phase of the cycle. From the position recordings finite screws between consecutive mandible positions were determined while the resultant force and moment at the corresponding instant during the movement were also calculated. The parameters of the finite screw and the resultant wrench were determined and used to verify that these two screw quantities alone completely define the mechanics of jaw-function in a simplified manner and thus assist in the longer term in the understanding of dysfunctional conditions.

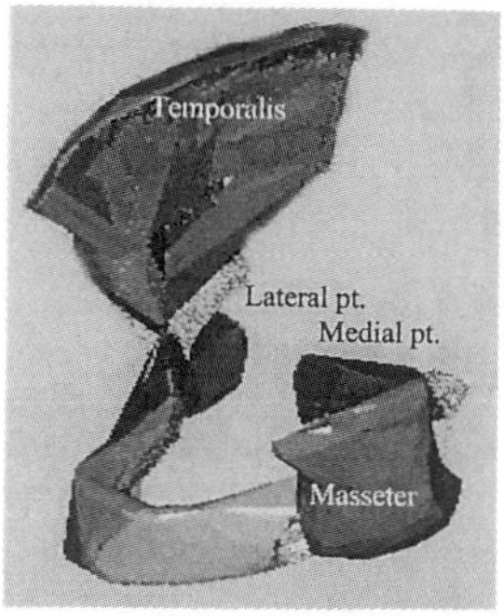

Figure 3. Partially reconstructed computer model of mandible with masseter, temporalis, lateral and medial pterygoid muscles.

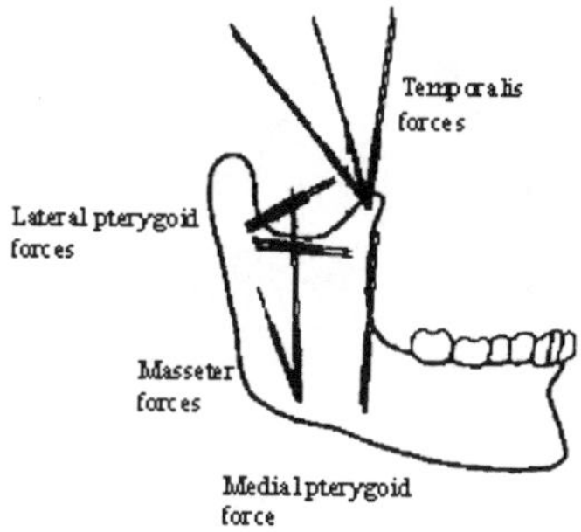

Figure 4. Lines-of-action for eight forces on each side at stationary start position during closing phase

3 Theory of Screws

3.1 Finite screw axis

A movement between two finitely displaced positions of any rigid body can be represented by the unique axis about which it rotates and along which it translates to reach the second location. This axis is the finite screw axis (FSA) and can be determined from monitoring the vector displacement of three points in the body, $\boldsymbol{p}_1$, $\boldsymbol{p}_2$, $\boldsymbol{p}_3$, to a new location $\boldsymbol{q}_1$, $\boldsymbol{q}_2$, $\boldsymbol{q}_3$. The required parameters of the FSA are the rotation matrix, R, and the translation vector, $\boldsymbol{v}$, where the movement can then be expressed as

$$\boldsymbol{q}_i = R\,\boldsymbol{p}_i + \boldsymbol{v} \quad \text{for i} = 1, 2, 3. \tag{1}$$

and calculating R and $\boldsymbol{v}$ depends on an optimisation procedure [2] based on the measured quantities $\boldsymbol{p}_i$ and $\boldsymbol{q}_i$ and derived from the function, F:

$$\mathrm{F}(\boldsymbol{v}, R) = \frac{1}{3}\sum_{i=1}^{3}(R\boldsymbol{q}_i + \boldsymbol{v} - \boldsymbol{p}_i)^T(R\boldsymbol{q}_i + \boldsymbol{v} - \boldsymbol{p}_i) \tag{2}$$

The other parameters of the FSA including its location and direction cosines, and the magnitudes of the angular and linear displacements of the body as it twists about and translates along the FSA can also be determined, but this is not detailed here. An approach based on this technique is adopted in this work to determine the FSA parameters for every sampled jaw position during three open-close cycles and to trace the associated FSA pathways.

3.2 Wrench axis

When a system of forces is acting on a rigid body such as the mandible there is a unique resultant where the vectors of the resultant force and couple are collinear. This resultant lies on an axis that is known as the wrench axis. The location and associated parameters of the wrench axis completely encapsulates the kinetic effect of the individual muscle forces acting together to move the jaw.

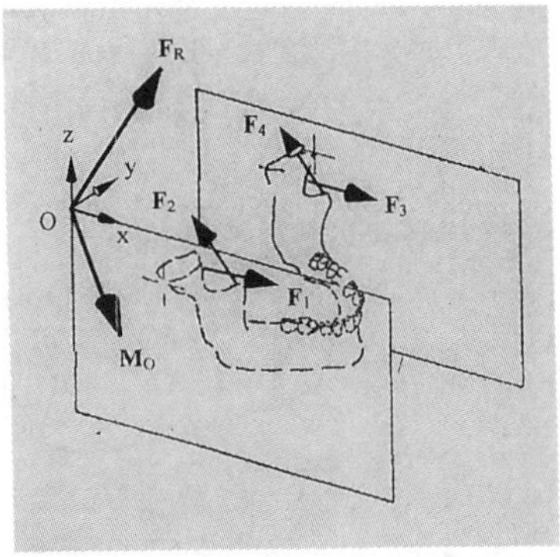

Figure 5. System of four forces acting on the mandible and the resultant force $\mathbf{F}_R$ and moment $\mathbf{M}_O$

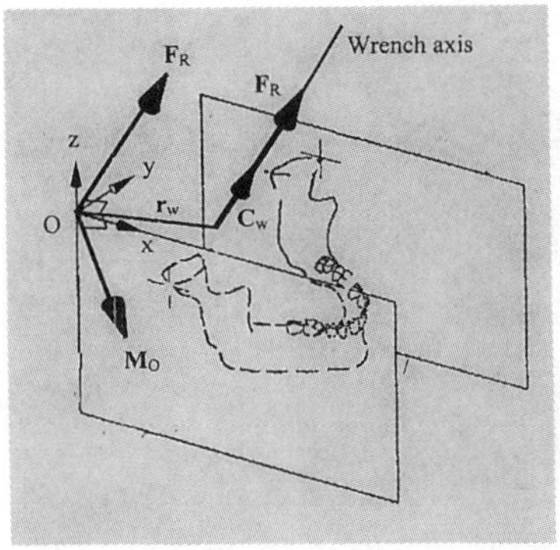

Figure 6. The wrench location with $\mathbf{r}_w$ the radius vector from O and $\mathbf{C}_w$ the couple at the wrench axis

The four forces $\mathbf{F}_i$ for i=1-4, in Figure 5, for example, representing four muscle forces, can be combined into a resultant force $\mathbf{F}_R$ and a resultant moment $\mathbf{M}_O$ at the origin to replace all the four forces. By moving $\mathbf{F}_R$ to a new location determined by the equations below, a unique combination of resultant force and couple is obtained where the corresponding vector directions are collinear resulting in a wrench with minimum couple [8] (see Figure 6).

Referring to Figure 6, the vector displacement, $\mathbf{r}_W$, from the origin, O, to the wrench axis location is given by the unit vector equation;

$$\hat{\mathbf{r}}_W = \hat{\mathbf{F}}_R \times \hat{\mathbf{M}}_O \tag{3}$$

where the magnitude $r_W = M_O \sin\gamma / F_R$ and the angle between $\mathbf{M}_O$ and $\mathbf{F}_R$, is γ, obtained from:

$$\cos\gamma = \hat{\mathbf{F}}_R \bullet \hat{\mathbf{M}}_O \tag{4}$$

By moving the resultant force to the new location the moment $\mathbf{M}_O$ changes to the couple $\mathbf{C}_W$ as shown in Figure 6 and its magnitude is given by:

$$C_W = \mathbf{M}_O \bullet \hat{\mathbf{F}}_R \tag{5}$$

The calculations in equations (3) to (5), formulated from the fact that at the wrench axis both the resultant force, $\mathbf{F}_R$, and the resultant moment, $\mathbf{C}_W$, are collinear, provide all the required information about the wrench axis. The ratio of the magnitudes of $\mathbf{C}_W$ and $\mathbf{F}_R$ is usually referred to as the pitch of the wrench axis and hence it can be treated as a 'screw ' in the same manner as the FSA. When the wrench axis representation of all the

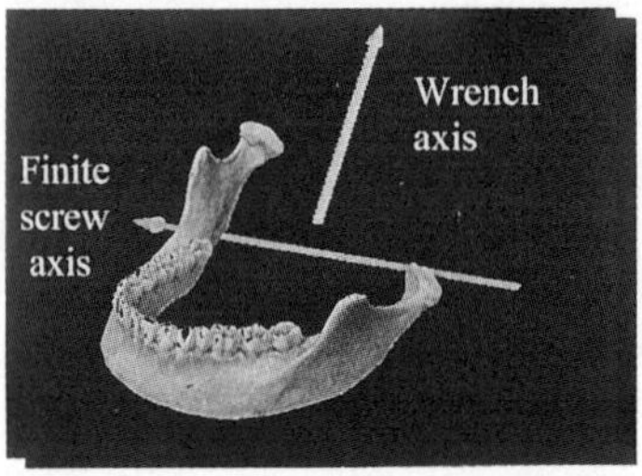

Figure 7. Mechanics of jaw-motion represented by the FSA and the wrench axis

forces is combined with the FSA representation of motion, a complete description of the mechanics of the mandible is obtained as illustrated in Figure 7 and there is sufficient information in these two screws to determine such mechanical property as the work done by the muscle forces.

4 Experimental results and discussion

Using the motion data and the segmented anatomical data from one human subject as described in section 2, three closing movements during open-close cycles were used to verify the FSA and wrench axis (WA) analysis. A computer simulation has been devised for tracing the FSA and WA pathways during the motion.

In Figure 8, a typical single closing cycle is shown with the FSAs plotted at the sampled time-intervals of 15ms. The WAs representing the resultant of all 16 putatively obtained muscles forces (see Figure 4) are also plotted during the motion showing little variation away from the mid-point between the two sagittal planes throughout the closing phase. Since the resultant wrench actuates the movement, the location of the WAs suggests that the closing motion is caused by the wrench generating a closing torque about the corresponding FSA. This appears to conform to the expected behaviour of a closing action.

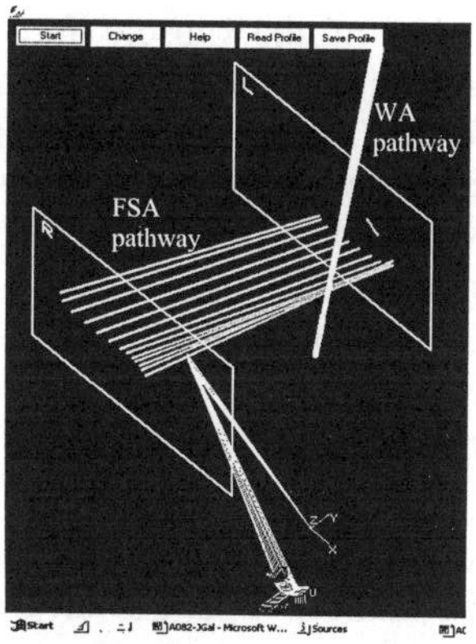

Figure 8. FSA and WA pathways for a closing cycle

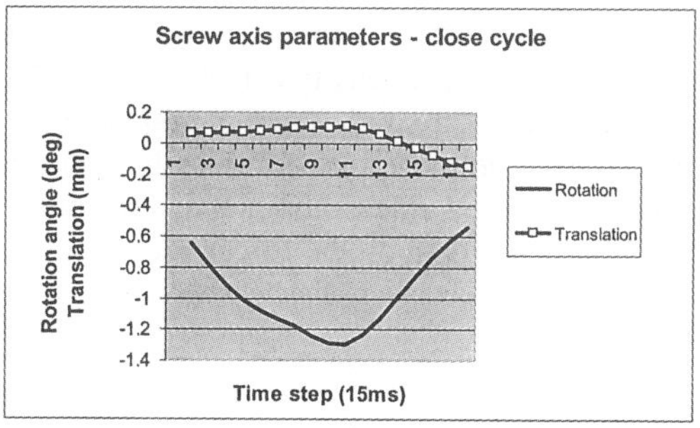

Figure 9. Parameters of the FSA for a typical closing movement; rotation and translation

The parameters of the FSA in a typical closing movement, Figure 9, show for each time step the small translation component along the FSA (±0.02mm maximum per step) with a relatively larger rotation about the FSA (±0.1deg per step) as expected.

Given that movement of the muscle insertion points during an open-close action is mainly in the planes parallel to the condyle planes as in Figure 8, the position of the wrench axis and the wrench force do not change very much during the movement. The wrench couple however does vary significantly with the jaw closing, reflecting the increase in effective torque due to the muscle forces. For the average of three closing cycles, the WA parameters of wrench force and wrench couple are shown in figure 10 while the orientation of the wrench axis in terms of angle from the vertical is shown in Figure11. The wrench force stays in the range 252N to 268N while the wrench couple varies from 34Ncm to 199Ncm over the three, recorded cycles.

The combination of the wrench and FSA parameters also allow the calculation of work or energy used in performing the closing movement of the jaw. This is not described in this paper.

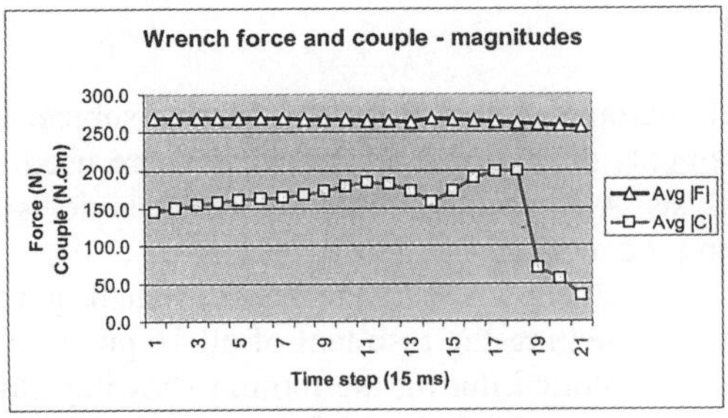

Figure 10 Wrench parameters averaged over three closing cycles (wrench force and couple)

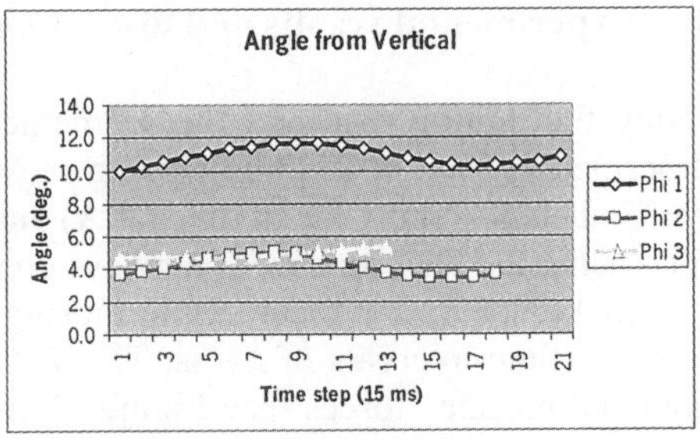

Figure 11 Orientation of the wrench axis from the vertical for three closing cycles

5 Conclusion

The application of screw theory in the form of screw axis and wrench axis for characterising the movement of the mandible and the action of the muscle forces, respectively, provide a unique and yet compact method of modeling the mechanics of jaw function. The results presented in this paper, based on recorded motion data and MRI anatomical data for a single human subject show that the screw theory approach to jaw-mechanics fully describes the complex behaviour of jaw movement and is the first step in modelling the forces due to the masticatory muscles. Further work will explore methods of obtaining better data on the forces present in the TMJ and contact areas.

References

1. JA Gal, LM Gallo, G Murray, SA Shahrestani, CW Johnson IJ Klineberg and S Palla. Moment arm parameters related to the Mandibular Finite Helical Axis, J Dent Res, Vol 79, Special Issue (Abs), 2000, p 80.
2. Gallo L.M., Airoldi G.B., Airoldi R.L. and Palla S., Description of mandibular finite helical axis pathways in asymptomatic subjects, J Dent Res, 1997, Vol. 76, No 2, pp.704-713.
3. S.G, Rugg, R.J. Gregor, B.R. Mandelbaum and L. Chiu. In Vivo Moment Arm Calculations at the Ankle Using Magnetic Resonance Imaging (MRI), J Biomech, 1990, Vol. 23, No. 5, pp. 495-501.
4. E. Kazazoglu, M. R. Heath, A. M. Ferman and G. R. Davis: Recording mandibular movement: technical and clinical limitations of the sirognathograph. J Orofac Pain, 1994, Vol. 8, pp. 165-177.
5. S.K. Boyd, J.L. Ronsky. Instantaneous moment arm determination of the cat knee, j Biomech, 1998, 31, pp. 279-283.
6. Koolstra J.H. and van Eijden T.M.G.J., The jaw open-close movements predicted by biomechanical modelling, J Biomech, 1997, Vol. 30, No 9, pp. 943-950.
7. Van Ruijven I.J. and Weijs W.A., A new model for calculating muscle forces from electromyograms, Euro J Appl Physiol (1990) 61:479-485.
8. Phillips J., Freedom in machinery: introducing screw theory, Vol. 1, 1984, Cambridge University Press.

BIOMECHANICAL AND VALIDATION CONSIDERATIONS FOR AN ADVANCED COCHLEAR IMPLANT ELECTRODE DESIGN

C. G. TREABA[a,b] AND L. ZHANG[b]

[a]*School of Aerospace, Mechanical & Mechatronic Engineering,*
The University of Sydney, NSW 2006, Australia
[b]*Cochlear Limited and the CRC for Cochlear Implant and Hearing Aid Innovation,*
14 Mars Road, Lane Cove NSW 2066, Australia

A spiral position of a cochlear implant electrode, closer to the nerve cells in the center of the cochlea, should result in reduced threshold currents and increased dynamic range of auditory stimulation. Patient psychophysics studies using various electrode configurations have confirmed these hypotheses. An intracochlear electrode array has been developed for safe yet simple implantation, resulting in peri-modiolar placement, close to the auditory nerve cells. Key components of this development have been the study of the biomechanics of cochlear electrode insertion, concept design and evaluation of alternatives, and validation of the preferred design choice. The new electrode array is made pre-curved, conforming to the inner curvature of the average cochlea after insertion. This design concept is based on a stylet introducer, which holds it almost straight and imparts the assembly appropriate properties for insertion. This paper explains the biomechanics of the insertion of the new electrode array, with emphasis on comparisons with other existing or considered cochlear electrode designs. Biological safety aspects of the validation of this electrode, leading to its regulatory approval and product launch, are also presented.

1 Introduction

Cochlear implant systems are used to aid patients having a hearing deficiency. The normal function of the cochlea is to convert mechanical vibration from the middle ear into electrical potentials, which stimulate the auditory nerve located in the center of the spiral (modiolus). A cochlear implant bypasses this natural function, receiving sound through a sound processor and transforming it into electrical pulses, which are transferred through an electrode array to stimulate the auditory nerve cells in the modiolus. An intra-cochlear electrode array ('electrode') consists of an array of stimulator sites (e.g., plates or rings) connected to insulated wires, and moulded in an insulator matrix (e.g., silicone rubber).

This paper summarizes the biomechanical and validation considerations for a new cochlear implant electrode design, implemented in the Contour Electrode produced by Cochlear Limited of Sydney, Australia. The main biomechanical considerations in electrode design are biocompatibility and biomechanical safety of implantation of the electrode array in the human cochlea. Other validation requirements include strength of welds and mechanical integrity of the electrode array. Finally, we present a summary of the feedback received since the Nucleus® 24 Contour™ (trademarks of Cochlear Limited) implant has been released.

2 Materials and Methods

The first requirement for an electrode array designed for chronic implantation inside the human cochlea is that it is biocompatible. That is, the implant electrode must not be rejected by the human body and shall not generate a severe body reaction during its lifetime function (e.g., unacceptable levels of tissue growth), as judged by expert histologists.

Materials proven biocompatible for implantation in the cochlea (inner ear) include platinum (Pt), platinum-iridium alloys (Pt-Ir), Parylene™ (trademark of Paratech, Inc.) and some medical grade silicone rubbers.

The second requirement for an intra-cochlear electrode array is that it can be inserted with minimal trauma, by trained surgeons. The safety of insertion is closely related to the mechanics of surgical insertion and location inside the cochlea resulting from desired functionality. Several studies have shown that placing the stimulator electrodes close to the auditory nerve reduces threshold stimulation amplitude and improves stimulus localisation [1, 2]. In the ideal implementation, the stimulator array takes a curved (spiral) shape, lying next to the modiolus after surgical implantation. By contrast, in order for the surgeon to be able to insert the electrode array in the scala tympani through a path drilled in the temporal bone, the electrode array needs to be, or be held, straight or only slightly bent. Design options developed to resolve these two conflicting requirements, are based on electrode arrays manufactured straight, then positioned (forced) to curve around the modiolus with the means of additional implements, and electrode arrays made curved, then straightened for insertion and released to a relaxed curve inside the scala tympani.

3 Electrode Design Platforms and the Contour Electrode

The first design option to locate an array of electrodes next to the cochlear modiolus is an electrode array made straight or slightly bent, then curved inside the cochlea using positioners. This option is more straightforward and generally much more efficient (cheaper) to manufacture than curved electrodes; however, this advantage usually comes at the expense of both short and long term trauma to the cochlear structures (Fig. 1).

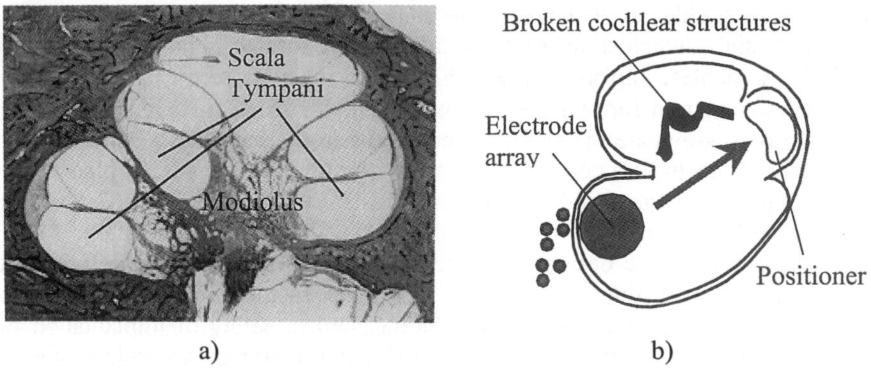

Figure 1. a) Cross-section through an adult human cochlea (courtesy of the CRC for Cochlear Implant and Hearing Aid Innovation, Melbourne); b) Potential damage with positioners

Acute and chronic (bone erosion) damage following insertion of electrodes with positioners [3, 4] can lead to increased tissue reaction, potentially resulting in loss of auditory nerve cells and poor recipient performance. A particular category of positioners consists of space-filling designs, which together with the electrode array occupy most (or all) of the available scala tympani volume [5]. Additional concerns with this design are, tissue reaction in the neighbouring chambers of the scala tympani, and significantly diminished possibilities for additional residual (natural) hearing.

The second design option for a peri-modiolar electrode array is an electrode made with a pre-determined curvature, which is held straight or slightly bent for insertion through the bone cavities, then released to its relaxed shape inside the scala tympani. Such an electrode is more difficult (expensive) to manufacture and requires specialised implements developed for its insertion. However, it is easy to insert and is the least invasive possible [4], with the resulting benefits in preservation of the cochlear structures (auditory nerve) and residual hearing of the recipient.

Several methods have been developed for temporarily holding the electrode straight for insertion, mostly consisting of rigid tubes of insertion tools [6], or flexible sheaths or dissolvable coatings [7]. An alternative design, which uses an internal component (stylet) to hold a pre-curved electrode array slightly bent for insertion, has been developed in a workable solution in Cochlear's Contour Electrode.

The Contour Electrode consists of an array of stimulator Pt electrodes, connected to the implant receiver-stimulator by individual Pt-Ir (90:10) alloy wires, insulated in Parylene™ (Fig. 2a). The electrode is made pre-curved with a lumen, with semi-circular Pt electrodes exposed to the inside of the curve. For insertion, the electrode is assembled with a malleable Pt stylet with graded annealing, which holds it slightly bent and is removed by the surgeon after insertion. Thus the structure inserted remains flexible, whilst it is the least invasive possible [8] (Fig. 2b).

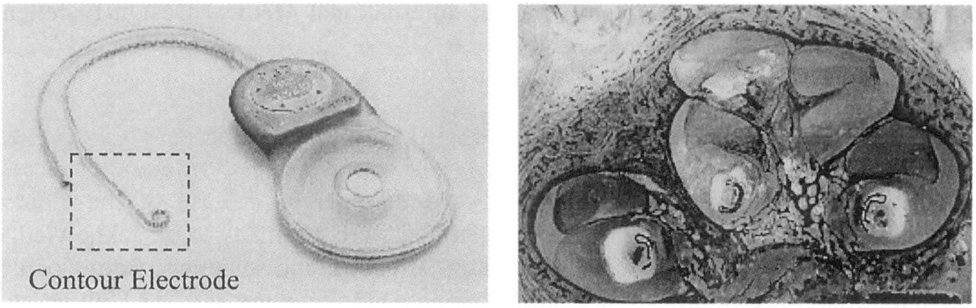

a) b)

Figure 2. a) Nucleus® 24 Contour™ implant (courtesy of Cochlear Limited, Sydney); b) Contour Electrode inserted safely – histological section (courtesy of the CRC for Cochlear Implant and Hearing Aid Innovation, Melbourne)

The biocompatibility requirement is satisfied in the Contour Electrode by the choice of biocompatible materials, all thoroughly proven by the established Nucleus™ technology [9, 10]: Pt, Pt-Ir, Parylene™, and four grades of silicone rubbers ('A-B mix', LSR-30™, adhesive RTV™, and elastomer HCRP-50™, all trademarks of Applied Silicone, Inc.).

The biomechanical safety requirement is satisfied by the Contour Electrode's design of insertion mechanism and overall structure. Once the electrode is inserted, the stylet is removed allowing the electrode to relax to a position close to the modiolus. The electrode is designed to return to a pre-determined, natural shape of the mean cochlear modiolus [11, 12], to ensure that there is no pressure on the cochlear structures. Simulated insertion tests in more than 100 temporal bones worldwide have proven that the Contour Electrode is safe to insert, by comparison with the proven Nucleus 24™ straight electrode array [13, 14].

In addition to these biomechanical requirements, to ensure maximum reliability of the welds, a simplified, robust welding method has been developed for the Contour Electrode. This involves only one contact resistance weld. The welding parameters were optimized to ensure intimate bonding between the Pt and Pt-Ir alloy materials, while maintaining low currents for maximum strength and fatigue resistance of the 25 micron Pt-Ir wires. The Contour Electrode assembly and connecting leads have passed extensive tests for resistance to stresses and fatigue as part of their validation testing prior to market release [15].

4 Discussion

The Contour Electrode array of Cochlear Limited has been developed to satisfy the essential requirements of biocompatibility, biomechanical safety of insertion, functionality, and reliability in use. The Contour Electrode is the result of evaluating a large number of different designs, and has a demonstrated record of safe insertion. By contrast, the wedging action of positioners, in particular the space-filling aspect of some designs, risk severe trauma to the cochlear structures, potentially leading to neural loss and deteriorating patient performance, also disabling any residual hearing. Clinical trials to date have confirmed that the functionality (i.e., location) objectives of the Contour Electrode design have been met [16].

The Contour Electrode implant was launched successfully by Cochlear Limited in early 2000, and has received the Australian Design of the Year 2000 Award in Australia and an international Silver Award in the 2001 International Medical Design Excellence Awards competition in the USA.

5 Acknowledgements

Contributors to the successful development and commercial release of the Contour Electrode by Cochlear Limited of Sydney, Australia, were too many to be acknowledged separately here. They are engineers, technicians, surgeons, and scientists, from Cochlear Limited, the CRC for Cochlear Implant and Hearing Aid Innovation and The University of Sydney, Australia, and Cochlear and cochlear implant clinics worldwide. We thank all of them for their contribution to the success of the Nucleus® 24 Contour™ implant.

References

1. Finley C. C., Wilson B. S., White M. W., Models of neural responsiveness to electrical stimulation. In J. M. MILLER and F. A. SPELMAN (eds.) *Cochlear Implants: models of the electrically stimulated ear.* ed by Miler J. M. and Spelman F. A. (Springer-Verlag, New York, 1990) pp.135-60.
2. Shepherd R. K., Hatsushika S., Clark G. M., Electrical stimulation of the auditory nerve: the effect of electrode position on neural excitation. *Hear Res* **66** (1993) pp.108-20.
3. Richter B., Aschendorff A., Lohnstein P., Husstedt H., Nagursky H., Laszig R., Alteration of fine cochlear structures after implantation of a new perimodiolar electrode array prototype. *European Academy of Oto-Rhino-Laryngology* **255** (1998) pp.92.
4. Tykocinski M., Cohen L. T., Pyman B. C., Roland T. J., Treaba C. G., Palamara J., Dahm M. C., Shepherd R. K., Xu J., Cowan R. S., Cohen N. L., Clark G. M., Comparison of electrode position in the human cochlea using various perimodiolar electrode arrays. *Am J Otol* **21** (2000), pp.205-11.

5. Seldon H. L., Dahm M. C., Clark G. M., Crowe S., Silastic with polyacrylic acid filler: swelling properties, biocompatibility and potential use in cochlear implants. *Biomaterials* **15** (1994) pp.1161-69.
6. Treaba C. G., Xu S.-A., Xu J., Clark G. M., Pre-curved electrode array and insertion tool. *Suppl Ann Otol Rhino Laryng* **166** (1995) pp.438-41.
7. Parker J. L., Treaba C. G., Use of bioresorbable polymers in cochlear implants and other implantable devices, U.S. Pat. no. 5,653,742 (1997).
8. Parker J. L., Treaba C. G., Cohen L. T., Tykocinski M., Saunders E., Gibson P., Darley I., Dadd F. B., Garg R., and Zhang L. C., Nucleus® 24 Contour™ Electrode with Stylet. Presented at the 6th International Cochlear Implant Conference, Abstract published in *Conference Proceedings, CI2000 6th International Cochlear Implant Conference* (University of Miami, 2000) pp.205.
9. Clark G. M., Shepherd R. K., Patrick J. F., Black R. C., Tong Y. C., Design and fabrication of the banded electrode array. *Ann New York Acad Sci* **405** (1983) pp.191-201.
10. Black R. C., Hannaker P., Dissolution of smooth platinum electrodes in biological fluids. *Appl Neurophys* **42** (1979) pp.366-74.
11. Hatsushika S., Shepherd R. K., Tong Y. C., Clark G. M., Funasaka S., Dimensions of the scala tympani in the human and cat with reference to cochlear implants. *Ann Otol Rhinol Laryngol* **99** (1990) pp.871-76.
12. Ariyasu L., Galey F., Hilsinger R., Byl F. M., Computer-generated three-dimensional reconstruction of the cochlea. *Otolaryngol Head Neck Surg* **100** (1989) pp.87-91.
13. Shepherd R. K., Clark G. M., Pyman B. C., Webb R. L., Banded intracochlear electrode array: evaluation of insertion trauma in human temporal bones. *Ann Otol Rhinol Laryngol* **94** (1985) pp.55-9.
14. Richter B. Aschendorff A., Lohnstein P., Husstedt H., Nagursky H., Laszig R., The Nucleus Contour Electrode array: a radiological and histological study. *Laryngoscope* **111** (2001), pp.508-14.
15. Cochlear Limited, Internal Validation Reports.
16. Saunders E., Cohen L., Aschendorff A., Shapiro W., Walzman S., Stecker M., Richter B., Tykocinski M., Roland T., Marangos N., Laszig R., Cowan R., Psychophysical measures and NRT thresholds as a function of electrode position in the cochlea: results of a multi-centre study using the Contour Electrode array. Presented at the 5th European Symposium on Paediatric Cochlear Implantation, Abstract published in *5th European Symposium on Paediatric Cochlear Implantation* (University of Antwerp, 2000) pp.70.

DYNAMICS OF CRICKET AND SQUASH BALL IMPACTS WITH SHEEP'S EYE

K.M. COOK, L.L. KOSS AND D. GOMEZ

Department of Mechanical Engineering, MonashUniversity, Clayton, Victoria, Australia 3800

This study examined static compression properties of sheep eyes and dynamic forces on the eye due to impact of squash and cricket balls with the eye. Dynamic forces were measured by using an accelerometer attached to the non- impact side of an impacting ball and dropping the ball onto the eye from two-drop heights. Bio- dynamicists may wish to employ this data. in their models for eye injury.

1 Introduction

Previous investigators [1,2,3,4,5,6,7] reported on eye injuries that occurred during playing of cricket and squash; data regarding eye static properties and dynamic forces were not available for assistance in the analysis of such injuries. Thus, an investigation was undertaken to determine mechanical properties of eyes. In vitro human eyes are not available for study or that from cadavers being easily available for testing but the medical faculty advised that sheep eyes are similar to human eyes and are an alternative for study. The main caveats being that the sheep eyes should be less than one day old and should not be frozen. The investigation determined compressive stiffness of sheep eyes and impact forces on the eye due to collisions between squash and cricket balls and sheep eyes.

2 Description of Tests

A specially designed rig was employed to carry out the static compression stiffness tests on three sheep eyes. Adding weights, 50 gm increments, to a 15 mm diameter steel indenter, manually increased the force acting on an eye and simultaneously the deflection of the indenter was measured using a displacement gage. The slope of the force deflection curve gives the static compressive deflection stiffness.

A two- piece men's cricket ball and a yellow dot (extra slow) squash ball were employed for the ball impact tests. The cricket ball had a mass of 160 gm and a diameter of 70.05 mm and the mass of the squash ball was 29 gm and had a diameter of 40.3 mm. The sheep's head was held in place using the same rig as for the static stiffness tests and the balls were dropped from a height of 350 mm or 1000 mm onto a sheep's eye. The fall of was not controlled, as a guide would interfere with the free fall. The accelerometer that was attached to the non-impact side of a ball was unguided. A two- channel spectrum analyzer, AND 3525, measured the voltage output from the accelerometer attached to the back of the balls and displayed the time history of the voltage. The force amplitude was calculated by using the accelerometer sensitivity, 10 mV/g, and the mass of a ball. The impulse delivered to the eye is calculated by measuring the area under the force- time history. The balls sometimes impacted the orbit of the eye prior to impacting the eye itself; the eye protrudes from the orbit and thus the ball comes in contact with the eye although the orbit is impacted initially. Data from this type of impact was also recorded in Table 2, results from the ball -eye impact tests.

3 Experimental Results

The compression tests on three different eyes gave similar data for deflection under load i.e., a load of 2 kg gave a deflection of 15 mm; a plot of deflection vs. stiffness for one eye is given in Figure 1. The stiffness and average stiffness of all three eyes are given in Table 1.

Table 1. Compressive stiffness of sheep eyes

Eye number	Least squares fitted stiffness N/mm
1	1.23
2	1.26
3	1.47
Average	1.32

Cricket ball stiffness was measured at 192 N/mm [8] and is considerably higher than that of the sheep eye and during impact of a ball the eye would give the greater deflection.

The rig used to measure impact forces between ball and eye is shown in Figure 2. Two ball drop heights were employed giving impact velocities of 2.62 and 4.42 ms^{-1}. These velocities would be much slower than that of 150 kph occurring in ball play. The impact force was determined by measuring deceleration of the ball using an accelerometer on the non-impact side of the ball and then multiplying this deceleration by the ball mass to give force. A typical force time history is shown in Figure 3.

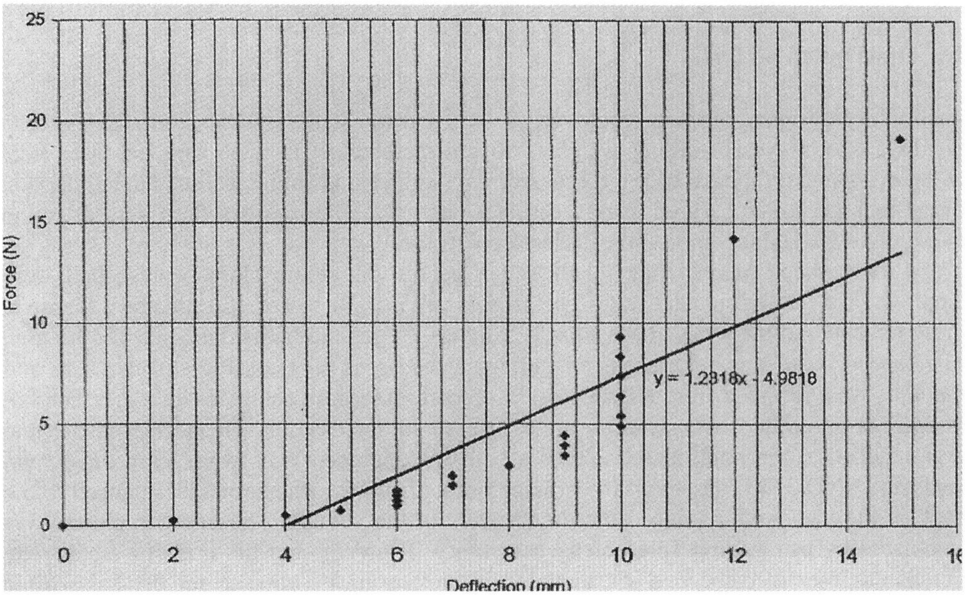

Figure 1. Static force- deflection curve for an eye.

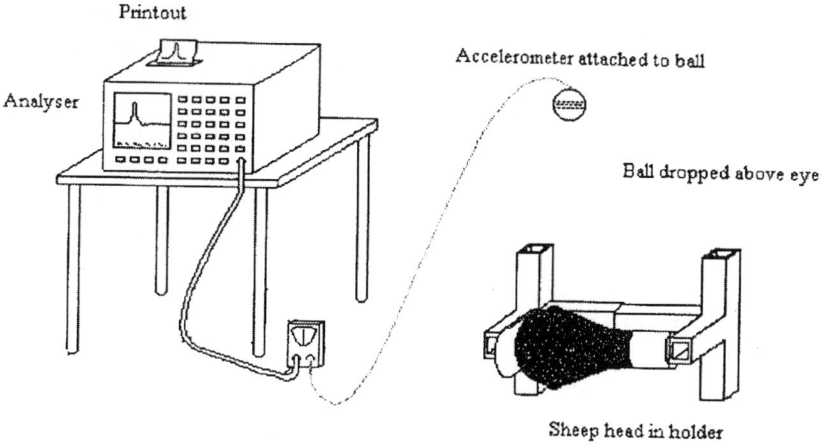

Figure 2. Schematic diagram of rig used to measure impact force on an eye. Accelerometer is located on non-impact side of ball.

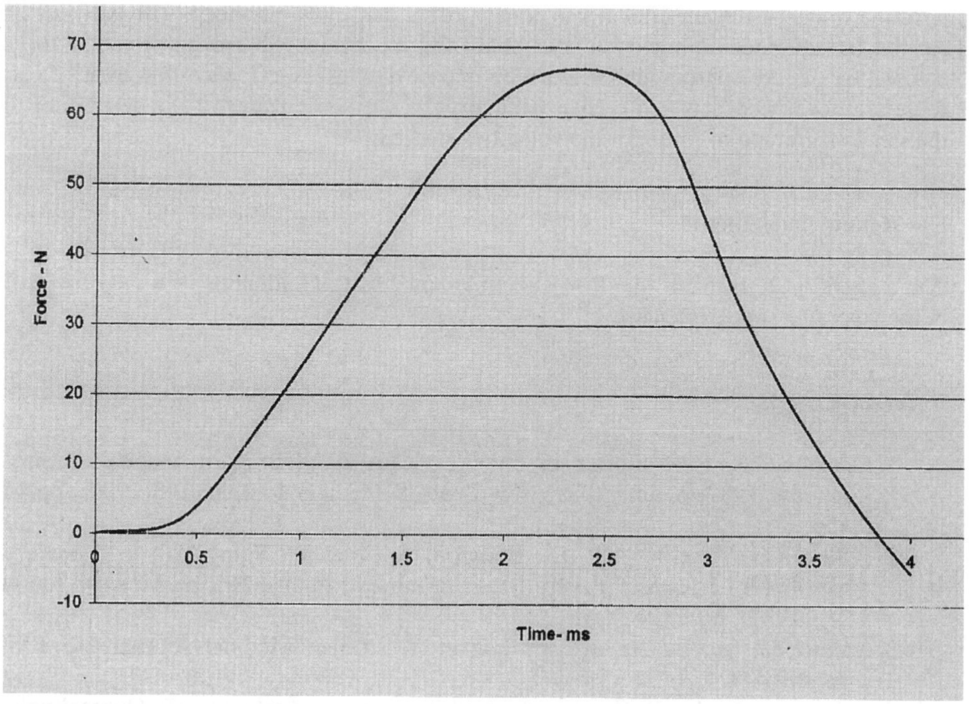

Figure 3. Force time history of impact of squash ball with sheep's eye, impact velocity is 4.42 ms^{-1}.

Table 2. Force and impulse associated with ball impacts with the eye.

Ball type	Impact velocity ms^{-1}	Position of impact	Average peak force – N	Average impulse- Ns
Cricket ball	2.62	orbit	251.0	2.30
Cricket ball	2.62	eye	275.9	2.19
Cricket ball	4.42	Orbit & eye	495.1	2.63
Cricket ball	4.42	eye	382.5	2.47
Squash ball	2.62	eye	20.0	0.040
Squash Ball	4.42	eye	30.3	0.052

The force and impulse delivered to the eye from cricket ball impact is much greater than that from the squash ball and is due to the larger mass of the cricket ball. The squash ball used in the tests was in the "cold" state and as a squash ball is played in a game its temperature increases. The temperature increase is the source of a reduction of the rubber loss factor, a decrease in the modulus of elasticity and the warming of the air inside the ball. Thus, the ball becomes more "bouncy" and possibly has a marginally lower stiffness. The impact force between the ball and the eye would be less if the stiffness decreased.

4 Conclusions

This paper has presented data for static compression stiffness of sheep eyes and data for dynamic impact between cricket and squash balls, and eyes' of sheep. The stiffness data can be incorporated into models that study the mechanics of eye deformation for the calculation of force, stress and strain. The impact data gives evidence that high levels of force are developed between cricket balls and the eye while squash balls do not develop the same magnitude of force for the velocities studied.

5 Acknowledgements

The authors wish to thank Mr. Ron Laemmle and Mr. P. Goldenberg for their assistance in the laboratory experiments.

References

1. Auben, N., Eye injuries in indoor cricket at Wellington hospital: a survey January 1987 to June 1989, New Zealand Medical Journal, 103(1990), pp.454-456.
2. Barrell,G. V., Cooper,P .J., Elington, A. R. et al., Squah ball to eyeball: the likelihood of squash players incurring an eye injury, British Medical Journal, 283(1981), pp.893-895.
3. Coroneo, M. T., An eye for cricket, Australian Medical Journal 142(1985), pp.469-471.
4. Elman, M. J., Racket-sports ocular injuries, Archives of Ophthalmology 104(1986), pp.1453-1454.
5. Genovese, M. T., Lenzo, N. P., Lim, R. K., et al,. Eye injuries among pennant squash players and their attitudes, The Medical Journal of Australia 153(1990, pp. 655-658.

6. Jones, N. P. and Tullo,A. B., Severe eye injuries in cricket, British Journal of Sports Medicine(1986), pp. 178-179.
7. Fong, L.P., Sports related eye injuries, The Medical Journal of Australia 160(1994), pp. 560-561.

AN ANLYSIS OF THE STRENGTH OF THE SURGICAL TOOL/TISSUE INTERFACE OF WAVE PROFILED SURFACES

GRANT H. ISKOV, JONATHAN VINCENT

School of Engineering and Industrial Design,
University of Western Sydney, Penrith 2747, Australia
E-mail: j.vincent@uws.edu.au

RICHARD L. OLLERTON

School of Quantitative Methods and Mathematical Sciences,
University of Western Sydney, Penrith 2747, Australia

An investigation into the surgical instrument/tissue interface of 2-Dimensional and 3-Dimensional wave profiled surfaces. Pairs of aluminium-profiled blocks were machined with teeth of differing size and shape. These were used to grip leather samples under a constant pneumatic pressure. The sample was extracted at a velocity of 50 mm/min until either the grip or the sample failed. The load at which the grip failed, the maximum load generated, and the type of failure was recorded. A quantitative relationship between wave amplitude and release load was established.

1 Introduction

Surgical graspers apply a gripping force to improve the manipulation of tissue during medical procedures. Such graspers are often characterised by sharply profiled and opposing surfaces that hinge from extended arms or linkages to provide a mechanical advantage.

Little is known about the effect of the profiled surfaces on grip security and the resulting tissue damage. Traditionally, the manipulation and skill of the surgeon have compensated for the limitations of surgical instruments. However, with the adoption of laparoscopic techniques, the surgeon has become sensually dislocated from the instrument/tissue interface. This has resulted in as increase of tissue traumas during laparoscopic procedures. For instance, there is an increased likelihood (13% - 40%) of perforating the gall bladder during laparoscopic cholecystomy over standard open cholecystomy [2]. In the future, surgeons may be further dislocated by the use of robots to perform laparoscopic surgery.

The aim was to investigate the surgical instrument/tissue interface of 2-Dimensional and 3-Dimensional ("egg carton") wave profiled surfaces. Wave profiles are being investigated in light of investigations by Marucci *et.al* [1] that concluded that wave shaped teeth result in significantly less tissue trauma than sharp teeth.

With the development of improved gripping surfaces it is hoped that the grip security/tissue trauma balance can be optimised, and contribute to new developments in the design of surgical instruments. This will provide benefit to patient and surgeon alike, as well as potentially improving the bedside manner of robots in surgery. Although the work has a medical focus, there is no reason why the research cannot be applied to non-medical applications, such as rubber handling in manufacture or cloth manipulation in the textile industry.

2 Procedure

The grip of different surface profiles were tested and compared. This involved machining pairs of aluminium blocks to the desired profile. These were then mounted in a pair of pneumatic grippers. A 25mm×40mm leather strip folded over a hook, and both ends were threaded between the profiled blocks.

The gripper was actuated and the leather sample was pulled from the profiled blocks at a rate of 50 mm/min, while the tensile load was recorded. The tensile test continued until the controller interpreted a sample failure (no load) or the extension of the sample reached a preset limit. The same tensile test was performed five times for each profile.

The tensile testing system used was the *MTS Mini-Bionix 858 Testing Machine* controlled by a *MTS 458.20 Micro Console*. Mounted to the base of the MTS was a 250 Newton S-bend load cell atop which was attached the *MTS light tensile load pneumatic grippers*. The pneumatic grippers applied a clamping force of approximately 48 Newtons at 414 Pa of regulated nitrogen pressure.

3 Wave Profile Design & Manufacture

All profiles were machined from aluminium using a *Bridgeport* Computer Numerical Control (CNC) mill with a *Heidenhein* controller. The Heidenhein controller provides the use of the programming language *Conversational Programming Language* that allows the manipulation of variables with basic mathematical and trigonometric functions.

3.1 Two-dimensional profiles

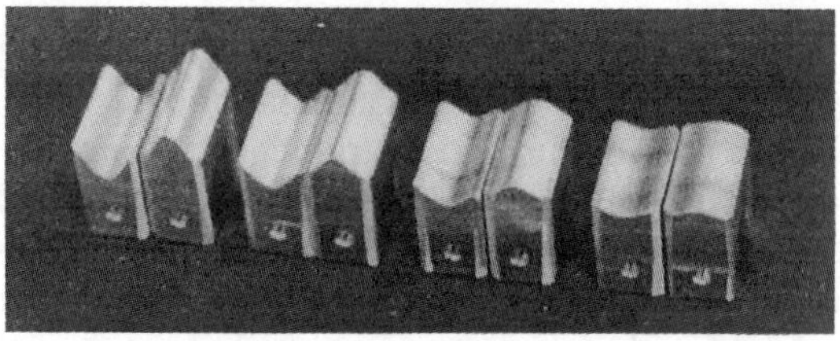

Figure 1. 2-dimensional sinusoidal profiles. (2, 1.5, 1 and 0.5 mm)

The surfaces are defined by the following equation (bearing in mind that the Heidenhein controller for the Bridgeport CNC mill works in degrees, not radians).

$$y = A\cos\left(\frac{360x}{8}\right)$$

Where: A is the amplitude, and was given values of 0.5mm, 1mm, 1.5mm, and 2mm, and x was between 0mm and 8mm.

Both opposing halves of the profile were milled from the same billet of aluminium. Because the milling is done in 2 dimensions we were able to use the radius compensation functions of the *Heidenhein* controller.

2.2 Three-dimensional "egg carton" profiles

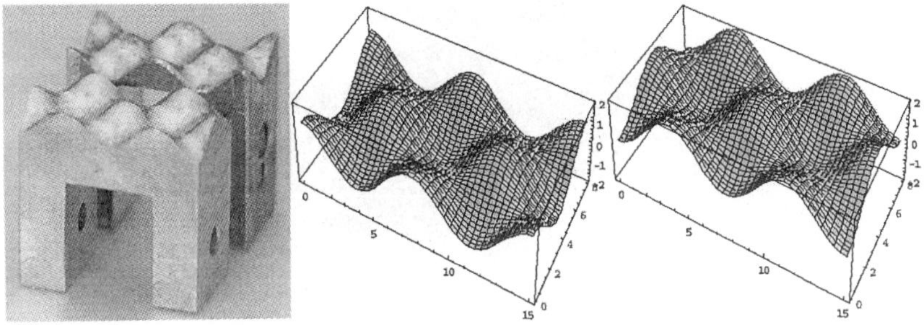

Figure 2. 3-dimensional sinusoidal opposing profiles (1 mm amplitude)

The equation for the surface is:

$$z = A\cos(Bx)\cos(Cy)$$

Where: A= amplitude
B=360/(wavelength in x-direction)
C=360/(wavelength in y-direction)

The *Heidenhein* controller is not capable of radius compensation in three dimensions, so this had to be calculated within the program.

Equations were therefore derived to calculate the path of the centre of the spherical cutting tool, so that its outer edge would cut the required profile. This was achieved by finding the unit normal vector at each point on the surface. This was then multiplied by the tool radius and added to the locating vector of the point on the surface. This gave the following equations.

$$X = x + \frac{RDA\sin(Bx)\cos(Cy)}{\sqrt{[DA\sin(Bx)\cos(Cy)]^2 + [CA\cos(Bx)\sin(Cy)]^2}}$$

$$Y = y + \frac{REA\cos(Bx)\sin(Cy)}{\sqrt{[DA\sin(Bx)\cos(Cy)]^2 + [CA\cos(Bx)\sin(Cy)]^2}}$$

$$Z = A\cos(Bx)\cos(Cy) + \frac{R}{\sqrt{[DA\sin(Bx)\cos(Cy)]^2 + [CA\cos(Bx)\sin(Cy)]^2}}$$

Where: A = amplitude
B = 360/(wavelength in x-direction)
C = 360/(wavelength in y-direction)
D = 2π/(wavelength in x-direction)
E = 2π/(wavelength in y-direction)
R = tool radius
X , Y, Z = position of tool bit
x, y, z = point of desired surface

Only one value of amplitude (1mm) for the 3-dimensional profiles was machined.

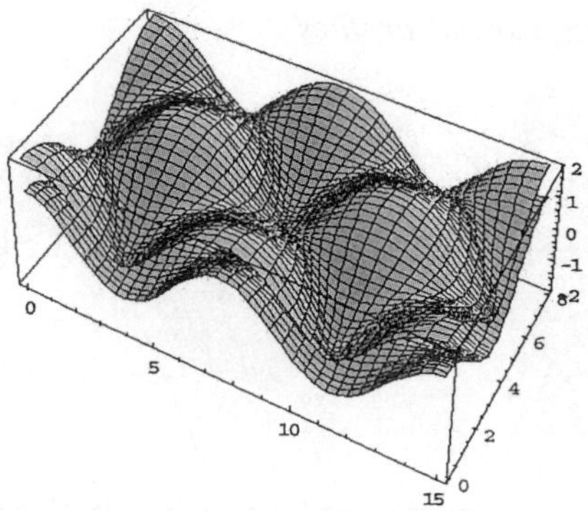

Figure 4. The 3-dimensional "egg carton" profile, showing the required final surface and the radius compensated path for the cutting tool's centre

4 Results

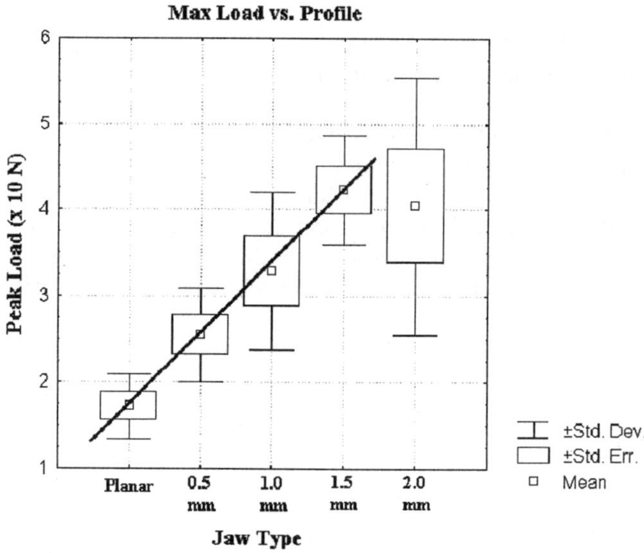

Figure 5. The variation of the peak load of 2-dimensional profiles of different amplitude.

A graph of the release load vs the profile amplitude for the 2-dimensional wave profiles can be seen in Figure 5. Excluding the 2 mm wave profile, there appears to be a linear relationship between the mean values of each of the data sets. This relationship is given by,

$$P = 17A + 17.5\,,$$

Where P is the peak load (in N) and A is the amplitude (in mm).

Figure 6 compares the peak load of 2-dimensional and 3-dimsional profiles with the same 1-mm amplitude. It illustrates that there is no clear improvement to grip security with 3-dimendional profiles over 2-dimensional profiles. However, there may still be the advantage that 3-dimensional profiles provide simultaneous grip security both longitudinally and laterally.

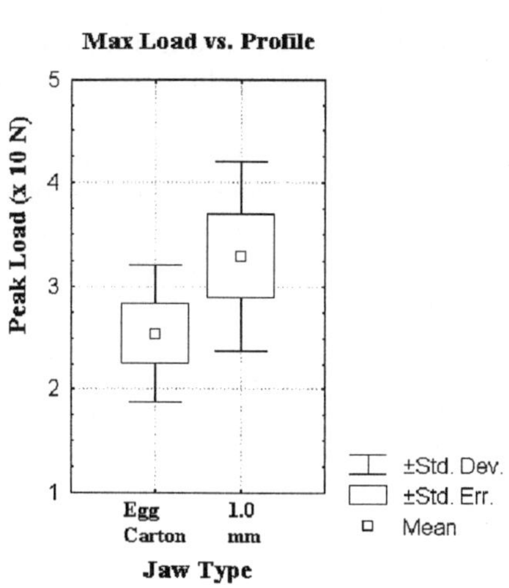

Figure 6. A comparison between the peak load of a 3-dimensional "egg carton" profile and a 2-dimensional profile, both with a 1mm amplitude.

Difficulties were encountered in securing the profiles of higher amplitude. Firstly, a slight hinging of the profiles in the direction of strain was apparent. Also, a slight separation of the gripper arms was also observed. Both contributed to the profiles not being oriented in parallel. Poorly secured profiles resulted in 'peaky' results (Fig 7) and premature grip failure. It is suggested that unparallel (poorly secured) profiles may limit peak grip loads and also exert, impulsive, high and localised pressures upon the tissue.

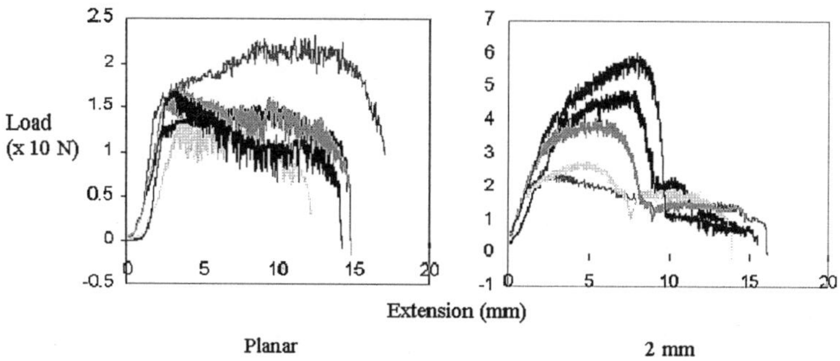

Figure 7. The 2mm amplitude profiles exhibited 'peaky' results are due to difficulties in securing the blocks sufficiently; the results using planar (0 mm amplitude) profiles are more reliable.

5 Conclusions

This investigation analysed the strength of the surgical tool/tissue interface of wave-profiled surfaces.

During the investigation, procedures for the design and manufacture of 2-dimensional and 3-dimensional wave profiled blocks were developed.

The results suggest that unsecured, or unparallel profiles will withstand reduced tensile loads and may exert high, localised pressure upon the tissue sample.

It is also demonstrates that for wave profiles, of the same wavelength, there is a linear relationship between the amplitude and the maximum tensile load that can be held when gripping the same sample tissue types.

Results indicate that 3-dimensional wave profiles do not have improved grip security over 2-dimensional wave profiles. However, 3-dimensional wave profiles do warrant further investigation, in light of inaccuracies identified with the experimental data and their potential to grip against lateral and longitudinal forces simultaneously.

There are many areas of the surgical instrument/tissue interface that warrant further investigation. This includes investigation into the load bearing properties of profiles when the tissue is pulled parallel to the axis of wave rotation (Fig. 8), the effects of offsetting the profiles (Fig. 9), and the load bearing properties of wave profiled teeth with tissue sample thickness compensation (Fig. 10).

Figure 8. Longitudinal profiles

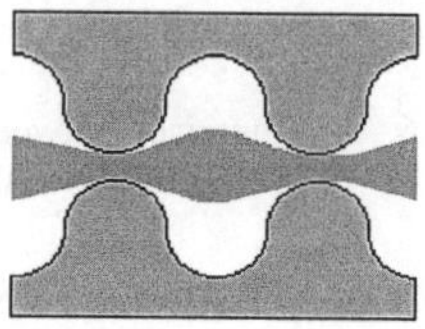

Figure 9. Offset profiles

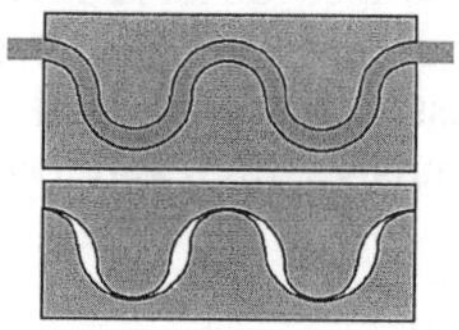

Figure 10. Sample thickness compensated profiles

Acknowledgments

Thanks must be extended to Damian Marucci and William Walsh of the Orthopaedic Research Laboratories at UNSW for the provision of testing equipment and instruction. We also thank Paul Toner and Terry Wolfe for their assistance in the manufacture of the profile blocks. Finally, thank you to Robert McDonald for his assistance with researching.

References

1. Marucci D. D., Cartmill J. A., Walsh W. R. and Martin C. J., Patterns of Failure at the Instrument-Tissue Interface. *Journal of Surgical Research* **93** (2000) pp 16–20.

2. Marucci D. D., Shakeshaft A. J. Cartmill J. A., Cox M. R., Adams S. G. and Martin C. J., Grasper Trauma During Laparoscopic Cholecystectomy, *Submitted for Publication*

3. Cartmill J. A., Shakeshaft A. J., Walsh W. R. and Martin C. J., High Pressures are Generated at the Tip of Laparoscopic Graspers. *Aust. N.Z. J. Surg.* **69**, (1999) pp 127-130.

MECHANICAL ANALYSIS OF AN ATHLETE'S TIBIA TO DETERMINE STRESS FRACTURE RISK LOCATIONS: PREPARATION FOR FEM

MELANIE FRANKLYN[#] AND BRUCE FIELD[##]

[#] *Accident Research Centre, Monash University and* [##] *Department of Mechanical Engineering, Monash University, Clayton, Victoria 3168, Australia*

E-mail: Melanie.Franklyn@general.monash.edu.au and Bruce.Field@eng.monash.edu.au

Stress fractures, which are a common injury in athletes and the military, are a result of applying repetitive cyclic loading to bone. They are commonly diagnosed in the tibia, either at the midshaft or at the junction between the mid and distal thirds of the tibia. Although much is known clinically on the diagnosis and etiology of stress fractures, the mechanical factors that contribute to their development are not as well understood.

Using a gait analysis lab with a full marker system, values of the forces and limb positions were obtained from an aerobic athlete running across the plate. These values were used in conjunction with supplementary information obtained from cross-sectional CT scans taken at regular intervals along the athlete's tibia to perform a mechanical analysis of the bone. The objective of this analysis was to determine the location and approximate magnitude of the highest stresses and which load or combination of loads produced them. This paper will discuss the results generated from the mechanical analysis and their clinical significance. In addition, the application of the gait data to a Finite Element (FE) Model will be discussed in the conference presentation.

1 Introduction

Stress fractures, which are a common injury in athletes (such as runners [6] and ballet dancers [4]), as well as the military [1, 7] are a result of applying repetitive cyclic loading to bone at loads below the fracture threshold. They are commonly diagnosed in the tibia, and in athletes, either at the midshaft or at the junction between the mid and distal thirds of the tibia [2].

There is much clinical knowledge on the diagnosis and etiology of stress fractures. In particular, refined medical imaging techniques allow accurate diagnosis [6] and clinical symptoms are well recognised. Furthermore, it has been demonstrated that associated risk factors such as demography (age, gender and race), anatomy, physical fitness and bone density influence the likelihood of suffering from a stress fracture [2]. Prior research has linked biomechanical factors (such as external rotation of the hip [5] and tibial width [7]) to stress fracture incidence. In addition, some finite element work, which has been linked to experimental data, has been completed on the tibia of rabbits [2, 3]. Nevertheless, there has been little investigation into the mechanical factors that contribute to stress fracture development, consequently, the relationship between biomechanical factors and stress fractures is not as well understood.

Although there is a significant amount of literature on the walking gait cycle [12], particularly in the analysis of prosthetics and neuromuscular disease, there is less literature on the gait analysis of running. To the authors' knowledge, gait systems have not previously been used for analysing the etiology of stress fractures from a mechanical perspective. Consequently, it is the objective of the current research to perform a mechanical analysis of the tibia using both full gait data with the corresponding CT scans from an athlete. This continuing investigation aims to pinpoint possible locations of stress fracture development from both a mechanical analysis (which is presented here) as well as a finite element analysis (which will be presented at the conference).

2 Methods

2.1 Data Acquisition

The data from an athlete running was captured using 8 infrared cameras to track motion trajectories of 25mm retroreflective markers. A single force plate was used to measure one cycle of the athlete's stance phase. The plate used was a piezoelectric Kistler plate, type 9281B (Kistler, Winterthur, Switzerland). The Vicon 512 data capture system and software (Oxford Metrics, Oxford, UK) and the Bodybuilder software (Oxford Metrics) was used to process the data. Vicon 512 initially measured and reassembled the motion data (ie. the raw marker trajectories from the 8 cameras was captured and reconstructed into a single 3D trajectory for each marker), while the Bodybuilder software was used to view the trials (as stick figures), process and export the data.

The athlete, who was a 30 year old female aerobics instructor, was accustomed to high-impact exercise consisting of aerobics and running 6-8 times a week. Her bone density was above the mean density for the same age and sex as ascertained by a bone density scan conducted prior to the experiment. The athlete performed twenty running trials, half with the left and half with the right foot impacting on the plate. In addition, the trials were completed with the athlete running from each side of the plate. This was undertaken to ensure that there would be some trials with complete sets of data, as the markers were thereby viewed from various camera angles.

The trials were viewed using Bodybuilder, and those that were missing significant marker trajectory data were discarded. Consequently, eight trials remained that were deemed to contain a complete set of data, and a representative trial was chosen from this group for analysis. The raw data from this trial was then processed and exported. The final data consisted of the three force components (F_x, F_y, F_z), the three plate moments (M_x, M_y, M_z) and the centre of pressure (COP) of the foot on the plate (X_{cp}, Y_{cp}). The vertical centre of pressure (Z_{cp}) was obtained from the plate dimensions. The Kistler force plate is illustrated by Figure 1.

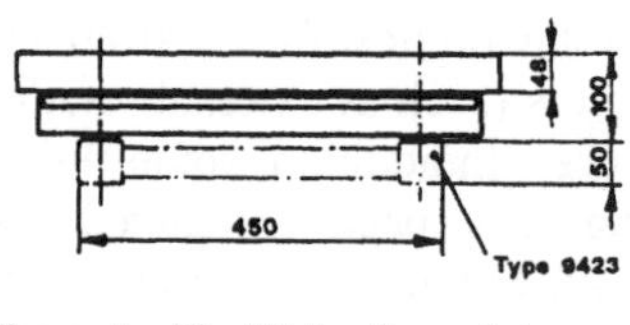

Figure 1. The Kistler force plate.

Thirty-nine cross-sectional CT images of the athlete's tibia were obtained at 10mm intervals from the knee to the ankle using a General Electric Pace CT scanner. A scout image (showing the positions of the cross-sectional slices) in addition to a lateral image was also obtained. After scanning, the images were digitised to TIF format using the Microsoft Global Lab Image frame grabber (Version 8, Marlboro, MA, US). The UTHSCA Image Tool for Windows (Version 2.0, The University of Texas Health Science Centre, San Antonio) was then used for displaying the digitised images so that measurements of the profile could be taken. Figure 2 depicts the overall process of data capture and processing.

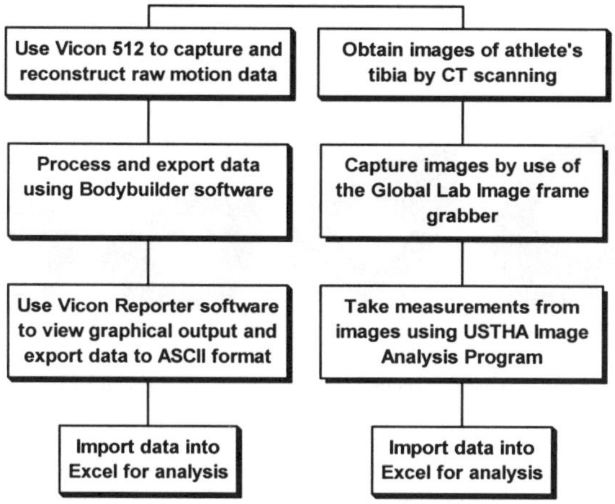

Figure 2. Overview of the data acquisition and processing.

2.2 Data Analysis

Graphs of the three force components and moments against time were plotted. Using the three force components and the joint positions (obtained from the marker data), the magnitude and direction of the resultant forces (F_R) were calculated for each time interval. The time step that corresponded to the highest bending moment on the tibia (the combination of a high resultant force and a large angle between the resultant force and the neutral axis of the tibia) was then used for the analysis. Four cross-sectional images were then selected from the CT scans, as illustrated by Figures 3 and 4. Figure 3 also shows the coordinate system used. Clinically, stress fractures are located midshaft (50% level) or at the junction of the middle and distal thirds of the tibia (33% level), hence these levels were chosen in the study [4]. To give a reasonable view of the tibial mechanics, two other sections were chosen (at the 23% and 67% levels), one each side of the first two sections.

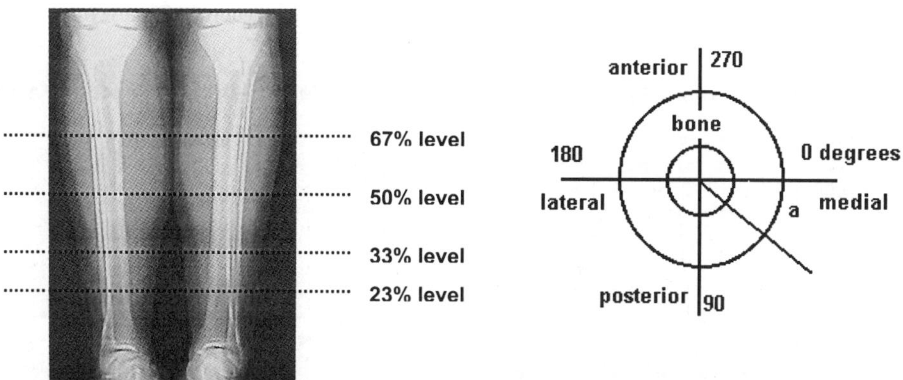

Figure 3. Scout film displaying the tibial CT slices scanned and (left) and a diagram showing the angle (a) around the slice.

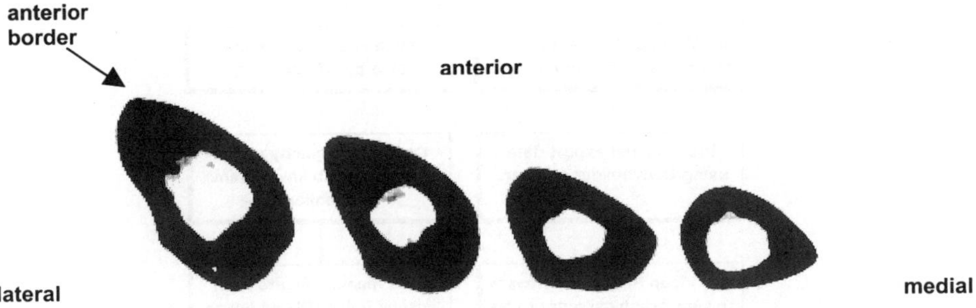

Figure 4. Images of the cross-sections at the 67%, 50%, 33% and 23% levels (left to right).

Cross-sectional properties were calculated numerically using a Fortran code written by the first author. This code was based on a modified version of the algorithm developed by Nagurka and Hayes [9]. The Fortran program was designed to read input data derived from the Image Analysis Program. The cross-sectional properties that were required for the analysis included the area, centroid and section modulus.

3 Results

Considerable literature exists on the analysis of the walking cycle and the familiar double-humped pattern of the vertical ground reaction force (F_z) is well known. However, the literature pertaining to running gait is not as well documented and furthermore, there is greater variability from subject to subject, with the shape of the ground reaction force primarily depending on running style [8]. Nevertheless, it is clear that two types of ground reaction force (GRF) curves exist for running gait: one displaying a double peaked graph while the other showing either an absence or minute initial peak following heel contact [8, 10]. The initial peak is the breaking phase while the latter phase is termed propulsion. In the current research, four sets of the complete data displayed the double peak characteristic while four sets contained a minute peak. These graphs, which are illustrated in Figure 5, are similar to that reported in the literature [10]. A representative trial was chosen from the eight complete sets of data and is shown in Figure 6.

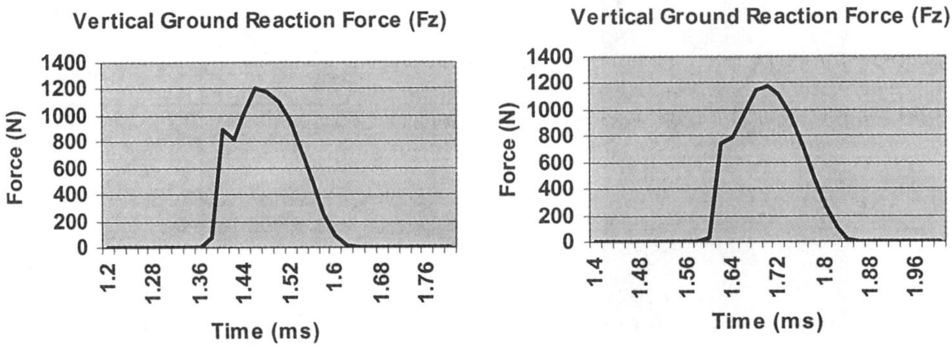

Figure 5. The two types of curves obtained for running gait in this experiment (a) double peaked and (b) almost single peaked.

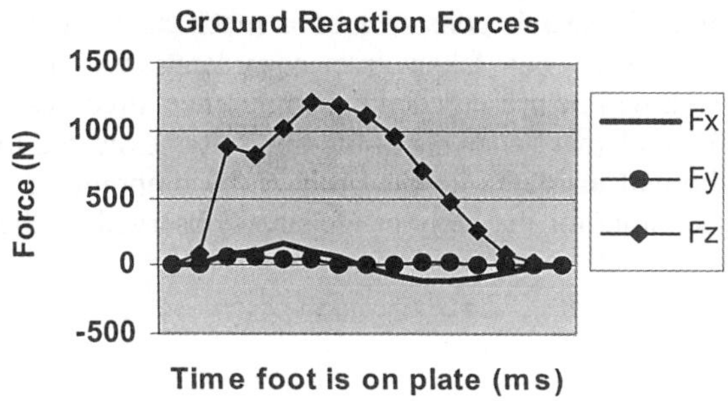

Figure 6. Graph illustrating the ground reaction forces against time.

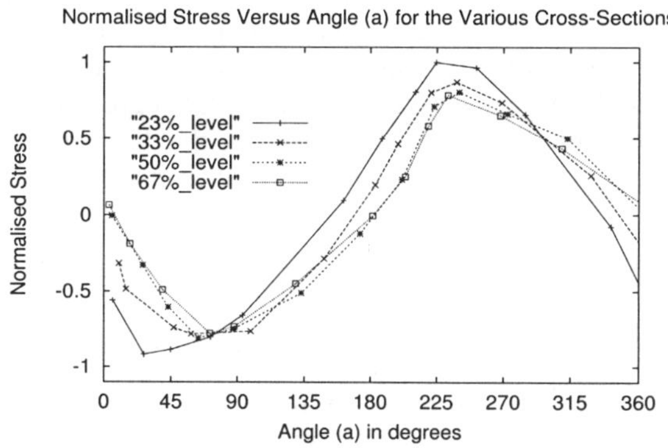

Figure 7. The resultant stresses at the chosen cross-sections.

4 Discussion

In the current research, the ground reaction force was calculated to be posterior to the cross-sectional centroids, hence it can be deduced that the highest tensile stresses should be located anterior to the centroid and the highest compressive stresses posterior to the centroid. In fact, Figure 6 indicates that during the stance phase, the resultant force is very nearly vertical. Hence when the knee is ahead of the ankle, the force is posterior to the tibia. The magnitudes of the peak compressive and tensile stresses are similar for a given cross-section (Figure 7), however, cortical bone is weaker in tension [11]. As the anterior border is the region of highest tensile stress, it can be concluded that this region most susceptible to stress fracture development. This result is consistent with the clinical data on athletes, which has shown that stress fractures are usually found on the anterior border of the tibia. Some stress fractures have also been observed clinically on the anteriomedial border at the 33% level, however, the current results so far do not clearly distinguish this region as being susceptible to stress injury.

From the results in Figure 7, it is clear that the stress magnitudes illustrated in the curves increase from proximal to distal tibia. This would suggest that the distal end of the tibia is more prone to stress injury. It has been demonstrated clinically that stress fractures in athletes generally occur at the 50% and 33% levels of the tibia [4]. What is not as apparent from the results so far is which mechanical parameters segregate these two levels from other lower tibial regions as sites of stress fracture development. For example, the current research consisted of a mechanical analysis of the tibia where material properties were not taken into consideration. The second part of this investigation, which will be presented at the conference, involves using the gait data from the athlete in a finite element analysis of the tibia. It is expected that this further analysis will provide additional insights into the problem by pinpointing which component(s) of the force are responsible for the high tensile stress observed on the anterior border of the tibia.

5 Conclusions

A mechanical analysis of the tibia using force plate data from an athlete running has demonstrated that the anterior border is the most susceptible region for stress fracture development in running athletes, as it is the location of highest tensile stress. Furthermore, cross-sections at the distal end of the tibia display slightly higher stress magnitudes (both in tension and compression) than those at the proximal end, suggesting that the distal end is more likely to result in this injury. It is still unclear at this stage why certain cross-sectional levels, such as the 33% level, are segregated from adjacent areas as key locations for stress fracture injury. It is expected that the next stage of this research, a finite element analysis of the tibia, will provide further insights and explanations for this phenomenon.

6 References

1. Beck T. J. Ruff C. B. et al., Dual-energy x-ray absorptiometry derived structural geometry for stress fracture prediction in male US Marine Corps recruits. *J. Bone and Min. Res.* **11, 5** (1996) pp. 645-653.
2. Burr D. B., Bone, exercise and stress fractures. *Ex. and Sports Sci. Rev.* **25** (1997) pp 171-194.
3. Burr D. B., Milgrom C. et al., Experimental stress fractures of the tibia. *J. Bone Joint Surg.* **72B, 3** (1990) pp. 370-375.
4. Devas M., *Stress fractures* (Medical Division of Longman Group Limited, New York, 1975).
5. Finestone A., Shlamkovitch N., Eldad A., et al. Risk factors for stress fractures among Israeli infantry recruits. *Milt. Med.* **156, 10** (1991) pp. 528-530.
6. Fredericson M, Bergman G. et al., Tibial stress reaction in runners. Correlation of clinical symptoms and scintigraphy with a new magnetic resonance imaging grading system. *The Am. J Sports Med.* **23, 4** (1995) pp. 472-481.
7. Giladi M. C., Milgrom C., Simkin A. et al., Stress Fractures and tibial bone width. A risk factor. *J. Bone Joint Surg.* **69B** (1987) pp. 326-329.
8. Munro C. F., Miller D. I. and Fuglevand A. J., Ground reaction forces in running: a reexamination. *J. Biomech.* **20, 2** (1986) pp. 147-155.
9. Nagurka M. L. and Hayes W. C., An interactive graphics package for calculating cross-sectional properties of complex shapes. *J. Biomech.* **13**, (1980) pp. 59-64.
10. Ounpuu S., The biomechanics of walking and running. *Clin. in Sports Med.* **13, 4** (1994) pp. 843-863.
11. Pattin C. A., Caler W. E., and Carter D. R. Cyclic mechanical property degradation during fatigue loading of cortical bone. *J. Biomech.* **29, 1** (1996) pp. 69-79.
12. Rose J. and Gamble J. G., *Human walking* (Williams and Wilkins, Baltimore, 1994).

MEASUREMENT OF BONE DENSITY AND BONE MINERAL CONTENT USING ULTRASONIC WAVES

E. Azmy, and R.N.Ibrahim
Department of Mechanical Engineering, Monash University
P.O.Box 197, Caulfield East, Victoria 3145, Australia
Email: Raafat.Ibrahim@eng.monash.edu.au

Measurement of bone density for women is of great importance especially for certain age, as by aging bone density and percentage of bone mineral content decrease causing weakness in bones. There are several techniques for bone density measurements but the most important one is the ultrasound waves, which are widely used, in medical applications. When voltage is applied, high frequency longitudinal waves are produced and measurements are made of the ultrasonic attenuation. In this research work, samples are taken and positioned between. two broadband ultrasonic transducers, one of them transmits short pulses of broadband ultrasound and the other used as a receiver, and a relation-ship between ultrasonic attenuation and frequency is established. In this paper, the relationship between attenuation coefficient and bone density is determined. Also, a correlation between bone density and percentage of mineral content is established.

1 Introduction

Bone consists of a hard, dense outer casing of cortical bone surrounding the medullary cavity that contains a less dense network of bone fibers, trabecular bone. The composition of cortical and trabecular bone remains relatively constant among healthy adults with aging, with only 2% increase in mineral and a 2% decrease in water in elderly bone compared to young, adult bone. [1,2] Changes can occur in the overall density of bone due to a wide variety of reasons other than normal age-related regression of bone and red marrow. Metabolic bone diseases are characterized by general change in bone mass.[1]

Bone tissues are elastic, and can be classified into two types depending on macroscopic and microscopic features. In particular, cancellous bone, being a two-component material comprised of a hard spongy core with cavities filled by soft tissue, has both strong anistropy and inhomgeneity and is therefore especially difficult to analyze. [3]

Ultrasonic waves is one of the most accurate systems for measuring bone density, propagation of ultrasonic wave in bone was studied by Biot where he found that there are two longtudinal waves, one is fast which corresponds to the wave in the solid and fluid moving in phase, while the other is a slow speed wave corresponding to motion out of phase.[4-6]

The propagation speeds and attenuation of the fast and slow waves were measured and compared quantitatively. The measured speeds of the fast and slow waves are in good agreement with estimated results, and the waves were confirmed to correspond to the two waves predicted by Biot. [7]

The goal of this study is to evaluate bone density and percentage of bone mineral content using acoustic parameters of ultrasonic waves. This goal can be achieved by evaluating the attenuation coefficient for each corresponding ultrasonic frequency, measuring porosity and its corresponding density for the bone sample for various demineralization periods, where the samples are demineralized in Nitric acid.

Using these data and another data produced through previously known equations, we can find an equation to calculate bone density directly and another one to calculate percentage of bone mineral content using the deduced bone density.

2 Experimental work

Sample preparation: Seven transverse slices of bovine bone samples were cut parallel to the plane of the tibial condyles. Each sample was machined and polished to make it uniform and smooth. Only well aligned structure was used. The trabeculae are oriented almost perpendicularly to tibia condole articulator surfaces. The soft tissue filling between the trabeculae was removed by boiling, and the spaces were saturated with distilled water. The samples were decalcified in 100 mL of Nitric Acid.

Bone density is determined by saturating samples by water and decompressing them to remove air bubbles. If all pores are interconnected and if no closed cells are present, the density of the frame material can be determined from apparent density of the porous material. The porosity β of each specimen is determined from the weights of sample saturated with water and that when water is drained from pores and replaced by air. Mineral content was calculated as the ash weight divided by the dry weight. Apparent ash density (ρ_s, kg/m^3) was calculated as the mineral content times the apparent dry density.

The schematic diagram of the ultrasonic system shown in (fig.1) was used to measure the acoustic parameters (longitudinal velocity and ultrasonic attenuation) along the length of the major alignment of the trabeculae.

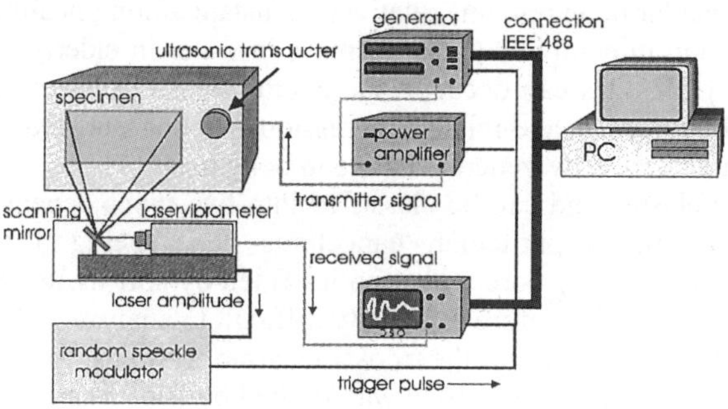

Figure1 Experimental Setup

Ultrasonic measurements were conducted using a substitution method. A broadband ultrasonic pulse was received with and without the sample interposed between the two transducers.

To measure the attenuation at each frequency, the following equation is used:

$$\alpha(f) = \frac{(20 \log \frac{V_2}{V_1})}{d} \tag{1}$$

where $\alpha(f)$ is the attenuation coefficient in db/cm at frequency f, V_1 is the peak amplitude without sample, represents the echo of the ultrasonic pulse transmitted through water, and V_2 is the peak amplitude with sample; d is the sample thickness in cm.

The attenuation increases non-linearly with frequency, the high frequency component is more attenuated than the low frequency component when an ultrasonic pulse passes through tissue.

$$\alpha(f) = \alpha_o f^n \qquad (2)$$

where α_o is the attenuation coefficient factor, and n is the frequency power factor.

3 Experimental results & analysis

The porosity and bone density have been measured for seven samples of bovine cancellous bone. The samples were cylindrically shaped with a diameter of 2 cm and varying lengths. The measured results are plotted in (fig.2). Bone tissues were decalcified in Nitric Acid to for different periods to measure porosity, porosity was found to increase as demineralization time increase as plotted in (fig.3).

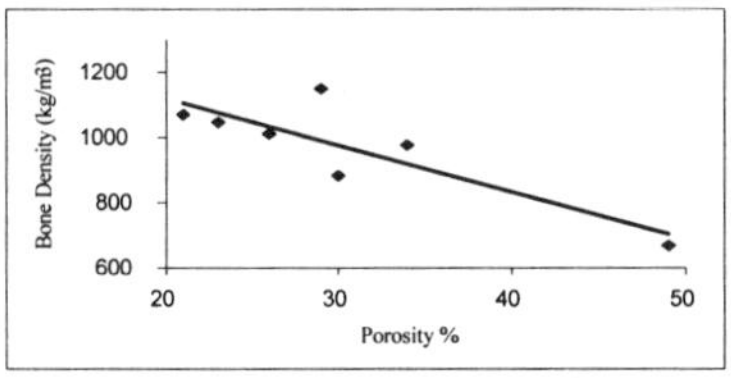

Figure 2 Porosity vs Bone density

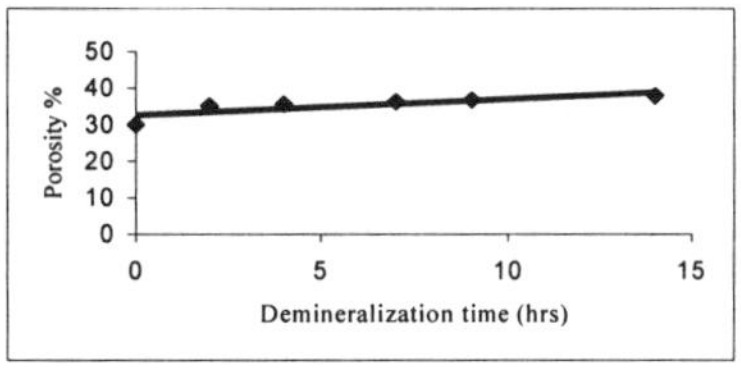

Figure 3 Relation between Demineralization time and porosity

For different frequencies of ultrasonic waves, attenuation coefficient was measured at various decalcification periods. Results are in table 1. Values of bone density and corresponding BMC% were measured and plotted in figure 4.

Table (1-a) Relation between frequency &attenuation coefficient

Frequency (MHz)	Attenuation Coefficient of decalcified bones (db/cm)				
	Fresh bone	2 hours	4 hours	7 hours	9 hours
0.8	12	10	8	4.8	4
1	11.5	11	10	9	8
1.5	14	13	11	10	8.5
2	19	16	14	13	10.5
2.2	17	17	16	12	11

Table (1-b) Relation between frequency and attenuation coefficient

Frequency (MHz)	Attenuation Coefficient after 14 hours of decalcification (db/cm)
1	2.2
2	6.5
3	7
4	10
5	11

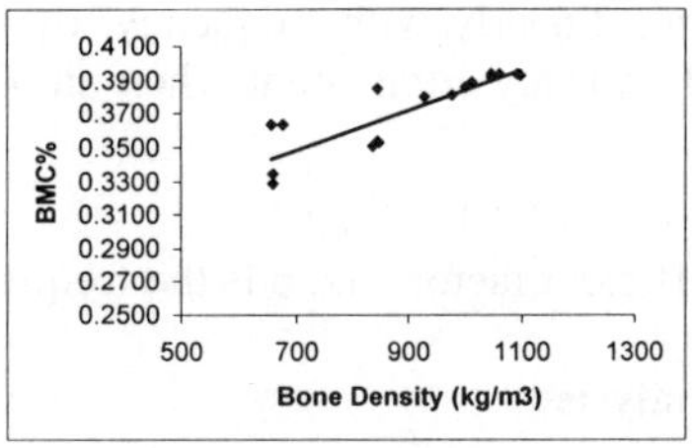

Figure 4 Bone density vs. Bone Mineral Content%

The goal of this study is to evaluate the bone density and Bone Mineral Content % from the measured measured acoustic parameters of ultrasonic waves (f, v_1 & v_2). To achieve this goal using equation 2, (α_o and n) must be expressed in terms of bone density (ρ).

3.1 Calculating the attenuation coefficient and the frequancy power factor

From equation 2, $\alpha = \alpha_o$ at f=1MHz, so using table 1, we can find α_o at (f= 1) for various periods of demineralization.

Also, (n) is found algebraically, from equation 2

$$n = \frac{\ln \alpha - \ln \alpha_o}{\ln f}$$

From this equation, we can find a value of ***n*** for different values of ***f*** and $\boldsymbol{\alpha_o}$ at all demineralization periods.

The values of attenuation coefficient factor α_o and frequency power factor n against demineralization time are listed in Table 2.

Table 2: Attenuation coeff. & frequency power factor against demineralization time

Dem. time (hrs)	α_o	n
0	11.5	0.3
2	11	0.386
4	10	0.46
7	9	0.8
9	8	0.81
14	2.2	0.941

3.2 Establishing the relationship between α_o, ρ & n

Since we have a relationships between "α_o and demineralization time" (table 2), "porosity and demineralization time" (figure 3) and "porosity and bone density" (figure 2), therefore, to deduce this relation, two steps must be followed:

Step one: Deducing a relation between β and α_o at same demineralization time :
From (fig.3) & Table 2, we can find the required relation, listed in Table 3.

Step two: Also, from (fig.2) and Table 3, a relation between α_o and ρ is deduced and listed in Table 4.

This relation is plotted in (fig.5), fitted and an equation relating α_o and ρ is deduced as follows:

$$\alpha_o = -0.00204\,\rho^2 + 3.79\,\rho - 1746 \qquad (3)$$

Table 3. Attenuation coeff. & Porosity

α_o	β %
11.5	30
11	35.14
10	35.58
9	36.2
8	36.684
2.2	37.786

Table 4. Attenuation coeff. & Bone density

α_o	ρ
11.5	964
11	891
10	885
9	876
8	869
2.2	853

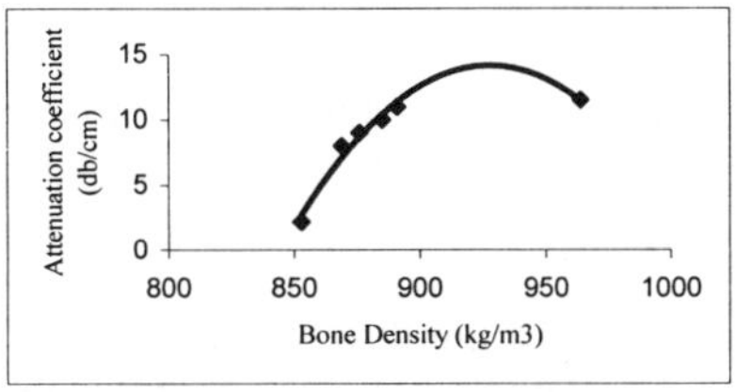

Figure 5 Attenuation coefficient vs Bone density

We will follow the same steps as in case of α_o and ρ, having a relation between "n and demineralization time", "porosity and demineralization time" and "porosity and bone density", therefore, to deduce this relation, two steps must be followed:

Step one: Deducing a relation between β and n at same demineralization time:

From (fig.2) & Table 2, we can find the required relation, listed in Table 5.

Step two: Also, from (fig. 1) and Table 5, a relation between n and ρ is deduced and listed in Table 6.

Table 5. Frequency power factor & porosity

n	β %
0.3	30
0.386	35.14
0.46	35.58
0.8	36.2
0.81	36.684
0.941	37.786

Table 6. Frequency power factor &density

n	ρ
0.3	964
0.386	891
0.46	885
0.8	876
0.81	869
0.941	853

This relation is plotted in (fig.6), fitted and equation relating n and ρ is deduced

$$n = 1.08 * 10\text{-}4\ \rho^2 - 0.202\,\rho + 95.1 \qquad (4)$$

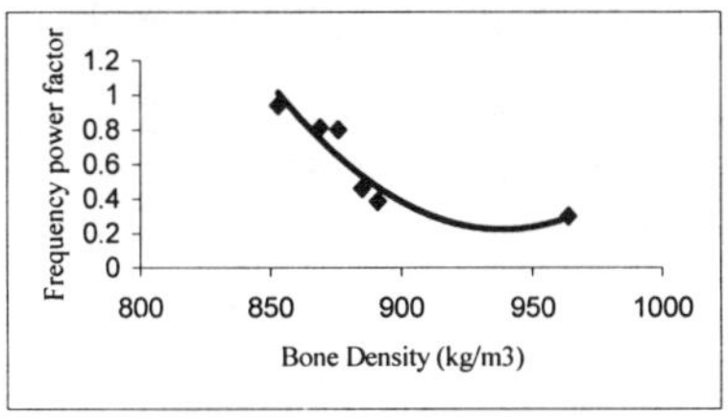

Figure 6 Frequency power factor vs Bone density

By substituting equations 3 & 4 in eqn. 2,

$$\alpha(f) = (-0.00204\ \rho^2 + 3.79\ \rho\ -1746) * f^{(1.08 * 10-4\ \rho 2 - 0.202\rho + 95.1)} \quad (5)$$

This is the required relation between attenuation coefficient and bone density at constant frequency.

Values of figure 4 are fitted to find an equation relating BMC% and bone density.

$$BMC\% = 0.0001\ \rho + 0.2661 \quad (6)$$

The bone density observed from Eqn. 5 is substituted in Eqn. 6 to deduce BMC%.

4 Discussion

Ultrasonic waves were previously studied in the matter of bone density evaluation; it was proved that they are one of the most accurate devices for the job. Ultrasonic waves were used in this study to find a direct equation to evaluate bone density and BMC% in a bone sample. These equations can be easily used given the frequency of the emitted wave from transducer and its corresponding attenuation coefficient through the bones.

Therefore, bone density can be calculated for any bone sample using these two parameters.

Of course, applying these two equations needs calibration to include the effect of tissues and fats in the human body on the results.

5 Conclusion

It was shown through this study how an acoustic parameter of ultrasonic waves (frequency) was used in evaluating bone density and BMC% of a bone sample. Knowing the frequency (f) of the wave entering the bone and following its attenuation through the bone (α_o), then substituting them in the given equations with other measured bone parameters, we were able to deduce a direct equation to calculate bone density and using this density we were able to calculate BMC%.

References

1. http://bonedensity\bonedensitometry.html
2. http://bonedensity\diagnosis.html
3. P.Laugier, P. Drion, A.M. Laval-Jeantet, and G. Berger, "In vitro assessment of the relationship between acoustic properties and bone mass density of the calcaneus by comparison of ultrasound parametric imaging and QCT", Bone, vol. 20, pp. 157-165, 1997
4. W. Xu, and J.J. Kaufman, "Diffraction correction methods for insertion ultrasound attenuation estimation", IEEE Trans. Biomed. Eng., vol. 40, pp. 563-5770, 1993
5. F.J. Fry and J. E. Barger, "Acoustical Properties of the human Skull", J. Acoust. Soc. Amer., vol. 63, pp, 1576-1589, 1978
6. M.A. Biot, "Theory of propagation of elastic waves in a fluid-saturated porous solid. I. Low-frequency range" J. Acoust. Soc. Am. 28, 168-178, 1956
7. A. Hosokawa and T. Otani "Ultrasonic wave propagation in bovine cancellous bone", J. Acoust. Soc. Am. 101 (1), January 1997.

A THERMOMECHANICAL APPROACH TO THE FORMULATION OF CONSTITUTIVE LAWS FOR GRANULAR MATERIALS

I.F.COLLINS
Department of Engineering Science, School of Engineering, University of Auckland, New Zealand..
E-mail:i.collins@auckland.ac.nz

A general method of constructing theoretical models describing the compaction and shearing of sands, soils, powders and other granular materials is described. The procedure is based upon the laws of thermo-mechanics, so that the models automatically obey the Second Law inequalities, and there is no need to appeal to "quasi-thermodynamic" stability postulates. For illustration the theory is couched in terms of the variables commonly used for tri-axial tests. It is shown that existing models can be made more realistic in a systematic way by modifying the initial dissipation function to include non-associated flow rules and shear as well as volumetric hardening. This new approach to model-building also pin-points some shortcomings of some existing procedures.

1 Introduction

Much of the observed mechanical behaviour of sands, soils, powders, aggregates and other granular materials, can be described within the continuum framework of rate-independent elastic/plastic materials. The irreversible, plastic processes may consist of "frictional effects" due to particle rearrangement by sliding or rolling, the accompanying asperity deformations, or granular crushing at sufficiently high isotropic stress levels. Many elastic/plastic models have been proposed to describe such behaviour, starting with the simple, perfectly-plastic Mohr-Coulomb model. This was followed by the family of volumetric hardening Cam-Clay models, based on conceptual ideas of energy dissipation and dilatancy, and the similar, but largely empirically based models, such as those of Lade and coworkers [4]. There is currently much interest in the development of more complicated models, such as those using kinematic hardening concepts to model the growth of anisotropy in sands and powders, and the use of multi-surface models to simulate the response of granular materials to cyclic loading histories. There is also much research into the formulation of incrementally non-linear (hypo-plastic) models.

Nearly all these theories use the classical procedure of constructing the elasticity law, yield condition, flow rule and hardening rule as four separate postulates. However modern developments in thermo-mechanics have thrown fresh light in a number of areas of irreversible continuum mechanics [5], and in particular, in the case of rate independent materials, have provided a procedure whereby these four postulates can be deduced from knowledge of the material's free energy and dissipation functions. An important consequence of this for geo-materials, is that when describing "frictional behaviour" the dissipation function has to depend on stress, and the resulting flow rule is necessarily non-associated [1,2]. In this paper we review a number of classical models in the light of these developments, and as a result establish a family of new critical state models which offer a number of advantages over the extant models.

2 The thermo-mechanical procedure for establishing constitutive laws

If the deformation is isothermal, the rate of working of the applied stresses is the sum of the rate of change of the free energy function Ψ, and the rate of dissipation function Φ, all these functions being defined per unit volume, so that:

$$\sigma_{ij}\dot{e}_{ij} = \dot{\Psi}(e_{ij}, e^p_{ij}) + \Phi(e^p_{ij}, \dot{e}^p_{ij}) \qquad (1)$$

where the free energy depends on the total and plastic strains. The dissipation function depends on the plastic strain and possibly other state variables and, if the material is rate-independent , it will also be a homogeneous function of degree 1 in the plastic strain rates. The function Φ is non-negative in value and vanishes with the plastic strain rates. Application of the standard procedures [5] shows that:

$$\sigma_{ij} = \partial\Psi / \partial e_{ij} \text{ and } \pi_{ij} = -\partial\Psi / \partial e^p_{ij} = \partial\Phi / \partial\dot{e}^p_{ij}, \qquad (2)$$

where π_{ij} is the dissipative (thermodynamic or generalized) stress. Since the derivative $\partial\Phi / \partial\dot{e}^p_{ij}$ is homogeneous of degree zero in the plastic strain rates, the five ratios of these tensor components can be eliminated from the six equations (2) to give the yield condition in π- space. In can also be shown that the plastic strain-rates are always given by the normal flow rule in this space [1,5], but that this normality is lost when the yield condition is expressed in true stress space if the dissipation function depends on any of the stress or total stain variables.[1]. If we assume that the elastic and plastic behaviour are de-coupled [1,2], the free energy function can be written as the sum of two functions $\Psi_1(e^e_{ij})$ and $\Psi_2(e^p_{ij})$, where e^e_{ij} is the elastic strain. Equations (2) now simplify to

$$\sigma_{ij} = \partial\Psi_1 / \partial e^e_{ij} = \rho_{ij} + \pi_{ij}, \qquad (3)$$

where $\rho_{ij} = \partial\Psi_2 / \partial e^p_{ij}$ is the shift stress, which has the effect of translating the yield surfaces without change of shape, as in linear kinematic hardening models [1,5]. The rate of working of the applied stresses is hence the sum of three terms:

$$\sigma_{ij}\dot{e}_{ij} = \sigma_{ij}\dot{e}^e_{ij} + \rho_{ij}\dot{e}^p_{ij} + \pi_{ij}\dot{e}^p_{ij} \qquad (4)$$

It is important to appreciate that the dissipation rate is the inner product of the plastic strain rate with the dissipative stress, not the true stress, and it is hence necessary to distinguish between the rate of plastic work and dissipation rate:

$$\dot{W}^p = \sigma_{ij}\dot{e}^p_{ij}, \text{ but } \Phi = \pi_{ij}\dot{e}^p_{ij}, \qquad (5)$$

which are only equal when the shift stress is zero.

3 A hierarchy of models for granular materials

Here we will discuss a number of one and two-dimensional models, beginning with the standard ones. For simplicity we will use the standard notation used for the analysis of triaxial test data. The isotropic and shear invariants of the true effective stress will be denoted by (p',q), and of the dissipative stress by (π',τ).

3.1 *Isotropic compression*

Here, the only non-zero strains are the volumetric components., and the only effective stress is the effective pressure p'. We will assume these strains to be logarithmic, so that, provided the solid phase is incompressible, they can be expressed in terms of the specific volume v, porosity n or voids ratio e. For example:

$$e_v = \ln(v_o / v), e_v^e = \ln(v^p / v), \text{and } e_v^p = \ln(v_0 / v^p) \tag{6}$$

where v^p is the "plastic specific volume", defined to be the specific volume of the element after it is unloaded from the current state to the initial reference state . We could replace the strains in the general formulation by any of these standard volume fraction measures, but here, for simplicity we will continue with the standard strain formulation.

Since the dissipation function has the dimensions of stress / time and is non-negative, it must, in the case of isotropic compression, be of the form $\pi_c(e_v^p)\left|\dot{e}_v^p\right|$, so that

$$\pi'\dot{e}_v^p = \Phi = \pi_c(e_v^p)\left|\dot{e}_v^p\right|, \tag{7}$$

where $\pi_c(e_v^p)$ is the hardening function. It follows that the yield condition in dissipative stress space is $\pi' = \pm\pi_c(e_v^p)$. It is clear that if we want to model a material with no strength in tension, we must choose the free energy function so that the shift stress is also $\pi_c(e_v^p)$, so that $p' = \pi' + \pi_c$. Thus in true stress space, yield occurs when $p' = 0$ or $p_c(e_v^p)$, where $p_c = 2\pi_c$ is the standard normal consolidation pressure. This situation is illustrated in Fig.1.

3.2 *Coulomb Friction Model*

Consider now a material whose dissipation function is proportional to the plastic shear strain-rate and the mean effective pressure $Mp'\left|\dot{e}_\gamma^p\right|$, so

$$\pi'\dot{e}_v^p + \tau\dot{e}_\gamma^p = \Phi = Mp'\left|\dot{e}_\gamma^p\right|. \tag{8}$$

This is the simplest model, which attempts to model Coulomb's law of dry friction. From (8) it follows that the shear component is $\tau = Mp'$, and the volumetric plastic strain rates are identically zero. This dissipation function is only meaningful if the plastic volume strains are zero, since, otherwise, according to (8), any plastic volume change can occur without producing any energy dissipation. If we transfer now to true stress space, by simply equating the true and dissipative stress (no shift stress), we deduce the classical Coulomb Model but with an incompressible flow rule (Fig 1):

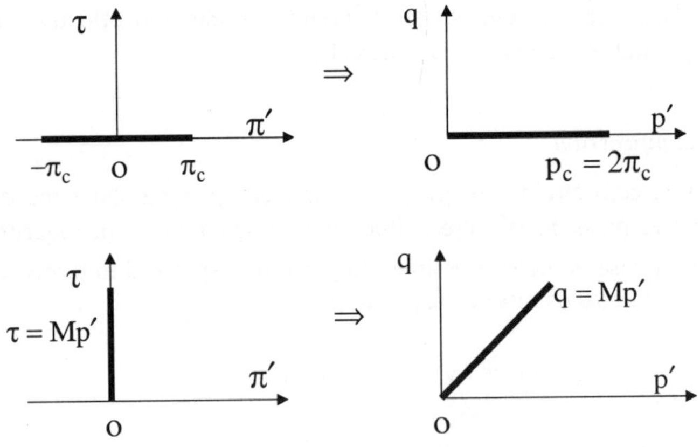

Figure 1. Transformation from dissipative stress space to true stress space for (a) isotropic compression, (b) pure friction.

$$q = Mp' \quad \text{and} \quad \dot{e}^p_v = 0 \tag{9}$$

3.3 Original Cam-Clay

Although this model is the precursor of many important developments, it is a pathological example. The dissipation function is the same as in 3.2. The standard argument is then to equate the plastic work-rate to this dissipation function: $p'\dot{e}^p_v + q\dot{e}^p_\gamma = Mp'|\dot{e}^p_\gamma|$, set up a differential equation for the plastic potential q=q(p'), by noting that $dq/dp' = -\dot{e}^p_v / \dot{e}^p_\gamma$, and integrating to derive the potential function : $q = Mp'\ln(p'_c/p')$. A normal or associated flow rule is then assumed so that the yield surface and potential are identical. This argument is flawed in many respects. It fails to draw the distinction between dissipation rate and plastic work, it assumes a normal flow rule, whereas in fact the thermo-mechanical functions uniquely determine *both* the flow rule and yield condition. Most importantly of all however is, as noted above, the fact that this simple "Coulomb" dissipation function can only be valid if the volumetric strain rates are identically zero. The simple model in 3.2 is the true model associated with this dissipation function. Whilst satisfying Drucker's postulate, the original Cam Clay model violates the Second Law of Thermodynamics.

3.4 Modified Cam Clay

This model stems from a dissipation function, which is proportional to the isotropic consolidation, volumetric hardening function and is the natural extension of that in section 3.1.

$$\pi'\dot{e}^p_v + \tau\dot{e}^p_\gamma = \Phi = \pi_c(e^p_v)\{(\dot{e}^p_v)^2 + M^2(\dot{e}^p_\gamma))\}^{1/2} \tag{10}$$

The dissipative stresses are obtained by using (2):

$$\pi' = {\pi_c}^2\dot{e}^p_v / \Phi \text{, and } \tau = M^2{\pi_c}^2\dot{e}^p_\gamma / \Phi \tag{11}$$

so that eliminating the strain rates between (10) and (11), we obtain the elliptical yield surface in dissipative stress space:

$$(\pi'/\pi_c)^2 + (\tau/M\pi_c)^2 = 1 \qquad (12)$$

and finally in true stress space, by introducing the same shift stress as in section 3.1.

$$(p' - p_c/2)^2/(p_c/2)^2 + q^2/M^2(p_c/2)^2 = 1\text{, or } q = M[p'(p_c - p')]^{1/2} \qquad (13)$$

The flow rule is necessarily associated, since the dissipation function does not involve the true stresses – see Figs 2(a) and (b).

3.5 *A new model with a non-associated flow rule*

The above examples suggest a general procedure for developing a hierarchy of models, by first constructing the yield surface and the, necessarily normal, flow rule in dissipative stress space. These are then transformed to true stress space, either by using a shift stress, or a "shearing deformation "as in 3.2, or both. For example consider

$$\Phi = \pi_c(e_v^p)\{(\dot{e}_v^p)^2 + M^2\xi^2(\dot{e}_\gamma^p)^2\}^{1/2}\text{, where } \xi = (\alpha + (1-\alpha)(p'/\pi_c)) \qquad (14)$$

This dissipation function reduces to that for modified Cam Clay when the constant $\alpha = 1$, but otherwise this function depends on p' and so leads to a model with non-associated flow rule. The plastic volumetric strain is still zero on the critical state line $q = M p'$, but this is no longer the line of maxima on the yield surfaces. The equation of the yield surfaces is:

$$q = M(\alpha + (1-\alpha)(2p'/p_c))[p'(p_c - p')]^{1/2} \qquad (15)$$

which are illustrated in Fig 3 (c) for the case where α =0.5.

3.6 *A new model with volumetric and shear hardening*

In order to model the peak stress behaviour of over-consolidated clays and dense sands in drained and undrained tests it is necessary to depart from the volumetric hardening assumption of the classical critical state models. Nova [6] suggested a model which depended on a linear combination of volumetric and shear strains, whilst Lade [4] has assumed that, as in metals, the yield surfaces are curves of constant plastic work. Here we prefer to follow a suggestion of Krenk [3] and assume that the hardening parameter is a weighted average of the volumetric and shearing plastic work, and write the increment in the consolidation pressure as:

$$\delta p_c = (p'\delta e_v^p + \beta q \delta e_\gamma^p)/(\lambda - \kappa) \qquad (15)$$

where λ and κ are the slopes of the consolidation and elastic loading lines in the classical $e_v^p - \ln(p')$ plots, and β is the weighting parameter. When β = 0, we recover the classic volumetric hardening model. Changing the hardening law in this way does not change the shape of the yield curves or the nature of the flow rule, but it does significantly effect the stress-strain loading paths. The major features of the observed stress-paths, including peak stress behaviour, and the "kink" in the undrained stress path for dense sands, can be reproduced with this mode

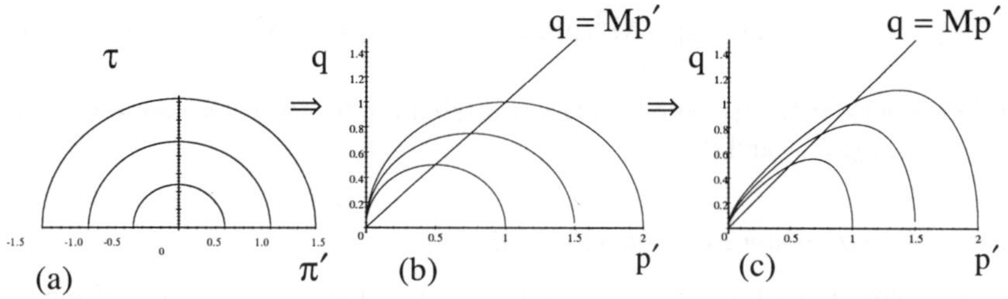

Figure 2. (a) Ellipses in dissipative stress space, (b) Modified Cam Clay (c) Non-associated model.

4 Conclusions

In this paper, it has been shown that a more rigorous use of thermodynamic concepts and methodologies can throw new light on the existing theories of the mechanics of granular materials. In particular it has been shown that the original Cam Clay model is in violation of the Second Law. More importantly, however, it provides a systematic procedure for developing new models. This has been indicated with brief discussions of an extended model exhibiting non-associated flow behaviour, and another which includes shear as well as volumetric hardening. These two extensions can be combined into a composite model which just adds two new model parameters, but which adds considerable realism to the material simulation. Full details will be given elsewhere.

References

1. Collins, I.F. and Houlsby, G.T. Application of thermomechanical principles to the modelling of geotechnical materials. *Proc. Roy. Soc. Lond. A* **453** (1997) pp 1975-2001.
2. Houlsby, G.T. and Puzrin, A.M. A thermomechanical framework for constitutive models for rate-independent dissipative materials.*Int.J.Plasticity*,**16**(2000),pp1017-1047.
3. Krenk, S. A characteristic state plasticity model for granular materials. *Mechanics of Granular Materials*, ed. by N.A.Fleck and A.C.F.Cocks, Kluwer. (1996) pp 83-94.
4.Lade, P.V. Effects of voids and volume changes on the behaviour of frictional materials. *Int.J.Num.Anal.Meth.Geomechanics*,**12** (1988),pp 351-370.
5.Maugin, G.A. *The thermomechanics of nonlinear irreversible behaviors*. World Scientific Series on Nonlinear Science,(1999)
6. Nova, R. On the hardening of soils. *Archives of Mechanics*, **29** (1977) pp 445-458.

ELASTIC-PLASTIC TENSOR OF THE TANGENT MODULUS BASED ON ASYMMETRIC NONQUADRATIC YIELD FUNCTION

VU T. NGUYEN*, ZHIPING CHEN** AND PETER F. THOMSON*

*School of Physics and Materials Engineering - Monash University
P.O. Box 69M Monash University, Vic. 3800 – Australia
Email: vu.nguyen@spme.monash.edu.au

**Boeing - Hawker De Havilland
226 Lorimer Street,Port Melbourne, Vic 3207
Email: Zhiping.Chen@Australia.Boeing.com

A nonquadratic yield function (Yld96) for aluminium alloy sheet, which simultaneously accounts for the anisotropy of uniaxial yield stress and *r*-values, was developed by Barlat *et al.*[1]. This model was based on isotropic hardening and was recently extended to account for the difference between tensile and compressive yield strength[2], resulting in predicting earing profiles in good agreement with experimental results. Isotropic hardening, however, is mainly applicable to monotonic loading. For problems with stress reversal, it fails to predict the material behaviour due to the Bauschinger effect. In this paper, an elastic-plastic tensor of the tangent modulus for plane stress problems based on Yld96 is derived, which includes the isotropic and kinematic hardening to account for the Bauschinger effect. It provides a convenient means for numerical analysis of plasticity such as the finite element analysis.

1 Introduction

Sheet metal forming processes have been employed widely in the automotive, aerospace, packaging and electronic industries. The analysis of such forming processes therefore plays a crucial role in reducing the product development time and final product cost. Various approaches have been used to study sheet metal forming processes, including experimental, analytical and numerical methods. With the advent of high-speed computers numerical methods, notably the Finite Element Method (FEM), have become powerful means for analysis of sheet metal forming. FEM input however requires material models or constitutive equations, which can be formulated at microscopic or macroscopic level to account for material characteristics.

The macroscopic approach is more phenomenological, but more attractive at the design stage of forming processes because it is numerically efficient[1]. Along this direction, several anisotropic yield functions associated with work hardening have been proposed. The quadratic criterion due to Hill [3] has been useful to explain the phenomena associated with anisotropic plasticity particularly for steel, followed by a number of non-quadratic yield functions, which can be used to improve the description of the yield of aluminium alloys (see Barlat *et al.*[1]).

Recently, Barlat *et al.*[1] proposed a nonquadratic yield function - Yld96 suitable for aluminium alloy sheet, which simultaneously accounts for the anisotropy of uniaxial yield stresses and *r*-values. This model was based on the isotropic hardening assumption and recently extended to account for the difference between tensile and compressive yield strength[2], which predicted earing profiles in better agreement with experimental results. Isotropic hardening, however, is mainly applicable to monotonic loading. For problems with stress reversal, it fails to predict the material behaviour due to the Bauschinger effect. In this paper, an elastic-plastic tensor of the tangent modulus for plane stress problems based on Yld96, which includes the isotropic and kinematic hardening to account for the Bauschinger effect is derived. The full description of Yld96 in 3 dimensional stress space is given by Barlat *et al.*[1]. Yld96 in plane stress subspace is summerised in the next section.

2 Yld96 in plane stress subspace

In what follows, it is assumed that yielding is independent of hydrostatic pressure and deformation is incompressible. The axes (denoted as x, y, z), which represent the mutually orthogonal axes of anisotropy, are aligned with the initial rolling, transverse and normal directions of the sheet. The components of the isotropic plasticity equivalent (IPE) deviatoric stress tensor s_{ij} are defined with respect to the components of Cauchy stress σ_{ij} in the matrix form as

$$\mathbf{s} = \mathbf{L}\boldsymbol{\sigma} \tag{2.1}$$

where the vector notation to express $\boldsymbol{\sigma}$ and $\mathbf{s}$ are

$$\boldsymbol{\sigma} := [\sigma_{xx}\ \sigma_{yy}\ \sigma_{xy}]^T, \qquad \mathbf{s} := [s_{xx}\ s_{yy}\ s_{xy}]^T \tag{2.2}$$

L is the 4th order linear operator, which can account for different kinds of material symmetry such as monoclinic, hexagonal, cubic etc. (Karafillis and Boyce[4]). For orthotropic symmetry, in the plane stress 3x3 square matrix representation, **L** is given by

$$\mathbf{L} := \begin{bmatrix} (c_2 + c_3)/3 & -c_3/3 & 0 \\ -c_3/3 & (c_3 + c_1)/3 & 0 \\ 0 & 0 & c_6 \end{bmatrix} \tag{2.3}$$

The yield function Yld96 is defined as

$$\varphi = \alpha_1 |s_2 - s_3|^a + \alpha_2 |s_3 - s_1|^a + \alpha_3 |s_1 - s_2|^a = 2\sigma_e^a \tag{2.4}$$

with a = 6 and a = 8 recommended to represent the behaviour of BCC and FCC materials, respectively. σ_e is a measure of the magnitude of the yield surface. $s_{i\ (i=1,\,2,\,3)}$ are the principal deviatoric stresses of s_{ij} given by

$$s_{1,2} = \frac{s_{xx} + s_{yy}}{2} \pm \sqrt{\left(\frac{s_{xx} - s_{yy}}{2}\right)^2 + s_{xy}^2} \tag{2.5}$$

Because the tensor s_{ij} is traceless, the third principal value of s_{ij} will be

$$s_3\ (\equiv s_{zz}) = -(s_1 + s_2) = -(s_{xx} + s_{yy}) \tag{2.6}$$

In Eq.(2.4), $\alpha_{k\ (k=1,\,2,\,3)}$ are defined as

$$\begin{aligned} \alpha_1 &= \alpha_x \cos^2\theta + \alpha_y \sin^2\theta \\ \alpha_2 &= \alpha_x \sin^2\theta + \alpha_y \cos^2\theta \\ \alpha_3 &= \alpha_{z0} \cos^2 2\theta + \alpha_{z1} \sin^2 2\theta \end{aligned} \tag{2.7}$$

where $\alpha_{z0} = 1$, θ is the principal direction measured counter-clockwise from the x-axis determined by

$$\theta = \tan^{-1} \frac{(s_1 - s_{xx})}{s_{xy}} \tag{2.8}$$

The material coefficient c_k in Eq.(2.3) and the quantities α_x, α_y, α_{z0} and α_{z1} in Eq.(2.7) describe anisotropic properties[1] which can be calculated from seven test results, i.e. tensile yield stresses at 0^0, 45^0 and 90^0 from the rolling direction (RD), balanced biaxial yield stress measured from the bulge test, and the *r*-values at 0^0, 45^0 and 90^0 from the RD.

3 Elastic plastic stress strain matrix

Before introducing the elastic plastic matrix ($\mathbf{C_M}^{ep}$) based on mixed-hardening, which takes simultaneously into account the isotropic and kinematic hardening, it is necessary to derive the elastic plastic matrix of Yld96 based on isotropic ($\mathbf{C_I}^{ep}$) and kinematic ($\mathbf{C_k}^{ep}$) hardening.

3.1 Isotropic hardening elastic plastic stress strain matrix $\mathbf{C_I}^{ep}$

In addition to the stress vector form, the components of the strain tensors are collected in the vector form as:

$$\boldsymbol{\varepsilon} := [\varepsilon_{xx}\ \varepsilon_{yy}\ \varepsilon_{xy}]^T; \qquad \boldsymbol{\varepsilon}^e := [\varepsilon^e_{xx}\ \varepsilon^e_{yy}\ \varepsilon^e_{xy}]^T; \qquad \boldsymbol{\varepsilon}^p := [\varepsilon^p_{xx}\ \varepsilon^p_{yy}\ \varepsilon^p_{xy}]^T \tag{3.1}$$

where $\boldsymbol{\varepsilon}$, $\boldsymbol{\varepsilon}^e$ and $\boldsymbol{\varepsilon}^p$ are total, elastic and plastic strain respectively. The elastoplastic equations can be formulated in the vector form as follows:

$$\boldsymbol{\varepsilon} = \boldsymbol{\varepsilon}^e + \boldsymbol{\varepsilon}^p \tag{3.2}$$

$$\boldsymbol{\sigma} = \mathbf{C}^e\, \boldsymbol{\varepsilon}^e \tag{3.3}$$

$$d\boldsymbol{\varepsilon}^p = d\lambda \boldsymbol{\sigma}' \tag{3.4}$$

$$Q = \varphi - k, \qquad k = 2\sigma_e^{\ a} \tag{3.5}$$

where $\mathbf{C}^e$ is the symmetric elastic constitutive matrix for plane stress, $d\lambda$ is a scalar function and k defines the hardening law in terms of the effective plastic strain. We have set for convenience:

$$\boldsymbol{\sigma}' = \partial\varphi/\partial\boldsymbol{\sigma} := [\partial\varphi/\partial\sigma_{xx}\ \partial\varphi/\partial\sigma_{yy}\ \partial\varphi/\partial\sigma_{xy}]^T \tag{3.6}$$

The calculation of the normal to the yield surface $\boldsymbol{\sigma}'$ is summarised in the appendix A.

For plastic loading, the plastic consistency condition requires that the stress point must remain on the yield surface, which implies that $dQ \equiv 0$. By differentiating Eq.(3.5) and assuming a unique relationship between σ_e and the effective plastic strain ($d\varepsilon_p$), we obtain:

$$dQ = \boldsymbol{\sigma}'^T d\boldsymbol{\sigma} - 2a\sigma_e^{\ a-2} H_p\, \boldsymbol{\sigma}^T\boldsymbol{\sigma}'\, d\lambda = 0 \tag{3.7}$$

where $H_p = d\sigma_e/d\varepsilon_p$. The expression for the stress increments is obtained by time differentiation of Eq.(3.3) in addition with the use of Eqs.(3.2) and (3.4), i.e.

$$d\boldsymbol{\sigma} = \mathbf{C}^e d\boldsymbol{\varepsilon} - \boldsymbol{\eta} d\lambda \tag{3.8a}$$

where $$\boldsymbol{\eta} = \mathbf{C}^e\boldsymbol{\sigma}' \tag{3.8b}$$

Substituting Eq.(3.8a) into Eq.(3.7) yields

$$d\lambda = \frac{\boldsymbol{\eta}^T d\boldsymbol{\varepsilon}}{N_I} \tag{3.9a}$$

where $$N_I = \boldsymbol{\eta}^T\boldsymbol{\sigma}' + 2a\sigma_e^{\ a-2} H_p\, \boldsymbol{\sigma}^T\, \boldsymbol{\sigma}' \tag{3.9b}$$

Finally, substituting $d\lambda$ from Eq.(3.9) into Eq.(3.8a) results in:

$$d\boldsymbol{\sigma} = \mathbf{C_I}^{ep} d\boldsymbol{\varepsilon} \tag{3.10}$$

where $$\mathbf{C}_I^{ep} = \mathbf{C}^e - \frac{1}{N_I}\boldsymbol{\eta}\boldsymbol{\eta}^T \tag{3.11}$$

3.2 Kinematic hardening elastic plastic stress strain matrix $\mathbf{C_k}^{ep}$

In order to account for the Bauschinger effect, kinematic hardening and back stress are necessarily introduced into the Yld96 formulation. Let B_{ij} be the second order invariant back-stress tensor, $\bar{\sigma}_{ij}$ be the reduced stress tensor, and $\bar{s}_{ij}$ be the reduced IPE deviatoric stress tensor. In plane stress subspace, the vector notation of which can be expressed as:

$$\mathbf{B}:= [B_{xx}\ B_{yy}\ B_{xy}]^T, \qquad \bar{\boldsymbol{\sigma}}:= [\bar{\sigma}_{xx}\ \bar{\sigma}_{yy}\ \bar{\sigma}_{xy}]^T, \qquad \bar{\mathbf{s}}:= [\bar{s}_{xx}\ \bar{s}_{yy}\ \bar{s}_{xy}]^T \tag{3.12}$$

With this notation, the following relationships can be drawn

$$\bar{\boldsymbol{\sigma}} = \boldsymbol{\sigma} - \mathbf{B}, \qquad \bar{\mathbf{s}} = \mathbf{L}\bar{\boldsymbol{\sigma}} \tag{3.13}$$

The yield function becomes:

$$\varphi = \alpha_1|\bar{s}_2 - \bar{s}_3|^a + \alpha_2|\bar{s}_3 - \bar{s}_1|^a + \alpha_3|\bar{s}_1 - \bar{s}_2|^a = 2\sigma_e^a \tag{3.14}$$

where $\bar{s}_k$ are the principal values of $\bar{s}_{ij}$. Assuming that only kinematic hardening takes place, the effective stress σ_e remains constant during plastic flow. The consistency condition now has the form:

$$dQ = \frac{\partial\varphi}{\partial\boldsymbol{\sigma}}^T d\boldsymbol{\sigma} + \frac{\partial\varphi}{\partial\mathbf{B}}^T d\mathbf{B} \tag{3.15}$$

Employing Prager's hardening rule and the use of Eq.(3.4) lead to:

$$d\mathbf{B} = c\, d\boldsymbol{\varepsilon}^p = c\, \boldsymbol{\sigma}' d\lambda \tag{3.16}$$

In Eq.(3.16), c is a material constant. It should be noted that:

$$\frac{\partial\varphi}{\partial\boldsymbol{\sigma}} = \frac{\partial\varphi}{\partial\bar{\boldsymbol{\sigma}}} \qquad \text{and} \qquad \frac{\partial\varphi}{\partial\mathbf{B}} = -\frac{\partial\varphi}{\partial\bar{\boldsymbol{\sigma}}} \tag{3.17}$$

Substituting Eqs.(3.16) and (3.17) into the consistency condition (3.15) results in:

$$dQ = \boldsymbol{\sigma}'^T d\boldsymbol{\sigma} - c\, \boldsymbol{\sigma}'^T \boldsymbol{\sigma}'\, d\lambda = 0 \tag{3.18}$$

Solving for $d\lambda$ from Eqs.(3.18) and (3.8a) gives:

$$d\lambda = \frac{\boldsymbol{\eta}^T d\boldsymbol{\varepsilon}}{N_k} \tag{3.19a}$$

where

$$N_k = \boldsymbol{\eta}^T\boldsymbol{\sigma}' + c\, \boldsymbol{\sigma}'^T\, \boldsymbol{\sigma}' \tag{3.19b}$$

Finally, substituting $d\lambda$ from Eq.(3.19) into Eq.(3.8a) results in:

$$d\boldsymbol{\sigma} = \mathbf{C}_k^{ep} d\boldsymbol{\varepsilon} \tag{3.20}$$

where

$$\mathbf{C}_k^{ep} = \mathbf{C}^e - \frac{1}{N_k}\boldsymbol{\eta}\boldsymbol{\eta}^T \tag{3.21}$$

3.3 *Mixed-hardening elastic plastic stress strain matrix* $\mathbf{C}_M^{ep}$

Assuming that the plastic strain increments can be further split into two parts[5], the isotropic hardening part $d\boldsymbol{\varepsilon}^{pI}$, and the kinematic hardening part $d\boldsymbol{\varepsilon}^{pk}$,

$$d\boldsymbol{\varepsilon}^p = d\boldsymbol{\varepsilon}^{pI} + d\boldsymbol{\varepsilon}^{pk} \tag{3.22}$$

The parameter M , $0 \le M \le 1$, is introduced to represent the relative amounts of isotropic and kinematic hardening, i.e.

$$d\boldsymbol{\varepsilon}^{pI} = M\, d\boldsymbol{\varepsilon}^p \qquad \text{and} \qquad d\boldsymbol{\varepsilon}^{pk} = (1-M)\, d\boldsymbol{\varepsilon}^p \tag{3.23}$$

In this case, the consistency condition derived from Eq.(3.5) has the form

$$dQ = d\varphi - dk = 0 \tag{3.24}$$

It is necessary to formulate $d\varphi$ and dk. The expression for $d\varphi$ can be employed from Eq.(3.15), i.e.

$$d\varphi = \frac{\partial\varphi}{\partial\boldsymbol{\sigma}}^T d\boldsymbol{\sigma} + \frac{\partial\varphi}{\partial\mathbf{B}}^T d\mathbf{B} \tag{3.25}$$

Prager's hardening rule, Eqs.(3.23) with (3.4) gives:

$$\mathbf{dB} = c\, \mathbf{d\varepsilon}^{pk} = (1\text{-}M)\, c\, \boldsymbol{\sigma}'d\lambda \tag{3.26}$$

Substituting for **dB** from Eq.(3.26) into Eq.(3.25) and using Eq.(3.17) yields:

$$d\varphi = \boldsymbol{\sigma}'^{T}\mathbf{d\sigma} - (1\text{-}M)\, c\, \boldsymbol{\sigma}'^{T}\boldsymbol{\sigma}'\, d\lambda = 0 \tag{3.27}$$

dk is now derived from Eq.(3.5), assuming a unique relationship between the effective stress σ_e and the isotropic hardening part of the effective plastic strain ($d\varepsilon_p{}^I$).

$$dk = 2a\sigma_e{}^{a-1} H_p{}^I\, d\varepsilon_p{}^I \tag{3.28a}$$

where $$H_p{}^I = d\sigma_e/d\varepsilon_p{}^I \tag{3.28b}$$

From the plastic work associated with the isotropic hardening part, one can show that:

$$d\varepsilon_p{}^I = M\, (1/\sigma_e)\, \boldsymbol{\sigma}^T\boldsymbol{\sigma}'\, d\lambda \tag{3.29}$$

Substituting for $d\varepsilon_p{}^I$ from Eq.(3.29) into Eq.(3.28a) yields

$$dk = M\, 2a\sigma_e{}^{a-2} H_p{}^I \boldsymbol{\sigma}^T\boldsymbol{\sigma}'\, d\lambda \tag{3.30}$$

The expression of dQ is now explicitly derived from Eqs.(3.24), (3.27) and (3.30) as

$$dQ = \boldsymbol{\sigma}'^{T}\mathbf{d\sigma} - (1\text{-}M)\, c\, \boldsymbol{\sigma}'^{T}\boldsymbol{\sigma}'\, d\lambda - M\, 2a\sigma_e{}^{a-2} H_p{}^I \boldsymbol{\sigma}^T\boldsymbol{\sigma}'\, d\lambda \tag{3.31}$$

The symmetry of $\mathbf{C}^e$ and the use of Eqs.(3.8a) and (3.8b) lead to

$$dQ = \boldsymbol{\eta}^T\mathbf{d\varepsilon} - \boldsymbol{\eta}^T\boldsymbol{\sigma}'d\lambda - (1\text{-}M)\, c\, \boldsymbol{\sigma}'^{T}\boldsymbol{\sigma}'\, d\lambda - M\, 2a\sigma_e{}^{a-2} H_p{}^I \boldsymbol{\sigma}^T\boldsymbol{\sigma}'\, d\lambda = 0 \tag{3.32}$$

Solving for $d\lambda$, i.e.

$$d\lambda = \frac{\boldsymbol{\eta}^T \mathbf{d\varepsilon}}{N_M} \tag{3.33}$$

where $$N_M = \boldsymbol{\eta}^T\boldsymbol{\sigma}' + (1\text{-}M)\, c\, \boldsymbol{\sigma}'^{T}\boldsymbol{\sigma}' + M\, 2a\sigma_e{}^{a-2} H_p{}^I\, \boldsymbol{\sigma}^T\, \boldsymbol{\sigma}'\, d\lambda \tag{3.34}$$

Finally, by substituting for $d\lambda$ from Eq.(3.33) into the stress increments (3.8a) we obtain

$$\mathbf{d\sigma} = \mathbf{C}_M{}^{ep}\mathbf{d\varepsilon} \tag{3.35}$$

where

$$\mathbf{C}_M^{ep} = \mathbf{C}^e - \frac{1}{N_M}\boldsymbol{\eta}\boldsymbol{\eta}^T \tag{3.36}$$

4 Conclusion

An elastic-plastic stiffness tensor of tangent modulus for plane stress problems based on Yld96, which includes nonlinear isotropic and linear kinematic hardening to account for the Bauschinger effect was derived. Prager's kinematic hardening rule was employed in this paper but not restricted to.

It should be noted that if M = 1, i.e. only isotropic hardening takes place, $N_M = N_I$ (Eq.3.9b), Eq.(3.36) simplifies to Eq.(3.11). If M = 0, i.e. only kinematic hardening occurs, then $N_M = N_k$ (Eq.3.19b). In this case Eq.(3.21) is recovered from Eq.(3.36).

References

1. Barlat, F., Maeda, Y., Chung, K., Yanagawa, M., Brem, J.C., Hayashida, Y, Lege, D.J., Matsui, K., Murtha, S.J., Hattori, S., Becker, R.C. and Makosey, S. (1997), Yield function development for aluminium alloy sheets, *J. Mech. Phys. Solids*, v.45, pp. 1727-1763.

2. Yoon, J.W., Barlat, F., Chung, K., Pourboghrat, F. and Yang, D.Y. (2000), Earing predictions based on asymetric nonquadratic yield function, *Int. J. Plasticity*, v.16, pp. 1075-1104.
3. Hill, R. (1948), A theory of the yielding and plastic flow of anisotropic metals, *Proc. R. Soc. Lond.*, A193, pp.281-297.
4. Karafillis, A.P. and Boyce, M.C. (1993), A general anisotropic yield criterion using bounds and a transformation weighting tensor, *J. Mech. Phys. Solids*, v.41, pp. 1859-1886.
5. Chen, W.F. and Han, D.J. (1988), *Plasticity for structural engineers*, Springer-Verlag, New York.

Appendix A

The quantity $\boldsymbol{\sigma}' = \{\partial\varphi/\partial\boldsymbol{\sigma}\}$ is obtained through the chain rule, i.e.

$$\frac{\partial\varphi}{\partial\boldsymbol{\sigma}}^{T} = \frac{\partial\varphi}{\partial\tilde{\boldsymbol{\eta}}}^{T}\frac{\partial\tilde{\boldsymbol{\eta}}}{\partial\mathbf{s}}\frac{\partial\mathbf{s}}{\partial\boldsymbol{\sigma}} \tag{A1}$$

where $$\tilde{\boldsymbol{\eta}} := [s_1\ s_2\ s_3\ \theta]^T \tag{A2}$$

$$\frac{\partial\varphi}{\partial\tilde{\boldsymbol{\eta}}} = \begin{bmatrix} a[\alpha_3(s_1-s_2)|s_1-s_2|^{a-2} - \alpha_2(s_3-s_1)|s_3-s_1|^{a-2}] \\ a[-\alpha_3(s_1-s_2)|s_1-s_2|^{a-2} + \alpha_1(s_2-s_3)|s_2-s_3|^{a-2}] \\ a[-\alpha_1(s_2-s_3)|s_2-s_3|^{a-2} + \alpha_2(s_3-s_1)|s_3-s_1|^{a-2}] \\ (\alpha_x-\alpha_y)\sin 2\theta\,(-|s_2-s_3|^a + |s_3-s_1|^a) - 2(\alpha_{z0}-\alpha_{z1})\sin 4\theta\,|s_1-s_2|^a \end{bmatrix} \tag{A3}$$

$$\frac{\partial\tilde{\boldsymbol{\eta}}}{\partial\mathbf{s}} = \begin{bmatrix} \frac{1}{2}+\frac{s_{xx}-s_{yy}}{4r} & \frac{1}{2}+\frac{s_{xx}-s_{yy}}{4r} & \frac{s_{xy}}{r} \\ \frac{1}{2}-\frac{s_{xx}-s_{xy}}{4r} & \frac{1}{2}+\frac{s_{xx}-s_{xy}}{4r} & -\frac{s_{xy}}{r} \\ -1 & -1 & 0 \\ \frac{1}{s_{xy}(1+\tan^2\theta)}\left(\frac{\partial\tilde{\eta}_1}{\partial s_{xx}}-1\right) & \frac{1}{s_{xy}(1+\tan^2\theta)}\frac{\partial\tilde{\eta}_1}{\partial s_{yy}} & \frac{1}{s_{xy}(1+\tan^2\theta)}\left(\frac{\partial\tilde{\eta}_1}{\partial s_{xy}}-\tan\theta\right) \end{bmatrix} \tag{A4}$$

where

$$r = \sqrt{\left(\frac{s_{xx}-s_{yy}}{2}\right)^2 + s_{xy}^2} \tag{A5}$$

The components of the 4th row of Eq.(A4) are obtained from the relationship given by Eqs.(2.5) and (2.8), which results

$$\frac{\partial\theta}{\partial s_{\alpha\beta}} = \frac{1}{1+\tan^2\theta}\left(\frac{\partial(\tan\theta)}{\partial\tilde{\eta}_1}\frac{\partial\tilde{\eta}_1}{\partial s_{\alpha\beta}} + \frac{\partial(\tan\theta)}{\partial s_{\alpha\beta}}\right), \quad \begin{matrix}\alpha = x, y, z \\ \beta = x, y, z\end{matrix} \tag{A6}$$

Finally, from Eq.(2.1) we have

$$\frac{\partial\mathbf{s}}{\partial\boldsymbol{\sigma}} = \mathbf{L} \tag{A7}$$

DERIVATION OF MOHR-COULOMB CRITERIA FROM RIGID –PLASTICITY THEORY

GRAHAM J. WEIR AND ROGER M. YOUNG

Industrial Research Ltd, PO Box 31 310, Lower Hutt, New Zealand
E-mail: g.weir@irl.cri.nz

Mohr-Coulomb Theory (MCT) assumes that the normal to the plane of failure lies in the plane containing the maximum and minimum principal stresses. This assumption can be derived from Elasto-Plasticity Theory (EPT). Both MCT and EPT predict the same orientation of the failure planes for associative failure theories, but this orientation is independent of the elastic parameters in EPT. In this paper we present a new (second-order) Rigid-Plastic Theory (RPT) of failure and show that it also agrees with EPT for associative failure theories, but it can predict failure angles different from those predicted by EPT for non-associative failure theories. The experimental results on the orientation of failure planes in unconsolidated sands are consistent with the predictions of RPT.

1 Introduction

Failure in solids is often characterised by the existence of well-defined failure surfaces, or failure planes. Observation of the orientation of these failure planes in shear-cells, laboratory tests or geological structures is used to infer properties of the state of stress in the solid at the time of failure.

Perhaps the most widely used theories for the interpretation of the state of stress in these failed solids are those of Coulomb proposed in 1773 and Mohr proposed in 1900 [1]. This early work predicted that the orientation of the failure planes depended only on the state of stress at the time of failure. However, in the early 20th century, workers such as Taylor and Quinney [5] pointed out that because of work hardening (and softening), failure depended on both the state of stress and also on the past history of deformation in the solid. In order to relate the failure of a solid to its stress and deformation history, plasticity theory was used.

A natural plasticity theory coupling stress and deformation in solids is elasto-plasticity theory (EPT), which assumes the solid has both elastic and plastic properties. The orientation of failure planes in EPT has been derived by Rice [2] and others [3] [4]. The key factors in EPT are a yield condition defining plastic behaviour, an incremental plastic flow law, and an elastic compliance tensor relating stress to elastic strains.

While EPT has wide acceptance [6], there are several areas of application where it encounters conceptual and quantitative difficulties. First, not all materials have an elastic behaviour. For example, unconsolidated sands are not elastic in behaviour, but can still show well defined failure surfaces. However, when the elastic compliance tensor is set to zero in EPT, the algorithm for calculating the orientation of the failure surface becomes singular. Second, for certain important classes of behaviour in EPT, such as that expressed by a Mohr-Coulomb yield and flow law, the orientation of the failure surfaces is independent of the elastic properties of the solid, but the existence of the elastic properties is vital in the calculational algorithm.

Hence there is a case for constructing a theory of failure surfaces in solids which applies in the absence of elasticity. However, given the success of EPT, such a theory may need the equivalent of an elastic compliance tensor, since this structure is vital in EPT. It is not obvious how this can be achieved by considering the failure surfaces in EPT across which first order derivatives of velocities can be discontinuous. However, there is a chance if second order derivatives of the velocity are discontinuous across the failure surface.

In this paper, we present and analyze a general rigid plastic theory (RPT) in which stress and first order derivatives of velocity are continuous across a failure surface, but higher order derivatives need not be continuous. We show that a Hessian structure appears in the analysis, and that this Hessian is analogous to the elastic compliance tensor in EPT. However, this Hessian is described completely within our RPT by the plastic behaviour of the system, and is not a freely defined structure, as in EPT. Finally, we show how this second order RPT imposes severe conditions on the failure properties of a Mohr-Coulomb system, and that these constraints for an associative theory correspond to earlier assumptions made by Coulomb and Mohr.

2 Second order rigid plasticity theory

The following assumptions are made in the basic model:

1. An infinitesimal surface element exits across which stress and first order derivatives of velocity are continuous, but first order derivatives of stress and second order derivatives of velocity may be discontinuous.
2. A solid in a wholly plastic state exists on either side of this surface element.
3. Away from the surface element, Newtons law of motion holds; the solid satisfies a yield condition F =0; and a flow rule holds.
4. A hardness function applies to the solid describing damage due to the deformation history of the solid.
5. Failure first occurs when the hardness function is at a maximum.

The corresponding yield condition for the solid is

$$F(\sigma,\kappa)=0$$

where σ is stress tensor, and κ is the hardness parameter.

The flow rule is

$$\varepsilon = \lambda\frac{dG}{d\sigma}$$

where ε is the strain rate, λ the plastic parameter, and G the flow function.

The hardness parameter is assumed to satisfy a first order incremental law

$$d\kappa = h(\sigma)d\lambda$$

An (infinitesimal) characteristic surface for the equations above is a surface element on which stress, velocity, plastic parameter and hardness are defined, but the equations above cannot be used to extend these surface values from the surface element. Suppose that a system of coordinates is aligned to the characteristic surface, with "0" being normal to the characteristic surface, and α, β, γ, δ lie in the surface. Since tangential derivatives of all quantities within the surface are continuous at the surface, it is only the normal derivatives of the stress, velocity and hardness which cause difficulty in extension from the surface.

Taking the normal derivatives of the equations above shows that the following matrix system follows

$$\frac{d^2G}{d\sigma_{\alpha\beta}d\sigma_{\gamma\delta}}d_0\sigma_{\gamma\delta}+\frac{dG}{d\sigma_{\alpha\beta}}\frac{d_0\lambda}{\lambda}=\text{known}$$

$$\frac{dF}{d\sigma_{\alpha\beta}}d_0\sigma_{\alpha\beta}-\Theta\frac{d\lambda}{\lambda}=\text{known}$$

and so the implied determinant of the above matrix must be zero, for the surface element to be part of a characteristic surface.

Consequently, the characteristic surface condition in this second order RPT involves the Hessian of the flow function G, and first order derivatives of the flow function G and yield function F with respect to the stress components, as well as the hardness function Θ. Hence Θ can be solved for, and then maximised to find the orientation of the characteristic surface in stress space.

It has been shown how these steps can be achieved in an isotropic [8] plastic system. An isotropic system is one in which G and F are functions only of the three principal stress invariants. An isotropic Mohr-Colomb flow function G satisfies by definition $G_2 = 0$, and $G_1 > 0 > G_3$. Note also that Coulomb flow is an example of a **linear** flow rule for which all elements $G_{ij}=\delta^2G/\delta\sigma_i\delta\sigma_j$ vanish. For a linear flow rule and in the principal stress coordinate system [1] it may be shown [8].

$$-\Theta=h_1\frac{\left(n_2^2F_3+n_3^2F_2\right)\left(n_2^2G_3+n_3^2G_2\right)}{n_2^2n_3^2}+h_2\frac{\left(n_1^2F_3+n_3^2F_1\right)\left(n_1^2G_3+n_3^2G_1\right)}{n_1^2n_3^2}$$
$$+h_3\frac{\left(n_1^2F_2+n_2^2F_1\right)\left(n_1^2G_2+n_2^2G_1\right)}{n_1^2n_2^2}$$

where

$$h_1=\frac{2(G_2-G_3)}{(\sigma_2-\sigma_3)},\; h_2=\frac{2(G_1-G_3)}{(\sigma_1-\sigma_3)}\text{ and }h_3=\frac{2(G_1-G_3)}{(\sigma_1-\sigma_2)}$$

and where (n_1, n_2, n_3) are the components of the vector normal to the characteristic surface in the principal stress coordinates, and for example, (G_1, G_2, G_3) are the derivatives of the flow function G relative to the principal stresses (σ_1, σ_2, σ_3). The principal stresses are ordered according to $\sigma_1 \geq \sigma_2 \geq \sigma_3$. The ratios $\mu_i = 1/h_i$, for example, are effectively shear moduli.

Suppose that $G_2=0$ but none of (n_1, n_2, n_3) is zero, and maximise Θ with respect to n_1 and n_3. Then we obtain $\sum h_i n_i^2 = 0$ and for Coulomb flow (for which μ_1, μ_2, μ_3 are non-negative) this equation cannot be satisfied. Hence the assumption that none of the (n_1, n_2, n_3) are zero is incorrect, and so at least one of these components must be zero. This component must be n_2, since with $G_2=0$, Θ is singular for any other component being zero. Therefore maximising Θ reduces to maximising

$$-\frac{\left(n_1^2F_3+n_3^2F_1\right)\left(n_1^2G_3+n_3^2G_1\right)}{n_1^2n_3^2}$$

subject to $n_1^2+n_3^2=1$. The solution is

$$\frac{n_1^2}{n_3^2}=\tan^2\theta_c=\sqrt{\frac{F_1G_1}{F_3G_3}}$$

where θ_c is the angle between the failure plane and the principal stress. If the functions G and F are both of Mohr-Coulomb type,

$$F=\frac{1}{2}(\sigma_1-\sigma_3)+\frac{1}{2}(\sigma_1+\sigma_3)\sin\phi$$

$$G=\frac{1}{2}(\sigma_1-\sigma_3)+\frac{1}{2}(\sigma_1+\sigma_3)\sin\psi$$

and

$$\cos 2\theta_c=-\frac{\sin\left(\dfrac{\phi+\psi}{2}\right)}{\cos\left(\dfrac{\phi-\psi}{2}\right)} \tag{1}$$

The characteristic surface orientation therefore depends on both the flow and yield functions. For associated flows, F = G, and Equation (1) reduces to the well-known Mohr-Coulomb expression $\theta_c=\pm\left(\frac{\pi}{4}-\frac{\phi}{2}\right)$.

Experiments by Vardoulakis [7] show that unconsolidated sands behave approximately as non-associative rigid-plastic systems, with Mohr-Coulomb flow and yield functions. Hence the expressions above should apply to such systems. Equation (1) has previously been derived by Vardoulakis using a different argument, and shown to be in good agreement with this experimental data. Equation (1) is different from that predicted by EPT, but the difference is not significant for the parameter values in the experiments of Vardoulakis.

3 Discussion

A general second order theory of rigid plasticity has been developed. The theory for the orientation of characteristic surfaces in stress space has been outlined. These characteristic surfaces should correspond to failure planes in physical systems.

Our results show that it is possible to develop a second order RPT, and that for Mohr-Coulomb flow functions, the predictions from this RPT are in agreement with experimental results for failure plane orientation in unconsolidated sands. Equation (1) shows that the failure plane orientation depends on both the flow and yield properties of the physical system, and for associative theories in which flow and yield functions are equal, then the classical results or Mohr and Coulomb are rederived.

For Mohr-Coulomb flow functions $(G_2 = 0)$ and any isotropic yield function, the failure plane must contain the intermediate principal stress direction, a result which is typically assumed in Mohr-Coulomb theory. In such systems, the dependence of the failure plane orientation on the intermediate stress enters at most through the parametric dependence of the yield function – no derivatives with respect to the intermediate stress appear.

References

1. Jaeger J. C. and Cook N. G. W., Fundamentals of Rock Mechanics (*Methuen, London, 1969*).
2. Rice J. R., The localization of plastic deformation. In W. T. Koiter (ed.) *Theoretical and Applied Mechanics*, 14th IUTAM Congr. (North Holland Publi., 1976) pp. 207-220.
3. Rudnicki J. W. and Rice J. R., Conditions for the localisation of deformation in pressure-sensitive dilant materials *J. Mech Phys Solids*, 23 (1975) pp. 371-394.
4. Ottosen N. S. and Runesson K., Properties of discontinuous bifurcation solutions in elasto-plasticity, *Int. J. Solids Structures*, 27(4), 1991) pp. 401-421.
5. Taylor G. I. and Quinney H., The plastic distortion of metals, *Phil. Trans. Roy. Soc.*, London, A. 230 (1931) pp. 323-362.
6. Vardoulakis I. and Sulem J., Bifurcation Analysis in Geomechanics (Chapman & Hall, Glasgow, 1995).
7. Vardoulakis I., Shear band inclination and shear modulus of sand in biaxial tests, *Int. J. Numer. Anal. Methods Geomech*, 4 (1980) pp. 103-119.
8. Weir G. J. and Young R. M., The second order characteristic condition for plastic flow, submitted *Int. J. Plasticity*.

IMAGE TENSOR ANALYSIS – A NUMERICAL TOOL FOR DATA ACQUISITION AND CONSTITUTIVE ANALYSIS OF HISTORIC CONTINUA

P.F. BRITTON AND J.G. LOUGHRAN

School of Engineering, James Cook University, Townsville, Australia

Image measurement based on the Fast Fourier Transform convolution theorem has been combined with a finite difference algorithm to produce image tensor analysis (ITA), a numerical tool for experimental determination of deformation for historic continua.
ITA allows the deformation within an image-compatible material body to be quantified using digital video footage taken from experimental events. Hence the deformation output from computational modelling of the same event can be directly compared to the measured deformation field given by the image-based numerical solution.

The current research application of ITA aims to provide experimental deformation data and characterisation of the constitutive behaviour of the fibro-porous media prepared sugar cane. This paper outlines the numerical procedure of ITA and details the possible post processing applications with respect to both data acquisition and constitutive modelling.

1 Introduction

Many industrial processes nowadays are being modelled by nonlinear finite element techniques. High-speed computers coupled with advances in solving techniques, contact mechanics and constitutive behaviour are enabling insight to be brought to bare on problems which until recently have been considered intractable. The crushing of sugar cane between grooved counter-rotating rollers to extract the high sucrose liquid is one such example. Here, a considerable advance has been made both in understanding the mechanics of the multiphase problem and in the computational algorithms required to solve the highly nonlinear fully coupled large strain problem [5,6,7]. While agreement with mechanical outputs such as roll load and torque are considered satisfactory from a trend perspective [1] further improvements require more detailed knowledge of the constitutive behaviour of the solid phase. To date, the deformation field within the solid phase continuum under process conditions has not been quantified; hence no kinematic data exists for validation of computational solutions.

Image tensor analysis (ITA) is an image measurement technique being developed specifically to calculate kinematic quantities in a two dimensional plane, based on the theory of continuum mechanics. While particle imagery techniques have been applied widely to Newtonian fluids [4], the concept of applying such techniques to single and multi-phase solid mechanics problems would appear relatively novel. The procedure aims to extend the understanding of the rolling of prepared cane, as well as to provide validation and assessment of the critical-state constitutive model that is being used to describe the response of the solid phase to a known stress state.

2 The ITA Procedure

The procedure of ITA involves taking video footage of the spatial and temporal deformation within a process experiment, such as extracting liquid from sugar cane. This footage is decomposed into a series of adjacent images, separated by known time intervals. A particle-tracking algorithm is employed to track points in the media and store the location of each point spatially at all time. This temporal location data forms the deformation function (χ).

Finite differencing of the deformation function is then conducted to obtain the spatial gradients of location with respect to a reference configuration. As a tensor, these differentials form the deformation gradient (F). This is perhaps the most fundamental of kinematic quantities in continuum mechanics and forms the basis for calculation of the right Cauchy-Green deformation tensor, which describes the stretch of a body [8].

In order to provide direct comparison with finite element modelling (FEM) output, further tensor algebra is necessary to obtain the desired mode of strain [8]. Figure 1 shows a flowchart of the numerical procedure.

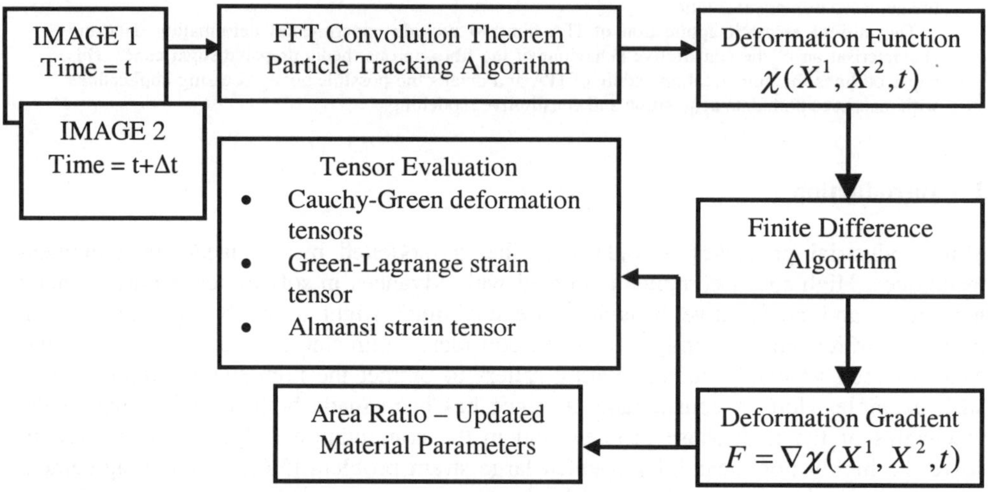

Figure 1. Flow chart illustrating the ITA numerical procedure.

2.1 *The Particle Tracking Algorithm*

The particle-tracking algorithm uses a pair of adjacent images from a series, as input. On the first of the images, the initial body point locations are identified, and this configuration is then chosen as the reference configuration, i.e. the configuration of the body points in the X^1, X^2 space at the time datum. A mathematical transformation is required to describe the deformation in terms of some other reference configuration [8].

Sub-regions of analysis are placed centrally over the initial body point locations. The Fast Fourier Transform convolution [3,4] is employed to determine the displacement of the numerical pixel pattern in each sub-region relative to the centre, from one image to the next. The pixel displacements are scaled to a physical length using a pixel to metre scaling factor. The resulting physical displacements are added to the initial location of each body point, and its new location in X^1, X^2 space is stored in multi-dimensional arrays. It is this temporal location data that is the deformation function. The sub-region associated with each body point is then relocated to the new location of that body point, and the procedure repeated.

For real-time tracking it is necessary, in each iteration, to disregard the first image from the previous pair and gather the next image in the series, hence following the same body point through space and time. Alternatively, the assumption of spatially steady-state can be made indicating that any body point at a certain location will behave as any other would if it were at that location. With this assumption the deformation function can be obtained directly from a single image pair. Note that steady-state analyses conducted on all relevant image pairs for a given experiment will produce the complete transient solution of deformation of the material body.

2.2 *The Finite Difference Algorithm*

A finite difference algorithm is required to obtain the deformation gradient [8]. The partial differentials of displacement with respect to the reference configuration are approximated using central, forward and backward difference techniques involving second order Taylor's Series expansions,

$$F = \nabla \chi \tag{1}$$

In component form,

$$\begin{aligned} F_{\alpha}^{k} &= X_{,\alpha}^{k} \\ &= {}^{\partial}X_{\alpha}[\chi_{\kappa}^{k}(X^{1}, X^{2})] \end{aligned} \tag{2}$$

where **κ** refers to the reference configuration and $\alpha, k = 1, 2$.

The Right Cauchy-Green deformation tensor is given by

$$C = F^{T} F \tag{3}$$

or

$$C_{\alpha\beta} = F_{\alpha}^{k} F_{\beta}^{m} g_{km} \tag{4}$$

where g_{km} are the covariant metric components in the selected co-ordinate system and $\beta,\ m = 1, 2$.

Further manipulation of the Right Cauchy-Green deformation tensor will yield the desired strain tensor field for validation of finite element output, i.e.

$$E_{gl} = \frac{1}{2}(C - I) \tag{5}$$

for Green-Lagrange and

$$E_{A} = F^{T} E_{gl} F \tag{6}$$

for Almansi.

2.3 *ITA Output*

The output of the ITA procedure results in both kinematic data and local material parameters. Principal strain components and directions are the eigenvalues and eigenvectors of the particular strain tensor. For example, we can solve

$$\det[E_{gl} - \lambda I] = 0 \tag{7}$$

for λ_i, the square of the principle stretches.

The deformation function (χ) allows material streamlines to be displayed. Velocities and accelerations are determined by taking the time derivatives of (χ).

$$v_i = \frac{d\chi_i}{dt}; a_i = \frac{d^2\chi_i}{dt^2} \tag{8}$$

At any time (t) the square root of the third invariant of the deformation gradient is the inverse ratio of elemental area to that of the same element in the reference configuration. Here we define the area ratio:

$$A_R = \frac{1}{\sqrt{\det[F]}} \tag{9}$$

Initial material parameters such as local density are determined based on the experimental material sample. Hence, for a later time (t), the local density at a body point can be evaluated.

$$\rho_t = A_R \rho_{initial} \tag{10}$$

This can also be applied to saturation, void ratio, porosity, etc.

2.4 *Experimental Considerations*

Image tensor analysis requires a pair or series of still images as input. Each image is read as a large pixel matrix, the elements of which represent one of the 256 grey-scale shades, at each x-y pixel location. It is therefore evident that the quality of data acquired from the footage is heavily dependent on the image quality. The FFT convolution relies on pixel colour (or matrix element) contrast in order to successfully determine the displacement of a body-point from one image to the next. Hence, depending on the material in question, samples may require seeding with particles of appropriate size, shape and colour. For results shown here, ITA has returned correlation coefficients [4] commonly above 0.85. Such accurate particle tracking results can be contributed to the absence of particle loss from within a subregion, a problem often encountered in velocimetry analyses of Newtonian fluids.

2.5 *Error Sources*

The two sources of error involved in the ITA procedure are, in order of decreasing significance:

- Camera position: Accurate results in the 2-dimensional plane of interest depend highly on the condition that the major axis of the camera is perfectly orthogonal to the subject. Careful precision must been taken in regard to the fabrication and implementation of the camera mounting.
- Finite difference procedure: The error associated with the second-order Taylor's series expansions can be determined for any grid size used for analysis in ITA. This error is decreased only through grid-size reduction. However, for many materials there exists a lower limit to grid size, depicted by the representative elementary volume [2].

No error exists in the tensor manipulation, however, finite difference error must be considered in all calculations in order to arrive at an error in the output variables.

Finally it is noted that the procedure involved in determination of deformation is based on the theory of classical continuum mechanics. Hence the ITA strain field is not applicable to discrete systems.

3 The Current Application

Computational modelling of the rolling of prepared sugar cane to extract the liquid sucrose requires modelling of the partially saturated / saturated fibro-porous media which is subjected to large strain deformation under evolving boundary conditions. The material is two-phase and in practice exhibits distinctive bi-modal fibre distribution and transversely isotropic mechanical properties [5]. Conventional grooved roll geometry only further increases the difficulty of the problem by creating a three-dimensional strain field that is not easily quantified.

Application of constitutive material models associated with porous media mechanics [6] has enabled some progress in the area, despite the lack of model validation beyond experimental load and torque data. Implementation of viewing panels and camera mountings to both a uniaxial test cell (with varying groove geometry) and a two-roll experimental mill has allowed collection of visual data that captures deformations due to rolling and grooving. Applying the numerical techniques in this paper enables the deformation field to be measured directly, and can be utilised for parameter estimation within a given constitutive model as well as a direct validation tool for FEM output.

3.1 Treatment of Rolling Data

Rolling data is captured via a digital video camera attached to the experimental apparatus as indicated by Figure 2(a). A three-column grid of points is tracked through an image series or a single image pair (for steady-state analyses). Post-processing of results is conducted by constructing a mesh depicted by the path of the centre column of the analysis grid. Figure 2(b)-(d) display results from a steady-state analysis conducted on footage taken from a 12kg prepared-cane rolling experiment.

4 Conclusions

The application of experimental image measurement is a powerful tool for acquisition of 2D displacement data for dynamic process experiments involving complex geometry. Knowledge of motion streamlines (χ) provides a basis for determination of spatial tensor output concurrent with output from finite element models. The technique is termed Image Tensor Analysis and it provides a basis for validation and intelligent tuning of constitutive models in regard to correct kinematic material behaviour.

5 Acknowledgements

This work is being funded by the SRDC of Australia under grant number JCU020. This support is gratefully acknowledged.

References

1. Adam C. J., Application of computational porous media mechanics to the rolling of prepared sugar cane (1997), James Cook Univesity, Australia, PhD Thesis.
2. Bear J. Dynamics of Fluids in Porous Media (Amerian Elsevier, New York, 1972).
3. Buchhave P., Particle Image Velocimetry–Status and Trends, Experimental Thermal and Fluid Science (1992) 5: 586-604.

4. Flannery, B. P. Numerical recipes in Fortran, 2nd Edition (Cambridge University Press, New-York, 1992).
5. Loughran, J.G. and Adam C.J., Properties of prepared cane for computational crushing models, Proc. Aust. Soc. Sugar Cane Technol., (1998) 20: 307-312.
6. Owen, D.R.J., Zhao S.Y. and Loughran J.G., Application of porous media mechanics to the numerical simulation of the rolling of sugar cane, Engineering Computations, (1995) 12: 281-302.
7. Owen, D. R. J. et al. Neto, E. A. d.S., Zhao, S. Y., Peric, D. and Loughran, J. G., Finite element simulation of the rolling and extrusion of multi-phase materials – Application to the rolling of prepared sugar cane, Computer Methods in Applied Mechanics and Engineering. V151, Issue 3-4, (1998) pp. (479-495).
8. Truesdell C. The Elements of Continuum Mechanics (Springer-Verlag, New York Inc., 1966).

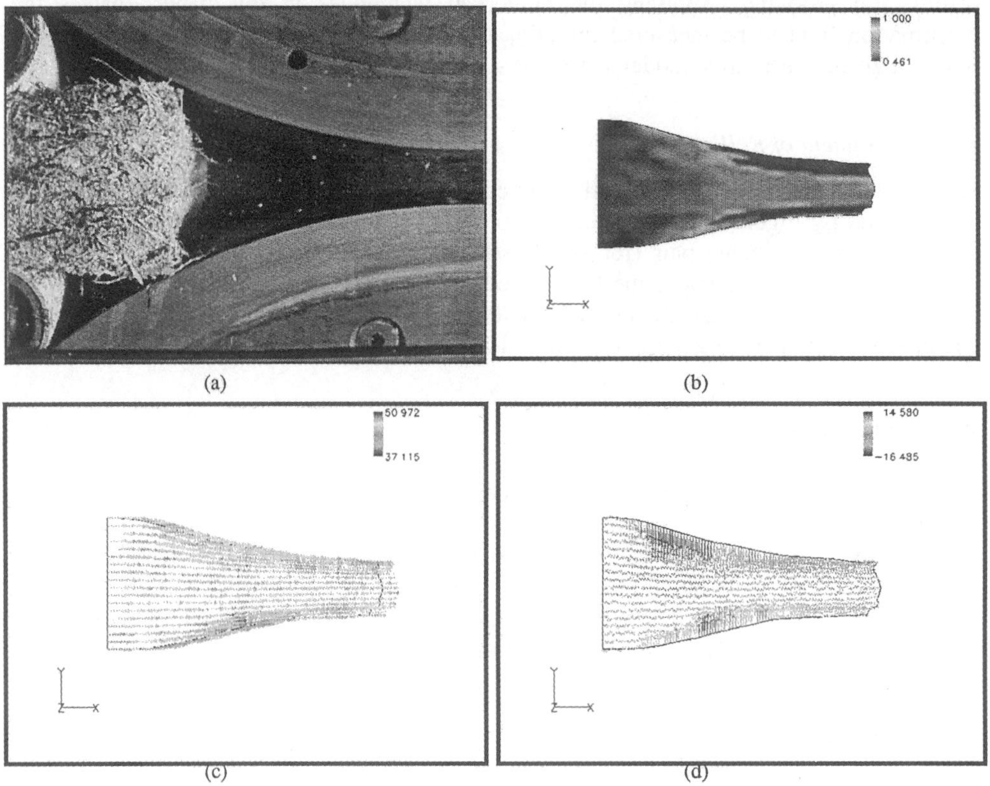

Figure 2. Response of the ITA Model: (a) Experimental set-up; (b) Measured contours of instantaneous saturation; (c) Vector plot of solid-phase X velocity; (d) Vector plot of solid-phase Y velocity.

A MECHANISM OF THERMAL FRACTURING OF CEMENTICIOUS MATERIALS

HAN CHUAN KHOR AND ARCADY V. DYSKIN
Department of Civil and Resource Engineering, The University of Western Australia, 35 Stirling Hwy, Crawley, WA 6009
E-mail: khchuan@cyllene.uwa.edu.au

Thermal fracturing (spallation) is a method of breaking rocks and concrete. It is based on heating the surface (usually with open flame), utilising thermal stresses to develop small chips (spalls) at the surface. The method is cheap and easy to use, however, its practical applications are restricted by the relatively low pace of the breakage and by sensitivity to the material properties and microstructure. Thermal spallation only occurs when the temperature is kept just below the melting point of the particular material. The process is also very sensitive to the presence of water in materials such as rock and concrete. It has been found that thermal fracturing is controlled by the internal permeability of the material in materials with low permeability, water in the pores and pre-existing cracks cannot escape during heating, evaporates and drives the crack in uncontrollable manner. In this case, the sample fails in splitting without any spallation. The difference between these two failure modes is also controlled by the material's tensile strength.

1 Introduction

As available resources get depleted, economically viable alternatives to current mining techniques are required. This paper analyzes one such alternative, the thermal fracturing method.

Several investigations have been conducted on the use of thermal fracturing for both structural and mining purposes. In the structural field, experiments have shown that the applied heat weakens the concrete and further increase of heat results in fracturing of the concrete structure [2,9]. In the mining industry, thermal fracturing has been conducted using flame jets; rocket-type burners using fuel mixed with oxygen or air to produce a high velocity flame. This technique has been used successfully in drilling and cutting of spallable rock such as taconite and granite in the mining and quarrying industries since the early 1950's [3]. The major difficulty in investigating the technique is with respect to the high heterogeneity of the rock mass. Rock samples taken from the test site usually cannot be regarded as representative since the scale of heterogeneity is relatively high, whilst dived field experiments are expensive. In this situation, it is advantageous to use artificial model materials with controlled microstructure to investigate the main features of the process and enable mathematical modeling. Cementicious materials will be used in this study to simulate the rock mass.

2 Thermal spallation

The idea behind the thermal fracturing method is to utilize the mechanical effect of heating (thermal energy) to fracture rock by producing thermal spallation. The main mechanism of rock spallation is compressive thermal stresses acting parallel to the free surface. This stress grows cracks that separate layers of the rock surface. The layers then buckle and spall off, leaving new surfaces. The higher the temperature, the more rapid the fracture process is, however it is essential that the temperature be kept below the melting point otherwise spallation would not occur [4, 6, 10]. Spalling behavior in rock appears be dependent on the material's thermal expansion and shear strain characteristics. Inelastic behavior of the rock inhibits failure by thermal spalling. In rocks that exhibit good spalling behavior, a thin heat layer with high thermal gradient forms during spalling [8].

Studies on explosive spalling have shown that, non-linear temperature distribution across a concrete section lead to compression failure [7]. Investigations have been conducted to study the effect of transient high temperatures on high strength concrete. These results show that some samples will fail explosively at temperatures of 320-360°C, pulverizing the concrete into small particles [2]. Extensive testing of concrete suggests that steam pressure below the surface, as a result of "moisture clogging", causes explosive spalling to occur [5]. Water-cured specimens were found to be significantly more susceptible to explosive spalling then moist cured specimens. Formation of a denser, less permeable layer of hydration products close to the surface assisted high pore pressures. The enhancement of the pressure from frictional resistance to the flow of vapour in the capillaries was identified as a major factor. This would further increase the internal stresses in the concrete.

3 Experiments

These experiments are designed to investigate the mechanism of thermal spallation under direct fire. Two sets of flame jet testing were done in this research. One is equipped with a flame jet that operates on oxygen and natural gas, while the other is a flame jet that operates on acetylene and oxygen. The prepared samples measure 200mm x 200mm x 80mm, with the composition shown in Table 1.

Table 1: Summary of the material composition.

Mix Type Material	Grain Size (μm)	1	2	3	4	5	6
Cement	40	1	1	1	1	1	1
Silica sand	300	-	2	2.5	-	-	1
Silica flour	150	-	-	-	3	-	-
Garnet	200	-	-	-	-	3	-
Gravel	4.6 mm	-	-	-	-	-	2
Water/Cement ratio		0.3	0.47	0.47	1	0.58	0.58

Fig. 1a shows the apparatus in the flame jet experiment. Fig. 1b shows the sample after spallation using the acetylene/oxygen flame jet. The types of fracturing obtained are summarized in Table 2. Three types of fracturing were observed depending upon the sample composition: extensive explosive spallation, non-explosive spallation and cracking perpendicular to the heated surface. The first two types of fracturing are the sought spallation types, differing in speed and extent. The last type is the one to be avoided since no surface cutting is produced.

Figure 1: The flame jet apparatus (a) and a sample produced by the oxygen/acetylene jet (b).

4 Properties of the samples

In order to investigate the mechanisms of thermal fracturing, several basic properties of the samples were examined such as the tensile and compressive strength, and the permeability. The strengths were measured in accordance with Australia Standard AS1012.10 [1] with results of the testing shown in Table 2 (N.B. tensile strength was measured by the Brazilian Test). The permeability was estimated using each component material's permeability by applying the method of weighted average to obtain the permeability value for each sample.

Table 2: The types of fracturing resulted from the flame experiments and properties of the samples.

	Sample 1	Sample 2	Sample 3	Sample 4	Sample 5	Sample 6
Type of Spallation	Extensive Explosive Spallation, Cracks	Cracks	Cracks	Cracks	Non- Explosive, Spallation, Cracks	Explosive Spallation, Cracks
Compressive Strength (MPa)	53.9	50.7	37.5	16.3	45.0	57.7
Tensile Strength (MPa)	2.38	1.89	1.97	0.80	2.07	2.71
Permeability (m/s)	1.54E-12	1.53E-04	1.68E-04	1.60E-05	5.16E-04	7.09E-02
Characteristic Length (m)	2.94E-02	2.66E+02	3.76E+02	5.79E+01	8.18E+02	7.34E+03

5 Mechanisms of thermal fracturing

The experimental results suggest that the tensile strength is one of the parameters that control the type of fracturing. High tensile strength appears to be factor that promotes the spallation process. Permeability is another parameter related to the fracture mechanism. Very low permeability correlates with extensive spallation.

In order to explain the results of the experiments we need to consider possible mechanisms involved in thermal fracturing. Fig. 2 shows the stress fields that result from surface heating [4].

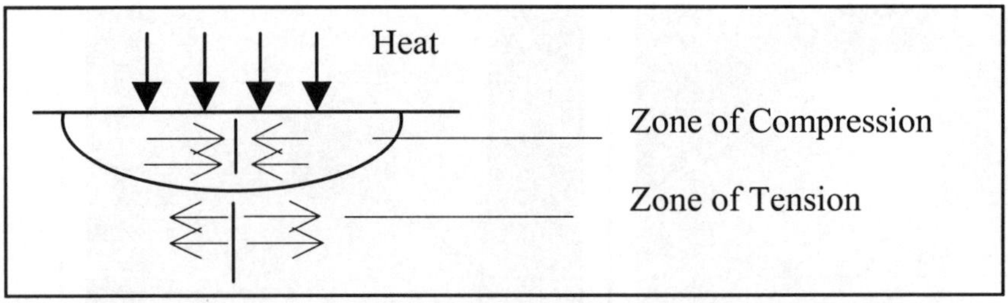

Figure 2: Compression and tension zones induced by surface heating.

The high temperature region under the surface is compressed. This component is balanced by the tensile zone beneath the compressed zone. The presence of two types of stressed zones makes possible two types of fracturing – compressive zone cracking and tensile zone cracking.

a) Compressive Zone Cracking

In the compressive zone, the compressive stress, increasing with heating to the free surface, make internal cracks that grow parallel (Fig. 3). As the cracks grow, they form a thin layer that can buckle under compression, thus forming the spalls.

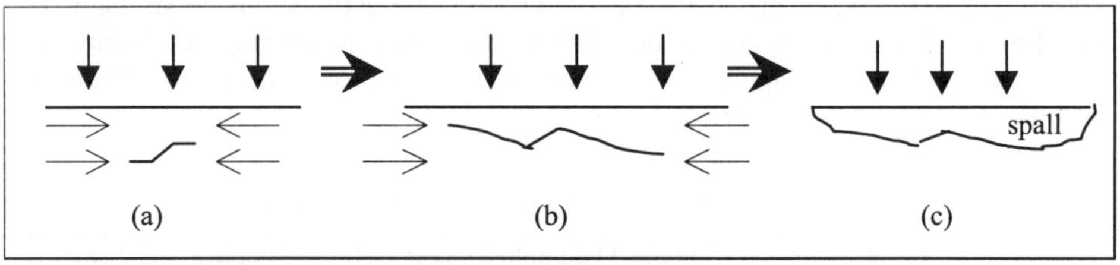

Figure 3: Growth of internal cracks parallel to free surface in compression. (a) initial stage of crack growth; (b) further crack growth under increasing compression; (c) spall formation.

b) Tensile Zone Cracking

Another fracture mechanism is the formation of tensile cracks in the zone of tension. These cracks grow perpendicularly to the surface and disable the spallation process. The obvious controlling parameter for the appearance of these cracks is the tensile strength. High tensile strength prevents these cracks from initiating, clearing the way for the spallation process.

The cementicious materials tested are characterised by the presence of water in pores. Evaporation of heated water in pores can generate cracks. This can considerably enhance the spallation process, however for this to occur, the vapour pressure must be sufficient to drive the cracks. The main factor reducing the pressure is water penetration into the surrounding material. This can be quantified by the characteristic distance that water vapour is able to travel during the time necessary to build up pressure in the crack. If this distance is comparable to the crack size, only a small portion of the water will be able to escape, so the driving pressure will not be affected. As a consequence the vapour will be able to drive the crack. However, if the distance is large compared to the crack size, a significant amount of water will escape, thus inhibiting this mechanism of crack growth.

The characteristic distance l is related to the subsidiary permeability (permeability of material without fracture networks) k_s as follows.

$$k_s = \frac{l^2 \eta}{Pt} \tag{1}$$

where η is the dynamic viscosity , P is the vapour pressure and t is the time for the very first crack developed.

The pressure P can be estimated, as the pressure required for crack propagation. Assuming the crack to be disk-like, of radius a, the criterion for crack propagation can be expressed in the form (eg, [10]),

$$K_{1C} = 2P\sqrt{a/\pi} \tag{2}$$

Here K_{IC} is the fracture toughness of the material (the resistance of the material to crack growth). Combining the above equations, one has

$$l \approx \left(\frac{K_{1C} k_s t}{2\eta}\right)^{1/2} \left(\frac{a}{\pi}\right)^{-1/4} \tag{3}$$

Assuming that $\eta \approx 3.0 \times 10^{-5}\ kg/ms$ and $K_{IC} \sim 1MPa\sqrt{m}$ (e.g. [11]) one obtains

$$l \approx \left(\frac{k_s t}{6.0 \times 10^{-5}}\right)^{1/2} \left(\frac{a}{\pi}\right)^{-1/4} \tag{4}$$

From the aspect ratio analysis, the thickness of the spalls collected from samples involved in explosive thermal spallation is on average, 9 mm. This suggests that for explosive type spallation, the thickness of the spalls is with respect to rate of heat penetration. If the temperature is increasing rapidly, the spalls' are expected to be thinner. This is due to less time being available for the heat to penetrate further in the sample.

Crack growth is another phenomenon observed in the thermal fracturing process. When the temperature is rising slowly, two possible phenomena might be observed. Firstly, if temperature increases relatively slowly compared to the heat conductivity of the sample, then macro cracks might form. . If crack growth has occurred before spallation then spallation will not take place. Secondly, if spallation occurs, the thickness of the spalls is expected to be higher as the heat will penetrate further in the sample and create a deeper boundary. Samples with fine aggregates (cement with silica flour) shows the best resistance to heat, as the time needed for cracks to develop is longer.

6 Discussion

In the experiments, three samples showed thermal spallation while the other three developed macroscopic cracks, splitting the sample. There were two types of thermal spallation observed; explosive and non-explosive spallation. Comparison with the values of tensile strength for different samples, suggest that high tensile strength is a necessary condition for thermal spallation, the role of high tensile strength being the suppression of splitting cracks. Another important parameter controlling the spallation type is material permeability. Low permeability leads to explosion-like spallation because, in this case, water vapour does not have time to escape from cracks into the bulk material. Instead, vaporisation of the water further drives the cracks. Observations also show that temperature and the rate, in which it is applied, also significantly influence the thermal spallation process. The spall area is larger but shallower when a sample is under a lower temperature flame. The reverse is also true: when the sample is heated with a higher temperature flame; the spall area is smaller but deeper.

Acknowledgements

The authors greatly appreciate the assistance of Dr. E Sahouryeh. The authors would like to acknowledge the support of the UWA Research Grant 2001.

References

1. *Australian Standard*, Methods of Testing Concrete- Determination of Indirect Tensile Strength of Concrete Cylinders, AS1012.10 – 1985.
2. Castillo. C & A.J. Durrani, Effect of transient high temperature on high strength concrete. *ACI Materials Journal V.*87, *No.* 1 (1990) pp. 47-53.
3. Cook N.G.W. and V.R. Harvey, An Appraisal of Rock Excavation by Mechanical, Hydraulic, Thermal and Electromagnetic Means. *Advances in Rock Mechanics Volume I, Part B* (1974) pp. 1599-1615.
4. Germanovich L.N., Thermal spalling of rock, *Advances in Fracture Research, Proceedings of the Ninth International Conference of Fracture,* 1-5 *April* 1997, *Sydney, Australia Vol* 6(1997) pp. 2771-2782.
5. Harmathy T.Z., Effect of Moisture On The Fire Endurance of Building Elements, *ASTM Special Technology Publication* (1965).
6. Preston F.W. and H.E.White, Observations on spalling. *J. Am. Ceram. Soc.*, 17,137 (1938).
7. Saito H., Explosive Spalling of Prestressed Concrete in Fire. *Fire Resistance of Prestressed Concrete Proceedings of a Symposium at Braunschweig* (1965).
8. Somerton W.H., *Thermal Properties and temperature-related Behaviour of Rock/Fluid Systems* (1992).
9. Sullivan P.J.E. and Zaman, A.A.A, Explosive Spalling of Concrete Exposed to High Temperatures, *Proceedings of the First International Conference on Structural Mechanics in Reactor Technology, Berlin* (1971) p H1/5.
10. Thirumalai K., Process of thermal spalling behavior in rocks- an exploration study. *Rock Mechanics –theory and practice, Proceedings Eleventh Symposium on Rock Mechanics held at California* (1969).
11. Cherepanov G.P., *Mechanics of Brittle Fracture* (1979).

THREE-DIMENSIONAL SOLUTIONS FOR STRESS CONCENTRATORS IN TRANSVERSALLY ISOTROPIC PLATES

ANDREI KOTOUSOV
Department of Mechanical Engineering, Monash University, P.O. Box 31, VIC 3800, Australia.
E-mail: Andrei.Kotousov@eng.monash.edu.au

CHUN-HUI WANG
Aeronautical and Maritime Research Laboratory, Defence Science and Technology Organisation, 506 Lorimer Street, Fishermans Bend VIC 3207, Australia.
E-mail: Chun.Wang@dsto.defence.gov.au

By adopting Kane and Mindlin's assumption that the through-the-thickness extensional strain is uniform through the plate thickness, a generalised plane-strain theory is developed for transversely isotropic plates. The three-dimensional governing equations are successfully reduced to two coupled equations in the two-dimensional space. With the new theory, explicit solutions of the three-dimensional stresses, especially the through-the-thickness component, around a circular hole and a U-shaped notch in a transversely isotropic plate of arbitrary thickness are derived. The analytical solutions are verified by comparing with finite element results.

1 Introduction

Plane-stress and plane-strain solutions are applicable to plates with vanishing thickness or infinite thickness, respectively. Furthermore, it is known that solutions obtained using the plane-stress solutions do not satisfy all the governing equations [1], although these solutions are reported to be good approximations for thin plates. However, there is no generally accepted criterion for identifying what thickness would qualify as plane-stress or plane-strain. Consequently there is a significant level of empiricism in deciding whether a particular plate could be treated as "thin" or "thick".

The most significant difference between plane-stress solutions and plane-strain solutions of a plate containing a notch is the through-the-thickness stress, although the in-plane stresses obtained from plane-strain theory have been found to agree well with approximate three-dimensional solutions [2,3] and finite element solutions [4-6]. In the case of elastic deformation, this out-of-plane constraint has only minor effect on the in-plane stress distribution. However, the out-of-plane stress can have a significant impact on plastic deformations and failure of notched components. For example, in the delamination or interlaminar matrix cracking of notched composites [7] the through-the-thickness stress plays the key role in the failure process. However, due to the lack of analytical tools, the through-the-thickness tensile stress is inevitably ignored in quantifying the delamination strength of composites.

Adopting the generalised plane-strain theory advanced by Kane and Mindlin [8] for isotropic plate, Kotousov and Wang [9] obtained an exact solution for the three-dimensional stress distributions around a circular hole, and approximate, explicit, solutions for V-shaped notched with a circular tip. The generalised plane-strain theory, which assumes that all extensional strains are uniform through the plate thickness, has been exploited by Yang and Freund [10] in studying the three-dimensional stress distribution near a crack tip.

The purpose of this article is to extend the Kane and Mindlin's generalised plane-strain theory to transversely isotropic materials and then apply the new theory to develop explicit solutions of the three-dimensional stresses near a circular hole or a non-circular hole with a circular tip.

2 Generalised plane strain theory for transversely isotropic plates

The basic assumption of the generalised plane-strain theory [8] is that the through-the-thickness extensional strain is uniform in the thickness direction. In the case of an elastic plate bounded by planes $z = \pm h$, where x and y represent the in-plane Cartesian coordinates, the displacement field takes the following form

$$u_x = u_x(x, y) \quad , \quad u_y = u_y(x, y) \quad , \quad u_z = \frac{z}{h} w(x, y) \cdot \tag{1}$$

It is clear that equation (1) implies that lines normal to the mid-plane of the plate in the undeformed state remain normal to the mid-plane in the deformed state and that these lines experience uniform extensional strain $w(x, y)/h$, where $w(x, y)$ is the out-of-plane displacement of the surface $z = h$ of the plate.

To facilitate the following analysis the stress results are defined as

$$(N_{xx}, N_{yy}, N_{zz}, N_{xy}) = \int_{-h}^{h} (\sigma_{xx}, \sigma_y, \sigma_{zz}, \tau_{xy}) dz \qquad (R_x, R_y) = \int_{-h}^{h} (\tau_{zx}, \tau_{zy}) z dz \tag{2}$$

It is seen that the N_{xx}, N_{yy}, and N_{xy} are the usual forces per unit length, and N_{zz} is $2h$ times the average transverse normal stress. The parameters R_x and R_y are components of "pinching" shear [8], playing a role similar to that of the transverse shear force in the corresponding equilibrium equations of flexible plates.

The equilibrium equations, in the absence of body forces and inertia effects, can now be expressed in terms of stress resultants and "pinching" shears, by substituting equation (1) into the three-dimensional field equations and then integrating with respect to z between the limits $\pm h$ [8],

$$N_{\alpha\beta,\beta} = 0 \qquad R_{\alpha,\alpha} = N_z \tag{3}$$

where the stress-free boundary conditions along the plate surfaces, *i.e.*, $\sigma_{zz} = \tau_{zx} = \tau_{zy} = 0$ at $z = \pm h$, have been used. The Greek indices (α, β) denote (x, y), and the comma the partial differentiation with respect to coordinates. Here and in the following discussion the summation rule applies for repeated indices.

Making use of the generalised Hooke's law for transversally isotropic medium (the plate is isotropic in the $x - y$ plane), the stress resultants can be expressed in terms of the displacement components,

$$N_{xx} = 2h(c_{11}\varepsilon_{xx} + c_{12}\varepsilon_{yy} + c_{13}\, w/h)$$

$$N_{yy} = 2h(c_{12}\varepsilon_{xx} + c_{11}\varepsilon_{yy} + c_{13}\, w/h)$$

$$N_{zz} = 2kh(c_{13}\varepsilon_{xx} + c_{13}\varepsilon_{yy} + c_{33}\, w/h)$$

$$N_{xy} = 2h\frac{c_{11} - c_{12}}{2}\varepsilon_{xy} \tag{4}$$

$$R_x = c_{44}\frac{2h^2}{3}\frac{\partial w}{\partial x}$$

$$R_y = c_{44}\frac{2h^2}{3}\frac{\partial w}{\partial y}$$

where the constants $(c_{11}, c_{12}, c_{13}, c_{33}, c_{44})$ are the elastic constants for the transversally isotropic material.

The equilibrium equations (3) can be automatically satisfied by introducing a function Φ, similar to the Airy stress function,

$$N_{xx} = \frac{\partial^2 \Phi}{\partial y^2}, \quad N_{yy} = \frac{\partial^2 \Phi}{\partial x^2}, \quad N_{xy} = -\frac{\partial^2 \Phi}{\partial x \partial y}$$

(5)

Denoting the mean in-plane stress resultant by $N = (N_{xx} + N_{yy})/2$, the normal stress resultant N_{zz} can be expressed in terms of w and N as,

$$N_{zz} = 2\nu_z\left(1 + \frac{E_z}{\nu_z}\frac{w}{N}\right)N$$

(6)

The following relation between w and N results if R_x, R_y and N_{zz} are eliminated from the equilibrium equations (3),

$$\nabla^2 w - \frac{3}{h^2}\frac{E_z}{G_z}w = \frac{3}{h^2}\frac{\nu_z}{G_z}N$$

(7)

The system of field equations is completed by enforcing strain compatibility, resulting in

$$\nabla^2\left[N - \frac{\nu_z E}{1-\eta^2}w\right] = 0 \quad , \quad \eta = \nu_z\sqrt{E/E_z}$$

(8)

Equations (7) and (8) furnish a set of coupled equations for the mean in-plane stress N and the out-of-plane displacement w, while the normal stress resultant N_{zz} can be found from (6). The boundary conditions are normally given in terms of the in-plane stress resultants $N_{\alpha\beta}$, which are determined separately by the stress function (5),

$$N_{\xi\varsigma} = \overline{N}_{\xi\varsigma} \quad , \quad N_{\varsigma\varsigma} = \overline{N}_{\varsigma\varsigma} \quad , \quad \omega_{,\varsigma} = 0$$

(9)

where the barred quantities are known quantities and co-ordinate system is a right-hand system with ξ measuring along the boundary and ς in the direction of outward normal. The last boundary condition for the function w just means the absence of the transverse shear stress on the boundaries of the body under consideration.

The coupled equations (7) and (8) can be de-coupled, resulting that both the displacement w and the stress-resultant function Φ obey two separate governing equations,

$$\nabla^4 w - \kappa^2 \nabla^2 w = 0$$

(10)

$$\nabla^6 \Phi - \kappa^2 \nabla^4 \Phi = 0$$

(11)

with $$\kappa = \frac{1}{h}\left[\frac{3E_z}{\mu(1-\eta^2)}\right]^{1/2}$$

It is clear that the present theory recovers the theory for isotropic materials [9] as a special case. Furthermore, it is also clear that both plane-stress ($\nabla^4\Phi = 0, \nabla^2 w = 0$) and plane-strain theories ($\nabla^4\Phi = 0, w = 0$) are special cases of the generalised plane-strain theory, therefore the solutions to be presented later will correctly recover the both plane-stress and plane-strain solutions as limiting solutions.

3 General Solutions for an Annulus

Now we present a general solution of the stresses in an annulus of arbitrary thickness. The inner and outer radii are denoted as R_i and R_o, and the thickness is $2h$. The arbitrary traction along the inner and outer edges can be expressed in terms of Fourier series. For simplicity, only the results pertaining to cosine series will be presented; the results for the sine series can be readily obtained using the same method. The arbitrary traction having an angular variation can be written in form

$$N_{rr}(R_i,\theta) = \sum_{n=0}^{\infty} \overline{N}_{rr}^{an}\cos(n\theta), \qquad N_{rr}(R_o,\theta) = \sum_{n=0}^{\infty} \overline{N}_{rr}^{bn}\cos(n\theta)$$

$$N_{r\phi}(R_i,\theta) = \sum_{n=0}^{\infty} \overline{N}_{r\theta}^{an}\sin(n\theta) \qquad N_{r\phi}(R_o,\theta) = \sum_{n=0}^{\infty} \overline{N}_{r\theta}^{bn}\sin(n\theta)$$

(12)

$$R_\theta(R_i,\theta) = 0 \qquad R_\theta(R_o,\theta) = 0$$

where the bared quantities ($\overline{N}_{rr}^{an}, \overline{N}_{r\phi}^{an}, \overline{N}_{rr}^{bn}, \overline{N}_{r\phi}^{bn}$) are constants representing the magnitude of the boundary traction.

The solution of equation can be obtained by the method of variation of parameters, as

$$\Phi_n = -\frac{1}{n}\left[r^n\int_r^b \xi^{1-n}N_n(\xi)d\xi + r^{-n}\int_a^r \xi^{1+n}N_n(\xi)d\xi\right] + A_5 r^{-n} + A_6 r^n$$

(13)

where ξ is merely the dummy variable of integration and

$$N_n(r) = \frac{\nu_z E}{1-\eta^2}\left[(1-\eta^2)(A_1 r^{-n} + A_2 r^n) + A_3 K_n(\kappa r) + A_4 I_n(\kappa r)\right]$$

(14)

here $K_n(\cdot)$ and $I_n(\cdot)$ are the modified Bessel functions of nth order.

Thus, we have six constants A_{1-6} in the stress function Φ (equations (13) and (14)) and six boundary conditions (12), so the stress function can be fully determined. Using the Fourier representation of the boundary conditions and the superposition principle we also can obtain solution for an arbitrary loading on the edges of the hole.

3.1 An Infinite Plate with a Circular Hole Undergoing Extensional Deformations

As example of the theory presented in the preceding section, consider an infinite plate of thickness $2h$ with a circular hole of radius R, which is subjected to a remote tensile stresses $\overline{N}^{\infty}$ parallel to the mid-plane of the plate.

After algebraic manipulations with the stress function (13) and boundary conditions (12) we find for the stress concentration coefficient

$$K_{T,\text{tension}} = \frac{N_{\phi\phi,\max}}{\overline{N}^{\infty}} = 1 + \frac{2\kappa R[\kappa R(1-\eta^2)K_1(\kappa R) + 2K_2(\kappa R)]}{[4\eta^2 + (1-\eta^2)\kappa^2 R^2]K_1(\kappa R) + 2\kappa R(1-\eta^2)K_2(\kappa R)}$$

(15)

Now define an out-of-plane constraint factor,

$$C_z = \frac{N_{zz}}{2\nu_z N} = 2 + 2\frac{E_z}{\nu_z}\frac{w}{N}$$

(16)

Referring to equation (16) the above defined constraint factor is equal to 0 in the plane stress limit and 1.0 in the plane strain limit. Based on the above solutions, final expression of out-of-plane constraint factor becomes

$$C_{z,\text{tension}} = \frac{\frac{8}{3}r^2 K_2(\kappa r)}{\kappa R^3(1-\eta^2)[K_1(\kappa R) + K_3(\kappa R)] + 4\eta^2 r^2 K_2(\kappa r)}$$

(17)

3.2 U-shaped Notch with a circular tip

Consider U-shaped notch with a round tip subjected to tensile loading. The out-of-plane constraint factor can be written as, omitting the details of derivations,

$$C_{z,\text{tension}}(t) = \frac{1}{(1-\eta^2)(1+\kappa R)e^{\kappa t} + \eta^2}$$

(18)

where $t\,(=r-r_0)$ denotes the distance ahead of the notch tip. The above solution is graphically shown in Fig.1 together with finite element results [5] obtained for isotropic plate.

4 Conclusions

Adopting Kane and Mindlin's assumption that the through-the-thickness extensional strain is uniform through the plate thickness, a generalised plane-strain theory has been developed for transversely isotropic plates in which the three-dimensional governing equations are successfully reduced to two coupled equations in the two-dimensional space. With the new theory, exact solutions are obtained for the three-dimensional stresses in a transversely isotropic plate containing a circular hole subjected to biaxial tension. Furthermore, a first order approximate solution has also been derived for a U-shaped notch with a circular tip. The theoretical solutions have been verified by comparing with finite element results.

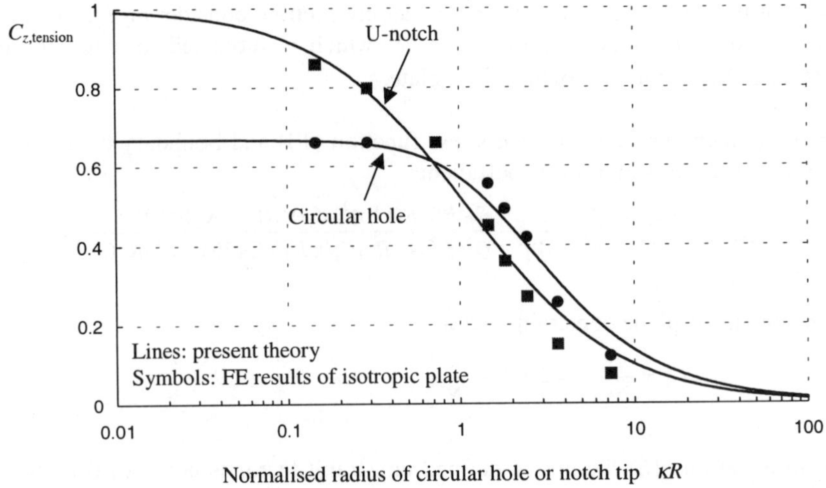

Figure 1 The maximum constraint factor for the cylindrical hole and deep U-notch in an infinite plate under uniaxial tension ($\eta = 0.3$).

References

1. Timoshenko, S. P. and Goodier, J. N. *Theory of Elasticity*, 3rd Edition, McGraw Hill (1970), pp.274-277.
2. Sternberg, E. and Sadowsky, M.A. Three-dimensional solution for the stress concentration around a circular hole in a plate of arbitrary thickness, *J. of Applied Mechanics* **16** (1949) pp. 27-38
3. Young, C. K. and Sternberg, E. Three-dimensional stress concentration around a cylindrical hole in a semi-infinite elastic body, *J. Applied Mechanics* **33** (1966) pp. 855-865.
4. Folias E.S. and Wang, J.J. On the three dimensional stress field around a circular hole in a plate of arbitrary thickness, *Computational Mechanics* **6** (1990) pp. 379-391.
5. Li, Z., Guo W. and Kuang Z. Three dimensional elastic stress fields near notches in finite thick plates, *Int. J. Solids and Structures* **37** (2000) pp. 7617-7631.
6. Li, Z. and Guo, W. Three dimensional elastic stress fields ahead of blunt V notches in finite thickness plate, *Int. J. Fracture* **107** (2001) pp. 53-71
7. Smith, P. A. Carbon fibre reinforced plasticity—properties, *Comprehensive composite materials, Vol.2*, A. Kelly and C. Zweben ed., Elsevier, Oxford, (2000) pp.107-150.
8. Kane, T.R. and Mindlin, R.D. High frequency extensional vibrations of plates, *Journal of Applied Mechanics* **23** (1956) pp. 277-283
9. Kotousov, A. and Wang, C. H. Three-dimensional stress constraint in an elastic plate with a notch, *Int. J. Solids and Structures* (2001) (in press).
10. Yang, W. and Freund, L. B. Transverse shear effects for through-crack in an elastic plate, *Int. J. Solids and Structures* **21** (1985) pp. 977-994.

CALCULATION OF THE STRESS INTENSITY FACTOR FOR SMALL-TO-MEDIUM CRACKS WITH APPLICATIONS TO SHAPE OPTIMIZATION PROBLEMS

ANDREI KOTOUSOV and RHYS JONES

DSTO Centre of Expertise in Structural Mechanics, Mechanical Engineering Department, Monash University, P.O. Box 31, Monash University, Victoria, 3800, Australia
E-mail: Andrei.Kotousov@eng.monash.edu.au

This paper presents a new and relatively simple engineering method for calculating the stress intensity factors for small-to-medium cracks emanating from a notch under arbitrary loading. The formulation can be used in calculating the fatigue life of notched components as well as in the shape optimisation problems with durability constraints. Several examples are considered to demonstrate the advantages of the present method in comparison with both existing approximate approaches and finite element techniques.

1 Introduction

Fatigue failures invariably initiate at some form of a geometrical discontinuity. It is therefore not surprising to find that a great deal of fatigue related research has been devoted to studying the behaviour of notched specimens. The fatigue life prediction of notched components can be approached using fatigue crack growth analysis. To characterize fatigue crack growth rate we require a knowledge of the stress intensity factors for a crack emanating from the notch.

The stress intensity factor for edge cracks is commonly expressed as:

$$K = F\sigma\sqrt{\pi l} \quad , \tag{1}$$

where l is the crack length, σ is the applied stress and F is a magnification factor.

In the absence of an analytical solution for F, its determination requires extensive numerical computations for each particular notch profile, crack length and loading condition. Consequently, only a few numerical stress intensity factor solutions for edge notch-emanating cracks are available in the literature. This has led to numerous attempts to obtain approximate solutions [1-11]

It has been indicated [1] that some of these approximate methods might not be sufficiently accurate and that when used in fatigue life calculations they can result significant errors. Moreover, the existing approximate approaches have, as a rule, been developed and validated for particular types of loading.

The aim of this paper is to develop a new and relatively simple method for calculating the stress intensity factors for small-to-medium cracks at notches under arbitrary loading. The method is based on the assumption that cracks occurring in similar stress fields with similar local geometries should produce similar stress intensity factors. In the present paper the similarity of the local geometry and stress fields is achieved by considering the ancillary problem of a crack emanating from a hole having the same radius as the notch-tip in the stress field corresponding to the (initial) notch configuration and loading conditions.

The proposed method is simple and transparent and can be used for numerous practical applications, viz: fatigue life calculations, and shape optimisation problems with durability constraints. For the last class of problems we require a knowledge of the stress intensity factors associated with cracks, of variable lengths, located at all points along the boundary being optimised. In such optimisation problems the present method can be very effective, because of its analytical nature, in comparison with direct finite element techniques.

2 Method

In this section we will develop an approximate method for calculating the stress intensity factors for an edge crack, with a length up to the order of the characteristic dimension of a local structural detail. In this paper we will define a crack to be small when $l \ll \rho$, medium when $l \sim \rho$ and long $l \gg \rho$, where ρ is the local radius at the point of interests. In this work we will only considered small and medium length cracks.

Whilst for many structural components the fatigue life is dominated by the time taken in growing from a "small"-to-a "medium" length crack it should be noted that there are instances when this is not true. This tends to occur in lightly loaded structures with large initial defects.

To achieve the similarity we consider the ancillary problem of a crack emanating from a hole having the same radius as the notch-tip, see Figure 1. We also need to determine the stress field associated with the (uncracked) notch as shown in Figure 1a).

Figure 1. Initial and ancillary geometries

The stress distribution for several types of holes in an infinite body can be obtained analytically based, for example, on the complex variable method or numerically.

After the replacing the notch geometry and the determining of the stress field the stress intensity factor $K_i = \{K_I, K_{II}\}$ for the edge crack can be written in the following form [12,13]

$$K_i = -\frac{\sqrt{2\rho}}{\sqrt{b-1}}\left\{s_i(b) + \frac{\pi}{n}\sum_{m=1}^{n} q_i(t_k)M_i(b,t_m)\right\} \tag{2}$$

where $b = 1 + l/\rho$, $M(\cdot,\cdot)$ is the kernel of an integral equation, $q(\cdot)$ is the solution of the integral equation and function $s(\cdot)$ is dependent on the loading applied.

Comparison of the present method with known analytical and numerical results [1,7,9,12,14] shows a very good agreement for the "small-to medium" crack lengths and will be omitted here.

3 Shape Optimisation Problem

Fatigue life extension strategies, such as shape reworking, are being increasingly used to restore the airworthiness and availability of ageing structural components. Often reworking of components will be needed to remove service induced cracking while still having the requirement to minimise local peak stresses. For some rework cases cracks cannot be fully removed and in structural optimisation work undertaken to date the effect of such a flaw, after reworking, has not been accounted for. Hence the interplay of an initial optimal rework shape (determined for the case when no crack was present) and an existing crack is an important issue, which needs to be addressed, and which to date has received little attention [15-18].

Unfortunately, current optimum design tools do not readily lend themselves to treating fracture strength as a design constraints. The reason for this is that generally the majority of the life of a component is used up in growing from an initial non-detectable (semi-elliptical) flaw to a complex flaw size, which is detectable. Thus at each stage of the life of the component it is necessary to analyse a complex 2-D or 3-D flaw under arbitrary loading. Due to the singularity along the crack (flaw) front this requires a fine numerical mesh. This in turn means a large number of degrees of freedom and the resultant structural optimisation problem becomes extremely time-consuming requires large computer resources and is computationally inefficient. Nevertheless, Mattheck, Burkhardt and Bethge [19-22] have used finite element tools together with biological growth algorithms to obtain an optimum shape so as to extend the crack initiation phase of a component.

In a damage tolerant design/analysis initial flaws are considered to be present at each of the "hot spots" in the structure and these initial flaws are allowed to grow. At each stage in the life of the structure the current crack length(s) are evaluated and the associated stress intensity factors K are computed. These values of K are then used to evaluate the residual strength and crack growth rate associated with each crack. At no stage in operational life of the structure should the residual strength fall beneath limit load.

Although this paper focuses on the problem of designing to ensure a specific residual strength it should be noted that, as commented on by Visick [23], mechanical repairs can not usually be expected to restore the original fatigue life of the parent structure. As a result it is generally accepted that that, to ensure continued airworthiness, a scheduled inspection process is essential. Consequently, for an optimal repair design, or a structural modification, the problem of choosing the geometry to optimise the inspection intervals also needs to be considered.

Stress-based optimisation is now offered in several commercial structural packages, e.g. ASTROS, Genesis, NASTRAN, ANSYS and others. However, the problem of shape optimisation with durability constraints is still in under development. The problem is made particularly challenging since it must allow for cracks at any arbitrary point along the boundary being optimised. In this section we will attempt to illustrate how the approximate solutions discussed previously can be used to meet this challenge.

To this end let us to first consider the problem of shape optimisation of a hole in an infinite plate subjected to arbitrary remote bi-axial loading. In this case a stress based optimisation, which minimises the maximum stress in the body, predicts the optimal shape to be an ellipse with the ratio of principle axis equal to the ratio of the principal remote stresses G. P. Cherapanov [24].

In this study we will limit the possible (hole) shapes to be elliptical and assume that all cracks are normal to the edge of the hole. Within this design space we will optimise the shape of the hole so as, for a range of given crack sizes, to have the maximum residual strength.

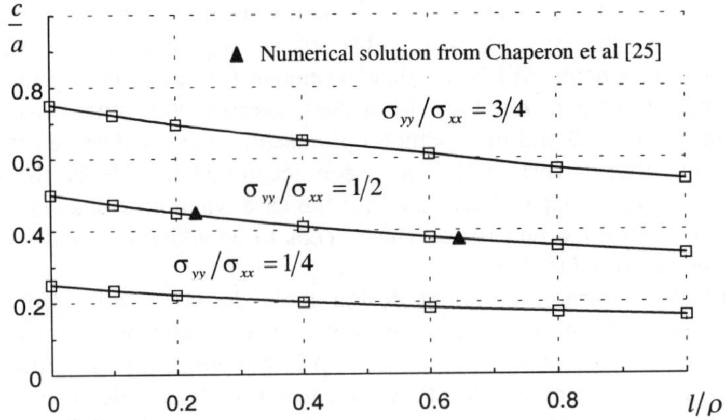

Figure 2. The optimal (fracture strength) shape as a function of the length of the pre-existing edge crack for various ratio's of the principle remote stresses. Here c/a is the ratio of major axis of the ellipse and l/ρ is the non-dimensional length of the pre-existing crack.

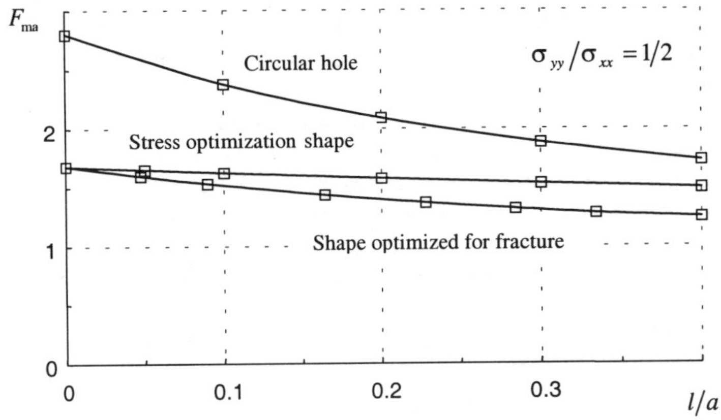

Figure 3. The maximum stress intensity factor magnification F_{max} as a function of the length of the initial crack for a circular hole, the stress optimised shape and the fracture strength optimised shape.

The resultant optimal shapes, as a function of the size l of the (pre-existing) edge crack, is shown in Figure 2. for various rations of principle remote stresses. Here ρ is the minimum radius curvature of the optimised hole. As can be seen from Figure 2 the comparison with the numerical solutions for the cases $\sigma_{yy}/\sigma_{xx} = 1/2$ given by Chaperon and Jones [25] is quite good. A comparative solution for other σ_{yy}/σ_{xx} ratios are not available.

The maximum stress intensity factor magnification coefficient as a function of the length of the initial defect (crack) is shown in Figure 3 for a circular hole, the stress optimised shape, and the fracture strength optimised shape. It is readily seen that the optimised shapes provide significant reduction for the stress intensity factor over all of the range pre-existing crack lengths considered. As expected the fracture strength optimisation did not produce a significant reduction in K for small cracks. This is because the stress intensity factor for small cracks is not sensitive to the local geometry detail and can be found as a production of the normal stress and factor 1.12.

The difference in the stress intensity factor for cracks emanating from a stress optimised geometry and a fracture strength optimised geometry become significant in the "medium" crack size regime, see Figure 3. Consequently, the optimised shape will depend on the size of the initial flaw. For problems associated with fatigue life optimisation the growth rate, and hence the life, is a function of the stress intensity factor raised to a positive power, typically 3 for structural aluminiums and steels. This means that the differences in F, and hence K, produce even larger differences in growth rate and consequently in fatigue life. We thus see that if the initial flaw size is small and the critical crack length lies in the medium crack length regime then stress optimised shapes may not result the optimum fatigue life. Similarly if the initial flaw size is not small then the stress optimised shape will not result in the optimum fatigue life.

Conclusion

The paper presents a simple method for calculating of the stress intensity factors for an edge crack at a notch and the results compared with other existing approaches. We have also shown the potential for this method to be used for shape optimisation with static strength constraints. Indeed, we have shown that the resultant fracture optimised shapes are in good agreement with those obtained using other numerical approaches. This method is being further developed account for non self-similar crack growth.

Acknowledgement – This work was undertaken as part of the DSTO Centre of Expertise in Structural Mechanics research task on "Structural optimisation with damage tolerance constraints". The authors wish to specially acknowledge the support of Dr Manfred Heller.

References

1. Kujawski, D. Estimation of stress intensity factors for small cracks at notches. Fatigue and Fracture Engineering Materials and Structures. 1991; 14 (10): 953-965.
2. Broek, D. The propagation of fatigue cracks emanating from holes. National Aerospace Laboratory, Report NLR TR 72134 U. Amsterdam, 1972.
3. Smith, R. A. and Miller K. J. Fatigue cracks at notches. International Journal of Mechanical Science. 1977; 19: 11-22.
4. Lucas, P. Stress intensity factors of small cracks at notches. Engineering Fracture Mechanics. 1987; 26: 471-473.
5. Karlsson, A. and Backlund, J. Summary of SIF graphs for cracks emanating from circular holes. International Journal of Fracture. 1978; 14: 585-596.
6. Benthem, J.P. and Koiter, W. T., 1973, Asymptotic approximations to crack problems. In: Mechanics of Fracture (Edited by G.C. Sih). Noordhoff; Leyden, 1973; 5: 131-178.
7. Newman, Jr. An improved method of collocation for the stress analysis of cracked plates with various shaped boundaries. NASA TN D-6376, 1971.
8. Schijve, J. The stress intensity factor of small cracks at notches, Fatigue of Engineering Materials and Structures. 1982; 5 (1): 77-90.
9. Nisitani, H. Solutions of notched problems by body force method. In: Mechanics of Fracture (Edited by G.C. Sih). Noordhoff; Leyden, 1978; 5:1-68.
10. Murakami, Y. Two semi-infinite solids bonded by a bond of arbitrary shape subjected to tensile load. In: Stress intensity factor handbook. Japan: Society of Material Science, 1986.

11. Ibrahim, R.N. and Kotousov, A. Eccentricity correction for the evaluation of fracture toughness from cylindrical notched test small specimens. Eng. Fracture Mechanics. 1999; 64: 49-58

12. Tweed, J. and Rooke, D.P. The distribution of stress near the tip of a radial crack at the edge of a circular hole. International Journal of Engineering Science. 1973; 11:1183-1195

13. Kotousov, A. and Jones, R. The stress intensity factor of a radial crack at the edge of a circular hole under shear loading. J. Theor. and App. Fracture Mechanics. 2001 (in print)

14. Hsu, Y.C. The infinite sheet with cylindrical hole under inclined tension or in-plane shear. International Journal of Fracture. 1975; 11(4): 571-581

15. Chaperon, P., Jones R., Heller, M., Pitt, S. and Rose, F. A methodology for structural optimization with damage tolerance constraints. J. Eng. Failure Analysis. 2000; 7: 281-300.

16. Abdi, R. El, Touratier, M. and Convert, P. Optimal design for minimum weight in a cracked pressure vessel of a turboshaft. Communications in Numerical Methods in Engineering. 1996;12: 271-280.

17. Bethge, K. and Mattheck, C. Fatigue testing of a shape-optimized circular hole in a plate under tensile and bending loads. Int. J. Fatigue. 1990;12(6): 489-492.

18. Cheng, G. D. and Fu, B. Optimization of continuum with crack. IUTAM Symposium on Structural Optimization. 1988. Melbourne, Australia.

19. Mattheck, C. and Moldenhauer, H. An intelligent CAD-method based on biological growth. Fatigue Fracture Engng Mater. Structures. 1990; 13(1): 41-51.

20. Mattheck, C., Erb, D., Bethge, K. and Begemann, U.Three-dimensional shape optimization of a bar with a rectangular hole. Fatigue Fracture Engng Mater. Structures. 1992;15(4): 347-351.

21. Mattheck, C. Design and growth rules for biological structures and their application to engineering. Fatigue Fract. Engng. Mater. Struct. 1990; 13(5): 535-550.

22. Mattheck, C. and Burkhardt, S. A new method of structural shape optimization based on biological growth. Int. J. Fatigue. 1990; 12(3):185-190.

23. Visick, J. Damage tolerance/maintenance of airworthiness - An operators view point. International Conference on aircraft Damage Assessment and Repair. 1991. Melbourne, Australia, August 1991: 15-22.

24. Cherapanov, G.P. Inverse problems of the plane theory of elasticity. PMM. 1974;38(6):963-979.

25. Chaperon, P., Jones, R., Heller, M., Pitt, S. and Rose, F. A methodology for structural optimisation with damage tolerance constraints. Eng. Failure Analysis. 2000; 7: 281-300

STRESS CONCENTRATION FACTORS OF WELDED TUBE-TO-TUBE T-JOINTS IN THIN-WALLED (T<4mm) SQUARE HOLLOW SECTIONS

F. R. MASHIRI[1], X. L. ZHAO[1], P. GRUNDY[1] AND L.W. TONG[2]
[1] *Department of Civil Engineering, Monash University, Clayton, VIC. 3800, AUSTRALIA*
[2] *Department of Building Engineering, Tongji University, Shanghai, 200092, P.R. CHINA.*

The fatigue design of tubular nodal joints in the latest IIW and CIDECT fatigue design guidelines is based on the hot spot stress method. Stress concentration factors (SCFs) form a significant part of the design of welded connections using the hot spot stress method. This paper describes the determination of experimental stress concentration factors in thin-walled tube-to-tube T-joints made up of square hollow sections, due to in-plane bending moment in the brace. The effect of induced axial load and induced bending moment in the chord on experimental SCFs is studied. The experimental SCFs are compared to the SCFs derived from existing parametric equations. It is found that the experimental SCFs are significantly lower than the SCFs from existing parametric equations. Observations are made in this paper which form a preliminary part of future research into SCFs of thin-walled tubular connections.

1 Introduction

The conventional method used for fatigue design of construction details is the classification method. In the classification method the fatigue design curve relates the nominal stress range determined by simple beam theory to the number of cycles to failure [3,5,11]. The other method that can be used for fatigue design is the hot spot stress method. The hot spot stress method is relatively new, and is primarily used for tubular welded nodal joints [1,2,4,5,7,14]. The hot spot stress method uses the maximum stress concentration factors at the "hot spot" (where fatigue crack initiation occurs) to convert nominal stress range to the hot spot stress range. In the hot spot stress method SCF determination is therefore vital. In the latest IIW [7] and CIDECT [14] fatigue guidelines, stress concentration factors are calculated using parametric equations within a given validity range. The validity ranges of parametric equations in IIW [7] and Zhao et al [14] do not cover most of the thin-walled tubular joints made up of cold-formed square hollow sections [6], which characteristically have a higher chord width to thickness ratio.

This paper describes the determination of stress concentration factors (SCFs) in thin-walled tube-to-tube T-joints made up of square hollow sections, due to in-plane bending moment in the brace. The SCFs are derived from hot spot stresses that are deduced from experimentally measured strains. The effect of induced axial load and induced bending moment in the chord on experimental SCFs is determined. The experimental stress concentration factors (SCFs) are compared to the SCFs derived from existing parametric equations. It is found that the experimental SCFs are in general significantly lower than the SCFs from existing parametric equations.

2 Experimental Determination of Stress Concentration Factors

Table 1 shows the tube sizes and corresponding joint parameters for the thin-walled tube-to-tube T-joints, which were simply supported with a span of l_0. Measurement of strains for determination of SCFs was carried out in ten connections, with different non-dimensional parameters β, τ and 2γ. See Figure 1 and Table 1 for the meaning of symbols used to define non-dimensional parameters. Hot spot stresses in welded nodal joints made up of square or rectangular hollow sections were measured at the hot spot locations, lines A to E [13], shown in Figure 1, where crack initiation occurs under fatigue loading, using foil gauges consisting of 5 strain-sensitive grids mounted on a common backing. For each connection, strains were measured at four different load levels, taken to be below the maximum elastic moment of the connection [9]. Hot spot stresses can be determined by the linear or the quadratic extrapolation methods. Previous research [10,13] has shown that the stress distribution at hot spot locations in nodal joints made up of rectangular or square hollow sections, is highly non-linear and is therefore better represented by quadratic extrapolation.

Table 1: Connections, non-dimensional parameters and total stress concentration factors

Name	Chord $h_o x b_o x t_o$ (mm)	Brace $h_1 x b_1 x t_1$ (mm)	Non-dimensional Joint Parameters				Experimental SCF_{tot} (Average Quadratic)			
			$\beta = b_1/b_0$	$2\gamma = b_0/t_0$	$\tau = t_1/t_0$	$\alpha = 2l_0/b_0$	Line E	Line B	Line C	Line D
D3D1	100x100x3	50x50x3	0.50	33	1.0	12.0	-	9.3	12.0	7.8
D3D2	100x100x3	50x50x1.6	0.50	33	0.5	12.0	-	5.6	7.1	5.1
D3D4	100x100x3	35x35x3	0.35	33	1.0	12.0	-	4.6	12.7	7.7
D3D5	100x100x3	35x35x1.6	0.35	33	0.5	12.0	-	3.7	5.9	5.8
D6D1	75x75x3	50x50x3	0.67	25	1.0	12.0	-	3.2	8.4	4.2
D6D2	75x75x3	50x50x1.6	0.67	25	0.5	12.0	-	2.5	8.3	1.6
V3V5	100x100x3	60x60x3	0.60	33	1.0	12.0	5.1	-	11.6	7.3
V3V4	100x100x3	30x30x2	0.30	33	0.7	12.0	1.8	-	4.4	5.1
V6V1	70x70x3	50x50x3	0.71	23	1.0	12.9	1.8	-	3.3	3.4
V6V2	70x70x3	40x40x2	0.57	23	0.6	12.9	2.4	-	5.5	3.6

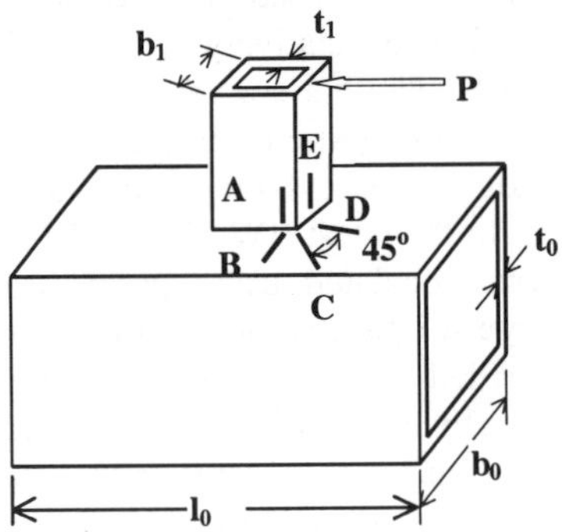

Figure 1: Lines A to E, Tube-to-tube T-joints

The stress concentration factor is defined as the ratio between the hot spot stress and the nominal stress applied to the connection. For the case of in-plane bending in the brace, the stress concentration factor on a hot spot location in a tube-to-tube T-joint is;

$$SCF_{tot} = S_{hs} / \sigma_{m1} \qquad (1)$$

where S_{hs} is the hot spot stress at a predetermined line (A to E), σ_{m1} is the nominal stress due to in-plane bending in the brace, and SCF_{tot} is the total stress concentration factor resulting from the effect of bending moment in the brace as well as the induced bending and axial stresses in the chord. The total SCFs obtained for the ten different connections, at some of the hot spot locations, A to E, where strain measurements were taken, are shown in Table 1. Measurement of strains was carried out mainly in the chord, where the highest stress concentrations are expected [13].

3 Influence of Induced Bending and Axial Force in Chord on Total SCFs

The experimental measurement of strain at the hot spot locations, A to E, results in the determination of SCF_{tot} as shown in equation 1. The values of stress concentration factors due to in-plane bending moment in the brace, SCF_{m1}, given in IIW [7] and Zhao et al [14], do not include any influence of induced stresses in the chord. In order to allow comparison of the SCFs determined experimentally to those in IIW [7], the total SCFs determined experimentally in equation 1 need to be corrected to remove the influence of induced stresses in the chord and thereby determining the corresponding value of SCF_{m1} for each experimentally determined SCF_{tot}. The total hot spot stress at any hot spot location for the case shown in Figure 1 [13] is;

$$S_{hs} = \sigma_{m1} \cdot SCF_{m1} + \sigma_{a0} \cdot SCF_{a0} + \sigma_{m0} \cdot SCF_{m0} \quad (2)$$

where, SCF_{a0} is the stress concentration factor due to nominal axial stress in the chord, SCF_{m0} is the stress concentration factor due to nominal in-plane bending stress in the chord, SCF_{m1} is the stress concentration factor due to nominal in-plane bending stress in the brace, σ_{a0} is the nominal axial stress in the chord, σ_{m0} is the nominal in-plane bending stress in the chord, and σ_{m1} is the nominal in-plane bending stress in the brace.

The stress concentration factor due to axial force in the chord, SCF_{a0}, is equal to the stress concentration factor due to in-plane bending in the chord, SCF_{m0} [7]. Equations (1) and (2) can be rewritten as:

$$SCF_{tot}/SCF_{m1} = \left(S_{hs}/\sigma_{m1}\right)/SCF_{m1} = 1 + [\lambda] \cdot \left(SCF_{m0}/SCF_{m1}\right) \quad (3)$$

where the multiplier $\lambda = \left(\sigma_{mo} + \sigma_{a0}\right)/\sigma_{m1}$. The multiplier in equation (3), λ, ranges between 0.03 and 0.26 for the connections and test set-ups used in this investigation. The values of SCF_{m0} obtained from parametric equations in IIW [7], are much smaller compared to SCF_{m1} for a given T- or X-joint especially for larger 2γ values which are characteristic of thin-walled sections. The axial force or bending moment in the chord also causes negligible stress concentration in the brace and the values of SCF_{m0} in the brace lines A and E are taken as zero [7]. The effect of induced loading in the chord also causes negligible stress concentration in the chord along line B, and the value of SCF_{m0} in the chord along line B can be taken as zero [7]. Since current fatigue guidelines [7,14] show that the influence of bending moment and axial force in the chord causes a negligible effect on SCF along lines A, B and E, therefore, SCF_{m0}=0, and hence SCF_{tot}=SCF_{m1} for lines A, B and E. Bending moment and axial force on the chord however cause stress concentrations along lines C and D. The SCFs due to bending moment or axial force on the chord are identical and are given as parametric equations in IIW [7]. Using the existing parametric equations from IIW [7] the ratio SCF_{m0}/SCF_{m1}, in equation (3) can be determined and hence the ratio SCF_{tot}/SCF_{m1}. The existing parametric equations for determining, SCF_{m0} and SCF_{m1} are used with the assumptions that the relative magnitude between SCF_{m0} and SCF_{m1} in welded thin-walled tubular connections is similar to that predicted by existing parametric equations. Using the existing parametric equations, the ratio SCF_{m0}/SCF_{m1} for the connections under investigation range from 0.03 to 0.31 for lines C and D. The resulting ratios SCF_{tot}/SCF_{m1} for lines C and D, are all approximately equal to 1.0 with the largest ratio being 1.07. These results show that the effect of induced bending and induced axial forces in the chord has very little influence on the total stress concentration factor, SCF_{tot} of the connections under investigation.

4 Effect of τ and β on Stress Concentration Factors

The values of SCF_{m1} determined experimentally and from parametric equations are shown in Table 2. The SCFs from existing parametric equations in IIW [7] and Zhao et al [14] are estimated with the assumption that, the current trend in SCFs as depicted by the existing parametric equations remains the same for thin-walled connections, where some of the joint parameters lie outside the validity ranges. The tested connections whose non-dimensional parameters lie outside the validity range in IIW [7] are marked with asterisks in Table 2.

Table 2: Comparison of Experimental SCF with SCF from parametric equations

Name	**IIW SCF_{ml}**				**Experimental SCF_{ml}** (Average Quadratic)				$\frac{Experimental\ SCF_{ml}}{IIW\ SCF_{ml}}$			
	Hot Spot Location, Line				**Hot Spot Location, Line**				**Hot Spot Location, Line**			
	A/E	**B**	**C**	**D**	**A/E**	**B**	**C**	**D**	**A/E**	**B**	**C**	**D**
D3D1*	12.1	22.7	30.5	17.5	-	9.3	11.9	7.7	-	0.41	0.39	0.44
D3D2*	12.1	14.2	19.1	10.9	-	5.6	7.07	5.0	-	0.40	0.37	0.46
D3D4*	6.7	12.5	14.0	15.5	-	4.6	12.7	7.7	-	0.37	0.91	0.49
D3D5*	6.7	7.8	8.7	9.7	-	3.7	5.9	5.8	-	0.47	0.67	0.59
D6D1	11.8	14.7	19.0	9.4	-	3.2	8.3	4.0	-	0.23	0.44	0.43
D6D2	11.8	9.2	11.8	5.9	-	2.5	8.2	1.5	-	0.27	0.69	0.26
V3V5*	15.4	26.9	36.4	16.3	5.10	-	11.5	7.1	0.33	-	0.32	0.44
V3V4*	5.1	6.6	6.9	9.5	1.80	-	4.4	5.1	0.35	-	0.64	0.53
V6V1	11.1	12.3	15.1	7.8	1.80	-	3.2	3.2	0.16	-	0.21	0.41
V6V2	9.5	8.9	12.2	6.9	2.40	-	5.5	3.5	0.25	-	0.44	0.51
									Mean = 0.43			
									Stan. Dev. = 0.16			
									COV = 0.37			

4.1 Effect of τ on SCF

For given values of β and 2γ, SCF increases with τ, for both experimentally and parametric equation [7,12,13] derived SCFs, see Figures 4(a) to 4(c). For two connections with the same β and 2γ values, the connection with a larger τ value has a larger brace wall thickness and hence a larger brace elastic section modulus. For a given load under bending, the connection with a larger τ value therefore has a smaller nominal stress, $\sigma = M/Z$. A load of a given magnitude produces hot spot stresses that are almost equal on two connections with the same β and 2γ as reported in Mashiri et al [10]. As such the ratio of hot spot stress to nominal stress (i.e SCF) is higher for the connection with a larger τ value.

4.2 Effect of β on SCF

Current trends in SCFs given in IIW [7] and Zhao et al [14] show that for a given value of τ and 2γ, the SCFs along a hot spot location A to E increases up to a certain value of β before decreasing again, see Figures 4(a) to 4(c). Some of the SCFs determined from experimental measurement of strain show a similar trend in SCFs for the range of β values investigated. The measured SCFs along lines B, C and D (Figures 4(a)-(c)) show a similar trend in SCFs to that predicted by the existing parametric equations in IIW [7] and Zhao et al [14], except for the measured SCFs along line C for τ=1.00, see Figure 4(b).

The different trend along line C for τ=1.00 has been found to be associated with reduced non-linearity of strains measured in the connections. Mashiri et al [10] described the non-linearity of measured strains as a ratio of the SCF determined using the quadratic extrapolation (SCF_q) to the SCF determined using the linear extrapolation method (SCF_l).

Different trends in SCFs for different values of β have also been given by previous researchers [7,12]. Whereas the SCFs given by IIW [7] increase up to a certain value of β

before decreasing again as shown in Figure 4(a) to 4(c), the SCFs given by Soh and Soh [12] continuously increase for the range of β values between 0.2 and 0.75, see Figure 4(d). Future work is needed to verify the influence of β in thin-walled tubular joints.

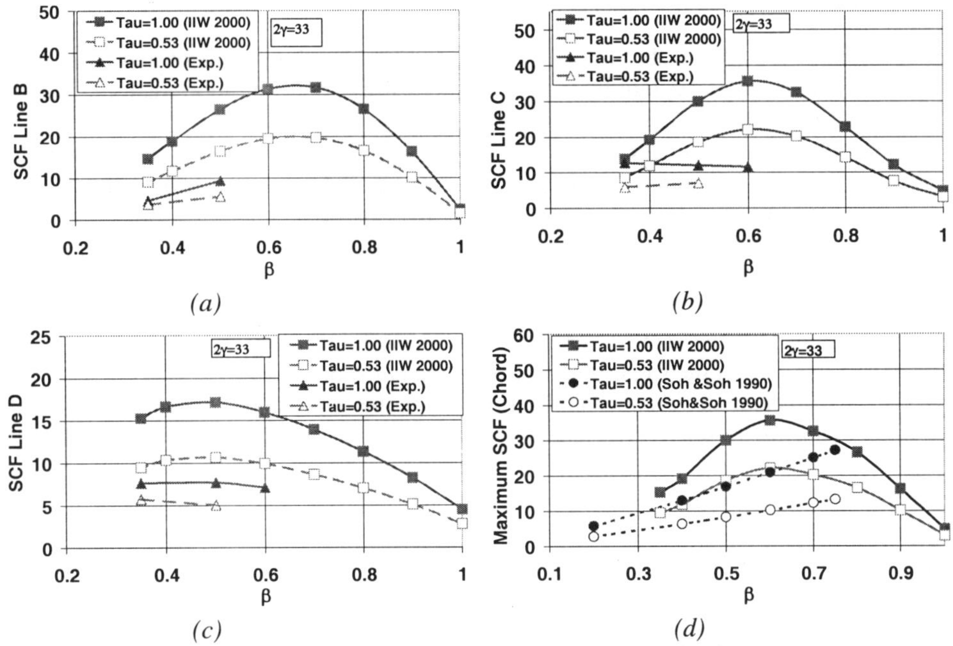

Figure 4: (a) Effect of β and τ on SCF, line B; (b) Effect of β and τ on SCF, line C; (c) Effect of β and τ on SCF, line D; (d) Comparison of SCF trend from [7] and [12]

5 Comparison of Experimental SCFs with SCFs from Parametric Equations

The ratios of the experimental SCFs to the parametric equation SCFs vary between 0.2 and 0.9, with a mean and coefficient of variation of 0.43 and 0.37 respectively, see Table 2. This shows that the experimental SCFs are significantly lower than the predicted SCFs from existing parametric equations. Table 2 also shows that the ratios of experimental SCFs to parametric equation SCFs for connections whose non-dimensional parameters lie within the validity range in IIW [7] are similar to the ratios of those connections whose non-dimensional parameters lie outside the validity range. This shows that the existing parametric equations for SCFs overestimate the SCFs of thin-walled tubular T-joints even when they are within the validity range of the existing parametric equations in IIW [7]. The variability in the ratio of experimental SCFs to the parametric SCFs shows that the experimental SCFs may be influenced by weld leg length for welded thin-walled tubular connections. Measured average weld leg lengths ranging between 5.8mm and 7.1mm have been found in these thin-walled welded connections, compared to a nominal value of 4.2mm (3mm throat thickness) recommended by AS4100-1998 [11] and EC3 [5]. Maddox *et al* [8] showed that for a tubular joint geometry and loading considered, increasing the leg length moves the weld toe into a lower stress region. Measurements of weld profiles carried out in this investigation and reported in Mashiri *et al* [10] have shown that the welds obtained for these thin-walled tube-to-tube with thicknesses less than 4mm are oversized. Future work also needs to be undertaken (e.g. using boundary element analysis) to check if the restriction of the first point of extrapolation to not less than 4mm from the weld toe does not actually remove the measurement of stress from the region of steep gradient for thin-walled joints less than 4mm.

6 Conclusions

1. It has been shown that the influence of the induced bending moment and axial force in the chord have negligible effect on total stress concentration factors, for the connections and setups used in this investigation.
2. Experimental stress concentration factors at hot spot locations in the chord, for connections with the same β and 2γ values increase with an increase in τ value.
3. Stress concentration factors derived from the experimentally determined hot spot stresses show dependence on the β value for connections with the same τ and 2γ value similar to that in IIW [7] for some of the connections tested. In other connections however, the dependence of stress concentration factors on β value for connections with the same τ and 2γ values, were found to be greatly influenced by the non-linearity of the stress distribution at the hot spot location.
4. The experimental stress concentration factors in the thin-walled tube-to-tube T-joints were found to be relatively smaller than the predicted stress concentration factors from the parametric equations, due to the oversized welds associated with thin-walled welded connections.

7 Acknowledgements

The authors would like to thank Monash University, Department of Civil Engineering staff for their help with the experiments. This project was funded by ARC and CIDECT. The first author was also supported by a Monash Postgraduate Publications Award.

References

1. API 1991, "Recommended practice for planning, designing and constructing fixed offshore platforms", Nineteenth Edition, August 1991, Washington, USA
2. AWS 1998, ANSI/AWS D1.1-98, American Welding Society, Miami, USA
3. CSA 1994: CAN/CSA-S16.1-M94, Canadian Standards Association, Canada
4. Department of Energy, 1990, "Offshore Installations: Guidance on design, construction and certification", Fourth Edition, London, HMSO, UK.
5. EC3 1992, Eurocode 3: ENV 1993-1-1, European Committee for Standardisation.
6. Hancock G.J. 1999: *Advances in Steel Structures Vol. I*, Editors Chan S.L. and Teng G.J., Elsevier Science Ltd, pp. 25-37
7. IIW 2000: Fatigue Design Procedures for Welded Hollow Section Joints, IIW Doc. XIII-1804-99, Eds. Zhao X.L. & Packer J.A., Abington Publishing, Cambridge, UK.
8. Maddox S.J. Wylde J.G. and Yamamoto N. 1995, Int. Conf. on Offshore Mech. and Arctic Eng., Vol. III, American Society of Mechanical Engineers, pp. 127-134
9. Mashiri F.R., Zhao X.L., Grundy P. 1999: *Advances in Steel Structures Vol. II*, Editors Chan S.L. and Teng G.J., Elsevier Science Ltd, pp. 983-990
10. Mashiri F.R., Zhao X.L., Grundy P. 2000, CIDECT Project No. 7T, Interim Report No. 3, Department of Civil Engineering, Monash University, Australia, May 2000.
11. SAA 1998, *Australian Standard AS 4100-1998*, Standards Association of Australia.
12. Soh A.K. and Soh C.K. 1990, *J. of Constr. Steel Res.*, No. 15, pp. 173-190
13. van Wingerde A.M. 1992: *Heron*, Vol. 37, No. 2, pp. 1-180
14. Zhao X.L., Herion S., Packer J.A., Puthli R., Sedlacek G., Wardenier J., Weynand K., Wingerde A., and Yeomans N. 2000, "Design Guide for Circular and Rectangular Hollow Section Joints under Fatigue Loading", Verlag TUV Rheinland, Germany.

SHAPE OPTIMISATION OF FATIGUE PRONE HOLES IN THE WING PIVOT FITTING OF THE RAAFS F-111 AIRCRAFT

M. BURCHILL, M. MCDONALD, M. HELLER AND K.C. WATTERS

Airframes and Engines Division, DSTO, PO Box 4331, Melbourne VIC 3001, Australia

Shape reworking has historically been used to remove cracks and reduce the peak stress at critical locations in the F-111 in service with the RAAF. Optimal reworks for the most critical fuel flow vent hole fatigue locations in the F-111 wing pivot fitting have been determined using a recently-developed finite-element-based gradient-less shape optimisation procedure, and typically provide a 35% reduction in peak elastic stresses as compared to current rework shapes. These optimal rework shapes were developed using both 2D PAFEC and 3D MSC.NASTRAN finite-element models. The final robust optimal rework shapes provide a low peak stress over a range of conditions. The numerical strain results have been validated with a recent full-scale static wing test. It is anticipated that the stress reductions achieved will provide a basis for extending recommended inspection intervals, thus reducing costs and increasing aircraft availability until the planned withdrawal date of 2020. It is planned that the RAAF will commence modifying F-111 fleet wings in 2001.

1 Introduction

The wing pivot fitting (WPF) is a primary structural component in the F-111 wing and has been the site of fatigue cracking in service and structural failures during cold proof load testing (CPLT). The upper and lower plates of the WPF, Figure 1, are high strength low fracture toughness D6ac steel with integrally forged stiffeners, which are welded together to form a box structure which transmits concentrated wing loads through a pivot pin. The main locations at which cracks have occurred are geometric features in the WPF upper plate stiffeners, known as fuel flow vent holes (FFVH) and stiffener runouts (SRO). In terms of the FFVHs, the most frequent location of cracking has been FFVH 13, however cracks have also been detected at FFVHs 11, 12 and 14, shown in Figure 2.

CPLT is a periodic proof load (-3.0g to +7.33g) test program performed on the F-111 that confirms that no flaw above a small critical size is present in the structure, upon completion. The CPLT causes localised yielding in areas of high stress concentration, such as the FFVHs, leaving residual stresses upon unloading. The CPLT induced large tensile residual stresses have been found to be the main factor driving FFVH fatigue cracks. Cracks are removed either by grinding or the local geometry is reworked to a predetermined shape, which aims to both remove the damaged material and attempts to reduce the peak stress. The current rework shapes, consisting of straight lines and circular arcs [1] were determined by trial-and-error, and have not prevented further cracking.

This paper discusses the development of the best possible rework shapes for the critical FFVH locations using gradient-less shape optimisation procedure implemented with the PAFEC and MSC.NASTRAN finite element (FE) solvers. These optimal reworks are designed to remove damaged material from the boundary and provide significant elastic stress reductions over a range of conditions. Which is anticipated to result in lower crack growth rates, reduced maintenance costs, increased aircraft availability and making it possible for the aircraft to reach its planned withdrawal date of 2020.

2 Gradient-less finite element shape optimisation method

The gradient-less optimisation method used here [2,3] involves iterative FE solutions to minimise the peak stress, by altering the local geometry to achieve a constant stress along the boundary. The general approach, is based on an analogy with the growth behaviour of biological structures [4], whereby material is added to regions of high stress and/or removed from regions of low stress, (refer to Heywood [5] for another similar technique). This can in principle achieve a free-form shape that produces a constant stress on the free surface. The amount of material to be added or removed at any point on the boundary is taken to be directly proportional to the difference between the local tangential stress and a suitable reference value. This process is then repeated iteratively until the boundary tangential stress is constant, or nearly constant to a prescribed tolerance. The boundary movement for any given position *i* on the boundary is given by :

$$d_i = \left(\frac{\sigma_i - \sigma_{th}}{\sigma_{th}} \right) ls \qquad \text{-(1)}$$

where d_i is the outward normal boundary movement, σ_i is the tangential stress at position *i*, σ_{th} is a non-zero threshold stress, *ls* is a scaling factor. This method is highly suitable for implementation with any standard FE analysis code that permits the node movement after the completion of an analysis. A common outcome in using a constant σ_{th} value shape optimisation, in constrained geometry is the formation of sharp corners, as shown in Figure 3. For practical problems, such sharp corners would be highly sensitive to remote loading and geometric misalignment. During this work considerable effort has been made to create robust optimal shapes that achieve a significant stress reduction over a range of conditions, by preventing the formation of sharp corners.

3 Fuel flow vent hole optimisation analysis and results

Preliminary optimal reworks for FFVH 13 were developed using a fully automated 2D version of the gradient-less shape optimisation method implemented to interface with the PAFEC FE code [2,3]. The FE model used is shown in Figure 4 [6], where focus for the optimisation was on the lower inboard corner, due to current code and model limitations. Significant mesh distortion was avoided using a convenient PAFEC element generation feature named PAFBLOCK, which automatically redistributes the internal mesh at the end of each iteration. In this work, the formation of sharp corners was prevented by the use of a simple three node averaging algorithm, where the minimum radius of curvature is dependent on the local element size.

Currently stress analysis of the WPF is performed using a 3D FE model, Figure 2, with the MSC.NASTRAN solver. Using standard h-type elements, the mesh refinement is suitable for capturing the stiffness of the structure and far field strains. In regions of high stress gradients, higher order p-type elements are used to capture the detail stress distributions. When developing the *final* optimal reworks the preliminary optimal PAFEC FFVH 13 shape was used as an initial shape. This two-staged process reduced the overall effort, the preliminary profile was developed efficiently in 200-300 iterations using the automated procedure, so that the 3D optimisation could be completed in far fewer (20-25) iterations. Only limited automation was achieved, due to a combination of complex 3D geometry, multiple optimisation regions (one for each stress peak) and the intermittent need for re-meshing. Robust optimal shapes, i.e. no sharp corners, were achieved by using a non-uniform stress threshold in Equation 1. As shown in Figure 3, the stress distribution about a constrained optimal hole is only marginally degraded by the removal of sharp corners. Additional work has found that sharp corners can be avoided by stipulating a non-uniform stress threshold, typically a polynomial of order 4, with the turning point nominally on the location of peak stress and/or geometric constraint [6]. Increasing the polynomial order reduces the robustness by giving a smaller radius of curvature and visa-versa. The final robust optimal and traditional profiles for FFVH 13, are shown in Figure 5, and the associated hoop strain results are presented in Figure 6. Similar optimal shapes were developed for FFVHs 11, 12 and 14.

For the most fatigue critical location of the lower inboard corner of FFVH 13 the robust optimal shape provided a very significant peak elastic strain reduction of 50 % and 35% compared to blueprint and traditional intermediate rework respectively, Figure 6. Similar results were achieved for the other FFVHs [6] as shown in the stress contour plots in Figure 7. The reductions in peak elastic strain have been confirmed with a full scale wing fatigue strain survey, results of which are also shown in Figure 6. The final robust optimal profiles have also been successfully manufactured into a damage tolerance and durability test wing to validate crack growth rate predictions used to set inspection intervals.

4 Conclusions

The robust optimal reworks determined for the most critical FFVHs, typically provide a 35% reduction in peak elastic stresses as compared to current rework shapes. Full scale strain surveys have confirmed the finite-element-based gradient-less shape optimisation results. It is anticipated that the stress reductions predicted will be sufficient to provide a basis for extending the recommended inspection intervals for FFVHs in the F-111 WPF. Implementation of the proposed reworks to the RAAF F-111 fleet is due to start in 2001, and is expected to provide significant cost savings and increase aircraft availability.

References

1. Keays, D.W., Molent, L. and Graham, A.D.(1992) *F-111 Finite Element Analysis Rework of Fuel Flow Vent Hole #13*, DSTO-TM-557, DSTO, Melbourne, Australia.
2. Kaye, R. and Heller, M. (1997) *Structural Shape Optimisation by Iterative Finite Element Solution*, DSTO-RR-0105,DSTO, Melbourne, Australia.
3. Waldman, W., Heller, M. and Chen, G. (2000) *Optimal free-form fillet shapes for tension and bending*, (accepted for publication in the Journal of Fatigue).
4. Thompson, D.W. (1917) *On growth and form,* Cambridge University Press, London [also abridged edition. Ed. Bonner, J.T. (1961) Cambridge University Press, London]
5. Heywood, R.B. (1945) Photo-Elasticity and Design Problems (Part II), *Aircraft Engineering*, **17**(August) 226-228.
6. Heller, M., Burchill, M., McDonald M. and Watters, K.C. (2000) *Shape Optimisation of Critical Fuel Flow Vent Holes in the F-111 Wing Pivot Fitting*, DSTO-TR-1120, DSTO, Melbourne, Australia.

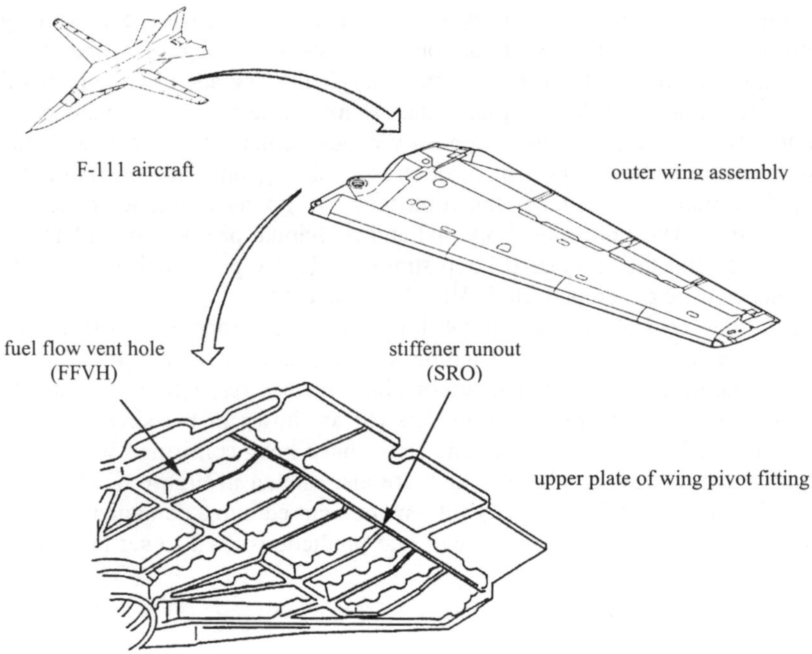

Figure 1. F-111 wing pivot fitting, showing details of fuel flow vent holes and stiffener runouts.

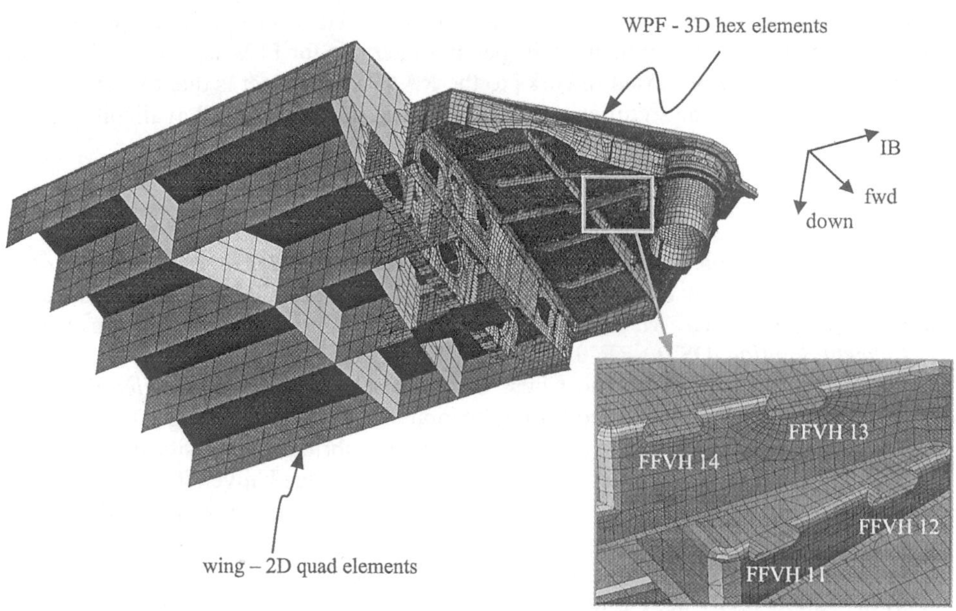

Figure 2. MSC.NASTRAN wing pivot fitting finite element model (upper section), showing detail of fuel flow vent holes 11, 12, 13 and 14 in blueprint condition.

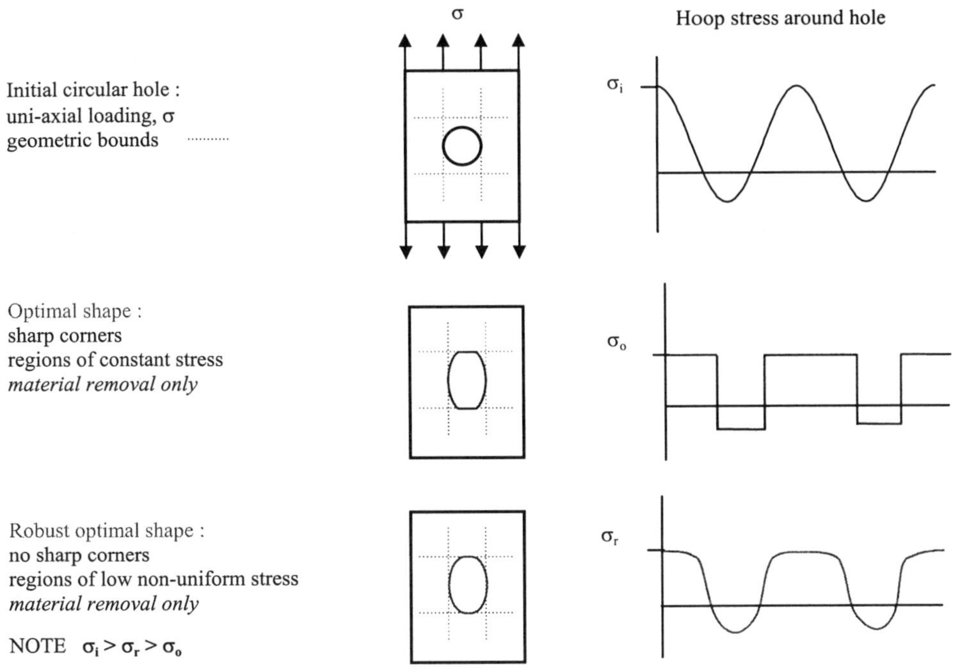

Figure 3. Developing a robust optimal profile for a constrained hole in a large plate under uni-axial loading, σ

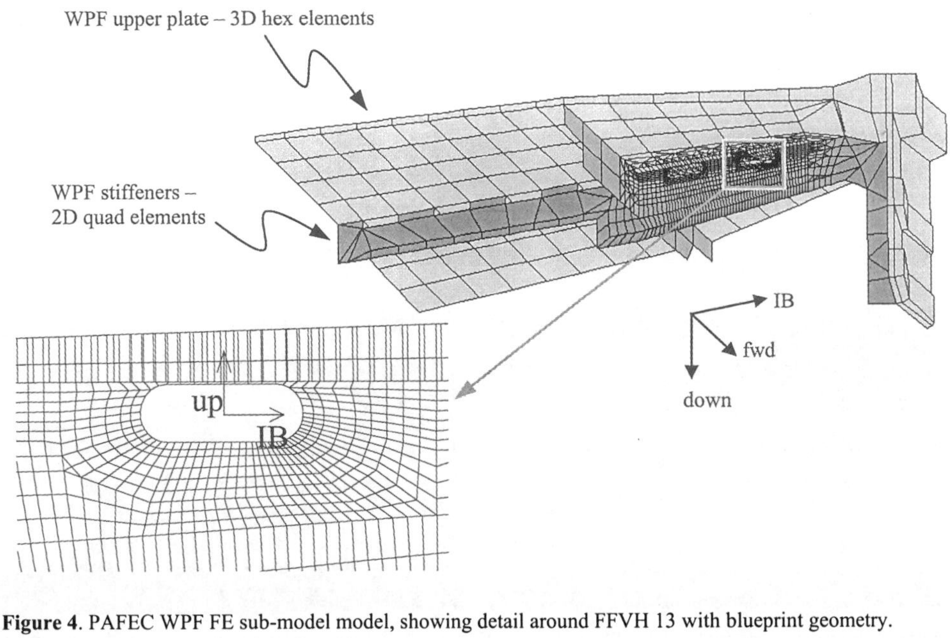

Figure 4. PAFEC WPF FE sub-model model, showing detail around FFVH 13 with blueprint geometry.

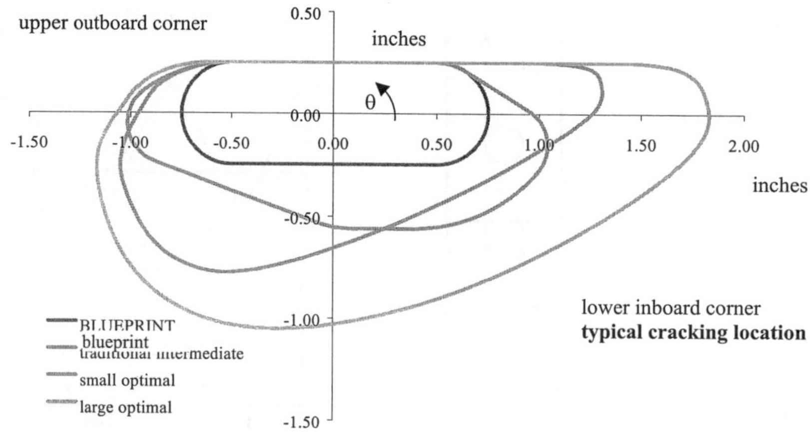

Figure 5. FFVH 13 profiles – blueprint, traditional and robust optimal rework shapes.

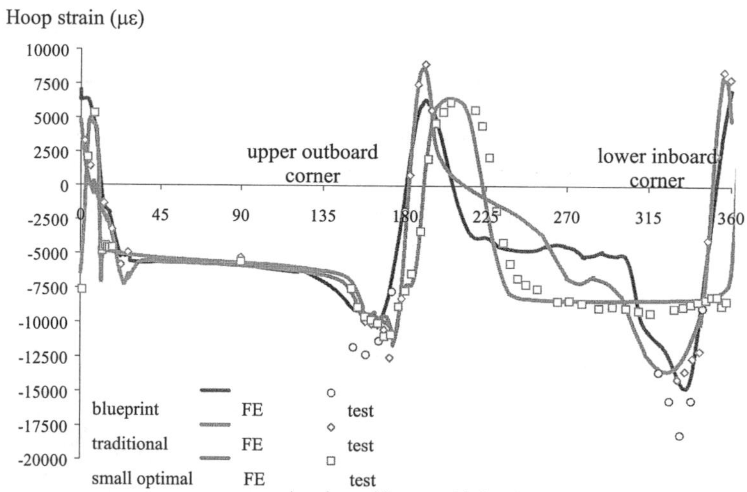

Figure 6. Scaled elastic hoop strain FE and test results around FFVH 13 rework shapes, at +7.33g CPLT load.

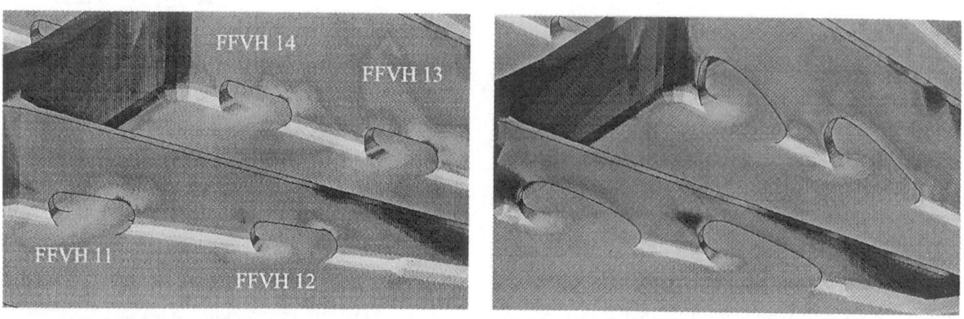

Figure 7. Stress contour plots with FFVHs 11, 12, 13 and 14 in blueprint and optimal condition, at +7.33g CPLT load.

FATIGUE CRACK GROWTH IN CARBON STEEL DUE TO REPEATED THERMAL SHOCK

B. KEREZSI, J.W.H. PRICE AND R. IBRAHIM

Department of Mechanical Engineering, Monash University, Caulfield Campus, East Caulfield, 3145, Victoria, Australia,
E-mail: Brian.Kerezsi@eng.monash.edu.au

Repeated thermal shock loading is common in many industrial situations including the operation of pressure equipment found in thermal power stations. Thermal shock can produce a very high stress level near the exposed surface that eventually may lead to crack nucleation. Further crack growth under the influence of repeated thermal shock is a very complex phenomenon due to both the transient nature of the highly non-linear thermal stresses and the strong influence of the environment. This paper describes an experimental analysis of crack growth in heated carbon steel specimens exposed to repeated thermal shocks using cold water. Analysis of the effect of steady state primary loads on the growth of the cracks is isolated using a unique test rig design. Environmental effects due to the aqueous nature of the testing environment are found to be a major contributor to the crack growth kinetics.

1 Introduction

The growth and arrest of cracks due to repeated thermal shock loading is of interest in industrial applications where predictions can allow decisions to be made on the necessity of planned inspections and component replacements. Current methods for predicting repeated thermal shock (RTS) crack initiation lifetime and growth rates rely on standards and codes of practice such as the ASME Boiler and Pressure Vessel Code [1,2], and British Standard BS7910 [3], both of which use models based on isothermal fatigue tests and simplified stress profiles. The conservatism of these codes when analyzing RTS cracking is of concern, particularly when predicting crack growth rates after initiation has been observed. An over-estimation of this growth rate can lead to premature replacement of components that otherwise may have been left in service. A particular reference to the conservatism of the ASME code can be found in work by Czuck et al [4], where growth at the tip of a crack exposed to cyclic thermal shock loading was found to be an order of magnitude less than that predicted by using the code.

Cracking due to repeated thermal shock (RTS) is a particular problem in thermal power stations. Thermal shocks in boiler equipment are an inevitable side effect of normal operation with start-up and shut-down procedures thought to be especially damaging. Cyclic operation of traditionally base-loaded units only increases the severity of the problem. In this work, tests that closely simulate the real life conditions of thermal shock encountered in thermal power station pressure components are completed. These tests are intended to develop data that will provide for realistic determination of lifetime to crack initiation and crack growth rates for service components as well as allow for an estimation of the conservatism of the current codes.

Many influential external factors that are present in thermal power plant equipment are ignored in existing codes. Foremost is the combined effect of external primary loads and the environment in which the crack is growing. The external loads can be a direct result of the pressure or mechanical loading of the components, the effect of which is to open any cracks, exposing them to the environment. The environment is dependent on the process in which the component is being used, which in the case of thermal power station equipment is often aqueous in nature (including chemical treatment to control pH and oxygen levels) and will modify conditions at the crack tip.

In this paper, selected results from crack growth tests using a unique test-rig arrangement are presented. Comparisons of the actual results with empirical prediction methods from current design codes are made.

2 Experimental Technique

The testing completed in this investigation has been carried out on a thermal fatigue test rig that has been purpose built for the investigation of crack initiation and growth due to repeated thermal shock loading. Consisting of a convection furnace, static loading structure and quenching system it allows for the monitored growth of cracks for a wide variety of component geometries.

The key advantages of this rig over those used in previous studies are:

- The component is heated by convection, which means that there are no unwanted heat effects at the crack tip as may be the case for induction or resistance heating.
- The component is quenched by room temperature pH and O_2 controlled water.
- The specimen size is representative of typical industrial components.
- Large specimen size permits multiple simultaneous experiments.
- An unloaded "control" specimen can be used.
- Approximately one-dimensional conditions exist at any one crack because of the unique specimen design.

A thorough analysis of the development of the test rig and specimen design, including a review of previous trends in the experimental investigations of thermal shock cracking can be found in [5].

Tests conducted with this rig have concentrated on identifying the effects of environment and primary loads on crack initiation and growth during repeated thermal shock. This was completed by simultaneously testing sets of two low-carbon steel, flat plate specimens (grade AS 1548-1995 [6]). One specimen is subjected to a 90MPa (13ksi) uniform tensile stress and the other left unloaded. The specimens, with a combination of 0.25mm and 0.1mm radius notches machined into the quenched faces, were fitted vertically in the furnace. The upper specimen temperature was limited to 370°C to remove any creep effects. Dissolved oxygen levels (D.O.) were varied between tests, while the pH was held steady at around 8.0. The first set of tests used fully oxygenated tap water with a D.O. level of around 8ppm. This water was vigorously pre-boiled in the second set of tests, driving off excess oxygen and reducing the D.O. level to around 2ppm.

Each thermal cycle consisted of a slow heat to a set temperature of between 290°C and 370°C, followed by a 7-second water quench. Cycle time was around 15 minutes. Thermocouples were used to record the temperature profiles of the specimens during testing. The corresponding elastic stress profiles were then generated using the theory of elasticity [7]. A weight function technique as outlined in Section XI of the ASME Boiler and Pressure Vessel Code [2] was then used to calculate the resulting stress intensity factor profiles. A typical stress intensity factor profile plot for a 7s thermal shock from 370°C with 90MPa primary load is presented in figure 1.

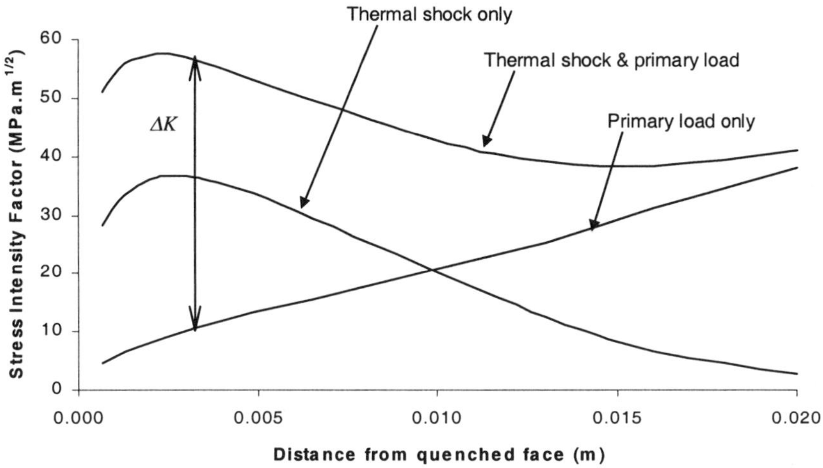

Figure 1: Maximum stress intensity factor profiles during 7s shock from 370°C, with and without 90MPa primary load.

After each period of 500 thermal cycles, the specimens were removed from the furnace and investigated for cracking at a low magnification (10 to 60x). A rough estimate of crack initiation lifetime was defined as when a full-face hairline crack was visible at the base of the notch. Extrapolations of the long crack growth data were used to further refine when initiation was to have occurred. Any subsequent crack growth after initiation was measured on each side of the specimen. Through crack depth was taken as the average of these two measurements.

3 Test Results

The number of thermal shock cycles required for crack initiation in the notched specimens is shown graphically in figure 2. Unlike the authors' previous work [5], where elastic stress amplitude was used, the crack initiation lifetime has been plotted against the Neuber pseudostress amplitude (S'_a). The pseudostress is defined as the local strain amplitude at the notch (ε_a) multiplied by Young's Modulus (E).

$$S'_a = \varepsilon_a E$$

When the behaviour of the material remote from the notch is predominantly elastic, the pseudostress amplitude can be related to the nominal stress amplitude (S_n) and the local stress amplitude (σ_a). The final relationship, outlined in work by Prater and Coffin [8], is reproduced below:

$$S'_a = \frac{(k_f S_n)^2}{\sigma_a}$$

Here k_f is the notch fatigue factor, determined from a notch sensitivity analysis as used in [5]. A maximum value of 5.0 has been established for k_f in accordance with a "worst case notch" analysis. Use of this equation also requires knowledge of the cyclic stress strain relationship for the material. Prater and Coffin [8] provide data for carbon steel that has been adapted for use in this investigation.

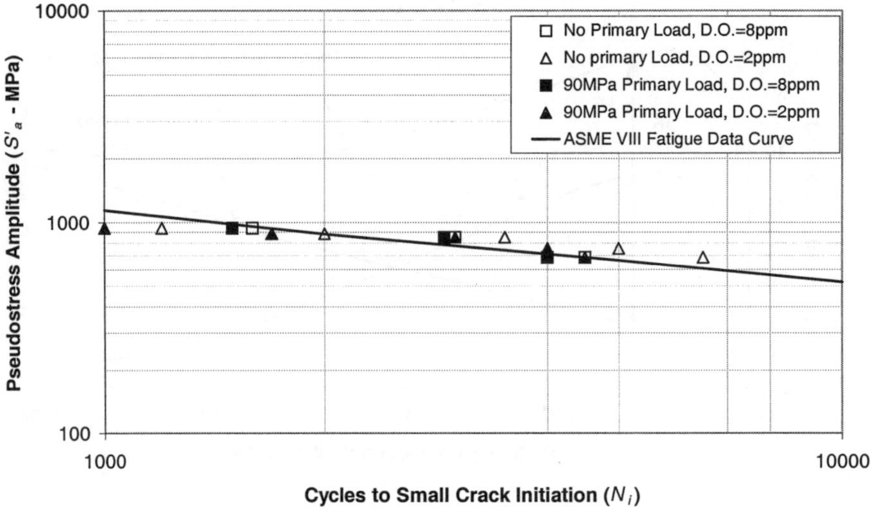

Figure 2: Effect of primary load and dissolved oxygen Level (D.O.) on small crack initiation lifetime.

Figure 2 also shows a curve based on ASME *fatigue* data. This curve is slightly different to the published *design* curve as shown in Section VIII of the ASME Code [2]. The design curve has a built-in conservatism of 2 times on stress and 20 times on lifetime on the fatigue data, whichever is the most conservative.

Crack growth results are shown on figure 3. Stress intensity factors have been plotted against the crack growth rate using the *R*-ratio to categorise the data. Both water and air environment crack growth curves for carbon steel, taken from the ASME Boiler and Pressure Vessel Code [2], are also plotted for comparison.

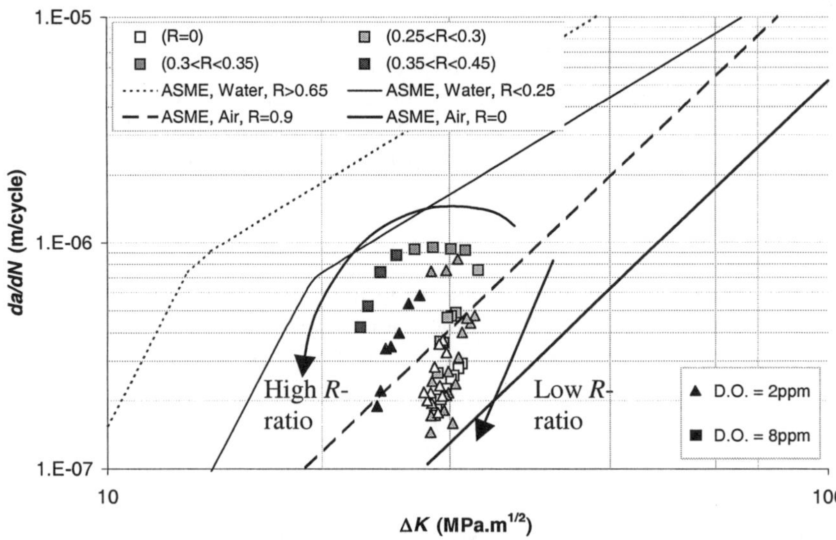

Figure 3: Crack growth rates as a function of stress intensity factor. Squares correspond to measurements made during testing with D.O.≈8ppm, triangles correspond to D.O.≈2ppm. The large arrows indicate only trends for individual crack growth – no curve fitting is implied.

4 Discussion

4.1 Crack Initiation

As shown in figure 2 and reported in [5], the application of a primary load displays little or no effect on crack initiation lifetime within the limitations of experimental variance. The same can be said for reducing the dissolved oxygen level from 8ppm to 2ppm.

In all of the test cases, the ASME *fatigue* curve seems to provide a good approximation of the relation between number of cycles to crack initiation and notch pseudostress amplitude. However, any attempt to use the ASME curves to predict crack initiation should be approached with care. This is because the ASME fatigue curves are based on uniaxial strain controlled testing of small cylindrical specimens. Failure in one of these specimens is defined as when complete fracture occurs. It is then assumed that the failure of these small specimens is equivalent to the initiation of a small crack in a larger structure.

Referring to ASTM E 8M – 96 [9], the standard small cylindrical specimen size for tensile tests is 12.5mm in diameter. Depending on the stress amplitude during fatigue testing and the presence of mean stresses, a small amount of crack growth will occur before specimen failure, probably in the order of a few millimetres. This means however that no exact "initiation" size can be known to allow comparison with experimental data from large specimens. Therefore, any correlation between the ASME code and the experimental work presented here must remain purely qualitative. Namely, the Neuber pseudostress developed at a notched carbon steel specimen exposed to RTS can be related to number of cycles to initiation in a linear manner similar to that shown by the ASME fatigue curve. Reciprocally this means the ASME fatigue curve can be used as a guideline for determining approximate times to develop small (<1mm) deep cracks in carbon steel exposed to RTS.

4.2 Crack Growth

A number of observations can be made from the crack growth data obtained by relating the change in stress intensity factor ΔK to the propagation rate *da/dN*.

- For all of the plotted points, the ASME water curves provide a conservative result (see figure 3). For some points this conservatism is more than an order of magnitude.
- Using the results of the ΔK profiles shown in figure 1 and the trend of the data of figure 3, it would seem that deceleration of all observed cracks is occurring.
- In specimens without steady state primary loading, crack growth rates begin decreasing immediately after crack initiation.
- In contrast, after initiation, cracks in the specimens with large steady state primary loading experience an increase in crack growth rate as the R-ratio increases. Then, after a period of rapid growth (at around 1 mm/1000 cycles), the rate decreases rapidly. It is possible that this high rate of crack growth, followed by the quick drop off may correspond to the start of an environmentally influenced diffusion controlled crack growth.

5 Conclusions

Results obtained from a test program developed to simulate repeated thermal shock conditions produced in operating thermal power station have shown:

1. The application of a primary stress had little or no affect on crack initiation lifetime during repeated thermal shock below the creep range. A reduction of dissolved oxygen levels in the quenching water from 8ppm to 2ppm also did not significantly affect initiation times. ASME Boiler and Pressure Vessel Code, Section VIII, Division 2 [1] fatigue data may be used as a guideline for determining approximate times to develop small (<1mm) deep cracks in notched carbon steel specimens exposed to repeated thermal shock.
2. Environmental and primary load interaction is highly influential in the growth of thermal shock cracks. Cracks with a low *R* ratio (no primary stress) show signs of rapid deceleration and cracks with high *R* ratio show signs of deceleration after a period of environmentally enhanced growth. The ASME provided crack growth curves for a water environment are conservative in all cases.

6 Acknowledgements

This work has been completed with the assistance of an Australian Research Council grant with contributions from HRL Technology Ltd, Optima Energy, Western Power, Pacific Power and the Electric Power Research Institute of USA. The Monash University Postgraduate Publications Award has also provided assistance.

References

1. ASME. ASME Boiler and Pressure Vessel Code, Section VIII, Division 2, ASME, New York (1998)
2. ASME. ASME Boiler and Pressure Vessel Code, Section XI, ASME, New York (1998)
3. BSI. BS 7910, *Guide on methods for assessing the acceptability of flaws in fusion welded structures*, BSI, London (1999)
4. Czuck, G., Mattheck C., Munz, D. and Stamm, H., Crack growth under cyclic thermal shock loading, *Nuclear Engineering and Design* **84** (1985) pp. 189-199.
5. Kerezsi, B.B., Kotousov, A.G. and Price, J.W.H., Experimental apparatus for thermal shock fatigue investigations, *International Journal of Pressure Vessels and Piping* **77** (2000) pp. 425-434.
6. Australian Standards. AS 1548-1995, *Steel plates for pressure equipment*, Standards Australia, Homebush, NSW (1995).
7. Timoshenko S.P. and Goodier J.N., *Theory of Elasticity*, 3rd edition, McGraw Hill, London (1982).
8. Prater T.A. and Coffin L.F., The use of notched compact-type specimens for crack initiation design riles in high temperature water environments, In: *Corrosion Fatigue: Mechanics, Metallurgy, Electrochemistry and Engineering*, ASTM STP 801 (1983) pp. 423-444.
9. ASTM. ASTM E 8M-96, *Standard test method for tension testing of metallic materials [metric]*, ASTM, Philadelphia (1996).

STUDIES IN THE MECHANICS OF DUCTILE AND BRITTLE FRACTURE

B.S. GLASS AND M. P. WEST
University of Wollongong, **Wollongong, NSW 2522 Australia**

Monotonic load tests of plain and notched cylindrical tensile specimens have been made using a variety of aluminium, brass, and steel materials. It is shown, as would be expected, that many of these normally ductile materials may exhibit brittle fracture in notched bar tests, depending on the notch geometry. The apparent notch toughening effect was measured as the ratio of nominal breaking stress of notched versus un-notched specimens. Notch toughening is then used as a parameter to describe the relative ductility or brittleness of a particular fracture, given that the material is normally ductile. Elastic-plastic analyses of the two types of notched specimens were carried out using highly detailed finite element modelling, to better understand the mechanics of deformation processes in the regions of high plastic strain. Good agreement was achieved between numerical analyses and experiments, with finite element analysis replicating the notch-toughening behaviour found in load tests. These results are used to demonstrate that the continuum plasticity model consistently predicts monotonic failure curves (applied load versus axial deformation) for notched metallic specimens, whether the failure is brittle or ductile. The ductility or brittleness of the failure mode are shown to depend on notch geometry and material characteristics, including yield strength, hardening exponent, and fracture ductility of a plain bar in uniaxial tension.

1 Introduction

The objective of this paper is to examine details of fracture behaviour of notched test specimens made from five different metallic materials. In particular, we try to understand the conditions by which, in monotonic loading, a material may fail in ductile fracture in one notched condition, and brittle in another. The load-deformation behaviour up to failure is also of interest, as it shows apparent notch toughening for some materials, but not others.

The materials chosen are: (a) free-cutting brass, (b) 6061-T651 aluminium, (c) 4340 steel, (d) 1080-O (high purity) aluminium, and (e) gray cast iron. For each material, uniaxial tension tests using 12.85 mm diameter cylindrical specimens were made. Stress-strain curves are shown in Figure 1. The tests were carried out on a 100 kN servohydraulic testing machine in manual load control, with a 25 mm gauge length strain extensometer. The basic tensile properties range from strong and tough (4340 steel), to soft and ductile (1080 aluminium), as well as inherently brittle (gray cast iron).

2 Notched Specimens

Two types of notched specimens were made, based on the original tensile bar geometry: (a) a transverse hole 6.35 mm diameter was drilled through the centre of tensile bars; (b) a 90° circumferential V-notch was machined into tensile bars, with nominal depth of 3.175 mm and root radius equal to 0.10 mm. The respective test specimens are shown in Figures 2 (a)-(c). The notched specimen dimensions were uniform for all of the five materials.

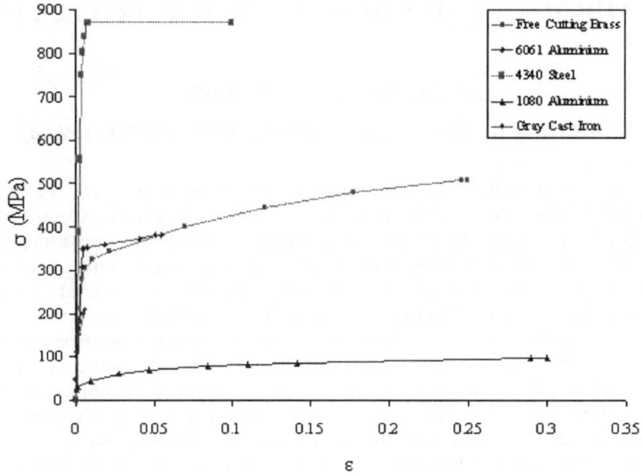

Figure 1. True stress-true strain curves for free-cutting brass, 6061 aluminium, 4340 steel, 1080 aluminium, and gray cast iron.

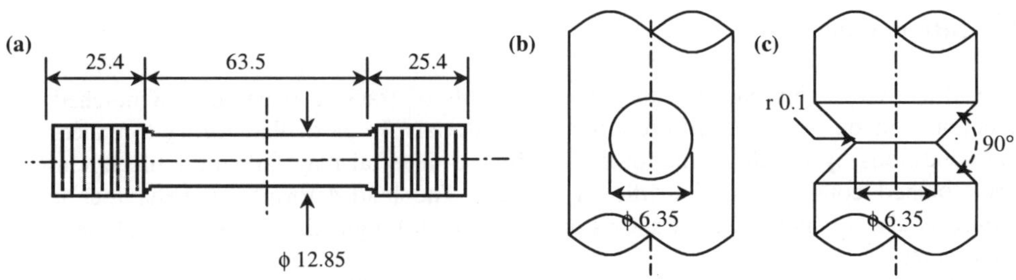

Figure 2. (a) Plain tensile bar; (b) Transverse hole geometry; (c) V-notch geometry.

3 Test Program

For each of the five materials, tensile load tests were made on the notched bars, with load-deformation curves recorded using the 25 mm clip gauge in each test. To aid in interpreting the results, the fracture load was divided by the least cross-sectional area of each specimen, giving the nominal engineering stress on the critical section at failure. Table 1 shows the results, along with the material ultimate tensile stresses from the plain bar tests. The free-cutting brass, 6061 aluminium, and 4340 steel specimens all showed ductile fracture surface appearances for the transverse hole notch type, and brittle appearance for the V-notch. 1080 aluminium was ductile, whereas cast iron was brittle (as well as for the plain bar). When considered along with the fracture surface appearances, the data of Table 1 are remarkable, insofar as the reduced elongation associated with brittle fracture in the free-cutting brass, 6061 aluminium, and 4340 steel specimens is accompanied by substantially higher net section stress at fracture, compared to the plain bar. The transverse hole notch led to little, if any, toughening for these materials. In the case of 1080 aluminium, all specimens were ductile, with little sign of notch toughening.

Table 1. Failure load, nominal stress and deflection data.

Specimen Type	Plain			V-Notch			Transverse Hole		
Material	P (N)	σ_o (MPa)	δ (mm)	P (N)	σ_o (MPa)	δ (mm)	P (N)	σ_o (MPa)	δ (mm)
Free-Cutting Brass	53000	408.677	14.3427	23350	718.441	0.2056	26200	391.91	0.97043
6061 Aluminium	46900	361.64	2.69336	21250	653.828	0.38242	25200	376.952	0.55923
4340 Steel	113000	871.329	0.39064	54800	1686.11	0.34952	49000	732.962	0.4112
1080 Aluminium	9750	75.1811	17.2704	3950	121.535	2.42608	-	-	-
Gray Cast Iron	27100	208.965	0.3084	11000	338.452	0.07813	-	-	-

Several interesting questions arise from these tests: what is the mechanism of notch toughening? Why does 1080 aluminium not show brittle fracture for the V-notch bar? Can finite element analysis based on conventional plasticity theory (von Mises yield surface with associated flow rule) predict the load-deformation behaviour up to fracture for all the cases tested?

4 Finite Element Analysis

A numerical study was conducted using Abaqus software from HKS Inc., chosen for its elastic-plastic large strain capabilities. Finite element models were generated using Femcad 2000 software [1]. The V-notch specimens were modelled using 8-node quadrilateral axisymmetric elements; the mesh is shown in Figure 3. A series of mesh configurations was tested for this specimen, with successively finer element subdivisions near the notch root, until the final stress field and limit load results converged. Note (see detail, Figure 3) that the initial notch root radius of 0.10 mm was carefully modelled in the finite element analysis. The bar with a transverse hole was modelled for 3-D analysis as shown in Figure 4, using 20-node hexahedral and 15-node pentahedral elements. Sufficient detail in the notch region was included in the finite element mesh to reveal accurately the stress concentration effects at the hole. The analysis was carried out for each of the two notched specimen types, for each of the five materials, using as input data, the respective uniaxial true stress-true strain properties obtained in the plain bar tests. A plain bar was modelled as well for each material, as a control on the finite element results.

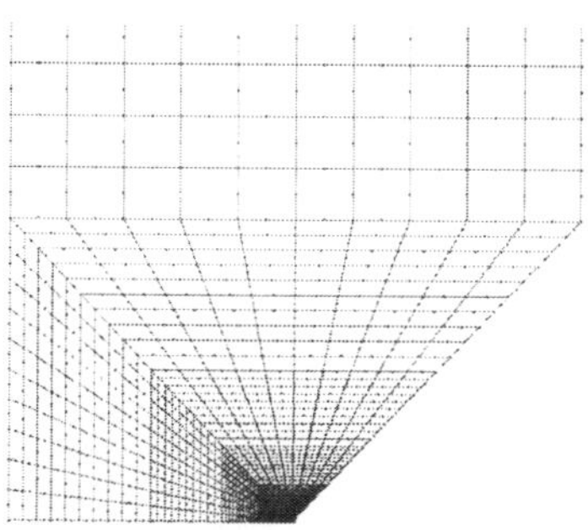

Figure 3. V-notch finite element model.

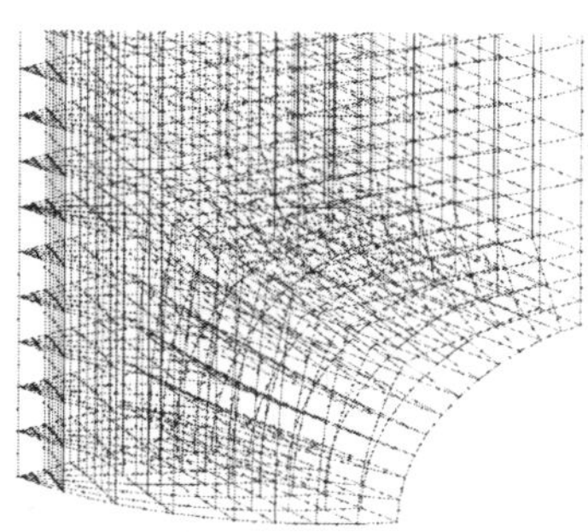

Figure 4. Transverse hole finite element model.

In the Abaqus elastic-plastic computations, automatic load increments were used, until the calculation terminated due to failure to achieve convergence in the strain field for the final load step. This condition is interpreted as the limit load state, representing incipient failure. The load-deformation characteristics obtained from finite element analysis are shown in Figures 5 (a)-(f) and 6 (a)-(b), with the experimental results redrawn for comparison. It is clear that the numerical model, incorporating the uniaxial stress-strain properties for each of the materials with carefully chosen finite element mesh design, does accurately predict ductile or brittle failure in each case. The continuum plasticity concept appears therefore to be sufficient to explain the range of fracture behaviour observed experimentally. The question remains as to what details of the stress and strain fields that arise during loading to failure can explain the observed ductility or brittleness. For this problem, the incremental results generated during the finite element analysis were examined, with the presumption that conditions controlling plastic flow in the notch region ultimately determine the fracture mode.

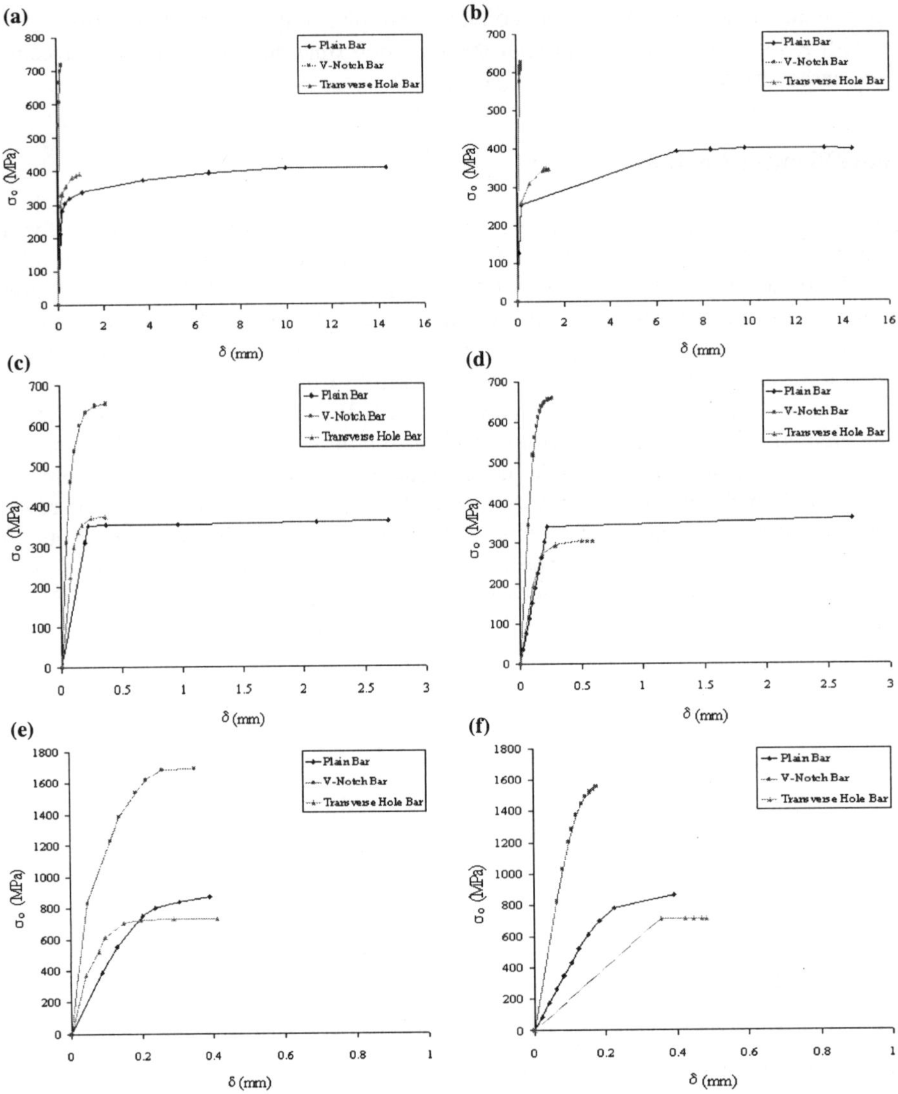

Figure 5. Nominal engineering stress-deflection curves: Free-cutting brass, (a) experimental, (b) analysis; 6061 aluminium, (c) experimental, (d) analysis; 4340 steel, (e) experimental, (f) analysis.

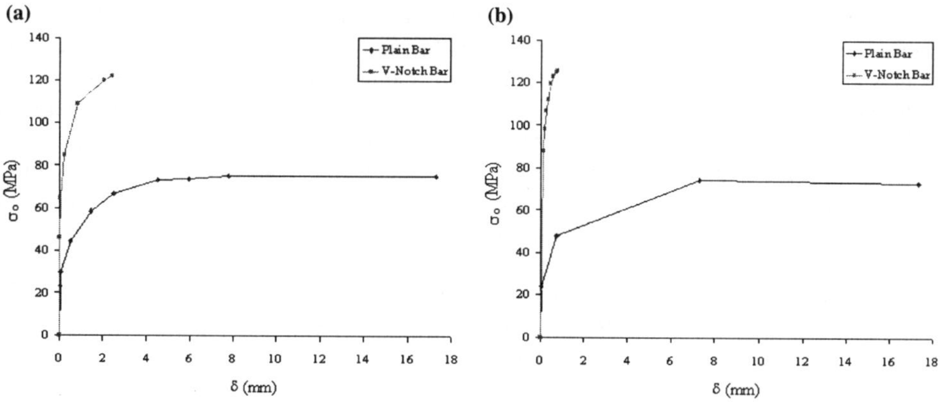

Figure 6. Nominal engineering stress-deflection curves: 1080 aluminium, (a) experimental, (b) analysis.

The transverse hole notch, which (except for cast iron) typically shows ductile fracture in tests, does not present any particular difficulty, as the computed stress field shows low triaxiality, with uniform spreading of the plastic zone as loading is increased (Figure 7 shows results for brass). The V-notch case shows high stress triaxiality, and the plastic zone remains confined to a small region at the root of the notch, despite the early appearance of axial stresses reaching ultimate tensile strength (Figure 8 shows results for brass). For the 1080 aluminium bar, the extent of the plastic zone is greater, as shown in Figure 9. This is what could be expected, given the low yield strength of this material, relative to the intensity of stress concentration and the small size of the specimen. As a consequence, the plastic strain spreads rapidly through the cross section, opening up the notch root and removing the sharp geometric discontinuity required to maintain a state of high triaxiality there.

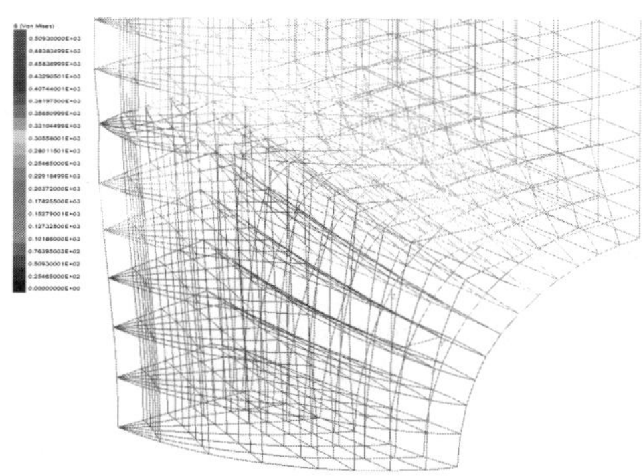

Figure 7. Transverse hole, free-cutting brass (von Mises stress).

5 Summary and Conclusions

It is clear from the results obtained that the continuum plasticity model is effective in consistently predicting the monotonic failure curves of notched metallic specimens. The true stress-true strain curve obtained from a uniaxial test of a plain bar provides sufficient material characteristics to effectively model the behaviour of a metallic material with different notch geometry, whether the failure is brittle or ductile. The correlation between the elastic-plastic analyses obtained from finite element analysis and the experimental results clearly show that the brittle or ductile behaviour of a material is dependant on notch geometry and material properties such as yield strength, hardening exponent and fracture ductility. Although the effects of necking are not modelled, the ductile failure was predicted to the point where necking would occur, hence accounting for the slight variation found between the results of the analyses and experiments.

The results of the analyses of the V-notched specimens clearly show that a distinct notch-toughening effect is present for those materials where brittle fracture occurred. For these materials, the observed state of stress across the fracture surface show a high stress triaxiality, with very little associated plastic strain. In comparison the plain and transverse hole specimens show considerable plastic strain before failure occurs, with a hydrostatic tensile stress state significantly lower than that for the V-notch specimen. In the case of 1080 aluminium, the plastic region is more dispersed away from the root of the notch, and hence in the V-notch case, a lower stress triaxiality and higher associated plastic strain are present across the fracture surface in comparison to the other materials tested.

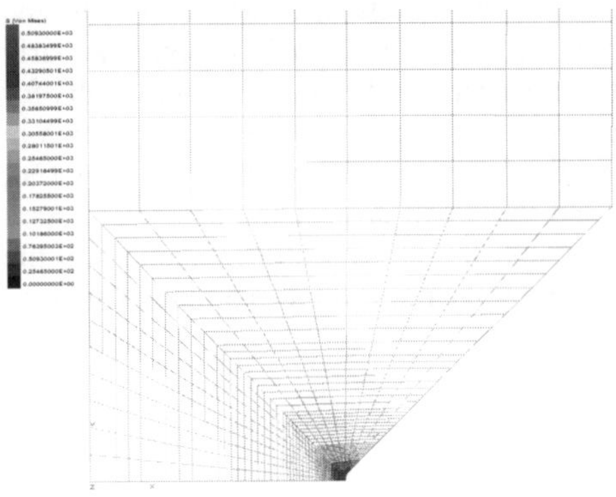

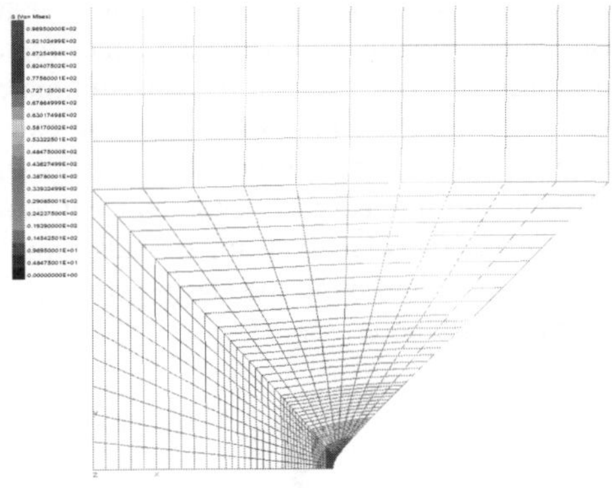

Figure 8. V-notch, free-cutting brass (von Mises stress). **Figure 9.** V-notch, 1080 aluminium (von Mises stress).

6 Acknowledgements

Support for this research has been provided by Rail Infrastructure Corporation and the Co-operative Research Centre for Welded Structures.

References

1. Glass, B.S., *FEMCAD 2000 Finite Element Mesh Modelling Software*, 1st Ed., University of Wollongong, 2000.

CRACK-PARTICLE INTERACTIONS IN COMPOSITE MATERIALS WITH PARTICULATE-MODIFIED TOUGH INTERLAMINAR LAYERS: FEA STUDY

D. STEVANOVIC, S. KALYANASUNDARAM, A. LOWE

Department of Engineering,
Faculty of Engineering and Information Technology,
The Australian National University, Canberra, ACT 0200, Australia
E-mail: dejan.stevanovic@faceng.anu.edu.au

P.-Y. BEN JAR

Department of Mechanical Engineering,
University of Alberta, Edmonton, Alberta T6G 2G8, Canada

A numerical analysis was conducted, using previously obtained experimental results, to establish basic toughening mechanisms and fracture behaviour of an interlayer-toughened composite material, containing particulate modified toughened layers. The aim of the analysis was to examine the interaction between the particles and the crack tip during crack propagation. Good agreement with the experimental data was found.

1 Introduction

Interlayer toughening of composite materials has been receiving significant attention from materials scientists in the past two decades [1]. Some of the recent developments in these materials involved matrix resins blended with thermoplastic particles for the interlayer material [2]. The major problem in assessing full fracture behaviour of these materials is believed to be associated with complicated stress distributions induced by the presence of particles in the interlayer. An effective approach to solve this problem can be the application of finite element analysis (FEA) for assessing the stress/strain distributions inside the interlayer, together with simulating crack-particle interaction using elastic-plastic crack propagation FEA algorithms. The main aim of this study is to examine interlaminar fracture processes under mode I and mode II loading of an interlayer-toughened composite material using FEA.

2 Finite Element Analysis

This study was undertaken by modelling a vinyl-ester(VE)/glass fibre composite material, interleaved with the matrix resin blended with ABS (poly(acrylonitrile-butadiene-styrene)) particles. It was found in our previous studies [3] that by varying the ABS particle concentration, a non-linear trend of fracture toughness was established, caused by chemical reaction of the vinyl-ester resin with ABS. Besides, as a result of the chemical reaction, 7% ABS blend was established as a transition point in the interlayer material chemical composition and mechanical properties. The blends beyond the transitional ABS content were established as two-phase materials composed of a matrix, having properties of the transitional blend (i.e. 7% of ABS), and particles of neat ABS. This hypothesis has been used in this study for modelling mode I (DCB) and mode II (ENF) fracture toughness specimens. The specimens were modelled as two linear elastic orthotropic layers, representing 20 unidirectional VE/glass composite plies (10 in each half), joined with a single elastic-plastic layer of VE/15% ABS blend, representing the tough interlayer. The interlayer was represented as an elastic-plastic matrix with a strain-hardening tensile stress-strain curve while the ABS particles were modelled as an elastic perfectly plastic material, and thus in good agreement with the obtained experimental results. The properties of the orthotropic layers and the tough interlayer (matrix and particles) are given in Table 1.

Table 1. Material properties applied in the model

	Orthotropic layers		Interlayer matrix	ABS particles
E_1 [GPa]	38.15	E [Gpa]	2.95	2.54
E_2 [GPa]	8.85	ν	0.35	0.35
G_{12} [GPa]	4	σ_{yield} [MPa]	34.2	35.6
G_{23} [GPa]	3.1	ε_{yield} [%]	1.4	1.6
ν_{12}	0.3	$\sigma_{failure}$ [MPa]	49.25	35.6
ν_{21}	0.07	$\varepsilon_{failure}$ [%]	2.8	~20
ν_{23}	0.42			

For this study the crack tip opening displacement (CTOD) was used as a fracture criterion to define position of new crack tip in each step of the analysis. The critical value of CTOD required for the analysis was assessed by employing an FEA approach. For this purpose, a plane strain finite element model was developed which represented a single end notch bend (SENB) specimen, for mode I fracture toughness measurement of plastics, containing a sharp pre-crack of 1 mm in length. The modelled SENB specimen was loaded to the load level required for fracture (experimentally obtained) and the CTOD was measured 20μm behind the sharp crack tip, being 8.22 μm for the interlayer matrix. All numerical models are developed and solved using fracture mechanics FEA code, FRANC 2DL [4]. For the purpose of defining the direction of the crack propagation in each step, the software used the $\sigma_{\theta\text{-max}}$ criterion, where the crack propagates in a direction perpendicular to the direction of maximum tensile stress.

2.1 Mode I models

The initial plane strain model was developed to simulate interaction between a single ABS particle and a crack that approaches the particles out of its symmetry line. The modelled geometry mimicked the geometry of the DCB specimen in every way except the length, which was made to cover 20mm in front of the initial crack, making the whole model 70mm long. This reduced the number of elements in the FE model without a reduction in solution accuracy. The initial crack was introduced by a gap of 20μm in thickness, having a semi-circular blunt tip, positioned in the mid region of the specimen (i.e. inside the interlayer). This simulated an aluminium foil, usually used to simulate initial crack within a DCB specimen. This is depicted in Figure 1 (a). In order to obtain additional and more accurate data on crack/particle interactions, a multi-particle model was developed. However, due to the rationalization of the employed number of degrees of freedom, incorporation of more than one particle in the DCB model would produce a numerically inefficient model.

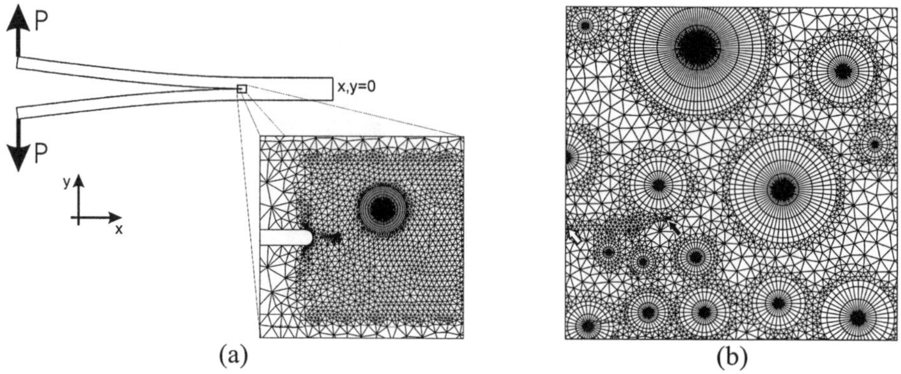

Figure 1. Mode I crack propagation models: (a) DCB configuration with a single particle; (b) Mesh of the multi-particle model with a developing crack (the crack mouth and tip are marked with a white and black arrow, respectively)

Therefore, a simple tensile model was developed, illustrated in Figure 1(b). The model represents a small section of 200X200μm under uniaxial tension in Y (i.e. vertical) direction. A random particle distribution was generated by employing an algorithm which used a uniform random number generator (RNG), generating a number of random circular particles within the area of given boundaries. A crack marked with white and black arrow in Figure 1(b) was generated and grown with the crack tip elements of 2μm. The applied boundary conditions simulated mode I loading by fixing the bottom edge of the model in Y direction and left edge in X direction.

2.2 Mode II models

The initial model mimicked the geometry of the ENF specimen. It was built from approximately 5500 plane strain second order elements, containing a line of gap elements which served to simulate a sharp crack. The gap elements were one-dimensional line elements possessing very low shear and very high compressive stiffness in order to obtain a proper load transfer from the upper to the lower half-arm of the crack. This model was loaded up to 0.8KN as illustrated in Figure 2(a).

This level of loading simulated fracture initiation (i.e. the first non-linear point at the load-displacement curve recorded during the test). However, since the region in front of the crack tip was of the main interest for this study a rationalization of the initial model had to be conducted. A small region surrounding the crack tip, 1x0.6mm for 200μm interlayers or 1x0.9mm for 500μm interlayers, was isolated containing realistically a modelled blunt crack tip and particles in its vicinity. The boundary conditions presented in Figure 2(a) were chosen to obtain an equivalent stress and strain distribution as that of the initial ENF model. The crack faces were fixed in the Y direction, while a displacement of 0.015mm in was applied in opposite directions along X axis, on each half-arm of the crack. The level of the applied displacement was a result obtained from the initial ENF model at the point 0.2mm behind the crack tip, since this point was chosen to be the left boundary of the equivalent shear model. The final mesh of an equivalent shear model of a specimen with 500μm interlayer is depicted in Figure 2(b).

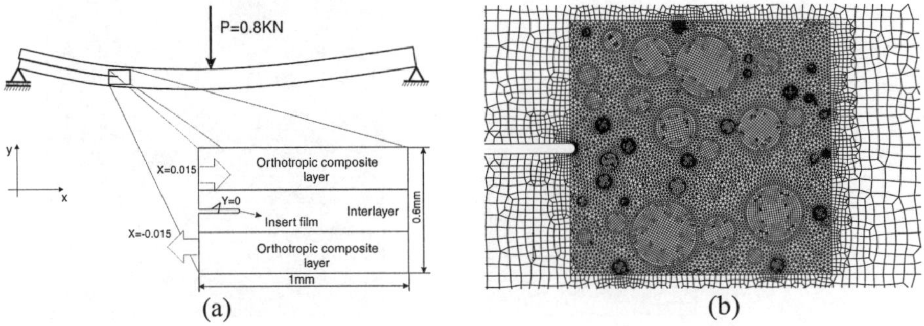

Figure 2. (a) DCB configuration with a single particle; (b) Mesh of the multi-particle model with a developing crack (the crack mouth and tip are marked with a white and black arrow, respectively) – 500μm interlayer

3 Results and Discussion

3.1 Mode I models

The typical fracture behaviour of the interlayer material, observed during this study, is illustrated in Figure 3(a), which depicts mode I crack propagation through a 500μm interlayer. Firstly, the main crack propagates through the interlayer without having any contact with the fibre-layer interface, proving a good interfacial bonding strength. This is the first prerequisite for the full exploitation of the interlayer toughening potential. Secondly, together with the main crack some secondary micro-cracking is quite visible, while on the other hand, there was clearly no particle debonding. Thirdly, the crack path was altered by the presence of particles. All these findings were the subject of a further examination using the numerical models used in this study.

The initial single-particle model showed that the particle can attract the propagating crack and that stress and strain levels were sufficient to cause damage in the vicinity of the matrix/particle interface. This finding was further explored with the multi-particle model. The crack was initiated manually to a length of 20μm and followed by automatic crack propagation, conducted in 12 steps. The average mesh size was 6μm, which was found to be sufficient for the purpose of this analysis. The obtained results are presented in Figure 3(b), as the effective plastic strain contour plot. High strain zones around the propagating crack, visible in Figure3 (b), illustrate the step-like nature of the analysis where each zone represents the position of the crack tip during the crack propagation, consequently resulting in high residual strains. The presence of the particles altered the crack path by attracting the crack. At the same time, the particles and surrounding matrix were subjected to high straining, sufficient to cause micro-cracking. This was clearly observed during the experiments. Since the ultimate plastic strain of the matrix material was only 1.1%, the illustrated contour plot suggests that failure of the matrix material close to the particles is quite possible. Interesting finding is nevertheless strain levels inside the particles, which are exceeding 20%, even in a zone not directly in contact with the crack tip. Obviously, failure of the particles is also possible, making the strain distribution in the whole region around the crack favorable for development of secondary cracks. During the crack propagation, it was clearly attracted by the larger particle, despite being under influence of several small particles during the initial crack extensions.

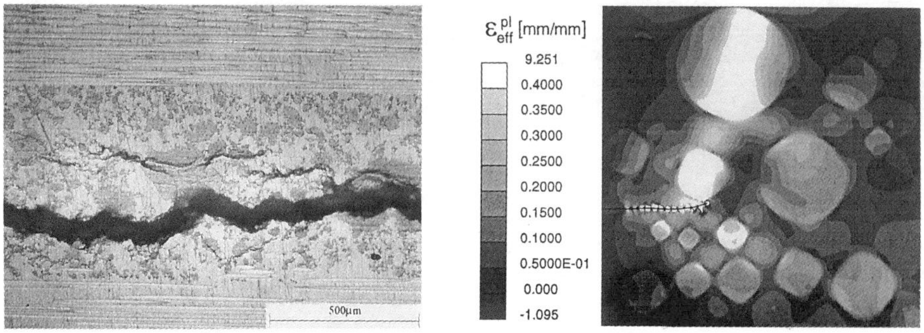

(a) (b)

Figure 3. (a) An experimental micrograph of mode I fracture within an interlayer – crack kinking and multi-cracking are evident; (b) Multi-particle model: equivalent plastic strain contour plot. Crack path was altered and terminated at the interface of a large particle, while interaction between particles is also visible

High strain zones among the particles are developed, suggesting that interactions between particles exist in addition to already observed crack/particle interactions. These particle-particles strain interactions make the whole crack process zone much larger, supporting possibility for secondary crack development.

3.2 Mode II ENF and shear models

The micrograph in Figure 4(a) contains some specific features that describe mode II fracture behaviour of the material. Firstly, the main crack (black arrow) was initiated from the film starter tip and propagated governed by the tensile failure mode. After reaching the interlayer/fibre interface crack propagated along the interface, but interacting with nearby particles through micro-cracking (white arrow). Significant yielding of the layer material is also visible (dark shaded regions).

The experimental findings were further examined by a multi-particle equivalent shear model of a 200μm interlayer, presented in Figure 4(b). The presented image represents the shear strains only in the yielded elements of the interlayer matrix. The crack propagated from the film insert tip, until it was terminated at a particle interface. The crack path was not altered by the near-by particle as it was found for the mode I loading. However, more important for this study is strain distribution above the propagating crack. Obviously, under the influence of the crack driving force (i.e. the maximum tensile stress) the particles were subjected to severe deformation, causing their straining and resulting in development of high strain zones among the particles. Based on the level of recorded strain the particles can cause micro-cracking during the propagation of the main crack towards the interface.

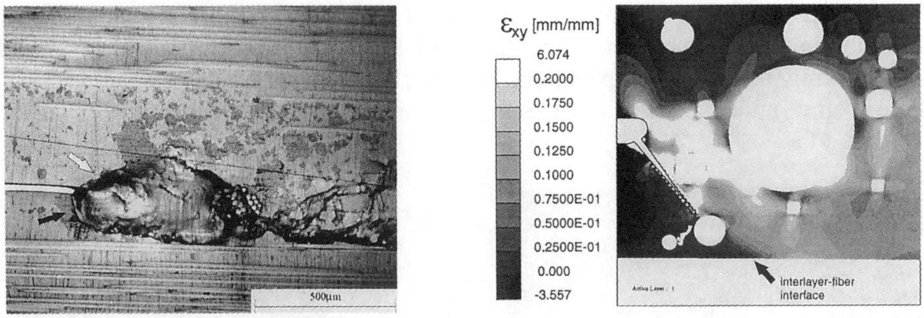

(a) (b)

Figure 4. (a) An experimental micrograph of mode II fracture within an interlayer – the main crack development (black arrow) is accompanied by numerous micro-cracks; (b) Multi-particle model with 200μm interlayer: shear plastic strain contour plot. Crack path was not altered but significant straining among particles indicates possibility of micro-cracking .

It can be assumed that this scenario is also active during the subsequent interfacial failure and the hackle formation, which was observed at the micrograph. This involved various unrecoverable deformations induced upon the matrix material among the particles, including micro-cracking and development of localized plastic deformations. Figure 4(b) suggests that the particle that stopped the crack was not subjected to severe deformation, as it was the case with the particles above the crack.

4 Conclusions

In summary, a methodology is presented to simulate mode I and II delaminations in interlayer toughened composite materials by using FEA approach. It was shown that during crack propagation under both loading modes, through region filled with particles, significant straining was induced in the matrix among the particles and in the particles alone. This suggested that particles influenced stress distribution within the interlayer and raised possibility for micro-cracking among them. The presented methodology, combined with experimental findings, assisted in assessing toughening mechanisms within these materials and in better understanding their fracture behaviour.

References

1. Ishai O, Sela N. Interlaminar fracture toughness and toughening of laminated composite materials: a review. Composites 1989;20(5):423-435.
2. Sue HJ, Jones RE, Garcia-Meitin EI. Fracture behaviour of model toughened composites under Mode I and Mode II delaminations. Journal of Materials Science 1993;28:6381-6391.
3. Stevanovic D, Lowe A, Kalyanasundaram S, Jar P-YB Chemical and mechanical properties of vinyl-ester/ABS blends. Polymer, to be submitted.
4. James MA. A plane stress finite element model for elastic-plastic mode I/II crack growth. PhD thesis, Department of Mechanical and Nuclear Engineering, Kansas State University, 1998.

COMPARISON OF SHOT PEENING AND LASER SHOCK PEENING TECHNIQUES IN ENHANCING FATIGUE LIFE OF 7XXX SERIES ALUMINIUM ALLOYS

Q. LIU, S. A.BARTER AND G. CLARK

Airframes and Engines Division, Aeronautical and Maritime Research Laboratory, DSTO, Melbourne, Australia

E-mail:qianchu.liu@dsto.defence.gov.au

A novel mechanical surface treatment using laser-induced shock waves has been applied to a 7xxx series aerospace aluminium alloy to improve the fatigue life. Laser shock peening (LSP) is applied by using a high energy pulsed laser to create a high amplitude stress wave or shock wave on the surface to be treated. This stress wave propagates into the material, causing the surface layer to yield and plastically deform, and developing a residual compressive stress. The residual compressive stress from laser peening extends much deeper than that from shot peening. The fatigue behaviour has been investigated at different applied stresses and is compared with those of conventionally shot peened specimens. The fatigue life of laser-shocked specimens was found to be higher than that following shot peening, particularly at extremely high stress (450MPa). This large increase in fatigue life appeared to be the result of the deeper residual compressive stress and the smoother surface produced in the laser treated specimens.. For this aluminium alloy, the laser shock surface treatment does not appear to create any microhardness modifications. Therefore, hardening does not appear to be responsible for the fatigue life extension.

1 Introduction

The fatigue life of a component can be improved by using surface enhancement methods such as shot peening, laser shock peening or other techniques. These methods have been widely applied in industry with some of them becoming an industrial standard; this is particularly true for the shot peening process.

Each method has its own advantages and limitations. Conventional shot peening is a traditional cold working process in which the surface of a component is bombarded with small media called shot made of hardened cast-steel, conditioned cut-wire, glass or ceramic beads applied at a relatively high velocity (40-70m/s). The surface is peened to full coverage, and then, usually, for the same time again ("200%" coverage). Each shot acts as a peen-hammer and impacts the exposed surface producing a small indentation. This results in a localised stretching of the surface along with cold working to a depth of about 120 ~ 400μm. A large number of overlapping indentations will generate an even layer of stretched metal, which because of the constraint of the unstretched metal beneath it is held in compression resulting in a residual compressive stress field. As nearly all fatigue and stress corrosion failures originate at the surface of a component, residual compressive stress induced by shot peening readily provides an increase in component fatigue life. However, shot peening may leave an undesirable surface finish such as laps, folds, embedded shot fragments and roughened surfaces, and research by DSTO [1] has highlighted the negative impact of this surface damage on fatigue life. The net effect depends on the balance between the beneficial residual stresses and the surface damage; in steels, the surface damage is usually minor, whereas in aluminium alloys, the surface damage can have greater effect. In practice, unless an abusive process is used, the overall outcome is an improvement in fatigue life.

The LSP process uses a high intensity, short pulse duration laser to create the impact event, which plastically deforms the material and creates the residual stress state. Laser shock peening or processing (LSP) has a number of advantages over the conventional shot peening technique. LSP can create a deeper residual compressive stress - up to 1.5mm in the case of Al alloys [2]. There is little or no cosmetic change to the surface of the metal. Even very thin sections can be laser processed without deformation or spallation of the part. In some component geometries which feature small holes and small slots, it is very difficult for conventional technology such as shot peening or cold expansion to be used due to limited accessibility, whereas LSP can easily reach these areas by using an optical fibre as a delivery system. In addition, the laser radiation is highly controllable and the process is readily amenable to automation. Therefore, it is expected that LSP technology will potentially be an important method for increasing fatigue life in developing modern aircraft, and repairing and maintaining "aging" aircraft.

This study briefly reports the investigation on the extension of fatigue life of aluminium alloy 7075 by using LSP and to compare its effects with a conventional mechanical treatment (glass bead peening). The details of the work will be reported elsewhere.

2 Experimental Procedures

2.1 Material and Test Specimen

The material used for this study is a wrought 7075-T7451 aluminium alloy of composition Al-0.08Si-1.50Cu-0.15Fe-2.55Mg-5.65n-0.06Ti-0.21Cr with static tensile yield and ultimate stresses of 435 and 505MPa respectively. Elongation at failure is 13%.

The specimens used were cylindrical dog-bone types, with a reduced gauge diameter of 12.0mm (d) and length of about 90mm (L) and grip region with a diameter of 24.0mm (D); total specimen length was about 190 mm (see Fig.1).

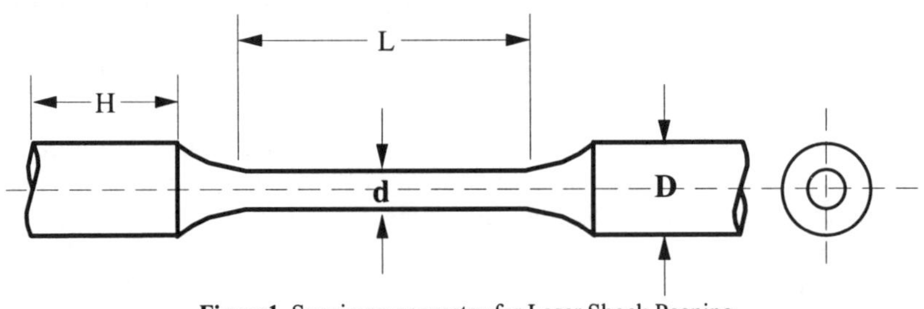

Figure1. Specimen geometry for Laser Shock Peening

2.2 Surface treatments by Laser shock peening (LSP) and Glass bead peening

The laser used in this study is the Q-switched high energy-pulsed neodymium (Nd)-glass phosphate laser with a wavelength of 1.054μm and a pulse duration of 15 to 25ns (nanoseconds). The repetition rate is 0.125 to 0.25Hz. The laser beam spot size is between 5.5mm and 9mm in diameter, which depends on the density of the spot. The LSP configuration was 2 beams, and 50 J/pulse.

This laser was used to treat the gauge part of the specimen (L in Fig.1). Because of the round geometry, the specimens were rotated steadily during the LSP treatment. After LSP the blank paint sacrificial coating was removed using acetone. The specimen surfaces were LS peened by *LSP Technology Inc.*, USA. In this study, the material at the gauge area of the specimen was peened using LSP. However, in order to prevent the specimen breaking at the outside of the gauge area (L in Fig.1), particularly in the grip areas, the rest of the specimen was treated by conventional shot peening (SP).

Glass bead peening was performed at DSTO with a coverage rate of 200% and Almen intensity of 10A using 0.25 mm glass beads.

2.3 Fatigue Test Procedure

Fatigue tests were carried out in a MTS 100kN, digitally controlled testing machine, under load control mode, using a spectrum representative of a fighter aircraft used for F/A-18 aircraft full-scale fatigue testing at AMRL. The loading spectrum consists of 13,475 turning points per program, with one program equal to 324.92 Simulated Flight Hours (SFHs).

2.4 Measurement of Crack Length and Morphology Analysis

The quantitative fractographic analysis capability developed by DSTO was used to measure fatigue crack lengths, based on the progression marks on the fracture surfaces. The details are referred to [3,4,5].

The fracture morphology and microstructure (including dislocations) of laser-peened, shot peened and unpeened specimens were analyzed using a scanning electron microscope. The surface roughness for both glass bead peened and laser shock peened specimens were measured using a laser surface profiler.

3 Results and Discussion

3.1 Microhardness and Fracture Mechanisms

The microhardness depth profile was measured by digital microhardness machine from the peened surface up to the unpeened area. The results show that the laser shock surface treatment does not create any microhardness modifications. This result indicates that the hardening may not be responsible for fatigue life extension of the studied material. The surface treated by laser shock peening is much smoother than that produced by traditional shot peening. For example, the averaged roughness on the glass peened surface (R_a) is 3.1 μm but the averaged roughness in the LSP specimen surface is only 0.2 μm.

An SEM examination of the broken test specimens showed that the crack initiation almost invariably commences at the surface of glass bead peened specimens (arrowed in Fig.2), because peening introduces some surface damage such as surface laps, folds, and embedded glass fragments lying on or just below the surface, which can promote fatigue cracking [6,7]. Therefore, in order to minimize the surface damage, the peening process should be well controlled. However, because of the relatively smooth surface in the LSP specimens, the crack initiation was usually just subsurface in this study (arrowed in Fig.3).

3.2 Fatigue behaviour

3.2.1 Laser shock peened

Fatigue testing was conducted under spectrum loading at the different applied stress levels. The fatigue crack growth results are presented in Fig. 4. The effect of residual compressive stress in reducing crack growth rate is more marked with decreasing applied stress. For example, the fatigue life at a peak load stress of 360MPa is about 8 times that seen at 420 and 450MPa, and 3 times that at 390MPa. This means that the beneficial effect of residual compressive stress induced by LSP processing is more significant at lower applied stress. In addition, the retardation is maintained over a depth range up to about 0.9mm in these tests, which is expected.

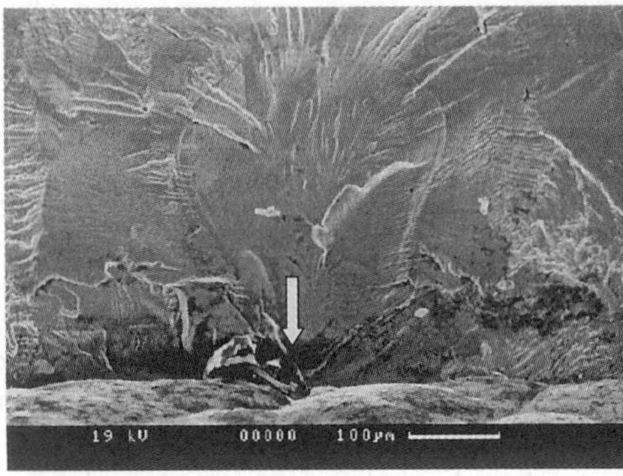

Figure 2 SEM fractography of fatigue origin for the laser peened specimens (arrowed)

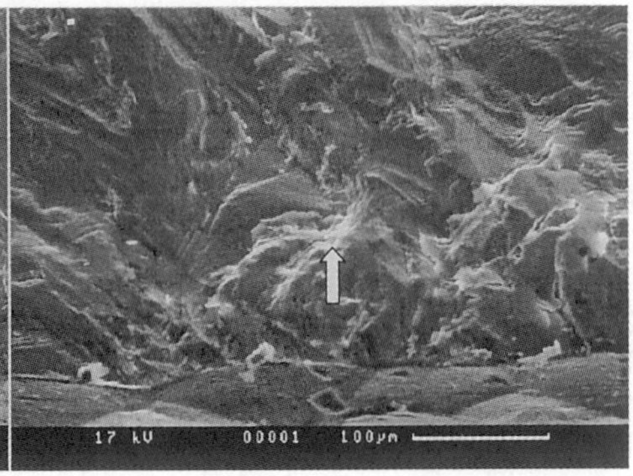

Figure 3 SEM fractography of fatigue origin for the shock peened specimen (arrowed)

3.2.2 Glass bead peened

The fatigue crack growth results for glass bead peening are plotted in Fig.5. The characteristic retardation due to the introduction of residual compressive stress occurred over up to a depth of about 300 μm, slightly less than was observed for the LSP case. The fatigue life enhancement was not as good as that with LSP; this is believed to be because of the shallower residual compressive stress and enhanced crack initiation resulting from surface damage such as surface folding, laps or glass bead fragments. Like LSP specimens, the beneficial effect of residual compressive stress depends on the applied stress. Previous research [1,7,8] on 7050 aluminium alloy has shown that the benefits of peening aluminium alloys are maximized when applied service stresses are low. At much higher applied stresses, peening appears to be less effective, because the residual compressive stress might be washed out by the deformation caused by the test loading. This conclusion is applicable to both LSP and SP specimens in this study.

3.3 Comparison of laser shock and glass peening

It is well known that the cyclic behaviour of mechanical components depends on the metallurgical, mechanical and geometrical surface states of the materials involved. Many experimental and practical results have proven that fatigue life is vastly improved by shot peening (SP) or other mechanical surface treatment methods. However, as the peening process creates a certain amount of surface damage which may promote fatigue cracking, the main beneficial effect of peening is therefore crack growth retardation and also its effect on crack initiation. This is well supported by fractographic studies showing impeded or arrested crack growth in the surface layers and an acceleration in the layers below the surface (Fig.2).

From Fig.6, both of SP and LSP produce a significant improvement in fatigue life for the studied material over the etched specimens (the reference surface condition). With decreasing applied stress the improvement of fatigue life is very obvious. Compared to glass bead peening, laser shock peening (LSP) further improves the fatigue life of a material although the experimental data for LSP specimens has more scatter at some applied stresses. The beneficial effect is believed to be associated with the deeper residual compressive stresses and better surface quality, which are expected to influence crack growth favourably.

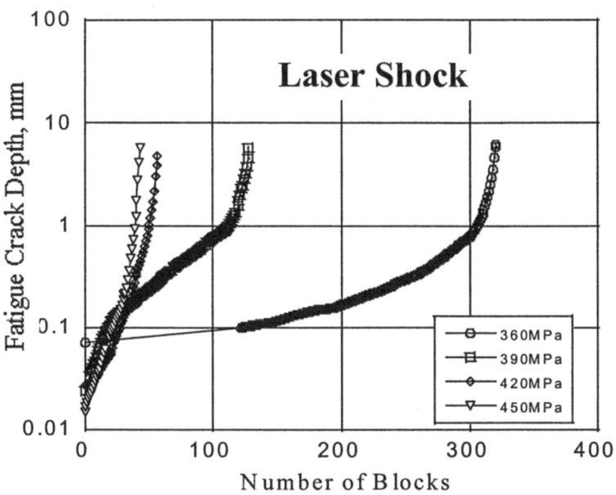

Figure 4 Fatigue crack growth length is measured by using fractographic method from the broken surfaces of the LSP specimens.

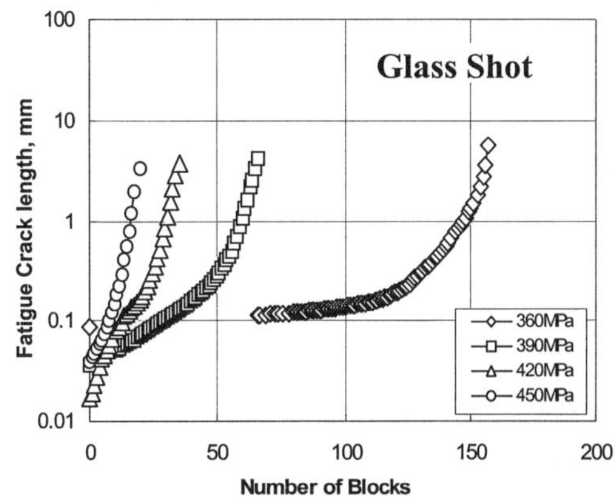

Figure 5 Fatigue crack growth length is measured by using fractographic method from the broken surfaces of the glass bead peened specimens.

4 Concluding Remarks

The effect of the surface treatments by laser shock peening (LSP) and conventional shot peening (glass bead peening) on a 7075-T7351 aluminium alloy has been studied. Both LSP and SP produce an improvement in fatigue life for the material studied, due to the introduction of residual compressive stress. However, LSP gives further improvement, particularly at high applied test stresses (e.g. 420 and 450MPa) since LSP is very effective at introducing deeper residual compressive stresses compared to SP, and because it leaves a better surface condition. Under spectrum loading, a great improvement in fatigue life was made for LSP specimens over the baseline (etched) and glass bead peened specimens.

The LSP process does not appear to produce any detrimental microstructure changes. Fractographic and microhardness studies seem to confirm that the residual compressive stress induced by LSP causes the enhanced fatigue life of the tested material.

The LSP method is a very promising technique, capable of enhancing the service fatigue life of a component, particularly for high-value components such as fatigue-critical aircraft components although the cost at the moment is relatively high. The cost of surface treatment by this method should reduce in the near future as the technique is introduced more widely, and as it is further improved by research.

However, it is essential to note that like conventional mechanical peening processes the quality of the LSP processing should be well controlled in order to obtain the best results. Otherwise, this new technique may carry some risks. Studies to determine any technological risks associated with the technique are being carried out by DSTO.

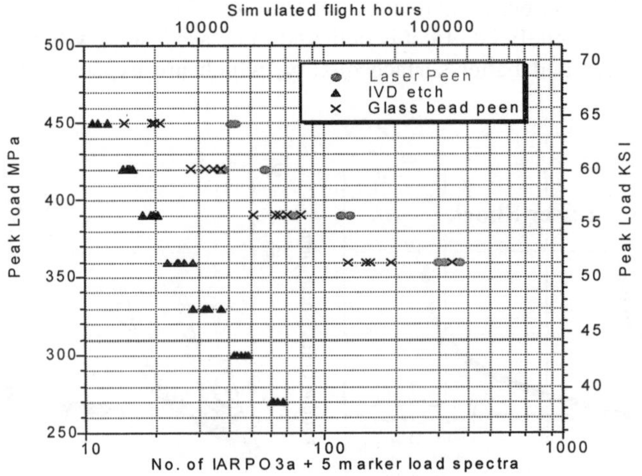

Figure 6 Fatigue tests on 7xxx series aluminium alloys for etched, glass bead peened and laser shock peened specimens.

References

1. Sharp, P.K, Clayton, J.Q. and Clark, G., The fatigue resistance of peened 7050-T7451 aluminium alloy-Repair and re-treatment of a component surface, *Fatigue Fract.Engng Mater. Struct.*, **17**(1994), pp.243-252.
2. Clauer, A. H. and Lahrman, D.F., Laser shock processing as a surface enhancement process, *Inter. Mech. Engng Congress & Exposition on the Durable Surfaces Symp.*, Nov.5-10, 2000, Orlando, FL, USA.
3. Goldsmith & Clark, Analysis and interpretation of aircraft component defects using quantitative fractography, *Quantitative Methods in fractography* (1990), American Society fort Testing and Materials, PA, pp.52-68.
4. Barter, S.A. and Houston, M.I., Examination of methods to control the depth of surface removal on the F/A-18 499 bulkhead, DSTO-DDP-0192, Defence Science and Technology Organisation, Aeronautical and Maritime Research Laboratory, 1997.
5. Barter,S.A. and Price, J., Effect of Surface Preparation Treatments on Fatigue Life of 7050 Aluminium Alloy, Structural Integrity and Fracture 2000, G. Heness (Ed.), Australian Fracture Group Inc. and University of Technology, Sydney, Australia, pp.140-153, 2000.
6. Mutoh, Y., Fair, G.H., Noble, B. and Waterhouse, R.B., The effect of residual stress induced by shot peening on fatigue crack propagation in two high-strength aluminium alloys, *Fat.Fract.Engng.Mater.Struct.*, **10**(1987), pp.261-272.
7. Clayton, J.Q. and Clark, G., The effect of steel shot and glass bead peening treatments on the fatigue resistance of 7050-T76351 aluminium alloy, Proc. Australian Fracture Group, *Fracture Mechanics in Engineering Practice*, Melbourne, (1988), pp.44-51.
8. Clark, G. and Clayton, J., Effectiveness of peening treatments in improving fatigue resistance of 7050 aluminium alloy under constant amplitude and spectrum loading, *Aust.Surface Engng Conf.*, Tech.Search, University of Adelaide, 12-14 March, 1991, B11.

R-CURVE METHOD OF FRACTURE TOUGHNESS DETERMINATION USING SEMI-CIRCULAR BEND SPECIMEN

M.D. KURUPPU

Curtin University of technology, PMB 22, Kalgoorlie, WA 6430, Australia

Email: M.Kuruppu @wasm.curtin.edu.au

M. SETO AND T. FUNATSU

Research Center for Deep Geological Environments, AIST, Tsukuba Central 7, 1-1-1 Hogashi, Tsukuba, Ibaraki, Japan

Email: m-seto@aist.go.jp

An experimental compliance method and a three-dimensional acoustic emission source location method were simultaneously used to determine the location of the crack tip in fracture toughness tests of Kimachi sandstone. Tests were done using the semi-circular bend specimen under three-point loading. Both methods gave agreeable results for the advancing crack length and they were used to construct the R-curve. The method enables the determination of size-independent fracture toughness of the rock material using a single specimen.

1 Introduction

Fracture toughness is an important material property that is related to the critical level of the stress intensity factor or fracture strength, as well as to the strain energy release rate or the fracture energy. In rock fracture mechanics, fracture toughness is used as an index of fragmentation and for numerical modelling purposes. A standard method for the fracture toughness determination of rock has been formulated [1]. The standard method uses two specimen configurations, namely the chevron bend specimen and the short rod specimen [2]. However, it is preferable to use three complimentary specimens for the fracture toughness determination of layered rocks from a single block of core (Figure 1). Moreover, materials such as concrete, asphalt and some large-grained rock tend to develop large process zones during crack propagation. A new specimen that can be reliably used in the presence of large non-linear deformations is required. Since geomaterials are weak in tension, tests should preferably be done with compressive loading where tensile fractures are induced. The semi-circular bend (SCB) specimen introduced by Chong and Kuruppu [3] is shown in Figure 1. It is core based and the preparation requires very little machining. It can be prepared from typical rock core, concrete, asphalt or any other material by splitting a circular disc into two halves followed by introducing a flat or a chevron notch. It is tested under three-point bending and is particularly suitable for measuring fracture toughness of materials undergoing substantial non-linear deformation before failure [4]. Some of the other alternative tests that have been developed are the modified ring test [5] and the centre-notched Brazilian disc [6].

Sub-critical and post-critical crack growth invariably occurs during fracture toughness tests. Measuring crack growth is often required for applications such as the determination of crack growth resistance, and evaluation of the remaining strength of

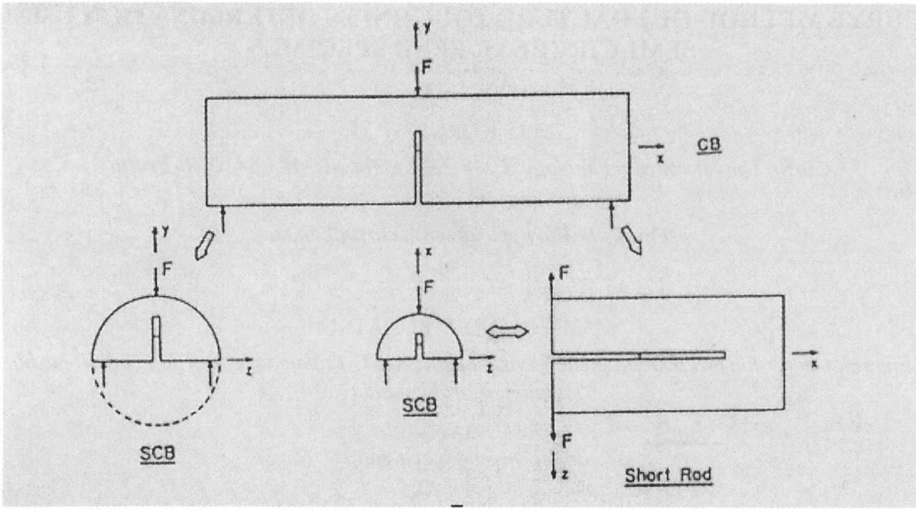

Figure 1. Core based fracture toughness specimens

rock structures containing cracks. The crack growth in test specimens can be conveniently measured using a suitable compliance method, and often employing the same equipment that is used for testing. Alternatively, an acoustic emission (AE) method can be used to determine the acoustic source locations associated with crack growth. In this investigation, the crack growth resistance in SCB specimen is measured as a function of crack extension. The instantaneous crack length is measured using compliance as well as a three-dimensional acoustic emission source location method developed at the National Institute of Advanced Industrial Science and Technology in Tsukuba, Japan.

2 Experimental methods

Most tests were done using the SCB specimen. However, chevron bend bar [1] and single edge-cracked round bar in bending [7] were also used for comparison and verification of some of the results. The testing program covered the following areas:

1. Fracture toughness of Kimachi sandstone was determined using 100 mm diameter SCB specimens. The crack growth resistance curve, also known as K-resistance curve and R-curve, was constructed using the fracture resistance and the crack growth.
2. Fracture toughness was determined using the ISRM standard method by employing a sample of chevron bend bar specimens.

As Kimachi sandstone is a sedimentary layered rock, test specimens were prepared such that the notch was aligned with one of the three principle orientations known as the divider, the arrester and the short transverse. Specimens of divider orientation yield average and representative properties of many different layers. Hence, the present investigation concentrated on measuring the fracture toughness in that orientation although some tests were done in other orientations as well. Specimens were prepared by slicing rock cores while noting the direction of bedding. Each core disc was then split into two halves producing two specimens having almost identical properties. Finally, a straight notch was introduced using a thin circular saw having a cutter thickness of 0.3 mm. The specimens were oven dried at 60^0C for a few days and all dimensions were recorded prior to testing. Other parallel tests showed that moisture content has a very significant effect on the fracture toughness [8]. Major mechanical properties of the rock considering the average values in all directions are: Young's modulus of 6.5 GPa; Poisson's ratio of 0.21; compressive strength of 28 MPa and tensile strength of 4 MPa.

SCB specimens made of Kimachi sandstone were tested using a MTS servo hydraulic test frame. The sizes were approximately 100 mm diameter and 25 mm thickness with a crack length to radius ratio, a/R, of 0.5. Specimens were tested under three-point bending while measuring the load, the load-point displacement and the crack opening displacement using a load cell and other gauges [9]. The loading fixture allows the support rollers to rotate and move apart slightly as the specimen is loaded minimising frictional resistance. The rollers were initially positioned against stops that set the span length and were held in place by low-tension springs. This fixture helps to achieve proper alignment in the load transfer system. A crack opening displacement (COD) gauge was attached to the specimen using knife-edges glued across the mouth of the notch. These knife-edges support the gauge arms and serve as the displacement reference points. The load-line displacement was measured by taking the average reading of two linear variable differential transducers (LVDT) that were placed between the top and the bottom loading platens.

12 AE sensors were glued to the surfaces of each specimen in order to determine the three dimensional acoustic source locations associated with the fracture process. In order to test the accuracy of the AE source location system, a calibration was carried out prior to fracture testing. This was done by deliberately producing a series of acoustic events on the surface of the specimen and comparing with the locations returned by the AE system. Most errors are less than 5 mm which is the diameter of each sensor, except that the system is unable to locate the position of any source close to the notch surface accurately. This is not of great concern as the objective is to determine the crack-tip location. During testing, the location where most AE events occurred was taken as the advancing crack tip. Specimens were loaded to failure under COD control at a constant rate of 0.06 mm/min. At least one partial unloading was done before reaching the peak load and a number of partial unloading-reloading cycles were done in the post-peak region. The high stiffness of the test frame enabled the test to be proceeded until the specimens could sustain no appreciable load.

3 Evaluation of fracture toughness and K-resistance curve

A compliance-based method was also used to determine the crack length. This method can be conveniently used, as it only requires a calibration of the specimen compliance, which can be done either numerically or experimentally. In this case an experimental calibration of the SCB specimen was done.

3.1 Elastic unloading compliance and crack length relation

The non-dimensional elastic compliance can be expressed as

$$C' = E' D C$$

where E' = E (Young's modulus) for plane stress and $E' = E/(1 - \upsilon^2)$ for plane strain. D is the diameter and C is the elastic compliance. Plane strain conditions were assumed although it may not strictly prevail away from the mid-thickness of the notch tip. A number of SCB specimens having a range of non-dimensional crack length, a/R, were tested by loading in three-point bending. Load versus load-line displacement was recorded in the elastic range for each of them. The relation between the non-dimensional compliance and the crack length is given by

$$C' = 1366\,\alpha^3 - 867\,\alpha^2 - 51.9\,\alpha + 129.4$$

where α is the crack length to radius ratio. This relationship is expected to be independent of material properties and specimen size.

3.2 Fracture toughness of Kimachi sandstone

Fracture toughness was determined for several samples of Kimachi sandstone using a procedure similar to level I and level II testing of ISRM standard method and the values are given in Table 1. The average value of level II fracture toughness determined using chevron bend bar specimens according to ISRM standard method is 0.99 $MPa\sqrt{m}$ [8]. In contrast the level II fracture toughness determined using SCB specimen as given in Table 1 is only about 0.69 $MPa\sqrt{m}$. This difference can be attributed to the size effects. Assuming that the fracture toughness is 0.99 $MPa\sqrt{m}$ the minimum size requirement using the formula suggested in [10] is 125 mm. Such a large-sized specimen could not be used due to the practical limitations. The crack growth resistance curves were constructed using two SCB specimens from the same sample as given in Table 1 and in the divider orientation. The K-resistance values were determined using the partial unloading cycles and following a procedure similar to the level II fracture toughness evaluation. The crack extension was measured as the average of values given by compliance and AE methods. Figure 2 shows R-curves derived using 2 specimens. Both curves reach a maximum limit of 1.0 $MPa\sqrt{m}$ that agrees with the value given by the ISRM standard method. Furthermore, this result agrees with that published by Matsuki et al for the same material [11]. It is observed that sub-sized specimens can be successfully employed to determine the fracture toughness by making use of the R-curve method.

Table 1. Fracture toughness of Kimachi sandstoneAttention:

Orientation	No. of specimens	Level I fracture toughness ($MPa\sqrt{m}$)	Level II fracture toughness ($MPa\sqrt{m}$)
Divider	6	0.45	0.65
Arrester	6	0.48	0.69
ST	9	0.41	-

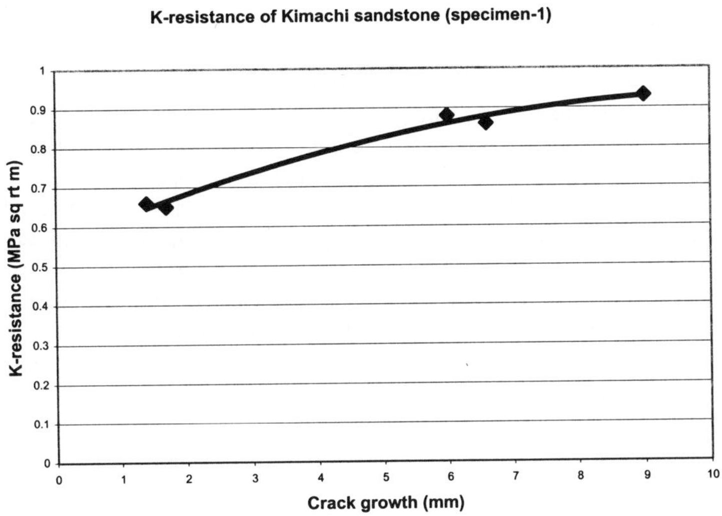

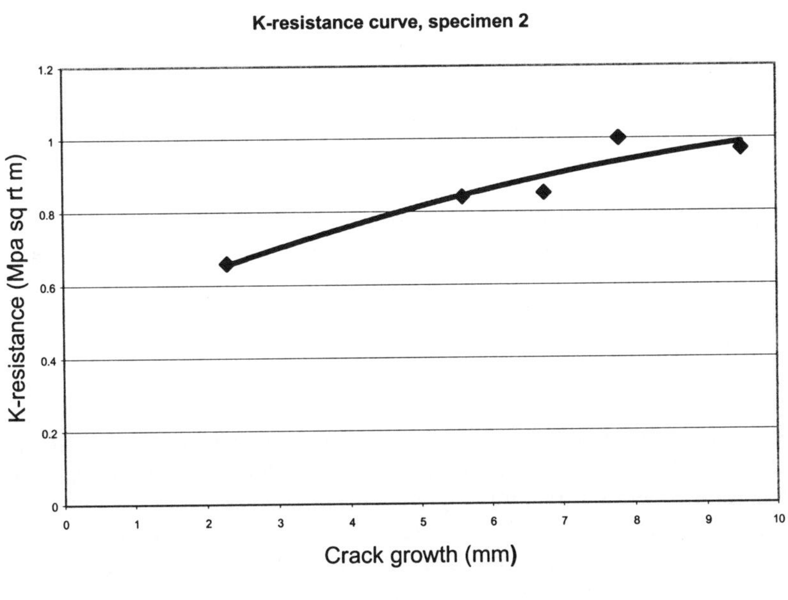

Figure 2. K-resistance curves of Kimachi sandstone (using SCB specimen).

4 Conclusions

Mode I fracture toughness of Kimachi sandstone was measured using 100 mm diameter and 25 mm thick SCB specimens. The growth of cracks was measured using an acoustic emission 3D source location system. The crack growth was also determined using the elastic unloading compliance method at regular intervals of each test. The AE method and the compliance method yielded crack length values that agreed within 5%, and both methods proved to be reliable. Following a procedure similar to the analysis of level II fracture toughness, the crack growth resistance was measured using the unloading-reloading cycles following the peak load. This result in combination with crack growth data enables the construction of K-resistance curve also known as R-curve. The procedure can be used to determine the R-curve using a single specimen and its limiting value yields the true fracture toughness. The R-curve gave a limiting value of 1.0 $\text{MPa}\sqrt{\text{m}}$ for a crack growth of order 10 mm in 100 mm diameter SCB specimens.

Acknowledgements

We thank Manabu Utagawa and Jae-Joon Song for the assistance in experimental and numerical work reported here.

References

1. Ouchterlony F., Suggested methods for determining the fracture toughness of rock, *Int. J. Rock Mech. Min. Sci.* **25** (1988) pp. 71-96.
2. Barker, L.M., A simplified method for measuring plane strain fracture toughness, *Eng. Frac. Mech.* **9** (1977) pp. 361-369.
3. Chong K.P. and Kuruppu, M.D., New specimen for fracture toughness determination of rock and other materials, *Int. J. Fracture* **26** (1984) pp. R59-R62.
4. Lim, I.L., Johnston, I.W., Choi, S.K. and Boland, J.N., Fracture testing of a soft rock with semi-circular specimens under three-point bending, *Int. J. Rock Mech. Min. Sci.* **31** (1994) pp. 185-197.
5. Thiercelin, M., Fracture toughness under confining pressure using the modified ring test, in *Rock Mechanics: 28th U.S. Symp. Rock Mech.* (Balkema, 1987), pp.149-156.
6. Fowell, R.J., Hudson, J.A., Xu, C., Chen, J.F. and Zhao, X., Suggested method for determining Mode I fracture toughness using cracked chevron-notched Brazilian disc (CCNBD) specimen, *Int. J. Rock Mech. Min. Sci.* **32** (1995) pp. 57-64.
7. Ouchterlony, F., A new core specimen for the fracture toughness testing of rock (SveDeFo Report DS1980:17, Swedish Detonic Research Foundation, Stockholm.
8. Kuruppu, M.D. and Seto, M., Determination of fracture toughness of rock under in-situ conditions using semi-circular specimen, *10th Int. Conf. Fracture* (2001) Dec.
9. Seto, M. Kuruppu, M.D. and Funatsu, T., Fracture toughness testing of rock at elevated temperatures and pressures, *Proc. DC Rocks*, (ARMA, 2001), To appear.
10. Chong, K.P., Kuruppu, M.D. and Kuszmaul, J.S., Fracture toughness determination of layered materials, *Eng. Frac. Mech.* **28** (1987) pp. 43-54.
11. Matsuki, K., Hasibuan, S.S. and Takahashi, H., Specimen size requirements for determining the inherent fracture toughness of rocks according to the ISRM suggested methods, *Int. J. Rock Mech. Min. Sci.*, **28** (1991) pp. 365-374.

INFLUENCE OF ADHESIVE REINFORCEMENT ON THE MODE II FRACTURE TOUGHNESS OF A BONDED JOINT

M. S. FORTE AND N.J. PAGANO

Air Force Research Laboratory, Materials Directorate Wright-Patterson AFB, OH 45433-7750 USA E-mail: mark.forte@wpafb.af.mil, nicholas.pagano@wpafb.af.mil

The influence of adhesive reinforcement on the Mode II fracture toughness of an end notch flexure (ENF) specimen is investigated with a computer model based on a variational theorem that has been shown to accurately predict stress fields and energy release rates for cracked laminates. A large radius axisymmetric fracture mechanics model was developed and used to determine the energy release rates for midplane cracking in aluminum specimens and in aluminum bonded specimens with varying amounts adhesive reinforcement. The energy release rates, which were determined from the present model, were used to determine the fracture toughness for specimens in a systematic experimental program and were compared to closed form analyses from the literature. The amount of reinforcement in the adhesive layer has been shown to have a significant effect on the Mode II adhesive fracture toughness and cracking behavior.

1 Introduction

Aerospace grade epoxy adhesives used in both original component fabrication and repair applications have various forms of carrier materials, each of which can have a dramatic effect on adhesively bonded joint strength and durability. Film adhesives are supplied in a tape form, which gives the advantage of bondline thickness uniformity making them easy to apply. Films can be supplied in unsupported or supported forms, in which case, the adhesive is cast into a mesh carrier cloth. The carrier cloths are available in either random mat, knit woven or square patterns in dacron, nylon or glass fabrics. Carrier cloths are placed in film adhesives during manufacture mainly for handling control during subsequent part fabrication; however, they also provide bondline thickness control of a cured bonded joint. The adhesive supported with the knit woven carrier gives the highest stiffness while the random mat provides superior handling characteristics [1]. A commercially available film adhesive with a tight knit carrier also provides a degree of electrical isolation between metal and graphite composite adherends to reduce galvanic corrosion [2].

In this paper, we will study the use a large radius axisymmetric damage plane strain model [3] to determine the energy release rates for midplane cracking in aluminum specimens and in aluminum bonded specimens with varying amounts of adhesive reinforcement. The adhesive chosen for this study is EA 9396 because it represents a state of the art epoxy based adhesive, which has been well characterized and is currently used as a repair material in current Air Force aircraft [4]. This system can be cured at room temperature to minimize thermal curing stresses, and fibers can be readily added to the resin system due to the low resin viscosity. Previous efforts using this resin system has shown that $\mathcal{G}_{IC}$ can be increased (up to 50%) or decreased (up to 43%) depending on the type and form of reinforcement [5, 6, 7].

2 Method

To accurately model the behavior of the ENF specimen in the current test program, a plane strain model that explicitly models the adhesive layer and the midplane traction free crack was developed using the Large Radius Axisymmetric Damage Model (LRAM) [3]. LRAM was developed as a convenient way to approximate the thermoelastic stress field and energy release rates of flat laminated bodies subjected to various idealized cracks. The formulation employs the idea that the stress field in an axisymmetric cylinder approaches that in a long flat coupon as the radius to thickness ratio approaches infinity provided the flat coupon stresses are independent of the length coordinate (x). In the limit as the ratio of radius R to laminate thickness T approaches infinity, the gradient of the hoop strain through the wall thickness approaches zero, generating a stress-strain field equivalent to a flat composite coupon under a uniform axial strain. Using this methodology, an axisymmetric cylinder with $R >> T$ is used to represent a flat laminate.

The model is generated by subdividing the body into regions consisting of concentric shells of constant length and satisfying the Reissner variational equation [8] with an assumed equilibrium stress field within each region. That is, we set

$$\delta H = 0 \qquad (1)$$

where

$$H = \int_V F dV - \int_{S'} \tilde{T}_i U_i dS \qquad (2)$$

$$F = \tfrac{1}{2}\tau_{ij}\left(U_{i,j} + U_{j,i}\right) - W_c\left(\tau_{ij}, e_{ij}\right) \qquad (3)$$

In these equations, W_c is the complementary energy density; τ_{ij} and U_i are the stress and displacement components, respectively, in Cartesian coordinates; e_{ij} are the engineering free expansional or non-mechanical strains; V is an arbitrary volume enclosed by S; S' is the part of S on which one or more traction components are prescribed; and $\tilde{T}_i$ are the Cartesian components of the prescribed tractions. A comma followed by a subscript denotes a derivative with respect to the indicated coordinate and the summation on repeated letter subscripts, but not superscripts, is implied. Reissner [8] has shown that the variational Eq. (1) is equivalent to the field equations and boundary conditions of linear elasticity theory in a homogeneous body provided one applies the variational operator δ to both stresses *and* displacements. The theory was extended to heterogeneous laminates and cylinders by Pagano [9, 10].

The stress field is assumed such that σ_θ and σ_ξ are linear in r within each region, while the forms of σ_r and $\tau_{r\xi}$ are chosen to satisfy the axisymmetric equilibrium equations of linear elasticity. The regions are selected such that the thermoelastic properties are constant and the boundary conditions do not change character on any of the bounding surfaces (i.e., traction and/or displacement components prescribed or continuity prescribed) within each region. The number of regions, in particular in the radial direction, can be increased in order to improve solution accuracy. Since the field equations within each region reduce to a system of algebraic equations and linear ordinary differential equations with constant coefficients, the general form of the solution for any of the dependent variables $f(\xi)$ is expressed by

$$f(\xi) = \sum_i A_i e^{\lambda_i \xi} + f_p(\xi) \tag{4}$$

within each region, where A_i are constants, λ_i are eigenvalues of a determinant and $f_p(\xi)$ is a particular solution, which in the present case is a simple polynomial.

3 Contemporary Fracture Toughness Models

Several models are available in the literature to predict Mode II fracture toughness. Data reduction has been performed [13] using a beam theory formulation to obtain the following mode II energy release rate for composites:

$$G_{IIc} = \frac{9a^2 P_c^2}{64 E_x^b b^2 h^3} \tag{5}$$

where a is the crack length, P_c is the critical load, E_x^b is the modulus of elasticity in bending, b is the specimen width and h the specimen height.

A higher order beam theory based on second degree displacement (in-plane) in the thickness direction using Reissner's variational principle was developed for homogeneous, orthotropic materials [14]. Both the in-plane normal stress and the interlaminar shear stress distribution exactly satisfy the equilibrium equations of classical theory of elasticity. With this approach the constitutive relations are determined directly without the need for shear correction factors. The expression for energy release rate is as follows:

$$G_1 = \frac{9\bar{a}^2 P^2}{64 E_1 b^2 h}\left\{1 + \frac{2}{\lambda \bar{a}} + \frac{1}{60\lambda^2 \bar{a}^2}\left[\lambda^2\left(\frac{E_1}{G_{13}}\right) + 60\right]\right\} \tag{6}$$

where $\bar{a} = \dfrac{a}{h}$ and $\lambda = 4\sqrt{\dfrac{14}{5}\left(\dfrac{G_{13}}{E_1}\right)}$.

4 Experimental/Analytical Correlation

4.1 Materials, Specimen Configuration

In this study, 2024-T3 aluminum adherends are selected for the end notch flexure testing to eliminate adherend failures and isolate the failures to the adhesive layer. The properties of the aluminum used in the analysis are E=72.4 GPa and ν = 0.33. Proper surface preparation of the aluminum adherends with an acid etch and priming with BR-127 was conducted to ensure that the failure is contained in the adhesive (cohesive failure) rather than in the interface or the adherend. The end notch flexure specimens were fabricated in several steps. The adhesive was debulked and was staged after the fibers were added. The adhesive was placed between two primed aluminum adherends and cured in a press for 1 hour at 65°C. Shims were used to achieve a desired 0.203 mm bondline thickness. The specimens were tested under ambient temperature and moisture conditions.

The percentages of fibers in the adhesive for the present study were selected to be nominally 10, 20 and 40 percent by volume because these amounts are achievable in the selected fiber type, allowing a variation over a large range of fiber volume fractions. The fiber volume percentages will help to establish limits in the amount of added reinforcement that is beneficial to adhesive durability and toughness. The specific reinforcement material selected for this study was plain weave fiberglass cloth all with the same silane sizing, which is the most common reinforcement for polymer matrix composites [11]. The in-plane material properties were determined by testing laminates of various thicknesses for each adhesive type, and the out of plane material properties were determined using the micromechanical 3-phase concentric cylinder model NDSANDS (N Directional Stiffness AND Strength) [12].

4.2 LRAM model of ENF specimen

An LRAM analysis of the ENF specimen was developed and refined to accurately model the thermoelastic stresses and displacements. Our initial effort was to develop a model without an adhesive and to compare the results from the present analysis with beam theory and Whitney's formula, which also neglect the adhesive layer. The initial crack length was 25.4 mm, and the specimen half thickness was 25.6 mm. Axial displacement was constrained at one point, and at the crack face, we assumed a smooth closed crack with no friction (continuity of both w and σ_z and τ_{yz}=0).

An investigation to study the effect of layer refinement on the predicted energy release rate was conducted. A comparison of the Whitney analysis and Beam Theory results with the present analysis for 14 and 5 layers, assuming a unit load P, are shown in Figure 1. The minimum layer thickness used in the 14-layer all aluminum model corresponded to the 0.038 mm thick region which is less than one half of the adhesive thickness (0.102 mm); the layer thickness from the midplane were 0.038, 0.102, 0.254, 1.27, 4.267, 8.534 and 12.802 mm respectively. LRAM proved to converge rapidly for the energy release rate predictions. The 14- layer model, which was refined towards the stress singularity at the midplane was used in subsequent calculations.

The LRAM model was modified to include the adhesive layer with the minimum layer thickness equal to one half of the adhesive thickness (0.102 mm). A comparison of energy release rate using the 14-layer model corresponding to the adhesive with 10% fiber volume was made to the Whitney model and normalized to the Beam Theory model as shown in Figure 2. Neither the Whitney nor the Beam Theory models account for the adhesive layer. Compared to the all aluminum specimen, the addition of the adhesive layer in the present analysis caused an increase in the energy release rate over the entire crack length range investigated because the specimen is more compliant. At small crack lengths, the energy release rates for all models diverge from Beam Theory. The difference in energy release rate becomes less pronounced as the crack length increases.

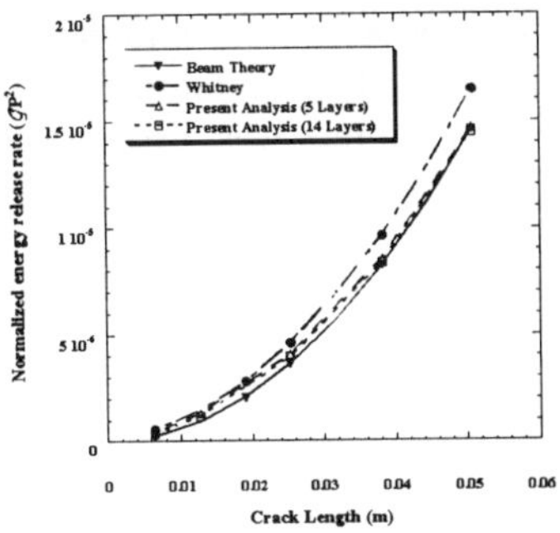

Figure 1. ENF LRAM layer refinement.

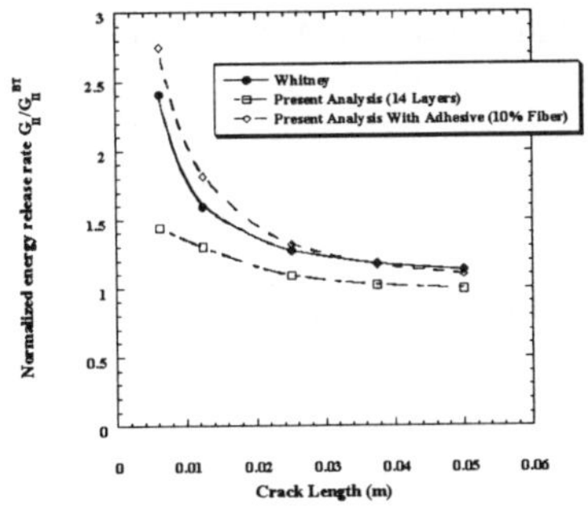

Figure 2. ENF analysis comparison.

4.3 Data Fitting

LRAM was used to predict energy release rates for all of the adhesives. The values generated from the analysis were based on a 0.203 mm bondline, a span to depth ratio of 5 to 1, and a model that contains 14 layers. For each adhesive/fiber combination, a calculation of $\mathcal{G}_{IIC}$ was made based on 3 specimens except the neat resin, which is comprised of only 1 specimen. The present analysis $\mathcal{G}_{IIC}$ comparison is provided in Figure 3. The neat resin specimens exhibited the highest $\mathcal{G}_{IIC}$. The addition of fibers at 10% and 20% fiber volume reduced the $\mathcal{G}_{IIC}$ by 40% while the 40% fiber volume resulted in a reduction of nearly 24%. As the fiber volume percentage in the bondline increased, the data scatter decreased.

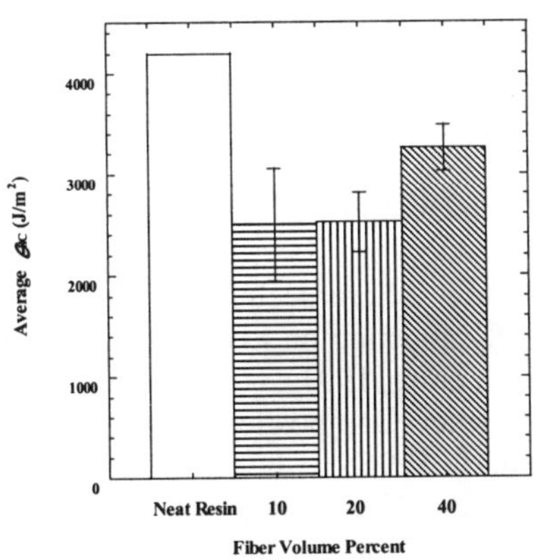

Figure 3. Present analysis $\mathcal{G}_{IIC}$ comparison

5 Discussion

An analytical and experimental effort was conducted to improve the understanding of how the carrier in an epoxy adhesive influences the fracture behavior of the adhesive. A plane strain analysis was used to determine the energy release rates for midplane cracking in aluminum bonded specimens with varying amounts of adhesive reinforcement. The experimental effort showed that the Mode II energy release rate was also degraded as a result of the fibers in the bondline.

Acknowledgements

We thank Greg Schoeppner for his assistance in with the LRAM analysis specimen development.

References

1. Handbook of Adhesives, FM-73 Film Adhesive, American Cyanamid Engineered Materials Department, 1990
2. Handbook of Adhesives, FM-300 Film Adhesive, American Cyanamid Engineered Materials Department, 1990
3. Schoeppner, G.A. and Pagano, N.J. "Stress Fields and Energy Release Rates in Cross-ply Laminates", *Int. J. Solids Struct.*, Vol. 35, No. 11, 1998, pp. 1025-1055
4. Askins, D.R., "Characterization of EA 9396 Epoxy Resin for Composite Repair Applications," Wright Laboratory's Materials Directorate Report No.WL-TR-92-4060, October 1991
5. Forte, M.S., "Fracture Toughness of EA 9396 Epoxy Resin," Wright Laboratory's Materials Directorate Evaluation Report No. MLS 97-005, January 1997
6. Forte, M.S., "A Controlled Study of the Effects of Bondline Reinforcement On the Fracture Behavior of a Brittle Epoxy Adhesive," Ph.D. Dissertation, May 1999
7. Forte, M.S., Whitney, J.M., and Schoeppner, G.A., "Influence of Adhesive Reinforcement on the Mode I Fracture Toughness of a Bonded Joint," submitted for publication in *Composites Science and Technology*, 1999
8. Reissner,E., "On a Variational Theorm in Elasticity", *Journal of Mathematical Physics*, 29(1950), 90-95
9. Pagano, N.J., "Stress Fields in Composite Laminates", *Int. J. Solids Struct.*, Vol. 14, 1978, pp. 385-400
10. Pagano, N.J., "Axisymmetric Micromechanical Stress Fields in Composites", Proceedings 1991 IUTAM Symposium on Local Mechanics Concepts for Composite Material Systems, Springer, Verlag, 1991
11. Jang, B.Z., "Advanced Polymer Composites: Principles and Applications," ASM International Publisher, January 1996
12. Pagano, N.J., and Tandon, G.P., "Elastic Response of Multi-directional Coated-fiber Composites," *Composites Science and Technology*, 31, (1988), pp. 273-293
13. Russell, A.J., and Street, K.N., "Factors Affecting the Interlaminar Fracture Energy of Graphite/Epoxy Laminates," *Proceedings of the 4th International Conference on Composite Materials*, Elsevier, North Holland, pp. 279-286, 1984
14. Whitney, J.M., "Analysis of the End Notch Flexure Specimen Using a Higher Order Beam Theory Based on Reissner's Principle," *Proceedings of American Society for Composites, 3rd Technical Conference*, Technomic Publishing, Lancaster, PA, pp. 103-112, 1988.

EVALUATION OF CONSTRAINT PARAMETERS IN MIXED-MODE FRACTURE

HAMID REZA DAGHYANI

Associate Professor of Amirkabir University of Ttechnology: Tehran, 15914, Iran

Email: hamid@cic.aku.ac.ir

ARASH NIKKHAAH

Postgraduate Student

SHAHRIAR FARIBORZ

Associate Professor of AmirkabirUuniversity of Technology

In the present study, a boundary layer model was proposed to analyze the effect of constraint parameters on mixed mode fracture toughness. A new definition of crack-tip constraint parameter, Q was presented to establish a unique relationship between the fracture toughness and the constraint parameter developing finite element models. The T-stress was also related to Q considering the far field mixity parameters employing modified boundary layer analyses.

1 Introduction

The J-Q theory in fracture mechanics provides a framework as two parameters allowing the toughness locus to be measured and utilized in engineering applications. Several research works have proposed that the J-integral and a constraint parameter, Q, are sufficient to characterize the full range of near-tip stress states [4,5]. Within these approaches, J and Q have separate roles: J sets the size scale over witch large stresses and strains develop, while Q scales the near-tip stress distributions relative to a reference high triaxiality stress state. Though, since the direction of a crack may change during propagation, the classical definition of Q breaks down in mixed-mode loading condition.

The J-Q evaluation of mode-I crack-tip fields was first derived by O'Dowd and Shih [4,5] developing a boundary layer model for a cracked body under both the small scale yielding (SSY) and the large scale yielding (LSY) conditions. The two-parameter evaluation of crack tip constraints was later studied by Ayatolahi [2] to define Q for the mixed-mode loading in the direction of crack growth in a finite size specimen. In the present study, the shortcoming of the previous studies was investigated and a new definition for Q in mixed-mode fracture was introduced to provide a unique relationship between the fracture toughness, J_C and constraint parameter, Q.

2 Constraint parameters in Mode-I fracture

Based on the research work reported by O'Dowd and Shih [4,5], the stress and strain fields for mode-I loading at the crack-tip are given by:

$$\frac{\sigma_{ij}}{\sigma_0} = f_{ij}\left[\frac{r}{(J/\sigma_0)}, \theta; Q\right] \quad (1)$$

$$\frac{\varepsilon_{ij}}{\varepsilon_0} = g_{ij}\left[\frac{r}{(J/\sigma_0)}, \theta; Q\right] \quad (2)$$

where σ_{ij} and ε_{ij} are the crack tip stress and strain components, respectively, σ_0 and ε_0 are the materials yield stress and yield strain, respectively, and r and $\grave{e}$ are the cylindrical coordinate system oriented at the crack tip. The relationship between the stress fields in a finite cracked body and a reference condition (*SSY, T=0*) can be expressed as:

$$\sigma_{ij} = \left(\sigma_{ij}\right)_{\text{Ref}} + Q\sigma_0\delta_{ij} + \left(\sigma_{ij}\right)_{\text{asymptotic power series}} \quad (3)$$

where, δ_{ij} is the Kroncker's delta.

O'Dowd and Shih [4,5] showed that $Q\sigma_0$ represents the difference in hydrostatic stress ahead of the crack tip between the reference field and the finite cracked body. Two definitions of Q proposed by O'Dowd and Shih [4,5] are:

$$Q = \frac{\sigma_m - \sigma_m^{SSY,T=0}}{\sigma_0} \text{ at } \quad \theta = 0, \quad r = 2J/\sigma_0 \quad (4)$$

$$Q = \frac{\sigma_{\theta\theta} - \sigma_{\theta\theta}^{SSY,T=0}}{\sigma_0} \text{ at } \quad \theta = 0, \quad r = 2J/\sigma_0 \quad (5)$$

where, σ_m is the mean stress and $\sigma_{\theta\theta}$ is the hoop stress. The difference between Equations (4) and (5) is less than 0.1, which is negligible. At small crack tip deformation states, finite cracked bodies experience SSY conditions and Q remains almost zero. At large crack tip strain fields, once LSY conditions prevail, the hydrostatic stresses at the crack tip are substantially less than those in the SSY field at the same applied J-value. This difference leads to negative Q values when the specimen deviates from the SSY condition.

The single-parameter fracture mechanics is based of the assumption that fracture toughness is a material constant. With J-Q theory, however, the critical J value for a given material depends on Q:

$$J_c = J_c(Q) \quad (6)$$

This fracture toughness is no longer a single value; rather it is a nonlinear function of Q. Thus, the critical fracture toughness, J_c as a function of Q is predicted by [4,5]:

$$\frac{J_c}{J_0} = \left[1 - \frac{Q\sigma_0}{\sigma_C}\right]^m \quad (7)$$

where J_0 is the critical fracture toughness when Q = 0, σ_C is the cleavage stress at a characteristics micro-structural distance, r_C, ahead of the crack tip [7] ($\sigma_C \approx 3.5\,\sigma_0$) and m is a material constant. Equation (7), predicts that fracture toughness depends on Q for mode-I fracture. However, the approach is sensitive to the material parameters such as σ_C and m, which are difficult to evaluate.

3 Constraint Parameters in Mixed-Mode Fracture

The constraint parameter, Q in the mixed-mode fracture was studied by Ayatolahi [2] developing a modified boundary layer (MBL) model where the Q was defined as follows:

$$Q_{\theta\theta} = \frac{\sigma_{\theta\theta} - (\sigma_{\theta\theta})_{ref}}{\sigma_0} \text{ at } \quad \theta = \theta_0, \quad 1 \leq \frac{r\sigma_0}{J} \leq 5 \tag{8}$$

where θ_0 is the direction of crack growth in a finite cracked size specimen. This definition was investigated in the present study and was modified presenting a new definition of Q In mixed-mode fracture. To evaluate the effect of various parameters on Q, a MBL model was developed using the ABAQUS [1] finite element software. The analyses were carried out considering a boundary domain around the crack tip where with the variation of stress intensity factors (K_I and K_{II}) and T-stress the conditions of finite cracked body and reference model were simulated [2,4,5]. The T-stress is the non-singular term of the stress field ahead of the crack tip defined by Williams [9]. The new definition for Q was presented in the present work in the mixed-mode loading as:

$$Q_{\theta\theta} = \frac{[\sigma_{\theta\theta}]_{\max} - [(\sigma_{\theta\theta})_{ref}]_{\max}}{\sigma_0} \text{ at } 1 \leq \frac{r\sigma_0}{J} \leq 5 \tag{9}$$

In Equation (9) Q was defined based on the stress fields in the direction of crack growth of a finite size model and a reference model separately. The direction of crack growth was considered normal to the maximum hoop stress, $\sigma_{\theta\theta}$. The difference between Equations (8) and (9) in Q influences to the relationship of fracture toughness and constraint parameter Q, discussed further.

First, the Ritchi [7] fracture criteria was considered where the relationship between the critical stress and fracture toughness was given by:

$$\sigma_c = \sigma_0 \left(\frac{J_c}{\alpha\sigma_0\varepsilon_0 I_n r_c} \right)^{\frac{1}{n+1}} \tilde{\sigma}_{\theta\theta,Q}\left(\theta = \theta_Q^*, M^p\right) + Q\sigma_0 \tag{10}$$

where n and á are the strain hardening exponent and a material constant in the Ramberg-Osgood deformation theory of plasticity, respectively. I_n is an integration constant, which is a function of n [3,6], θ_Q^* is the direction of crack growth in a structure with $Q \neq 0$ and M^p is the near tip mixity parameter [8,10]. For the reference model with a fracture toughness equal to J_0 and $Q = 0$, the Ritchi criteria becomes:

$$\sigma_c = \sigma_0 \left(\frac{J_0}{\alpha\sigma_0\varepsilon_0 I_n r_c} \right)^{\frac{1}{n+1}} \tilde{\sigma}_{\theta\theta,Q=0}\left(\theta = \theta_{Q=0}^*, M^p\right) \tag{11}$$

where $\tilde{\sigma}_{\theta\theta}$ is a dimensionless function [3,6]. Using Equations (10) and (11) and eliminating r_C, then J_C is related to J_0 by:

$$J_C = J_0 \left(\frac{1 - Q\sigma_0}{\sigma_C} \right) N(Q, M^P) \tag{12}$$

$$N(Q,M^P)=\left[\frac{\tilde{\sigma}_{\theta\theta,Q}(\theta=\theta^*{}_Q,M^P)}{\tilde{\sigma}_{\theta\theta,Q=0}(\theta=\theta^*{}_{Q=0},M^P)}\right]^{\frac{1}{n+1}} \tag{13}$$

where $N(Q,M^P)$ depends on the difference between the crack growth direction in the finite cracked structure and the reference model.

Our numerical analyses show that N(Q,Mp) may be considerably less than unity. For instance, N(0.8,0.8) is 0.44. Therefore, the previous definitions of Q may not alone establish a unique relationship between the fracture toughness of a finite cracked structure and the reference model in mixed-mode fracture, thus, another parameter should be defined. Now, using Equation (9) one can write [8]:

$$\tilde{\sigma}_{\theta\theta,Q}(max)=\tilde{\sigma}_{\theta\theta,Q=0}(max)\Rightarrow\tilde{\sigma}_{\theta\theta,Q}\left(\theta=\theta_Q^*\right)=\tilde{\sigma}_{\theta\theta,Q=0}\left(\theta=\theta_0^*\right) \tag{14}$$

Considering Equations (13) and (14), then $N(Q,M^P)=1$, In this case, with the new definition of Q in Equation (9), which we call it Q^N, a unique relationship between the fracture toughness of a finite size specimen and a reference structure is established.

4 Relationship between the *T*-stress and the Q^N parameter

By employing the definition of Q^N in mixed-mode fracture, the relationship between *T*-stress and Q^N can be obtained using MBL model. The model was first compared to the results of O'Dowd and shih [4,5] in pure mode-I, as well as the results of Ayatolahi [2] in pure mode-II, which were correlating well with

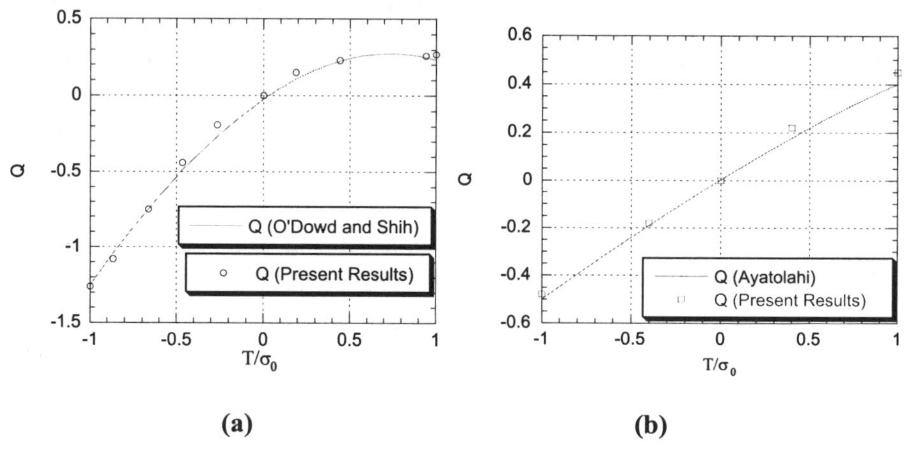

Figure 1: The relationship between *T* and *Q* for a) mode-I and b) Mode-II.

their results, as shown in Figure 1. Figure 2 shows the relationship between the Q^N-parameter and *T*-stress for a Ramberg-Osgood and linear elastic power law hardening materials with $n = 10$ and $\acute{o}_0 = 420$ MPa. The results were obtained based on the far field mixity parameter [8,10], M^e. As shown in Figure (2), by increasing the ratio of mode-I to mode-II in a cracked structure, Q was decreased, which means the crack tip constraint was relieved.

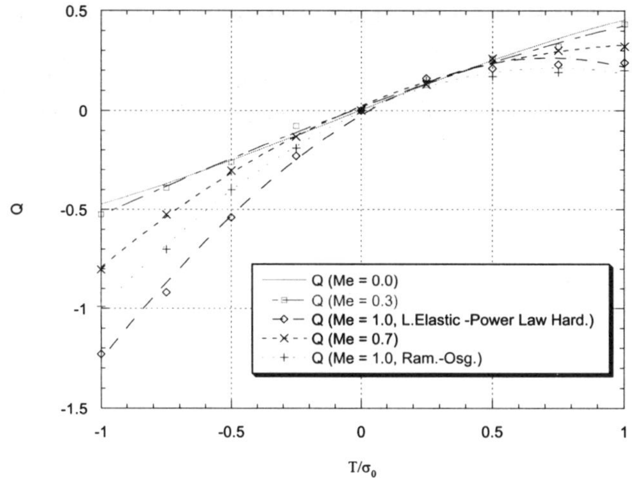

Figure 2: The relationship between T and Q^N

5 Conclusion

A new definition of Q-parameter, Q^N, was presented, which is more convenient in characterizing the mixed-mode fracture. By using Q^N, a unique relationship was defined between the fracture toughness and the constraint parameter, Q in mixed-mode. Thus, the fracture toughness of a finite size structure can be related to that one of a reference model with two parameters J_C and Q^N. The relationship between the Q^N-parameter and T-stress was also studied.

References

1. ABAQUS/Standard User's Manual, Ver. 5.7-5, Habbitt, Karlsson and Sorensen Inc.
2. Ayatolahi, M. R., "Geometry and Constraint Effect in Mixed-Mode Fracture", PHD dissertation, University of Bristol, 1998
3. Hutchinson, J. W., "Singular Behaviour at the End of a Tensile Crack in a Hardening Material", *J. Mech. Phys. Solids*, 16, 1968, pp. 13-31.
4. O'Dowd, N. P. and Shih, C. F., "Family of Crack Tip Fields Characterized by a Triaxiality Parameter, I- Structure of Fields", *J. Mech. Phys. Solids*, 39, 1991, pp. 989-1015.
5. O'Dowd, N. P. and Shih, C. F., "Family of Crack Tip Fields Characterized by a Triaxiality Parameter, II- Fracture Applications", *J. Mech. Phys. Solids*, 40, 1992, pp. 939-963.
6. Rice, J. R. and Resengren, G. F., "Plane Strain Deformation Near a Crack Tip in a Power-Law Hardening Material", *J. Mech. Phys. Solids*, 16, 1968, pp. 1-12.
7. Ritchie, R. O., Knott, J. F. and Rice, J. R., "On Relationship between Critical Tensile Stress and Fracture Toughness in Mild Steel", *J. Mech. Phys. Solids,* 21, 1973, pp. 395-410.
8. Shih, C. F., "Small Scale Yielding Analysis of MixedMode Plane Strain Crack Problems in Fracture Analysis", ASTM STP 560, 1974, pp. 187-219.

9. Williams, M. L., "On the Stress Distribution at the Base of a Stationary Crack", J. App. Mech., 24, 1957, pp. 109-114.
10. Zhu, X.K., Chao, Yuh J., "Constraint Effects on Crack-Tip Fields in Elastic Perfectly Plastic Materials", J. Mech. Phys. Solids, 49, 2001, pp. 363-399.

LOAD INTERACTION EFFECTS ON FATIGUE CRACK GROWTH

H. HOSSEINI-TOUDESHKY AND M.R. MOFAKHAMI

Aerospace Engineering Department, Amirkabir University of Technology (Tehran Polytechnic)
Tehran-Iran
E mail: hosseini@cic.aku.ac.ir

Crack growth prediction of components subjected to spectrum loadings is an important issue in fatigue life assessment of industrial components. Effects of load interaction cause acceleration and/or retardation in fatigue crack growth rate. Disregarding of this phenomenon may lead to the nonconservative life prediction in some cases. In this paper two methods for load interaction consideration are used. The first model is Chang-Willenborg that is a modified Willenborg model and the second one is a new model based on the compressive residual stresses at the crack tip and original Willenborg model. In the development of the new model the required residual stresses at the crack tip are obtained from an elastic-plastic finite element analysis. Also the Original Rainflow Counting Method and Modified Rainflow Counting Method have been used to prepare the spectrum loading data. Using the above mentioned cycle counting methods and load interaction models, the crack growth of the plates are obtained for two different flight spectrum loading. Finally the results obtained from different methods are compared with those obtained from experiments presented by the others.

KEY WORDS: Spectrum loading, Load interaction effect, Retardation, Acceleration, Fatigue crack growth, Fatigue life.

1 Introduction

Fatigue life assessment of cracked panels subjected to random loading is an important issue in design of many structures such as aircrafts. Effects of load interaction cause acceleration and/or retardation in fatigue crack growth rate of the components. Tensile overloads cause retardation in crack growth and compressive loads in tension-compression load cycles reduce the retardation effect, which is caused by a tensile overload [1]. Therefore to estimate more accurate life these phenomenon could be considered in the fatigue modeling. If the loading profile consist compressive loading and reduction in retardation due to the compressive loading is not considered, the obtained results maybe too non-conservative.

In this paper two methods for load interaction consideration are used. The first model is the Chang-Willenborg method that is a modified Willenborg model and the second one is a new model based on the compressive residual stresses at the crack tip and the Original Willenborg model. Also the Original Rainflow Counting Method and Modified Rainflow Counting Method have been used to prepare the spectrum loading data.

In the following sections the cycle counting methods and the modified Willenborg model are explained.

2 Counting method

To prepare the spectrum loadings in fatigue crack growth rate equations, two cycle counting methods of original rainflow [6], and modified rainflow [2] are used. Original Rainflow Counting Method (ORCM) is commonly recognized an efficient cycle counting method in fatigue life prediction, however some stress-strain response cannot be counted by this method [2].

The Modified Rainflow Counting Method (MRCM) is an efficient method with no such shortcoming. In this method two types of load points arrangement which shown in Figure 1 are considered, and for each of them different rules for cycle counting are sugested [2]. The algorithm used in the developed computer program is independent of the loading points arrangement [5].

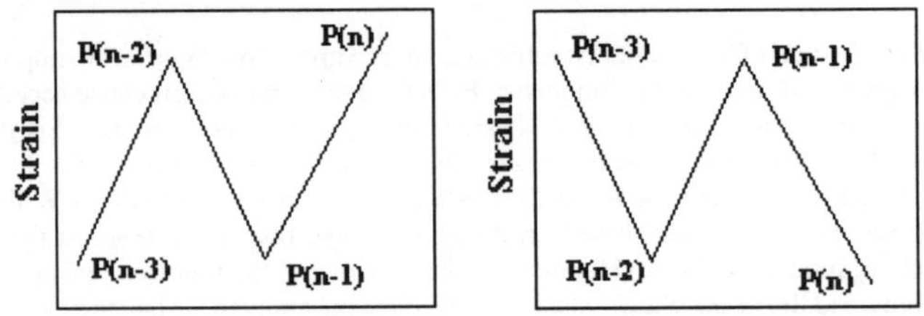

Figure 1. Two forms of loading points arrangement [2].

3 Fatigue Crack Growth Rate Equation

A fatigue crack growth program is developed to predict crack growth using several crack growth rate equations such as Paris, Forman and Modified Walker equation [3]. The Modified Walker equation can be expressed as follows:

$$\frac{da}{dN}=C\left[\frac{\Delta K}{(1-R)^{1-m}}\right]^{n} \tag{1}$$

where C, n and m are the fatigue crack growth rate constants, ΔK is the stress intensity factor range and R is the stress ratio. If $\Delta K < \Delta K_{th}$ then crack growth rate must be set to zero (ΔK_{th} is the threshold of ΔK) where:

$$\Delta K_{th}=(1-R)\,\Delta K_{th0} \tag{2}$$

and ΔK_{th0} is the threshold value of ΔK for R=0.

4 Load Interaction Models

To consider the load interaction effects in fatigue crack growth predictions two models based on the Generalized Willenborg model are used. In these models tensile overload retardation effects are accounted by Willenborg Model [4].

4.1 Chang Model

In this model, which is a combination of the Generalized Willenborg model and Chang acceleration scheme [3], when effective stress ratio is negative the Chang acceleration scheme calculates the compressive load acceleration effect as follows:

$$\frac{da}{dN}=C\left[(1-R)^{\bar{q}}(K_{max})\right]^{n} \tag{3}$$

$$\bar{q}=\left[\ln(r)/\ln(1-R)\right]/n \tag{4}$$

where r is the ratio of the crack growth rate at a specific negative stress ratio to its R=0 and n is the fatigue crack growth rate exponent determined from the R=0 crack growth data, however to simplify the calculation procedure an average of q can be used.

This model is applied for the reduction of the tensile overload retardation effect caused by the compressive load immediately following the tensile overload using an effective overload retardation zone size concept [3].

$$(Z_{Ol})_{eff} = (1 + R_{eff})(Z_{Ol}) \tag{5}$$

where Z_{ol} is the plastic zone size produced by the tensile overload and R_{eff} must be obtained from the Generalized Willenborg model.

4.2 New Model

The proposed Chang model assumes that the crack tip plastic zone associated with the underload is smaller than the monotonic plastic zone associated with the overload [8]. But this is an assumption then with respect to the crack-tip residual stress theory the Willenborg model is modified in the following manner. In the new model to obtain a relation between the residual stress and crack growth rate using the concept that residual stress is a static quantity [7], then the effect of the residual stress can be applied to the mean value of the load cycle. This residual stress will affect the following load cycle.

To perform the new model, calculation of the residual stress at the crack tip under different load cycles using finite element method is the first step. For this propose center-cracked panels with the geometry shown typically in Figure 2, with several crack lengths under loading with different stress ratio were analysed. The residual stresses in each case were determined using elastic-plastic analysis.

These analyses were performed using the finite element method. In Figure 3 a typical finite element mesh of the model is shown. Also for example the residual stresses at the crack tip for the crack length equal to 0.45 inch are shown in Figure 4.

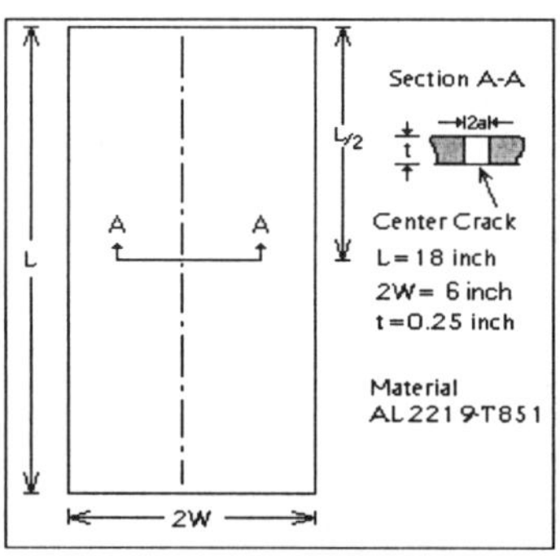

Figure 2. Center-cracked panel.

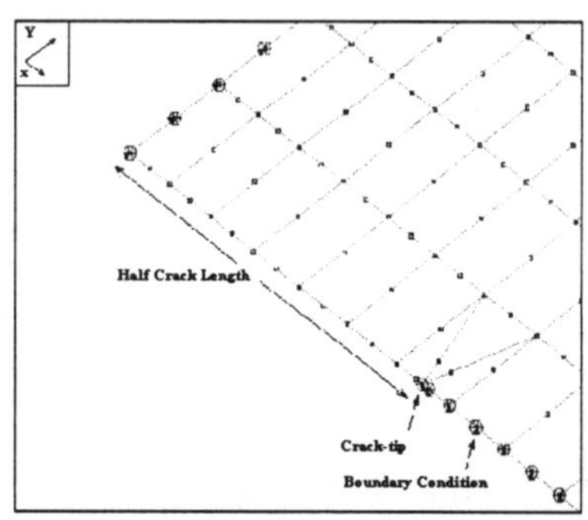

Figure 3. A typical finite element mesh.

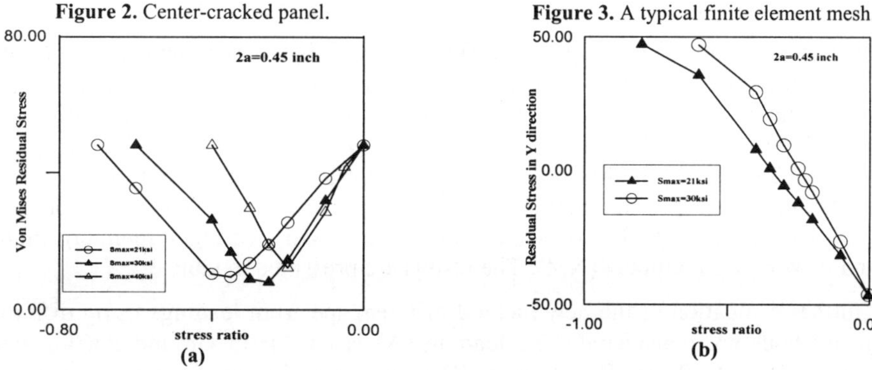

Figure 4. a- Von Mises Residual Stress **b-** Local Residual Stress in Y direction.

Using the results of the elastic-plastic analysis it is possible to determine values of residual stresses at the crack tip for each crack length. In Figure 4 variations of the residual stresses at the crack tip versus stress ratio for various maximum stresses are shown for a typical crack length. Figure 4-b shows that the residual stresses in Y-direction versus stress ratio varies linearly. The slopes of these curves were calculated using a least square method for various stress ratios and different crack lengths. The obtained results were fitted in the following equation to predict the residual stresses (S_i+1), at the crack tip of the Al 2219-T851 plates.

$$S_i = \frac{\sigma_{\max_i}}{21} * R * \left[3.69 - 0.118(\frac{a}{w})^{-1} + 0.0015(\frac{a}{w})^{-2} \right] \tag{6}$$

where a is half of the crack length and w is half of the width of the panel.
The application of this model is as follow. It assumes that the retardation effect of the tensile overloads and the effect of the compressive underloads can be superimposed linearly and the reduction in retardation effect is considered with the accelerating in crack growth of the following cycles. In this procedure at first R_{eff} is calculated from the willenborg model then with respect to the equation (6) this value of R_{eff} is increased by the same value of the increase in residual stress at the crack tip. This new effective stress ratio is named R_{eff}. With respect to this fact that the acceleration effect of the compressive underload is damped after several cycles it is assumed that this effect can be applied until the half of the crack length is equal to the value of a_r ($a_r = a_{ol}+[1-(\sigma_{max})_i/(\sigma_{max})_{ol}]Z_{ol}$). However to simplify the calculation procedure, an average value of a_r for a specific overload can be used.

5 Crack-Tip Stress-Intensity Factor and Instability Criteria

The following stress intensity factor is used in calculations:

$$K = f(g)\sigma\sqrt{\pi a} \tag{7-a}$$

$$f(g) = \left[1 - 0.025(a/w)^2 + 0.06(a/w)^4\right]\sqrt{Sec(\pi a/2w)} \tag{7-b}$$

The instability criteria is the value of the critical stress intensity factor (K_c), which normally is presented by the experimental results [3].

6 Evaluation of the models

The required Parameters and material properties are listed in Table 1 [3].

To evaluate the Original Willenborg (OW) model, fatigue crack growth results of a panel shown in Figure 2 under three types of loading profiles are predicted. In these cases the results obtained from the Original Willenborg model are compared with those obtained with No-load-Interaction effects (NLI) and with the results obtained from the experiments in Table 2 [3]. In the NLI cases the compressive loads are set to zero.

Then, fatigue crack growth under several loading profiles containing the compressive loading are calculated with No-Load-Interaction effect (NLI), Chang/Willenborg model (CW) and New Modified model (NM). The results are presented in table 3.

For further evaluation of the new method different spectrum loadings using different counting methods are considered. The loading (A) is an Air to Ground loading with design limit stress of 30 ksi. The loading (B) is an Air to Ground loading with design limit stress of 40 ksi and loading (C) is an Instrumentation and Navigation loading with design limit stress of 40 ksi. The numerical values of these loadings are available in reference [1]. For example the power spectral density of the loadings(A) and (B) are shown in Figure 5.

Table 1- Material constants and parameters required for walker equation.

Fatigue crack Rate constant	**Parameters**
C = 8.367e-10	$K_c = 65$ ksi(in)$^{1/2}$
N = 3.64	$\sigma_{ty} = 48$ ksi
M = 0.60	$K_{th} = 2.5$ ksi(in)$^{1/2}$
q = 0.30	

Table 2 Evaluation of the Original Willenborg model.

1.1.1.1.1. est	Loading Profile	σ_{max} (ksi)	σ_{min} (ksi)	$\sigma_{max,ol}$ (ksi)	$\sigma_{min,ul}$ (ksi)	N_{II} (cycle)	N_{II} (cycle)	a_I-a_f (inch)	N_T (cycle)	N_P/N_T	
										.2 NLI	OW
T1	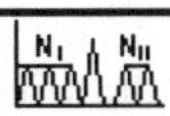	20	0	30	0	2500	To Failure	0.148-f	15182	0.85	0.88
T2	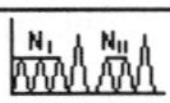	20	0	30	0	2500	2500	0.149-f	16623	0.75	0.86
T3	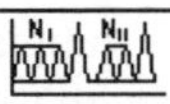	20	6	40	0	2500	2500	0.155-f	225090	0.11	0.25

NP: Predicted Life. NT: Test Life.

Table 3 Evaluation of compressive loads effects in fatigue crack growth.

1.2.1.1. est	Loading Profile	σ_{max} (ksi)	σ_{min} (ksi)	$\sigma_{max,ol}$ (ksi)	$\sigma_{min,ul}$ (ksi)	N_I (cycle)	N_{II} (cycle)	a_I-a_f inch	N_T (cycle)	N_P/N_T		
										NLI	CW	NM
T4	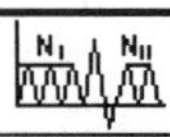	20	0	30	-6	2500	Failure	0.145-f	13460	0.98	1.01	1
T5	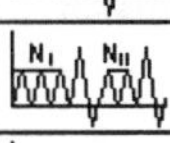	20	0	30	-6	2500	2500	0.150-f	36083	0.34	0.39	0.37
T6	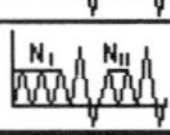	20	0	40	-12	2500	2500	0.160-f	22850	0.33	1.1	0.88

NP: Predicted Life. NT: Test Life.

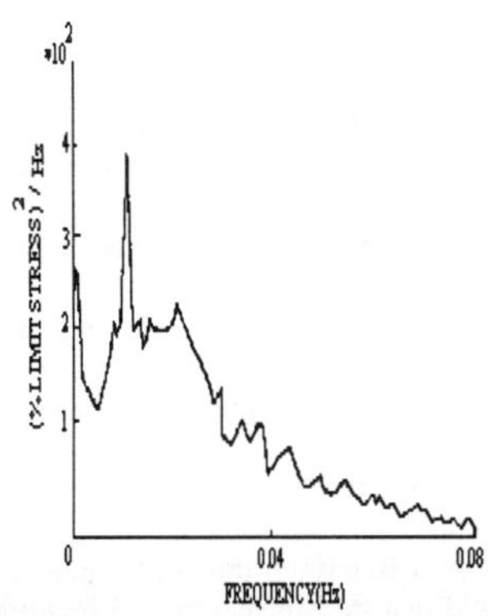

Figure 5. PSD of the loadings(A),(B).

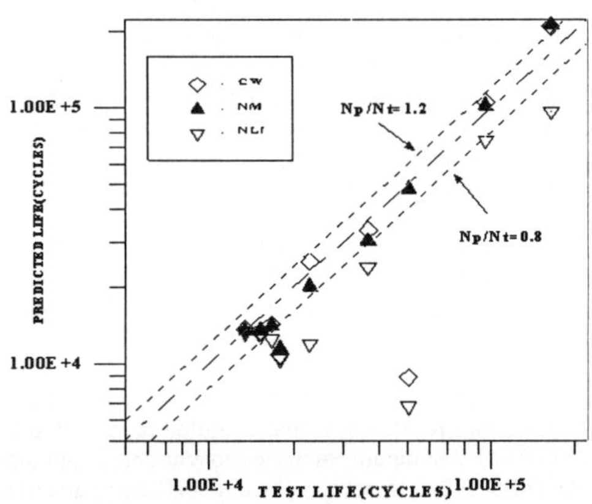

Figure 6. Predicted Life versus Test Life.

The results of fatigue crack growth calculations using the New Modified model (NM), Chang/Willenborg model (CW) and with No-Load Interaction effect (NLI), under these spectrums are presented in Table 4. The predicted lives for different cases using above methods versus experimental results are shown in Figure 6.

Table 4 Results of fatigue crack growth under spectrum loading.

Test	a_i-a_f inch	N_T (cycle)	N_P/N_T					
			ORCM			MRCM		
			NLI	CW	NM	NLI	CW	NM
A	0.144-f	95642	0.78	1.01	1.06	0.79	1.10	0.98
B	0.153-f	36367	0.66	0.91	0.83	0.65	0.92	0.76
C	0.150-f	164738	0.57	1.26	1.28	0.59	1.27	1.29

NP: Predicted Life. NT: Test Life.

7 Conclusion

The fatigue life assessment results using different methods for different cases presented in Tables 2 to 4 show that the scatter of the results obtained from the New Modified model is smaller than those obtained from the Chang/Willenborg model. But the results of the new model are more conservative than the results obtained from the Chang/Willenborg model. Also when the load interaction effects are not considered in the fatigue crack growth prediction the scatter of the obtained results is much more than those which consider the effects of the load interactions. It is also concluded that when the load interaction effects are not considered the results for fatigue crack growth prediction under different spectrums loading are not affected significantly by the cycle counting methods. However the Original Rainflow Counting Method has some shortcomings in comparison with the Modified Rainflow Counting Method but there is no much differences between the results of fatigue crack growth prediction for these two cycle counting methods.

References

1. J. B. Chang, " Round-Robin Crack Growth Predictions on center cracked Tension Specimens under Random Spectrum Loading", Methods and Materials for Predicting Fatigue Crack Growth under Random Loading, ASTM STP 748, American Society for Testing and Materials, 1981, p 3-40.
2. Nie Hong, "A Modified Rainflow Counting Method ", Int.J.Fatigue, No.6, 1991,p465-469.
3. J. B. Chang,. M. Szamossi and K-W. Liu, "Random Spectrum Fatigue Crack Life Predictions with or without considering Load Interactions", ASTM STP 748, American Society for Testing and Materials, 1981, p 115-132.
4. B. Farahmand ," Fatigue and Fracture Mechanics of High Risk Parts , Application of LEFM & FMDM", International Thomson Publishing,1997.
5. M.R. Mofakhami, " Fatigue Crack Growth Prediction Under Irregular Loading", Msc Thesis, Amirkabir University of Technology (Tehran Polytechnic), 2000.
6. Annual Book of ASTM Standard, Standard Practices for Cycle Counting in Fatigue Analysis, American Society for Testing and Materials, E10-49-85, 1995.
7. D. Broek, "The Practical Use of Fracture Mechanics ", Kluwer Academic Publishers, 1989.
8. A.F. Liu, "Structural Life Assessment Methods", ASM International, 1998.
9. J. B. Chang, R. M. Engle and J. Stolpestad , "Fatigue Crack Growth Behavior and Life Predictions for 2219-T851 Aluminum Subjected to Variable Amplitude Loading",Fracture Mechanics , 13th conference ,ASTM STP 743, American Society for Testing and Materials,1981, p 3-27.

A DAMAGE IDENTIFICATION APPROACH BASED ON STRESS WAVE PROPAGATION

ZHONG-QING SU[1], LIN YE[1,*], XIONG-ZHU BU[1,2], YIU-WING MAI[1]

[1]Centre for Advanced Materials Technology (CAMT)
School of Aerospace, Mechanical and Mechatronic Engineering
The University of Sydney, NSW 2006, Australia
[2]Nanjing University of Science and Technology, 210094, P.R.China
E-mail: ye@aeromech.usyd.edu.au

XIAO-MING WANG

CSIRO Infrastructure Systems Engineering, Highett, VIC 3190, Australia

An approach for structural damage diagnosis employing the wavelet transform technique is proposed in the paper. A damage identification system with built-in piezoceramic transducer is developed in correlation with a stress wave propagation model. In the system, signal-processing techniques in the time-frequency domain are applied to reduce noise interference and facilitate the acquisition of diagnostic information from acquired raw data via signal processors. In addition, the capability of this system is investigated through experiments considering a beam pre-treated with various degrees of damage, which aims at simulating the damage growth. It is shown that the damage position and degree in a flexural beam can be detected quantitatively and efficiently by this approach.

1 Introduction

During the past two decades, increasing demands for effective and economical detecting techniques for structural damage identification have been recognized. Considerable efforts have been made for the early stage diagnosis of structural damage in aerospace, mechanical and civil engineering. In particular, detection methods based on elastic waves propagation have been attracting great interests from both researchers and engineers, regarded as one of encouraging solutions to damage diagnosis.

Signal processing is a key point to make damage identification systems success and applicable. As known, traditional spectrum or time series analysis is incapable of achieving good solutions both in time and frequency domain simultaneously, and much useful information may be consequently lost. However, recent work demonstrates that the wavelet analysis is able to overcome this problem successfully by decomposing acquired signal into time domain as well as frequency domain synchronously. Combined with this technique, the elastic wave-based identification approaches have ability to extract more detailed and accurate information from raw signals, which therefore make quantitative damage evaluation more feasible and reliable.

In the past few years, researches on the application of wavelet analysis for damage identification have made significant progress. For Example, Sung and Oh[1] (2000) developed an impact monitoring technique; Kim and Kim[2] (2000) reported their damage identification system for a defective structure; Samuel *etc*[3] (1997) accomplished the detection via wavelet for a rotorcraft planetary gear system.

This paper reports one of our recent ongoing work for damage identification, aiming at developing a damage identification system incorporated with built-in piezoelectric transducers. With this system, the presence, location and degree for a pre-treated

* Corresponding author.

transverse cut-shaped damage with increasing depth, simulating the damage growth, in a typical Euler-Bernoull beam, can be identified promptly and accurately. Scarcely could the similar work touching on continuously variable damage be found so far.

2 Fundamentals of Wavelet

A brief background about wavelet analysis employed in this study is introduced in this section. More detailed knowledge can be found in [4].

As for any selected wavelet transform function Ψ(t), the Continuous Wavelet Transform (CWT) of arbitrary sampling signal *f*(t) can be defined as[4]:

$$W(a,b)=\frac{1}{\sqrt{a}}\int_{-\infty}^{+\infty} f(t)\cdot\Psi^*(\frac{t-b}{a})\cdot dt \tag{1}$$

where a is the scale parameter and b is the translation parameter, corresponding to frequency and time, respectively. $\Psi^*(t)$ denotes the complex conjugate of $\Psi(t)$. Further, the total energy of a signal can be obtained as:

$$E=C\cdot\int_{-\infty}^{+\infty}\int_{-\infty}^{+\infty}\left|W(a,b)\right|^2\cdot\frac{da\cdot db}{a^2} \tag{2}$$

thereby, the wavelet coefficient $W(a,b)$ reflects the distribution of wave energy over the time and frequency domain.

Practically, it is unnecessary to calculate wavelet coefficients at every scale, therefore the Discrete Wavelet Transform (DWT) is innovatively put forward. Herein, continuous a and b are discretized by dyadic parameters, with which signal can be broken down into low-resolution components, i.e. Detail and Approximation, by selecting different dyadic scales[4].

3 Theoretical Model

Consider a uniform Euler-Bernoulli beam with a defect. The motion equation for traveling wave can be simplified as:

$$EI\cdot\frac{\partial^4 w}{\partial^4 x}+\rho A\cdot\frac{\partial^2 w}{\partial^2 t}=p(x,t) \tag{3}$$

where $w(x,t)$ is the transverse wave displacement. And the solution of Eq.3 can be written as the sum of four series of flexural wave components[5]:

$$w(x,t)=[\sum A_1e^{-ikx}+\sum A_2e^{ikx}+\sum A_3e^{-kx}+\sum A_4e^{kx}]\cdot e^{i\varpi\cdot t} \tag{4}$$

where k denotes the wavenumber.

A transverse cut with increasing depth, as depicted in Fig.1, is made to simulate the damage growth in a metal beam. We hereby assume the damage begins at the left side and ends at the right side:

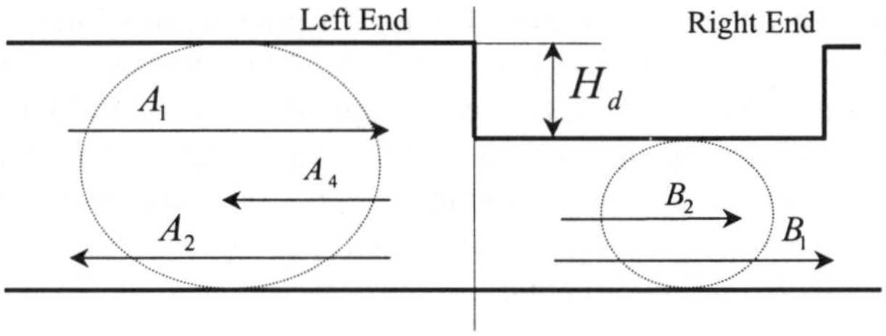

Figure 1. Damage model.

Upon localizing Eq.4 at left end with continuity conditions, the reflection coefficient at the left end can be yielded as:

$$R = \frac{A_2}{A_1} = \frac{-iI_1^2k_1^4 + 2I_1I_2k_1^3k_2 + 2iI_1I_2k_1^2k_2^2 - 2I_1I_2k_1k_2^3 - iI_2^2k_2^4}{I_1^2k_1^4 + 2I_1I_2k_1^3k_2 + 2I_1I_2k_1^2k_2^2 + 2I_1I_2k_1k_2^3 + I_2^2k_2^4} \tag{5a}$$

$$I_1 = \frac{l \cdot H^3}{12}, \; I_2 = \frac{l \cdot (H - H_d)^3}{12} \tag{5b}$$

where l and H denote the width and height of the cross section of beam, respectively. While H_d represents the depth of the damage.

Based on Eq.2, the amplitude ratio of two waves in spectrogram can be proved to be equivalent to the integration ratio of their corresponding wavelet coefficients along the scale axis at specific times, where more detailed derivation can be found in [7]. Therefore, the depth of damage can be produced through solving Eq.5 after introducing the reflection coefficient. The calculated relationship between reflection coefficient and damage degree (H_d / H) is drawn in Fig.2.

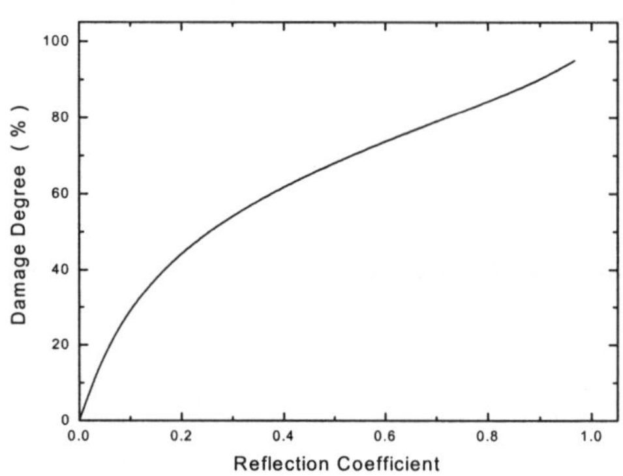

Figure 2. Relationship between reflection coefficient and damage degree.

4 Identification System Configuration in Experiments

The diagnosis system developed in this work consists of the signal transducer, signal acquisition unit and central control unit (CCU). A *PI*© piezoceramic element

(*20mm×10mm×1mm*) is chosen as the signal sensor. Signal acquisition is implemented via *Agilent*© VXI instrument (*HP E3242A* and *HP E1437A*) through IEEE488 bus; CCU performs both as controller and as interpreter, supported by *NI* Labview© and Matlab© platform. A simple yet practical force hammer is employed for impact excitation. In Fig.3, a sketchy pictorial of an aluminum beam for experiment (Cross Section: *10mm×10mm*, Young's Modulus: *71.2Gpa*; Density: *2.69g/cm³*) is given. The depth of simulated defect varies from *2mm* to *8mm*.

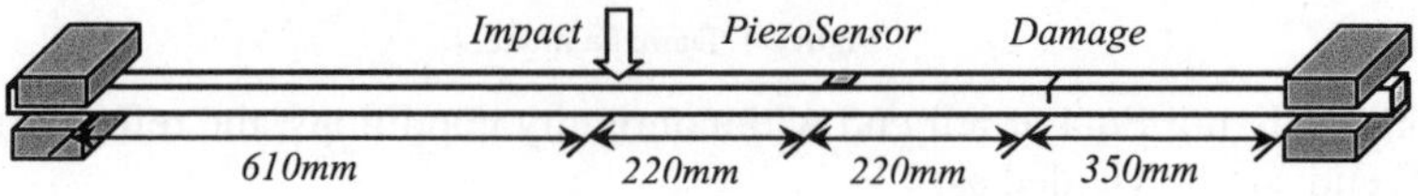

Figure 3. Specimen in experiment.

5 Results and Discussion

Wave velocities for different modes are necessary in this work, with which the different wave modes and reflected waves due to damage or boundaries can be distinguished. Further, the damage position can be determined. The velocities measured by experiment and calculated values are listed in Table.1.

Table 1. Wave Velocities List.

Wave Mode	Measured	Calculated
Longitudinal Wave (m/s)	6365	6298
Transverse Wave (Shear Wave) (m/s)	3205	3161

The signals for different damage severity are recorded with sampling rate of *20.48MHz*. As typical examples, the signals for *4mm*-depth damage and *6mm*-depth damage are comparatively shown in Fig.4 and Fig.5, respectively.

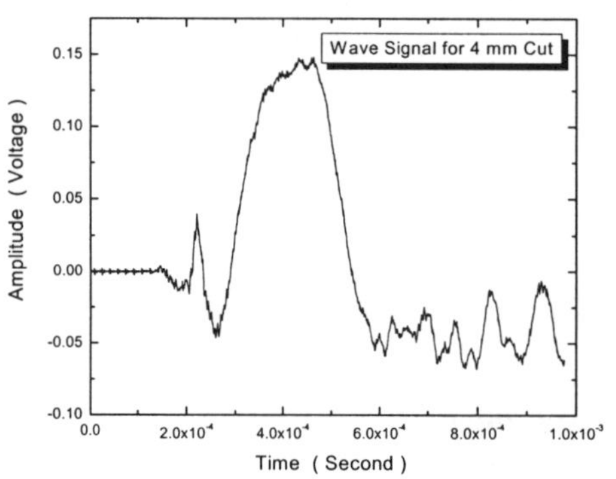

Figure 4. Sampling signal for *4mm*-depth damage.

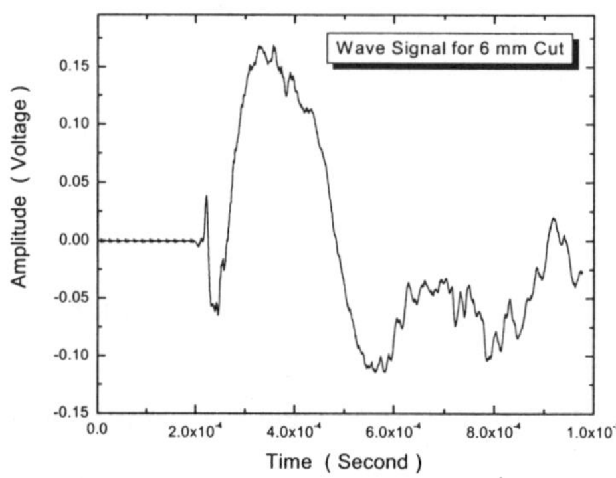

Figure 5. Sampling signal for *6mm*-depth damage.

Hardly can we extract any useful information directly from these raw signals. However, spectrogram analysis based on wavelet transform offers great advantages in detecting singularity intuitively. Here, both DWT and CWT analyses are performed. *Daubechies* function at level 4 (*db4*) is chosen as the transform function, which is demonstrated effective in the following signal processing. In particular, the details of sampling signal at level 8 via DWT and 3D spectrogram via CWT for *6mm*-depth damage, as a paradigm, are displayed in Fig.6 and Fig.7, respectively.

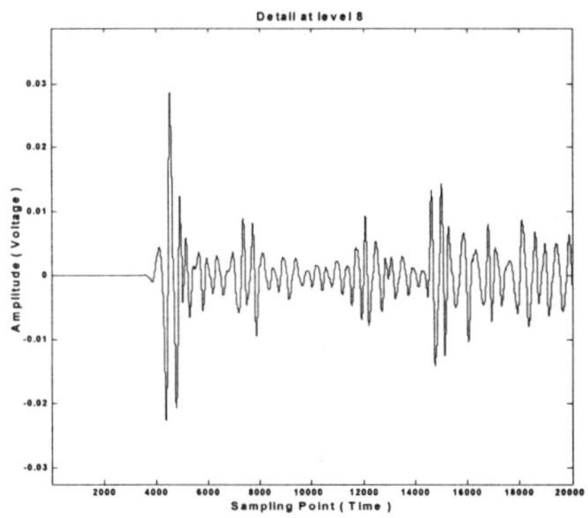

Figure 6. Detail at level 8 for *6mm*-depth damage.

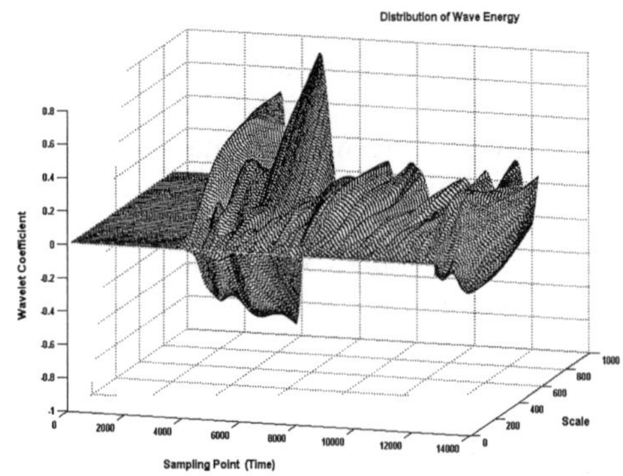

Figure 7. 3D CWT spectrogram for *6mm*-depth damage.

In Fig.6, the wavelet detail at level 8 deploys the incident wave due to the excitation and reflected waves from the damage and boundaries clearly. Time difference between waves can be used to calculate the damage location. Meanwhile, in Fig.7, several energy ridges can be observed, which provides a possibility to fulfill the quantitative damage identification with the aid of Eq.5 and Fig.2. The damage diagnostic results achieved by this identification system and the actual situations are compared in Fig8 and Fig.9,

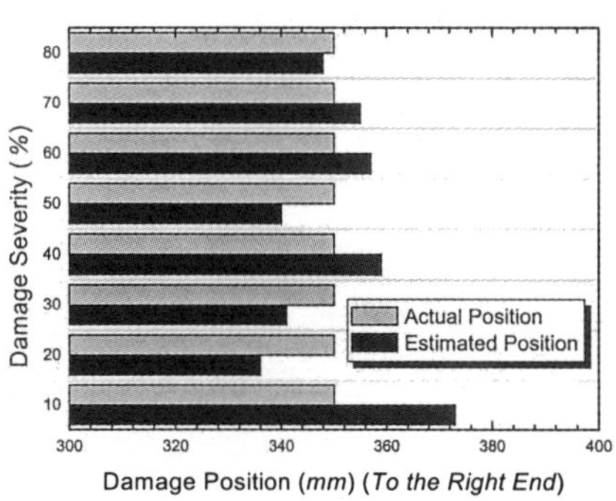

Figure 8. Capability for damage location estimation.

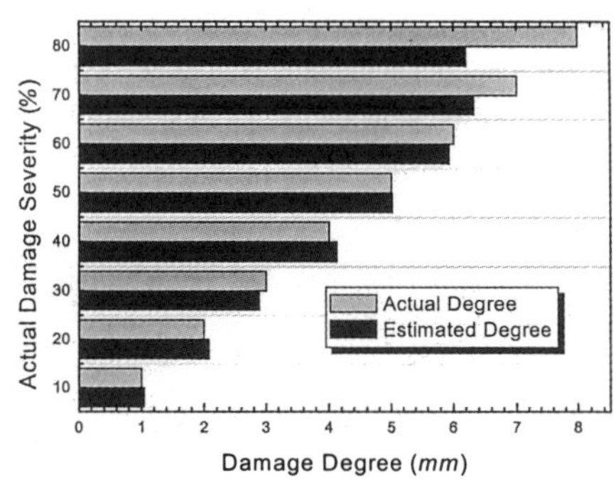

Figure 9. Capability for damage degree estimation.

Both the estimation errors for damage location and degree diminish as the damage severity increases, however, the latter reach the minimum when damage approaches *5mm-6mm* approximately, then expand dramatically. This is attributed to the inapplicability of hypotheses and continuity condition for Euler-Bernoulli beam theory when the damage severity is beyond *65%* of the entire cross section.

6 Conclusion

An effective damage identification approach based stress wave propagation has been developed. A theoretical model with a cut-shaped damage vertical to beam axis with variable depth is established via Euler-Bernoulli beam theory, which is demonstrated experimentally its feasibility for damage diagnosis. Continuous and discrete wavelet transforms are performed for signal processing. Quantitative identification can be consequently fulfilled and facilitated. Furthermore, The experiment results imply the detection accuracy depends on the severity of damage. But only when the damage is very severe (over *65%*), will the unacceptable error take a clear effect. In practice, such severe damage is not the concern of structural health monitoring. Accordingly, this detection scheme can be applied for common damage identification purpose. More results about the diagnosis for complicated damage in laminated composite plates have not yet been addressed in this paper and will be covered in coming literatures.

References

1. Sung D.U., Oh J.H. and Kim C.G., Impact monitoring techniques for smart composite laminates. *Proceedings of the second Asian- Australasian conference on composite materials*. 18-20 Aug 2000, Korea pp.1123-1128.
2. Kim Y.Y. and Kim E.H, A new damage detection method based on a wavelet transform, *Proceedings of 18th IMAC*, Feb. 7-10, 2000, San Antonio, Texas, U. S. A. pp. 1207-1212.
3. Samuel P., Health monitoring/damage detection of a rotorcraft planetary geartrain system using piezoelectric sensors, *SPIE*, **3041** pp. 44-53.
4. Chan Y.T., Wavelet Basic. (Kluwer Academic Publishers, Boston, 1995).
5. Doyle J.F., Wave propagation in structures. (Springer-Verlag, New York, 1997).
6. Zhou Z., Noori M. and Amand R.St., Wavelet-based approach for structural damage detection, *Journal of Engineering Mechanics*, **7** (2000) pp. 667-683.
7. Su Z.Q., Ye L. and Bu X.Z. *etc*, Quantitative assessment of structural damage by impact response, (to be submitted).

DAMAGE DETECTION IN BOLTS UNDER SERVICE LOAD BY STRESS WAVE ATTENUATION

D.W. BARKE AND W.K. CHIU*

Department of Mechanical Engineering. Monash University, Clayton Victoria 3800 Australia

** Author to whom Correspondence Should be Addressed.*

S. FERNANDO

Ajax Technology Centre. Ajax Fasteners, Mills Rd Braeside Victoria 3195 Australia

Preliminary work has shown that damage in a short round bar can be quantified by determining the attenuation of an ultrasonic stress wave. A bar provides a waveguide for the impulse to travel and has a simple geometry. A shoulder or other geometric feature will reflect the excitation, which would change the recorded response. To investigate the possibility of applying this technique in situ, three bolts were also loaded under conditions of varying levels of uniaxial tension, in a large steel block. The aim being to observe the effect of an alternative load path and a mean stress on the attenuation of an acoustic stress wave and to compare this with an untensioned bolt. A linear relationship between attenuation and crack depth was observed.

1 Introduction

Preliminary work suggests that it is possible to detect damage in a short structure by exciting an ultrasonic stress wave at one end and recording the response of the structure at the opposing end. Mostafa et al.[1] have successfully applied this technique to tensile test specimens, in which they induced failure by fatigue crack growth. Attenuation of response has also been used to quantify the extent of fatigue damage in a short concrete compressive test specimen [2]. In both of these tests and the present authors' preliminary work, the structure is a waveguide which provides only one path for an acoustic stress wave to travel through.

Components commonly have shoulders and other geometric features, which will cause reflection and refraction, altering response to an impulse. In situ testing also requires that the structure be tested without disassembly and therefore alternative paths may exist for waves to travel through. Such a path for the wave, such as the body in which a component is mounted could significantly reduce the sensitivity of proposed non destructive testing by pulse transmission.

Changes in response of a bolt under tension, due to the introduction of damage, are investigated in this paper. The effect of a surrounding structure, in which the bolt is mounted, on response through which waves can travel is also observed.

2 Response Quantification

By quantifying the magnitude of the fluctuating signal the effect of damage can be compared. It was chosen in this paper to measure overall strength of signal as a root mean square (RMS) response. For discretely sampled data, the root mean square (RMS) voltage is defined by equation 1, which is suggested by Vary [3] as an appropriate measure of stress wave energy.

$$V_{RMS} = \sqrt{\frac{1}{t_{k\cdot tof} - t_{ktof}} \int_{tof}^{k\cdot tof} V^2 \cdot dt} \tag{1}$$

Where:
V_{RMS} = the root mean square voltage
V = the voltage across the piezoelectric patch
t_{tof} = the time of flight of the fastest wave in the structure
$t_{k\cdot tof}$ = k times t_{tof}, where k is an integer. In this study k is set to be 3 to avoid the influence of reflected waves on the RMS response.

2.1 Definition of Attenuation

For the purposes of comparison between the response of the undamaged structure and the response of the damaged structure, to an ultrasonic impulse, the attenuation of the signal was calculated. The definition of attenuation is defined by equation 2, a definition similar to that used by Fei, Zhu and Cui [4].

$$\alpha = \ln\left(\frac{A_{RMSo}}{A_{RMS}}\right) \tag{2}$$

where:
A_{RMSo} is the root mean square response of the undamaged model or component
and
A_{RMS} is the root mean square response at a known damage level.

3 Effect of Mean Tension and Alternate Wave Paths on Attenuation

Vary [3] suggests that attenuation of ultrasonic stress waves may be used to determine the load bearing capacity of composite components. He also suggests that the technique may be applicable to other waveguides. However, such results may be inapplicable when a bolt has its design load applied. Furthermore, a bolt, in service, will be mounted in a surrounding structure, which may limit the ability of the stress wave to detect damage in a bolt.

To establish the effect of increasing the degree of damage in a bolt on impulse attenuation, in the presence of a surrounding structure and a tensile load, three tests were undertaken on mass produced Class 4.6 M12x75 set screws. To quantify strain levels in the bolts, a reduced section was machined into the middle of each bolt to accept a 350Ω Strain Gauge, as depicted in Figure 1. A 10x10 mm piezoelectric patch was glued to the head of the bolt and 5x5mm patch glued to the end, both bonded with an epoxy resin, for mechanical strength and durability, and a conductive cement, to complete the electrical circuit. Both positive and negative leads were bonded with the same conductive cement used to locate the patch.

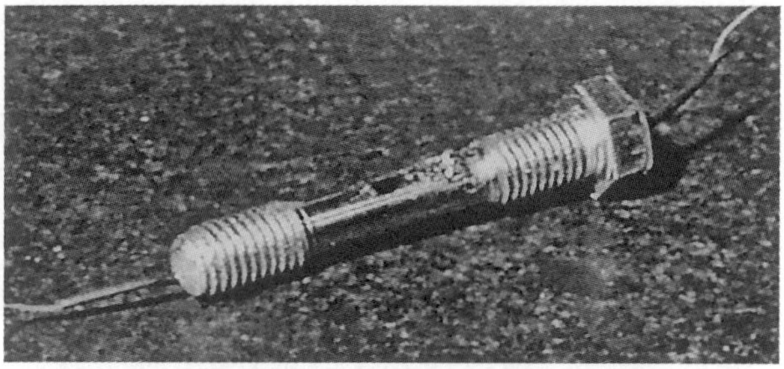

Figure 1. A reduced shank bolt made from a Class 4.6 Screw, with a strain gauge attached to the shank and piezoelectric patches attatched to either end.

Figure 2. A Strain Gauged and Piezoelectrically Patched Bolt, in Tension in the Steel Test Block. Holes in the block provide location for the bolt, access for strain gauge wires and access for cutting the bolt by passing a handsaw blade through the hole.

The piezoelectric patch attached to the bolt head was excited with a single cycle of a square wave, of frequency 1MHz and magnitude 10 Volts (ground to peak) which was generated with a Hewlett Packard 33210A Function Generator. Each response was acquired from the receiving patch with a Tektronix TDS210 digital oscilloscope, which was downloaded directly to computer. The Root Mean Square of the response voltage was then evaluated.

To understand the behaviour of the bolt outside the test block, a bolt was cut with a 0.45mm wide handsaw and the response recorded whilst this bolt was sitting on aerated foam. The characteristic response of the actuator-bolt-sensor system, in its undamaged state, to an impulse was recorded and then a shallow cut made in the centre of the reduced section. The response to an impulse of the same dimensions was recorded at this damage state. The process of cutting the bolt by small increments and recording response was repeated until the cut was approximately 3.5mm deep.

A bolt was mounted in the test block, shown in Figure 2, and tensioned manually to a strain of 95$\mu\varepsilon$. The bolt was not retensioned or unladen throughout the test. The characteristic response, to an impulse, of the undamaged bolt was recorded and the bolt then cut in small increments, whilst in tension, recording the response as the cut depth was increased.

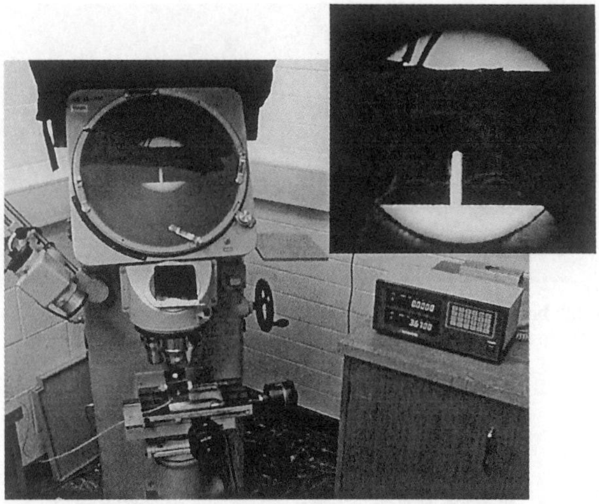

Figure 3. The Optical Comparator is used to measure the depth of cut in the bolt. The table on which the specimen rests has a digital coordinate recorder, the display of which can be seen in the right of the figure. Also shown is a detail view of the projection. The strain gauge glued to the bolt can be seen in the top of the projection.

A third bolt was fixed in the test frame and tensioned to 1100με by turning the nut whilst holding the head of the bolt with spanners, in the usual fashion. This bolt was characterised and damaged in the same manner as the previous tests.

In each test, a guide was used to ensure a cut perpendicular to the bolt axis and in the same location at each step of the process. Where the bolt was mounted in the test block, a piece of equal angle steel was clamped to the side of the block to guide the handsaw. For testing outside the block a cut was made halfway through a short piece of pipe in an appropriate location to provide a guide for the saw. This section was placed over the shank of the bolt and held in place with a nut on the thread of the bolt. The method of cutting a bolt is artificial and in service, a crack may not be as clean as a sawcut.

Cut depth was measured using an Optical Comparator which projects light past the test specimen, through a magnifying lens and onto a screen with cross hairs, as depicted in Figure 3. The table of the comparator, on which the specimen rests, has two digital gauges enabling position to be measured in terms of x and y coordinates to an accuracy of 0.01mm or greater.

3.1 Attenuation with Cut Depth in a Bolt in Tension

The average of five response samples from the undamaged bolt is plotted in Figure 4. The analytically predicted time of flight, t_{tof}, was 13.2μs and this did not change when loading conditions were altered.

Root Mean Square voltage was evaluated between 13.2 and 39.6μs, which is three times t_{tof}. The second arrival of the longitudinal wave at the sensor patch will be two times t_{tof} after it arrives the first time. The portion of the response reaching the sensor, after 39.6μs will contain waves reflected from thread roots, cracks (where they exist) and other geometric features present. Therefore the RMS evaluation does not include this part of the response.

Attenuation increases with cut depth in Figure 5. It can be seen here that the behaviour of the unladen bolt and the bolt laden to its design tension are similar, indicating that the mean tension and the effect of the surrounding structure are negligible. When a bolt was laden in tension at a working strain of 1000με it was observed that the change in attenuation with increasing crack depth of both it and the unladen bolt were similar, although not identical.

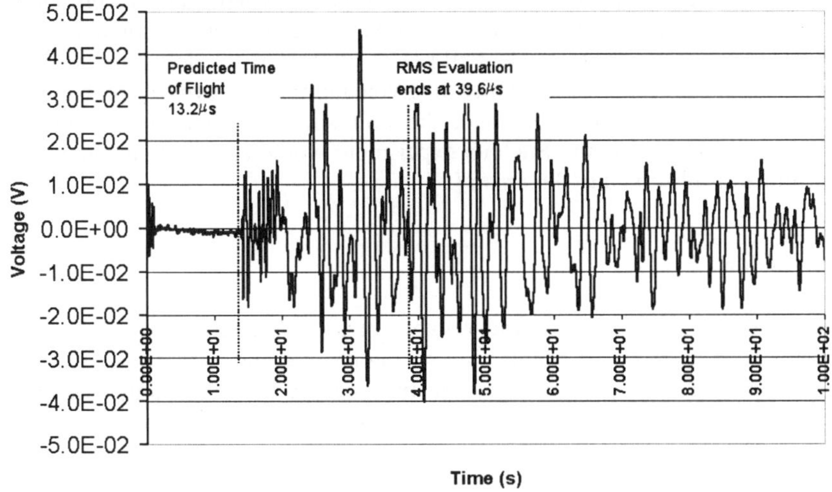

Figure 4: The Average of Five Voltage Response Samples of a Bolt In Tension, Excited by a 10V 1MHz Single Square Burst.

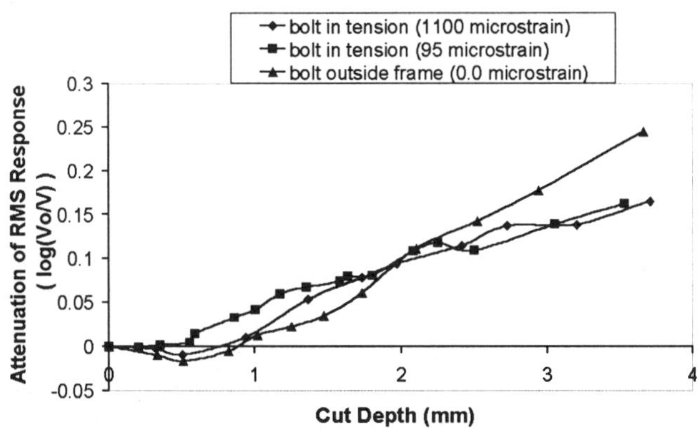

Figure 5: Effect of a Tensile Load and Cut Depth Upon Attenuation, with Comparison to an Unladen Bolt, with Increasing Cut Depth. Results From the Finite Element Model Have Been Included for the purposes of Comparison.

4 General Discussion

It was observed that attenuation of response in a bolt occurs with crack growth and that this occurs in the presence of a mean tension. The presence of alternate paths through which the wave might travel or an energy sink, such as a large structure, did not limit the effectiveness of pulse attenuation non destructive testing.

Guidelines prescribing the minimum amount of time over which the root mean square response should be evaluated in are required. This should include the location of bounds, which define the earliest point in time at which the RMS could be evaluated and the latest point at which it could be evaluated. Such an experiment requires explicit definition of the noise levels present, or a means of artificially introducing noise to a response.

It is strongly recommended that a test be completed where standard production bolts which have piezoelectric patches bonded to either end have cracks fatigued into them by cyclic uniaxial tension. These cracks will not be as even as the defect induced in the test specimens of the present work. Once a given number of cycles are completed and attenuation measurements are recorded, the bolt should then be pulled to failure, to determine its residual strength. Such an experiment would establish characteristic curves for bolt failure, as observed by Peterson [5].

5 Conclusion

Pulse attenuation is an effective and reliable means of determining damage levels in components such as bolts. It enables in situ monitoring and estimation of damage size. Such techniques are not prevented by the presence of a surrounding structure, which potentially provides a path for a wave to travel through and a medium in which a wave might be absorbed.

Piezoelectric patches provide a means of exciting ultrasonic stress waves in a structure. Magnitude of response, by taking a root mean square of the signal provides an efficient and effective means of determining crack depth. Noise also alters signals obtained and work is required to select appropriate ways of recording the true signal obtained from the excitation. Work is also required to establish characteristic curves of stress wave attenuation of fatigue damage to bolts.

References

1. Mostafa, I., Hailu S., Welsch G.E., Hazony D., and Halford, G.R.. Ultrasonic pulse transmit-receiver method for detecting and monitoring of fatigue damage. In *Fatigue and Fracture Mechanics* **29** *ASTM STP 1332* ed. by T.L. Panontin and S.D. Sheppard (American Society for Testing and Materials, West Conshohocken, PA. 1999). pp. 642-657.
2. Suaris, W. and Fernando, V., Detection of Crack Growth in Concrete from Ultrasonic Intensity Measurements. *Materials and Structures* **20** (1987) No.117 pp. 214-220.
3. Vary, A., The Acousto-Ultrasonic Approach. In *Acousto-Ultrasonics: Theory and Application* ed. by Duke, J.C. (Plenum Press, New York, New York, 1988) pp. 1-21.
4. Fei, G.T., Zhu Z.G., and Cui, P., Ultrasonic attenuation study of defects in Aluminium under the action of constant stress *Materials Science,* **31** (1996) pp. 2725-2728
5. Petersen, R.E., Stress Concentration Factors (John Wiley and Sons, New York, New York 1987).

ON DAMAGE ASSESSMENT OF IMPACTED CERAMICS VIA 3-D X-RAY COMPUTED TOMOGRAPHY

J.M. WELLS, W.H. GREEN, AND N.L. RUPERT
US Army Research Laboratory, Weapons & Materials Research Directorate, APG, MD 21005-5069
E-mail: jwells@arl.army.mil

The recent application of x-ray computed tomography has enabled new 2-D and 3-D in situ diagnostic and visualization capabilities in both the pre- and post-mortem damage assessment of impacted ceramic targets. Examples of such capabilities presented include: (a) 2-D and 3-D asymmetrical mesocracking in TiC, (b) residual tungsten alloy penetrator distribution in TiB_2, (c) subsurface cracking in body armor ceramic panels and (d) cracking of monolithic ceramic tiles contained within a cast or welded and hot isostatic pressed metal encapsulation. These results have enabled, and will continue to provide, damage descriptions that are far more accurate and revealing than previous approaches and that affect more efficient and detailed visualization and modeling of the various complexities present in practical armors. This innovative nondestructive CT technique provides a vital ballistic damage assessment approach which has applicability to other physical damage situations such as quasi-static fracture, fatigue and corrosion as well.

1 Introduction

Visualization and assessment of *in situ* ballistic damage internal to opaque metal and ceramic target materials is a difficult and challenging task. Furthermore, interrogation of the impacted target materials to obtain comprehensive knowledge of the size, shape and location of various damage elements nondestructively is highly desirable. Figure 1 [1,2] shows a summary comparison of several NDE technologies and their relative scale of detectable features. The two types of x-ray techniques of interest here, namely a shadowgraph or digital radiograph, DR, and the computed tomography, CT, slice are shown schematically in figure 2. A DR is a 2-dimensional (2-D) projection of a 3-dimensional (3-D) object. Both volumetric shape and spatial information is lost with a DR but retained with CT slices. Recently, the present authors have successfully demonstrated innovative 2-D and 3-D nondestructive x-ray computed tomography, XCT, techniques for visualizing the comprehensive internal mesoscale (> 200 m) damage in opaque ceramics following ballistic impact [3-6]. These nondestructive CT analyses are not conducted in real time during the ballistic event but rather are conducted in a post-mortem fashion. When an impacted ceramic remains essentially intact, these CT techniques can be applied and digital images may be reconstructed to show the resulting internal cracking in considerable detail on *any* 2-D planar section or 3-D volumetric perspective desired. While the original articles contain appropriate further details, it is our intention here to briefly provide an updated overview of the progress and capabilities demonstrated in the application of this technology to ballistic damage assessment to date.

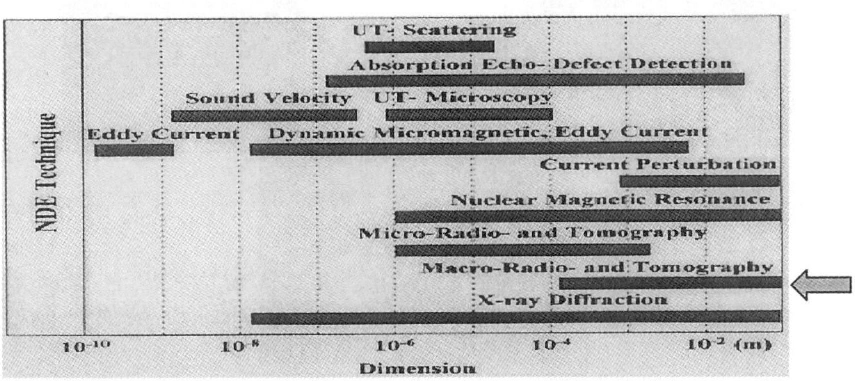

Figure 1. Dimension scale of features detectable by various NDE Methods [1,2]

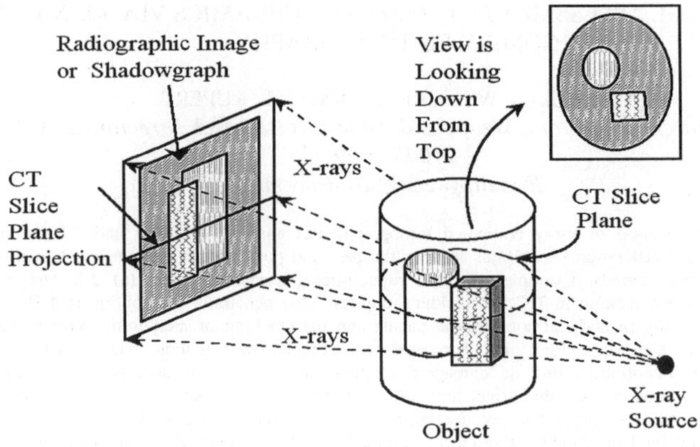

Figure 2. Schematic of x-ray Computed Tomography Techniques

These virtual representations of internal impact damage features are a significant advance in the diagnostic damage visualization and assessment tools now available to the ballistic research community.

2 Mesocracking Damage Visualization

A 3-D solid reconstruction of the original CT scan data permits the visualization of the mesocracking details internal to the impacted sample on any arbitrary virtual planar section. Such damage, with features including conical, laminar and radial cracking, is clearly evident in figure 3 on both a vertical and a horizontal planar section of the original TiC disk segment. Furthermore, a "point cloud" reconstruction technique has also been demonstrated (see figure 4) to visualize the comprehensive details of the *isolated* impact damage in 3-D for this same TiC ceramic target. These later images are particularly noteworthy in that the previous visual obstruction related to the opacity of the TiC ceramic in figure 3, has been eliminated, thus allowing the entire asymmetric mesocracking morphology to be visualized in total. Considerable detail including the bifurcation of the radial cracking on the right hand side becomes readily apparent when viewed with a rotating 3-D solid image of this point cloud.

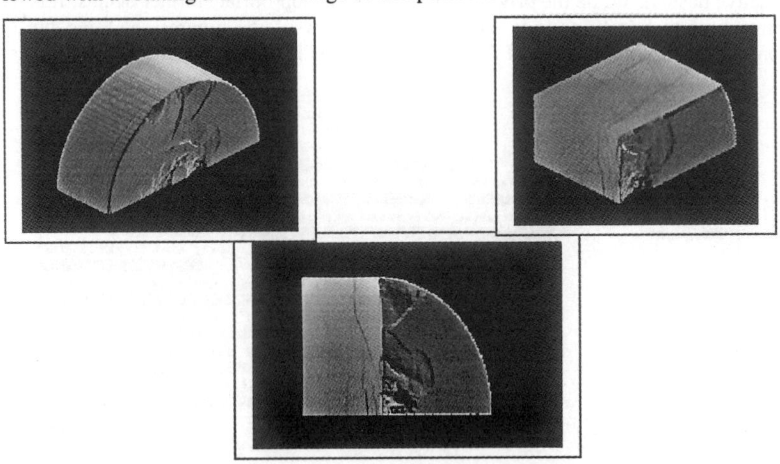

Figure 3. 3-D Solid Visualization of mesocracking on vertical and horizontal virtual sections of a 72 mm diameter x 25 mm thick TiC disk segment following interface defeat ballistic impact. [4]

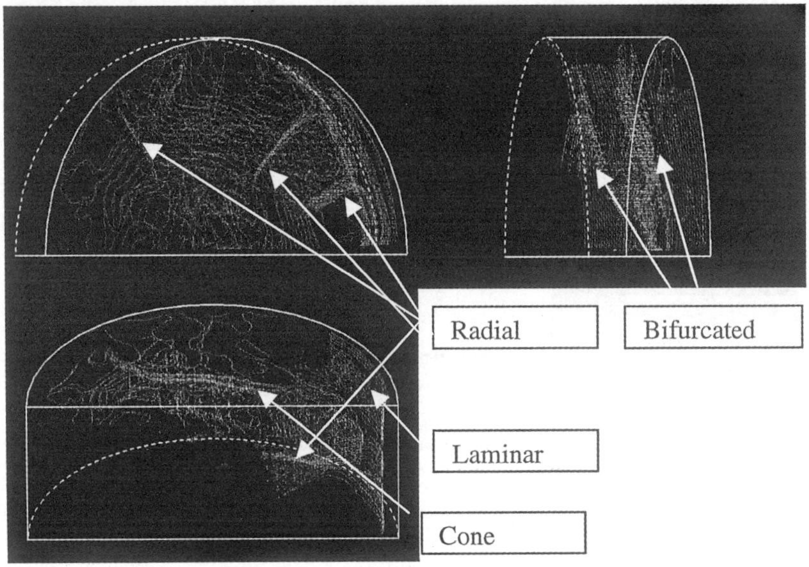

Figure 4. XCT Point Cloud virtual image of isolated 3-D mesocracking damage morphology in a ballistically impacted TiC ceramic target.

3 Residual Penetrator Material in TiB_2 Ceramic Target

When a subscale long rod penetrator overcomes the initial resistance of an encapsulated armor ceramic and penetration into the armor ceramic occurs, residue material from the fractured penetrator often remains inside the ceramic along the penetration cavity. This residual penetrator material can be clearly visualized in situ, on arbitrarily selected virtual planar sections of a 72 mm diameter x 25 mm thick TiB_2 disk, as the white (high-density) areas shown in figures 5b and 5c [5,6]. Considerable mesocracking is also observable in the TiB_2 ceramic throughout the volume radially outward from the penetrator in these same sectioned virtual images.

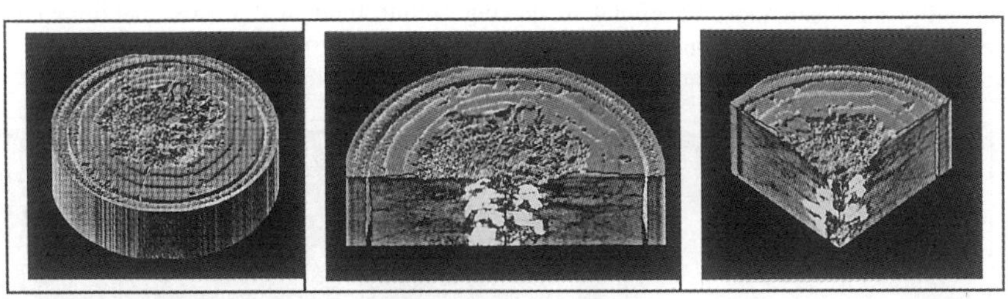

Fig. 5a. 3-D solid image of 72 mm dia. TiB_2 disk

Fig. 5b. 3-D solid image of ½ of the TiB_2 disk

Fig. 5c. 3-D solid image of ¼ of the TiB_2 disk

4 Metal Encapsulated Armor Ceramic Samples

4.1 Ceramic tiles before encapsulation

Ceramic materials of various sizes were procured and are being evaluated for ballistic penetration resistance in the metal encapsulated configuration. CT scanning of as-received SiC ceramic tile samples was conducted and has revealed no mesoscale defects in the as-manufactured condition. Examples of defect-free DR and CT scans for one of these as-received 10cm x 10cm SiC tiles are shown in figure 6.

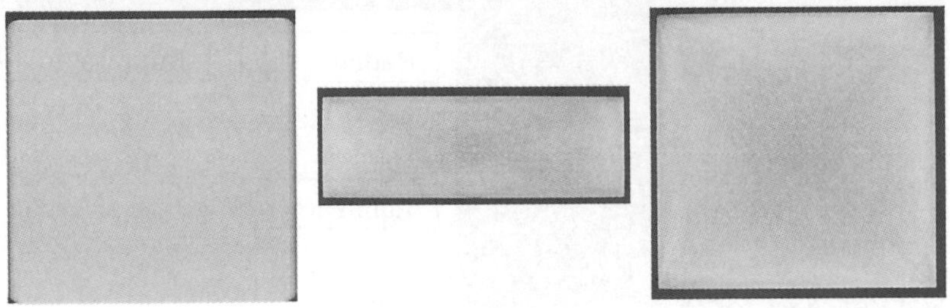

(a) (b) (c)

Fig. 6 Digital Radiograph (a) and Orthogonal XCT Scans (b & c) showing absence of defects in typical as-received SiC tile (10 cm x 10 cm x 1.2 cm).

4.2 As-Fabricated Ceramic Encapsulated Assemblies

Samples of ceramic armor tiles have been individually encapsulated by two methods: (1) cast encapsulation of SiC ceramics in a discontinuously reinforced aluminum (DRA) metal matrix composite (MMC) surround material [7] and, (2) welding and hot isostatic pressing (HIP) in a titanium alloy, Ti-6Al-4V [8]. CT scanning of these encapsulated samples was conducted and has shown several defects in the as-fabricated assemblies. Significant defect conditions were detected and accurately located within encapsulated assembles including: severe cracking in both the SiC tiles and in the DRA MMC material, corner and center cracks in two Al_2O_3 tiles (but not the lower density SiC third tile shown in figure 7b) of one welded/HIP sample, unclosed partial weld seams after HIP'ing, and cases of tears in the Ti-6Al-4V cover plates of additional encapsulated samples. Examples of some of these as-fabricated defects are shown in figure 7. Such defect detection capability has been most beneficial in making architectural and process modification decisions for further specimen fabrications as well as in providing a baseline damage condition prior to ballistic impact for individual specimens.

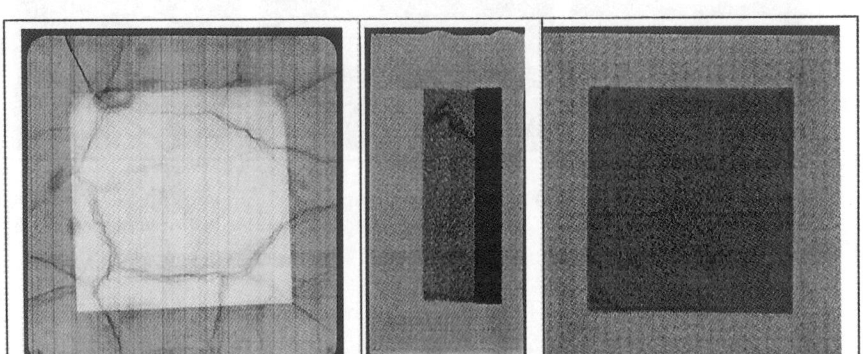

Figure 7. Defects shown via Digital Radiograph (a & b) and CT Scan (c) in the as-cast DRA MMC/SiC (a) and in Weld/HIP Ti-6Al-4V/Al_2O_3/SiC specimens (b & c).

5. Impacted Encapsulated Ceramic Tiles In Situ

Following impact by a sub-scale long rod tungsten alloy penetrator, selective CT scans were conducted to examine the ballistic damage in the encapsulated 25 mm thick SiC ceramic tile. Two CT scans are shown in figure 8 which reveals the existence and location, but decreasing severity, of the interior damage as the rear face is approached.

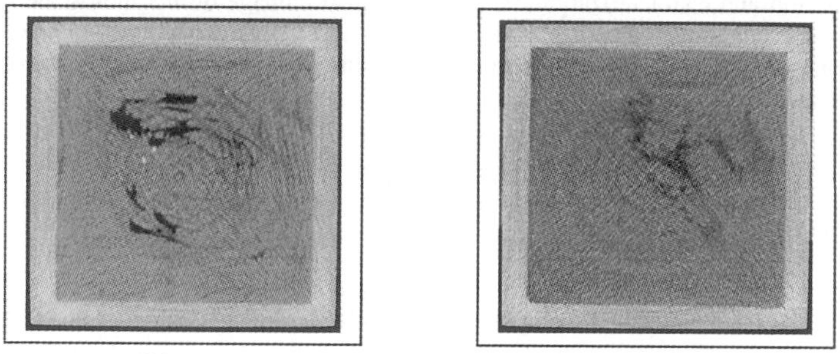

(a) **(b)**

Figure 8. Damage shown via CT Scans in impacted Weld/HIP Ti-6Al-4V/SiC encapsulated specimens, (a) 13mm & (b) 3mm from rear face.

Selected CT scans from initial work on small caliber ballistic impact of polymeric encapsulated ceramic tiles in rigid and flexible body armor specimens are displayed in figure 9.

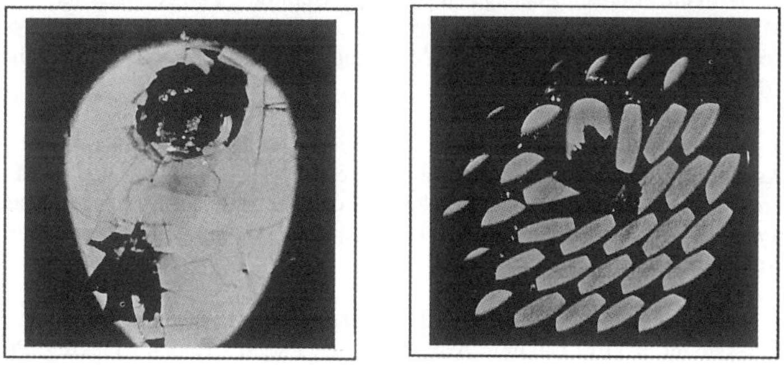

Figure 9. Defects shown via CT Scans of impacted rigid (left) and flexible (right) body armor ceramic specimens.

5 Discussion

This paper presents several recent and striking examples demonstrating both the feasibility and the actual application of x-ray computed tomography to the damage assessment of armor ceramics. Mesoscale internal defects caused by either sample fabrication or by ballistic impact have been detected, located and visualized in situ by 2-D and 3-D CT techniques with spatial resolution of ~200 m and within ceramic sample volumes of 50 cm^3 and larger. Traditional conical, radial and laminar cracking modes were observed in TiC and TiB_2 ceramics. Residual penetrator material was observed and mapped in TiB_2 ceramic. Several forms of as-fabricated defects were observed in early designs of encapsulated armor ceramic tiles that lead directly to architectural design changes. Ballistic damage of impacted encapsulated samples has been captured and visualized as well.

Other armor applications of the CT damage assessment techniques described above are possible and are being explored as part of the ongoing research of the authors. Some of these applications include CT examination of helmets and individual body armor components. Ultimately, the CT methodology may likely evolve into a standard armor quality assurance practice as well as a ballistics research and model development and verification tool. This unique nondestructive method of comprehensive 2-D and 3-D visualization and characterization of in situ physical damage on the mesoscale in armor target materials offers exciting new diagnostic tools and capabilities to the ballistic community and to other physical damage situations such as quasi-static fracture, fatigue and corrosion.

References

1. Kim, J and Liaw, P.K., The Nondestructive Evaluation of Advanced Ceramics and Ceramic-Matrix Composites, November 1998, (vol. 50, no. 11), JOM-e

2. P. Holler et al., eds., *Nondestructive Characterization of Materials* (1988).

3. Green, William H. and Wells, Joseph M., Characterization of Impact Damage in Metallic/Nonmetallic Composites Using X-ray Computed Tomography Imaging, AIP Conference Proceedings 497, pp. 622-629, 1999.

4. Wells, J. M., Green, W. H., and Rupert, N. L., Nondestructive 3-D Visualization of Ballistic Impact Damage in a TiC Ceramic Target Material, Proceedings MSMS2001, 2nd Intn'l Conf. on Mechanics of Structures, Materials and Systems, 14-16 February 2001, University of Wollongong, Wollongong, NSW, Australia, pp 159-165.

5. Green, W. H., Doherty, K. J., Rupert, N. L., and Wells, J. M., Damage Assessment in TiB2 Ceramic Armor Targets; Part I - X-ray CT and SEM Analyses, Proceedings MSMS2001, 2nd Intn'l Conf. on Mechanics of Structures, Materials and Systems, 14-16 February 2001, University of Wollongong, Wollongong, NSW, Australia, pp 130-136.

6. Rupert, N. L., Green, W. H., Doherty, K. J., and Wells, J. M., Damage Assessment in TiB2 Ceramic Armor Targets; Part II - Radial Cracking, Proceedings MSMS2001, 2nd Intn'l Conf. on Mechanics of Structures, Materials and Systems, 14 - 16 February 2001, University of Wollongong, Wollongong, NSW, Australia, pp 137-143.

7. Wells, J. M., Green, W. H., and Rupert, N. L., Pre-Impact Damage Assessment Using X-ray Tomography of SiC Tile Encapsulated in Discontinuously Reinforced Aluminum Metal Matrix Composite, Proceedings ACUN-3, International Composite Conference, 6-9 February, University of New South Wales, Sydney, NSW, Australia, pp549-555.

8. Bruchey, W., Horwath, E., Templeton, D., and Bishnoi, K., System Design Methodology for the Development of High Efficiency Ceramic Armors, Proc. 17th Intn'l Symp. on Ballistics, v.3, Midrand, S. Africa, March 23-27,1988, pp167-174.

SOAKED BEHAVIOUR OF EPS GEOFOAM IN GEOTECHNICAL APPLICATIONS

A. KEATING, C. J. LEO AND Y. ZOU

University of Western Sydney, Locked Bag 1797, South Penrith DC NSW 1797, Australia
E-mail: aaron_spurs@yahoo.com

Expanded Polystyrene (EPS) geofoam is a thermoplastic material (of synthetic origin) used increasingly as an ultra-lightweight fill in geotechnical applications. Its use extends from replacing the low load bearing, highly compressible subgrade soil in road pavements to as a backfill body in the construction of retaining walls and embankments. In these and other sub-surface constructions, the material is subjected to varying moisture conditions. It has been observed in field studies that as much as 9 % moisture pickup by volume occurred in blocks lying permanently under the ground level for 9 years. Moisture impinges on geotechnical design in that it not only increases the unit weight of the EPS geofoam, but may significantly affect the mechanical properties as well. As little by way of experimental evidence is available to form appropriate conclusions, the authors have undertaken a study to investigate moisture effects on the mechanical behaviour of EPS. In the study, EPS samples were prepared and carefully soaked in distilled water for various lengths of time ranging from two weeks to one year. At appropriate points in time, batches of thirty soaked samples were withdrawn and subjected to uniaxial and triaxial compression tests to assess the mechanical behaviour. Preliminary findings show that the volumetric moisture content (VMC) increases with the duration of soaking. It appears that moisture has little effect on the material elastic stiffness or its yield stress, but causes an increase in the plastic modulus.

1 Introduction

In geotechnical applications, Expanded Polystyrene (EPS) geofoam is typically used as a sub-grade and backfill material. Although essentially considered to be hydrophobic, EPS is still known to absorb small quantities of moisture over time. For instance, hydrostatic pressure develops in and around the geofoam blocks immersed below the groundwater level, and drives the moisture into the material. Vaporized water is also absorbed into and condenses in the material under the influence of thermal gradients. Studies done by the Norwegian Road Research Laboratory research have found that the moisture content of the material varies between 9 % in highly wetted regions to 1 % in drier parts [1]. In over twenty years of roadway insulation experience in Alaska, water absorption in EPS was found to be roughly 6 % [4]. BASF [2] has similarly undertaken research into the moisture absorption characteristics of their own product, revealing that the amount of water absorbed after 7 days was in the vicinity of 0.5% to 1%, and after 28 days 1% to 3 %.

The amount and the rate of moisture ingress will depend on a number of controlling factors including the density, surface area and thickness of the EPS, together with the phase of the water (whether liquid or vapor), the submersion depth and length of exposure time. While some data is available on the rate of moisture absorption, its effect on the mechanical properties of EPS such as strength and stiffness are less clear, the prevailing view being that these properties are not affected by the presence of moisture in and around EPS. A long-term study has been initiated by the authors to gather experimental data about the effects on the mechanical properties due to the presence of moisture. This paper presents some of the preliminary findings.

2 Test Program

Uniaxial and triaxial compression tests were performed on dry and moist EPS specimens. A total of 150 uniaxial tests have been conducted over a period of twelve months with thirty EPS specimens tested for each of the 2, 4, 12, 26 and 52 week soaking periods. Triaxial testing of moist EPS was also conducted for confining pressures of 0, 10 and 20 kPa. These tests were completed for 2 and 52 week soaking periods along with dry EPS tests, five for each confining pressure.

RMAX of the Pacific Dunlop Group manufactured and supplied all EPS material used in this investigation. The tested material has a density 20 Kg/m^3 (RMAX M class) [5] because it represents the typical density used for geotechnical applications.

Care was taken to ensure that each test specimen was prepared and handled in exactly the same way to avoid introducing any more variables than there already were. The EPS specimen was cut with a hot-wire cutter to a height of 50 mm from a long core of 50 mm nominal diameter. The dimensions were measured and the specimen weighed using a balance accurate to 0.01g. Each batch of about 240 specimens were then held in 160L water filled drums in a fully submerged state and observed for lengths of time ranging from 2 weeks up to 1 year. The water used was distilled so as to remove the majority of the impurities and avoid effects from contamination. At the appropriate times, the specimens were retrieved from the drums, again being handled exactly in the same fashion, and tested in the laboratory. Before any test, the specimen was inspected for any signs of damage, with the ones considered damaged discarded. After this, the specimens were pat dried for 30 seconds using a paper towel and re-weighed. The difference between the weights before and after soaking accounts for the amount of moisture, which has been absorbed by the specimen.

The loading frame employed in the uniaxial and triaxial testing was the Wykeham-Farrance Tritech 50 kN compression machine. A constant deformation rate of 0.5 mm/min was applied throughout the loading and continued until 20 % strain. The values of the sought after properties: elastic modulus, yield stress and plastic modulus of the EPS specimen were evaluated after plotting the data in an electronic spreadsheet. Figure 1 shows schematically the definitions of these parameters [6].

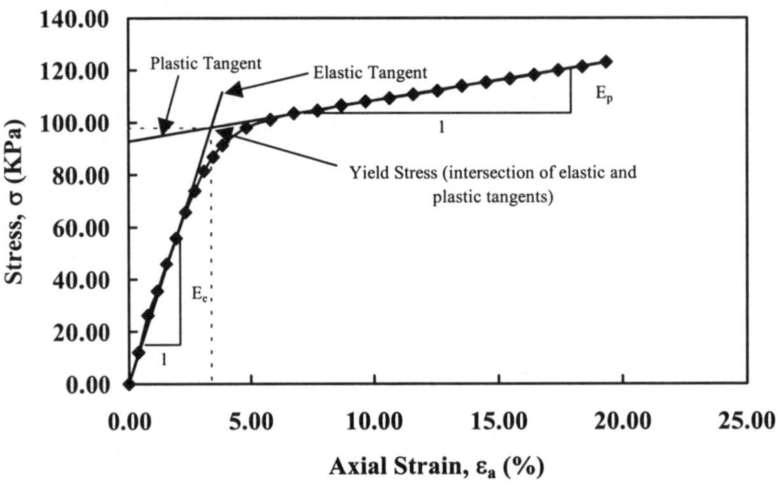

Figure 1 Definitions of Parameters

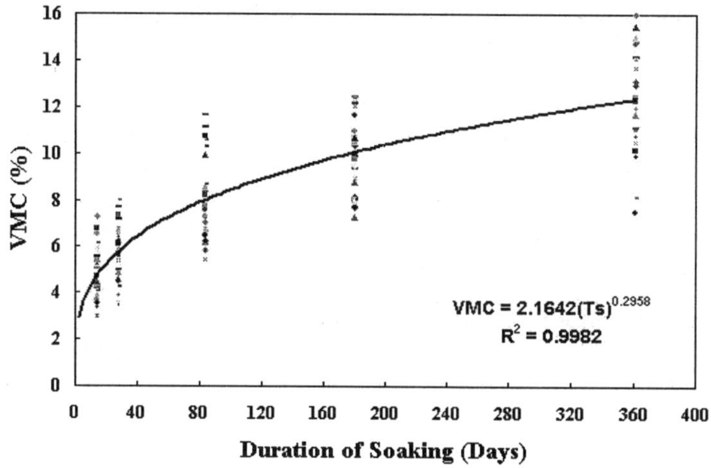

Figure 2 VCM versus Duration of Soaking

3 Test Results

In this paper, the moisture content refers to the volumetric moisture content (VMC). VMC denotes the volume of water within the EPS cells compared to the total specimen volume. This is a better measure of the moisture content of the material than the calculation based on the weight of water compared to weight of the specimen. Figure 2 shows the VMC against the duration of soaking in confined uniaxial compression test. The VMC clearly increases over time in a power-like curve, rapidly in the beginning but more gradually at later times. Because the results for VMC were quite scattered, it was useful to refer to the average VCM of a certain soaking period. Each colored symbol in the figure represents a specific test sample and this representation also applies in subsequent figures.

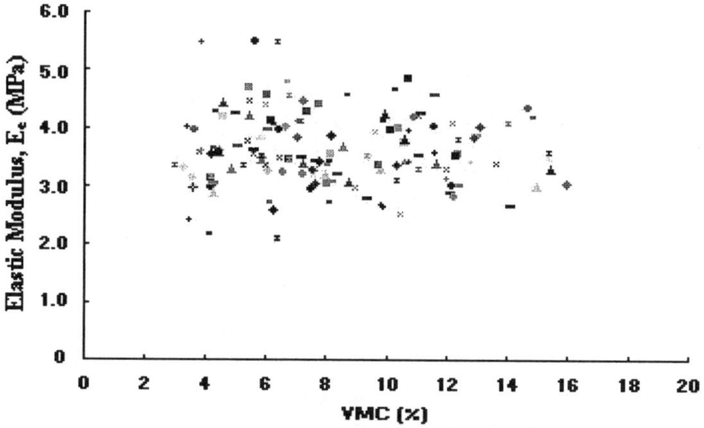

Figure 3 Elastic Modulus versus VMC

Figure 3 and 4 show the elastic modulus and yield stress against the duration of soaking, respectively for the confined uniaxial tests. The scatter in these plots suggests that that there is little dependency between these parameters and the VMC. The effect of moisture on these two properties appears to be negligible.

Figure 5 shows an approximately linear increase in the plastic modulus with VMC under uniaxial loading, which suggests that the plastic modulus of moist EPS is higher than that of dry EPS.

A summary of the triaxial test results including the mean elastic modulus, plastic modulus, yield stress and VMC is given in Table 1. It is also quite evident from referring to the values in the table that the elastic modulus and the yield stress do not experience any discernible change for different VMC as well as confining stress. Due to a lack of space, the plots are not shown.

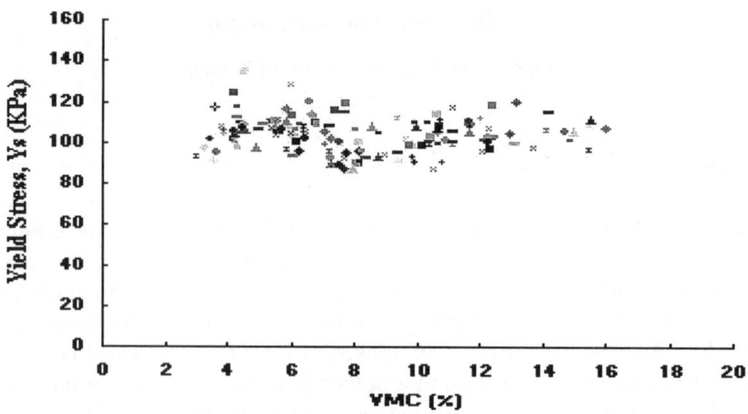

Figure 4 Yield Stress versus VMC

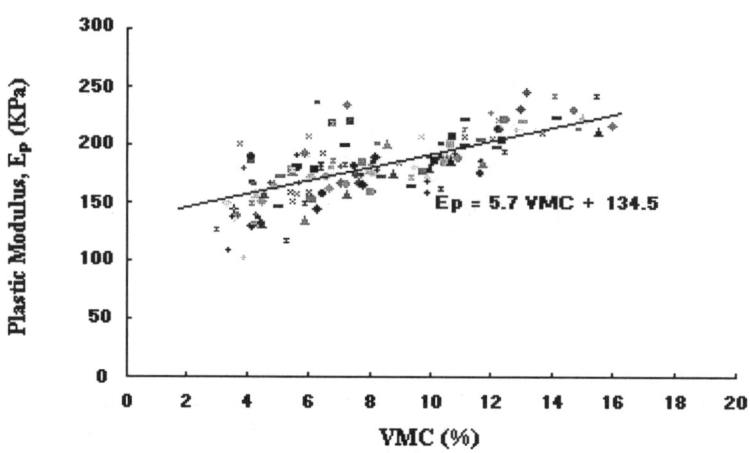

Figure 5 Plastic Modulus versus VMC

Table 1 Summary of Triaxial Test Results

Confining stress (kPa)	Ts[1], (weeks)	Ee[2] (MPa)	Ep[2] (kPa)	YS[4] (kPa)	VMC (%)
0	0	3.42	169	88	0
10	0	3.68	171	90	0
20	0	3.45	169	89	0
0	2	3.56	166	90	4.2
10	2	3.55	169	86	4.2
20	2	3.17	185	81	4.2
0	52	3.65	206	87	12.3
10	52	3.74	211	84	12.3
20	52	3.61	220	84	12.3

1 Ts – time of soaking 2 Ee – elastic modulus
3 Ep – plastic modulus 4 YS – yield stress

4 Discussion of Test Results

It is found that the VMC increases with duration of soaking, rapidly at first and then less gradually at later times. As the EPS is a closed cell structure mostly made up of empty space, during the manufacture, small punctures and imperfections in the wall structure are inadvertently formed. Additionally, service conditions probably exacerbate cell wall cracking, which are randomly distributed throughout the entire body of EPS and formed countless numbers of micro-conduits for the water to pass through. Therefore even though the material is considered to be hydrophobic, some moisture will permeate through the material under hydrostatic pressure applied for a certain length of time. Short-term exposure of EPS to moisture is expected to cause relatively small amounts of moisture absorption because of its hydrophobic qualities but longer exposure (several months or years) results in comparatively large amounts of moisture absorption, in volumetric terms.

In accordance with Norwegian and German field experience [2][3], the VMC was found to approach a point of equilibrium over the duration of soaking when no more moisture may be admitted into the EPS. This is due to the entire cell structure of the material being effectively saturated under normal hydrostatic pressure conditions. During experiments, a number of EPS specimens were split open to reveal the extent of the moisture ingress. Moisture was visibly observed to exit from within and around the cells across the entire cross section of the specimens. The non-zero rate of increase of moisture at the end of 1 year suggests that saturation has not occurred yet in our tests.

The results appear to validate for the most part the contention that the presence of moisture in EPS does not affect the mechanical properties. This is true of the elastic modulus and the yield stress. However, experimental evidence points to some dependency between the plastic modulus and the VMC. The authors suggest that the increase in plastic stiffness is caused by the build up moisture within the EPS replacing the air in the cell structure of the EPS. This effect is felt only at large deformation because initially, it is possible for the moisture to escape into other micro-conduits under applied loading, but eventually a stage is reached where there are few or no more micro-conduits for the moisture to escape into. While air is readily compressible, water is not and the effect results in additional stiffness or increased modulus as the incompressible water being squeezed under loading has little else to escape into.

The results from the triaxial tests generally not only collaborate with those of the uniaxial tests, they also show that the influence of the confining pressure is negligible, i.e. different levels of confinement do not change the assessment of the moisture effects. This is consistent with the findings from Zou and Leo [6] who have reported that confining pressure has little effect on most mechanical properties of dry EPS specimens.

In the instance of cyclic wetting and drying which occurs regularly in soils, the complexity of the problem increases many folds. Whether this phenomenon may only be a temporary short-term nature where if the VMC were to return to its original value, the plastic stiffness correspondingly follow as well, would be of interest. Cyclic wetting and drying is not considered in this investigation due to its complexity. It may feature in future investigations since it is an important consideration.

5 Conclusion

The VMC increases with respect to the duration of soaking, following a power-like curve where the rate of increase is high initially but decreasing gradually to a small value at larger times. Preliminary data based on 1 year data and uniaxial testing appears to support for the most part the contention that the presence of moisture in EPS does not affect the mechanical properties. This is true for the elastic modulus and yield stress, but this assessment may not be accurate for plastic modulus. Experimental data shows that plastic modulus increase with increasing VMC. The same results are generally repeated under triaxial testing, where the test data further suggests that there is little dependency between the measured parameters and the confining pressure. Further testing at larger times are needed to confirm the observed trends. The data gathered so far suggests that saturation or equilibrium point appears to take longer than one year to achieve.

References

1. Aaboe, R. 13 Years of Experience with Expanded Polystyrene as a Lightweight Fill Material in Road Embankments." *Publication No.61, Norwegian Road Research Laboratory*, NRRL, Oslo, Norway, (1987) pp. 21-7.
2. BASF, Plastics Styropor® Expanded Polystyrene, Styrenic Polymers, *Technical Information CD-ROM* (1998): TI 101, TI 110, TI 120e, TI 141, TI 220, TI 610 TI 800 and TI 1-801e.
3. Horvath, J. S. *Expanded Polystyrene (EPS) Geofoam: An Introduction to Material Behaviour* (1993).
4. Negussey, D. *A Report to the Society of the Plastics Industry*. Department of Civil and Environmental Engineering, Syracuse University (1996).
5. R MAX, Pacific Dunlop Group, Isolite® Expanded Polystyrene, *Technical Data*.
6. Zou, Y. and Leo, C.J. (1998) 'Laboratory studies on the engineering properties of expanded polystyrene (EPS) material for geotechnical applications', *Proceedings of the 2nd International Conference on Ground Improvement Techniques*, pp 581-588, 7-9 Oct, Singapore.

CONSOLIDATION OF FINE-GRAINED COMPRESSIBLE SOILS BY VERTICAL DRAINS BASED ON THE FREE STRAIN APPROACH

C.J LEO
Department of Civil Engineering, University of Western Sydney
Locked Bag 1797, South Penrith DC, NSW 1797, Australia
E-mail: c.leo@uws.edu.au

Semi-analytical solutions of free strain consolidation of fine-grained soils by vertical drains are developed in this paper. The solutions obtained are semi-analytical in the sense that the problem is first solved analytically in the Laplace transform domain and the corresponding values in real time and space are then obtained by numerical inversion of the Laplace transform solutions. Solutions are obtained for the distribution of excess pore pressure and degree of consolidation subjected to a single step incremental loading.

1 Introduction

The use of vertical drains has long been regarded as an efficient way to accelerate the consolidation of soft compressible fine-grained soils subjected to surface loading. Experience over several years testifies to the ability of this method to alleviate undesirable post-construction settlement and associated creep problems. The implementation of vertical drain technology is often aided by a theoretical model used to make assessment of the likely rate of soil consolidation and dissipation of excess pore pressure. Reference is often made to the theoretical solutions for equal-strain consolidation developed by Barron [1] and by Hansbo [2]. Recently, Leo [3] also developed a more general analytic solution for equal-strain consolidation that can be conveniently evaluated on an electronic spreadsheet. However, reference to free-strain consolidation, which may be more appropriate for a flexible foundation is rare and no simple solution is available.

A free-strain analytic solution should be attributed to Yoshikuni and Nakanodo [5]. This solution is often mistaken as an equal-strain solution, but a closer examination will reveal it to be a free-strain solution that combines the effects of vertical and radial drainage, and a well with no well smear. Although analytic in form, the solution is considered unwieldy and more difficult to evaluate than the solutions of [1], [2] or [3]. The problem lies in finding the roots of the Fourier-Bessel function equations, which clearly require some kind of iterative procedures to solve. At the practical level the solution can be regarded as a semi-analytic rather than an analytic one.

This paper introduces a simpler semi-analytic solution of free-strain consolidation with an approach that uses an integral transform, the Laplace transform. This technique eliminates the time-variable via the Laplace transform and then solves the free-strain problem analytically in transform space, yielding a solution expressed as a series expansion of the sine functions. The efficient and highly recommended inversion algorithm of Talbot [4] is then used to give the final solutions in real space and time.

2 Development of Solutions

In the field, this technology requires the installation of a network of vertical drains in the consolidating fine-grained soil layer. For the purpose of analysis, consider a typical consolidating cylindrical cell with a vertical side-wall within the network of vertical drains (Figure 1). The center of the cylindrical cell has a drainage well of radius r_w surrounded by a zone of remoulded (smeared) soil bounded by an outer radius r_s which is in turn surrounded by the undisturbed soil extending to a radius of influence r_e.

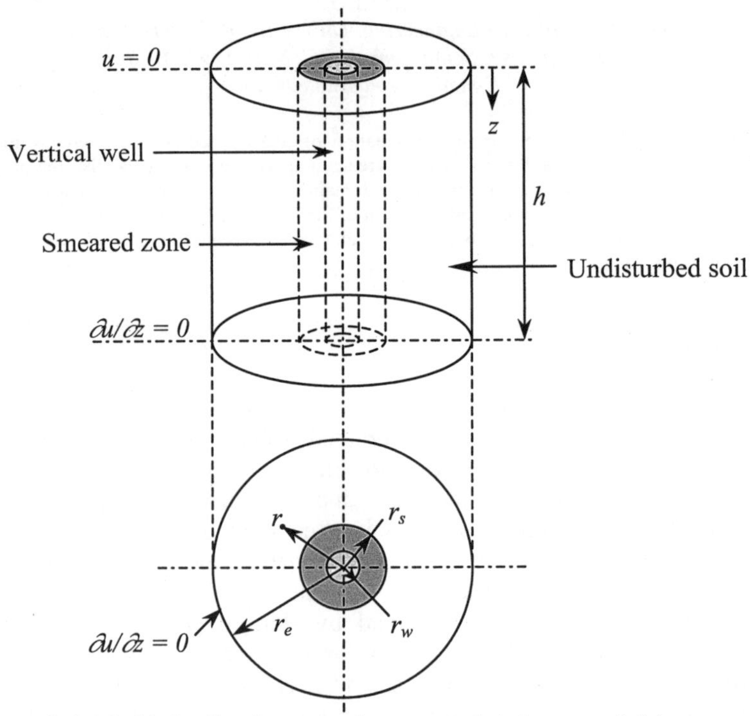

Figure 1. A typical cylindrical cell unit representing a vertical drain surrounded by smeared zone and undisturbed soil.

The soil has a layer thickness h, it is freely draining at the top and impermeable at the bottom. The continuity equation in the consolidating soil gives [1],

$$\frac{k_h}{\gamma_w}\left(\frac{\partial^2 u(r,z,t)}{\partial r^2}+\frac{1}{r}\frac{\partial u(r,z,t)}{\partial r}\right)+\frac{k_v}{\gamma_w}\frac{\partial^2 u(r,z,t)}{\partial z^2}=-\frac{\partial \varepsilon_z(r,z,t)}{\partial t} \tag{1a}$$

where $u(r,z,t)$ is the excess pore-water pressure, t is the elapsed time, k_h is the horizontal soil permeability, k_v is the vertical soil permeability in the soil, γ_w is the unit weight of the fluid, r is the radial distance from the center of the cell, $\varepsilon_z(r,z,t)$ is the vertical strain. A typical cell would be very much constrained on its vertical sides due to the presence of surrounding neighbouring cells. The strain is related to the effective stress as follows,

$$\varepsilon_z(r,z,t)=m_v\left[q(t)-u(r,z,t)\right] \tag{1b}$$

where m_v is the coefficient of volume compressibility of the soil, $q(t)$ is the spatially uniform applied surface loading. It is to be noted that in the case of free strain, no arching is assumed and thus the effect of the applied surcharge on the soil directly corresponds to the surface loading. To solve the free strain consolidation problem, it is useful to introduce a finite Fourier sine transform defined by,

$$u(r,z,t)=\sum_{n=0}^{\infty}U_n(r,t)\sin\omega_n z \tag{2a}$$

$$q(t)=\sum_{n=0}^{\infty}Q_n(t)\sin\omega_n z \tag{2b}$$

where $\omega_n = (2n+1)\pi/(2h)$ is chosen so that the drainage boundary conditions (i.e. pore pressure) are satisfied rigorously in the z spatial direction. If equation (2) were substituted into equation (1) the following equation written for each of the Fourier terms results,

$$c_h\left(\frac{\partial^2 U_n(r,t)}{\partial r^2}+\frac{1}{r}\frac{\partial U_n(r,t)}{\partial r}\right)-c_v\omega_n^2 U_n(r,t)=\frac{\partial U_n(r,t)}{\partial t}-\frac{\partial Q_n}{\partial t} \tag{3}$$

where $c_h = k/(m_v\gamma_w)$, $c_v = k_v/(m_v\gamma_w)$ are the coefficients of consolidation in the horizontal and vertical directions respectively. The solution process is further facilitated by the introduction of the Laplace transform defined typically as

$$\overline{U}_n(r,s)=\int_0^{\infty}U_n(r,t)e^{-st}dt \tag{4}$$

where the superior bar on the left hand side indicates the transformed variable and s is the Laplace transform parameter. It follows after taking the Laplace transform of equations (2) and (3) that,

$$\overline{u}(r,z,s)=\sum_{n=0}^{\infty}\overline{U}_n(r,s)\sin\omega_n z \tag{5a}$$

$$\overline{q}(s)=\sum_{n=0}^{\infty}\overline{Q}_n(s)\sin\omega_n z \tag{5b}$$

$$c_h\left(\frac{\partial^2\overline{U}_n}{\partial r^2}+\frac{1}{r}\frac{\partial\overline{U}_n}{\partial r}\right)-c_v\omega_n^2\overline{U}_n = s\left(\overline{U}_n-\frac{U_{0n}}{s}\right)-s\left(\overline{Q}_n-\frac{Q_{0n}}{s}\right) \tag{5c}$$

in which U_{0n}, Q_{0n} are the Fourier components of the initial spatially uniform excess pore pressure and the applied surcharge respectively at $t = 0$. The assumption that all vertical loads are initially carried by the excess pore water pressure ensures that $U_{0n} = Q_{0n}$. Rearranging equation (5c) gives,

$$\frac{\partial^2\overline{U}_n}{\partial r^2}+\frac{1}{r}\frac{\partial\overline{U}_n}{\partial r}-\mu_n^2\left(\overline{U}_n-\phi_n\right)=0 \tag{6}$$

where,

$$\mu_n^2=\frac{\left(s+c_v\omega_n^2\right)}{c_h} \tag{7}$$

$$\phi_n=\frac{s\overline{Q}_n}{s+c_v\omega_n^2} \tag{8}$$

It may be noted that the general form of ϕ_n as given in equation (8) depends on the loading sequence which hitherto has not been addressed. In this paper, the loading is assumed to be a single incremental loading, that is $q = q_0$ when $t > 0$ thus,

$$\overline{Q}_n = \frac{4q_0}{s(2n+1)\pi} \tag{9}$$

Now, the general solution to equation (6) for the consolidating soil is,

$$\overline{U}_n = c_{1n} I_0(\mu_n r) + c_{2n} K_0(\mu_n r) + \phi_n \tag{10}$$

where I_0, K_0 are the modified Bessel functions of first and second kind, zero order respectively, and c_{1n}, c_{2n} are constants of integration the values of which are to be determined.

2.1 *Remoulded Zone*

As in [1], the consolidation in the remoulded (smeared) zone is ignored and the radial drainage equation gives,

$$\frac{\partial^2 u'}{\partial r^2} + \frac{1}{r}\frac{\partial u'}{\partial r} = 0 \tag{11a}$$

where u' is the excess pore water pressure in the smeared zone. A Laplace transform then yields,

$$\frac{\partial^2 \overline{u}'}{\partial r^2} + \frac{1}{r}\frac{\partial \overline{u}'}{\partial r} = 0 \tag{11b}$$

Let $\overline{u}' = \sum_{n=0}^{\infty} \overline{U}_n'(r,s) \sin \omega_n z$ and substituting into equation (11b) gives,

$$\frac{\partial^2 \overline{U}_n'}{\partial r^2} + \frac{1}{r}\frac{\partial \overline{U}_n'}{\partial r} = 0 \tag{11c}$$

where the general solution is,

$$\overline{U}_n' = c_{3n} \ln r + c_{4n} \tag{12}$$

and c_{3n}, c_{4n} are the as yet undetermined constants of integration.

2.2 *Well Resistance*

Ignoring consolidation and radial drainage, the equation of flow in the well is given by,

$$\frac{\partial^2 u_w}{\partial z^2} + \frac{2}{r_w}\frac{k_s}{k_w}\left(\frac{\partial u'}{\partial r}\right)_{r=r_w} = 0 \tag{13}$$

and taking the Laplace transform gives,

$$\frac{\partial^2 \overline{u}_w}{\partial z^2} + \frac{2}{r_w}\frac{k_s}{k_w}\left(\frac{\partial \overline{u}'}{\partial r}\right)_{r=r_w} = 0 \tag{14}$$

Let $\bar{u}_w = \sum_{n=0}^{\infty} U_{wn}(r,s)\sin\omega_n z$ and substituting into equations (14) while using equation (12) leads to,

$$-\omega_n^2 U_{wn} + \frac{2}{r_w^2}\frac{k_s}{k_w} c_{3n} = 0 \tag{15}$$

It only remains to determine the values of the coefficients c_{1n} and c_{2n} to complete the solution for the excess pore pressure.

2.3 *Solution of Excess Pore Pressure*

The values of the c_{1n} and c_{2n} are found by imposing the continuity conditions (excess pore pressure and radial flow) at the interfaces and the outer boundary. At $r = r_w$,

$$\bar{U}_{wn} = \bar{U}_n' = c_{3n}\ln r_w + c_{4n} \tag{16}$$

and after substituting into equation (15) yields,

$$c_{4n} = \Omega_n c_{3n} \tag{17}$$

where,

$$\Omega_n = \left(\frac{2}{\omega_n^2 r_w^2}\frac{k_s}{k_w} - \ln r_w\right) \tag{18}$$

At $r = r_s$, the continuity in radial flow and excess pore pressure result in,

$$c_{3n}\ln r_s + c_{4n} = c_{1n}I_0(\mu_n r_s) + c_{2n}K_0(\mu_n r_s) + \phi_n \tag{19}$$

$$-\frac{k_s}{\gamma_w}\frac{c_{3n}}{r_s} = -\mu_n\frac{k_h}{\gamma_w}\{c_{1n}I_0'(\mu_n r_s) + c_{2n}K_0'(\mu_n r_s)\} \tag{20}$$

Rearranging (20) gives $c_{3n} = \mu_n r_s \dfrac{k_h}{k_s}\{c_{1n}I_0'(\mu_n r_s) + c_{2n}K_0'(\mu_n r_s)\}$, which after substituting into (19) leads to,

$$a_n c_{1n} + b_n c_{2n} = \phi_n \tag{21}$$

where,

$$a_n = \mu_n r_s \frac{k_h}{k_s}(\Omega_n + \ln r_s)I_0'(\mu_n r_s) - I_0(\mu_n r_s) \tag{22a}$$

$$b_n = \mu_n r_s \frac{k_h}{k_s}(\Omega_n + \ln r_s)K_0'(\mu_n r_s) - K_0(\mu_n r_s) \tag{22b}$$

At the outer boundary, $r = r_e$, the radial flow is zero so that,

$$-\frac{k_h}{\gamma_w}\frac{\partial \bar{U}_n}{\partial r} = 0 = -\mu_n\frac{k_h}{\gamma_w}\{c_{1n}I_0'(\mu_n r_e) + c_{2n}K_0'(\mu_n r_e)\} \tag{23}$$

i.e.

$$e_n c_{1n} + f_n c_{2n} = 0 \tag{24}$$

where,

$$e_n = I_0'(\mu_n r_e) \qquad\qquad f_n = K_0'(\mu_n r_e) \tag{25}$$

Solving equations (21) and (24) simultaneously gives,

$$c_{1n} = \frac{f_n \phi_n}{(a_n f_n - b_n e_n)} \qquad\qquad c_{2n} = \frac{-e_n \phi_n}{(a_n f_n - b_n e_n)} \tag{26}$$

Hence the solution for the excess pore pressure is completely determined.

2.4 Degree of Consolidation

For the single stepwise loading, the overall degree of consolidation of the entire cell with respect to the excess pore pressure is defined as,

$$\overline{U} = \frac{1}{s} - \frac{\frac{1}{\pi(r_e^2 - r_s^2)h} \int_0^h \int_s^{r_e} \overline{u}(r,z) 2\pi r dr dz}{q_0} = \frac{1}{s} - \frac{1}{q_0} \sum_{n=0}^{\infty} \Phi_n \tag{27a}$$

where,

$$\Phi_n = \frac{2}{\omega_n h \mu_n^2 (r_e^2 - r_s^2)} \{c_{1n} [\mu_n r_e I_1(\mu_n r_e) - \mu_n r_s I_1(\mu_n r_s)] - c_{2n} [\mu_n r_e K_1(\mu_n r_e) - \mu_n r_s K_1(\mu_n r_s)]\} + \frac{\phi_n}{\omega_h h} \tag{27b}$$

3 Conclusion

Semi-analytic solutions of coupled free-strain consolidation by use of vertical drains have been developed in the present paper. The effort to find the solutions is not more difficult than required for the equal-strain analytical solutions once the inversion algorithm of Talbot [4] is used.

4 References

1. Barron R.A. (1948) 'Consolidation of fine-grained soils by drain wells. Trans. ASCE, Vol. 113, Paper No. 2346.
2. Hansbo S. (1981) 'Consolidation of fine-grained soils by prefabricated drains. Proc. of the 10th Int Conference on Soil Mechanics and Foundation Engineering, Stockholm, **3**, pp677-682.
3. Leo (2000), 'Equal strain consolidation of fine-grained compressible soils by vertical drains', Engineering Research Report, UWS
4. Talbot A. (1979). The accurate numerical integration of Laplace transforms. *J. Inst. Math. App.*, **23**, 97-120.
5. Yoshikuni H. and Nakanodo H. (1974), 'Consolidation of soils by vertical drains with finite permeability', Japanese Society of Soil Mechanics and Foundation Engineering, Vol. 14, No. 2, pp35-46.

RECENT DEVELOPMENTS OF MINING MACHINERY AND THE IMPROVEMENTS OF PRODUCTIVITY

M. D. KURUPPU

Curtin University of Technology, Locked Bag 22, Kalgoorlie, WA 6433, Australia

E-mail: M.Kuruppu @wasm.curtin.edu.au

Mobile fleet of mining equipment is one of the highest cost contributors in a mining operation. This paper describes various advancements of haulage methods that reduce the unit cost while improving safety of mining. Proper maintenance practices are of paramount importance to improve the reliability and the availability of equipment. The performance analysis of a few mines is given in relation to key performance indicators that describe the availability and the utilisation of equipment. Some benchmarks for equipment performance based on achievable targets are suggested and compared with the actual performance.

1 Introduction

Australia is a big producer of gold, steel, base metals, coal and many other minerals. The annual earning from the export of mineral resources stands at about 30 billion Australian Dollars. Considering the low prices paid for raw materials worldwide Australia's mining and metallurgical sector has been doing well and remains profitable thanks to continuous reduction of costs that improves the efficiencies of mining. Adequate rewards in most sectors of mining have attracted a highly skilled workforce. Ready availability of contractors in most mining towns has further improved the strategic flexibility of the industry.

The success of the Australian mining industry must be valued in comparison with the slow growth in other major mineral producing centres such as South Africa, Canada and the U.S. In order to remain competitive with the rest of the world, the Australian mining industry is adopting innovative methods to reduce costs. Some of the improvements relate to the introduction and the developments of new technology. For example, several Australian companies have developed world-class mine planning and information systems which are used by the mining industry worldwide. Others relate to the operational and the maintenance efficiency. These systems allow to improve the equipment availability and utilisation so that the productivity can be increased and costs reduced without compromising safety of mining.

2 Recent advances of fleet control and haulage

Some of the major new technologies that have been introduced with the aim of further optimising the operations and to improve safety are:

- Autonomous operation of equipment such as load haul dump (LHD) trucks in underground mines
- Computerised truck dispatch systems
- Trolley-assist for speeding up the haul trucks in open pits
- Equipment condition and performance monitoring

- Advanced mine planning and optimisation computer codes

Remote control systems for underground truck transportation are now commercially available. However, a number of research organisations are engaged in further developing these systems [1]. Computerised truck dispatch systems have been developed to minimise the waiting time of trucks during their travel between loading points and dumps. They usually make use of Global Positioning System (GPS) or field beacons to locate truck positions. A number of advanced dispatch control packages are available commercially. The trolley-assist mode of truck transportation uses external electrical power as and when required for hauling. The resulting faster truck speeds translate directly into higher productivity. Due to the high cost of diesel fuel in comparison to electricity, a number of large Australian surface mining operations are currently interested in this option. Most mining trucks are fitted with on-board computers and an array of sensors. Data is continually logged to determine the equipment condition, and its performance such as the load and the speed [2]. For example Caterpillar company has developed the VIMS™ system to monitor the condition of trucks [3]. Novel designs of trucks to minimise tyre wear and to increase equipment availability are also available [4].

3 Improvements in maintenance practices

It has been found that nearly 50 percent of the costs of a typical open-pit mining operation is maintenance related [5]. Therefore, improvements in maintenance management are likely to reap good benefits. Objectives of the maintenance function must closely follow objectives of the mining operation. A good maintenance organisation may have the following favourable features.

- Adequate maintenance infrastructure: Supply of all necessary facilities such as buildings, repair bays, repair equipment and maintenance personnel. Inventory of spares should be carefully planned to make sure that the required parts are readily available. In the case of remote operations (e. g. those in country Western Australia and some overseas locations) adequate stocks should be maintained depending on how long it would take to receive goods ordered.
- Planning and scheduling: All work must be planned and scheduled in advanced. For example, most preventive maintenance jobs can be planned. This ensures availability of all necessary resources at the time of performing maintenance. This function can make use of the computerised maintenance management systems (CMMS) that are widely available.
- Job order system: This is the main communication medium from the point of raising the job order by the maintenance supervisor until the job is completed and the job order sent to the financial division for costing.
- Co-operation with production/operation personnel: For the maintenance function to be successful, full co-operation must be obtained from the operations division. For example, equipment should preferably be taken away for maintenance only during scheduled work breaks. Furthermore, operator assistance is necessary to carry out minor maintenance functions such as checking fuel and lubrication levels, maintaining correct tyre pressures etc. Also proper education and training of operators will ensure minimum wear and tear of equipment during usage.

- Optimisation of maintenance: Optimum usage of available resources such as maintenance personnel, availability of spares and inventory control, and adopting a predictive maintenance philosophy will minimise overall costs and maximise the effectiveness.

4 Maintenance performance indicators

A good maintenance practice is favourably reflected in the availability of the fleet of equipment. The availability as well as the effective utilisation of equipment in turn contributes to the productivity of the operation. In this study, a set of key performance indicators (KPI) are used to measure the availability and utilisation of a number of operations. Moreover, the KPI s can be used as benchmarks in order to compare the relative performance of similar operations [6-8]. A suggested set of performance indicators is given in Table 1. Many companies have adopted these benchmarks and use them regularly for performance monitoring. The information required to determine these performance parameters can be obtained from equipment operating and downtime log, and mine work hours. Table 2 illustrates how a particular KPI can be used for performance evaluation and control [9]. Tables 3 and 4 give performance indicators for four operations. Three of them are mining operations situated in Western Australia and the other is a large quarry operation in Sri Lanka. The latter operation is also managed by an Australian company. Data was gathered over a period of one month.

Operation A is a large open pit gold operation working round the clock in two shifts a day. Its location is close to a major mining town having all necessary facilities. A contractor who owns equipment does mining. Most aspects of the operation have been optimised using the latest available technologies such as computerised equipment dispatch aided by GPS, and pit optimisation software. Equipment maintenance is performed to a high standard by well-trained staff.

Operation B is a fairly large underground gold operation, which uses only in-house resources and has no major contractor to work on the site. Attention is paid to equipment maintenance that is up to a high standard. Being an underground operation, care must be exercised in comparing efficiencies with those of surface mining operations although the benchmarks are applicable equally.

Operation C is a medium size open pit hard rock mining operation located far from major cities and towns. Mining is done by a company-owned fleet of equipment and using in-house labour. This operation seems to have difficulty in keeping trained staff. High labour turnover affects all its operations unfavourably. D is a quarry operation located in Sri Lanka. It is a 12-hour daily operation. The absolute utilisation is the same as the effective utilisation for this particular period and therefore not given.

The operation A has optimised its performance as evidenced by a mechanical availability of 87 percent and use of availability of 90 percent. A time analysis of the fleet of trucks shows that the equipment stand by time has been only 7 percent of the operating time. The low stand by time is mainly attributed to the advance truck dispatch system being used.

Operation B shows an overall mechanical availability of 76 percent and use of availability of 38 percent. These figures indicate that the company has a poor equipment utilisation strategy although it takes reasonably good care of the equipment. The fact that the spares and specialised repair facilities are not readily available due to the remoteness of the location is a likely reason for over equipping the mine. However, in this case the losses due to the under utilisation seems to be quite high.

Table 1. Factors for mining performance - definitions and benchmarks. Notation: W - operating hours, R - total maintenance hours, S - stand-by hours, T - total scheduled hours, C - calendar hours, m - maintenance man-hours, W_p - planned maintenance hours.

Performance factor and description	Definition	Benchmark for mining performance
Mechanical Availability: measure of efficiency of maintenance	W/(W+R)	exceeds 85 to 90 percent through machine economic life
Physical Availability: measures the operational availability	(W+S)/T	not applicable
Use of Availability: measures the efficiency of operation and planning	W/(W+S)	exceed 90 percent of total scheduled hours
Effective Utilisation: measure of the overall efficiency of operation based on total scheduled hours	W/T	not applicable
Absolute Utilisation: measure of the efficiency of operation based on calendar hours	W/C	not applicable
Maintenance Manpower: maintenance man-hours per operating hour	m/W	0.25 to 0.30
Maintenance Efficiency: planned maintenance hours as a percentage of total down time	W_p/R	80 to 85 percent

Table 2. Example of the use of the performance indicator

Information	Description
KPI Title	Use of Availability
Definition	The ratio of work hours and the equipment available hours
Purpose	To measure the efficiency of operation and planning
Relates to	Efficiency of operation
Target	>90%
Formula	Operating hours/total of operating and stand-by hours
Frequency	monthly
Person to measure	Operator, supervisor
Sources of data	Equipment time sheets, downtime records
Person to take action	Maintenance manager
Description of action	If this falls below 90% operating procedures, maintenance records, failure or failure mode etc. must be investigated and corrective action taken

Table 3. Performance indicators (percentage) for the three Australian operations.

Note: standard deviations given in brackets

	Mechanical Availability	Physical availability	Use of availability	Effective utilisation	Absolute utilisation
Operation A					
trucks	89 (5.5)	90 (5.1)	92 (3.2)	83 (6)	83 (6)
loaders	84 (9)	87 (7.2)	87 (14)	75 (15)	75 (15)
drills	85 (9)	85 (10)	84 (6)	72 (13)	72 (13)
overall	87 (7.5)	88 (7.3)	90 (8.2)	79 (11)	79 (11)
Operation B					
trucks	72 (12)	85 (8)	44 (9)	37 (8)	37 (8)
loaders	76 (6)	86 (6)	47 (15)	41 (12)	41 (12)
others	80 (10)	94 (3)	27 (11)	25 (10)	25 (10)
overall	76 (10)	89 (7)	38 (14)	34 (11)	34 (11)
Operation C					
trucks	80 (7)	87 (5)	65 (5)	56 (6)	55 (6)
excavator	94	95	93	88	87
others	93 (2)	96 (1.9)	65 (9)	62 (8.3)	61 (8)
overall	86 (8.6)	90 (6.3)	67 (10)	61 (11)	60 (11)

Table 4. Performance indicators for a quarry operation in Sri Lanka **(Operation D)**

Equipment	Mechanical availability (%)	Physical availability (%)	Use of availability (%)	Effective utilisation (%)
Loaders	96	95	91	90
Excavators	95	95	85	81
Air compressor	84	88	69	61
Crushing plant	73	82	59	49
Generator	95	97	67	65

Operation C has an overall mechanical availability of 86 percent and the use of availability of 67 percent. The figures show that the fleet of equipment is under utilised although the availability has been good. Only the excavator achieves the benchmarks with the mechanical availability of 94 percent and the use of availability of 93 percent.

Operation D has an overall mechanical availability of 90 percent and a use of availability of 78 percent. The mechanical and the physical availability as well as the utilisation are good for the fleet of mobile equipment, which is quite new. The availability of the 12-year old crushing plant is generally satisfactory although there is room to improve its utilisation. This analysis shows that the management should make every effort to increase the utilisation of the plant. It may require further analysis into operator efficiency, operator training, any capacity mismatch between the quarry operation and the crushing plant, and other operational difficulties. Being a 12-hour operation, some of the maintenance can be done during off-shift hours and it helps to keep the availability high whereas this is not applicable for operations A, B and C that are 24-hour operations.

The ratio of planned equipment maintenance to the total manpower expenditure for all equipment maintenance in operation A was 59 percent and that for operation C was 55 percent. Both figures are well below 80 percent, which is the benchmark. The companies appear to be doing all necessary time-based preventive maintenance tasks. However, there is definite room for improvement in predictive maintenance that will enhance the effectiveness of planned maintenance. Most operations are yet to embrace the full potential of predictive maintenance as well as other emerging strategies such as Reliability Centered Maintenance (RCM). An analysis of maintenance manpower revealed that for operation C, the ratio of maintenance man-hours per operating hour is 0.37 that is not far from the desired benchmark of 0.25. Similar data for other operations were not available.

5 Conclusions

Australian mining industry has been adopting new technologies that reduce costs and make operations more efficient. Some of these technologies are: computerised mine-planning systems, computerised maintenance management systems, and computerised truck dispatch systems. Several more technologies such as trolley-assisted haulage systems, and steep angle conveying systems are closely investigated by a number of large mining operations.

Four typical mining/quarrying operations were investigated to determine the equipment performance and to compare the performance indicators with some suggested benchmarks. Many of the operations were found to have achieved satisfactory performance in equipment availability. However, utilisation indicators were low for most operations. Furthermore, the unplanned equipment maintenance was found to be well above its target of 20 percent revealing the inadequate use of predictive maintenance as well as other emerging maintenance strategies.

References

1. CMTE, Automation of mining equipment program, Cooperative Research Centre for Mining Technology and Equipment, St. Lucia, Qld.
2. Golosinski, T.S., Advances in automation and robotization of open pit mining. In *Mining in the New Millennium*, ed. by Golosinski (Balkema, 2000) pp. 145-152.
3. Caterpillar Inc., (2000) Company product information at www.cat.com.
4. Liebherr Inc., (2001) Mining trucks at www.tirupati-international.com.
5. Tomlingson, P.D., Maintenance management information systems (Maclean Hunter, Chicago, 1985).
6. Wireman, T., Developing performance indicators for managing maintenance (Industrial Press, Inc., New York, 1998).
7. Holman, P., Driving down haulage costs. In *Proc. Annual general meeting of the Soc. For Mining, Metallurgy and Exploration*, Reno, (1993) Preprint No. 93-257.
8. Kuruppu, M.D. and Golosinski, T.S., Maintenance practices of mining machinery - A Western Australian Perspective. In *Mine Planning and Equipment Selection*, ed. by R.K.Singhal (Balkema, 1998) pp. 607-612.
9. Neely, A., Richards, H., Mills J., Platts, K. and Bourne, M., Designing performance measures: a structured approach. *Int. J. of Operations and Production Management.* **17** (1997) pp. 1131-1152.

ENERGY ABSORBING ROADSIDE POSTS

PAUL BRIOZZO & ANDREI LOZZI
School of Aerospace, Mechanical & Mechatronic Engineering
University of Sydney

This paper provides an overview of the energy and crush space required to provide survivable impacts between motorcars and energy-absorbing poles. In addition this paper considers some of the constraints that would apply if those poles or posts would be required to be remaining upstanding and serviceable after the crash. A simplified acceleration/ displacement pulse for the combined car and pole is used in all the modeling described here. This pulse is arrived at using results from full-scale tests and FEA simulations for impacts from 60 km/h and for a maximum acceleration of 20 g. These limitations and the shape of the pulse give us the total crush depth for any given collision energy. The depth into the car, at the point when 20 g is reached is arrived at using published analysis of full-scale car-to-pole tests and it is a function of the car's mass, width and length, and the pole diameter. The division of crush depth between the pole and car is then estimated using the assumption that the pole can be designed to crush from the time that the car reaches 20 g acceleration. The results show that the poles that bring about 20 g acceleration in the car can be provided only for a narrow range of car masses. A pole that suits medium sized cars will cause reduced acceleration to heavier cars, to the point that it may be overrun and demolishing by much heavier cars. The same pole will cause higher accelerations for cars that are lighter than the chosen mass, to the point that it may be equivalent to a rigid pole for very light cars.

1 Introduction

Rigid poles (and trees) because of their extreme rigidity absorb effectively no energy when collided onto by motorcars, consequently the car structure absorbs nearly all of the energy. Breakaway poles because of their built-in low resistance to transverse loads also absorb essentially no energy, but neither does the car. Collisions with such poles result in an uncontrolled car and a fallen pole, thereby leading to the possibility of secondary impacts by the moving vehicle or by the pole. Wrap-around energy-absorbing poles on the other hand are typically rolled onto the ground by the car, absorbing nearly all of the vehicle's energy. They are usually intended to cause low accelerations to the car, resulting in little damage to it, but as a consequence they require several meters of space to bring the car to a halt. A fourth type of pole, one that absorbs a significant amount of energy while remaining off the roadway and reasonably upright, would have properties somewhere between the rigid and the wrap-around type.

2 Modelling of Car-to-Pole Collisions

Wood et al [1] reviewed the results of 202 barrier tests and 19 pole impacts and derived a model for some of the main outcomes. The model showed how the volume of the car crushed by the impact is a function of parameters of the car and pole diameter. Wood's results showed a distribution about the line of best fit for the cars tested of ±30%, with the standard deviation being 15%. This distribution may be viewed as a combination of uncertainty in modelling and variations between cars. Wood simplified the force-intrusion pulse to two adjacent rectangles and referred to these as the first and second stages of the crash. These are shown on Figure 1a). The first stage generates low accelerations and is due to components ahead of the engine crushing. This initial displacement is roughly 300 mm and absorbs 10% of the total energy. Large oscillations occur in this region, but because of their insignificance in moderate to high-energy impacts we represent this region by a rectangle. Energy may be absorbed over the shortest distance if the force is maintained at the highest level over the whole of the crush depth represented by a rectangular pulse. It has not been possible to maintain this force constant. FEA simulations using LS-DYNA [2] with deformable poles show an achievable pulse is one where acceleration is constant for only 30% of the high-energy stage. This results in the same maximum acceleration but a displacement that is 60% longer than that for a rectangular pulse.

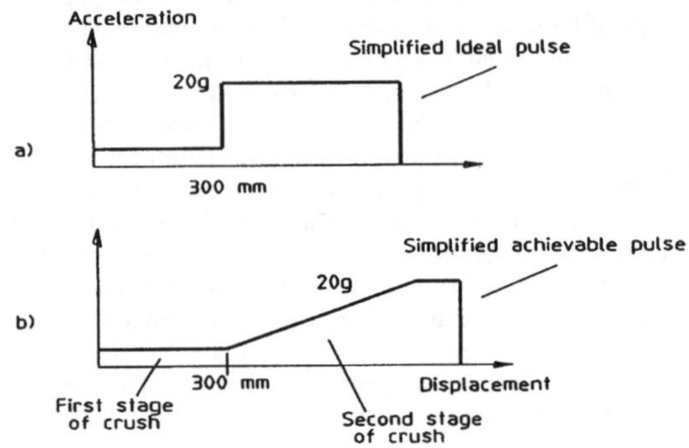

Figure 1 a) & b). Simplified acceleration/displacement pulses. a) A rectangular ideal outline, showing the two stages of crush. b) A simplified but achievable outline, with the same energy and maximum accelerations as in a).

3 Estimates of Intrusions and Accelerations

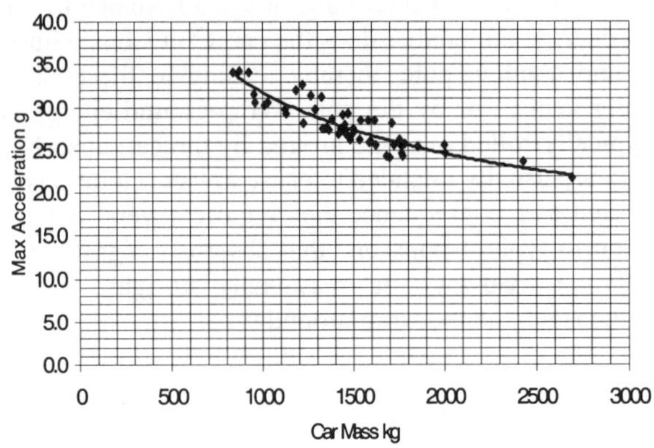

Figure 2 Maximum acceleration versus car mass for impacts from 60 km/h onto a 500 mm diameter pole.

Figure 2 provides a plot of maximum acceleration against car mass. For Figure 2 and all remaining figures the impacts are simulated from 60 km/h onto 500 mm diameter poles. The data points shown on this and following graphs are the result of applying the weight, width and length of 55 sedans of 10 years of age or less to Wood's analysis.

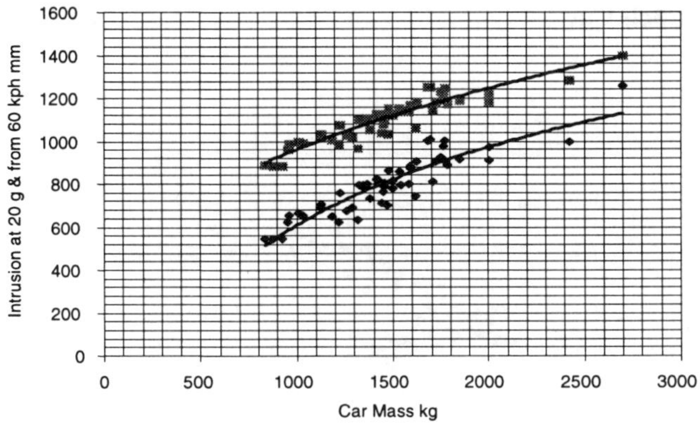

Figure 3 Car-to-pole intrusions for impacts from 60 km/h onto a 500 mm diameter pole (upper data), intrusion that generates 20 g acceleration (lower data).

The upper graph of Figure 3 shows car-to-pole intrusions resulting from impacts from 60 km/h. The lower graph is of the intrusion at the point when 20 g acceleration is reached during an impact from a high velocity. An acceleration limit of 20 g is used here as a representing an acceleration, which would be survivable. Figures 2 and 3 show that lighter cars can be expected to be stiffer than heavier ones. An acceleration of 20 g is reached with less intrusion in the lighter cars and the maximum acceleration is almost 100% higher in the lightest cars than the heaviest. An explanation is that a larger fraction of the mass of lighter cars is used in their structure than may be the case for heavier cars.

4 Collision Energy - Absorbed by the Car and Pole

Figure 4 shows the collision kinetic energy remaining in the cars at the point when 20 g acceleration is reached, from a 60 km/h impact. The heavier cars had initially double the energy due to their being double the weight of the lighter ones, but had only 40% remaining as opposed to 70% for the lighter cars, when 20 g was reached. We note then that there is the same amount of energy to be dissipated into the pole from the light cars vs heavy cars. Hence, for the lightest cars the resistive force provided by the pole has to be half that for the heaviest cars, with the crush depth into the poles double that for the heaviest cars. It is this fact that makes a general energy-absorbing stand-up pole difficult to achieve.

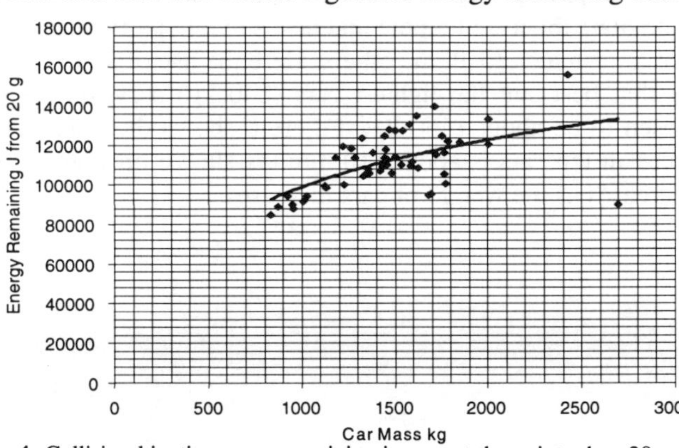

Figure 4 Collision kinetic energy remaining in cars at the point when 20 g acceleration is reached from a 60 km/h impact onto 500 mm diameter pole.

5 Pole Matched to Each Car

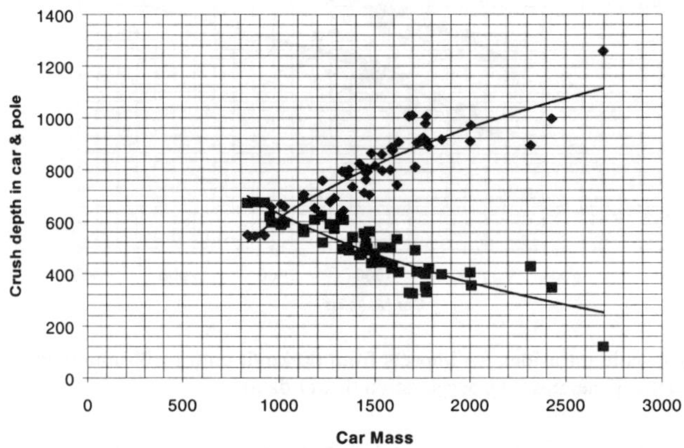

Figure 5 Crush depth in cars that generate 20 g acceleration (upper data) and displacement required of the poles (lower data) to maintain that acceleration

The lower graph on Figure 5 provides us with the depth of crush into a pole if the resistance provided by that pole could be tailored to limit the acceleration to 20 g in each car. The upper graph on Figure 5 gives us the crush depth into the cars that generates that maximum acceleration. If we sum the crush into the pole and car this figure provides us with the total crush that suits each car to come to a halt, which is nearly constant at 1300 mm. If one pole could be made to provide this function for the population of cars, then its stiffness would have to increase in equal proportions to the force that it would be subjected to during an impact. When calculating the data points on Figure 5 the complete crash pulse was assumed to be of the type described by Figures 1 b) and 2 b). This may be conservative but it has been our experience after many trials that it is difficult to make a pole or structure that deflects at anything like a constant force.

6 Poles Matched to Light and Medium Cars

It seems inescapable at present that the practical applications of the use of energy absorbing poles necessitate the use of a pole that best suit one particular weight of car and is to some degree be a mismatch to all the others. Here the use of just two type poles will be considered, one that best suit the lightest cars (840 kg) surveyed here and one for 'medium' cars (chosen to be about 1460 kg). Figure 6 shows the outcomes of using the softer pole. The crush in the cars is about 510 mm (as shown on Figures 3 & 5) for the lightest cars and decreases as the weight of cars increases.

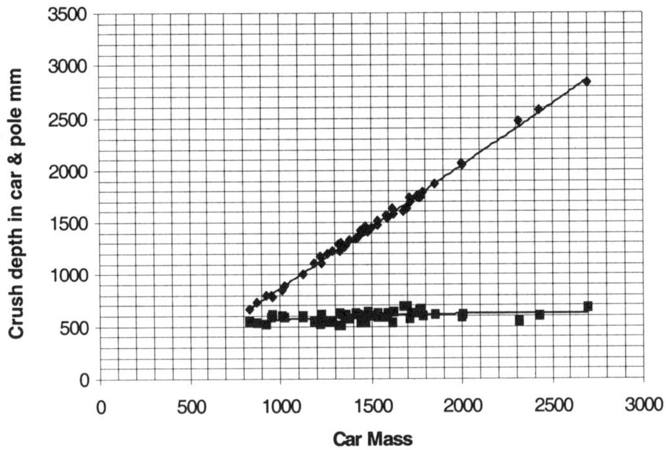

Figure 6 The crush depth caused by a pole that suits 840 kg cars, when impacted on by the whole population of cars. Pole displacement -upper points, car crush -lower points.

The displacement of the pole begins at about 700 mm for the lightest cars but becomes very large for the heaviest, to about 2500 mm. For medium size cars of 1460 kg, the pole displacement is 1600mm. If the pole were of a large diameter initially (say 800 mm) the pole may have remained up-standing after an impact with a very light cars, but its section would have been completely crushed by almost all other cars. Figure 7 shows the results of using a pole to suit medium size sedans. Here the car crush is essentially constant at about 800 mm. The pole crush is now much less beneficial for the lightest cars, being of only 100 mm. The pole displacement for medium cars is about 500 mm, which may be provided for by a suitable stand-up pole of 600 mm or more in diameter. Figure 6 indicates that even for a car of 1200 kg, which is of quite small mass, the pole crush has to be of the order of 1200 mm. In short, the pole that suits the lightest cars becomes a wrap-around pole to nearly all other cars, because it would be brought to the ground by all such impacts. Furthermore, it would be of such limited strength and rigidity that its use to carry signs or lights would find few applications.

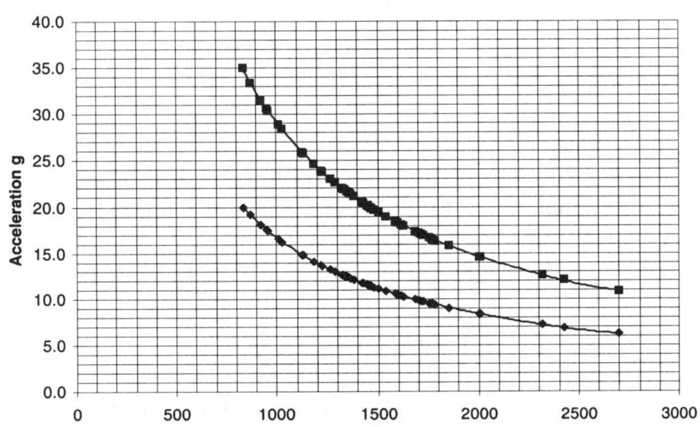

Figure 7 The accelerations that would result for the whole group of cars if either a pole that suits 1460 kg cars were used (upper data) or for the 840 kg cars (lower data).

Figure 7 plots the accelerations that would results to the whole range of cars if the light and medium poles were impacted on by all cars. This Figure indicates the accelerations that would accompany the crush shown on Figure 7. The upper curve shows the estimated accelerations if the mid-weight pole were to be used and the lower for the lightweight pole. When compared to Figure 2 the lighter pole would cause low accelerations to the whole range of cars, and that the medium pole would have only slight benefits to the lightest cars. The problem with the lightweight poles is that they would effectively be wrap-around poles, they do not meet our objectives stated in the Introduction of energy-absorbing stand-up poles. The current steel posts support large signs, which are typically heavy and generate large aerodynamic drag during storms.

7 Discussion

The Transport Research Board of the USA in their Report 350 [4] provide guidelines on the tests conditions and required performances for general support structures and breakaway utility poles. These guidelines have achieved broad support. They require that when impacted by vehicles with masses of 700 and 2000 kg no more than 20g acceleration is generated in the car. Figure 6 indicates the performance of a pole designed to suit an 840 kg car, and it shows that more than 2 meters of pole crush would be required to stop a car of more than 2000 kg. This crush space may be provided on wide new and refurbished roads (while maintaining the sign off the roadway) but in city roads the requirements of Report 350 are difficult to meet with the single post structure considered here.

8 Conclusions

In the development of less hazardous roadside signposts on city roads a number of contrary requirements have to be considered. There is a need for large traffic signs supported above roadways to provide directions to the traffic. The limited roadside area dictates that single posts of considerable strength support these signs. The high accelerations that accompany frontal collisions between motorcars and posts may be reduced but may not be rendered innocuous.

References

1. Wood, D. P., Doody, M., & Moomey, S., "Application of a Generalised frontal Crush Model of the Car Population to Pole and Narrow Object Impacts". SAE paper 930894, Society of Automotive Engineers, 1993.
2. LS-DYNA, Livermore Software Technology Corp, 2876 Waverley Way, Livermore, CA 94550.
3. Frontline System Inc. 1999. http://www.frontsys.com/.
4. Report 350, "Recommended Procedures for the Safety Evaluation of Highway features", National Cooperative Research Program, Washington, D.C., 1993.

DYNAMIC STRESS RESPONSE OF AN ORTHOTROPIC HOLLOW CYLINDER UNDER TORSIONAL IMPACT

G. LU [a], X. WANG [b*] X.D. HUANG [a] AND S.G. GUILLOW [a]

[a] *School of Engineering and Science, Swinburne University of Technology, Hawthorn, VIC., 3122, AUSTRALIA*

[b] *School of Civil Engineering and Mechanics, Shanghai Jiaotong University, Shanghai 200240, People's Republic of China*

In this paper an analytic equation for dynamic shear stress in an orthotropic hollow cylinder under torsional impact loading is derived using the eigen-function method. Mixed boundary conditions are assumed. Example calculations are presented applying this equation to determine stress time history response and distribution for two different orthotropic thick-walled cylinders under torsional impact.

1 Introduction

In many applied engineering fields, a notable problem is determination of dynamic shear stress under torsional impact loading conditions. This dynamic stress may affect the dynamic strength of the structure. Torsional impact loading can take place in many different engineering applications such as machine drilling, geological exploration drilling and structural bolt fastening. The material properties of the drilled plate, earth or rock may be isotropic or anisotropic. Previous research into theoretical solutions for structures under impact loads has mainly focused on radial impact [1-6]. General solution methods used for thick-walled cylindrical shell dynamic problems are : the integrate transform method [1], the ray method [3], the finite Hankel transform [4] and the eigen-function method [2]. To date the theoretical solution of thick-walled shells under torsional impact has seldom been considered. In reference [7], the author gives a theoretical solution for a hollow cylinder under torsional impact by making use of the finite Hankel transform. After analysis, the theoretical solution given in [7] was seen to be manifestly incorrect and no example calculations were given.

This paper develops analytic equations for dynamic shear stress in an orthotropic hollow cylinder with mixed boundary conditions under torsional impact loading by making use of the eigen-function method [2]. In our examples, we determine dynamic shear stress for two different orthotropic hollow cylinders under sudden torsion loads. By analysing the calculated results it was found that the solution has the properties of a wave. At the shear stress wavefront there is a strong discontinuity effect. The method given in this paper is concise and can be applied to practical problems.

2 Dynamic Equation and Solution Procedure for Torsional Problem

The assumed structure for an orthotropic hollow cylinder under torsional impact is shown in Figure 1. Considering the geometry and loading shown in Figure 1, this is seen to be an axi-symmetric plane strain problem. The dynamic equilibrium equation for the

*Corresponding author. E-mail: xwang@mail.sjtu.edu.cn

orthotropic hollow cylinder under torsional impact is expressed as

$$\frac{\partial^2 V}{\partial r^2}+\frac{1}{r}\frac{\partial V}{\partial r}-\frac{V}{r^2}=\frac{1}{C_L^2}\frac{\partial^2 V}{\partial t^2} \tag{1a}$$

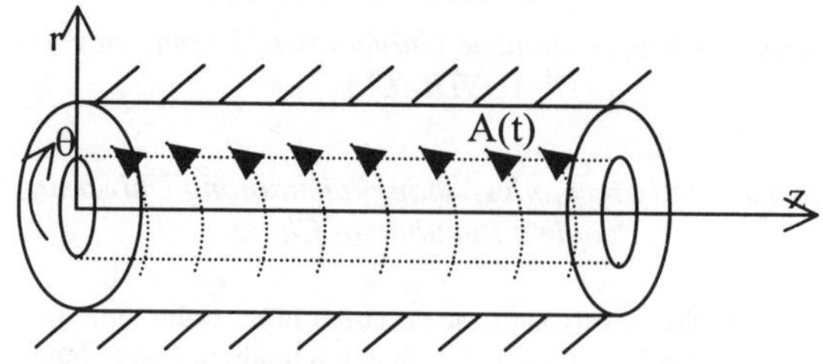

Figure 1. Coordinate system for the cylinder

The corresponding mixed boundary conditions is

$$\sigma_{r\theta}(a,t)=C_{66}\left[\frac{\partial V(r,t)}{\partial r}-\frac{V(r,t)}{r}\right]_{r=a}=A(t)\ ,\qquad V(b,t)=0 \tag{1c}$$

Initial conditions

$$V(r,0)=V_0(r),\quad \frac{\partial V(r,t)}{\partial t}=\dot{V}_0(r) \tag{1d}$$

In the above, r is the radial variable, a and b are respectively the internal radii and external radii of the hollow cylinder, $C_L=\sqrt{C_{66}/\rho}$ is the shear wave speed, $C_{66}=G_{r\theta}$ is the shear elastic constant in the r-θ plane and ρ is the material density.

Utilising eigen-function method [2], the general solution for dynamic equilibrium under torsional impact loading may be expressed as

$$V(r,t)=V_s(r,t)+\sum_{n=1}^{\infty}V_n(r)q_n(t) \tag{2}$$

where $V_S(r,t)$ satisfies the following quasi-static equation with inhomogeneous mixed boundary conditions

$$\frac{\partial^2 V_s(r,t)}{\partial r^2}+\frac{1}{r}\frac{\partial V_s(r,t)}{\partial r}-\frac{V_s(r,t)}{r^2}=0 \tag{3a}$$

$$C_{66}\left[\frac{\partial V_s(a,t)}{\partial r}-\frac{V_s(a,t)}{r}\right]=A(t)\quad ,\qquad V_S(b,t)=0 \tag{3b,c}$$

In the above formula, t is the time variable. Substituting equation (2) into equation (1) and utilising (3), the dynamic equation with homogeneous mixed boundary condition can be expressed as

$$\frac{d^2V_n(r)}{dr^2}+\frac{1}{r}\frac{dV_n(r)}{dr}+\left(k_n^2-\frac{1}{r^2}\right)V_n(r)=0 \tag{4a}$$

$$\left[\frac{dV_n(r)}{dr}-\frac{V_n(r)}{r}\right]_{r=a}=0\ ,\qquad V_n(b)=0 \tag{4b,c}$$

The time function $q_n(t)$ should satisfy the following equation (5) and the corresponding initial condition (1d)

$$\sum_{n=1}^{\infty}\left[\ddot{q}_n(t)+\omega_n^2 q_n(t)\right]V_n(r)=-\ddot{V}_s(r,t) \tag{5}$$

In equations (4) and (5), k_n and ω_n are respectively called the eigen-value and natural frequency and are related as follows

$$\omega_n = k_n \cdot C_L \tag{6}$$

Starting with equation (2), these equations are solved respectively for $V_s(r,t)$, $V_n(r)$ and $q_n(t)$ leading to the analytic solution of the elasto-dynamic equation (5).

Solving the quasi-static equation (3), we have

$$V_s(r,t)=\psi(r)A(t) \tag{7a}$$

where
$$\psi(r)=\frac{a^2}{2c_{66}}\left(\frac{r}{b^2}-\frac{1}{r}\right) \tag{7b}$$

Utilising eigen-value problem [2], for corresponding natural frequency $\omega_n (n=m)$ the general solution of the eigen-function $V_n(r)$ may be expressed as

$$V_n(r)=B_n C_1(k_n r) \tag{8a}$$

where
$$B_n=\left\{2\pi\int_a^b C_1^2(k_n r)rdr\right\}^{-\frac{1}{2}}=\frac{\sqrt{\pi}k_n J_a}{2}\left\{J_a^2-J_1^2(k_n b)\left[h^2+k_n^2\left(1-\frac{1}{k_n^2 a^2}\right)\right]\right\}^{-\frac{1}{2}} \tag{8b}$$

and
$$C_1(k_n r)=J_1(k_n r)Y_1(k_n b)-Y_1(k_n r)J_1(k_n b) \tag{8c}$$

The time function $q_n(t)$ can be determined as follows. The two sides of equation (5) are multiplied by eigen-function $V_m(r)$ and integrated across the cross-sectional area of the hollow cylinder, thus

$$\sum_{n=1}^{\infty}\iint_s\left[\ddot{q}_n(t)+\omega_n^2 q_n(t)\right]V_n(r)V_m(r)ds=-\int_s \ddot{V}_s(r,t)V_m(r)ds \tag{10}$$

Utilising the normalising property of the eigen-function, equation (10) is simplified as

$$\ddot{q}_n(t)+\omega_n^2 q_n(t)=\ddot{\varphi}_n(t) \tag{11a}$$

where
$$\varphi_n(t)=-2\pi\int_a^b V_s(r,t)V_n(r)rdr \tag{11b}$$

Thus the solution of equation (11a) satisfying the initial conditions may be expressed as

$$q_n(t)=q_n(0)\cos\omega_n t+\frac{t}{\omega_n}\dot{q}_n(0)\sin\omega_n t+\frac{1}{\omega_n}\int_0^t\ddot{\varphi}_n(\tau)\sin\omega_n(t-\tau)d\tau$$

$$=\left[q_n(0)-\varphi_n(0)\right]\cos\omega_n t+\frac{1}{\omega_n}\left[\dot{q}_n(0)-\dot{\varphi}_n(0)\right]\sin\omega_n t+\varphi_n(t)-\omega_n\int_0^t\varphi_n(\tau)\sin\omega_n(t-\tau)d\tau \tag{12a}$$

where
$$q_n(0)=2\pi\int_a^b V_0(r)V_n(r)rdr+\varphi_n(0)\ ,\quad \dot{q}_n(0)=2\pi\int_a^b \dot{V}_0(r)V_n(r)rdr+\dot{\varphi}_n(0) \tag{12c,d}$$

Substituting equations (7), (8) and (12) into equation (2) leads to an exact solution for the elasto-dynamic equation (1) of an orthotropic hollow cylinder under torsional impact:

$$V(r,t)=V_s(r,t)+\sum_{n=1}^{\infty}V_n(r)\varphi_n(t)+$$

$$\sum_{n=1}^{\infty}\left[\alpha_n\cos\omega_n t+\frac{1}{\omega_n}\beta_n\sin\omega_n t-\omega_n\int_0^t\varphi_n(\tau)\sin\omega_n(t-\tau)d\tau\right]V_n(r) \tag{13a}$$

where $$\alpha_n=2\pi\int_a^b V_0(r)V_n(r)rdr \quad \beta_n=2\pi\int_a^b \dot{V}_0(r)V_n(r)rdr \tag{13b}$$

From the expression of tangent displacement (13) and utilising physical and geometrical relationships, the dynamic shear stress of an orthotropic hollow cylinder may easily be expressed as

$$\sigma_{r\theta}=C_{66}\left[\frac{\partial V(r,t)}{\partial r}-\frac{V(r,t)}{r}\right] \tag{14}$$

3 Examples and Discussion

Assume that the orthotropic hollow cylinder is subjected to an average torsional impact load distribution such that

$$A(t)=\tau_0 e^{-\alpha t} \qquad t\geq 0^+ \tag{15}$$

When $\alpha=0$, A(t) represents a sudden torsion load; and when $\alpha\neq 0$, A(t) is an exponentially decaying load. Assuming that before impact the initial tangential displacement of the structure is at rest, the initial conditions are described in equation (1d). The corresponding solution of the quasi-static equation (7) may be expressed as

$$V_s(r,t)=\tau_0\psi(r)e^{-\alpha t} \quad , \quad t\geq 0^+ \tag{16}$$

Substituting equations (8) and (16) into (11b), we have

$$\varphi_n(t)=\tau_0 B_n D_n e^{-wt} \tag{17a}$$

where $$D_n=-2\pi\int_a^b (r)C_1(k_n r)rdr \tag{17b}$$

Utilising the initial conditions and substituting equation (17) into equation (12a), the corresponding time function may be expressed as

$$q_n(t)=\tau_0 B_n D_n Q_n(t) \tag{18a}$$

where $$Q_n(t)=\frac{-\alpha\omega_n\sin\omega_n t+\omega_n^2\cos\omega_n t+\alpha^2 e^{-\alpha t}}{\alpha^2+\omega_n^2} \tag{18b}$$

where $\psi(r)$, B_n, D_n, $Q_n(t)$ and $C_1(k_n r)$ are respectively shown in the equations (7b), (8b), (17b), (18b) and (8c). From equation (14) we obtain the corresponding analytical expression for dynamic shear stress which may be expressed as

$$\sigma_{r\theta}=-\frac{\tau_0 a^2}{r^2}e^{-\alpha t}+\tau_0 C_{66}\sum_{n=1}^{\infty}B_n^2 D_n Q_n(t)\left[k_n C_1'(k_n r)-\frac{C_1(k_n r)}{r}\right] \tag{19}$$

From equation (19) for dynamic shear stress, it is seen that the quantities reflecting material properties are C_{66} and $C_L=\sqrt{C_{66}/\rho}$. For different materials C_{66} varies and hence the shear stress wave speed and natural frequency of the material are also different. Thus the dynamic shear stress time response history and distribution will be different for various materials.

For convenience in the present calculations all variable will be expressed in a dimensionless form: $\sigma^*_{r\theta} = \sigma_{r\theta}/\tau_0$. Material constants are taken as : $C_{66} = G_{r\theta} = 7.1\,\text{GPa}$, $\rho = 5.0\times10^3\,\text{kg/m}^3$. In this paper, the response histories and distribution are calculated for the two kinds of a hollow cylinder: first, with thickness as $(b-a)/a = 20$,where $R^* = (r-a)/a$, $T^* = tC_L/a$ and second, with $(b-a)/a = 2$, where $R^* = (r-a)/(b-a)$, $T^* = tC_L/(b-a)$ under a sudden torsional impact load. -•- represents quasi-static solution.

In order to verify that the solutions obtained appear in wave property and accurate, the hollow cylinder with thickness as *(b-a)/a*=20 is first calculated and the response time is taken as $T^* \leq 20$. In the case, the response histories and distribution of shear stress in structure can not be interfered by the reflection of the shear stress wave at boundary.

Fig. 2 shows the shear stress time history response and distribution in a hollow cylinder with thickness *(b-a)/a*=20, under a sudden torsional impact load. It may be noted from this figure that the dynamic shear stress has its maximum value and is strongly discontinuous where the wavefront arrives. When the wavefront propagates from the point to external boundary, the stress at the point will gradually decay from the maximum to the quasi-static solution at the point.

Fig. 2b shows the radial distribution of dynamic shear stress after a sudden torsion impact load. It is seen from this figure that when the stress wavefront arrives at some point, the stress at the point has a maximum value and when the stress wavefront does not arrive at place, the stresses at those points equal to zero.

Fig. 3 shows shear stress time history response and distribution in a hollow cylinder with thickness *(b-a)/a*=2, under a sudden torsion impact load. The response time was taken as $T^* = 20$, which is much larger than the time required for the stress wave to propagate from the internal to external surface. Thus we must consider that in this case the response time history and distribution of torsional shear stress are affected by the reflected wave between internal and external surfaces. Except that the dynamic shear stress at the internal surface must satisfy the given boundary values, the stresses at other points appear to dramatically oscillate about the quasi-static solution.

References

1. Baker,W.E , Axisymmetric modes of Vibration of thin Spherical Shell. J. Acoust. Soc. Am. 33. (1961) pp.1749-1758.
2. Eringren.A. C.,and Suhubi, E.S., Elastodynamics.Vol.2, (Linear Theory) Academic Press. New York (1975).
3. Pao, Y. H. and Ceranoglu, A. N., Determination of transient responses of a thick-walled Spherical Shell by the ray theory. J. Appl. Mech.,ASME 45, (1978) pp.114 -122.
4. Cinelli, G., An extension of the finite Hankel transform and application. Int. J. Engng Sci. 3, (1965) pp. 539-550.
5. Cho, H., Kardomates, G.A., Valle, C.S. , Elastodynamic solution for the thermal shock stresses in an orthotropic thick cylindrical shell. J. Appl. Mech.,ASME 65, (1998) pp184-193.
6. Wang, X., Zhang, K. and Zhang, W., Theoretical solution and finite element solution for an orthotropic thick cylindrical shell under impact load., J. Sound and Vibration, 236, (2000) pp.129-140.
7. Cinelli,G., Dynamic Vibrations and Stresses in Elastic Cylinders and Spheres. J. of Appl. Mech., (1966) pp825-830.

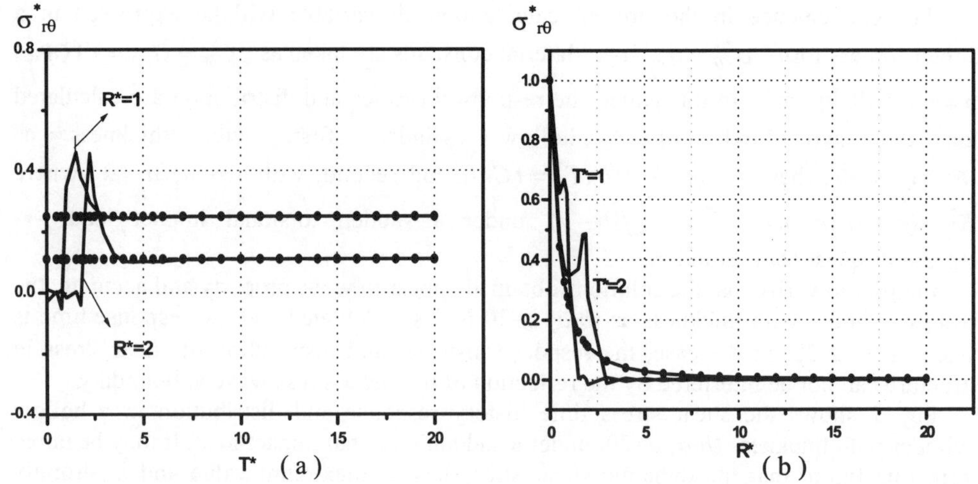

Figure 2. Shear stress time history response and distribution in hollow cylinder with thickness (b-a)/a=20, under sudden torsion impact load. $\sigma_{r\theta}^{*}=\sigma_{r\theta}/\tau_0$, $T^*=tC_L/a$, $R^*=(r-a)/a$. —•— quasi-static solution.

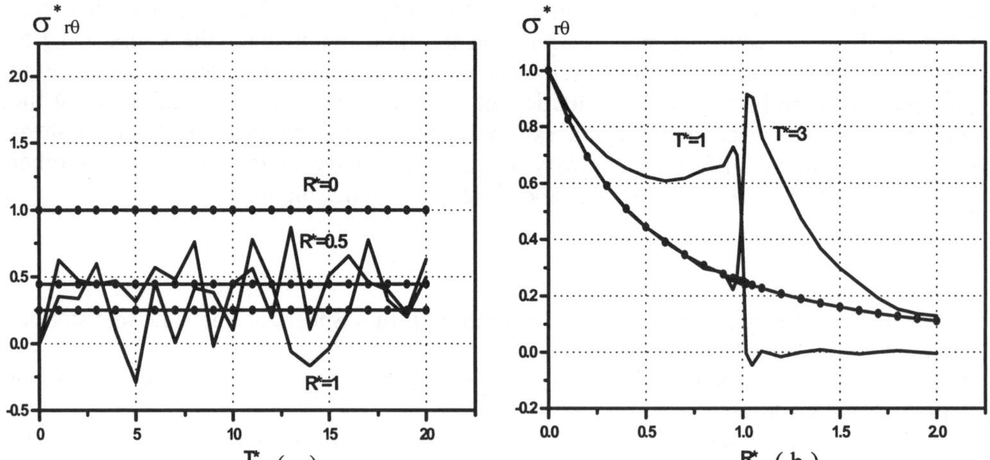

Figure 3. Shear stress time history response and distribution in a hollow cylinder with thickness *(b-a)/a*=2, under sudden torsion impact load. $\sigma_{r\theta}^{*}=\sigma_{r\theta}/\tau_0$, $T^*=tC_L/(b-a)$, $R^*=(r-a)/(b-a)$. —•— quasi-static solution.

AXISYMMETRIC PROGRESSIVE CRUSHING BEHAVIOUR OF CIRCULAR METAL TUBES

X. HUANG, G. LU AND S.R. GUILLOW

School of Engineering and Science, Swinburne University of Technology, Hawthorn, VIC. 3122, AUSTRALIA

E-mail: glu@swin.edu.au

A theoretical analysis of an axially compressed circular metal tube deforming in progressive axisymmetric folds is presented. This new model is developed based on our experiments and finite-element analysis. It helps explains our previous empirical finding that the non-dimensionalised mean crushing force P_m/M_0 is not proportional to $(D/h)^{0.5}$. An empirical formulae for the mean crushing force and half wavelength are established to best fit the numerical calculations. Comprehensive comparisons have been made among the present theoretical prediction, experiments and FEA results.

1 Introduction

The progressive crushing of thin-walled circular metal tubes under axial loading conditions has been studied extensively over the last two decades [1, 2] due to the common use of such tubes in energy absorption devices. The behaviour of a tube during crushing depends on both its geometrical parameters and material parameters. Relatively short and thick tubes normally exhibit an axisymmetric mode.

Alexander [3] was the first to model the axisymmetric collapse of a tube, the wall forming straight line elements between stationary hinges. Abramowicz and Jones [4,5] proposed an improved model with the tube wall bending in the meridianal direction into two oppositely curved arcs instead of a straight line. Grzebieta [6] further modified the meridianal profile, but adopted an equilibrium approach. In the above models the tube wall folded only outwards (or only inwards); but in fact observation shows it deforms both inwards and outwards. To account for this, Wierzbicki et. al. [7] introduced a parameter known as the eccentricity factor, η, which is the ratio of outward fold length to the total fold length. In this way, the occurrence and position of a second peak within each fold can be explained. Also, the force-displacement curve could be worked out, not just the average crush force. This work was further refined by Singace et. al. [8, 9].

Figure 1 shows a typical load-displacement curve for the axisymmetric collapse of a metal tube. Apart from the initial peak, which is governed by elastic-plastic buckling, the curve exhibits a characteristic fluctuation in axial force. Thus mean crush force, P_m, is a significant parameter for quantifying the behaviour of axially compressed tubes. All of the previous theoretical studies predict that the normalized mean crushing load (P_m/M_0) should be proportional to $(D/h)^{0.5}$. However, experimental observations by Guillow et. al. [10] indicate that functional dependence should be about $(D/h)^{0.33}$. Thus it seems that some important characteristic has been overlooked in previous studies.

The present study proposes a refined analytical model using an effective hinge length which varies with the wall thickness, based on experimental observations. The calculated results help explain the paradox between the experiments and the previous theories, and have good agreement with the experiments and FEA results.

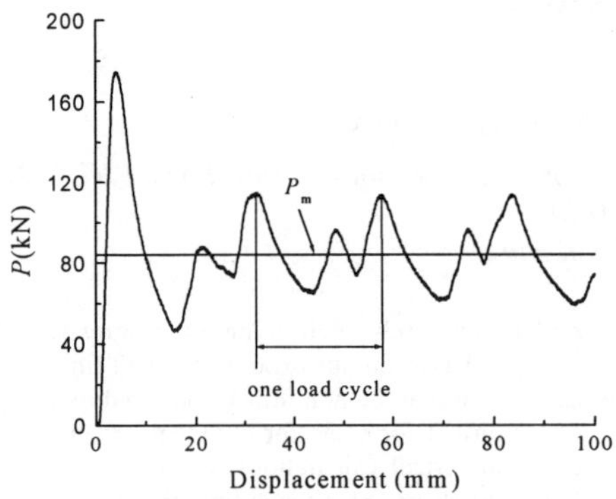

Figure 1 Load-deflection curve of an Aluminium tube with $D = 100$mm, and $h = 3.0$mm.

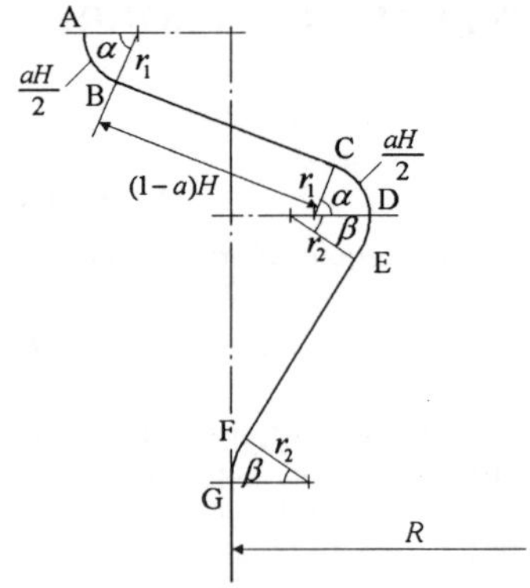

Figure 3 Instantaneous profile during crushing.

2 Crushing Model

Based on experimental observations, the assumed crushing process for one crushing cycle is demonstrated in Figure 2. Figures 2(a) to (c) show the first stage of the crushing cycle with one outwards fold (element 1) completed and the next inwards fold (element 2) starting to form. In the second stage, Figures 2(c) to (e), the inwards fold (element 2) is completed and the next outwards fold (element 3) starts to form. These two stages repeat successively as the tube progressively crushes; this leading to a periodic variation in the force-displacement curve with two force peaks per complete fold cycle. Figure 3 shows the instantaneous shape of the deformation zone with each element consisting of three segments : two of these are arcs of equal radius and the third a straight line which is tangent to the arcs and connects them together. By definition, the length of the arcs is equal to the effective hinge length, aH . H is the half plastic wavelength, which is an unknown parameter to be determined in the course of the analysis. The parameter a is $0 \le a \le 1$. When $a = 0$ the present model reduces to the well-known stationary hinge model; and when $a = 1$ it corresponds to a moving hinge model.

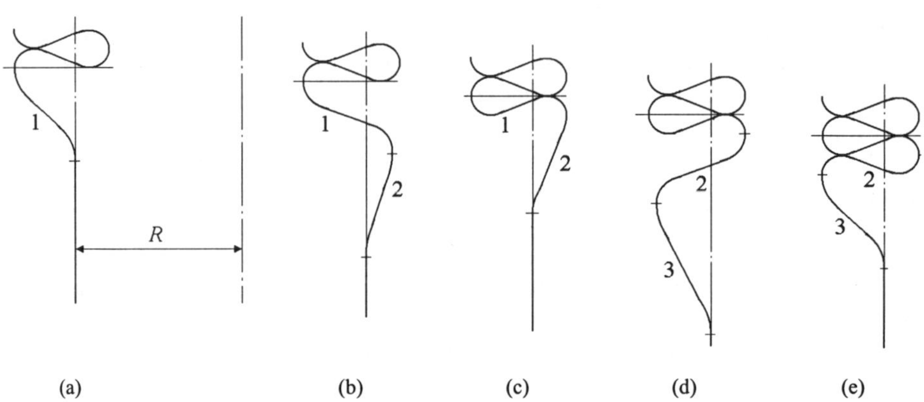

Figure 2 Successive deformation stages in one load cycle.

The instantaneous shape of the deformation zone (refer Figure 3) is defined by the instantaneous angles α and β. The instantaneous radii of the two arcs are

$$r_1 = \frac{aH}{2\alpha} \qquad r_2 = \frac{aH}{2\beta} \tag{1}$$

The relationship between angles α and β may easily be deduced from the eccentricity factor, η. Therefore for a given value of the eccentricity, this model is a one-dimensional system determined by either α or β. In the first half crushing cycle (Figures 2(a) to (c)), the limits for α and β are

$$\alpha_0 \le \alpha \le \alpha_f \text{ and } 0 \le \beta \le \beta_0 \tag{2}$$

In the second half crushing cycle (Figures 2(c) to (e)), the limits for α and β are

$$0 \le \alpha \le \alpha_0 \text{ and } \beta_0 \le \beta \le \beta_f \tag{3}$$

where $\alpha_0 \ne \beta_0$ while $\alpha_f = \beta_f$. The final angle, α_f or β_f, is obtained from the geometry of the fully crushed model. It is expressed by

$$\cos\alpha_f = \frac{a}{2(1-a)} \frac{1-2\sin\alpha_f}{\alpha_f} \tag{4}$$

Therefore the final angle, α_f or β_f, is determined by the parameter a.

The crushing distance in a complete crushing cycle is expressed by

$$\delta = \Delta(a)H \tag{5}$$

where $\Delta(a) = 2 - \frac{a}{\alpha_f}$.

3 Energy dissipation

Bending energy dissipation only occurs in the arc sections. In the first stage of collapse associated with outward folding (see Figure (3)), the rates of curvature are

$$\dot{\kappa} = \frac{2\dot{\alpha}}{aH} \quad \text{(sections AB and CD)} \tag{6a}$$

$$\dot{\kappa} = \frac{2\dot{\beta}}{aH} \quad \text{(sections DE and FG)} \tag{6b}$$

Hence the rate of bending energy dissipation is

$$\dot{E}_b = 4\pi R_0 M_0 (\dot{\alpha} + \dot{\beta}) \tag{7}$$

In the second stage of collapse associated with inward folding, the expression for bending energy rate is the same as eqn. (7). Integrating eqn. (7), the bending energy dissipation in a complete crushing cycle is

$$E_b = 8\pi R_0 M_0 A_1(a) \tag{8}$$

where $A_1(a) = \alpha_f$.

The rate of the membrane energy dissipation due to the circumferential strain is expressed as

$$\dot{E}_m = 2\pi R_0 N_0 \int_s \left| \frac{\dot{w}}{R_0} \right| ds \tag{9}$$

where $N_0 = \sigma_0 h$ is the fully plastic membrane force per unit length; and $\dot{w}$ denotes the radial velocity of the wall of the tube. From Figure 3, $\dot{w}$ is expressed as follows

$$\dot{w} = \frac{aH}{2\alpha^2}(\phi \sin\phi - 1 + \cos\phi)\dot{\alpha} \qquad \text{(section AB)} \qquad (10a)$$

where ϕ is an angular variable and $0 \le \phi \le \alpha$.

$$\dot{w} = \left[\frac{aH}{2\alpha^2}(1-\cos\alpha) - \frac{aH}{2\alpha}\sin\alpha - s\cos\alpha\right]\dot{\alpha} \qquad \text{(section BC)} \qquad (10b)$$

where s is a length variable and $0 \le s \le (1-a)H$

$$\dot{w} = \left[\frac{aH}{2\alpha^2}(1 - 2\cos\alpha + \cos\phi + \phi\sin\phi - 2\alpha\sin\alpha) - (1-a)H\cos\alpha\right]\dot{\alpha} \quad \text{(section CD)} \quad (10c)$$

The expression of the radial velocity for sections DE, EF and FG are the same as for CD, BC and AB respectively, except that the parameter α is replaced with β.

Substituting eqn's (10a-c) into (9), the rate of the membrane energy can be written as

$$\dot{E}_m = 2\pi N_0 H^2 \left[f(\alpha)\dot{\alpha} + f(\beta)\dot{\beta}\right] \qquad (11)$$

where the function $f(x)$ is

$$f(x) = \frac{a^2}{4}\frac{2x\cos x + 2x^2\sin x - 2x}{x^3} + \frac{a(1-a)}{2}\left(\frac{\cos x + x\sin x - 1}{x^2} + \cos x\right) + \frac{(1-a)^2}{2}\cos x \qquad (12)$$

Integrating eqn. (11) over a complete crushing cycle, the membrane dissipation energy is

$$E_m = 4\pi N_0 H^2 A_2(a) \qquad (13)$$

where $A_2(a) = \int_0^{\alpha_f} f(\alpha)d\alpha$

The total work done by the mean crushing load over a crushing cycle, $P_m\delta$, is equal to the energy dissipation,

$$P_m\delta = E_b + E_m \qquad (14)$$

Substituting eqn's (5), (8) and (13) into eqn. (14), the mean crushing load can be written as

$$P_m = \frac{8\pi R_0 M_0 A_1(a) + 4\pi N_0 H^2 A_2(a)}{\Delta(a)H} \qquad (15)$$

Here, the half plastic wavelength parameter, H, is as yet unknown. It can be obtained by minimizing P_m with respect to H, which produces

$$H = H(a)\sqrt{Dh} \qquad (16)$$

where $H(a) = \frac{1}{2}\sqrt{\frac{A_1(a)}{A_2(a)}}$. Substituting eqn. (16) into eqn. (15) and normalizing the mean crushing load with respect to M_0,

$$\frac{P_m}{M_0} = P(a)\sqrt{\frac{D}{h}} \qquad (17)$$

where $P(a) = 16\pi\frac{\sqrt{A_1(a)A_2(a)}}{\Delta(a)}$.

4 Determination of the effective plastic hinge length

The concept of an effective length for plastic hinges was first introduced by Nonaka [11] while investigating the plastic dynamic behaviour of beams. He suggested this length was equal to one or two depths of the beam depending on axial load. Based on experimental observations, Stronge and Yu [12] suggested an effective length of two to five times beam depth. Theoretical determination of the effective plastic hinge length is not as yet possible, as discussed in the literature [13]. It is best to estimate this length from experimental observations. In our experiments, the plastic hinge length was estimated at approximately $4h$ for a range of thicknesses. From eqn. (16) it is found that

$$\frac{D}{h} = \left(\frac{4}{aH(a)}\right)^2 \tag{18}$$

The above equation shows that a tube with a certain value of D/h has a corresponding dimensionless effective hinge length a. In other words, D/h can be determined inversely from the parameter a. For example, when $a = 0$, D/h tends to infinity. When $a = 0.6$, D/h equals to 36.8.

A computer program was written to calculate the coefficients in the above equations. For a given value of a the corresponding average crush load P_m / M_0, half wavelength H/h and D/h ratio were determined. Finally, from a best fit analysis, the following empirical relations were obtained

$$\frac{P_m}{M_0} = 35.84\left(\frac{D}{h}\right)^{0.41} \tag{19}$$

for the mean axial crush force, and

$$\frac{H}{h} = 1.78\left(\frac{D}{h}\right)^{0.36} \tag{20}$$

for the half wavelength.

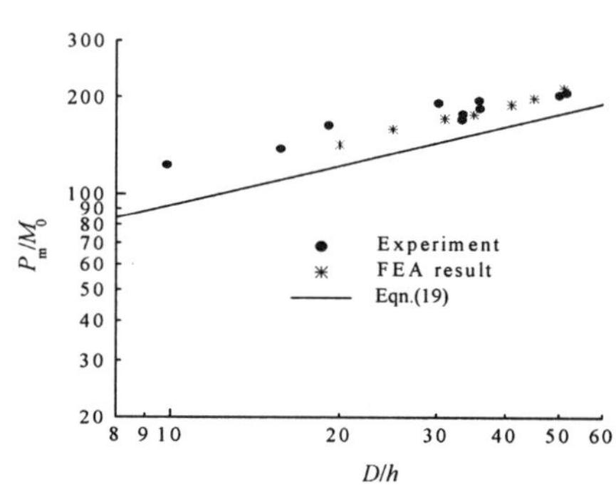

Figure 4 Comparison of the present predication of mean crushing force with experiments and FEA results.

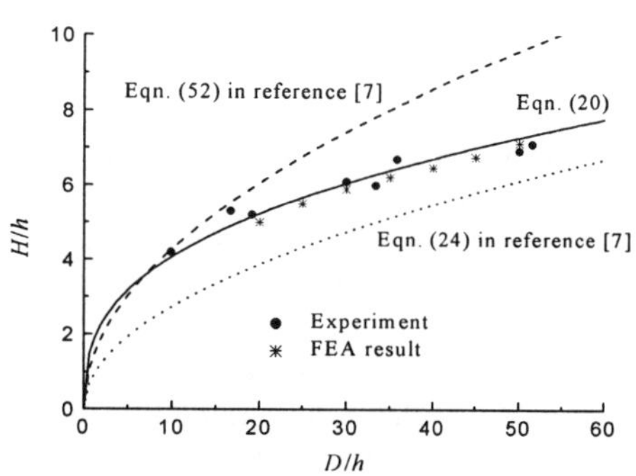

Figure 5 Comparison of the present predication of wavelength with experiments, FEA results and previous theories.

5 Comparison and Conclusion

The present model was verified against experimental test results and finite-element analysis using the commercial program ABAQUS. Comparisons of the calculated mean crush force and half wavelength with experimental data and FEA results for aluminium tubes are presented in Figure 4 and Figure 5. In calculating the full plastic moment, the flow stress σ_0 was assumed to be equal to the ultimate stress of the material due to the large strains involved. Figures 4 and 5 show broadly, a good agreement between theoretical predication, experiments and FEA numerical results.

The present model predicts that the non-dimensional crush force, P_m / M_0, is proportional to $(D/h)^{0.41}$, compared to $(D/h)^{0.5}$ in previous theories for axi-symmetric mode. Estimation of plastic wavelength, 2*H*, is much improved over previous models. Variation of non-dimensional load and geometry of the tube with D/h is well modeled.

References

1. Reid S. R., Plastic deformation mechanisms in axially compressed metal tubes used as impact energy absorbers. *Int. J. Mech. Sci.*, **35** (1993) pp. 1035-1052.
2. Magee C. L. and Thornton P. H., Design consideration in energy absorption by structural collapse. *SAE paper*, **780434** (1978).
3. Alexander J. M., An approximate analysis of collapse of thin-walled cylindrical shells under axial loading. *Quar. J. Mech. & App. Math.*, 13 (1960) pp. 10-15.
4. Abramowicz W. and Jones N., Dynamic axial crushing of circular tubes. *Int. J. Impact Eng.*, **2** (1984) pp. 263-281.
5. Abramowicz W. and Jones N., Dynamic progressive buckling of circular and square tubes. *Int. J. Imp. Eng.*, **4** (1986) pp. 243-269.
6. Grzebieta R. H., An alternative method for determining the behaviour of round stocky tubes subjected to axial crush loads. *Thin-Walled Struc.* **9** (1990) pp. 66-89.
7. Wierzbicki T., Bhat S. U., Abramowicz W. and Brodkin D., Alexander revisited - a two folding element model of progressive crushing of tubes. *Int. J. Sol. Struc.*, **29** (1992) pp. 3269-3288.
8. Singace A. A., Elsobky H. and Reddy T. Y., On the eccentricity factor in the progressive crushing of tubes. *Int. J. Sol Struc.*, **32** (1995) pp. 3589-3602.
9. Singace A. A. and Elsobky H., Further experimental investigation on the eccentricity factor in the progressive crushing of tubes. *Int. J. Sol. Struc.*, **33** (1996) pp. 3517-3538.
10. Guillow S. R, Lu G. and Grzebieta R. H., Quasi-static axial compression of thin-walled circular aluminium tubes. *Int. J. Mech. Sci.*, (2001).
11. Nonaka T. Some interaction effects in a problem of plastic dynamics, Parts 1-3. *ASME J. Appl. Mech.* **34** (1967) pp. 623-643.
12. Stronge W. J. and Yu T. X., *Dynamic Models for Structural Plasticity*, Chapter 5. Springer-Verlag (1993).
13. Su X.Y., Yu T. X. and Reid S. R., Inertia-sensitive Impact energy-absorbing structures Part I: effects of inertia and elasticity. *Int. J. Impact Eng.*, **16(4)** (1995) pp. 651-672.

BALLISTIC ASSESSMENT OF A LOW-COST Ti-6Al-4V TITANIUM ALLOY

S.J. CIMPOERU[1], S.J ALKEMADE[1], M. SZYMCZAK[2], N.L. RUPERT[3], W.H. GREEN[4] AND J.M.WELLS[5]

[1]*DSTO Aeronautical and Maritime Research Laboratory, PO Box 4331, Melbourne 3001, Australia*

[2]*Defence Research Establishment Varcartier, 2459 Boulevard Pie-Xi Nord, Val-Bélair QC G3J 1X5, Canada*

Army Research Laboratory, AMSRL-WM-TD[3]/MD[4]/MC[5], APG, MD 21005-5069, USA

Titanium alloy Ti-6Al-4V, produced via a new low-cost Electron-Beam Cold-Hearth Single-Melt Process, was ballistic limit tested against 12.7 mm projectiles. A brief overview is given of emergence of low-cost titanium as a real option for armour applications as well as the characteristics of the tested Electron-Beam Single-Melt Titanium. Simple plasticity-based analytical models are used to predict the ballistic limit for the non-deforming projectile case and to illustrate how important it is to understand the detail of target deformation and failure. The work provides essential background to future studies that will better characterise the ballistic deformation and damage in the target plate to further improve predictive modelling efforts

1 Introduction

The potential of Titanium as an armour material was noted as early as 1950 [1], but it is only in recent years that it has become an option for combat vehicles [2]. This is essentially because of a real drop in the price of Titanium metal in the 1990s, which arose because of the appearance of increasing quantities of scrap Titanium and Titanium sponge on the world market and a recognition that conventional high purity, high-cost aerospace Titanium grades are not necessarily required for armour applications. Unfortunately, low-cost Titanium armour has yet to become a reality. This has partly been because new commercial applications, e.g. Titanium Golf Club Heads [2], develop as the cost of Titanium falls, but also because the application of Titanium armour is still in its infancy and it's properties, particularly ballistic properties, are not well understood compared with the conventional armour materials that industry is more familiar with.

Titanium Armour is currently being used as a retrofit armour to replace existing vehicle components to lighten existing combat vehicles [2]. It also has been proposed as an applique armour for vehicles such as the ubiquitous M113 family of vehicles [3] and ultimately it is a favoured candidate for the structural armour of the next generation of combat vehicles.

The main Titanium alloy used for armour applications is Ti-6Al-4V. It has been recognised in recent times that the Military Specification for Titanium Armour [4], extra low interstitial Ti-6Al-4V, was expensive and not functionally required for many armour applications. Substantial work was therefore undertaken to characterise the ballistic performance of Ti-6Al-4V that had more relaxed compositional requirements [5-7]. A significant amount of work [8,9] has also recently been performed to evaluate whether the ballistic performance of Ti-6Al-4V can be enhanced by improving its thermo-mechanical processing and heat treatment, identifying which of these processes are of the greatest importance. A new Military Specification for Titanium Armour which allows for lower cost grades has also recently been promulgated [10].

The current paper evaluates the ballistic properties of Ti-6Al-4V that was manufactured using a new low-cost Electron-Beam Cold-Hearth Single-Melt Process [11]. Such a one-step process should have significant cost advantages over Ti-6Al-4V produced using conventional double or triple melting.

Overall, the key to producing low-cost Titanium armour, is the combination of new low-cost production methods, revised military specifications that allow for low-cost grades and a better understanding of what thermo-mechanical processing is essential for improved ballistic performance.

2 Electron-Beam Cold-Hearth Single-Melt Titanium

The Electron-Beam Cold-Hearth Single-Melt process uses an electron beam to remelt titanium sponge or scrap under vacuum. Ti-6Al-4V normally requires two vacuum arc furnace remelts to refine its composition for aerospace applications (or even three remelts for critical aerospace applications) [11]. While Cold-Hearth melting has been used in the past to substitute for one of the vacuum arc remelts and remove inclusions [11], substantial cost reductions could be achieved if a single Cold-Hearth remelt was used to produce Ti-6Al-4V for armour applications. Such a process will also be low cost because it uses low-cost raw materials, e.g. 100% scrap or sponge, and the final product can be rolled in conventional steel rolling mills. However, it is only recently that production experience has been obtained with plate from single melts of Ti-6Al-4V [11]. This paper further examines the ballistic properties of such plate.

The ingot met the compositional and mechanical property requirements of Class 2 Titanium Armour [10], with the fabrication, composition and mechanical properties of the ingot and plate being reported previously [11]. The Ti-6Al-4V plate that was tested had a maximum thickness of 25.4 mm and a measured hardness of 330 HV for 12.7 mm APM2 targets and 37-40 HRC for 12.7 mm SLAP targets.

Ti-6Al-4V is highly susceptible to adiabatic shear and its mechanical properties are highly anisotropic, and the latter may be attributed to both texture and banding of the microstructure [12]. The banding is parallel to the plane of rolling, is caused by variations in composition [12] and can be seen in the lower magnification image in Figure 1. The bands have a nominal maximum separation of approximately 0.25 mm, and would help contribute to any planes of weakness parallel to the plane of rolling.

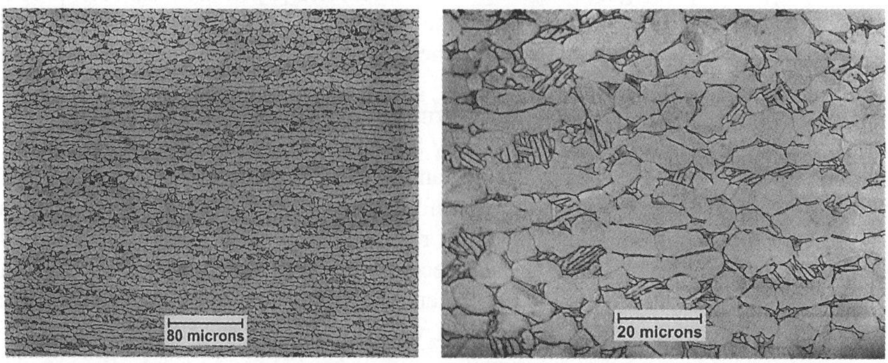

Figure 1. The microstructure of Electron-Beam Cold-Hearth Single-Melt Ti-6Al-4V at two magnifications. The light grains are α phase wheras the dark grains are transformed β phase. The rolling direction is across the page.

3 Ballistic Limit Testing

The ballistic performance of the plate had previously been assessed against 20 mm Fragment Simulating Projectiles and 30 mm APDS rounds [11] and found to compare favourably against the minimum standard specified in the Military Specification [10]. The current study examines the performance of the plate against 12.7 mm APM2 and 12.7 mm SLAP (Sabot Launched Armour Piercing) rounds. The performance of the former round against Titanium Armour is well known [4] wheras much less information is known about the performance of the latter projectile. Table 1 compares these two projectiles.

Table 1. Test Projectiles

Projectile	Core	Mass (g)	Length (mm)	Core Diameter (mm)
12.7 mm APM2	Hardened Steel 60-65 Rc	25.4	47.2	10.9
12.7 mm SLAP	Tungsten Alloy	23.2	37.0	7.7

Ballistic performance is commonly determined by measuring the ballistic limit or V_{50} velocity of a given thickness of material. As projectile velocity is increased there is a transition from projectiles being stopped (termed as partial penetrations) to projectiles perforating (termed as complete penetrations). The velocity at which 50% of impacts are partial penetrations and 50% of impacts are complete penetrations is the V_{50} or ballistic limit velocity. A V_{50} was obtained for three projectile/test temperature combinations at normal obliquity to the test target. Five partial and five complete penetrations were obtained for each test condition according to TOP 2-2-710 [13], with the method of DiDonato and Jamagin, Jr [14] being used to determine a precise V_{50} and its standard deviation. No witness sheet was used for the V_{50} determination for the APM2 tests, but this was not considered to have affected results in the present instance.

The 12.7 mm APM2 tests were performed with a standard 0.50 Cal. Browning HMG with a 45 inch long barrel at the DSTO Aeronautical and Maritime Research Laboratory (AMRL) wheras the 12.7 mm SLAP tests were performed with a 36 inch long 0.50 Cal. Mann barrel at the Canadian Defence Research Establishment Valcartier (DREV). Velocity measurements at AMRL were made with Oehler velocity screens wheras DREV velocity measurements were made with four orthogonal flash radiography stations, two in front of the target and two behind.

4 Results

Table 2 summarises the ballistic test results. The APM2 results are compared to the minimum ballistic performance requirement for the military specification [4], and it is found that the results compare favourably. The SLAP performance was determined at two temperatures: room temperature (RT), a nominal 20°C, and -40°C, a standard temperature for the assessment of protection at low temperatures. Greater ballistic performance was obtained with decreasing temperature, i.e. a 25 ms^{-1} increase in ballistic limit for a 60°C decrease in temperature. Overall, the plate was found to perform similarly against both APM2 and SLAP projectiles. This is an interesting co-incidence as the projectiles have significantly different characteristics as shown in Table 1.

Table 2. Test Results

Projectile	Test Location	Temperature (°C)	V_{50} (ms^{-1})	Standard Deviation (ms^{-1})	Spread (ms^{-1})
12.7 mm APM2	AMRL	RT	687	9.3	33
MIL-A-46077D [4]	-	RT	682	-	-
12.7 mm SLAP	DREV	RT	682	12.5	43
12.7 mm SLAP	DREV	-40	707	2.4	25

5 Discussion and Analysis

Titanium Armour has similar strength to armour steel and could be expected to have similar ballistic performance but with greater mass efficiency, i.e. it is nominally 60% of the weight of an equivalent thickness of steel. However, as mentioned earlier Ti-6Al-4V normally suffers from a spalling or discing failure mechanism on its back face because of the presence of planes of low fracture toughness which will reduce its ballistic performance and usually leads to the ejection of spall, though in the present instance, there was little behind-armour debris ejected. Figure 2(a) and (b) show the rear faces of targets penetrated by APM2 and SLAP projectiles, respectively.

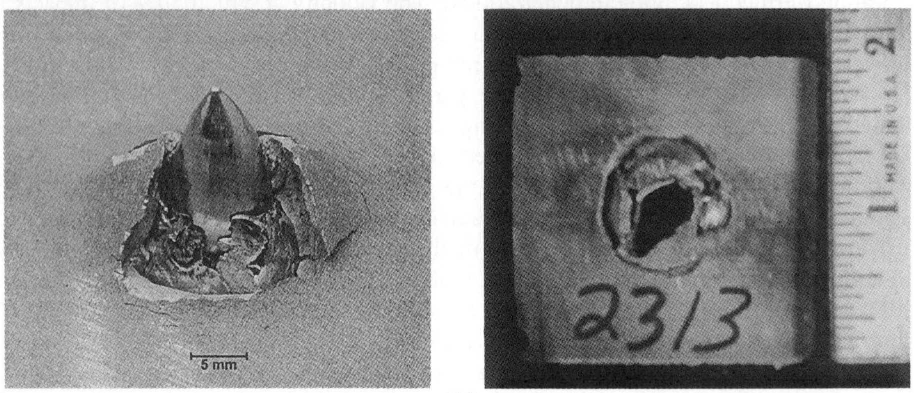

(a) (b)

Figure 2. The rear faces of typical targets perforated by: (a) 12.7 mm APM2 and (b) 12.7 mm SLAP projectiles.

The failure mechanisms developed during an armour/penetrator interaction will have a critical influence on ballistic performance. Simple plasticity-based analytical models are therefore used to study the mechanisms involved in the APM2 projectile penetration and predict its ballistic performance. This is possible as this projectile is non-deforming unlike the SLAP projectile, allowing the target failure mechanism to be readily analysed.

A first-order estimate of the ballistic limit can be made using the Taylor ductile hole formation failure (DHF) mechanism [15]. The work done in radially expanding a hole from zero diameter to diameter **D** in a plate of thickness **h**, $\mathbf{W_{DHF}}$, is

$$W_{DHF} = \frac{\pi}{2} D^2 Yh$$

where **Y** is an effective flow stress, which for Ti-6Al-4V of similar strength is 1685 MPa [16]. Provided this formula is used with care, reasonable lower-bound estimates can usually be made of the ballistic limit, though Ti-6Al-4V is normally an exception to this rule [16]. A ballistic limit of 793 ms^{-1} was obtained, which substantially overpredicts the experimental result as expected.

In the present case the rear of the APM2 targets actually failed by petalling, Figure 2(a), which can be treated as dishing, i.e. stretching and bending deformation on the rear of the target [15]. The petalling pattern is caused by the circumferential strains associated with dishing exceeding the failure strain of the material [15]. A better prediction can be obtained by combining the work done for a dishing mechanism for the rear of the target with the work done by a ductile hole formation mechanism for the front of the target [15]. Thus a thickness $\mathbf{h = h_o\text{-}0.5D}$ is considered to deform by ductile hole formation (where $\mathbf{h_o}$ is the original plate thickness) while a thickness $\mathbf{h = 0.5D}$ deforms by dishing [17]. The work of dishing, $\mathbf{W_D}$, can be predicted by

$$W_D = \sum_{i=1}^{n} \frac{\pi}{8} DY h_i \left(D + 2\pi h_i \right)$$

where $\mathbf{W_D}$ is obtained by summing the terms for **n** dishing plates, each of thickness $\mathbf{h_i}$, with $\mathbf{n = 1}$ in the present instance. However, since the Taylor equation normally assumes stretching on the rear of a target, the work of stretching or Thomson Work [15], $\mathbf{W_S}$, must be subtracted from the combination of $\mathbf{W_{DHF}}$ and $\mathbf{W_D}$.

$$W_S = \frac{\pi}{8} D^2 Yh$$

where again $\mathbf{h = h_o\text{-}0.5D}$. A ballistic limit of 714 ms^{-1} was predicted, which is a substantial improvement but again overpredicts the experimental result.

Closer inspection of the specimen revealed that the rear, dishing portion of the target was approximately one half projectile diameter in thickness and appeared to consist of two layers, as shown Figure 2(a) where an outer petal has broken away, with the layers forming along planes of weakness in the plane of rolling. The thickness of these dishing layers varied with orientation to the rolling direction, but they were typically between 1.5 mm and 3 mm in thickness. The previous calculations were therefore repeated but with the dishing thickness of $\mathbf{h = 0.5D}$ being split into two dishing layers of equal thickness, i.e. $\mathbf{h}$ = 2.7 mm, and $\mathbf{n = 2}$. A ballistic limit of 676 ms^{-1} was obtained, which is in excellent agreement with experiment, though the degree of agreement may to some extent be fortuitous.

While simple plasticity-based models enable first-order predictions to be made of ballistic performance, they also identify the importance of the various mechanisms to the energy absorption process and in this case quantitatively illustrate how planes of weakness parallel to the rolling direction contribute to target deformation and ballistic performance. Follow-on work will carefully characterise the ballistic deformation and damage in the target plate to further understand the nature of the deformation and ultimately improve predictive modelling efforts.

6 Conclusions

Ti-6Al-4V plate produced via a new low-cost Electron-Beam Cold-Hearth Single-Melt Process was found to meet the minimum requirements of the Titanium Armour military specification. It was also determined that while ballistic performance increased with decreasing temperature, the magnitude of this effect was not large. Simple plasticity-based analytical models were found to be capable of predicting the ballistic limit for a non-deforming projectile as well as demonstrate how important a precise understanding of target deformation and failure is for accurate predictions of ballistic performance.

Acknowledgements

The authors would like to thank M.G.H. Wells, formerly of the US Army Research Laboratory, for his efforts in initiating the joint work presented in the current paper.

References

1. Pitler, R. and Hurlich, A. Some Mechanical and Ballistic Properties of Titanium and Titanium Alloys, Watertown Arsenal Laboratory Report, WAL-TR-401/17, Watertown Arsenal Laboratory, MA, March 1950.
2. Montgomery J.S., Wells M.G.H., Roopchand, B. and Ogilvy J.W. Low-Cost Titanium Armors for Combat Vehicles, Journal of Metals, **49** (1997) pp. 45-47.
3. RMI Titanium Company, Titanium Armor, Product Brochure.
4. U.S. Military Specification. Military Specification, Armor Plate, Titanium Alloy, Weldable. MIL-A-46077D, 28 April 1978.
5. Gooch, W.A., Burkins, M.S., Ernst, H.-J., and Wolf, T.J., Ballistic Performance of Titanium Against Laboratory Penetrators with Aspect-Ratios of 10 or Greater, 15th International Symposium on Ballistics, Jerusalem, Israel, 21-24 May 1995, ADPA.
6. Gooch, W.A., Burkins, M.S., Ernst, H.-J., and Wolf, T.J., Ballistic Penetration of Titanium Alloy Ti-6Al-4V, Lightweight Armour Systems Symposium '95, The Royal Military College of Science, Shrivenham, Swindon, England, 28-30 June 1995.
7. Gooch, W.A., Burkins, M.S., and Frank, K. Ballistic Performance of Titanium Against Laboratory Penetrators, First Australasian Congress on Applied Mechanics, Melbourne, Australia, 21-23 February 1996, IE(Aust.), pp. 47-52.
8. Burkins, M.S., Hansen, J.S., Paige, J.I. and Turner, P.C. The Effect of Thermomechanical Processing on the Ballistic Limit Velocity of Extra Low Interstitial Titanium Alloy Ti-6Al-4V, ARL-MR-486, U.S. Army Research Laboratory, Aberdeen Proving Ground, MD, July 2000.
9. Burkins, M.S., Love W.W. and Wood, J.R. Effect of Annealing Temperature on the Ballistic Limit Velocity of Ti-6Al-4V ELI, ARL-MR-359, U.S. Army Research Laboratory, Aberdeen Proving Ground, MD, August 1997.
10. U.S. Military Specification. Detail Specification, Armor Plate, Titanium Alloy, Weldable. MIL-DTL-46077F, 29 Dec. 1998.
11. Wells, M., Burkins, M.S., Fanning, J.C. and Roopchand, B. The Mechanical and Ballistic Properties of Electron-Beam Cold-Hearth Single-Melt Ti-6Al-4V Plate, 9th World Conference on Titanium, St. Petersburg, Russia, 7-11 June 1999.
12. Woodward, R.L. Metallographic Features Associated with the Penetration of Titanium Alloy Targets, Met. Trans. A, **10A** (1979) pp. 569-573.
13. U.S. Army Test and Evaluation Command, Ballistic Tests of Armor Materials, TOP 2-2-710, Aberdeen Proving Ground, MD, 7 February 1984.
14. DiDonato, A.R. and Jamagin, Jr. M.P. Maximum Likelihood Estimates in Quantal Experiments, SIAM J. Appl. Math, **26** (1974) pp. 447-454.
15. Woodward, R.L. The Penetration of Metal Targets by Conical Projectiles, Int. J. Mech. Sci., **20** (1978) pp. 349-359.
16. Woodward, R.L. A Rational Basis for the Selection for Armour Materials, J. Aust. Inst. Metals, **22** (1977) pp. 167-170.
17. Woodward, R.L. and Cimpoeru, S.J. A Study of the Perforation of Aluminium Laminate Targets, Int. J. Impact Engng, **21** (1998) pp. 117-131.

EXTENSIONAL WAVE PROPAGATION IN PRE-STRESSED IMPERFECTLY BONDED INCOMPRESSIBLE LAYERED ELASTIC COMPOSITES

SOMSAK LEUNGVICHCHAROEN AND ANIL C. WIJEYEWICKREMA

Department of Civil Engineering, Tokyo Institute of Technology, O-okayama, Meguro-ku, Tokyo 152-8552, Japan E-mail: anil@cv.titech.ac.jp

In the present analysis harmonic extensional wave propagation in a pre-stressed, incompressible, symmetric layered composite is considered. The bi-material composite consists of two isotropic elastic materials. The imperfect interface is simulated by a shear spring type resistance model that can accommodate the extreme cases of perfect bonding or a sliding interface. Using the technique of the propagator matrix, the dispersion relation is obtained. Four possible cases of dispersion curves that depend on the limiting shear wave speeds of the bi-material composite are discussed. The effect of the imperfect interface is more pronounced for low wave numbers.

1 Introduction

Elastic wave propagation in infinite layers has previously been discussed in the monographs of Achenbach [1], Eringen and Suhubi [2] and Graff [4], where it has been shown that there are two uncoupled modes of wave propagation, viz. symmetric mode (extensional waves) and antisymmetric mode (flexural waves). The effect of finite primary deformations on waves in elastic layers have been studied by Ogden and Roxburgh [6] and Rogerson and Fu [7]. Motivated by the wide usage of layered composites in engineering practice, wave propagation in pre-stressed layered composites have been considered by Rogerson and Sandiford [8-10].

In general, layered composites are assumed to be perfectly bonded, although in practice due to defects in manufacturing or damage due to loading, the interface may be unable to perfectly transfer stresses due to displacement discontinuities at the interface. Such a case is considered in this paper where a shear spring type resistance model is used to simulate an imperfect interface. Both perfect bonding or a sliding interface can be accommodated by such a resistance model.

The basic theory and equations of incremental elastic deformations are given by Ogden [5]. The bi-material, symmetric composite is assumed to consist of isotropic incompressible elastic material. Four possible cases of dispersion curves that depend on the limiting shear wave speeds of the bi-material composite are discussed, for the case of extensional wave propagation.

2 Formulation of the problem

The layered incompressible composite shown in Fig. 1 consists of outer layers of identical isotropic material of thickness h and an inner layer of isotropic material of thickness $2d$.

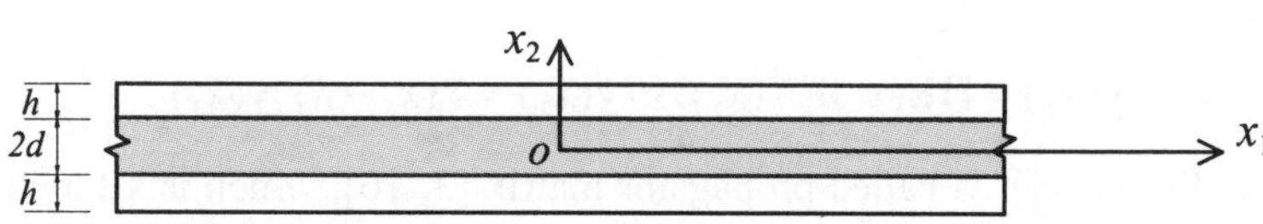

Figure 1. Geometry of symmetric layered composite.

2.1 Basic equations and harmonic wave solution

From the theory of incremental elastic deformations [5] the equations of motion, for pre-stressed, incompressible, elastic materials under plane strain conditions can be written as

$$\begin{aligned} &B_{1111}u_{1,11} + (B_{1122} + B_{2112})u_{2,21} + B_{2121}u_{1,22} - p^*_{,1} = \rho\ddot{u}_1 , \\ &(B_{1221} + B_{2211})u_{1,12} + B_{1212}u_{2,11} + B_{2222}u_{2,22} - p^*_{,2} = \rho\ddot{u}_2 , \end{aligned} \tag{1}$$

where B_{ijkl} are components of the appropriate fourth order elasticity tensor, u_i are the displacements superimposed upon a pre-stressed equilibrium state, ρ is the density of the material, p^* is a time-dependent pressure increment and comma indicates differentiation with respect to the spatial coordinates. The two stress increments σ_{21} and σ_{22} can be written as

$$\begin{aligned} &\sigma_{21} = B_{2121}u_{1,2} + (B_{2112} + \bar{p})u_{2,1} , \\ &\sigma_{22} = B_{2211}u_{1,1} + (B_{2222} + \bar{p})u_{2,2} - p^* , \end{aligned} \tag{2}$$

with $\bar{p}$ denote a static pressure.

The harmonic wave solution of Eq. (1) is expressed in the form

$$(u_1, u_2, p^*) = (A_1,\ A_2,\ kP)\ e^{kqx_2} e^{ik(x_1 - vt)} , \tag{3}$$

where the A_1, A_2 and P are unknown coefficients and the parameter q is determined by inserting Eq. (3) into Eq. (1), and the use of the incompressibility condition.

2.2 Propagator Matrix

The solutions of Eq. (3) may be used with Eq. (2) to establish that

$$\mathbf{y}(x_2) = \mathbf{HE}(x_2)\mathbf{a} , \tag{4}$$

where $\mathbf{a}$ and $\mathbf{H}$ are respectively, a vector of arbitrary constants and 4×4 matrix which are independent of position x_2, and $\mathbf{y}(x_2)$ and $\mathbf{E}(x_2)$ are a displacement-stress increment vector and a diagonal matrix given by

$$\begin{aligned} &\mathbf{y}(x_2) = (-iU_1(x_2),\ U_2(x_2),\ \frac{\tau_{21}(x_2)}{ik},\ \frac{\tau_{22}(x_2)}{k})^{\mathrm{T}} , \\ &\mathbf{E}(x_2) = \mathrm{diag}(e^{kq_1x_2}, e^{-kq_1x_2}, e^{kq_2x_2}, e^{-kq_2x_2}) , \end{aligned} \tag{5}$$

where $U_1(x_2)$, $U_2(x_2)$, $\tau_{21}(x_2)$ and $\tau_{22}(x_2)$ contain the x_2 dependent parts of the displacement and stress increments u_1, u_2, σ_{21} and σ_{22}, respectively.

The arbitrary constant vector $\mathbf{a}$ is eliminated from Eq. (5) by introducing the vector $\mathbf{y}(\bar{x}_2)$ at some location $x_2 = \bar{x}_2$ to yield

$$\mathbf{y}(x_2) = \mathbf{HE}(x_2)\mathbf{E}^{-1}(\bar{x}_2)\mathbf{H}^{-1}\mathbf{y}(\bar{x}_2) = \mathbf{P}(x_2 - \bar{x}_2)\mathbf{y}(\bar{x}_2) . \tag{6}$$

The matrix $\mathbf{P}(x_2 - \bar{x}_2)$ is called propagator matrix [3, 10], which is a function of wave number k, phase speed v and the distance $x_2 - \bar{x}_2$.

Equations (3)-(6) are equations for the outer layer. The equations for the inner layer are obtained from the previous equations by interchanging $q \to p$ and denoting quantities associated with the inner layer by a tilde, for example $B_{ijkl} \to \tilde{B}_{ijkl}$ and $\mathbf{y}(x_2) \to \tilde{\mathbf{y}}(x_2)$.

2.3 Mid Plane, Boundary and Interfacial Conditions

Due to the symmetric properties of extensional waves, only the upper half of the composite is considered. The mid plane conditions and the boundary conditions at the surface can be written as

$$u_2(0) = \sigma_{21}(0) = 0\,, \quad \sigma_{21}(d+h) = \sigma_{22}(d+h) = 0\,, \tag{7}$$

and the continuity conditions at the interface between inner layer and outer layer are

$$u_2(d) = \tilde{u}_2(d)\,, \quad \sigma_{21}(d) = \tilde{\sigma}_{21}(d)\,, \quad \sigma_{22}(d) = \tilde{\sigma}_{22}(d)\,. \tag{8}$$

The shear spring condition is introduced to the interface between inner layer and outer layer by

$$\sigma_{21}(d) = \frac{k_x \mu}{d}\{U_1(d) - \tilde{U}_1(d)\}\,, \tag{9}$$

where k_x and μ are the dimensionless spring constant and the Lamé parameter respectively.

2.4 Dispersion Relation

The propagator matrix with conditions in Eqs. (7)-(9) yield a system of four equations in four unknowns, from which the dispersion relation for extensional waves in an imperfectly bonded composite is obtained as

$$\left\{P_{3i}\tilde{P}_{i1}P_{4j}\tilde{P}_{j4} - P_{3i}\tilde{P}_{i4}P_{4j}\tilde{P}_{j1}\right\} + \frac{dk}{k_x \mu}\left\{P_{3i}\tilde{P}_{i1}P_{41}\tilde{P}_{34} + P_{4i}\tilde{P}_{i4}P_{31}\tilde{P}_{31} - P_{3i}\tilde{P}_{i4}P_{41}\tilde{P}_{31} - P_{4i}\tilde{P}_{i1}P_{31}\tilde{P}_{34}\right\} = 0\,, \tag{10}$$

where the repeated indices i, j are summed from 1 to 4.

Considering the extreme values of shear spring stiffness k_x i.e., $k_x \to \infty$ the left part of the above equation is the dispersion relation for a perfectly bonded interface which agrees with Rogerson and Sandiford [8], and when $k_x \to 0$ the right part is the dispersion relation for a fully sliding interface. Thus, both extreme cases can be recovered from Eq. (10).

3 Numerical Results

The same materials used in [8] are used in the present analysis for the layered composite. Table 1 shows the parameters of those materials of which α, β and γ are defined by

$$\alpha = B_{1212}, \quad 2\beta = B_{1111} + B_{2222} - 2B_{1122} - 2B_{1221}, \quad \gamma = 2B_{2121}. \tag{11}$$

Table 1. Material parameters

Material	α	2β	γ	ρ
1	2.0	3.5	1.5	1
2	4.0	5.0	1.0	1
3	4.05	5.4	1.8	1
4	3.52	1.76	0.22	1

For an incompressible, pre-stressed, elastic material the limiting wave speed v_L depends on the above materials parameters as follows,

$$\rho v_L^2 = \begin{cases} \rho v_{S_1}^2 = \alpha & ; \alpha \le 2\beta \\ \rho v_{S_2}^2 = 2\beta - 2\gamma + 2\sqrt{\gamma}\sqrt{\alpha + \gamma - 2\beta} & ; \alpha > 2\beta. \end{cases} \tag{12}$$

From Eq. (12) it is seen that there are four different cases of dispersion curves which depend on limiting wave speeds of bi-material composite as shown in Table 2. The effect of the imperfect interface is presented in Figs. 2-5 for the shear spring parameter $k_x\mu/d$ = ∞, 100, 25, 1, 0.001, 0 ; where σ_2 is Cauchy principal stress along the x_2 − axis.

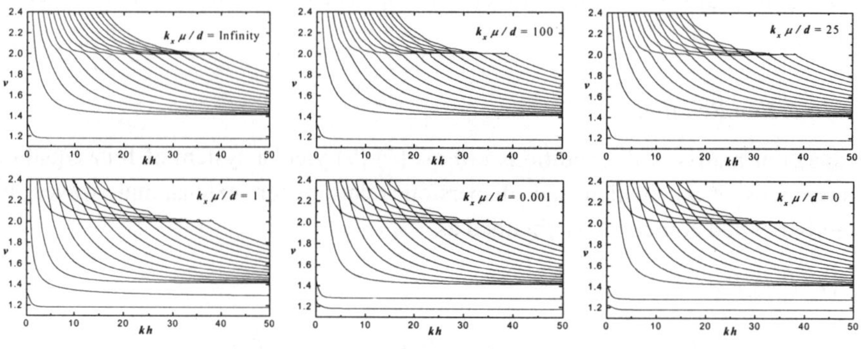

Figure 2. Dispersion curves for Case 1 (σ_2 = 2.5).

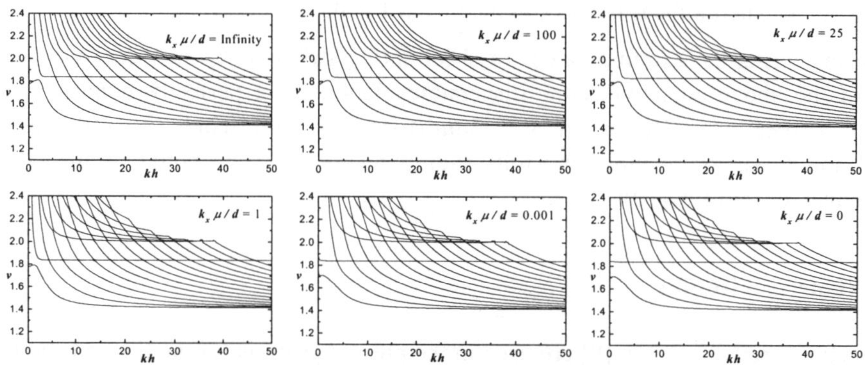

Figure 3. Dispersion curves for Case 2 (σ_2 = 1.8).

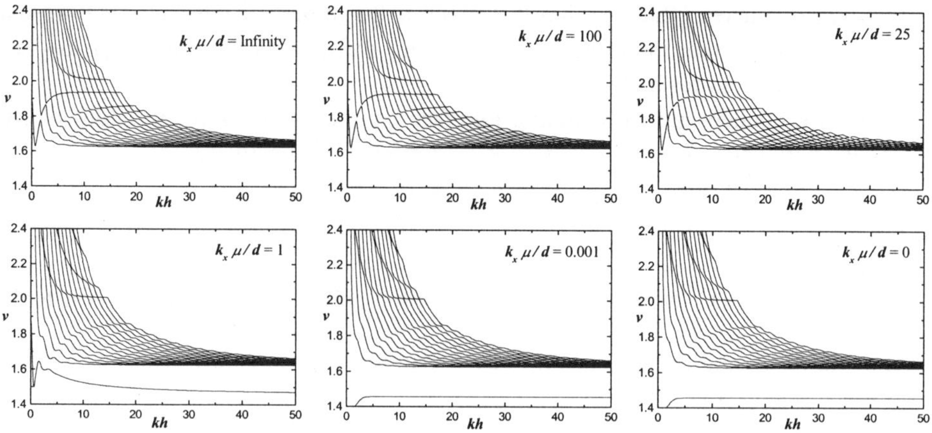

Figure 4. Dispersion curves for Case 3 (σ_2 = 1.0).

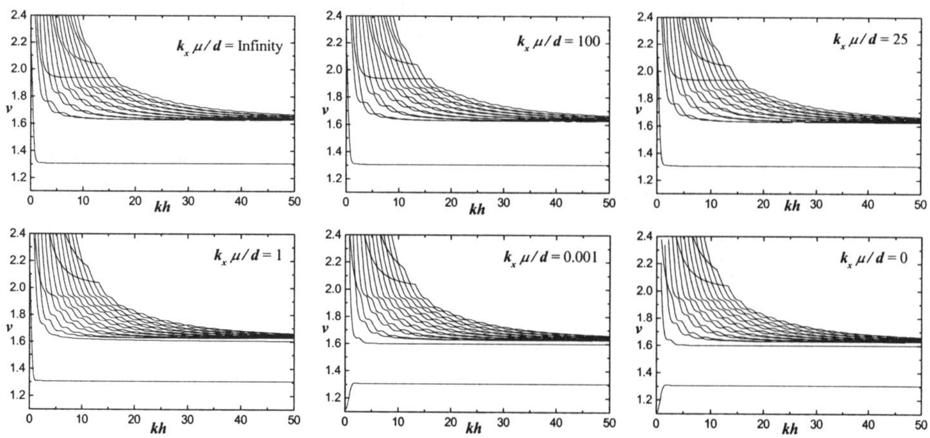

Figure 5. Dispersion curves for Case 4 (σ_2 = 0.5).

Table 2. Limiting velocities for large wave number for different material combinations

Case	Condition	Material		ρv_L^2		$\tilde{\rho}\tilde{v}_L^2$	
		o	*i*	$\rho v_{S_1}^2$	$\rho v_{S_2}^2$	$\tilde{\rho}\tilde{v}_{S_1}^2$	$\tilde{\rho}\tilde{v}_{S_2}^2$
1	$v \to v_{S_1}$ when $\alpha \le 2\beta$ and $\rho v_{S_1}^2 < \tilde{\rho}\tilde{v}_L^2$	1	2	2	–	4	–
2	$v \to \tilde{v}_{S_1}$ when $\tilde{\alpha} \le 2\tilde{\beta}$ and $\tilde{\rho}\tilde{v}_{S_1}^2 < \rho v_L^2$	2	1	4	–	2	–
3	$v \to \tilde{v}_{S_2}$ when $\tilde{\alpha} > 2\tilde{\beta}$ and $\tilde{\rho}\tilde{v}_{S_2}^2 < \rho v_L^2$	3	4	4.05	–	–	2.64
4	$v \to v_{S_2}$ when $\alpha > 2\beta$ and $\rho v_{S_2}^2 < \tilde{\rho}\tilde{v}_L^2$	4	3	–	2.64	4.05	–

Note : ***o*** : outer layer, ***i*** : inner layer

4 Conclusions

- Dispersion relation for extensional waves in an imperfectly bonded layered composite is obtained, where both perfectly bonded or fully sliding interfaces are accommodated in this dispersion relation.

- Dispersion curves which exhibit the effect of the imperfect interfaces are presented and it is found that the effect of shear spring stiffness, is quite high for low wave numbers and less for high wave numbers.
- Perfectly bonded and fully sliding interface can be assumed when values of shear spring parameter $k_x\mu/d > 25$ and $k_x\mu/d < 0.001$, respectively.

References

1. Achenbach, J. D., Wave propagation in elastic solids (North-Holland, Amsterdam, 1973).
2. Erigen, A. C. and Suhubi, E. S., Elastodynamics, Vol. 2: Linear Theory (Academic Press, London, 1975).
3. Gilbert, F. and Backus, G. E., Propagator matrices in elastic wave and vibration problems, *Geophysics* **31** (1966) pp. 326-332.
4. Graff, K. F., Wave motion in elastic solids (Clarendon Press, Oxford, 1975).
5. Ogden, R. W., Non-linear elastic deformations (Ellis Horwood, New York, 1984).
6. Ogden, R. W. and Roxburgh, D. G., The effect of pre-stressed on the vibration and stability of elastic plates, *Int. J. Engng. Sci.* **31** (1993) pp. 1611-1639.
7. Rogerson, G. A. and Fu, Y., An asymptotic analysis of the dispersion relation of a pre-stressed incompressible elastic plate, *Acta Mechanica* **111** (1995) pp. 59-74.
8. Rogerson, G. A. and Sandiford, K. J., The effect of finite primary deformations on harmonic waves in layered elastic media, *Int. J. Solids Structures* **37** (2000) pp. 2059-2087.
9. Rogerson, G. A. and Sandiford, K. J., Flexural waves in incompressible pre-stressed elastic composites, *Q. J. Mech. Appl. Math.* **50** (1997) pp. 597-624.
10. Rogerson, G. A. and Sandiford, K. J., On small amplitude vibrations of pre-stressed laminates, *Int. J. Eng. Sci.* **34** (1996) pp. 853-872.

MECHANICAL AND DYNAMIC PROPERTIES OF SPORTS BALLS

K.M. COOK, L.L. KOSS AND D. GOMEZ

Department of Mechanical Engineering, Monash University, Clayton, Victoria, Australia 3800

This study examined mechanical and dynamic properties of a range of balls used in many sports with a view towards relating these properties to physical injury due to impact of a ball and a person. The study examined impact forces, force durations, impulse, impact pressure and rebound heights for thirty different balls that included fourteen cricket balls. The data presented may be employed in models for ball- human impact studies.

1 Introduction

Injuries from cricket and other ball sports can occur in a number of ways that include ball impact with a person. Eye injuries in cricket have been reported by [1,2,3,4], squash ball-eye injuries have been reported by [5,6] and racquet sports and eye injuries have been reported by [7]. In the playing of indoor cricket occurrence of injuries are greater than that of outdoor cricket due to the confined area of play. The effects of ball characteristics on injuries are unknown due to the lack of previous publications in this field. Thus, the aim of this program was to determine mechanical and dynamic properties of a range of sport balls. Balls from cricket, baseball, squash, tennis, roller hockey, golf and table tennis were investigated by dropping them from several heights onto a force hammer, and a hard wooden surface.

2 Tests

Balls tested are outlined in Table 1 along with their material of manufacture, mass, diameter and seam diameter. These balls were dropped from 1, 2 and 4-metre heights onto wood to determine their coefficient of restitution and imprint impact area (maximum area of impact by using carbon paper on top of white paper). Force time histories of impacts were obtained by dropping the balls onto a PCB impulse sledge -hammer with a wood impact surface. The force time histories were stored on an AND 3525 analyzer for further analysis. A hard rubber impact surface was also used as well as the wooden surface. The impulse of the impacts was calculated by measuring the area under the force time histories. A typical force time history of an impact is given in Figure 1. Measured results were non-dimensionalised by using the following dimensionless parameters:

$\pi 1$= measured impact imprint area/frontal area of ball

$\pi 2$= impact pressure/($\rho v^2/2$)

The term "v" is impact velocity and the term "ρ" is ball density, both being in SI units.

Table 1. Ball used in the investigation and their properties.

Symbol	Sport	Material	Description	Mass (g)	Diameter (mm)	Seam Diameter (mm)
BS	Baseball	Leather etc.	Stitched ball (Old)	148	70.59	N/A
BU	Baseball	Plastic etc.	Unstitched ball (Old)	126	70.07	N/A
C4	Cricket	Leather etc.	4 Piece Ball (Old)	162	71.55	70.90
CC	Cricket	Composition	White Compo.	155	72.03	70.54
CD	Cricket	Composition	Dayglo	155	69.63	70.67
CEI	Cricket	Plastic etc.	Incrediball	143	70.65	74.30
CI	Cricket	Leather etc.	Indoor Ball (Old)	122	71.34	73.98
CM	Cricket	Leather etc.	2 Piece Ball (New)	160	70.05	71.56
CN4	Cricket	Leather etc.	4 Piece Ball (New)	165	72.25	73.05
CNI	Cricket	Leather etc.	Indoor Ball (New)	108	71.22	72.97
CNW	Cricket	Leather etc.	Women's Ball (New)	143	69.39	71.60
CP	Cricket	Plastic	Practice Ball	94	69.77	70.25
CRK	Cricket	Rubber etc.	K Practice Ball	153	68.13	66.93
CRT	Cricket	Rubber etc.	T Practice Ball	158	66.62	66.43
CS	Cricket	Plastic etc.	Club Safetyball	97	65.54	68.56
CW	Cricket	Leather etc.	Women's Ball (Old)	136	67.60	69.37
GP	Golf	Plastic	Toy Golfball	8	38.36	N/A
GT	Golf	Unknown	Golfball	46	42.51	N/A
HA	Roller Hockey	Plastic etc.	AGS Ball	88	66.57	N/A
HS	Roller Hockey	Plastic etc.	Street Attack Ball	57	66.63	N/A
NC	None	Foam	Swirly coloured	41	54.63	N/A
NF	None	Coated Foam	Soft Toy	78	62.90	N/A
NL	None	Plastic	Spiky/Lumpy	120	83.06	90.00 (Spikes)
NSL	None	Rubber	Large Superball	41	43.02	N/A
NSM	None	Rubber	Medium Superball	26	36.99	N/A
NSS	None	Rubber	Small Superball	9	24.83	N/A
PP	Ping Pong	Plastic	Ping Pong	2	37.31	N/A
SBD	Squash	Rubber	Blue Dot (Fast)	28	40.39	N/A
SYD	Squash	Rubber	Yellow Dot (Extra Slow)	29	40.31	N/A
TB	Tennis	Rubber etc.	Tennis Ball	56	61.02	N/A

N/A is non applicable.

2.1 *Impulse Results for 2 metre drop height*

Impulse data for collisions between balls and the wood -hammer head and rubber -hammer head for the two- metre drop height is given in Figure 2. The relation is given by

$$\text{Impulse (rubber head)} = 1.18 \text{ Impulse (wood head)} - \text{SI units.}$$

The experimental data is very close to a straight line as seen in Figure 2.

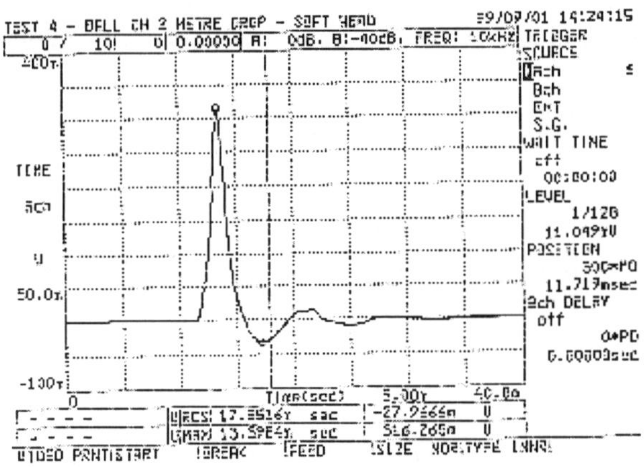

Figure 2. Measured force time history for a two metre drop of the CM ball onto the rubber force hammer head. Horizontal axis is time (40 ms) and peak vertical force is 1096 N.

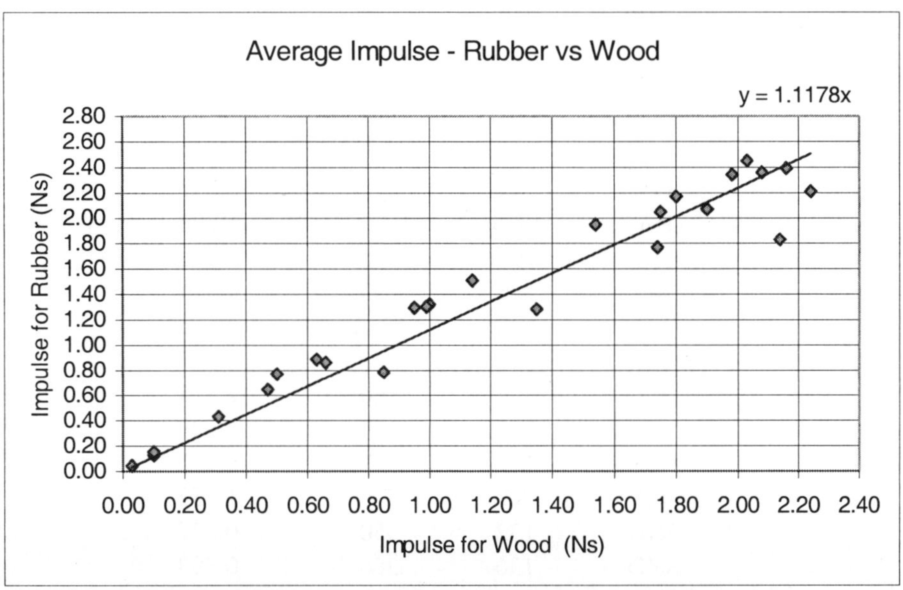

Figure 3. Measured impulse for ball drops onto rubber vs. measured impulse for ball drops onto wood. Each data point is the average of three drops.

2.1.1 Pressure and area data

Measured force, impact imprint area and pressure for the two- metre drop height onto the wood hammer head is given in Table 2 and in non-dimensional format in Figure 3. The non-dimensional relationship is given by

$$\pi 2=31.29\ \pi 1^{-1.619}$$

and the plot of the data appears hyperbolic in shape as shown in Figure 3. An outlier point exists for π_1 equal to 0.05 and is due to the NL ball, spiky-lumpy; this is a soft ball with 90 spikes sticking out from its surface. The outlier result is due to its unusual geometry.

Table 2. Impact force, impact imprint area and pressure for the two metres drop height.

Ball Type	Force N	Area mm^2	Pressure MPa
BS	2056	380	5.411
BU	551	980	0.562
C4	2713	210	12.919
CC	3458	140	24.700
CD	2765	300	9.217
CEI	753	780	0.965
CI	766	680	1.126
CM	2990	230	13.000
CN4	3449	272	12.680
CNI	1635	312	5.240
CNW	2522	212	11.896
CP	332	430	0.772
CRK	713	1150	0.620
CRT	820	1080	0.759
CS	756	340	2.224
CW	2889	190	15.205
GP	309	18	17.167
GT	2152	28	76.857
HA	1974	125	15.792
HS	790	300	2.633
NC	220	700	0.314
NF	390	1240	0.315
NL	142	278	0.511
NSL	444	160	2.775
NSM	312	120	2.600
NSS	189	60	3.150
PP	83	72	1.153
SBD	132	350	0.377
SYD	136	368	0.369
TB	273	860	0.317

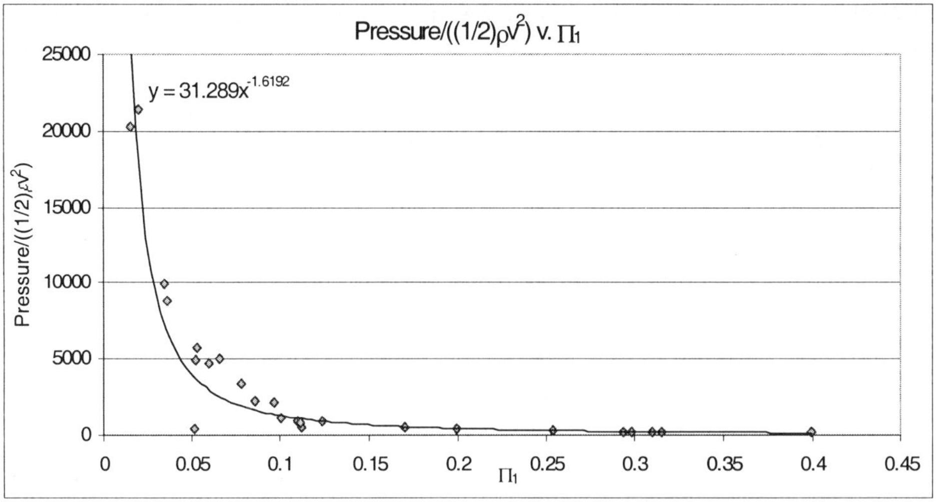

Figure 4 Dimensionless pressure ratio, π2, vs. dimensionless area ratio, π1, for ball impacts onto wood.

3 Conclusions

Experimentally determined mechanical and dynamic data for ball impacts onto wood and rubber surfaces have been presented for a variety of balls used in various sporting activities. The impulse developed between the balls and the two surfaces tested are about the same (within 12 %) for the same drop height although, the force time history for the impact on wood is much sharper than on rubber. Peak forces of up to 3000 N have been measured for impacts on the wood surface and up to 1100 N have been measured for impacts on the rubber surface. These impacts can "conceptually" correspond to ball impacts on to a stiff bone structure and on to a muscular structure, respectively. The data may be used in ball impact studies on human bodies.

4 Acknowledgements

The authors wish to thank Mr. Ron Laemmle and Mr. P. Goldenberg for their assistance in setting up the experiments.

References

1. Auben, N. , Eye injuries in indoor cricket at Wellington hospital: a survey January 1987 to June 1989, New Zealand Medical Journal, 103(1990),pp.454-456.
2. Jones, N.P. and Tullo, A.B., Severe eye injuries in Cricket, British Journal of Sports medicine ,(1986),pp. 178-179.
3. Coroneo, M.T., An eye for cricket, Australian Medical Journal 142,(1985),pp.469-471.
4. Corrigan, A.B., Cricket injuries, Australian Family Physician, 13(8), (1984),pp.558-562.

5. Barrell,G. V., Cooper, P.J., Ellington,A. R., et al, Squash ball to eyeball: the likelihood of squash players incurring an eye injury, British Medical Journal, 283(1981), pp. 893-895.
6. Genovese, M.T., Lenzo, N.P., Lim, R.K. et al,. Eye injuries among pennant squash players and their attitudes, The Medical Journal of Australia 153(1990), pp. 655-658.
7. Elman, M. J., Racket –sports ocular injuries, Archives of Ophthalmology, 104(1986), pp. 1453-1454.

Differential Diagnostics of Gear and Bearing Faults using Spectral Correlation

R. B. RANDALL AND J. ANTONI

School of Mechanical and Manufacturing Engineering
The University of New South Wales, Sydney 2052, Australia
E-mail: b.randall@unsw.edu.au

A case has been encountered in a critical helicopter gearbox, where a spalling fault in a bearing inner race manifested itself almost exclusively at discrete frequencies with a spacing equal to the speed of the shaft on which the bearing was mounted. A fault on a gear on the same shaft would produce the same pattern of frequencies, and so a method is sought to differentiate between the two situations. In the specific case of a fault in a bearing inner race, which passes through the load zone at a rate equal to the shaft speed, the signal is not periodic, but cyclostationary, because the rolling elements are in a different position on the rough fault surface for every revolution. Cyclostationary signals have the property that their autocorrelation function is periodic, so that their spectral correlation contains lines at discrete cyclic frequencies (α), even though it varies continuously in the normal frequency (f) direction. The power spectrum is represented by the line $\alpha = 0$, and the spectral correlation of stationary signals is zero for all other α. The spectral correlation of periodic signals contains discrete peaks in both directions. The proposed method thus uses the fact that at values of α representing multiples of the shaft speed (thus eliminating additive stationary noise), the spectral correlation will have non-zero values between the peaks that come from the periodic part of the combined signal. The paper illustrates the technique using both simulated and actual signals.

1 Introduction

In many machines, signals from gear and bearing faults largely occupy different frequency ranges, and can be separated on this basis. However, in high speed gearboxes, such as helicopter gearboxes, the gear-related signals from harmonics of gearmesh frequencies extend into the high frequency range where also bearing faults manifest themselves, and thus give problems by masking the often weak bearing signals. Where the signals are purely additive, it has been shown that self adaptive noise cancellation (SANC) can be used to separate the bearing from the gear signals on the basis that the gear vibration signals are phase-locked to shaft speeds, while bearing signals have a certain degree of randomness because of varying slip between the components [1].

However, a number of cases have been encountered where the bearing fault signal modulates the gearmesh signals, and they are thus no longer additive and separable by an adaptive filter. Most bearing fault frequencies are at non-integer multiples of the shaft speeds, and this also gives a possibility for separation in the frequency domain, but distributed inner race faults give a modulation at shaft speed (the rate at which the fault is passing through the load zone), and if they are relatively smooth, they do not give strong impacts at the ball-pass frequency. Faults on a gear on the same shaft would give a modulation at the same frequency, and the problem is to distinguish between them.

A possible solution resides in the fact that the gear signal is perfectly periodic, whereas the bearing signal, while being modulated at the shaft speed, would have a somewhat random carrier, because the rolling elements are on a different part of a rough surface for each revolution. A model of the expected type of signal is illustrated in Fig. 1.

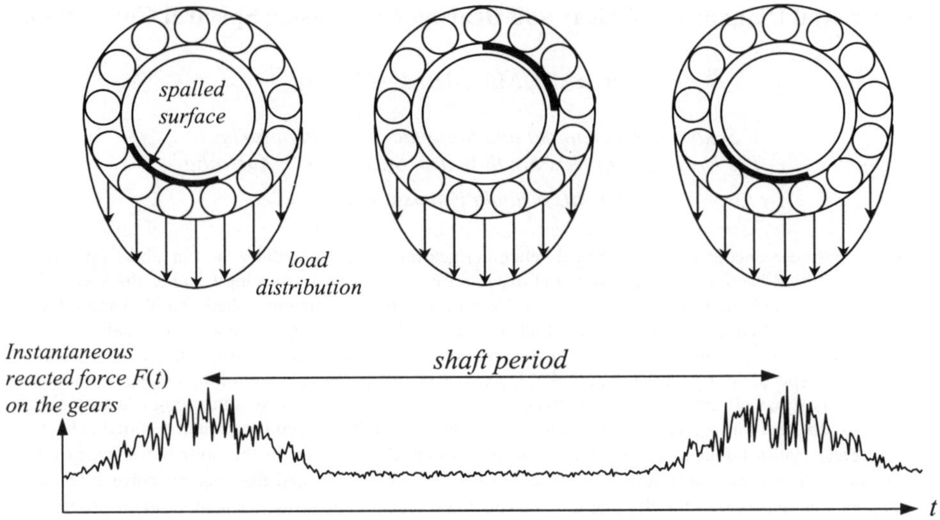

Figure 1: Illustrative scheme for the advanced spalled inner race defect.

Signals of this nature are known as "cyclostationary", as their autocorrelation function is not constant, as it would be for a stationary random signal, but varying periodically. This paper proposes a technique to separate this type of inner race bearing fault from a gear fault, based on the fundamental differences between periodic and cyclostationary signals.

2 Cyclostationary Signals

The bivariate autocorrelation function for a general process can be expressed as:

$$R_x(t,\tau) = E[x(t+\tau/2)x(t-\tau/2)] \tag{1}$$

For stationary signals this does not vary with t and can be represented as $R_x(\tau)$, and its Fourier transform is the power spectrum $S_x(f)$. For cyclostationary signals it is periodic with t, i.e. $R_x(t,\tau) = R_x\,(t+T,\tau)$ and thus a Fourier transform with respect to t will yield discrete frequency components at harmonics of the periodic frequency $1/T$. The two-dimensional Fourier transform of $R_x(t,\tau)$ yields the "spectral correlation density" (CSD) defined by:

$$S_x(\alpha,f) = \underset{t\to\alpha}{\Im}\,\underset{\tau\to f}{\Im}\left(R_x(t,\tau)\right) \tag{2}$$

This quantity has proven especially useful for recognising cyclostationary signals and consequently to distinguish them from stationary or periodic signals. Specifically, the spectral correlation density of cyclostationary signals is made up of a family of parallel spectra, continuous in the f direction but discretely distributed in the α direction, on all multiple of $\alpha = 1/T$ (Fig. 2.a). For stationary signals, only the value $\alpha = 0$ carries a spectrum, which actually is the power spectral density. On the other hand, for periodic signals, the spectral correlation density is composed of a set of discrete points as shown in Figure 2.b.

Based on the fact that gear signals are essentially periodic whereas bearing signals are better modelled as being cyclostationary, the spectral correlation density may be used for distinguishing between gear and bearing faults in complex systems where the two types of faults are likely to occur.

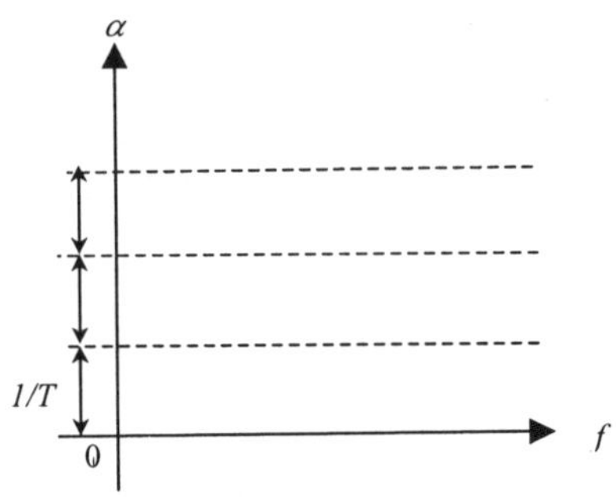

Figure 2.a: 2D structure for the spectral correlation density of cyclostationary signals.

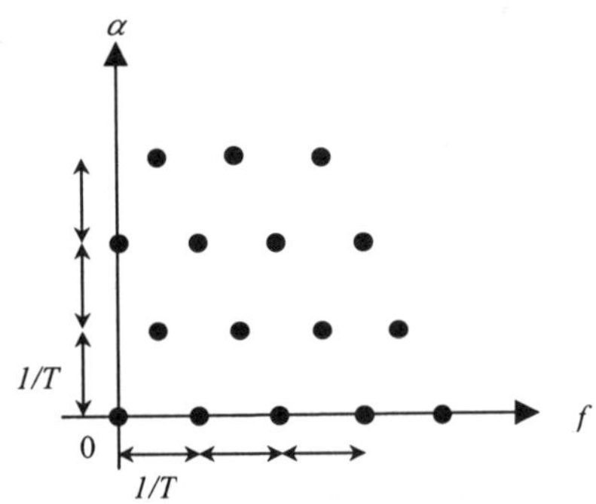

Figure 2.b: 2D structure for the spectral correlation density of periodic signals.

3 Detection scheme

According to the specific structures of the spectral correlation density as described above, the occurrence of an advanced spalled inner race defect (Fig.1) would give rise to non-zero parallel continuous spectral densities in the α-f frequency plane. A detection scheme is thus devised which measures the level of power in the spectral correlation density on those lines which indicate the presence of cyclostationary signals only. For instance, checking for power on the line defined by $\alpha = 1/T$ may be sufficient in practice. In addition, the points at f equal to a multiple of $1/T$ should be disregarded from the analysed line, as they are very likely to carry power from the periodic gear signal as well (Fig. 2.b). Indeed, one very convenient solution is to first remove all the periodic components from the vibration signal before examining it for cyclostationary components. This may be achieved by using the self-adaptive noise cancellation (SANC) technique discussed in [2] or other kinds of comb-filtering techniques.

4 Examples

For illustration, a typical vibration signal from a one-stage gearbox was simulated by amplitude modulating a gearmesh frequency with 6 harmonics. Some white noise was also added to produce a realistic SNR of 5. Figure (3.a) displays the vibration signal where only the modulation from the gear is present. Figure (3.b) displays the expected signal when an advanced spalled inner race defect has occurred. As can be seen from these figures, the two vibration signals are barely distinguishable in spite of the modulating functions being rather different.

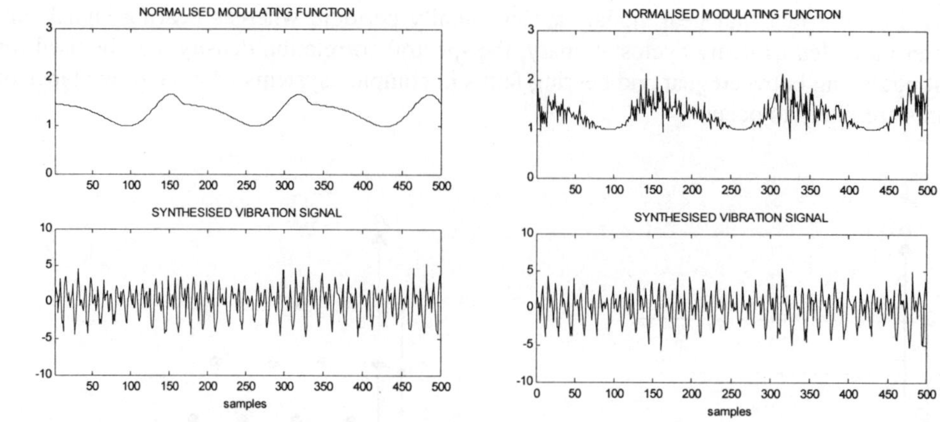

Figure 3: (a) Typical modulating function in the fault free case. (b) Typical modulating function in case of an advanced bearing fault. (c-d) Related vibration signals.

As matter of fact, the detection of the cyclostationary fault cannot be done from the power spectral densities either (Figs. 5.a-b), because the observed increase of 10 dB in power may just come from additional stationary noise. Nevertheless, the increase of 25 dB in the spectral correlation density (Fig. 5.b) taken for $\alpha = 1/T$ does necessarily come from a cyclostationary process and thus can be attributed to the bearing fault.

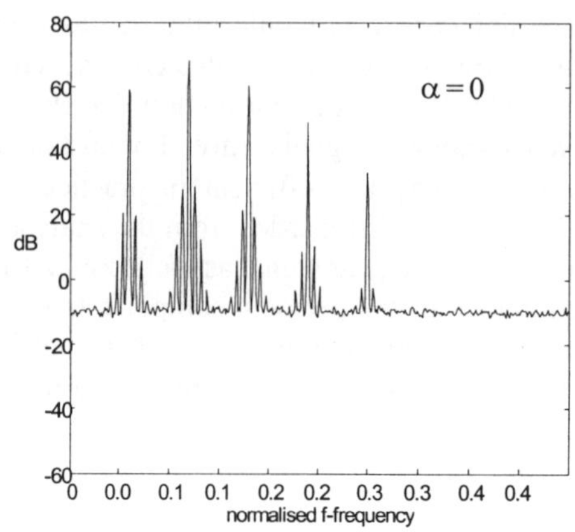

Figure 4.a: Power spectral density of the vibration signal in the fault free case.

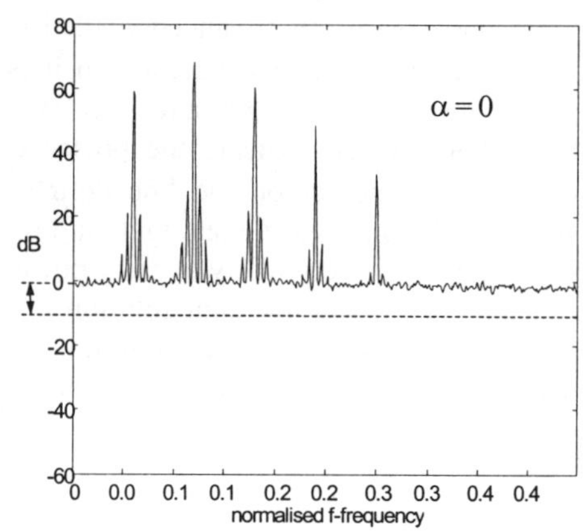

Figure 4.b: Power spectral density of the vibration signal in case of a fault.

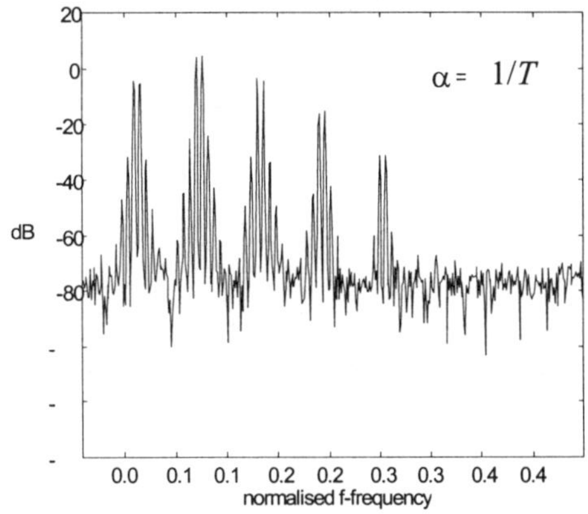

Figure 5.a: Cyclic power spectrum of the vibration signal for $\alpha = 1/T$ in the fault free case.

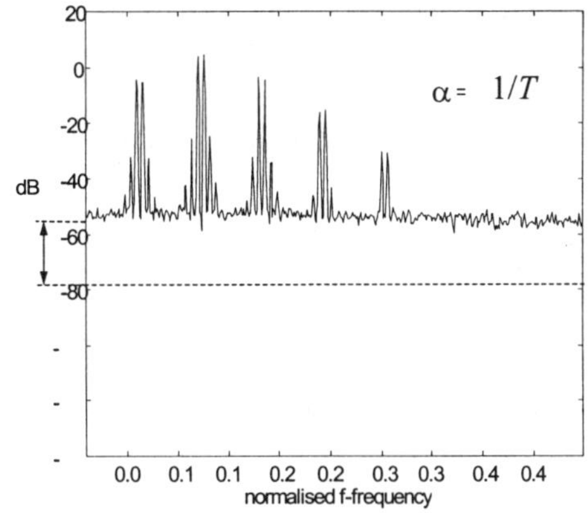

Figure 5.b: Cyclic power spectrum of the vibration signal for $\alpha = 1/T$ in case of a fault.

The next example is taken from an actual gearbox from a Black Hawk Helicopter SH-60, where evidence of a bearing fault of the type investigated in this paper was reported. Although some extra amplitude modulation was apparent on the time signals, there was no way to relate it either to a bearing or a gear fault. Indeed, the power spectral densities in the fault-free and the faulty cases were almost identical. Figure (6) shows the spectral correlation density taken for $\alpha = 1/T$, where again the presence of a bearing fault is undoubtedly evidenced by an overall increase in power. Note that in this case the discrete harmonics stemming from the periodic gear signals were first removed by the SANC.

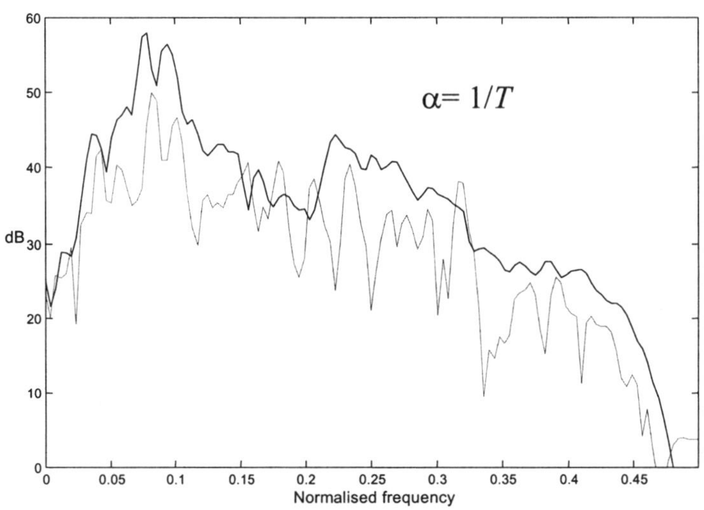

Figure 6: Cyclic power spectrum for $\alpha = 1/T$ after applying SANC. Dotted line: fault-free case. Full line: faulty case.

5 Conclusion

The cyclostationary properties associated with a bearing fault allow them to be separated from a (periodic) gear fault at the same frequency.

Acknowledgement

This work was supported by DSTO via the Centre of Expertise in Helicopter Structures and Diagnostics. The helicopter bearing signal was obtained from the US Naval Air Warfare Center, Aircraft Division.

References

1. Ho, D. and Randall, R. B., Optimisation of Bearing Diagnostic Techniques Using Simulated and Actual Bearing Fault Signals, *Mechanical Systems and Signal Processing*, **14** (5), September (2000), pp. 763-788.
2. J. Antoni and R. B. Randall, Optimisation of SANC for separating gear and bearing signals, *COMADEM Conference*, September 2001, Manchester.

STABLE AXISYMMETRIC ANEURISMS

J.W.H. PRICE

Mechanical Engineering, Monash University, PO Box 197, Caulfield East, Vic, 3145, Australia.

E-mail: john.price@eng.monash.edu.au

In the free axisymmetric blowing of hollow axisymmetric forms, a stable anuerism often develops which then travels through the form. This paper presents a procedure for carrying out experiments where by stresses and strains can be measured during axisymmetric free blowing. This paper explores the relationship between material properties and this behaviour to explain the conditions under which stable anuerisms can occur.

1 Introduction

Axisymmetric free blowing (AFB) is blowing of hollow axisymmetric forms where the word "free" indicates that the blowing is unrestrained by contact with solids or moulds. AFB is a technologically important process that describes the initial stages of the blowing processes used for making containers such as beverage bottles. In this case the material starts as an axisymmetric injection moulded preform (also sometimes called a parison) and then is formed through high strains at elevated temperatures in processes termed blow moulding (Ibrahim et al 1994). Another process where stable aneurisms are formed is glass blowing.

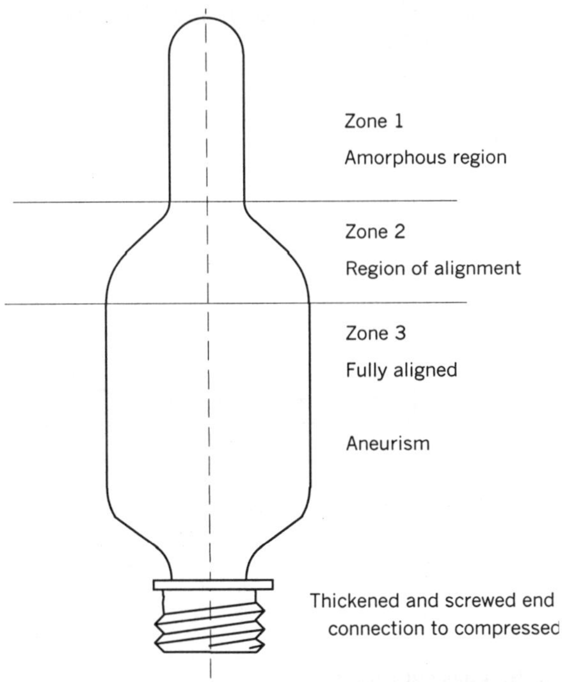

Figure 1 The formation of axisymmetric stable aneurism in a thermoplastic hollow form half way through a blow.

An aneurism is a bulging or dilation of a hollow form and is a term also found in medicine for the dilation of a blood vessel. A stable aneurism can be defined as a dilation which is stable because instead of bursting when it reaches its maximum strain condition, the aneurism progresses steadily and in stable process through the form. Only when the aneurism has progressed entirely through the form is it possible to increase the pressure and finally to burst the form. A very familiar case of this phenomenon is found when blowing up long party balloons.

Not all bulges or aneurisms are stable. For example, bulges on steel pipes occur where there is thinning of the material but these progress to burst (Price, 1993). The formation of stable aneurisms does not normally happen in steels or many other materials because the material properties are inappropriate. This paper explains why.

A procedure for carrying out controlled tests of thermoplastic axisymmetric free blowing has been developed by the author and the theory of this is discussed below. Results from one such experiment are used to explore the conditions under which a stable aneurism occurred.

2 Literature background

The nature of blowing axisymmetric forms from a material that has suitable properties is shown on Figure 1. The expansion of the form on applying internal pressure can be characterised as three zones 1, 2 and 3 (Erwin et al., 1983). Zone 1 corresponds to the state prior to major deformation or preform while Zone 3 can be described as a blown state. Zone 2 is a transition stage between amorphous and crystalline state and is stage of the formation of the aneurism.

The properties of the materials used in these processes are critical to their success. In some commercially important plastics such as Polyethylene Terephthalate, PET, the orientation in one direction affects the properties in each of the other principal directions. A typical stress strain curve for a polymer is shown on Figure 2. The amorphous sage is stage 1, alignment occurs in the "necking region" stage II and deformation of the highly anisotropic polymer occurs in stage III.

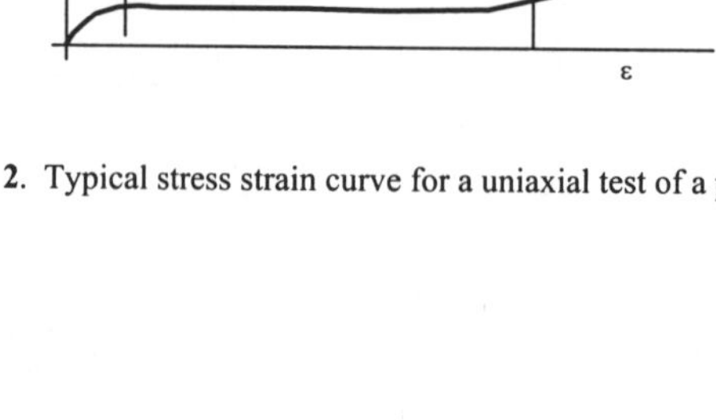

Figure 2. Typical stress strain curve for a uniaxial test of a polymer.

3 Theory

Nomenclature

h	Wall thickness at time t
p	Difference between the internal pressure and the external pressure
r_0	Average radius in the circumferential or hoop direction
r_1	Average radius in the meridional plane (averages of r_φ.)
Δt	Short time interval between two successive strain changes

θ	Circumferential or hoop direction
φ	Meridional angle
σ_θ	Hoop true stress
σ_φ	Meridional tangential true stress
ε_θ	Hoop logarithmic or true strain
ε_φ	Meridional tangential logarithmic or true strain
ε_h	Strain through wall

Space does not permit the presentation of the full derivation of the equations for stress and strain in an aneurism. The derivation of these equations was an original contribution of this work. For the axisymmetric free blowing situation, the through thickness stress can be set to 0 when compared with the much larger values stress in the plane of the material. Given this we can obtain the following simplified equations for the hoop strain:

$$\varepsilon_\theta = \lambda\sigma_\theta(1-\kappa),$$

$$\text{where } \kappa = \frac{\sigma_\varphi}{\sigma_\theta} = \frac{1/2}{(1 - \dfrac{r_0}{2r_1\sin(\varphi)})} \qquad (1)$$

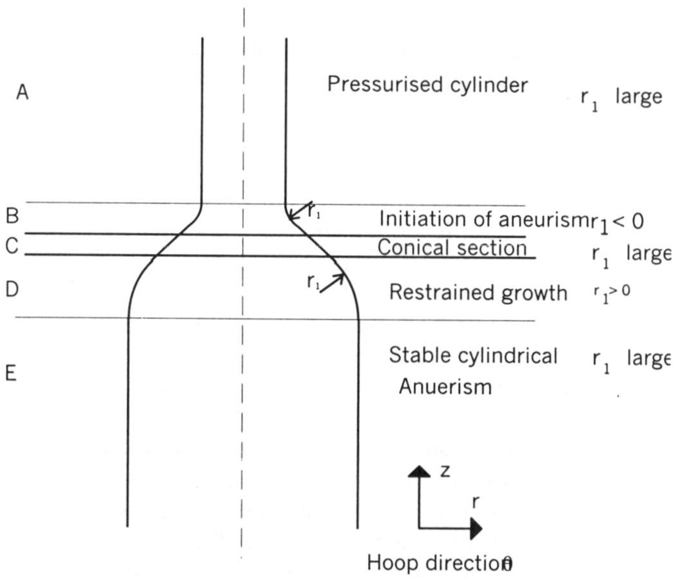

Figure 3. The regions for the formation of a stable aneurism

κ in equation 1 has a variety of possible values. These correspond to the various stages in the forming shown on Figure 3.

A. In the pure cylindrical case $r_1 >> r_0$ and $\varphi = 90°$ so $\kappa = 0.5$.

B. In the case where r_1 is negative and φ approaches 60°, κfalls below 0.5 and the meridional stress falls away. The values plotted in Figure 6 correspond to $r_1/r_0 =$ -100, -10, -1, -0.8 and returning.

C. Stage C is almost conical, $|r_1| >> r_0$ and $\varphi \neq 90°$. κ is approximately 0.5 in this case.

D. In D r_1 is positive and the meridional stress grows and can exceed the hoop stress. There is a point of singularity as $r_0/(2r_1\sin\varphi)$ approaches unity so meridional stress can grow to any value. The values of r_1/r_0 plotted on Figure 6 correspond to 10, 1, 0.9 and returning.

E. The shape is once again cylindrical and the stress state is similar to stage A. κ typically may have values as shown on Figure 4.

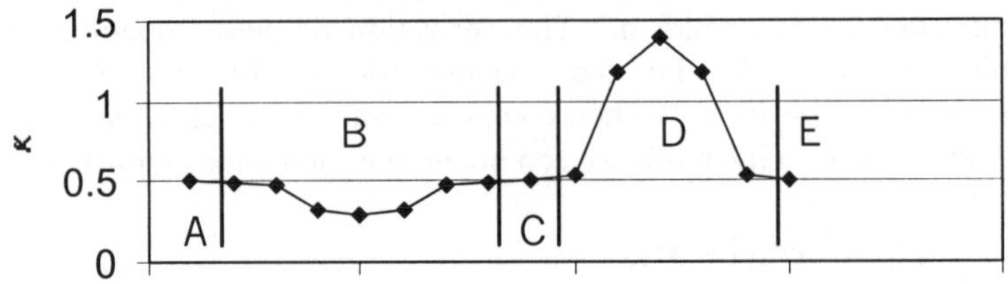

Figure 4. Possible values of κ in various regions of the aneurism

4 Experimental results

In order to confirm these hypotheses about the nature of stable aneurism, the results of experiments where stresses and strains were measured are considered. The manner in which deformation occurs is shown on Figure 5. The measurements of the deformation are conducted by measuring at a number of equally spaced lines drawn on the preform. The form is photographed at regular intervals, throughout the blowing and then the position of these lines is recorded. These equally spaced records of the deformation are termed "frames". When these measurements are combined with pressure measurements inside the form it is possible to calculate the hoop and meridional stress and strain at each stage of the forming process.

Examining Figure 5 and concentrating on the marked section defined by lines 13 and 14 it can be seen the section itself undergoes only very slow strain until between frames 4 and 7 it expands rapidly. After it has reached its stretched size the section only deforms slowly again.

Figure 6 shows the stress strain curves derived from the lines 13 and 14 in the test shown in Figure 4 for the hoop and the meridional directions. The hoop direction shows strain softening between frames 2 and 4 and then progressively strain hardens. The meridional direction shows no strain at all until frame 6 and then does most of its straining before frame 7. The material in the meridional direction requires about twice the stress to start straining in frame 6 when compared to the material in the hoop direction in frame 2.

After frame 8 in Figure 5 the material is in the post-aneurism state and the stresses in the hoop direction are about twice those in meridional direction. This is the stress state expected in cylinders.. Nevertheless some strain occurs in both directions in this state as the polymer continues to harden.

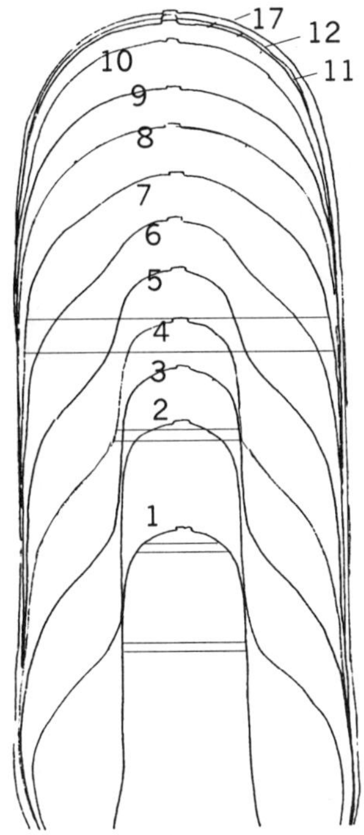

Figure 5. Deformation of the preform as captured in 17 separate photographic frames. The approximate positions of lines 13 and 14 are shown at frames 1, 4 and 7. Lines 2 and 3 are indicated on frame 1 only.

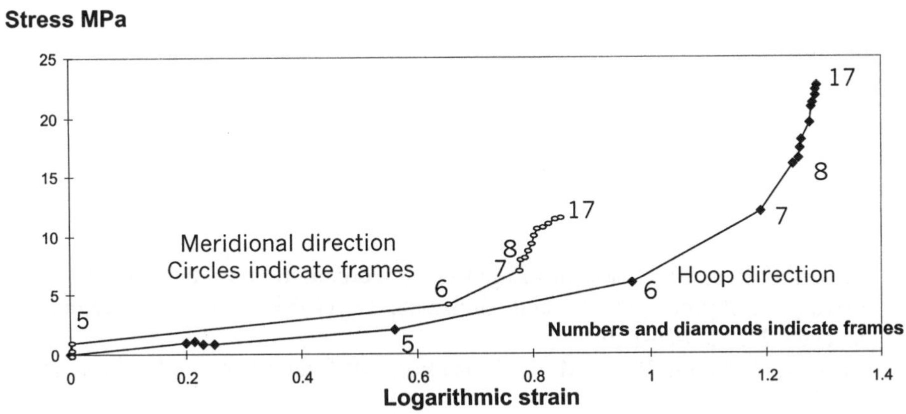

Figure 6 Stress strain curves derived for section 13-14 from Figure 6.

Figure 7 shows the relationship between the meridional stress and the hoop stress. This picture confirms the features shown in Figure 4. It is only when κ exceeds 0.5 that any significant straining occurs in the meridional direction.

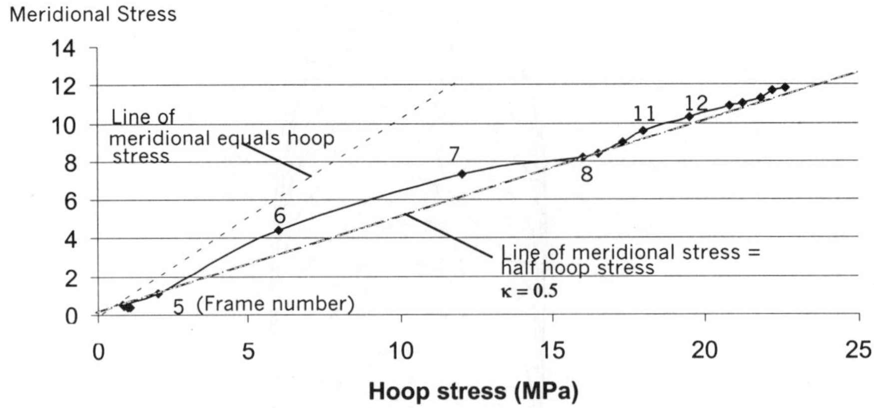

Figure 7. The relationship of meridional and hoop stress during the forming. Almost all the meridional strain occurs between frames 5 and 8.

5 Conclusions

A means of calculating the stresses and strains in a axisymmetric free blowing form has been developed. This permits the derivation of stress and strain curves throughout the blowing of stable aneurisms. It also permits the understanding of the key factor of the formation of a stable aneurism.

The key factor is that material must have a stress strain curve which displays a region of very low or negative strain hardening followed by a region of increasing strain hardening such as shown on figure 7. Such properties are typically found in long chain polymers.

Acknowledgments

The experimental work for this paper was conducted by Mr. A Walley in the course of his studies for a MEngSci with the assistance of Amcor research.

References

1. Crawford, R. J. (1981) Plastics Engineering, Pergamon Press, New York.
2. Erwin, L., Pollock M. A. and Gonzalez, H. (1983), Polymer Engineering and Science, **23**, 15, 826-829.
3. Haessly, W. P. and Ryan, M. E. (1993) Polymer Engineering and Science, **33**, 1279-1287.
4. Harms R. and Michaeli, W. (1994) Journal of Reinforced Plastics and Composites, **13**, 439- 447.
5. Ibrahim, R. N., Walley, A. P. and Lam, Y. C. (1994) Proceedings of International Symposium on Fracture and Strength of Solids, FEFG'94, **8**, 531-534.
6. Price, J.W.H. (1993) The criteria for failure of damaged boiler tubes, in Pressure Vessels and Pipework, Failures, Repairs and Life Assessment II, Conference Institute of Metals & Materials, Sydney, Australia.
7. Ryan, M. E. and Garcia-Rejon, A. (1994) ANTEC, 1824-1830.

ON CREEP STRAIN MEASUREMENT

K. ZARRABI AND Z. LU
The School of Mechanical and Manufacturing Engineering
The University of New South Wales
Sydney 2052, AUSTRALIA
E.Mail: k.zarrabi@unsw.edu.au

At the University of New South Wales a machine vision system is being developed for creep strain measurement at room and elevated temperatures. The system is based on a digital speckle correlation technique and it uses white light. This paper describes the development of this system and its successful application to measure creep strain at room temperature using a specimen made of nylon 66.

1 Introduction

The knowledge of the stress fields is important in: design, design evaluation or modification, failure analysis or life assessments of engineering components. In design and design evaluation/modification various stresses at critical points must be less than specified values usually given in design codes. In failure analysis the stress field is used to determine the critical defect size below which the component can perform its duty without undergoing any catastrophic failure. In life assessment, the knowledge of the stress field is used to estimate the time over which the component maintains its integrity.

There are three main approaches by which the stress field within a component can be determined, viz, theoretical, experimental, and hybrid. In the theoretical approach, the governing differential equations are set up on the basis of mathematical models consisting of the: material behaviour (constitutive equations), loading system, and component geometry using simplifying assumptions. The solving of the governing differential equations leads to the stress field determination. In the experimental approach, either the actual components or physical models of them are subjected to the operating or simulated loading conditions, and physical parameters such as displacements that are linearly proportional to stress components (in case of Hookean behaviour) are measured. To the authors' knowledge, there is no experimental technique to measure stress directly. This may be attributed to the fact that, as yet, the nature of stress is not completely understood. The complexity of the stress analysis problem decides which approach is the most economical to employ. It may be that a combination of both theoretical and experimental approaches, i.e., hybrid approaches, yields the most economical method. An example of a hybrid approach is where the experimental techniques are used as a means of obtaining the stress distribution on the component surface. These data are then applied as boundary conditions for solving the governing differential equations to give the stress field within the component.

While the development of powerful numerical methods of stress analyses, eg, the finite element method, has provided theoretical solutions for some complex stress analysis problems, the validity of the results depends on how well the assumed mathematical models match the reality. In particular the theoretical methods may not produce sufficiently accurate results for welded joints used in the high temperature pressure components These components are manufactured from expensive creep resistant steel alloys and used in the electricity generating power plants as well as chemical processing and refinery plants. Their sufficiently accurate life and integrity assessments can produce savings of many millions of dollars for the industry, each year.

Recently, experimental techniques have received a tremendous impetus from availability of new sophisticated instrumentations for data acquisition and processing that has created new experimental approaches for stress determination. While these techniques have, by and large, been successful for ambient temperature applications, their applications to the high temperature problems have been less successful.

The University of New South Wales (UNSW), the Australian Nuclear Science and Technology Organization (ANSTO) and the Pacific power International have commenced a collaborative research with the objective of developing an experimental technique for creep displacement/strain measurements at room and elevated temperatures. This research is being carried out in several stages. This paper describes the first stage of this project and its finds that is concerned with development of an optical measuring technique for creep displacement/strain measurements at room temperature.

2 Creep Strain Measurement System

A digital speckle correlation technique (DSCT) was developed for the measurement of creep strain. The DSCT was initiated and developed at the University of South Carolina [1-5] and then refined and extended by other researchers, e.g., Vendroux and Knauss [6-8], Vendroux [9], Jin, et al [10], Zarrabi [11], Zarrabi and Oreb [12] and Lu and Cary [13]. The DSCT has the advantage of being capable to measure large strains accurately as well as it is easy to use and has a low susceptibility to environmental vibration. In what follows, first the DSCT that was used for creep strain measurements will be described. Then the application of this system to measure creep strains at the room temperature using specimens made of nylon 66 will be detailed.

2.1 Digital Image Correlation Technique

In our technique, a speckle pattern was first created on the surface of a specimen. To this end, the surface of the specimen was uniformly covered by white paint and then sprayed by black paint. The principle of strain measurements using the DSCT was to track the displacements of a small sub-image of size of MxM pixels within the speckle region. The centre of this sub-image coincided with the point of interest. The original and some of the subsequently developed DSCTs relied on measuring both the displacement components and the displacement gradients of the sub-image. But to the authors' experience [11,12,14], the DSCT cannot measure the displacement gradients accurately in presence of creep deformation. In the present study, therefore, the measured displacements were fitted a bicubic spline surface in a piece-wise manner. Then, the spline surface was numerically differentiated to obtain strains at various times. The bicubic spline surface implemented a third-order polynomial that allowed both grey-level values and C^2 continuous gradients to be determined at any location within the domain over which the surface was defined.

The DSCT that was used for the present study is shown in Figs. 1 and 2. It consisted of a Data Translation® DT3157 PCI frame grabber, a PULNiX TM-1300 monochrome progressive CCD camera with a spatial resolution of 1300 x 1030 and 10-bit gray-level, a Navitar Zoom 7000 (F2.5/18-108MN) lens, a Pentium II-333 personal computer and a normal fluorescent tube as the light source. The CCD camera was mounted on a 3-Axis Macroblock Flexure Stage with a differential micrometer driver having a resolution of 0.001 mm. The assembly of the camera and Macroblock Flexure Stage was mounted on a tripod. This arrangement allowed the system to be set up and calibrated in-situ with relative ease (see later).

The principal of creep strain measurement using our DSCT is as follows. Using the DSCT, one measures in-plane displacements in the x and y-directions where the x and y-axes represent a local coordinate system attached to the specimen surface. As mentioned before, the measured displacements are then converted to strains by numerical differentiation. The displacement measurement is based on grabbing two speckled images of the object surface at two different time. The images are then digitised and numerically correlated to obtain the displacements at various points within the area of interest on the specimen surface during the time interval. This was achieved by maximizing the following correlation coefficient, C:

$$C = \frac{\sum_{ij}^{M}[f(x_i,y_j)-\bar{f}]*[g(x_i,y_j)-\bar{g}]}{\sqrt{\sum_{ij}^{M}[f(x_i,y_j)-\bar{f}]^2} * \sqrt{\sum_{ij}^{M}[g(x_i,y_j)-\bar{g}]^2}} \tag{1}$$

where $f(x,y)$ and $g(x,y)$ were gray-levels of the speckle patterns at two different time of interest, and $\bar{f}$ and $\bar{g}$ were the average values of the $f(x,y)$ and $g(x,y)$ respectively. MxM represented a square sub-image in the two speckle patterns the centre of which was the point at which the in-plane displacement and strain components were computed. The value for M was obtained from calibration tests (see later). The inclusion of $\bar{f}$ and $\bar{g}$ in equation (1) produced a sharper pick for the correlation coefficient function improving the accuracy of the measurements. The correlation coefficient (C) was optimised using a direct search method. This method first searched for the peak point of C along the x-direction. Then through this peak point it completed a second line search along the y-direction to discover the second peak point that was closer to the absolute maximum value of C. If the C value was not large enough, then this search algorithm was repeated. To increase the accuracy of the measurements to the sub-pixel level, the cubic spline interpolation function was used to compute the gray-levels of points fell between pixels. The initial search step was set to 2 pixels and then it reduced continually to a sub-pixel level until the desired accuracy is achieved. An object-oriented software using a C++ complier was developed to perform the required computations.

2.2 Specimen

A standard uniaxial tensile specimen was machined from nylon 66 material. At the mid-length and on the one side of the specimen a 'white-light speckle' pattern covering an area of 15x15 mm was created. On the other side of the specimen and at its mid-length (i.e., at opposite side to the 'white-light speckle' pattern) a 90-rossette strain gauge was mounted. The stain gauges were employed to check the accuracy of strain measurement using DIC technique within initial 2% as the strain gauges were capable of measuring strain up to 2%.

2.3 *Calibration*

Calibrations tests were carried out to determine: (a) the size of the sub-image MxM whose centre was the point of interest and (b) to relate displacements in units of pixel to that in unit of millimetre.

Two factors decided the size of the sub-image MxM. Firstly, each MxM sub-image in the two speckle patterns represented areas over which the strains were assumed to be constant. Therefore, to make measurement pertinent to a point on the specimen surface, M should be selected as small as possible. Secondly and on the other hand, the correlation algorithm was based on the assumption that the sum of the gray-levels over each sub-image of the size of MxM in the two speckle patterns was statistically unique. To satisfy this requirement, M should be sufficiently large. The following calibration tests were carried out to decide the value of M. In the first calibration test, two consecutive images of the speckled specimen surface were taken while keeping the specimen unloaded and the camera stationary. Ideally and for a correct value of M the two digitised 'speckle patterns' were identical and hence when used in conjunction with the numerical correlation algorithm described before should result in zero displacements and thus zero strains. However, because of various type of noise in the system the measured displacements were different from zeros. In measuring displacements, this test however indicated that the standard deviation of the error could be minimized to 0.03 pixels by setting M to 24 pixels or larger (Fig. 3). Therefore, in the correlation algorithm that was later used to determine displacements and strains on the specimen surface, M was set to 24 pixels.

Second sets of calibration tests were performed to relate displacements of the points within the sub-image MxM in units of pixel to that in unit of millimetre For this purpose, first an image of the speckle specimen surface was taken and digitised while keeping the specimen unloaded and the camera stationary. Then while the specimen was kept unloaded and stationary, the camera was moved relative to the specimen surface in the x and y-directions in steps of 0.001 mm and at each step an image of the speckled specimen surface was taken, digitised and correlated to the first image to determine the camera displacements in units of pixels. From these tests it was concluded that a displacement of 0.00153317 mm in the x-direction and 0.00153393 mm in the y-direction are equivalent to 1 pixel movement in each direction.

2.4 *Test Set-up and Results*

The specimen was subjected to a constant force of 82 KN in an Instron 5582 universal testing machine. Stains (ε_l) in the y (longitudinal) directions of the specimen at various time points were measured using three different methods, viz., (a) DSCT, (b) stain gauges and (c) the built-in extensometer in the Instron. To obtain strains in the longitudinal direction of the specimen using method (c), the expression $\varepsilon_l = \ln(\frac{l}{l_0})$ was used where l_0 and l are the initial and elongated lengths of the gauge length respectively. The measured stains using the above three methods are depicted in Fig. 3. Referring to Fig. 3, one may observe that: (1) the strain gauges failed after 1667 hours and at 2% strain, (2) The difference of strain values measured by the strain gauges and DSCT was 0.018% while that measured by the strain gauges and the Instron extensometer was 0.055%. After 1667 hours, the comparison of the strain values measured by the DSCT and the Instron extensometer showed a maximum difference of 0.028%. Therefore, it may be concluded that creep strains can be measured reasonably accurate using DSCT over a long period of time.

3 Conclusions

A digital speckle correlation technique was described that used white speckle pattern. It was shown that this technique could successfully and accurately be used to measure large creep strains at room temperature using specimens made of nylon 66.

Acknoweldgements

The Australian Research Council and the Pacific Power International financially support this research.

References

1. Peters WH and Ranson WF. Digital imaging techniques in experimental stress analysis. Opt. Eng. 1982;21; 427-432.
2. Sutton MA, Wolters WJ, Peters WH Ranson WF and McNeil SR. Determination of displacements using an improved digital image correlation method. 1983:1:3:133-139.
3. Sutton MA, Cheng M, Peters WH, Chao YJ and McNeil SR. Application of an optimized digital image correlation method to planar deformation analysis. Image Vision Computing. 1986:4:3:143-150.
4. Bruck HA, McNeil SR, Sutton MA and Peters WH. Digital image correlation using Newton-Raphson method of partial differential correlation. Experimental Mechanics. 1989:29:261-267.
5. Luo PF, Chao YJ, Sutton MA and Peters WH. Accurate measurement of three-dimensional deformation in deformable and rigid bodies using computer vision. Experimental Mechanics. 1993:33:123-132.
6. Vendroux G and Knauss WG. Submicron deformation field measurements: Part 1. Developing a digital scanning tunneling microscope. Experimental Mechanics. 1998:38:18-23.
7. Vendroux G and Knauss WG. Submicron deformation field measurements: Part 2 Improved digital image correlation. Experimental Mechanics. 1998:38:86-91.
8. Vendroux G and Knauss WG. Submicron deformation field measurements: Part 3. Demonstration of deformation determinations. Experimental Mechanics. 1998:38:154-160.
9. Vendroux G. Correlation: a digital image correlation program for displacement and displacement gradient measurements. GALCIT Report No. SM90-19. California Institute of Technology. 1990.
10. Jin GC, Bao NK and Chung PS. Advanced digital speckle correlation method for strain measurement and nondestructive testing. Proceedings of International Conference on Experimental Mechanics: Advances and Applications. Singapore, 4-6 December 1996.
11. Zarrabi K. A remote sensing transducer for measuring in-plane displacements. Non-destructive Testing, Australia, A Journal of Measurement Control & Testing. 1993:30:5.
12. Zarrabi K and Oreb BF. Current experimental techniques for high-temperature displacement/strain evaluation: an outline of available techniques. Proceedings of International Conference on Experimental Mechanics: Advances and Applications. Singapore, 4-6 December 1996.
13. Lu H and Cary PD. Deformation measurements by digital image correlation: implementation of a second-order displacement gradient. Experimental Mechanics. 2000:40:4:393-400.

14. Zarrabi K. Application of image analysis to obtain in-plane displacements. Proceedings of the International Conference on Data and Knowledge Systems for Manufacturing and Engineering. Hong Kong, May 2-4, 1994.
15. B.W. Smith, X. Li and W. Tong Error Assessment for Strain Mapping by Digital Image Correlation. Experimental Techniques July/August 1998

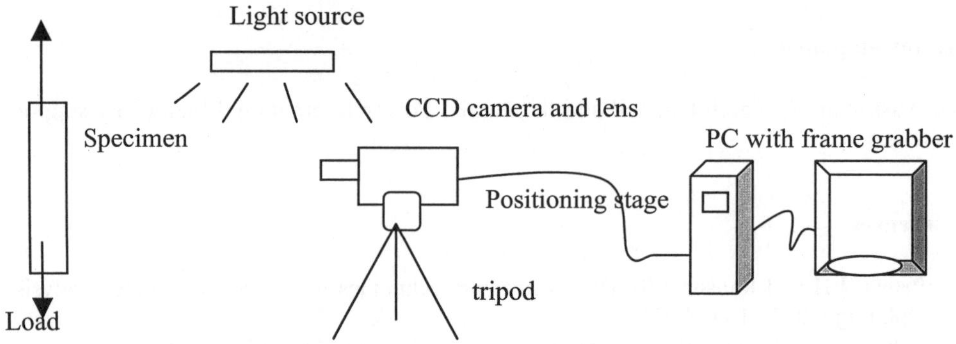

Figure 1 – Schematic details of the digital speckle correlation techniques (DSCT)

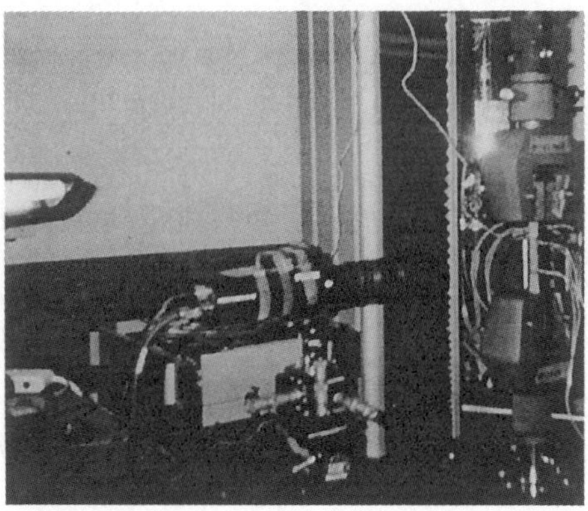

Figure 2- Arrangement of DSCT and Instron rig for creep strain measurement using nylon 66 specimens

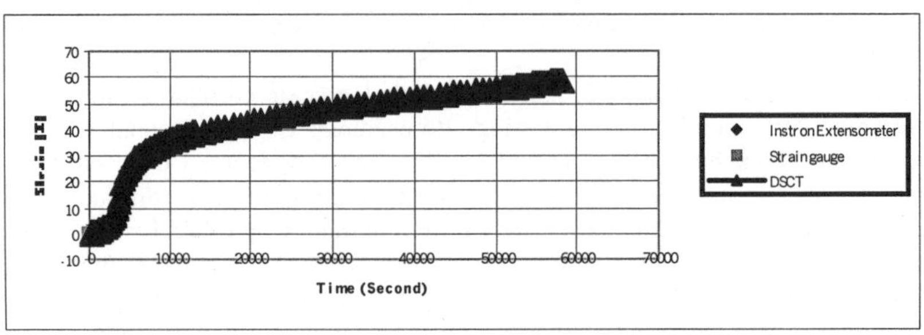

Figure 3 – Comparison of measured creep strains using various techniques

A HOLOGRAPHIC INTERFEROMETRY STUDY OF PRESSURE SENSITIVE ADHESIVE TAPES

R.B. HESLEHURST, O.T. KIEBOOM & R.C. CLARK
School of Aerospace & Mechanical Engineering, University College, University of New South Wales, Australian Defence Force Academy, Canberra, ACT

Pressure sensitivity adhesive tapes are become more common in many environmentally demanding applications. One such example is the use of paint film appliqués that have pressure sensitive adhesive backings applied to aircraft as paint schemes. A concern that exists with such paint scheme applications is durability of the adhesive and adhesion interface in the harsh environment of flight. A holographic interferometry study has been undertaken to study the effects of adhesive degradation and surface condition deterioration. The results show that poor surface conditions and bondline defects can be identified in holographic interferograms. Characterisation of several forms of bondline defects are presented and qualitatively analysed.

1 Introduction

Pressure sensitive adhesive (PSA) tapes are becoming more common in many environmentally demanding applications. One such example is the use of paint film appliqués that have pressure sensitive adhesive backings applied to aircraft as paint schemes. A concern that exists with such paint scheme applications is durability of the adhesive and adhesion interface in the harsh environment of flight.

A previous investigation of bonded joint effects using holographic interferometry (1) revealed that weak bondlines over an adhesion interface can be identified by:

- A global irregularity in the fringe pattern,
- A high fringe density in a local regions of the global fringe pattern,
- A non-uniform fringe pattern shape, and
- Fringe line discontinuities.

Aim. A holographic interferometry study has been undertaken to study the effects of adhesive degradation and surface condition deterioration. This study reviews the detectability of debonds and weak bonds in pressure sensitive adhesive tapes.

Scope. This study deals with artificially induced debonds on PSA tapes and simulated weak bonds that are produced in real life by, for instance, poor surface preparation or surface contamination.

2 Holographic interferometry test method

To measure the deformation of the PSA tape, the double-exposure white-light reflective holographic interferometry technique (2) was employed. The regions of interest in each panel (at the particular PSA tape covered areas) were covered with a holographic plate (Figure 1) and the first half exposure taken at approximately at 46 cm Hg vacuum load. The second half exposure was taken at approximately a 5 cm Hg lower vacuum load (41 cm Hg). The holographic plate was then developed and captured using the standard technique (2). A schematic of the holographic interferometry set-up is shown in Figure 2.

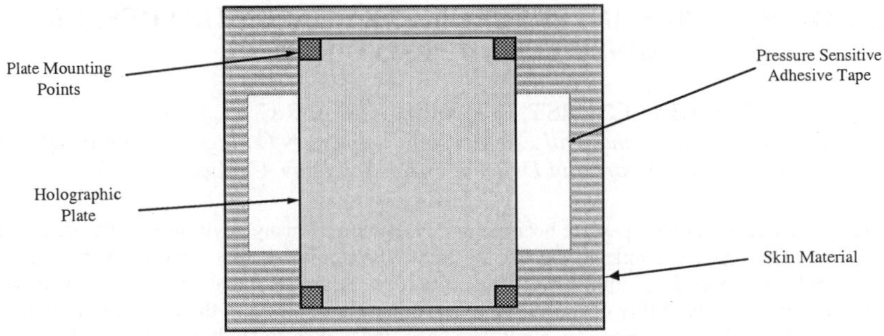

Figure 1: Position of the Holographic Plate on the Panel

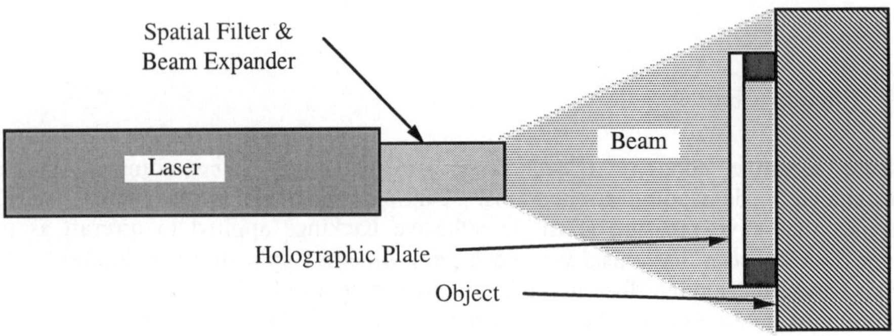

Figure 2: Holographic Interferometry Procedure Set-up

3 Specimen description and test setup

The test regions were prepared by using a section from a Boeing 737 rudder skin. The configuration of the rudder skin is six plies of graphite/epoxy composite material separated by 10.5 mm Nomex honeycomb core. Various pressure sensitive labels were applied to the skin surface and each had a different surface modification. Figure 3 shows the panel and four PSA tapes with their surface conditions indicated.

Figure 3: Honeycomb Rudder Skin Panel

The label and surrounding region were then covered with a small vacuum box, in order to create stresses in the skin material by applying a vacuum load. An initial vacuum (46 cm Hg, 18 inch Hg) was applied to the region and the first exposure of the holographic plate was taken. A portion of the vacuum was released (to 41 cm, 16 inch Hg) and the second exposure of the holographic plate taken. After development of the holographic plate the image was capture by CCD camera for analysis.

4 Interferograms

The Figures 4 to 10 show all the pertinent interferograms taken during the tests. Each interferogram is discussed in the follow on section.

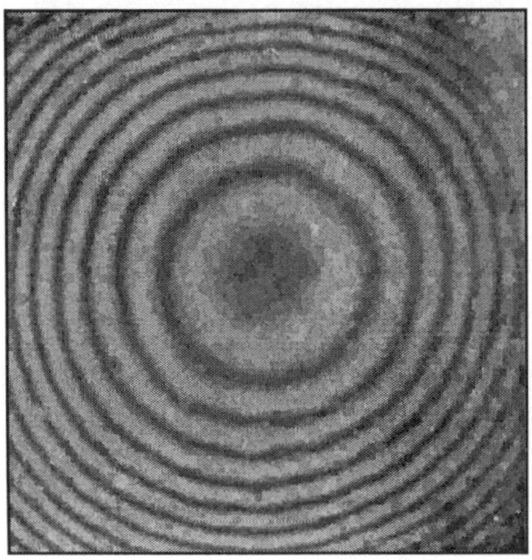

Figure 4: Response of Skin Material to Pressure Differential

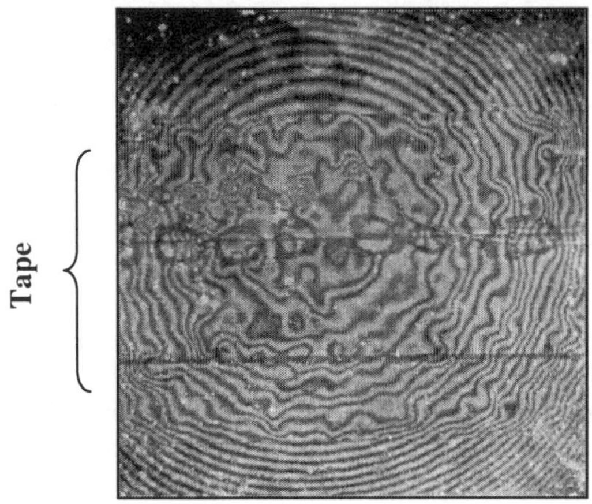

Figure 5: Good Bond

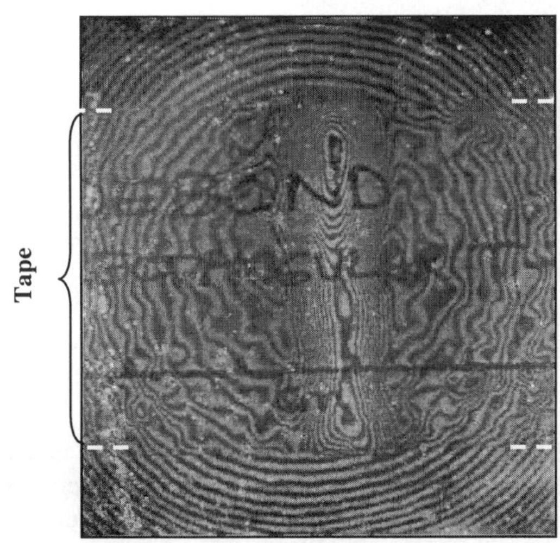

Figure 6: Circular Central Debond

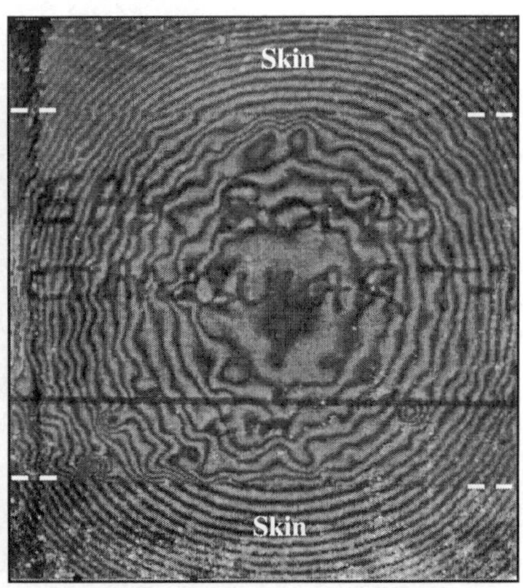

Figure 7: Circular Central Weak Bond

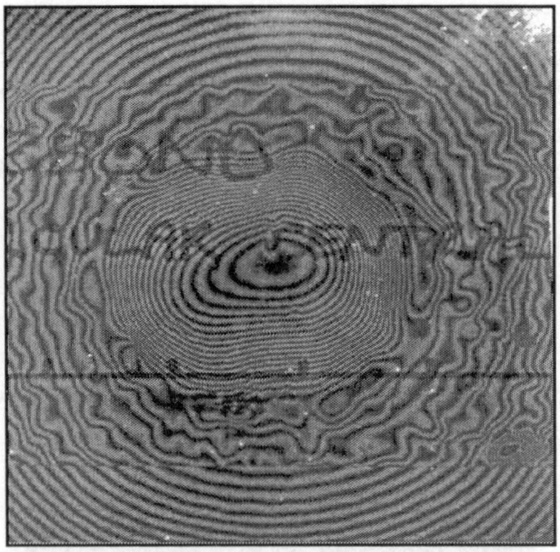

Figure 8: Rectangular Central Debond T-T-W

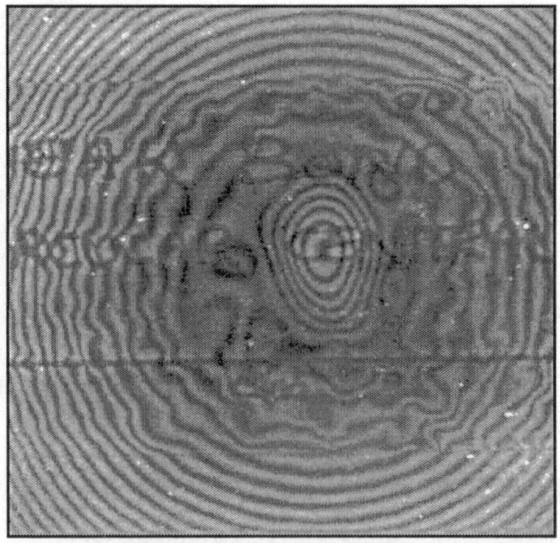

Figure 9: Rectangular Central Weak Bond T-T-W

5 Commentary on interferograms

Figure 4 shows the response of the skin material without the adhesive tape attached to the skin. The circular shape that the vacuum loaded skin material has in the centre causes the concentric fringe lines. Note that each fringe line represents a height contour. The distance between fringes becomes smaller away from the centre, indicating a domed skin surface in the middle. Taking a vertical cut through the centre of the interferogram and transposing the cross-section as height plot shows this dome effect for this interferograms and the following interferograms in Figure 10. In Figure 10 the gradient of the lines corresponds with the density of the fringe patterns.

In Figure 5, and in the subsequent figures, the dashed lines indicate the location of the tape in the interferogram. If the fringes in the interferogram are continuous over the edge of the tape, this means that there was no relative movement between the tape and the skin surface. Minor relative movement indicates good adhesion.

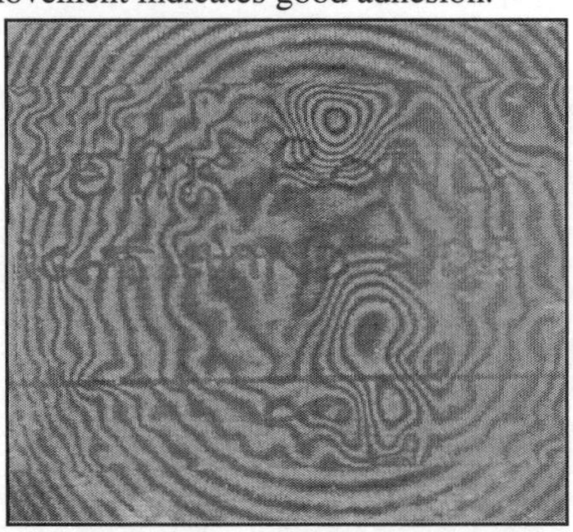

Figure 10: Weak Bond on right hand side

The adhesion between the PSA tape and the skin for the 'good bond' was achieved under the best possible surface conditions. However, in Figure 5 the fringe pattern is significantly different from that of Figure 4. The irregular trajectory of each fringe line over the PSA tape region in Figure 5 would indicate that the PSA quality is not uniform over the contact area. The overall fringe pattern still shows relatively distinguishable concentric lines. Several areas within the bonded area indicate local lack of adhesion uniformity. These characteristics are shown in Figure 11 as a symmetrical U-shape for the predominant concentric shape and the lack of smoothness caused by non-uniformity of the fringe pattern.

A circular debond of 23 mm shows the typical holographic interferometry fringe pattern in Figure 6. Since the disbond PSA tape region has a relatively smaller flexural rigidity than the surrounding material, it will tend to deform out-of-plane more. This is represented by the higher fringe density across the debonded region. A vertical cross-section of this region is also shown in Figure 11, the steeper lines near the centre indicating the higher fringe density and the transition of fringes on the PSA to fringes on the skin shows up in line as a change of gradient around fringe number 24.

A weakly bonded central region produced a holographic interferogram of Figure 7. The fringe pattern is very similar to the fringe patterns of the skin material response and the 'good bond' interferograms of Figure 4 and 5, respectively. However, the central region of the weak bond area shows a similar interferogram pattern to the debonded fringe pattern of Figure 6. The distinguishing feature of this weakly bonded region is the lower fringe density when Figure 6 (debonded) is compared to Figure 7 (weak bond). Apparently, the weak bond shows up as a less distinctive dome shape, which is understandable because of partially still active adhesion. A more definitive comparison can be seen by the vertical cross-sections of Figures 5 and 6 shown in Figure 11, where the central weak bond line is considerably less steep than the central debond line due to the lower fringe density of the weak bond.

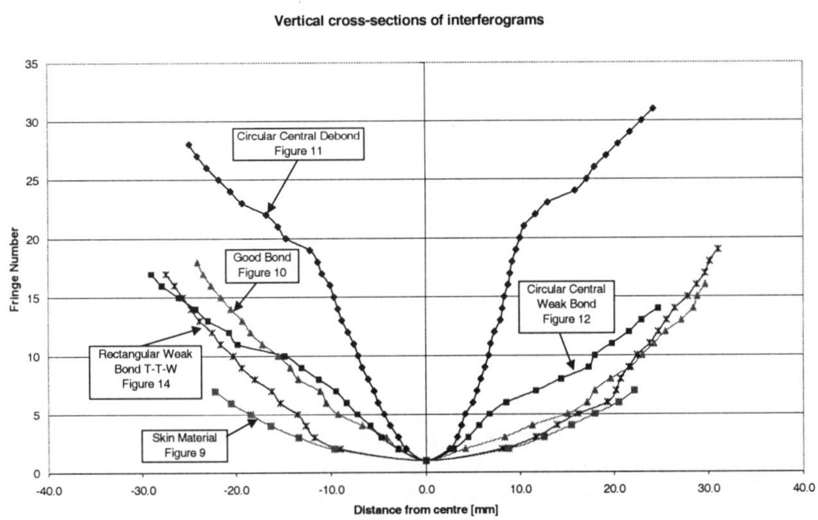

Figure 11: Vertical cross-sections of the interferograms

Another comparison of a debond and a weak bond is given in Figure 8 and Figure 9. The rectangular debond in Figure 8 shows up nicely, but the rectangular weak bond is more difficult to distinguish in Figure 9. The lack of fringe pattern changes in the weakly bonded rectangular through-the-width (T-T-W) interferogram, Figure 9, when compared with a 'good bond' fringe pattern, Figure 5, would suggest that no effective weakening of the bondline had taken place. The sensitivity of the PHITS process strongly supports that a lack of bonding in this region is actually present, as has also been reported by Heslehurst (1). Figure 11 underlines these observations with about equally smooth lines and similar curves for Figure 5 and 9, for a good bond and a T-T-W bond, respectively. Note that the T-T-W debond of Figure 8 is not represented in Figure 11, because a vertical cross section of the fringe pattern would result in an incomparable line.

As final experiment a piece of adhesive tape with a weak bond over the whole contact surface on the right hand side from the centre. The resulting interferogram is shown in Figure 10. Two areas with smooth and highly dense fringes may be seen on the weak bond side of the tape. Whilst the fringe patterns of the upper and lower regions of Figure 10 may suggest debonding, the fringe density indicates weak bonding. The characteristic fringe patterns of Figures 5 and 6 (debond and weak bond) provide the supporting evidence that Figure 10 is showing local weak bonding. The vertical asymmetry of this type of debond makes this interferogram also unsuitable for representation in Figure 11.

Summarizing, debonds can be detected easily and certainly in pressure adhesive tapes using holographic interferometry, as Figure 6 and Figure 8 prove. The weak bonds, however, are less easy to detect. This is logical as the difference in adhesion quality is smaller between a good and a weak bond and at significantly greater between a good bond and a debond. If we compare Figure 5 and Figure 7, the weak bond is quite easy to detect, but when compare Figure 5 and Figure 9; the difference is small, leaving room for doubt and/or misinterpretation.

In vertical cross-section diagrams, debonds are also quite distinguishable by their steeper U-shapes. Weak bonds show up in a vertical cross-section diagram as lying below the debond lines, but above the line for the skin-surface, and their curves are less smooth.

6 Follow-on studies

Good results were obtained in this investigation of poor surface conditions and bondline defects with holographic interferograms. The pressure adhesive tape that was used was porous, which is not representative for adhesive products used in the aerospace industry, such as paint film appliqués. Therefore, an additional investigation with airtight pressure adhesive tapes will be undertaken. Environmental conditioning of the PSA tapes will allow qualitative studies of free edge degradation. Variation in fringe line continuity at the free edges of the PSA tape has shown the same characteristic responses in several holographic interferograms.

7 Conclusions

Poor surface conditions and bondline defects in pressure sensitive adhesive tapes can be identified in holographic interferograms. Very clear images were obtained for debonds, which leave no room for misinterpretation of these results. The results for the weak bonds were not as clear, but there was still a good outcome. In addition, the potential to identify bondline degradation is indicated by several characteristic features of the weak bond regions. More extensive studies will allow better understanding of the weak bondline mechanism in adhesive system applications.

References

1. Heslehurst R.B. Ph.D. Thesis Chapter 7, UNSW, 1999.
2. Heslehurst R.B. (editor), "Portable Holographic Interferometry Testing System Operating Manual", School of Aerospace and Mechanical Engineering, University College, UNSW, ADFA, Canberra, 1996.

EFFECTS OF POLYMER PROPERTIES ON THE SURFACE FINISH AND CHIP FORMATION IN ORTHOGONAL CUTTING

KEQIN XIAO AND LIANGCHI ZHANG

School of Aerospace, Mechanical and Mechatronic Engineering
The University of Sydney, NSW 2006, Australia

This study aims to evaluate the machinability of polymers and understand the interaction between strain rate and temperature rise during orthogonal cutting. The materials used were high density polyethylene (HDPE) and low density polyethylene (LDPE). It was found that the average surface roughness and shear strength of both HDPE and LDPE decreases with increasing the cutting speed, demonstrating that the strain rate has a greater influence than that of the temperature rise. A relatively high mechanical strength is preferred in order to create a smoother surface of soft polymers with a low glass transition temperature and mechanical strength. It was revealed that the chip deformation mechanism varies with material properties.

1 Introduction

With the rapid growth of new polymers and their applications in engineering, machining of polymeric materials has become an increasingly important operation in manufacturing industry. Machining is also necessary when the quantity of components does not justify the cost of tooling for moulds or extrusion dies, or when a product needs a particularly high dimensional accuracy.

It has reported [1-3] that the surface roughness of a machined polymeric component is dependent on the cutting speed. Some studies have also tried to correlate the material proerties, such as glass transition temperature, melting temperature, viscosity and relaxation, to surface finish of polymeric workpieces. For example, Smith and Scattergood [4] considered the effects of polymer glass transition temperature on the surface quality of some amorphous thermoplastic polymers subjected to single-point turning. They claimed that the glass transition temperature of a polymer, T_g, is the most important parameter in controlling the surface roughness. If the temperature in the cutting zone exceeds T_g during machining, a smooth surface finish can be obtained, and the material would be removed in a ductile manner; otherwise, a rough surface is formed because the material had experienced a brittle fracture process. More recently, Carr and Feger [5] presented a more detailed analysis on the effect of material properties on the surface roughness of several polymers with different T_g and molecular weights, when subjected to single-point diamond machining. They explained their experimental results using the well-known Time-Temperature Superposition Principle of polymers. They discussed the minimum roughness of VESPEL polymer (DuPont, Wilmington, DE) against tool speed as a result of time/temperature interaction. For the polymethylmethacrylate (PMMA) polymer, they further concluded that cutting occurs in the melt since they failed to identify any effect of molecular weight on surface roughness.

The purpose of this paper is to investigate the machinability of two typical polymers, the high density polyethylene (HDPE) and low density polyethylene (LDPE), and understand the effects of cutting speed on surface roughness and chip formation in orthogonal cutting. The variations of surface roughness and material removal behavior will be discussed in relation to the long-range mobility of molecular segments and shear strength of the polymers and the temperature rise in the cutting zone.

2 Material preparation and properties

The raw granules of HDPE and LDPE were provided by Qenos Pty Ltd, Australia, and the plates for cutting and material property testing were produced by injection moulding at about 170°C. The material properties, *i.e.*, the tensile strength (σ_y), elongation at failure (ε_f), glass transition temperature (T_g), melting temperature (T_m) and density (ρ), as summarized in Table 1, were measured in the authors' laboratory. It can be seen that both the thermoplastics, LDPE and HDPE, are very ductile at room temperature.

Table 1. Material property of four polymers used in this study.

Materials	T_g [°C]	T_m [°C]	σ_y [MPa]	ε_f [%]	ρ [g/cm^3]
LDPE	-109.5	110.6	8	521	0.93
HDPE	-90.4	127.3	18	317	0.95

3 Orthogonal Cutting

The materials were cut at a cutting depth of 0.2mm using a CNC Minini M286 grinding machine at a series of cutting speeds. A fixed rake angle of +20° was employed. Cutting forces in three directions were recorded automatically using a KISTLER 3-Component dynamometer (Type 9257B). The surface roughness of a machined surface was measured using a profilometer (Mitutoyo, Surftest 402 Series 178) after ultrasonic and ethanol cleaning. The morphology of a machined surface was examined using an optical microscope and a scanning electron microscope (SEM) 505C at a voltage of 20 kv.

3.1 Surface finish

The dependence of surface roughness of HDPE and LDPE workpieces on the cutting speed is presented in Fig. 1. It can be clearly seen that the surface finish of both LDPE and HDPE are improved significantly with increasing the cutting speed, while HDPE has a lower surface roughness than LDPE at the same cutting conditions. Considering the fracture strength and mobility of molecular segments of these two materials, it seems to imply that for the extremely soft polymers a too low mechanical strength of the material is not preferred in order to produce a good surface finish. It is easier to understand if taking into account the deformation at the cutting zone. Both the cutting force and thrust force produce permanent plastic deformation on both the cutting surface and its immediate sub-surface, resulting in a rough surface. This is further confirmed in the following discussions about the dependence of the shear strength on the cutting speed and the observations of the damaged layer beneath the machined surface. It should be mentioned that if T_g of the polymer is far beyond room temperature, the surface finish may be improved when a high long range mobility of molecular segments and a low mechanical strength are achieved, as reported by Carr and Feger [5].

It can be seen from Fig. 2 that the cutting force, F_c, and thrust force, F_t, remain constant beyond a high impact peak where the cutting tool comes into contact with the workpiece. The cutting force is much larger than the thrust force. During the cutting of HDPE at a speed of 50 mm/min, the former reaches 3.4 N, while the latter is only 0.39 N. The shear strength is a good indication of the resultant influence from the strain rate and temperature rise in cutting, and can be calculated using the model developed by Ref. [6]:

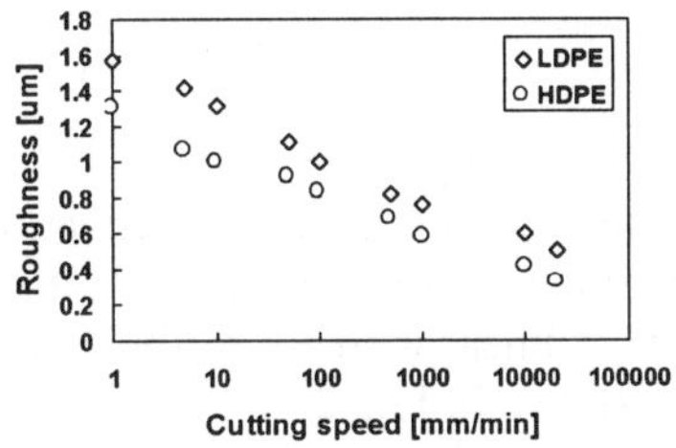

Figure 1. Surface roughness of LDPE and HDPE at different cutting speed.

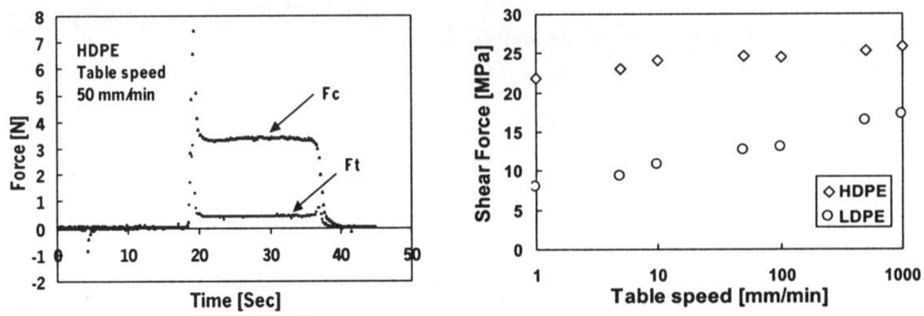

Figure 2. Left: recorded cutting force, F_c, and thrust force, F_t, during machining of HDPE. Right: Calculated shear strength of LDPE and HDPE as a function of cutting speed.

$$\tau_t = \frac{[(F_C \cos\phi) - (F_t \sin\phi)]\sin\phi}{A_C}$$

where τ_t is the apparent shear strength of the material on the shear plane, A_c the cross-sectional area of the uncut chip and ϕ is the shear angle given by

$$\tan\phi = \frac{r_c \cos\gamma}{1 - r_c \sin\gamma}$$

where γ is the rake angle the cutting tool, r_c is the cutting ratio of a_c, the undeformed chip thickness, to a_o, the chip thickness.

As shown in Fig. 2, the shear strengths of both LDPE and HDPE increase almost linearly with the logarithm of the cutting speed. According to the time-temperature superposition principle, it can be deduced that the increase in strain rate has a greater influence than the temperature rise in the cutting zone. The disturbance applied externally (cutting in this case) is faster than the intrinsic relaxation time of the material at the temperature in the cutting zone. Consequently, material strength increases. In relation to the change of surface roughness against cutting speed, it is again confirmed that a higher mechanical strength results in a better surface finish in very soft polymers with great mobility of molecular segments and low mechanical strength.

It is further noticed that the rate of increase in shear strength of the softer LDPE is faster than that of the relatively hard HDPE, showing the dependence of the strain rate effect on the material properties. Experimental work on metals [7] has shown that shear strength remains constant for a given material at different cutting speed and over a wide range of other cutting conditions. The significant shear strength variation of polymers with the strain rate (cutting speed) should be attributed to their intrinsic viscoelasticity. In polymers, the relationship between stress and strain rate can be described by Eyring's law [8], *i.e.*, yield stress increases linearly with the logarithm of strain rate, $\dot{\varepsilon}$.

$$\sigma_y = \frac{4kT}{\xi V^*}(Ln\frac{\dot{\varepsilon}}{A} + \frac{\Delta H}{kT})$$

where ξ is the stress concentration factor, V^* is the activation volume, k is Boltzmann's constant, ΔH is activation enthalpy and A is a constant. If it is assumed that the strain rate is proportional to cutting speed, a linear increase in shear strength with strain rate is also seen in the polymer cutting, as shown in Fig. 2.

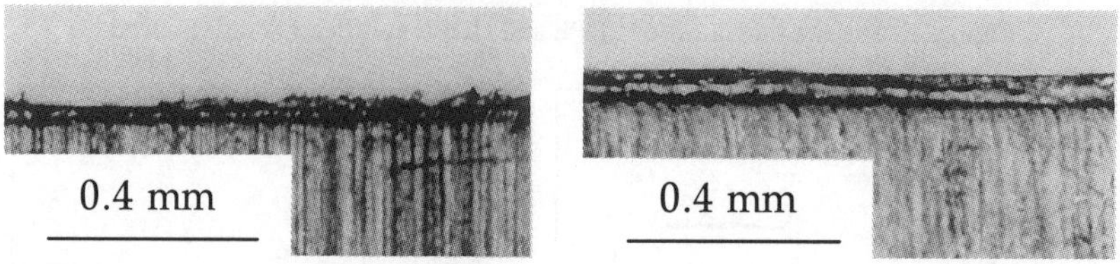

Figure 3. The damaged layer beneath the cut surface of LDPE at the cutting speeds of 1 mm (Left) and 20000 mm/min (Right).

A damaged layer beneath the cutting surface was found, and the side view is shown in Fig. 3. The thickness of the damaged layer in the LDPE decreases approximately from 80 μm at the cutting speed of 1 mm/min to 60 μm at the cutting speed of 20000 mm/min. This decrease is ascribed to an increase in mechanical strength of the material with the cutting speed, which is consistent with our previous discussions.

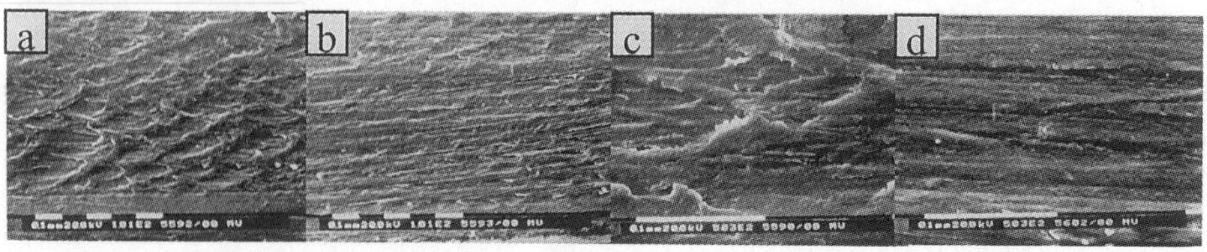

Figure 4. The machined surface morphology. (a) LDPE at 1 mm/min, (b) LDPE at 20000 mm/min, (c) HDPE at 1 mm/min and (d) HDPE at 20000 mm/min.

Morphologies on the cutting surfaces of LDPE and HDPE were examined by SEM and displayed in Fig. 4. The clear traces of plastic deformation on the cutting surfaces were observed in both the cases. However, severer deformation occurred at the low cutting speed. Moreover, the cut surface of HDPE is smoother than that of LDPE. These observations are in good agreement with the previous measurements of surface roughness and discussions on the variations of material properties with cutting speed.

3.2 Characteristics of chip formation

The study of chip formation mechanism is of fundamental importance, since it offers the understanding of surface roughness, material property-machinability relationship and other machining characteristics. For LDPE and HDPE polymers, continuous chips were formed over the range of cutting speeds used in this study, as shown in Fig. 5. At a low speed, chips are slightly curled. However, they become heavily curled with increasing the cutting speed. Comparing shapes of chips between two materials, it was found that chips of LDPE curl inwards or anticlockwise, while those of HDPE twist outwards or clockwise. Figure 6 shows the plastic deformation layers on the chip surfaces and shear deformation characteristics in chips. For LDPE, the thickness of the deformation layer on the chip-tool contact surface clearly increases with the cutting speed, which is about 20 μm at 1 mm/min but reaches about 80 μm at 20000 mm/min. This deformation layer is formed due to friction heating generated between the chip and the tool. Since the frictional heat increases with increasing the cutting speed, the thickness of the layer also increases. The material in the layer is plastically deformed in the direction opposite to the movement of the chip, which can be clearly seen from the morphology of LDPE chip (the middle one in Fig. 6). Consequently, a large residual stress forms in the layer at the direction of the material deformation, and makes the chips of LDPE curled anticlockwise. The material between the two plastic deformation layers of the LDPE chip is barely plastically deformed, as shown by the unchanged directions of the fine lines perpendicular to the chip-tool contact surface, marked on purpose before cutting. Thus, the deformation of a chip in the middle zone is essentially elastic, which is different from the case in cutting metals. Moreover. The chip above has two distinguished characteristics compared with the so-called 'continuous-flow type" chip found by Kobayashi [1]. Kobayashi's chip has no plastic deformation layer at the tool-chip interface and its thickness is equal to the depth of cut. In the present case, however, a clear plastically deformed layer exists as shown above. The average chip thickness varies with the increase in the cutting speed, from about 235 μm at the cutting speed of 1 mm/min to about 261 μm at 20000 mm/min when the depth of cut is fixed at 200 μm.

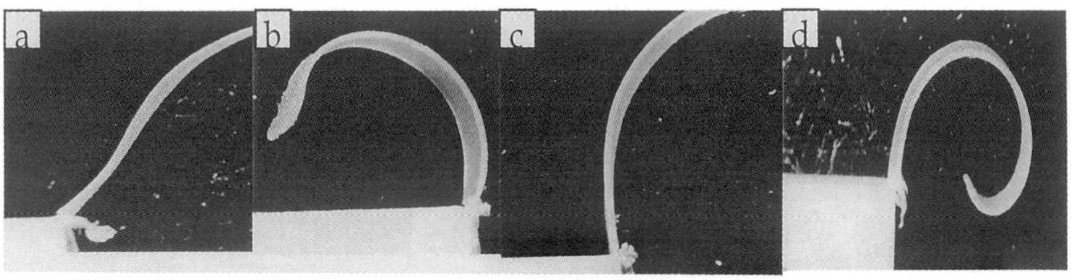

Figure 5. Continuous curly chips. (a) LDPE at 1 mm/min, (b) LDPE at 20000 mm/min, (c) HDPE at 1 mm/min and (d) HDPE at 20000 mm/min. The cutting tool moved from left to right.

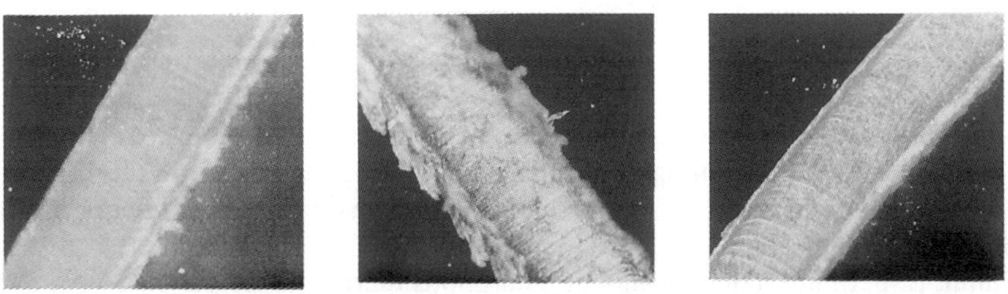

Figure 6. Shear deformation in chips. Left: LDPE at 1 mm/min, Middle: LDPE at 20000 mm/min, and Right: HDPE at 20000 mm/min. In these photos, the surface at the left hand side is the tool-chip interface. The cutting tool moved from left to right.

The deformation behavior of the HDPE chips is quite different from that of the LDPE's. In this case, a smoother chip edge and a thinner plastic layer beneath the tool-chip interface are identified in HDPE (Right in Fig. 6). The difference between the two materials is attributed to a lower melting temperature and shear strength of LDPE. Since the melting temperature of LDPE is only about 110°C, it is likely that the melting temperature of LDPE was reached at the tool-chip interface, resulting in a build-up-like morphology. The chip of HDPE is also different from Kobayashi's continuous shear chip [1] that has not got a plastically deformed layer. Further, a clear shear deformation in the area between two deformation layers is seen, similar to the common phenomenon in metal cutting. The shear deformation of the material on the shear planes is responsible for the clockwise curling of HDPE chips.

4 Conclusions

In this study, an orthogonal cutting was performed on LDPE and HDPE specimens, which are thermoplastics with very low glass transition temperatures and melting temperatures. Surface roughness and deformation behavior in chips were discussed in detail in relation to material properties and effect of cutting speed. The following conclusions can be drawn:

(1) Surface roughness of a machined HDPE specimen is better than that of a LDPE's because of the higher glass transition temperature and greater mechanical strength of the former material. With increasing the cutting speed, the surface finish of both the LDPE and HDPE are improved significantly.

(2) Shear strength was proposed as a criterion to evaluate the resultant influence of strain rate and temperature rise in the cutting zone. It was found that appropriate shear strength is preferred in order to produce a smooth surface of very soft polymers with low T_g or T_m and low mechanical strength.

(3) A special type of chips was found in the cutting of LDPE. The frictional heat between the tool and the chip was the major reason for the formation of the plastic deformation layer. The residual stress in the deformed layer bends the chips anticlockwise while shear deformation is restricted within the layer. A chip of HDPE is similar to a continuous chip in cutting a metal where shear deformation occurs across the chip thickness.

Acknowledgement

The authors appreciate the financial support of ARC and Sola through a SPIRIT grant.

References

1. Kobayashi, A. K. Machining of plastics, McGraw-Hill, Inc. New York, 1967.
2. Kobayashi, A. K. Machining of plastics: Part II, Modern plastics **43** (1963) pp. 109.
3. Kobayashi, A and Saito, K. On the cutting mechanism of high polymers, J. Polymer Sci. **58** (1962) pp. 1377-1395.
4. Smith, E. F. Single-point turning of amorphous thermoplastic polymers, M. S. thesis, North Carolina State University, Raleigh, NC, 1989.
5. Carr, J. W. and Feger, C. Ultraprecision machining of polymers, Precision Engineering, **15** (1993) pp. 221-237.
6. Boothroyd, G. Fundamentals of metal machining and machine tool, McGraw-hill book company, New York, (1975) pp. 74.
7. Wallace, P. W. An investigation on the friction between chip and tool in metal cutting, fellowship thesis, Department of Mechanical Engineering, University of Salford, England, 1962.
8. Bucknall, C. B. Toughened Plastics, Applied Science, London, 1977.

ANALYSIS OF DELAMINATION IN DRILLING COMPOSITE MATERIALS USING CORE DRILL

H. HOCHENG
Department of Power Mechanical Engineering, National Tsing-Hua University
Hsinchu, Taiwan, ROC
E-mail: hocheng@pme.nthu.edu.tw

C. C. TASO
Department of Automation Engineering, Tahua Institute of Technology
Hsinchu, Taiwan, ROC
E-mail: aetcc@thit.edu.tw

Delamination is one of the serious concerns during drilling. This paper presents a comprehensive analysis of delamination in use of core drill. The critical thrust force at the onset of delamination is predicted and compared with the twist drill. The well known advantage of industrial use of core drill is fundamentally illustrated. A guide for drill design can be developed based on the proposed models, and this approach can be extended to examine the effects of various drill or future innovative drill design

1 Introduction

Amongst all machining operations, drilling using twist drill is the most commonly applied method for generating holes for riveting and fastening structural assemblies. A lot of reference of the drilling of fiber-reinforced plastics reports that the quality of the cut surfaces is strongly dependent on drilling parameter [1-2]. The drill geometry is also considered the most important factor that affects drill performance [3]. Damage development and detection, new tooling design, and the influences of cutting conditions have been studied [4-5]. Doerr et al. designed the drill to cut materials toward the hole center and to shear at the hole edge [6]. Haggerty and Ernst found that "spiral" point drills performed much well than the conventional ones [7]. The influence of tool wear and the resulted increase of thrust were discussed [8-9]. Wu used multifaceted drills to reduce the thrust force [10]. Koenig et al. investigated the effect of processing variables on drilling damage [1] [11]. A general overview of the various possibilities for composites machining can be found [12].

Among the defects caused by drilling, delamination damages is the most critical. Drilling-induced delamination occurs both at the entrance and the exit planes of the workpiece. Investigators have studied analytically and experimentally in which delamination in drilling have been correlated to the thrust force during exit of the drill. The first analytical model to determine the critical thrust force of the drill was formulated by Ho-Cheng and Dharan [13]. They employed a linear elastic fracture mechanics method and solved for the critical thrust force that relates the delamination of composite laminates to drilling parameters and composite material properties. Jain and Yang developed a method for correlating the feed rate with the onset of delamination [14-15]. These works simplified the drilling thrust force by a single concentrated central load. Beside the twist drill, the effects of various drill geometry were rarely discussed in analytical fashion. This paper presents a comprehensive analysis of delamination in use of core drill. The critical thrust force at the onset of delamination is predicted and compared with the twist drill.

2 Delamination Analysis

In drilling of composite laminates, the uncut thickness to withstand the drilling thrust force decreases as the drill approaches to the exit plane. The laminate at the bottom can be separated from their interlaminar bond around the hole. In consideration of the expense of time and cost, an analytical approach is endeavored to embody the aspects of materials and process parameters in a model to predict the onset of delamination.

2.1 *Physical Model*

At the propagation of delamination, the drill movement of distance dX is associated with the work done by the thrust force F_A, which is used to deflect the plate as well as to propagate the interlaminar crack. The energy balance equation gives

$$G_{IC}dA = F_A dX - dU \quad (1)$$

where dU is the infinitesimal strain energy, dA is the increase in the area of the delamination crack, and G_{IC} is the critical crack propagation energy per unit area in mode I.

2.2 *Mathematical Analysis*

(I) Concentrated Load for Twist Drill [13]

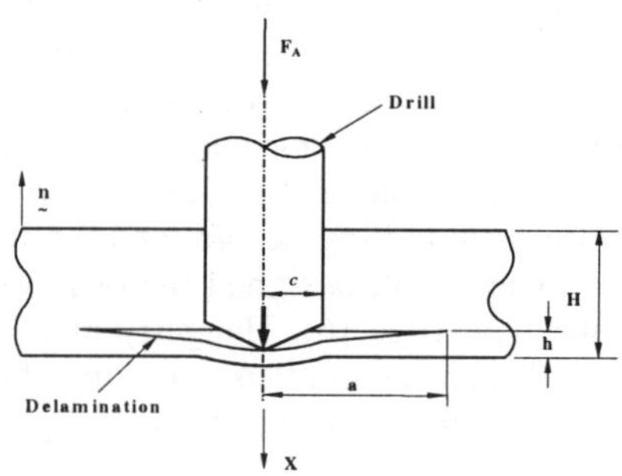

Figure 1 Circular plate model for delamination analysis (Twist Drill)

Fig. 1 depicts the schematics of delamination. In Fig. 1, the center of the circular plate is loaded by a twist drill of diameter d. F_A is the thrust force, X is the displacement, H is the workpiece thickness, h is the uncut depth under tool, and a is the radius of delamination. The isotropic bahavior and pure bending of the laminate are assumed in the model. For a circular plate subject to clamped ends and a concentrated load, the stored strain energy U is

$$U = \frac{8\pi M X^2}{a^2} \quad (2)$$

where

$$M = \frac{Eh^3}{12(1-v^2)} \quad (3)$$

and the displacement X is

$$X = \frac{F_A a^2}{16\pi M} \tag{4}$$

The thrust force at the onset of crack propagation can be calculated

$$F_A = \pi\sqrt{32 G_{IC} M} \tag{5}$$

(II) Distributed Circular Load for Core Drill

Fig. 2 depicts the schematics of a core drill and the induced delamination. Since the tool-work contact is located in a circular area, at bottom of drilling core the thrust force of the core drill is considered a downward area load (p_1) subtracted by an upward area load (p_2), as shown in Fig. 2. Let

$$p_2 = (1-\beta)^2 p_1 \tag{6}$$

where t is the thickness of core drill, and β is the ratio between thickness and radius of core drill (namely, $t = \beta c$)

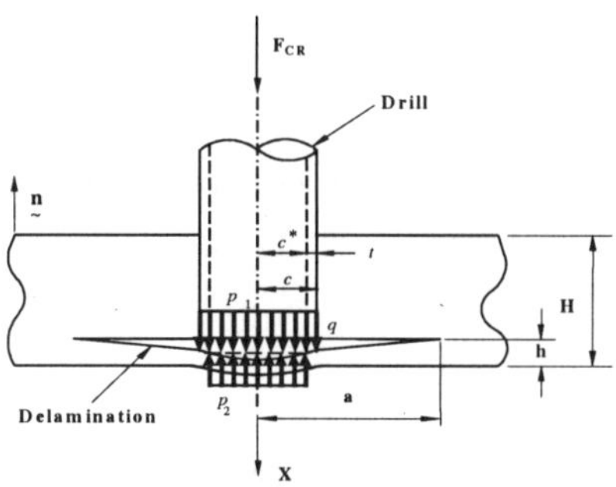

Figure 2 Circular plate model for delamination analysis (Core Drill)

The energy balance equation can be expressed as follows

$$\begin{aligned} G_{IC}.2\pi a &= (p_1\frac{dW_1}{da} - p_2\frac{dW_2}{da}) - (\frac{dU_1}{da} - \frac{dU_2}{da}) \\ &= \frac{p_1^{\,2}}{16\pi M}(a - \frac{c^2}{2a}) - \frac{p_2^{\,2}}{16\pi M}(a - \frac{c^{*2}}{2a}) \end{aligned} \tag{7}$$

where the subscripts 1 and 2 denote the variables for the downward force and the upward force, respectively.

For a laminate subject to an uniformly distributed load (Fig. 3), the following analysis is available [16]. The outer and inner deflection of a circular plate of radius a, which is clamped and subjected to a uniformly distributed load over a central circular area of radius c, is given by

(i) $0 \le r \le c$ (inner portion)

$$X = \frac{F_D}{64\pi M}[(4a^2 - 3c^2 + 4c^2\ln\frac{c}{a}) - 2r^2(\frac{c^2}{a^2} - 4\ln\frac{c}{a}) + \frac{r^4}{c^2}] \tag{8}$$

(ii) $c \leq r \leq a$ (outer portion)

$$X = \frac{F_D}{32\pi M}[2a^2 + c^2 - r^2(2+\frac{c^2}{a^2}) + 2c^2 \ln\frac{r}{a} + 4r^2 \ln\frac{r}{a}] \tag{9}$$

The stored strain energy is

(i) $0 \leq r \leq c$ (inner portion)

$$\begin{aligned} U_1 &= \pi \int_0^c M[(\frac{d^2X}{dr^2} + \frac{1}{r}\frac{dX}{dr})^2]rdr \\ &= \frac{{F_D}^2}{64\pi M}[\frac{2c^2}{3} - \frac{c^4}{a^2} + \frac{c^6}{2a^4} - \frac{4c^4}{a^2}\ln\frac{c}{a} + 4c^2 \ln\frac{c}{a} + 8c^2(\ln\frac{c}{a})^2] \end{aligned} \tag{10}$$

(ii) $c \leq r \leq a$ (outer portion)

$$\begin{aligned} U_2 &= \pi \int_c^a M(\frac{d^2X}{dr^2} + \frac{1}{r}\frac{dX}{dr})^2 rdr \\ &= \frac{{F_D}^2}{64\pi M}[2a^2 - 2c^2 + \frac{c^4}{2a^2} - \frac{c^6}{2a^4} - 8c^2(\ln\frac{c}{a})^2 + \frac{4c^4}{a^2}\ln\frac{c}{a}] \end{aligned} \tag{11}$$

The critical thrust force F_D at the onset of crack propagation can be calculated

$$F_D = \pi\sqrt{\frac{32G_{IC}M}{1-\frac{1}{2}s^2}} \tag{12}$$

where $s = c/a$. The comparison of F_D and F_A in Eq.(12) and Eq.(5) gives

$$\frac{F_D}{F_A} = \frac{1}{\sqrt{1-\frac{1}{2}s^2}} \tag{13}$$

as shown in Fig. 4.

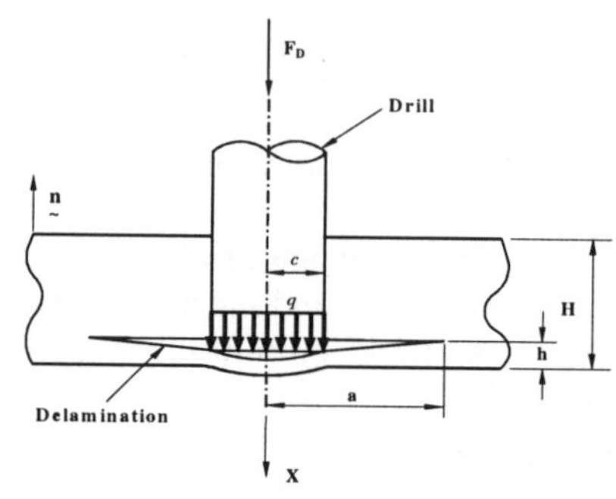

Figure 3 Circular plate model for uniform load

The critical thrust force F_{CR} at the onset of crack propagation can be calculated

$$F_{CR} = \pi \sqrt{\frac{32G_{IC}M}{\left\{[1-(1-\beta)^4] - \frac{1}{2}s^2[1-(1-\beta)^6]\right\}}} \tag{14}$$

The comparison of F_{CR} and F_A in Eq.(14) and Eq.(5) gives

$$\frac{F_{CR}}{F_A} = \frac{1}{\sqrt{\left\{[1-(1-\beta)^4] - \frac{1}{2}s^2[1-(1-\beta)^6]\right\}}} \tag{15}$$

as shown in Fig. 5.

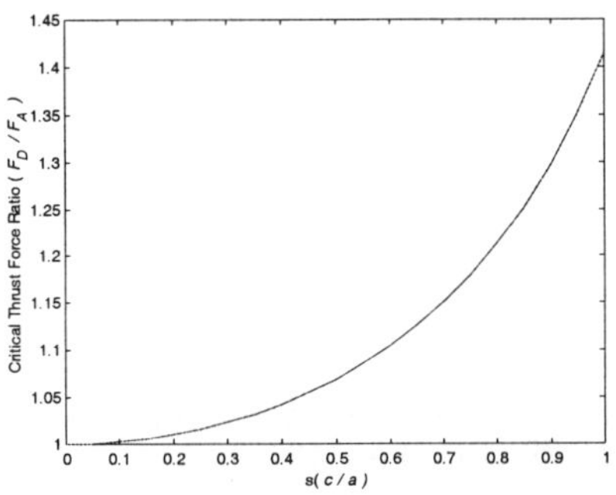

Figure 4 Critical Thrust Ratio between Uniform Load and Concentrated Load

3 Results and Discussions

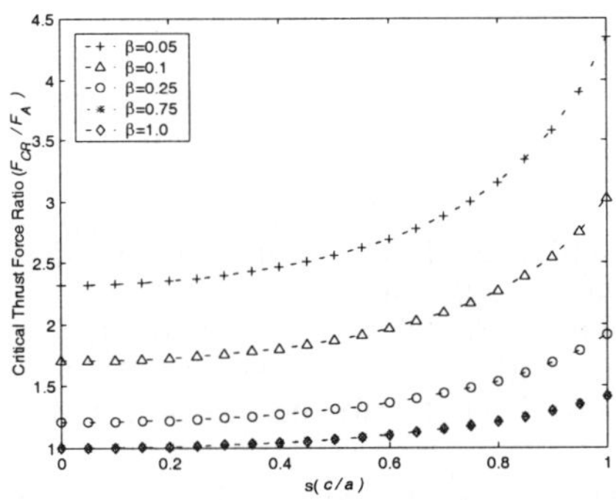

Figure 5 Critical Thrust Ratio between Core Drill and Twist Drill

Results for the predicted critical thrust force of the core drill is presented in Fig. 5. Since the total thrust force is distributed toward the periphery at a ratio of β (see Sec. 2.2 (II)) rather than concentrated at center, the drill is expected to be advantageous for allowing larger critical thrust force responsible for the onset of delamination. Fig. 5 illustrates the more the thrust force is distributed toward the periphery (smaller β), the higher becomes the critical threshold. One notices in Fig. 5, that $\beta = 1$ represents the fully uniform load, while the decreasing β approaches to the saw drill case.

4 Conclusions

A comprehensive analysis for delamination caused by twist drill and core drill in drilling of composite materials is presented in the study. The analytical results are obtained based on classical elasticity, linear elastic fracture mechanics and energy conservation law. Based on the proposed models, the core drill offers the higher critical thrust force than twist drill. The results agree with the industrial experience. The drill allows for larger critical thrust force can be operated at larger feed rate without delamination damage. A guide for drill design can be developed based on the proposed models, and this approach can be extended to examine the effects of various drill or future innovative drill design.

Acknowledgement

This work is partially supported by National Science Council, Taiwan, ROC, under contract NSC89-2212-E007-121.

References

1. Koenig, W., Wulf, C., Grass, P., and Willerscheid, H., Machining of Fiber Reinforced Plastics, *Ann. CIRP*. **34** (1985) pp. 538-548.
2. Komanduri, R., Zhang, B. And Vissa, C. M., Machining of Fiber Reinforced Composites, *Proce. Manufact. Compos. Mater.* **49/27** (1991) pp. 1-36.
3. Galloway, D. F., Some Experiments on the Influence of Various Factors on Drill Performance, *Trans. ASME* **79** (1957) pp. 191-237.
4. Caprino, G., and Tagliaferri, V.,Damage Development in Drilling Glass Fiber Reinforced Plastics, *Int. J. Mach. Tools Manufact.* **35** (1995) pp. 817-829.
5. Miller, J. A., Drilling Graphite/Epoxy at Lockheed, *Am. Mach. Auto. Manufact.* (1987) pp.70-71.
6. Doerr, R., Greene, E., Lyon, B. and Taha, S., Development of Effective Machining and Tooling Techniques for Kevlar Composites, Technical Report, No. AD-A117853 (1982).
7. Haggerty, W. A. and Ernst, H., The Spiral Point Drill – Self-Centering Drill Point Geometry, ASTE Paper No. 101, **58** (1958).
8. Koplev, A., Lystrup, A., and Vorm, P., The Cutting Process, Chips and Cutting Forces in Machining CFRP, *Compos.*, **14** (1983) pp. 371-376.
9. Veniali, F., Llio, A. Di, and Tagliaferri, V., An Experimental Study of the Drilling of Aramid Composites, *ASME J. E. Res. Tech.*, **117** (1995) pp. 271-278.
10. Wu, S. M., Shen, J. M., and Chen, L. H., Multifacet Drills, *Proceedings 14th National SAMPE Tech. Conf.*, (1982) pp. 456-463.
11. Koenig, W., Grass, P., Heintze, A., Okcu, F., and Schmitz-Justin, C., Developments in Drilling, Contouring Composities Containing Kevlar, *Prod. Eng.* (1984) pp. 56-61.
12. Komanduri, R., Machining Fiber-Reinforced Plastics, *ASME Mech. Eng.* (1993) pp. 58-64.
13. Ho-cheng, H., and Dharan, C. K. H., Delamination During Drilling in Composite laminates, *ASME J. Eng. Ind.* **112** (1990) pp. 236-239.
14. Jain, S., and Yang, D. C. H., Effects of Feedrate and Chisel Edge on Delamination in Composite Drilling, *ASME J. Eng. Ind.* **115** (1993) pp. 398-405.
15. Jain, S., and Yang, D. C. H., Delamination-Free Drilling of Composite Laminates, *ASME J. Eng. Ind.* **116** (1994) pp. 475-481.
16. Timoshenko, S., and Woinowsky-Keiger, S., *Theory of Plates and Shells*, 2nd ed., McGraw Hill, (1959) pp. 51-59, 376-377.

IMPROVING THE TRIBOLOGICAL AND MECHANICAL PROPERTIES OF STEEL COMPONENTS BY GRINDING

I. ZARUDI AND L. ZHANG

School of Aerospace, Mechanical and Mechatronic Engineering, The University of Sydney, NSW 2006, Australia

This paper studies a method of improving the tribological and mechanical properties of steel elements by grinding. It was found that a proper use of grinding heat and stress can create favorable microstructures and promote high wear and fatigue resistance. The thickness of the treated surface layer can be up to 1mm. An enhanced phase transformation, intensive dislocations and a uniform carbon distribution make the microstructure of the layer beneficial.

1 Introduction

In the current manufacturing practice, elements after a soft machining need to be transported to the surface hardening bay, cleaned and undergone the heat treatment. They are then discharged, moved back and incorporated into the production line for a final precision surface finishing.

On the other hand, the thermo-mechanical process in surface grinding often brings about undesirable changes of the mechanical properties generated by heat treatment. For example, the material removal in grinding produces considerable heat, and most heat conducts into the workpiece and often causes thermal damage. As a result, relevant studies on grinding technology in the last three decades have been trying to eliminate such damages by extracting the grinding heat from workpieces [1-3]. The above production procedure is inefficient because it wastes most grinding energy and makes the prevention of the treated surface structure difficult.

In the investigation on grinding-induced residual stresses in steel workpieces, Zhang and co-workers [4-6] found that phase transformation during surface grinding, which results in hardening characteristics, could be achieved by a proper selection of grinding conditions. Under the specified distribution of heat flux, convection process of coolants and constitutive models of work materials, the depth and nature of the martensite layer and distribution of residual stresses can be described quantitatively with respect to the material removal rate and coolant application method. Experimental studies [7-8] also reported that the grinding energy could be used to generate hard surface layer and the term "grind-hardening" was introduced [7]. However, a hardened layer does not necessarily mean beneficial, rather, it can be very harmful if the layer is associated with a poor microstructure [1-3].

The purpose of the present work is to understand the thermo-mechanical mechanism of surface heat treatment by grinding and find out a reliable method for creating suitable functional properties – wear and fatigue resistance of the ground component.

2 Experiment

The testing material selected was an initially tempered AISI4140 steel. Grinding was performed on a surface grinder, Minini Junior 90 CF CNC M286, with the grinding parameters specified in Ref. [9]. Coolant (compressed air) was applied at a distance *d* to the central line of the grinding wheel.

The properties of the ground surfaces and subsurfaces were characterized by a scanning electron microscope (SEM) 505 (JEOL) and a transmission electron microscope (TEM), CM12 (Philips).

Dry wear tests were carried out on a Plint-Cameron pin-on-disc machine. Two pin samples were held against a rotating disk of high-speed steel. Each test was run on a fresh track, with a sliding distance of 500m, normal stress between 1MPa and 5MPa and sliding velocity of 0.98m/s.

The fatigue experiments, using three-point bending, were carried out on an alternating bending machine "Servopulser" (Shimadzu Seisakusho Ltd.) at 10Hz. Flat samples were with a gauge length of 20mm and a cross-section of 2 x 15mm.

3 Results and Discussion

3.1 Micro-hardness and depth of grinding-induced layer

For convenience of the discussion below, we will call the martensite introduced by the conventional quenching the ordinary martensite (OM) and the hard layer generated by grinding the grinding-induced layer (GIL).

The depth of the GIL varied from 0.18mm to 1mm depending on grinding variables (see Ref. [9] for details). The micro-hardness of the GIL exceeds the original micro-hardness of the steel ($HV_{(300g)}$ =350) by 2.2 times and that of the OM ($HV_{(300g)}$ = 560) by 1.3 times. Minor fluctuations of micro-hardness happened across the layer.

3.2 Microstructures of the GIL and OM

The structure of the OM under an electron microscopy is shown in **Fig.1a**.

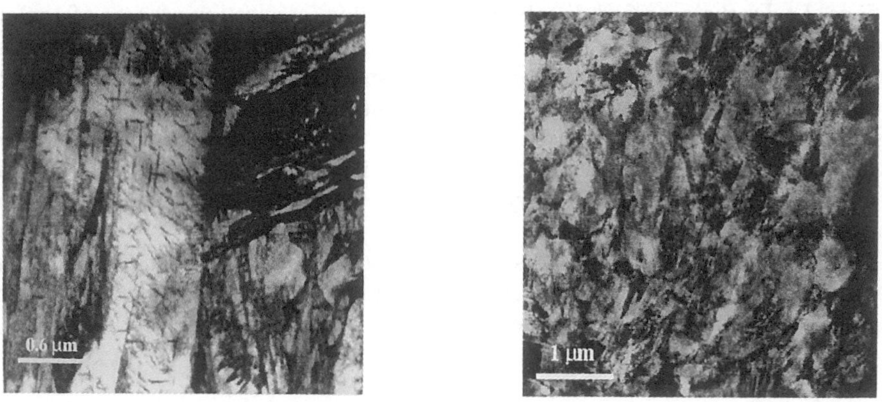

(a) (b)

Figure 1. The microstructures revealed by electron microscopy. (a) the OM. Note the large laths; (b) the GIL (grinding conditions: depth of cut = 70μm, infeed velocity = 0.1m/min). Note the refined martensite laths.

Martensite laths were formed on quenching with crystals being grouped in packets, having an average length of 2-3μm and a width of 0.3-0.5μm. The density of dislocations was high but carbides were not found.

In the GIL, however, the material of first 5-6 μm under the surface (**Fig.1b**) was composed of fine and nearly circular grains with a characteristic dimension of 0.2-0.5μm. The c/a parameter of the structure, according to TEM, was in the range of 1.02-1.03, indicating that the structure was a martensite with a great refinement of grains. Moreover, the dislocation structure of the martensite fragments was mainly cellular and no carbides observed.

In the deeper subsurface of the GIL, martensite laths became greater. For example, at the depth of 200μm from the surface the characteristic dimension of the laths was 0.8-1μm. Again no carbides were admitted and the dislocation structure of the sub-layer resembled that of the OM.

However, in the further deeper subsurface, at the depth of 400μm for instance, carbides could be easily located and the dislocation density was significantly lower.

3.3 Formation mechanism of the GIL

The above observation shows that the microstructures of the OM and GIL are rather dissimilar, and the microstructure in GIL varies along the depth of the layer.

The ordinary quenching has an appropriate soaking at a temperature above Ac_3 to allow phase transformation to take place.

The surface treatment by grinding entails the heating to above Ac_3 through the wheel-workmaterial interaction. It brings about an absolute maximum temperature at the surface with a sharply decreasing gradient along the depth, making the GIL form in a variable temperature domain. In addition, the stress and strain rate fields in the grinding zone change and in turn affect the GIL formation. When the temperature rise is sufficient to make a ferrite-austenite transformation, the stress field brings about the deformation of austenite at a high strain rate so that cellular and polygonal dislocation structures are developed. The fast cooling of the top layer does not allow the structural relaxation of dislocations and thus martensite inherits. The substructure generated in the austenite prevents the growth of martensite crystals and facilitates an increase in the number of nucleation centers. It causes the overall refinement of the martensite structure.

In the deeper subsurface, say at the depth of 200μm under the surface, the GIL structure is accomplished with a lower temperature rise and less deformation at a smaller strain rate. As a result, the dislocation density of the deformed austenite diminishes. Meanwhile, relaxation takes place, as the cooling time for this part of material is longer. The lower dislocation density in austenite results in larger martensite laths.

In the further deeper subsurface, at the depth of 400μm under the surface, for instance, the deformation in austenite becomes minor. In addition, the low cooling rate at this depth enables the formation of ferrite/bainate/martensite. Hence, both dislocations of low density and carbides appear.

The above analysis shows that the formation of GIL is governed by grinding variables and the structure formed is quite special, characterized by a very fine-grained martensite on the top sub-surface, a gradual growth of the grain-size along the subsurface depth, to a structure containing additional ferrite/bainite elements at the bottom. This structure promotes the beneficial mechanical properties of the GIL as to be discussed below.

3.4 Wear

The wear rate of the GIL is two orders lower than that of the OM, as shown in **Fig.2**. For the OM, transition from mild to sever wear happens when the normal stress applied varies from 2MPa to 3MPa. At the stress of 5MPa, adhesion takes place. However, the GIL always exhibits mild wear in the whole process of the wear testing.

The superior property of high wear resistance and low friction coefficient of the GIL is due to its special microstructure. In the mild wear regime, the GIL has a higher wear resistance because its hardness is much greater. In the severe wear regime, however, the wear resistance of a material depends largely on the stability of its microstructure [10-11]. Both grain refinement and polygonal dislocation arrangement enhance the thermal and mechanical stability of the structure. Since the energy consumption for changing the GIL structure is higher, the GIL possesses a better wear resistance.

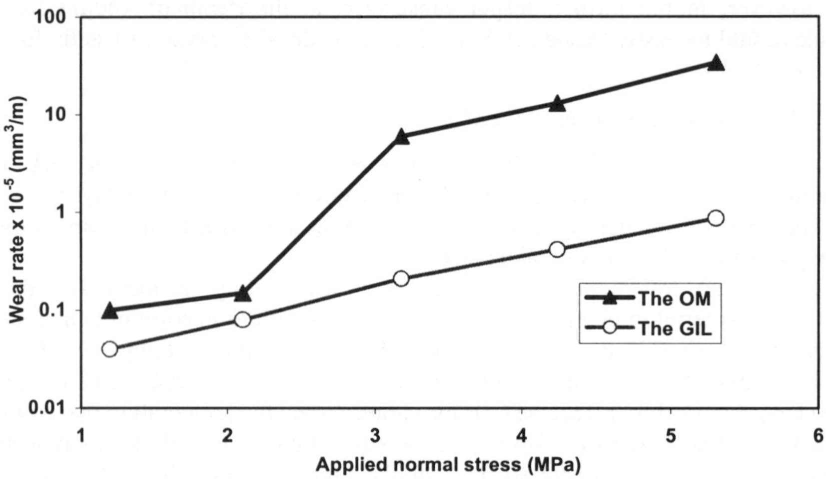

Figure 2. Wear of OM and GIL

Now consider the microstructural changes occurring during the wear tests. An extensive precipitation of carbides takes place in both the OM and GIL. However, the size and distribution of the carbides are different. In the OM, coarse plate-shaped carbides form throughout the laths. In the GIL, however, carbides are more dispersed and appear mostly inside the laths. The two obvious factors that caused the difference are: (1) The refined martensite laths in the GIL make its total boundary length much longer and thus reduced the carbide concentration along the boundaries. (2) The cellular and polygonal distribution of dislocations also promotes the carbide precipitation inside the martensite laths in the GIL.

The precipitation of carbides during the wear tests increases the hardness of both the OM and GIL, and thus improves their wear resistance. It is also advantageous from the point of view of structural changes because the formation of the tempered martensite increases the structural stability.

3.5 Fatigue

Fig.3 shows that the GIL has a higher fatigue resistance. At the highest bending stress of 1.75GPa in this study, its fatigue life is twice that of the untreated steel (US). At a lower bending stresses level, the improvement varies from 12% to 20%.

The topography of the fractured surfaces after fatigue experiments presents distinct characteristics of the GIL and US. The fractured surface of the US shows large brittle facets. The refinement of martensite structure in the GIL makes the facets much smaller. Since microcracks run mostly inside the martensite laths, networking of microcracks of individual martensite laths in the GIL is harder, because of its finer grains. In addition, the high compressive residual stresses in the GIL, which hinders the initiation and propagation of microcracks, play an important role in the improvement of the GIL's fatigue life. A more detailed investigation about the residual stresses is still being undertaken in the authors' laboratory.

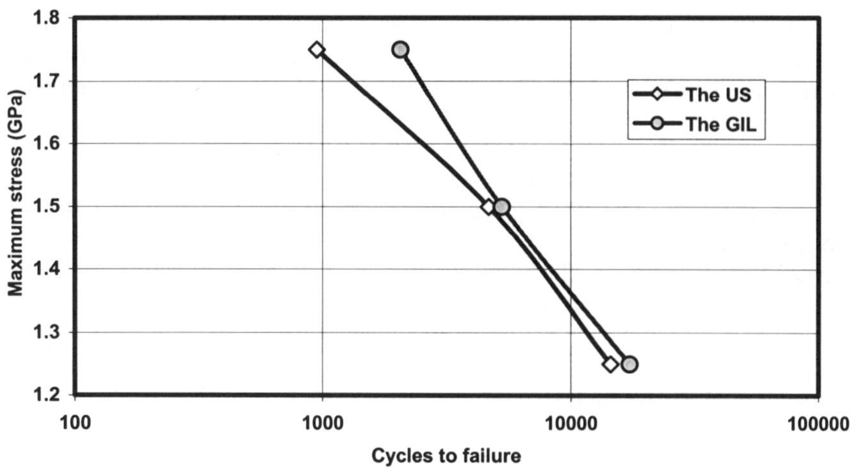

Figure 3. Fatigue life of OM and GIL.

4 Conclusion

The method of surface treatment by grinding leads to a grinding-induced layer, which has an advantageous microstructure and enhances the wear and fatigue properties of the ground components. Although the present study limits to one material, the mechanisms explored are applicable to a large class of quenchable steels whose heat treatment involves martensite phase transformation.

Acknowledgements

The authors wish to thank the Australian Research Council (ARC) for continuing support of this project and the Electron Microscope Unit of Sydney University for use of its facilities.

References

1. Malkin S., Burning limit for surface and cylindrical grinding of steels. *Annals of the CIRP* **27** (1978) pp. 233-236.
2. Shaw M. C., Vyas, A., Heat-affected zones in grinding steel. *Annals of the CIRP* **43** (1994) pp. 279-182.
3. Ihalainen E. K., Simple difractometrical procedure to determine the damaged surface layer in grinding of hardfield manganese steels. *Acta Polytechnica Scandinavica, Mechanical Engineering Series* (1981) pp. 1-24.
4. Zhang L. Mahdi M., Applied mechanics in grinding – Part IV. The mechanism of grinding induced phase transformation. *International Journal of Machine Tools & Manufacture* **35** (1995) pp. 1397-1409.
5. Mahdi M., Zhang L., Applied mechanics in grinding -Part VI. Residual stresses and surface hardening by coupled thermo-plasticity and phase transformation. *International Journal of Machine Tools & Manufacture* **38** (1998) pp. 1289-1304.
6. Mahdi M., Zhang L., Applied mechanics in grinding - Part VII: Residual stresses induced by the full coupling of mechanical deformation, thermal deformation and phase transformation. *International Journal of Machine Tools & Manufacture* **39** (1999) pp. 1285-1298.
7. Brinksmeier E., Brockhoff, T., Utilization of grinding heat as a new heat treatment process. *Annals of the CIRP* **45** (1996) pp. 283.
8. Brockhoff T., Grind-hardening: A comprehensive view. *Annals of the CIRP* **48** (1999) pp. 255-260.
9. Zarudi I. and Zhang L., Thermal analysis in surface modification by grinding. In: ACAM 2002 (Sydney, Australia, 2002).
10. Wang. Y. L. T., Liu J., Tribo-metallographic behavior of high carbon steel in dry sliding. III. Dynamic microstructure changes and wear. *Wear* **231** (1999) pp. 20-37.
11. Bernshtein M. L., Kaputkina L. M., Konyukova E. V., Nikishov N. A., Shakhkerimov R. N., Changes in the structure of thermo-mechanically hardened austenite under the effect of temperature and deformation. *Russian Metallurgy* **1** (1985) pp. 80-88.

THERMAL ANALYSIS IN SURFACE MODIFICATION BY GRINDING

I. ZARUDI AND L. ZHANG

School of Aerospace, Mechanical and Mechatronic Engineering,
The University of Sydney, NSW 2006, Australia

This study tends to investigate the reliability of the surface modification by grinding through understanding the heat flux formation and temperature distribution in the process. A method of capturing the heating zone movement by means of an optical microscope and a CCD camera was used to detect the temperature rise in the subsurface of a component subjected to grinding. The heating parameters, temperature field and stability of the method were evaluated. It was shown that a triangular heat flux gives the best fit for up-grinding. The contact grinding length can deviate significantly from the geometrical contact length due to thermal expansion of workpieces. It was also proved that stable grinding conditions might produce beneficial surface layers in ground component.

1 Introduction

The surface treatment by grinding allows an improvement of mechanical properties of components [1]. The beneficial microstructure is formed as a result of the special temperature-stress field in grinding. The reliability of the technique, to a great extent relies on the stability of the grinding temperature field.

A number of studies tried to evaluate the heat flux profile in conventional and creep-feed grinding [2-6]. A heat flux has been modeled in various ways, *i.e.*, described as uniform [2], triangular [3-8], square [9] and variable profiles [10].

The uniform heat flux model [2] led to an evident contradiction [10]. Rowe [9] proposed the parabolic heat flux distribution based on his experimental findings. However, a theoretical work of Zhang and co-authors [8] proved that a realistic heat flux should be neither rectangular nor parabolic. Because the surface heat generated has a close relationship with the local deformation of the workpiece material, the heat flux could be approximated by a triangular heat source [8, 11]. It was also confirmed that the temperature field could vary significantly with the location of the triangular apex [6-7]. However, their studies did not give rise to a correlation of flux distribution with real grinding conditions.

The experimental studies specified above used a thermocouple technique to measure the temperature rise in grinding [9,11]. Nevertheless, the method implies a number of limitations. It does not allow locating the position of the flux apex with a satisfactory resolution [12]. It results in an inadequate accuracy while grinding with large depth of cut. Furthermore, many passes are required if one needs to obtain the temperature distribution in the subsurface of a workpiece, which causes an additional problem because it is difficult to maintain identical wheel conditions in different grinding passes.

The present study aims to determine the heated subsurface zone by an optical method that can provide with higher accuracy and permits to avoid some drawbacks of the thermocouple technique.

2 Experiment

The testing material selected was a widely used steel, AISI4140. Grinding was performed on a surface grinder, Minini Junior 90 CF CNC M286, with the grinding parameters listed in **Table 1**.

A temperature measuring system composed of an optical microscope, a CCD camera and a video recorder were constructed as shown in **Fig.1a**. The wear flats of the grinding wheel were evaluated by means of Laser Scanning Confocal Imaging System MRC-600 and image analysis technique.

Table 1 Grinding conditions

Workpiece material	Steel 4140	Grinding width (mm)	15
Grinding wheel	A120MVAA	Table speed (m/min)	0.02, 0.05, 0.1, 0.2, 0.3
Wheel diam. (mm)	280	Depth of cut (μm)	35, 70, 105

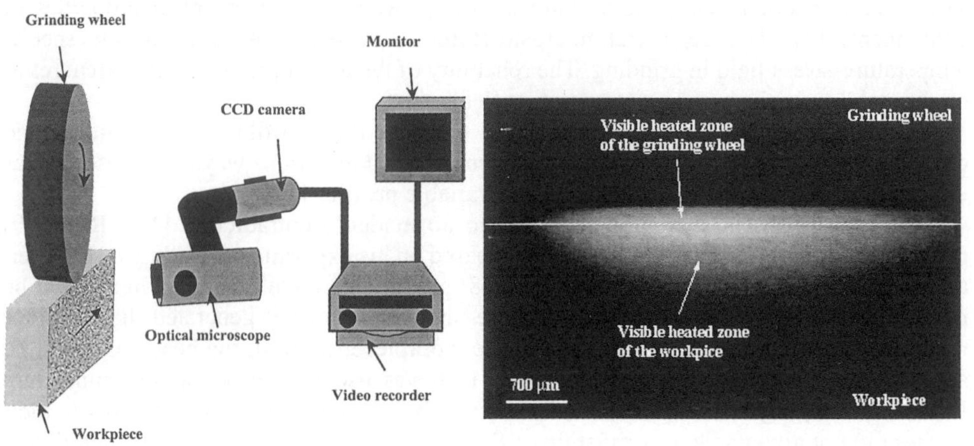

(a) (b)

Figure 1 (a) The temperature field recoding system ; (b) The heated zone captured by the camera. (Depth of cut = 70 μm. Infeed velocity = 0.1m/min).

3 Results and Discussion

3.1 Heated zone measured

The profiles of the heated zone in both the workpiece and the grinding wheel were asymmetric about the center of the grinding zone **(Fig.1b)**.

The workpiece material has a higher conductivity and thus has a much deeper heat penetration zone with the maximum temperature rise at its surface. An increase of the depth of cut resulted in a greater heated zone with its width and depth growing as detailed in

Table 2. This is reasonable because there was a higher energy input in grinding when the depth of cut increased. The infeed velocity affects the dimension of the heated zone in a more complicated manner. Initially, an increase of the infeed velocity leads to a larger heated zone (**Table 2**). However, when it increases further, the development of the zone size becomes non-monotonic. The shrinkage of the zone could be due to the decreasing heating time when the infeed velocity was large, because in this case the heat source on the workpiece surface moved with a faster speed and less amount of heat could actually conduct into the workpiece. At a low infeed velocity (e.g., from 0.02 to 0.1 m/min), the heated zone did not appear to be continuous but flashed (see **Table 2** for details). Physically, the flashing indicated that the heating was unstable and the temperature in the workpiece was frequently below the critical value. The stability of the heated zones increases with the rise of the infeed velocity and becomes steady as the velocity reaches 0.2-0.3 m/min.

3.2 Correlation between energy partitions and grinding wheel parameters

The energy partitions were determined by the following grain contact model proposed by Rowe et al. [12] :

$$R_{ws} = \frac{1}{1+\dfrac{k_{ge}}{\sqrt{r_0 v_s}} \times \dfrac{1}{\sqrt{(k\rho c)_w}}},$$

where k_{ge} is the effective thermal conductivity of the alumina grinding wheel, r_0 is the wear flat area of the alumina grain, v_s is the grinding wheel velocity, $\sqrt{(k\rho c)_w}$ is the thermal contact coefficient for the workpiece material. The wear flat areas of alumina grains determined by a confocal imaging system were in the range of 0.859 to 0.876 with the confocal depth, varying from 35μm to 350μm that significantly exceeds experimental depths of cut used in the study. The above measurement led to R_{ws}=0.864, 0.866, 0.869 for the grinding with the depth of cut equal to 35μm, 70μm, 105μm, respectively.

Table 2 Effect of grinding variables on the stability of the heated zone

Grinding regimes		Depth of the hardened layer, μm	Duration of flashing, s	Time between flashes, s	Width of the heated zone, mm	Depth of the heated zone, μm
Depth of cut; μm	Infeed velocity, m/min					
35	0.02	0	0	N/A	0	0
	0.05	130	0.06	0.33	1.73	140
	0.1	460	0.1	0.29	2.2	420
	0.2	500	0.15	0.09	2.4	460
	0.3	580	Pulsation	0	2.8	530
70	0.02	180	0.13	1.1	2.3	180
	0.05	350	0.14	0.24	2.60	400
	0.1	650	0.19	0.14	4.10	640
	0.2	880	Pulsation	0	5.20	840
	0.3	690	Continuous	0	4.50	720
105	0.02	300	0.056	1.09	2.45	230
	0.05	740	0.10	0.3	4.5	650
	0.1	830	Pulsation	0	5.30	800
	0.2	1000	Continuous	0	6.10	950
	0.3	900	Continuous	0	5.40	840

3.3 *Heated zone by theoretical prediction*

The well-known Jaeger's model of a moving heat source [13] was used in calculating the heated zone. The heat flux was modeled as a triangle.

The base of the heat flux was first assumed to be equal to the geometrical contact area and the heated zone evaluated. The heating, however, will lead to an expansion of the workpiece and grinding wheel. This expansion of the workpiece and the grinding wheel were assessed and the contact zone extended to the area where gaps between grinding wheel and workpiece became zero. Modified heated zone was then calculated based on the new contact length. The interaction method was applied till no differences in contact zone in two consecutive steps of iterations were detected.

The geometrical contact zones were close to the calculated when the infeed velocity and depth of cut were low. However, the deviation was quite pronounced for high infeed velocities and high depths of cut as the heating zone stretches beyond the geometrical, leading to a growth of the contact zone up to 2 times for grinding with 105 μm depth of cut and 0.2 m/min infeed velocity.

The apex position in the triangular heat flux was defined as:

$$l_a = \frac{2\zeta_a}{L_c},$$

where L_c is the length of the heat flux and ζ_a is defined by **Fig.2a**.

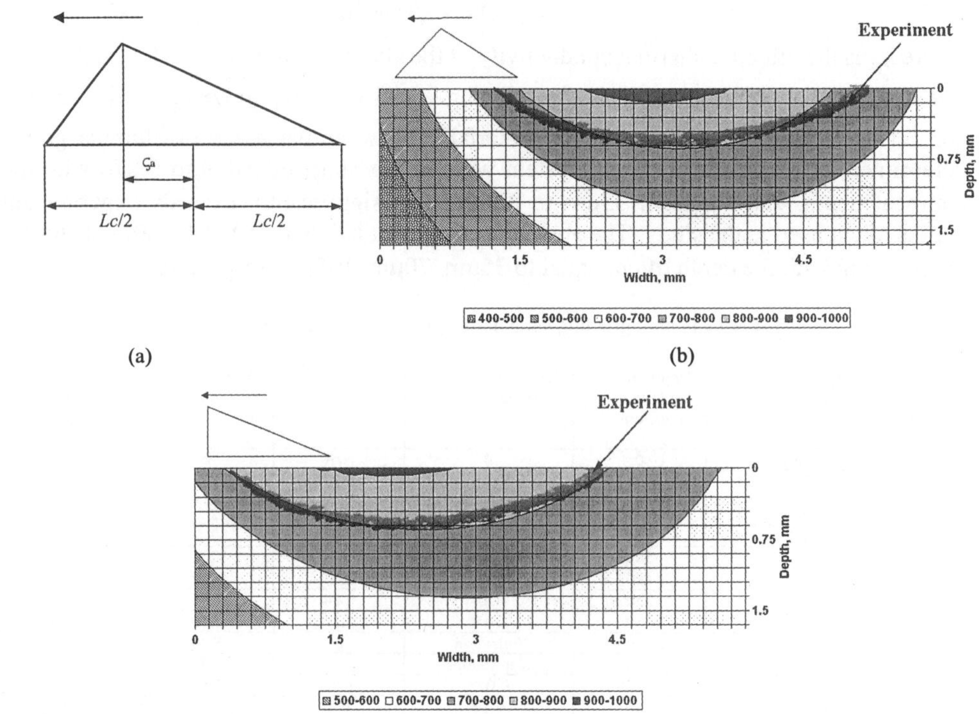

Fugure 2 (a) A theoretical triangular model of the heat flux; (b-c) a comparison of the experimental and theoretical temperature field (depth of cut = 70μm and infeed velocity = 0.1m/min) (b) $l_a = 0.25$; (c) $l_a = 1$.

Four types of triangle profiles with l_a=0.25, 0.5, 0.75 and 1 were considered in order to evaluate the best suitable profile with the experiment. As concluded in Ref. [1], the critical temperature rise responsible for phase transformation in the surface treatment of steel 4140 by grinding is 800°C. In the following analysis we will compare the heated zone profile experimentally measured at 800°C with the corresponding theoretical contours.

The results shown in **Figs 2b and 2c** indicate that l_a = 1 yields the best fit to experimental measurement. The decrease of l_a from 1 to 0.25 reduces the width of the heated zone but does not change its depth.

Experimental and theoretical parameters (width and depth) of the heated zones are in good agreement for high-infeed velocities (**Fig.3**). However, at low infeed velocities the deviation is quite large, especially at a low depth of cut. It could be attributed to the instability of the heated zone (**Table 2**), because the heated zones are unstable at low infeed velocities. The triangular heat flux used in calculations cannot accommodate this instability.

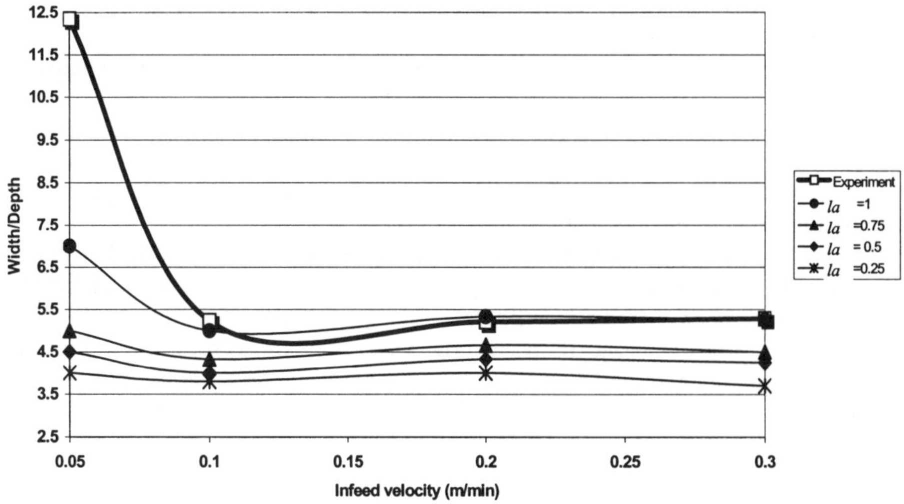

Figure 3 Effect of the grinding variables on the threshold of width/depth for experimental and theoretical heating zones (depth of cut = 70μm).

4 Conclusions

A method of capturing the heated zones in grinding was introduced in this study. The method permitted evaluating the temperature field in the subsurface of a component in a single grinding pass. In addition, it was found that

1. the triangular heat flux with l_a = 1 gives the best fit for all up-grinding conditions investigated.
2. the grinding contact length can deviate significantly from the geometrical if the grinding creates a sufficiently high temperature rise.
3. experimental and theoretical parameters of the heated zones are in good agreement under stable grinding conditions.

Acknowledgements

The authors wish to thank the Australian Research Council (ARC) for continuing support of this project and the Electron Microscope Unit of Sydney University for use of its facilities.

References

1. Zarudi I., Zhang, L.C., Mechanical property improvement of quenchable steel by grinding. *Journal of Materials Science* (submitted 2001).
2. Shaw M. C., Temperatures in cutting and grinding. *American Society of Mechanical Engineering* **146** (1990) pp. 17-24.
3. Lavine A. S., Malkin S., and Jen T. C., Thermal aspects of grinding with CBN wheels. *CIRP Annals* **38** (1989) pp. 557-60.
4. Guo C., Malkin, S., Analysis of energy partitions in grinding. *Journal of Engineering for Industry* **117** (1995) pp. 55-61.
5. Zhang L. C., Suto T., Noguchi H., Waida, An overview of applied mechanics in grinding. *Manufacturing review* **4** (1992) pp. 261-73.
6. Zhang L. C., Suto T., Noguchi H., Waida T, On some fundamental problems in grinding.(ed.), In First international conference on computer methods and experimental measurements for surface treatment effects, (Southampton Boston, Computational Mechanics Publications,1996) pp. 275-284.
7. Mahdi M., Zhang L., Applied mechanics in grinding Part-VI. Residual stresses and surface hardening by coupled thermo-plasticity and phase transformation. *International Journal of machine Tools & Manufacture* **38** (1998) pp. 1289-1304.
8. Zhang L. C., Suto, T., Noguchi, H., Waida, T., Applied mechanics in grinding Part II: Modelling of elastic modulus of wheel and interface forces. *International Journal of machine Tools & Manufacture* **33** (1993) pp. 254-255.
9. Rowe W. B., Black S. C., Mills B., Qi H. C., Morgan M. N., Experimental investigation of heat transfer in grinding. *Annals of CIRP* **44** (1995) pp. 329-32.
10. Jen T. C., Lavine A. S., A variable heat flux model of heat transfer in grinding: model development. *Journal of Heat Transfer* **117** (1995) pp. 473-478.
11. Guo C., Malkin S., Inverse heat transfer analysis of grinding, Part 2: Applications. *Journal of Engineering for Industry* **118** (1996) pp. 143-149.
12. Rowe W. B., Morgan M. N., Black S. C., Validation of thermal properties in grinding, *Annals of the CIRP* **47/1** (1998) pp. 275-279.
13. Jaeger J., Moving sources of heat and the temperature at sliding contacts. *Proc. of the Royal Society of NSW* **76** (1942) pp. 203-24.

SURFACE ENHANCEMENT TECHNOLOGIES FOR AIRCRAFT ENGINE COMPONENTS

W. ZHUANG AND B. WICKS
Airframes and Engine Division, Aeronautical and Maritime Research Laboratory
Defence Science and Technology Organisation, Melbourne, VIC 3207, Australia
E-mail: wyman.zhuang@dsto.defence.gov.au

Various mechanical surface enhancement technologies have been invented and developed to induce a protective layer of compressive residual stress at the surface of aircraft engine components where the operating loads are tensile dominated. The benefits of these surface enhancements are to prevent crack initiation, to retard propagation of small cracks and even to resist corrosion and wear damage. In this paper the literature on these technologies and their effects on component reliability and durability is reviewed, with an emphasis on Shot Peening (SP), Laser Shock Peening (LSP) and Low Plasticity Burnishing (LPB). The relative advantages of these three surface enhancement technologies as well as their limitations are identified and evaluated. The most important issues for these three technologies, and the relative merits of the resultant residual stress fields are presented and discussed with a view to enhancing engine component operating lives.

1 Introduction

For both airframes and engines fatigue and corrosion damage, often initiated from the surfaces of components, represents a threat to the safety and availability of aircraft (James 1991 and Ratwani et al 1997). For gas turbine engines in particular, the inspection requirements for fatigue and corrosion damage, the maintenance to replace critical components, and the rework required to remove fatigue and corrosion damage considerably increase operating costs. The down-time required to perform these inspections and repairs significantly impacts on aircraft airworthiness and readiness.

Various surface enhancement and modification technologies have been developed to induce a protective layer at the surface of aircraft engine components to enhance wear, corrosion and fatigue properties. Some of these technologies, such as plating, coating, ion implantation, heat treatment for surface hardening, and shot peening are well established and routinely applied to engine components in manufacturing and maintenance (Lessells and Brodrick in 1956, Levine 1978, Meetham 1978, Happ 1984, Bailey and Whalen 1990). However, these technologies have a number of limitations. For instance, shot peening induces only a shallow compressive residual stress layer (about 0.2 mm), with a large degree of cold work (strain hardening) that can cause damage to the microstructure of materials and promote accelerated relaxation of the beneficial compressive residual stresses (Prevey 2000). Although shot peening is routinely applied to engine components, engine manufacturers are unable to quantify the benefits and hence to allow the operators derive any benefit from extended component replacement intervals. In particular there is a lack of confidence in the development and sustainment of the residual stresses over long periods of service (Bailey and Whalen 1990).

Thus the risk that the compressive residual stress layer induced by SP will relax during engine operation prevents credit being taken for the benefits of the SP surface enhancement in increasing the safe operational life of engine components. At the same time there is a risk of surface damage from excessive peening which can actually reduce the component life. These concerns have lead to the development of advanced mechanical surface enhancement technologies, such as Laser Shock Peening (LSP) and Low Plasticity Burnishing (LPB). LSP was first developed in early 1970's as a fatigue enhancement surface treatment for metallic materials (Metz and Smidt in 1971, Fairand et al. 1972). Recently, based on the principle of roller and ball burnishing used originally for refinement of surface finishes, a new concept – Low Plasticity Burnishing was patented by Lambda Research in Cincinnati, Ohio (Prevey 2000). The published data has demonstrated that both LSP and LPB are able to introduce large and deep compressive residual stresses with low cold work, although they have very different features and capabilities that will be presented and discussed in this paper.

It is worth noting that there are other kinds of surface enhancement technology under research and development. It is impossible for this paper to exhaustively review and cover all of these technologies, and this paper highlights only the application of mechanical surface enhancement technologies. The technical features and induced residual stress fields will be presented and discussed and the paper will conclude with a statement of the future research requirements for this area, including the generation and relaxation of residual stresses, and the likely implications for gas turbine engine life assessment, life extension and repair.

2 Shot Peening (SP)

2.1 The Basic Shot Peeing Process

The shot peening process is a traditional and inexpensive surface enhancement method, and it is currently applied to many aircraft structures and engine components. The process is that hard shots in an air-blast repeatedly bombard the surface of the work-piece resulting in local plastic surface deformation. Since the plastically deformed surface layer tends to stretch/expand and the adjacent layer of elastic material beneath restrains the expansion there is a compressive residual stress field generated in the surface layer balanced by sub-surface tensile residual stresses.

2.2 The Effect of Shot Peening Control Parameters

Although shot peening is simple to apply, it is much more difficult to develop a model of the mechanism and the subsequent induced deformation and residual stress relaxation. It is equally difficult to quantify the SP benefits and the effect on the fatigue behaviour of shot peened components. In general, of the various important shot peening parameters (shot size, shape, mass, density, hardness, velocity, coverage, etc.), two parameters, the Almen intensity and the coverage are the most important for controlling the quality of the shot peening process.

2.3 Applications of Shot Peening to Aircraft Engines

One of the main merits of the SP process is that it can be applied to large and irregular surfaces of components at relative low cost. Because of the enhancement in fatigue life produced by the process, shot peening was applied initially to a variety of aircraft engine components, such as crankshafts, connecting rods, ring gears and so on in 1940's (Brookman and Kiddle 1946). A wide range of materials has since been SP treated, including steel, titanium and aluminium alloys relevant to the aircraft industry.

3 Laser Shock Peening (LSP)

3.1 The Basic Laser Shock Peening Process

Laser shock peening was first developed for enhancing the fatigue life of metal parts in the early 1970's, but the LSP process has only recently been applied to aircraft engine components for components such as fan blades and compressor blades. The main reason for the delay in applying the process has been the low repetition rate of the original pulsed Nd-glass laser system – about one cycle every eight minutes (Vaccari 1992). A new laser system developed by LSP Technologies, Inc in Dublin Ohio is a high-energy, pulsed neodymium-glass laser which reduces pulse duration to 15 to 30 nanoseconds and produces output energy per pulse of 200 joules or more (Clauser 1982).

The following expresses the basic process of Laser Shock Peening. A coat of black paint is first applied to the area of the component to be peened. A high energy pulsed laser is then used to vaporize the opaque coating into a plasma. As the plasma explosively expands, a powerful shock wave is generated that impacts and propagates into the components due to the presence of a transparent layer confining the energy outwards. The intensity of the shock wave travelling within the component can exceed two times the dynamic yield strength of the material at the surface and the shock wave then gradually disperses through the thickness (Clauer 1996). The shock wave penetrates a relatively deep surface layer to plastically deform the material to produce a relatively deep compressive residual stress layer.

3.2 The Control Parameters for Laser Shock Peening

The residual stress distribution induced by the LSP process is strongly dependent on the type and intensity of the process and the material properties. In principle, by increasing the intensity and exposure, the depth of the compressive residual stresses can be increased from 0.9 mm to 1.8 mm depending on the material (Clauer 1996).

3.3 Applications of Laser Shock Peening in Aircraft Engines

The LSP process has been successfully applied to enhance the fatigue resistance of engine materials such as Al7075-T7351 (Peyre et al 1996), IN718 (Prevey 2000), Ti-6-4 (Ruschau et al 1999) and Ti-6-2-4-6 (Clauer 1996). However, one of the significant barriers to the application of LSP is the slow processing rate, which is mainly due to the inability to remove the black coating rapidly for multiple shots.

4 Low Plasticity Burnishing

4.1 The Basic Low Plasticity Burnishing Process

Low plasticity burnishing (LPB) has been developed as an improved surface enhancement technology that can induce deep compressive residual stress layers reaching to 1.0 mm with a low degree of cold work (about 5 %), and a proportionate increase in the stability of the residual stress layers. The basic LPB process is that rather than using shots to bombard the component surface, a hydraulically floated ball is used to press and roll freely along the surface so that near-surface layers are deformed plastically. A series of overlapping passes are applied until the intended surface area has been treated. Based on Hertz contact mechanics, great depths of compressive residual stress can be induced by the LPB process. Because the rolling ball minimizes plasticity (no shearing due to sliding), less cold work is generated at the surface (Prevey 2000).

4.2 The Effect of Low Plasticity Burnishing Processing Parameters

There are several LPB processing parameters that affect the resultant residual stresses at the surfaces. The most important parameters are the LPB burnishing force, the burnishing ball diameter, the feed and speed. The effect of the LPB burnishing loads on residual stresses has been investigated (Prevey and Shepard 2001). Two levels of burnishing loading were chosen to represent the low and high LPB conditions. It was demonstrated that the compressive residual stresses in IN718 increase with the LPB burnishing load. As in the LSP process, increasing the burnishing load will increase the percentage of cold work, not necessarily increasing the depth of the compressive stresses for thin-wall structures.

4.3 Applications of Low Plasticity Burnishing in Aircraft Engines

LPB has been demonstrated as a cost-effective surface enhancement technology with process speeds comparable to those of conventional machining operations such as surface milling. It has been demonstrated that LPB can (i) improve fatigue strength of the engine materials, IN718 (Prevey at el 2000), Ti-6-4 (Prevey and Shepard 2000), and Al 7075-T6 (Prevey and Cammett 2000); (ii) increase resistance to foreign object damage (Prevey at el 2000); and (iii) eliminate corrosion damage (Prevey and Cammett 2000). The advantages of LPB have attracted attention from USAF and US NAVY. It is currently being investigated for application to both F404 (F/A-18 aircraft) and T56 (C-130 and P-3C aircraft).

5 Evaluation and Comparison

5.1 Residual Stress Profiles

In general, shot peening with small shots generates a low Almen intensity and a shallow depth of plastic deformation, resulting in shallower compressive residual stresses. A comparison of residual stress profiles induced by SP, LSP and LPB for engine material IN718 is shown in Figure 1 (Zhuang and Halford 2001). Figure 1(a) shows that, after the SP process, high compressive residual stresses are confined to a shallow surface layer (0.2 mm) while low tensile residual stresses spread deeper through the cross-section. Unfortunately, the degree of cold work can be exceptionally high (up to 30-40 %) at the surface, Figure 1(b).

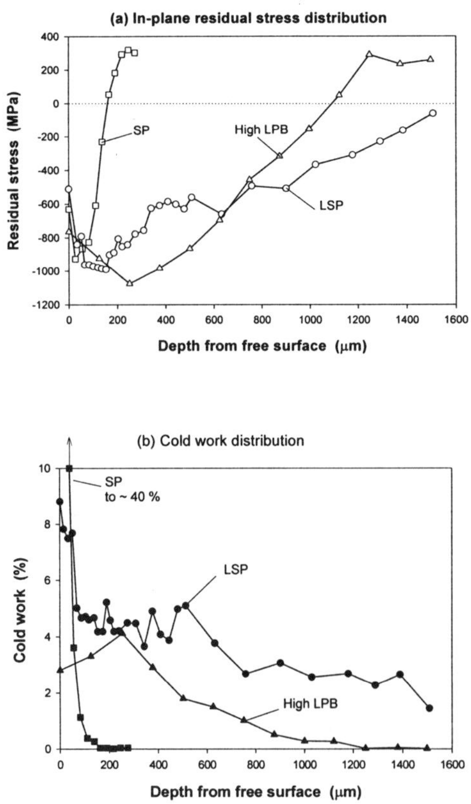

Figure 1. Residual stress and cold work distributions in IN 718 after SP, LSP, & LPB.

5.2 Comparison of The Features of Each Process

It is known that a deep and high-magnitude compressive layer at the surface (~ 1 mm+ by LPB and LSP) can significant resist both crack initiation and short crack propagation. Very low cold work (2-5 %) can improve both thermal stability and resistance to thermal and cyclic load relaxation of these residual stresses. Moreover, application of the process to industrial components requires a process which is adaptable to existing CNC machine tools, at speeds comparable to conventional machining operations such as surface milling. A comparison of the applicability of the three surface enhancement technologies is summarised in Table 1.

6 Conclusions

A review has been made of three mechanical surface enhancement technologies and their application to aircraft engine component life extension and repair. The two most advanced technologies, Laser Shock Peening and Low Plasticity Burnishing are described and compared with Shot Peening, a well established process. The advantages of the technologies are highlighted as well as the limitations.

References

1. Bailey, P. G., Lombardo, D. R., Popp, H. G. and Thompson, R. A., Full assurance shot peening of aircraft gas turbine engine components. *Proceedings of The Six International Conference on Shot Peening*, San Francisco, California, USA, (1996) pp. 320-327.
2. Brookman, J. G. and Kiddle, L., The prevention of fatigue failures in metal parts by shot peening. *The Failure of Metals by Fatigue*. Melbourne University Press (1946).
3. Clauer, A. H., Walters, C. T., and Ford, S. C., The effects of laser shock processing on the fatigue properties of 2024-T3 Aluminum. *Lasers in Materials Processing*, American Society for Metals, Metals Park, Ohio (1983), pp. 7-22.
4. Clauer, A. H., Laser shock peening for fatigue resistance. *Surface Performance of Titanium, Eds*. by J. K. Gregory, H. J. Rack, and D. Eylon, TMS, Warrendale, PA. 1996, 217-230.
5. Fairand, B. P., Wilcox, B. A., Gallagher, W. J. and Williams, D. N., *Journal of Applied Physics*. **3892** (1972) pp. XXX-YYY.
6. James, D., Ageing aircraft – The AATF present programme, *Proceedings of an international conference on Ageing Aircraft and Structural Airworthiness*, Washington, D. C., November 19-21, (1991) pp.15 –26.
7. Kobayashi, M., Matsui, T. and Murakami, Y., Mechanism of creation of compressive residual stress by shot peening. *International Journal of Fatigue* **20** (1998) pp. 351-357.
8. Levine, S. R., High temperature surface protection, *Surface Treatments for Protection*, The Chameleon Press Ltd, London SW18 4SG, (1978).
9. Meetham, C. W., The protection of compressor and turbine components of the aero gas turbine engine, in *Surface Treatments for Protection*, The Chameleon Press Ltd, London SW18 4SG, (1978).
10. Metz, S. A. and Smidt, F. A., *Journal of Applied Physics* **207** (1971).
11. Prevey, P. S., The effect of cold work on the thermal stability of residual compression in surface enhanced IN718, *Proceedings of the 20th ASM Materials Solutions Conference & Exposition*, St. Louis, MO, Oct. 10-12, (2000).
12. Ratwani, M. M., Koul, A. K., Immarigeon, J-P. and Wallace, W., *Aging airframes and engines*, AGARD-CP-600 Vol. 1, (1997).
13. Ruschau, J., John R., Thompson, S and Nicholas, T., Fatigue crack growth rate characteristics of laser shock peened Ti-6al-4V. *Journal of Engineering Materials and Technology*, **121** (1999) p. 321-329.
14. Smith, P. R., Shepard, M. J., Prevey, P. S. and Clauer, A. H., Effect of power density and pulse repetition on laser shock peening of Ti-6Al-4V. *Journal of Materials Engineering and Performance*, **9** (2000) pp. 33-37.
15. Vaccari, J. A., Laser shocking extends fatigue life. *American Machinist*, July (1992) pp. 62-64.
16. Zhuang, W. and Halford G., Investigation of Residual Stress Relaxation under Cyclic Load. Accept by *International Journal of Fatigue* (2001).

MODELLING OF THE ELASTIC DEFORMATION IN COLD ROLLING BY THE FINITE ELEMENT METHOD

Z. Y. JIANG, A. K. TIEU, C. LU AND W. H. SUN

Mechanical Engineering, Faculty of Engineering, University of Wollongong, Wollongong, NSW 2522, Australia

E-mail: jiang@uow.edu.au

The deformation zone in cold rolling includes elastic entry, elastic exit and plastic deformation regions. The elastic deformation zones have an effect on the lubricating film thickness and the accuracy of the final product. Based on the ABAQUS/Explicit code, the authors analyze cold strip rolling, and focus on modelling of the elastic entry region of a roll bite. The deformed elastic entry zone is obtained from the FE simulation and the effects of tension and rolling speed on the deformation are studied. The calculated rolling forces are close to the measured values. This model can predict the hump at the elastic entry zone of the strip, and contribute to the accuracy of the final product through a correct roll gap set up and improvement of the mixed film lubrication.

1 Introduction

The finite element (FE) technique has found wide application in metal forming processes, and in particular in the steel and aluminum industries. The finite element method (FEM) has been employed to solve many metal rolling problems, such as flat rolling [1], shape rolling [2, 3], edge-rolling [4], special shape steel rolling [5, 6] and so on. There are also some issues about the algorithms such as the modified Newton Raphson method [3] and the combination of the shifted incomplete Cholesky decomposition and conjugate gradient (shifted ICCG) method [7] used to improve the simulation processes. The variation in friction [8, 9, 10] at the strip-roll interface in the roll bite has also been investigated by experiments and modelling. The simulations by FEM show that the variation in friction has a significant influence on the calculated results. The rolling of steel involves a large deformation, and in general, the elastic deformation has been considered too small to be taken into account. The plastic deformation, therefore, is the main part that forms the desired shape and products. However, the elastic deformation affects the dimensional accuracy of the final products for cold strip rolling, and it also affects the lubricating film thickness (see Fig. 1). If there is a hump at the elastic entry zone, the film thickness will be reduced to h'-h''.

In cold rolling, the effect of the elastic entry and elastic exit zones (see Fig. 2) on mixed film lubrication and the prediction of the thickness of the strip produced is significant. In order to study the cold rolling process realistically, the elastic deformation zones at entry and exit should be considered in simulation. In this paper, the authors used the ABAQUS/Explicit package to analyze cold strip rolling, considering cases with and without tension, and they developed an FEM model to simulate the humps of the elastic entry zone. Based on this model, the deformed elastic entry zone can be obtained, and the effect of tension on the humps can also be quantified, which helps to improve the accuracy of the final product.

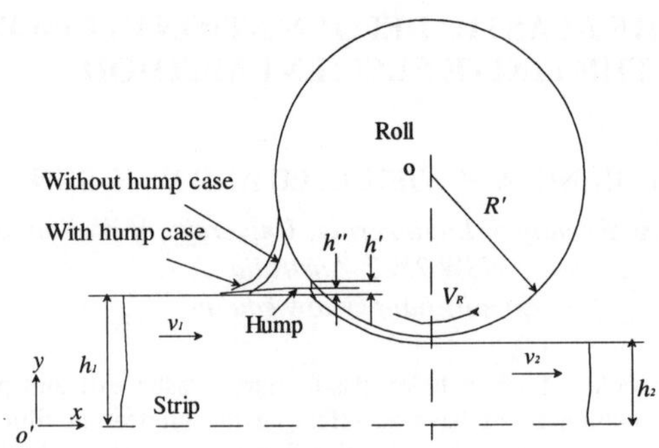

Figure 1. Effect of the hump at the entry zone on film thickness.

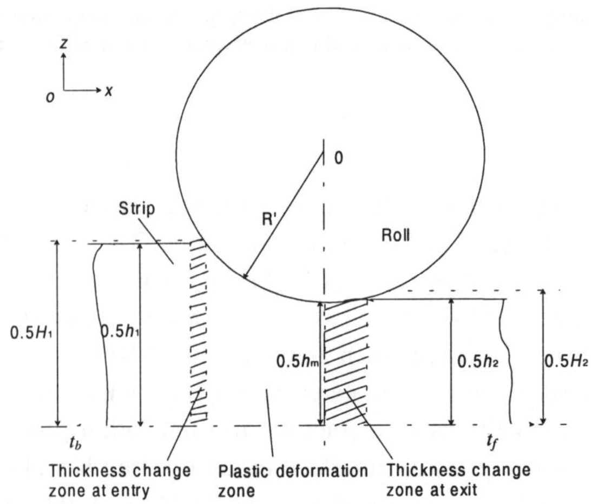

Figure 2. Strip rolling model with the thickness changes at entry and exit zones.

2 Theoretical models

2.1 *Virtual work principle*

For any virtual velocity δv_i, the statically admissible stress field ($\sigma_{ij} = \sigma_{ji}$) satisfies the following virtual work principle,

$$\int_V \sigma_{ij} \delta \dot{\varepsilon}_{ij} dV = \int_{S_p} \overline{p}_i \delta v_i ds \tag{1}$$

where σ_{ij} are the components of stresses, $\overline{p}_i$ is external force and

$$\delta \dot{\varepsilon}_{ij} = \frac{1}{2}(\delta v_{i,j} + \delta v_{j,i}) \tag{2}$$

2.2 *Newton-Raphson method*

The finite element models generated in ABAQUS are usually nonlinear and can involve from a few to many thousand unknown. In terms of these variables, the equilibrium equations can be obtained by minimising the following functional:

$$\phi = \int_V \sigma_{ij} \delta \dot{\varepsilon}_{ij} dV - \int_{S_p} \overline{p}_i \delta v_i ds \tag{3}$$

where the true velocity field v_i satisfies the minimum of the functional, that is

$$\frac{\delta\phi}{\delta\ v_i} = 0 \tag{4}$$

The equilibrium equations can be written symbolically as

$$F^N(u^M) = 0 \tag{5}$$

where F^N is the force component conjugate to the Nth variable in the problem and u^M the value of the Mth variable. The basic problem is to solve Eq. (5) for u^M throughout the history of interest.

Using the Taylor series about the approximate solution u_i^M , then neglecting the first order and second order differential, a linear system of equations is obtained and the approximation to the solution is then

$$u_{i+1}^M = u_i^M + c_{i+1}^M \tag{6}$$

and the iteration continues. c_{i+1}^M is the error of the solution.

The convergence of the Newton-Raphson method is measured by ensuring that all entries in F_i^N in c_{i+1}^N are sufficiently small.

3 Simulation Conditions

3.1 *FE mesh*

The FE mesh used in this simulation for the half thickness of the strip by ABAQUS/Explicit is shown in Fig. 3. The length of the strip was determined from the length of deformation zone, including the plastic zone, elastic entry zone and elastic exit zone. A fine mesh with CPE4R [11, 12] elements at the head and tail of the strip is used, and 3601 is the node number.

3.2 *Calculation conditions*

The rolling process parameters are as follow - initial thickness: h_1=2.2mm; finished thickness: h_2=1.6mm; friction coefficient: 0.05; diameter of rolls is 506.0 mm; average yield stress: 533 MPa; Poisson's ratio: 0.3; Young modulus: 2.1×10^5 MPa; rolling speed: 3.55 m/s. The strip is given an initial velocity in the longitudinal axial direction.

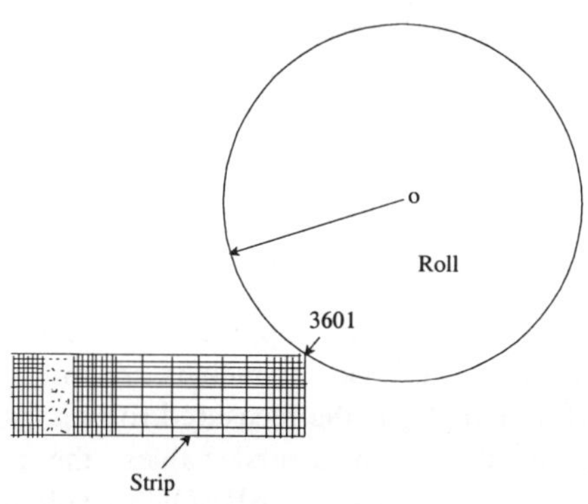

Figure 3. FE mesh used in the simulation for half strip.

4 Results

4.1 *Effect of tension on the hump of strip at the entry zone*

The hump at the elastic entry zone is shown in Fig. 4 for the case where there is no tension. It is seen that the change in thickness starts at a distance of about 0.51 mm from the entry section.

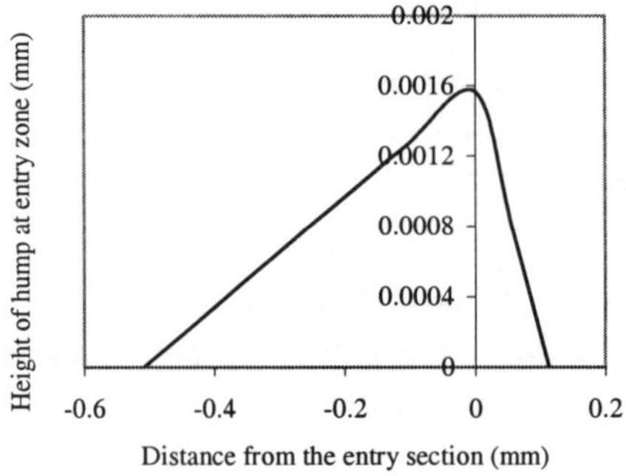

Figure 4. Height of the hump value of the strip at the elastic entry zone at rolling speed 3.55 m/s.

When tension is applied to the strip, the humps at entry zone are affected significantly. Fig. 5 shows the effect of front tension on the hump at the elastic entry zone. The hump decreases as the front tension increases.

The change in thickness of the elastic entry zone is shown in Fig. 6 for the case in which back tension is applied to the strip. The hump decreases and becomes negligible as back tension increases, but reduction in thickness becomes significant. The thickness of the strip at the elastic entry is reduced by applied back tension.

4.2 *Effect of rolling speed on hump*

For the case the initial thickness: h_1=1.6mm; finished thickness: h_2=1.0mm; friction coefficient: 0.04; roll diameter: 502.0 mm; average yield stress: 644 MPa; Poisson's ratio: 0.3; Young's Modulus: 2.1×10^5 MPa; rolling speed: 5.77 m/s, the effect of front tension on magnitude of the hump at the elastic entry zone is shown in Fig. 7. The magnitude of the hump decreases to less than 0.00158 mm (see Fig. 4) when the front tension increases for a rolling speed 5.77 m/s. Comparing the hump of the strip at rolling speeds of 5.77 m/s and 3.55 m/s (0.00147 mm) [12], it can be seen that the hump decreases when the rolling speed increases.

4.3 *Comparison of the calculated rolling force and measured value*

In order to verify the effectiveness of this approach, a comparison of calculated rolling force with the measured value was made, which is shown in Table 3. P_c is the value calculated by ABAQUS/Explicit, P_m is the measured rolling force. The calculated rolling forces are in good agreement with measured values; the maximum error is 8.28%. Simulation of strip cold rolling process by ABAQUS/Explicit package can predict the elastic entry and elastic exit zones, and the elastic recovery after rolling.

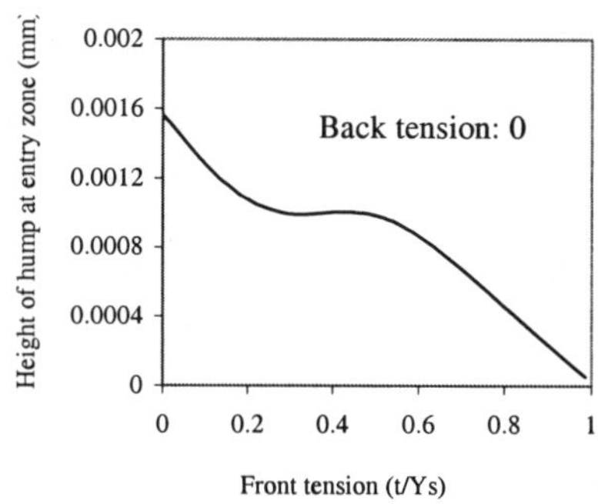

Figure 5. Effect of the front tension on the elastic entry zone at rolling speed 3.55 m/s.

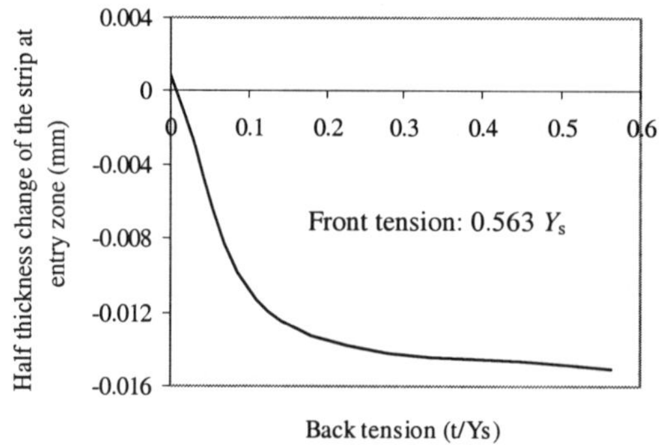

Figure 6. Effect of the back tension on the elastic entry zone at rolling speed 3.55m/s

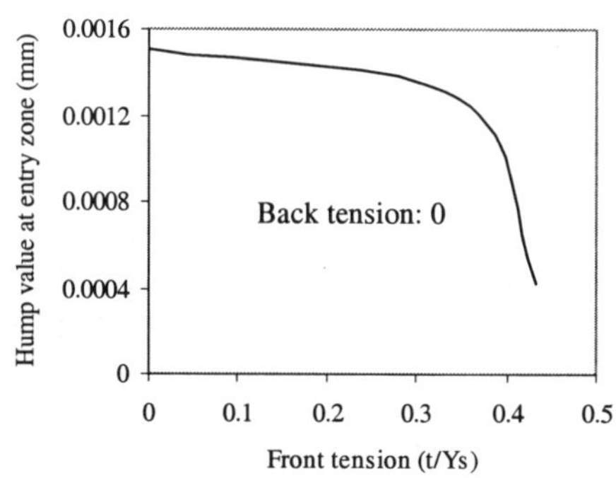

Figure 7. Effect of the front tension on the elastic entry zone at rolling speed 5.77 m/s.

Table 1. Comparison of calculated rolling force with measured value.

Y_s / MPa	h_1 / mm	h_2 / mm	R' / mm	P_c / MN/m	P_m / MN/m
533	2.2	1.6	253	6.97	7.60
644	1.6	1.0	251	8.31	8.10

5 Conclusions

Using the ABAQUS/Explicit package, the magnitude of the hump in the elastic entry and exit zones can be predicted for the practical cold rolling of strip. It is helpful to determine the film of lubrication in the roll bite and the elastic recovery of strip in cold rolling. The magnitude of the hump is reduced significantly by application of strip tension and decreases as the thickness of rolled strip decreases and the rolling speed increases. The calculated rolling forces are also compared with the measured values.

6 Acknowledgements

This work was supported by the New Researcher Grant, UoW and the Australian Research Council (ARC) IREX Award No. X00106473.

References

1. Kobayashi S., Oh S.I. and Altan T., *Metal Forming and Finite Element Method*, New York: Oxford University Press, 1989.
2. Mori K. and Osakada K., Simulation of Three-Dimensional Rolling by the Rigid-Plastic Finite Element method, In Pittman J.F.T., Wood R.D., Alexander J.M. and Zienkiewicz O.C. (eds), *Proc. Int. Conf. Numerical Methods in Industrial Forming Processes*, Swansea: Pineridge Press, 1982, pp747-756.
3. Liu X.H., *Rigid-Plastic FEM and Its Application in Steel Rolling*, Beijing: Metallurgy Industrial Press, 1994.
4. Huisman H.J. and Huetink J., Combined Eulerian-Langrangian Three-Dimensional Finite-Element Analysis of Edge-Rolling, *J. Mech. Working Technol.*, **21,** 1985, pp333-353.
5. Jiang Z.Y., Xiong S.W., Liu X.H., Wang G.D. and Zhang Q., 3-D Rigid-Plastic FEM Analysis of the Rolling of a Strip with Local Residual Deformation, *J. Mater. Proc. Technol.*, **79,** 1998, pp109-112.
6. Jiang Z.Y., Liu X.L., Liu X.H. and Wang G.D., Analysis of Ribbed-Strip Rolling by Rigid-Viscoplastic FEM, *Int. J. Mech. Sci.*, **42**(4), 2000, pp693-703.
7. Jiang Z.Y., Wu W.P., Thomson P.F. and Lam Y.C., Solution of the Equations of Rigid Plastic FE Analysis by Shifted Incomplete Cholesky Factorisation and the Conjugate Gradient Method in Metal Forming Processes, *J. Mater. Proc. Technol.*, **102**(1-3), 2000, pp70-77.
8. Lenard J.G., Friction and Forward Slip in Cold Strip Rolling, *Tribology Transactions*, **35**(3), 1992, pp423-428.
9. Liu Y.J., Tieu A.K., Wang D.D. and Yuen W.Y.D., Friction measurement in cold rolling, *J. Mater. Proc. Technol.*, **111**(1-3) (2001) pp142-145.
10. Jiang Z.Y., Tieu A.K., Modelling of Rolling of Strips with Longitudinal Ribs by 3-D Rigid Visco-Plastic Finite Element Method, *ISIJ International*, **40**(4), 2000, pp373-379.
11. Hibbitt, Karlsson & Sorensen, Inc., *ABAQUS/Standard User's Manual Vol. II (version 5.7)*, Pawtucket, 1997, 14.1.3-1.
12. Jiang Z.Y., Tieu A.K. and Lu C., Analysis of the elastic deformation regions in strip rolling, In Mori K., Simulation of Materials Processing Theory, Methods and Applications, Tokyo: A. A. Balkema Publishers, 2001, pp513-518.

INTELLIGENT OPTIMIZATION FOR UNSTEADY PROCESSES IN TANDEM COLD ROLLING

D. D. WANG AND A. K. TIEU
Faculty of Engineering, University of Wollongong, NSW 2522, Australia
E-mail: ktieu@uow.edu.au

GIOVANNI D'ALESSIO
Packaging Products, BHP Steel, Port Kembla, NSW 2505, Australia

W. Y. D. YUEN
Steel Research Laboratories, BHP Steel, Port Kembla, NSW 2505, Australia

The threading and acceleration processes of tandem cold rolling are modeled and dynamically simulated. Fluctuations in the processes are investigated theoretically and taken into account in the simulation, such as motor speed droop; speed dependent parameters during acceleration including friction coefficient between the rolls and strip, and oil film thickness in hydro-dynamically lubricated roll neck bearings. Strip thickness deviations under different threading conditions are compared and optimal threading speed is searched based on a heuristic optimization procedure. For acceleration process, the compensation coefficient of Automatic Gauge Control (AGC) is optimized to reduce off-gauge length of the strip. The simulation results correspond well with plant data and show that the off-gauge situation with optimal threading and acceleration parameters applied can be alleviated.

1 Introduction

The control of head and tail off-gauge is becoming an urgent requirement in the steel industry to reduce crop loss and produce high-quality steel products. As there are various kinds of fluctuations in unsteady states, and they are always accompanied by non-linearity and hysteresis, it is hard to accurately model and control these processes.

Although a few papers suggest possible approaches for the improvement of off-gauge, articles addressing a substantial solution for the problem are not easy to find. One United States Patent suggests a method for making delivery gauge corrections at low mill speed during threading and tailing out by varying the inter-stand tension [7]. Hennes reported the modernization of a 1970 tandem cold rolling mill [4]. The improvements addressed above are palliative rather than remedial for the head and tail off-gauge problem. Literature also shows the various ways in which the mill speed may influence the roll force and strip thickness in cold rolling. Sims described his experiment work for a range of mill speeds [9]. The results show that the variation of the coefficient of friction between the rolls and strip is the principal cause of roll force and thickness changes. Faizullin investigated the relationship between rolling speed and the thickness of the lubricating film in the backup bearings and the relationship between the rolling speed and the friction coefficient between work roll and strip [3]. Ogan addressed the speed influence on the output of continuous strip mills [6]. Yuen et al. presented a quantitative formula showing the variations of the friction coefficient with rolling speed [10].

This paper presents an endeavor to dynamically model, analyze and simulate both the threading and acceleration processes of a tandem cold rolling mill such that the parameters of these processes are optimum and off-gauge length of the rolled strip reduced accordingly.

2 Threading and Acceleration Processes

In the cold rolling process, a gauge deviation is produced between uniform (body) and end (head and tail) sections due to the change of rolling conditions. Before a process is well controlled, it is important to investigate the process first, then model and optimize it.

2.1 Threading Process

The threading process begins when the head of the strip enters the first stand of a tandem rolling mill and ends when the rolling mill accelerates. Due to the lack of front tension plus deviations of strip entry thickness and hardness, the whole threading process will produce fluctuations of some key process variables. When the strip enters Stand 1 of a typically five-stand mill, the back-tension is first established. The tension between Stands 1 and 2 does not exist until the strip head enters Stand 2. The process continues until the strip head goes into the coiler. During threading, the deviation of strip thickness is caused by several influence factors that can be inferred from Bland-Ford-Hill force calculation formula [1]. The back and front tensions are directly related to the roll force, which, in turn, influence the strip exit thickness.

2.2 Acceleration Process

Acceleration phase begins when the strip head enters the coiler, and ends when the rolling speed reaches the preset constant rolling speed. During acceleration, the strip thickness decreases as the mill is accelerated [5]. The main causes are as follows: (1) the oil film separating the roll neck and the chock changes with mill speed. This results in the change of roll gap position and exit thickness accordingly [8]; (2) the friction coefficient between rolls and strip decreases as the rolling speed increases [10]. This results in a decrease in roll force and the exit thickness of the strip; and (3) the tensions applied to the rolled strip on the entry and exit ends fluctuate. As the tensions cannot be returned to preset values as quickly as the hydraulic roll gap control, this leads to a deformation resistance on the rolled strip, and results in the strip thickness deviation.

3 Modeling of Dynamic Threading and Acceleration Processes

3.1 Modeling of Threading Process

The laws governing threading process include "constant mass flow" and "tension variation principle" [2]. The simulation starts from the moment the strip head enters the second stand. As described in equation (1), the exit tension t_i^{n+1} at time $(n+1)$ at Stand i is equal to the summation of the tension t_i^n at time (n) and product of time increment Δt and tension change rate $\frac{dt_i^n}{dt}$. In the following equations, E denotes the Young's modulus of the strip, and L is the distance between two neighboring stands, VH_{i+1}^n and Vh_i^n are the strip velocity at the entry of the $(i+1)^{th}$ stand, and at the exit of the ith stand, respectively. h_{i+1}^n and H_{i+1}^n are the strip thickness at the exit of the $(i+1)^{th}$

stand, and at the entry of the $(i+1)^{th}$ stand, respectively. v_i^n and v_{i+1}^n denote the strip velocities at the exit of the ith and $(i+1)^{th}$ stands, and f_i^n and f_{i+1}^n the forward slip ratios at the ith and $(i+1)^{th}$ stand, respectively.

$$t_i^{n+1} = t_i^n + \frac{dt_i^n}{dt}\Delta t \quad (1) \qquad \frac{dt_i^n}{dt} = \frac{E}{L}(VH_{i+1}^n - Vh_i^n) \quad (2)$$

$$VH_{i+1}^n = \frac{h_{i+1}^n}{H_{i+1}^n} v_{i+1}^n (1 + f_{i+1}^n) \quad (3) \qquad Vh_i^n = v_i^n (1 + f_i^n) \quad (4)$$

During steady state rolling, the strip thickness at the entry of the $(i+1)^{th}$ stand is equal to the strip thickness at the exit of the ith stand. However, a thickness time delay occurs during threading, and the time delay $\Delta\tau_i$ between the ith and $(i+1)^{th}$ stands is equal to the quotient of the inter-stand distance L_i and the strip velocity Vh_i^n at the exit of the ith stand as shown in equation (5). If the speed of a certain point on the strip at time τ^n is Vh_i^n, the time τ^{n+1} can be described as the summation of $\Delta\tau_i$ and τ^n which is the time at the nth moment. At time τ^{n+1}, the strip thickness at the exit of the ith stand is changed to h_i^{n+1} and the exit speed is changed to Vh_i^{n+1}. After a time period Δt, the distance from the point to the ith stand is calculated as per equation (7). If $L_i^n \geq L_i$, the point is going to enter the $(i+1)^{th}$ stand of the mill. The above procedure will be repeated between the $(i+1)^{th}$ and $(i+2)^{th}$ stands and until the strip head enters the coiler. In the simulation, the time step Δt =0.005 seconds is employed.

$$\Delta\tau_i = L_i / Vh_i^n \quad (5) \qquad H_{i+1}^{\tau} = h_i^{\tau - \Delta\tau_i} \quad (6)$$

$$L_i^{n+1} = \sum_{j=n}^{n+1} Vh_i^j (\Delta t + \tau^n) \quad (7) \qquad \Delta S_0 = \frac{1 + \frac{M}{K}}{1 + (1 - C)\frac{M}{K}} \cdot \Delta h \quad (8)$$

3.2 Modeling of Acceleration Process

The algorithms for acceleration simulation are similar to that used in the threading simulation. The main difference is that the AGC control loop is closed during acceleration, but not during threading [5]. The speed dependent parameters addressed in Section 2.2 are considered. During the rolling process, the rolling force P and the no-load roll gap S_0 may be measured at any time. Therefore, the exit thickness calculated from gauge-meter equation (8) can be used to replace the delayed gauge signal by measurement. In equation (8), *C* is a compensation coefficient, *K* the rolling mill elastic modulus and *M* the plastic modulus of the material being rolled.

4 Simulation and Optimization of Threading and Acceleration Processes

4.1 Flow Chart of the Simulation

The threading and acceleration processes are simulated on a PIII 500 computer. The flow chart for the threading simulation is outlined in Figure 1. The flow chart for acceleration simulation is similar to that for the threading. However, the initial parameters and simulation termination condition are different. For the threading, the front tensions at all stands are initialized to zero while the initial front and back tensions for acceleration simulation are from threading simulation. The threading simulation stops when the strip head enters the coiler while the acceleration continues until rolling reaches the full running speed.

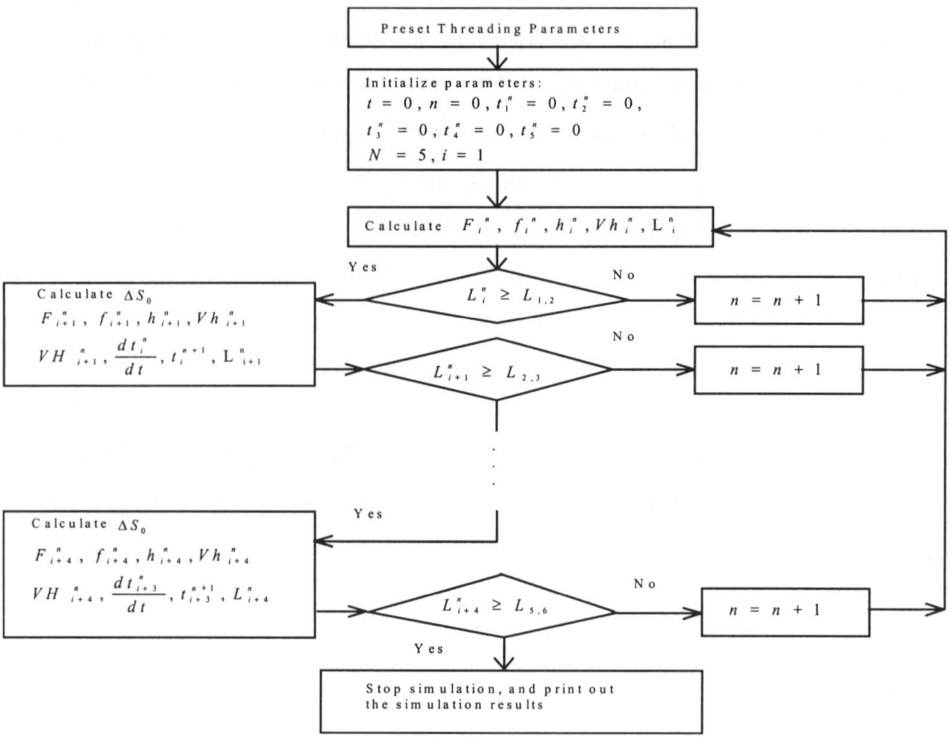

Figure 1 Flow chart of the simulation for threading process

4.2 Optimization of Threading and Acceleration Parameters

Although the unsteady processes are dynamic and can be difficult to control [5], it is still possible to optimize the process parameters so as to reduce off-gauge length of the strip. The threading parameters are usually determined based on experience. However, there is no guarantee whether the empirical threading parameters are optimal. A potential optimal set of threading parameters may exist in a local area of empirical parameters. An effort has been made to optimize the threading speed. The optimization is Genetic Algorithms (GA) based and the standard deviation of strip thickness at the exit of the last stand is employed as the optimization objective. The optimization for the acceleration process focuses on the AGC control compensation coefficient C in Equation (8). The optimization results for both threading and acceleration processes are reported in the following section.

4.3 Numerical Simulation Results

In the simulation, the strip dimensions are as follows: entry thickness 2.20mm, final thickness 0.273mm, and width 965mm. Other parameters are illustrated in Table 1.

The optimal threading speed for the fifth stand is determined from a practical speed range of 0.1m/s to 5m/s. The work roll speed of 0.966m/s is selected as the threading speed for the last stand. The threading speed for other four stands are determined based on constant mass flow principle. The threading force from the simulation is compared with the actual measurement from a steel plant and shown in Figure 2. It corresponds well to the measurement. The exit thickness variation at Stand 5 before and after optimization is compared and illustrated in Figure 3. It shows that the exit thickness off-gauge situation is improved with the optimized threading speed.

Table 1 The simulation parameters

	Stand 1	Stand 2	Stand 3	Stand 4	Stand 5
Scheduled Reduction Rates (%)	28.1	42.7	36.9	31.3	30.6
Scheduled Exit Gauges (mm)	1.582	0.906	0.572	0.393	0.273
Work Roll Diameter (mm)	594.66	578.82	556.0	591.0	590.64

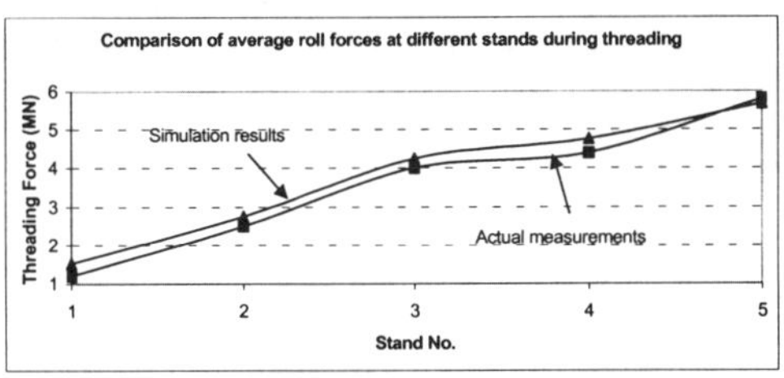

Figure 2 Comparison of threading forces from the simulation with actual measurements

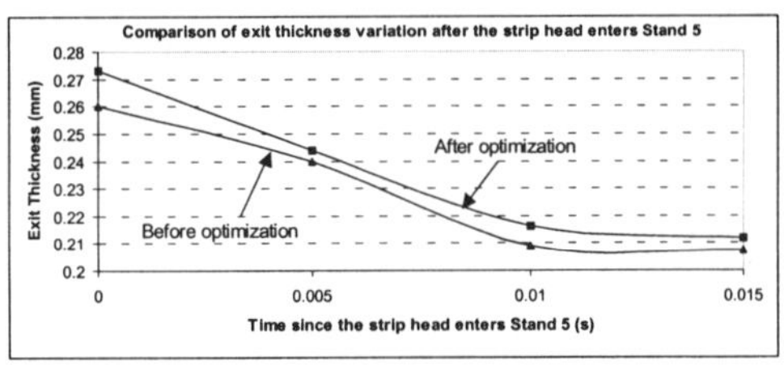

Figure 3 Comparison of exit thickness variation at Stand 5 before and after optimization

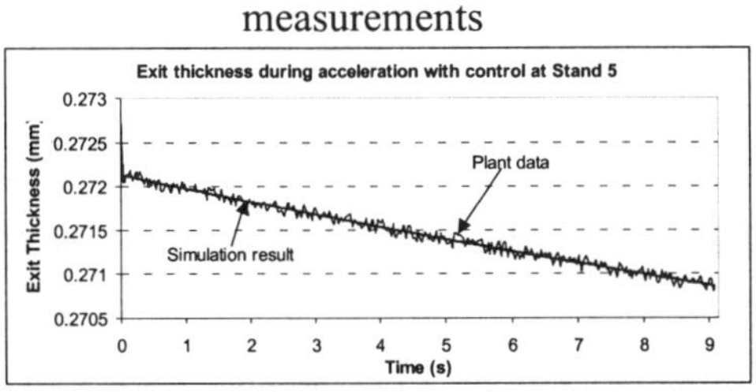

Figure 4 Exit thickness variation at Stand 5 during acceleration with AGC control

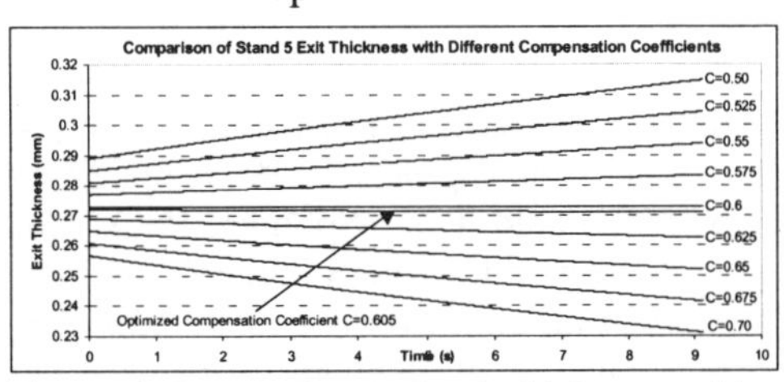

Figure 5 Comparison of exit thickness with different compensation coefficients

The acceleration simulation is conducted based on the acceleration rate 2 of m/s^2 at Stand 5. The simulation result matches well with the plant data shown in Figure 4. Acceleration process optimization is on the basis of Equation (8). The compensation coefficient C is usually determined by experience. When $C = 1$, the mill elastic deformation is completely compensated for, and the mill stand seems infinitely stiff. When $C = 0$, the no-load gap adjustment ΔS_0 is equal to the exit thickness error Δh , partial compensation is obtained, the mill stand becomes less rigid. So the compensation coefficient C is ideally a value between 0 and 1, which is determined based on the aforementioned GA optimization approach. The exit thickness at Stand 5 with different compensation coefficients is illustrated in Figure 5. The off-gauge situation is improved when the optimal compensation coefficient $C = 0.605$ is applied.

5 Conclusions

The mechanism of strip off-gauge in unsteady rolling processes is investigated. The Genetic Algorithms approach is employed to optimize the threading and acceleration parameters. The threading process of a tandem cold rolling mill is simulated and the result from the simulation matches well with the actual measurements from the plant. The optimal parameters for threading are searched by employing a GA based evolutionary optimization approach. The simulation results show that the thickness deviation with the optimal threading speed is minimized, and this will potentially reduce the off-gauge length of the strip during threading. The acceleration process of tandem cold rolling is also modeled and simulated, taking into account the speed dependent parameters including backup bearing oil film thickness and friction coefficient. The relationships between the final strip thickness and speed are reported. The simulation results show that AGC control provides an efficient correction of the speed effect, and they match the actual plant data well. When the optimized compensation coefficient is applied, the strip thickness deviation at the last stand is minimized, and the strip off-gauge situation is improved.

References

1. Bland, D. R., and Ford, Hugh, The Calculation of Roll Force and Torque in Cold Strip Rolling with Tensions. *Proc. I. Mech. E.*, **149** (1948), pp. 144 -153.
2. Fainberg, Yu. M., Laws Governing Variations in Tension When Rolling Flat Sections, *Stal* 8 (1963), pp. 637-638.
3. Faizullin, V. Kh., Effect of Transition Process in Continuous Cold Rolling on Thickness Variations of Strip. *STAL* 2 (1967), pp. 140-143.
4. Hennes, James M., Inland Steel's 80-in., 5-Stand Tandem Cold Reduction Mill. *Iron and Steel Engineer* 9 (1997) pp. 22-28.
5. McDonald, I. R., Spooner, P. D., Cockerell, R. A., Edwards, W. J., and Thomas. P. J., Dynamic Mill Simulator for Flat Metal Rolling Mills. *Proceedings of the first International Conference on Modeling of Metal Rolling Process* (1993), pp. 420-428.
6. Ogan, Ya. P., Influence of Speed Cycles on the Output of Continuous Strip Mills. *STAL* 1 (1969), pp. 39-47.
7. Peterson, Robert S., and Larsen, John A., Method for Rolling On-Gauge Head and Tail Ends of a Workpiece. *United States Patent*, Patent Number: 4998427 (1991).
8. Polukhin, V. P., *Mathematical Simulation and Computer Analysis of Thin-Strip Rolling Mill* (1975).
9. Sims, R. B., and Arthur, D. F., Speed-Dependent Variables in Cold Strip Rolling. *Journal of The Iron and Steel Institute* **11** (1952), pp. 285-295.
10. Yuen, W. Y. D., et al, Improved Scheduling and Setup for Reversing Cold Mills. *Proceedings of the 7th International Conference on Steel Rolling* (1998), pp. 778-783.

THE THERMAL EFFECT IN A MIXED-FILM LUBRICATED COLD ROLLING PROCESS

Y. J. LIU AND A. K. TIEU

Faculty of EngineeringUniversity of Wollongong, Northfields Avenue, Wollongong, NSW 2522, Australia

E-mail: yl25@uow.edu.au, Kiet_Tieu@uow.edu.au

A thermal model has been developed for mixed-film lubrication in cold rolling. The energy equation has been used to determine the temperature variation in the inlet, outlet and the plastic work zones. The temperature at the asperity contacts in the plastic work zone is also calculated. The influence of the thermal effect on the film thickness and pressure is significant and should be considered in a mixed film model. The model has been validated by experiment.

1 Introduction

In practice, rolling metals in the fully-hydrodynamic film regime, in which the surfaces are completely separated by a thin film of lubricant, minimizes friction and wear. However, the lubrication system must provide sufficient friction between the strip and the roll to draw the strip into the roll bite. These considerations generally require that the production of strip rolling operation runs in the mixed lubrication regime, in which part of the total interface pressure is provided by asperity contact at peaks and the remaining part by the pressurized lubricant in valleys. Significant progress has been made in the analysis of this lubrication regime recently [3,8,12]. The assumption of hydrodynamic lubrication is also adopted by other researcher [10], in which the contacting surfaces are fully separated by a thin film. Some researchers [5,7] have developed models which considered the effect of inlet & outlet zone under the fully hydrodynamic condition.

Wilson and Mahdavian [11] developed a thermal Reynolds equation to take into account viscosity variations across the lubricant film due to energy dissipation within the film under hydrodynamic conditions, but the dominant mode of heat transfer in the lubricant is conduction. Bhatt and Sengupta [1] developed a similar thermal Reynolds equation but took into account the effect of both conduction and convection on heat transfer in the hydrodynamic lubricant film. Lugt *et al.* [7] considered the thermal effect in his hydrodynamic model. It assumed that the deformation energy is entirely transformed into heat, whereas friction-induced heat generation and the conduction of heat into the rolls were neglected. The viscosity-pressure-temperature equation developed by Roelands [9] is used. Lin and Houng [5] analysed the thermal effect in the three zones namely hydrodynamic inlet, plastic work zone, and hydrodynamic outlet. Conduction heat was assumed to be the only mode of heat transfer in the three zones and the energy equation was used in these zones to calculate the temperature variations. All the work described above has been carried out for the fully hydrodynamic lubrication.

Liu, Tieu and Qiu [6] developed a different model to consider the effect of the inlet and outlet under mixed film lubrication but the thermal effect was not considered. It has been shown that the inlet zone has an influence on the accuracy. This paper concentrates on the thermal effects in the three zones, inlet, outlet and plastic work zones under mixed film lubrication.

2 Theoretical analysis

The lubricated zone is divided into three major zones: inlet zone, plastic work zone and outlet zone. Fully hydrodynamic lubrication prevails at the inlet zone; mixed-film lubrication occurs in the plastic work zone where the roll and strip surface are separated by the lubricant in some areas and the asperity contact in the remaining area; fully hydrodynamic lubrication takes place in the outlet zone. The whole rolling process is shown in Figure 1. Assumptions in the analysis are made as follows:

(1) The rolls deform elastically; (2) Conduction is assumed to be the only mode of heat transfer in the three zones; (3) The density of the lubricant is constant; (4) Reynolds equation for steady incompressible flow applies; (5) The lubricant is Newtonian with a viscosity η at pressure p_f and the average temperature T_m, given by $\eta = \eta_0 \exp[\alpha p_f - \delta(T_m - T_0)]$; (6) the flow of the lubricant is laminar and inertia forces are neglected; (7) The yield stress of material is not affected by the temperature in cold rolling.

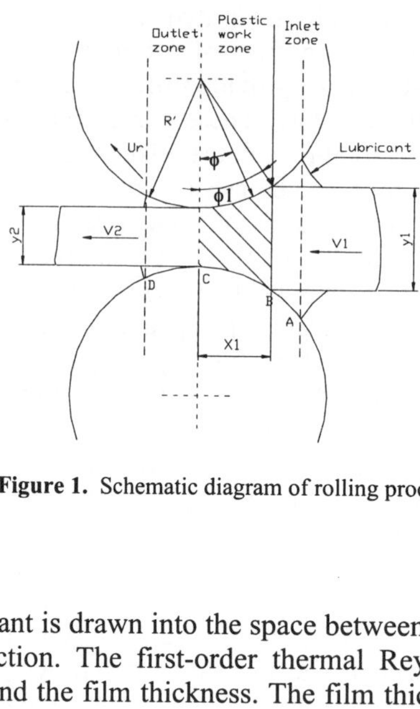

Figure 1. Schematic diagram of rolling process

2.1 Inlet zone

In the inlet zone, the lubricant is drawn into the space between the rolls and workpiece by a hydrodynamic wedge action. The first-order thermal Reynolds equation is used to analyse the film pressure and the film thickness. The film thickness in the inlet zone can be written from the geometry:

$$h = h_B + (x^2 - x_1^2)/2R' \tag{1}$$

The first-order Reynolds equation can be written as:

$$\frac{dp_f}{dx} = -6u_r \eta_0 \exp[\alpha p_f - \delta(T_m - T_0)].[1 + (1 + S_f).(y_2 / y_1)].\frac{(x^2 - x_1^2)/2R'}{[h_B + (x^2 - x_1^2)/2R']^3} \tag{2}$$

Since conduction is assumed to be the only mode of heat transfer in the inlet zone, the energy equation in the film is given as [5]:

$$\frac{\partial^2 \overline{T}_f}{\partial Y^2} = \frac{-P_1}{C}\{6[(1 + S_f)(1 - \varepsilon_e) + 1].(H - H_1) \times Y/H + [(1 + S_f)(1 - \varepsilon_e) - 1]\}^2 \exp[-E(\overline{T}_m - 1)] \tag{3}$$

$$\frac{\partial^2 \overline{T}_f}{\partial Y^2} = E_1[E_2^2 Y^2 / H^2 + 2E_2E_3Y/H + E_3^2] \tag{4}$$

where

$$\begin{cases} E_1 = (-P_1/C)\exp[-E(\overline{T}_m - 1)] \\ E_2 = 6[(1+S_f).(1-\varepsilon_e)+1].(H - H_1) \\ E_3 = (1+S_f).(1-\varepsilon_e) - 1 \end{cases} \tag{5}$$

The boundary conditions of the inlet zone are:

$$\overline{T}_w = 1; \quad \overline{T}_r = 1; \quad \overline{T}_f = 1 \quad \text{at} \quad X = \infty \tag{6}$$

The 'moving' boundary conditions for the roll and strip surfaces are given as [5]

$$\overline{T}_r = \overline{T}_f(X, 1/2) = P_3 \int_{\infty}^{x} (1/H)\frac{\partial \overline{T}_f}{\partial Y}\Big|_{Y=1/2} \times \frac{dX'}{(X'-X)^{1/2}} + 1 \tag{7}$$

$$\overline{T}_w = \overline{T}_f(X, -1/2) = \frac{P_2}{[(1+S_f)(1-\varepsilon)]^{1/2}} \times \int_{\infty}^{x} (-1/H)\partial \overline{T}_f / \partial Y\Big|_{Y=-1/2} \times \frac{dX'}{(X'-X)^{1/2}} + 1 \tag{8}$$

The fluid temperature $\overline{T}_f$ can be obtained directly by integrating Eq. (4) yielding

$$\overline{T}_f = E_1\left(\frac{E_2^2}{12}.\frac{Y^4}{H^2} + \frac{E_2E_3}{3}.\frac{Y^3}{H} + \frac{E_3^2}{2}Y^2 + C_1Y + C_2\right) \tag{9}$$

The fluid mean temperature is:

$$\overline{T}_m = \int_{-1/2}^{1/2} \overline{T}_f dY = E_1\left(\frac{E_2^2}{960H^2} + \frac{E_3^2}{24} + C_2\right) \tag{10}$$

2.2 Plastic work zone

The plastic work zone under the mixed film lubrication is governed by the equations in reference [8].

In the plastic work zone, the rolling condition is under mixed film condition. So the temperature in the lubricant and at the asperities contact should be calculated respectively.

Temperature in lubricant:

In the plastic work zone, Eq.(4) can still be used here, but E_2 and E_3 in Eq. (5) are written:

$$\begin{cases} E_2 = 6[(1+S_f)(1-\varepsilon_e)/\overline{Y} + 1].(H - H_1) \\ E_3 = (1+S_f)(1-\varepsilon_e)/\overline{Y} - 1 \end{cases} \tag{11}$$

In the rolling process, if part of energy of the plastic work is retained in the strip, the surface temperature of the strip can be written in non-dimensional form as:

$$\overline{T}_w = \overline{T}_{wb} + \gamma[\sigma_y/(c_w\rho_w T_0)]\ln \overline{Y} \tag{12}$$

γ is defined as the fraction of plastic work converted to heat.

The moving boundary condition for roll surface can be written as:

$$\overline{T}_r = P_3 \int_1^X \left(\frac{1}{H}\right)\frac{\partial \overline{T}_f}{\partial H}\Bigg|_{Y=\frac{1}{2}} \frac{dX'}{(X'-X)^{1/2}} + \overline{T}_{r_1} \tag{13}$$

Temperature at the contact between asperities:

The temperature increment at the point of (x, y) on either of two surfaces at any time t due to an instantaneous heat source of strength $\omega(x',y',t')dx'dy'dt'$ at x', y', and t' can be calculated by the following equation given by Carslaw and Jaeger [2].

$$dT = \frac{\omega(x',y',t').dx'.dy'.dt'}{4\rho c_p[\pi K(t-t')]^{3/2}} \times \exp\left[-\frac{(x-x')^2+(y-y')^2}{4K(t-t')}\right] \tag{14}$$

Eq. (14) can be simplified to a one-dimensional expression as follows:

$$dT = \frac{\omega(x',t').dx'.dt'}{2\pi k(t-t')} \times \exp\left[-\frac{(x-x')^2}{4K(t-t')}\right] \tag{15}$$

The heat flux in the contact area is:

$$\omega_a = \mu_a p_a V_s \tag{16}$$

V_s is a sliding speed – the difference in speed between the roll and the strip.
The heat generated is conducted into the two contact surfaces according to a partition factor $f(x',t')$, which represents the fraction of heat flux conducted into the roll surface with velocity u_r. The fraction of heat flux conducted into the strip surface with velocity u_w, then become $[1-f(x',t')]$. The expression for the temperature rise of the surfaces acted on by a single heat source at time t' is extended as follows.

Temperature rises of the roll and strip surface are:

$$dT_r = f(x',t')\frac{\omega_a(x',t').dx'.dt'}{2\pi k_r(t-t')} \times \exp\left\{-\frac{[(x-x')-u_r(t-t')]^2}{4K_r(t-t')}\right\} \tag{17}$$

$$dT_w = [1-f(x',t')]\frac{\omega_a(x',t').dx'.dt'}{2\pi k_w(t-t')} \times \exp\left\{-\frac{[(x-x')-u_w(t-t')]^2}{4K_w(t-t')}\right\} \tag{18}$$

The surface temperature rise produced by all the frictional heat sources can be obtained by integrating the above differential temperature over the contact heat source area x_1 from time $t'=0$ to time $t'=t$, and then discretizing the equations

$$\Delta T_r(x,t) = \sum_{l=0}^{L-1}\sum_{j=0}^{N} f(x',t')\frac{\omega_a(x',t')}{2\pi k_r(t-t')}.\exp\left\{-\frac{[(x-x')-u_r(t-t')]^2}{4K_r(t-t')}\right\}\Delta x'.\Delta t' \tag{19}$$

$$\Delta T_w(x,t) = \sum_{i=0}^{L-1}\sum_{j=0}^{N}[1-f(x',t')]\frac{\omega_a(x',t')}{2\pi k_w(t-t')}.\exp\left\{-\frac{[(x-x')-u_w(t-t')]^2}{4K_w(t-t')}\right\}\Delta x'.\Delta t' \tag{20}$$

L is the number of time increment, and N is the number of heat sources thorough the roll gap.
The temperature rise caused by the work of plastic deformation can be expressed as follows:

$$\Delta\overline{T}_w = \gamma[\sigma_y/(c_w\rho_w T_0)]\ln\overline{Y} \tag{21}$$

In order to calculate the temperature of the contact surfaces, the heat partition factors $f(x',t')$ must be solved by assuming that the surface temperatures at the asperity contacts are equal because there is no temperature jump across the interface. This yields:

$$\overline{T}_{rb} + \Delta\overline{T}_r(x,t) = \overline{T}_{wb} + \Delta\overline{T}_w(x,t) + \Delta\overline{T}_w \tag{22}$$

so the temperature at asperity contact can also be solved as follows:

$$\overline{T}_r = \overline{T}_{rb} + \Delta\overline{T}_r(x,t) \tag{23}$$

$$\overline{T}_w = \overline{T}_{wb} + \Delta\overline{T}_w(x,t) + \Delta\overline{T}_w \tag{24}$$

2.3 Outlet zone

In the outlet zone , the first-order Reynolds equation is used to analyse the film pressure and the film thickness. The film thickness in the outlet zone can be written from geometry:

$$h = h_C + x^2/2R' \tag{25}$$

The first-order Reynolds equation can be written as:

$$\frac{dp_f}{dx} = -6u_r\mu_0 \exp[\alpha p_f - \delta(T_m - T_0)].[1+(1+S_f)].\frac{x^2/2R'}{(h_C + x^2/2R')^3} \tag{26}$$

The energy equation in the outlet zone is the same as Eq. (4) and Eq. (5) except for the expressions for E_2 and E_3 .

$$\begin{cases} E_2 = 6(S_f + 2).(H - H_2) \\ E_3 = S_f \end{cases} \tag{27}$$

3 Results and discussion

Before the calculation, the time step interval in the roll bite must be chosen. With small time steps, the results of computation of temperature can differ from the steady state value. In the present study, 100 time steps is recommended to reach the steady state. Although more accurate results can be obtained with the more time steps and steps thorough roll gap, it will take a longer time to complete the calculation.

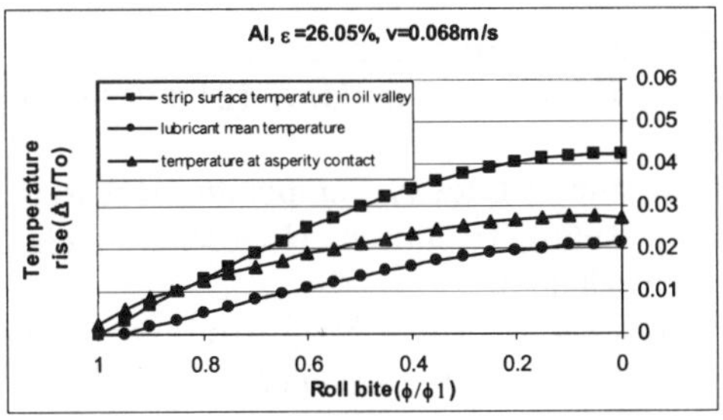

Figure 2. Temperature distribution in the roll bite

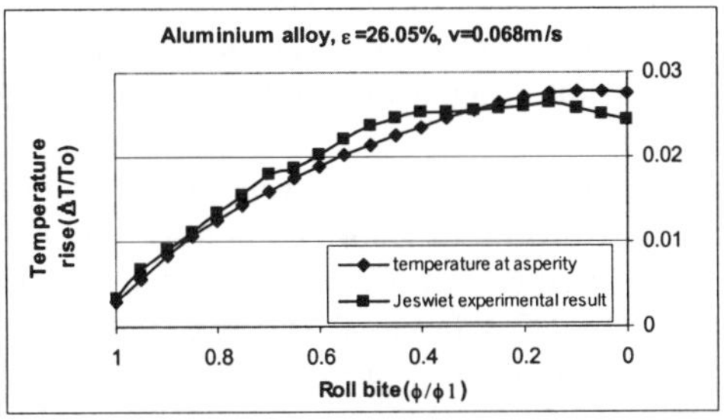

Figure 3. Comparison of temperature with experiment

Figure 2. shows the mean lubricant temperature, the strip surface temperature in the lubricant, and the temperature at asperity contacts. Temperature always increases from inlet to outlet with increasing plastic work. At the inlet area, the temperature at asperities caused by the work of friction is higher than strip surface temperature in the valley because frictional work plays more important role than plastic work when plastic deformation just starts, but the strip temperature rises as more plastic deformation occurs. Figure 3 shows the calculated results and experimental results [4] which compare well with each other.

4 Conclusions

The influence of thermal effects in inlet, outlet and plastic work zones has been discussed in the mixed film lubrication condition. The calculated results have been verified by the experimental data. The thermal effects in mixed film lubrication are significant and therefore should be considered in a mixed film model.

References

1. Bhatt V.K. and Sengupta D.K., Analysis of Thermal Effects in the Yield Phase of Hydrodynamic Lubricant Film in Plane Strain Forging. *Journal of Tribology*, **118** (1996) pp. 880-885.
2. Carslaw H. S. and Jaeger J. C., Conduction of Heat in Solids. Oxford (1959), Clarendon Press.
3. Chang D.F., Marsault N. and Wilson W.R.D., Lubrication of Strip Rolling in the Low-Speed Mixed Regime. *Tribology Transactions*, **39** (1996) pp 407-415.
4. Jeswiet J. and Zou, S., Temperature, Heat Flux and Conductivity in Bar Rolling. *Annals of the CIRP*, **41** (1991) pp 299-302.
5. Lin J.F. and Houng S.C., A Refined Model for the Thermal Hydrodynamic Lubrication Theory of High-Speed Rolling. Part 1: Effects of Operating Conditions. *Tribology International*, **24** No.3 (1991) pp 151-161.
6. Liu Y.J., Tieu A.K. and Qiu Z.L., The Influence of Inlet and Outlet Zone in Mixed-Film Lubricated Cold Rolling Process. *Proceedings of 2nd International Conference on Mechanics of Structures, Materials and Systems(MSMS 2001)*, 2001, Wollongong, Australia, pp 321-326.
7. Lugt P.M. and ten Napel W.E., The Influence of Elastic Deformation of the Roll and the Sheet in a Hydrodynamically Lubricated Cold Rolling Process. *Transactions of the ASME*, **117** (1995) pp 468-475.
8. Qiu Z.L., Yuen W.Y.D. and Tieu A.K. The Mixed-Film Lubrication Theory and Tension Effects on Metal Rolling Processes. *Journal of Tribology*, **121** (1999) pp 908-915.
9. Roelands C.J.A., Correlational Aspects of the Viscosity-Temperature-Pressure Relationship of Lubricating Oils. Ph.D. Thesis (1966), Technische Hogeschool Delft, The Netherlands (V.R.B., Groningen, The Netherlands).
10. Saxena S., Dixit P.M. and Lal G.K., Analysis of Cold Strip Rolling under Hydrodynamic Lubrication. *Journal of Materials Processing Technology*, **58** (1996) pp 256-266.
11. Wilson W.R.D. and Mahdavian S.M., A Thermal Reynolds Equation and Its Application in the Analysis of Plasto-Hydrodynamic Inlet Zones. *Journal of Lubrication Technology*, October 1974, pp 572-578.
12. Wilson W.R.D. and Chang D.F., Low Speed Mixed Lubrication of Bulk Metal Forming Processes. Tribology in Manufacturing Processes, ASME publication, CRTD-Vol. 30, Trib-Vol. 5, PED. **69** (1994), pp 159-168.

ELASTIC-PLASTIC FINITE ELEMENT ANALYSIS OF ASPERITY DEFORMATION UNDER COLD ROLLING CONDITION

B. MA, A. K. TIEU AND Z. JIANG

Faculty of Engineering, University of Wollongong, Northfields Avenue, Wollongong, NSW 2500, Australia

This paper presents an elastic-plastic finite element model for asperity deformation in cold rolling. The asperity is assumed to be spherical and an isotropic hardening material is applied. The emphasis of this simulation is placed on the relationship between contact pressure and actual area of contact. The influence of friction is investigated as well. Results and discussion of variations of main contact parameters are presented. Comparison with previous work shows that most results are in good agreement.

1 Introduction

The micro-contact between asperities on the surface of steel strip and the roll surface has become an important aspect in the research into rolling deformation and strip surface quality control. In order to simulate the whole surface contact accurately, it is essential to establish a model for a single asperity contact under the conditions of cold rolling. Prior to the cold-rolling operation, the hot strip is generally pickled, leaving a comparatively rough and broken surface. The roll surface is generally much harder and smoother than the metal surface, as shown in Figure1.

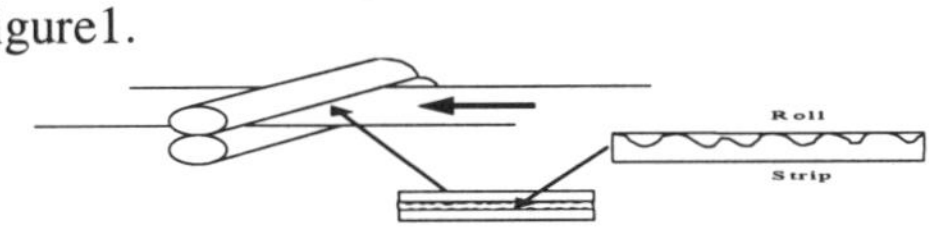

Figure 1. Illustration of interface between the rolls and strip in cold rolling

In the past, some researchers studied the case of a worn rough tool surface interacting with smooth work pieces, which is not applicable to cold rolling [1]. Wilson et al [6] and Sutcliffe [5] have investigated the effect of bulk plastic deformation on asperity flattening and proposed some models. However, the asperity was assumed to be of a sawtooth profile and the friction between the asperity and roll was not taken into account.

In this paper, the asperity was assumed spherical and the roll was considered rigid. The model was established to obtain the direct connection between contact parameters, such as rolling pressure, contact pressure, contact ratio and friction.

2 Modeling

2.1 Finite Element Meshing

The 3-node linear plane strain elements were used over the whole asperity, as it was found that this kind of element was effective in avoiding singular points for this case. The finite element mesh, partly illustrated in Figure 2, consists of 1220 elements comprising 736 nodes. The contact area was subdivided to obtain detailed contact information under large plastic deformation.

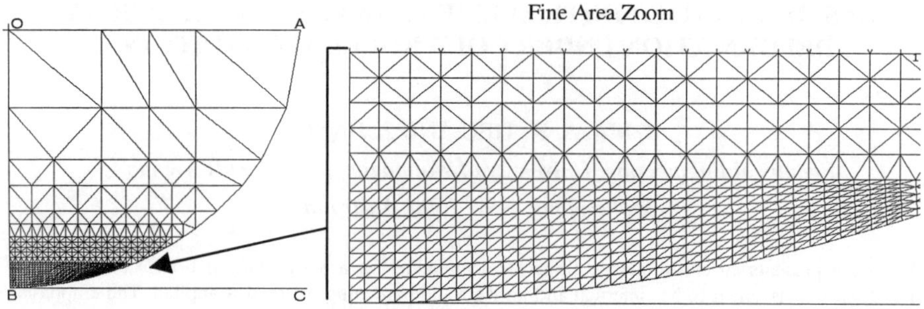

Figure 2. The finite element mesh for the asperity

The surface of the roll, which is harder and smoother than the material surface deformed in cold rolling, was assumed to be rigid and flat. It was defined as a rigid surface comprising 56 rigid elements in contact with the asperity.

2.2 Boundary Conditions

Because the deformation is symmetrical about the plane OB(Figure 2), half of the asperity was adopted to perform the simulation. Considering the asperity deformation process, plane OA was restrained on both the vertical and horizontal axes, while plane OB was restrained on the horizontal axis.

2.3 Strain hardening material properties

Mild steel, assumed to harden isotropically, was used to simulate the asperity deformation. The Young's modulus was 210GPa and Poisson's ratio was assumed equal to 0.3. The plastic properties were given by the relationship between log strain and true stress, shown in Table 1, where the initial yield stress is 168.72MPa.

Table 1. Plastic material properties

Log Strain	0	0.1	0.2	0.3	0.4	0.5
True Stress(MPa)	168.72	219.33	272.02	308.53	337.37	361.58
Log Strain	0.6	0.7	0.8	0.9	1.0	
True Stress(MPa)	382.65	401.42	418.42	434.01	448.45	

2.4 Plasticity Models

The finite element package ABAQUS was employed to perform the simulations. In the inelastic response models that are provided in ABAQUS, the elastic and inelastic response are distinguished by separating the deformation into recoverable (elastic) and non-recoverable (inelastic) parts. This separation is based on the assumption that there is an additive relationship between strain rates:

$$\dot{\varepsilon} = \dot{\varepsilon}^{el} + \dot{\varepsilon}^{pl}$$

Because isotropic hardening was assumed in this analysis, the von-Mises yield function was applied. In terms of the uniaxial yield stress plastic-strain relationship, the yield criterion reduces to:

$$\sigma_M = \sqrt{\frac{3}{2} S_{ij} S_{ij}} = \sigma^0$$

where σ_M is the von-Mises equivalent stress and σ^0 the material yield stress.

2.5 *Simulation*

The contact between the roll and asperity was simulated by incrementally applying a 20 percent reduction on the asperity under a variety of rolling conditions. Different friction conditions were applied to investigate their effects on the asperity flattening. Loading histories were reviewed to analyze the relationships among some important variables, such as the contact pressure, rolling pressure, contact length, contact ratio, etc.

The simulation was performed on a Sun E3000 computing server with four UltraSparcII CPUs and 512MB memory running at 4x167MHz. Typical cpu time was about 550 seconds.

2.6 *Verification*

To examine the suitability of the finite element mesh and modeling assumptions, the Hertz contact problem was compared with the classical solution. The comparison, which is shown in Figure 3, shows that the pressure between the two surfaces predicted by this FE model closely follows the classical elliptic distribution. In view of the favorable comparisons with analytical results, the modeling is considered appropriate for the above-mentioned simulation of asperity deformation.

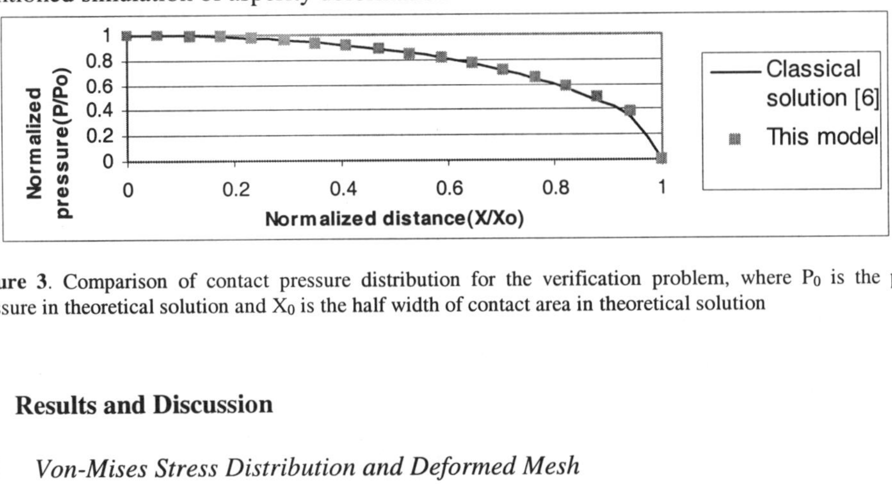

Figure 3. Comparison of contact pressure distribution for the verification problem, where P_0 is the peak pressure in theoretical solution and X_0 is the half width of contact area in theoretical solution

3 Results and Discussion

3.1 *Von-Mises Stress Distribution and Deformed Mesh*

Figure 4 shows the distribution of the von-Mises equivalent stress in the asperity after 20 percent reduction. It can been seen that the maximum stress (contour number 12) occurs near the edge of the contact area. This result corresponds to the analysis of contact pressure distribution and could be explained by material flow theory. During the process of asperity flattening, the material flows to the edge, which has the minimum restraint, and forms the so-called the 'Edge Effect'. The maximum plastic strain and von-Mises stress occurs therefore near the edge of contact area. Another result of the 'Edge Effect' is the deformation of the section MN to a non-circular profile. This effect is important for the study of multi-asperity deformation and the plastic deformation with lubricant.

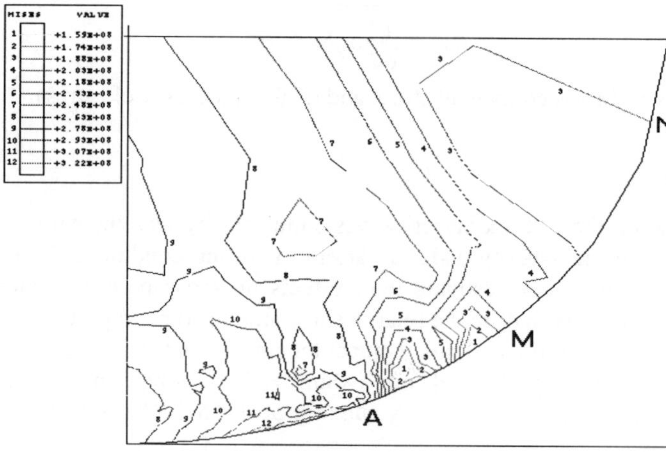

Figure 4 Contour of von-Mises stress distribution

3.2 *Contact Pressure Distribution*

3.2.1 *Contact Pressure Distribution versus Rolling Load*

Figure 5(a) shows the variation of the contact pressure distribution along the contact length as a function of rolling load. It is obvious that the deviation of the pressure profile changes significantly with the evolution of plastic deformation. Particularly, increasing the rolling load leads to a significant flattening of the contact pressure distribution, a shift of the maximum pressure toward the contact edge, and a remarkable growth of the contact length. For the cases with relatively small rolling load, such as curve 1 (Pa/Py=0.1) and curve 2 (Pa/Py=0.45) shown in Figure 5(a), the maximum pressure occurs at the center and the pressure distribution partly follows the Hertzian solution. The explanation for this result is that the spherical shape makes the stress concentrate on the centre and the plastic deformation is too small to form an edge flow. For the cases with large load, such as shown in curve 3 (Pa/Py=0.92), curve 4 (Pa/Py=1.04) and curve 5 (Pa/Py=1.15) (see Figure 5(a)), the edge flow is dominant and pressure distributions are flattened. These results are in agreement with previous analysis [4].

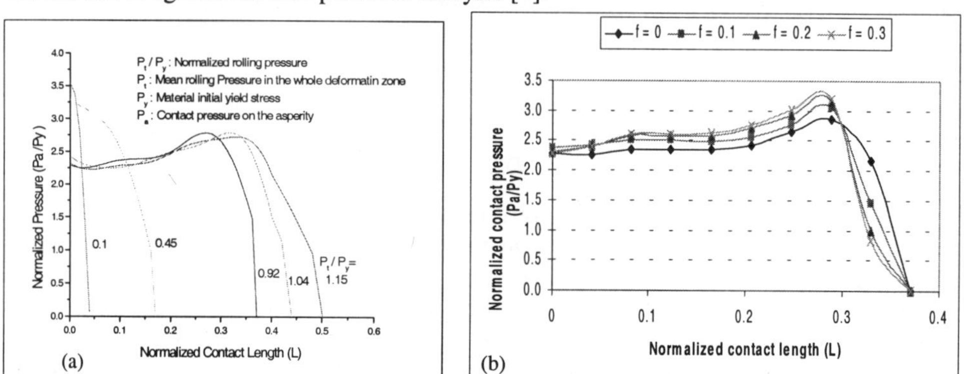

Figure 5 (a) Contact pressure distribution along contact length under various rolling loads; (b) Contact pressure distribution along contact length under various friction conditions. *Note: Normalized contact length is defined as the contact length along horizontal axis (see Figure 2) over the asperity radius

3.2.2 Contact Pressure Distribution versus Friction Condition

To investigate friction effects on the flattening of asperities, a Coulomb friction model was used in ABAQUS for the simulations. Figure 5(b) shows the contact pressure distributions obtained by giving a variety of friction coefficients. It shows that the contact pressure increases as the coefficient increases and the maximum pressure moves slightly toward the center as the friction increases. This is obviously caused by a higher deformation resistance and shear stress when the friction coefficient is increased.

3.3 Contact Ratio

The prediction of contact ratio, which is defined as the contact length along horizontal axis (see Figure 2) over the asperity radius, is the foundation for further modelling of the random surface deformation based on this asperity flattening model. The discussion on contact ratio, rolling pressure (pressure on the tool) and contact presure on asperity is presented respectively. Here, the maximum contact ratio obtained in the analysis is about 0.5, because the reduction was set to 20 percent.

3.3.1 Rolling Load versus Contact Ratio

Figure 6(a) shows the variation of the mean pressure on tool (Pm) with contact ratio A for a smooth contact case. For A between 0 and about 0.5, Pm increases almost linearly as the contact ratio increases. This result compared well with the previous work of Kimura[2] shown in Figure 6(a).

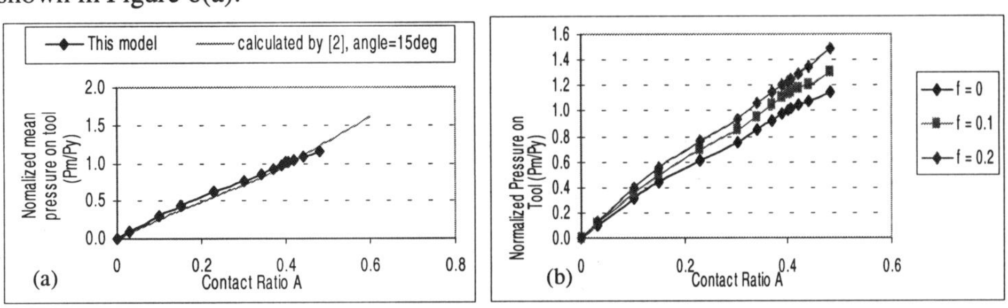

Figure 6 (a) Comparison of rolling pressure versus contact ratio with Kimura [2]; (b) Relationship of rolling pressure and contact ratio under various friction conditions

The friction effects on contact ratio and pressure on tool are shown in Figure 6(b). For the same contact ratio, the pressure Pm on the tool increases slightly with friction.

3.3.2 Contact Pressure versus Contact Ratio

Figure 7(a) shows the variation of mean contact pressure on the asperity (Pa) with contact ratio A. The curve can be separated into two sections: section 1 with A between 0 to 0.28 and section 2 with A between 0.28 to 0.5. Asperity contact pressure falls gradually in section 1, while contact pressure is independent of A in section 2. Comparisons with previous work of Kimura et al [2] are shown in Figure 7(a) as well. Kimura assumed the asperity section to be of trapezium profile. The variable angle in Figure 7(a) is the initial angle of the slope of an asperity. Two curves are given here for 15 degree and 25 degree slopes respectively. In the present model the asperity profile is assumed spherical, which can be initially treated as corresponding to 0 degree in slope angle. Contact pressure is therefore close to the curve with 15 degree in section 1. As the contact ratio A increases, the asperity is flattened. When A reaches 0.28, the asperity tangent angle is about 23 degree. Hence, the contact pressure follows the curve with 25 degree in section 2. The present model therefore agrees well with Kimura's results.

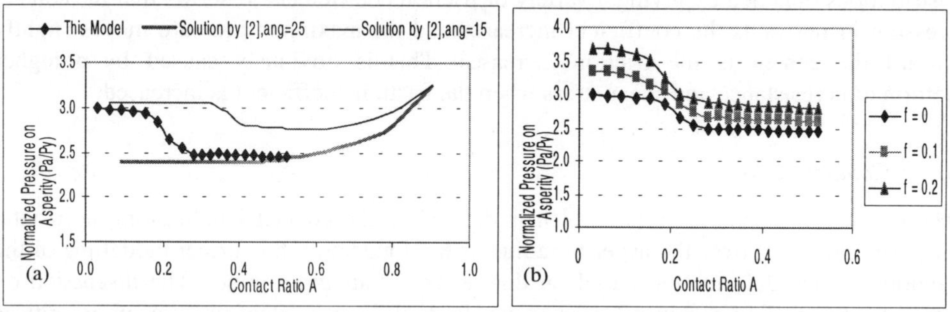

Figure 7 (a) Comparison of mean contact pressure versus contact ratio with Kimura et al [2] (b) Relationships of mean contact pressure and contact ratio under various friction conditions

Figure 7(b) shows how the relationship between contact pressure and contact ratio is influenced by friction coefficient. The contact pressure Pa increases at a same contact ratio by increasing the friction coefficient f.

4 Conclusions

(1) The present elastic-plastic FEM model for asperity deformation has been verified by the classical Hertz contact solution.
(2) Increasing the rolling load leads to a significant flattening of the contact pressure distribution and a shift of the maximum pressure toward the contact edge.
(3) Rolling pressure increases almost linearly with contact ratio from 0 to 0.5.
(4) Contact pressure experiences a slight reduction if contact ratio A is between 0 and 0.28 and is independent of contact ratio for A between 0.28 to 0.5.
(5) Friction affects the asperity flattening.

References

1. Keife, H. and Sj gren, C. , A friction model applied in the cold rolling of aluminum strips, Wear **179** , 1994, pp137-142
2. Kimura, Y. and Childs, T.H.C. , Surface asperity deformation under bulk plastic straining conditions, Int. J. Mech.Sci.Vol.**41,** 1999, pp283-307
3. Korzekwa, D.A., Dawson, P.R. and Wilson, W.R.D. , Surface asperity deformation during sheet forming, Int. J. Mech.Sci.Vol.**34**, No. 7, 1992, pp. 521-539
4. Kral, E.R., Komvopoulos, K., Bogy, D.B. , Elastic-Plastic Finite Element Analysis of Repeated Indentation of a Half-Space by a Rigid Sphere, J. App. Mech., Vol.**60**, 1993, pp829-841
5. Sutcliffe, M.P.F. , Surface asperity deformation in metal forming processes, Int. J. Mech.Sci. Vol. **30**, No.11, 1988, pp847-868
6. Wilson, W.R.D. and Sheu, S. , Real area of contact and boundary friction in metal forming, Int. J. Mech.Sci.Vol.**30**, No. 7, 1988, pp475-489

OPTIMIZING DESIGN OF CVC ROLL PROFILE

CHENG LU, A KIET TIEU AND ZHENGYI JIANG
University of Wollongong, Wollongong
NSW 2522, Australia
E-mail: chenglu@uow.edu.au

Continuously variable crown (CVC) system is an effective method for controlling strip flatness and profile. The key issue in the development of the CVC system is the design of the roll profile, which in this paper is based on a cubic function. Not only the desired variable crown can be obtained, but also the axial force acted on the rolls can be minimized.

1 Introduction

The flatness and profile of strip are mainly dependent on the configuration of the roll gap across the width of the strip. The shape of roll is one of the main factors affecting the shape of a roll gap. Work rolls with a specific crown are used for controlling the flatness and profile of strip in conventional flat rolling mills, as shown in Fig.1. However, some compensation methods, such as multizone cooling and rolling bending, are needed in order to meet the continuously varying rolling conditions.

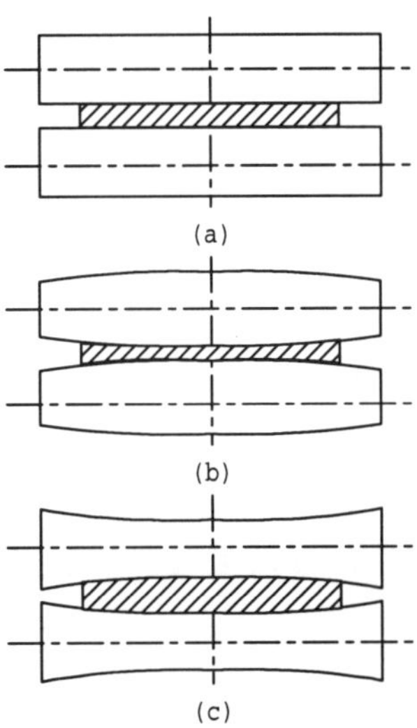

Figure 1. Conventional roll with crown

(a) Neutral crown; (b) Positive crown
(c) Negative crown

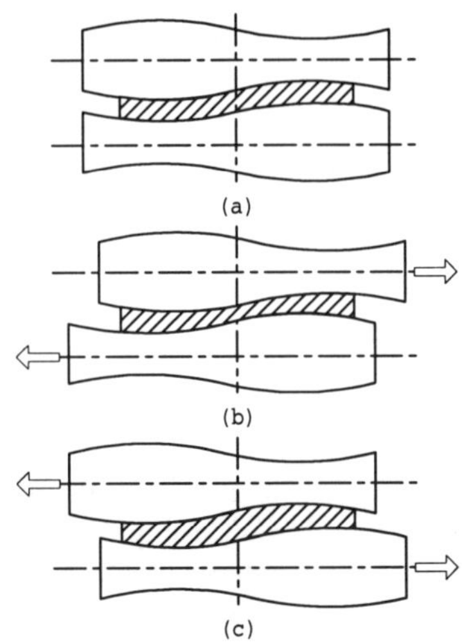

Figure 2. CVC roll with the roll shifting

(a) Neutral crown; (b) Positive crown
(c) Negative crown

SMS Schloemann-Siemag developed a continuously variable crown (CVC) system [1,2] and employed it in some cold rolling mill and hot rolling mill [3-6]. In CVC system both the upper and bottom rolls are ground to an S-shape and the axial shifting of rolls against each other produces a continuously adjustable roll gap contour. The performance of the CVC system is illustrated in Fig.2.

The key issue of CVC system is the principle for the design of the CVC rolls profile. Some research works have been done. The design of Zhang and Chen [7] attempted to minimise the diameter difference of CVC roll along the barrel. Jiang et al [8] designed a CVC roll with quintic curve to control quarter buckle. In this paper, a cubic curve is used as the CVC roll profile. The design of CVC roll profile can not only meet the desired adjusting range of the equivalent crown, but also minimise the axial force acted on the rolls.

2 Analysis of CVC roll profile

A cubic curve is used as the CVC roll profile, which has been confirmed to be effective in controlling the common shape defects including the centre buckle and edge buckle [8]. It is assumed that the origin of coordinates is located at the centre of left side of roll, as shown in Fig.3.

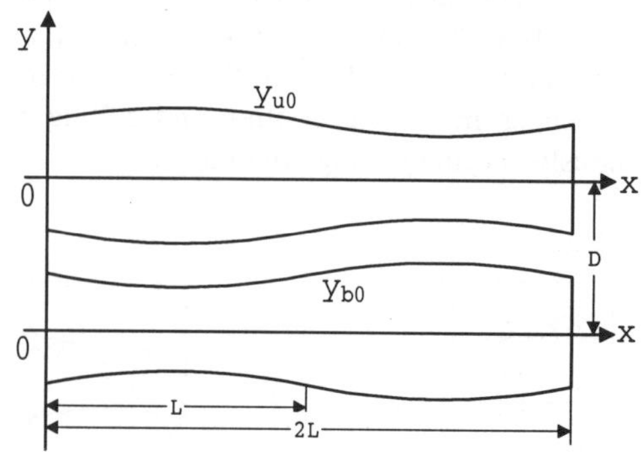

Figure 3. Coordinates of CVC roll profile

For the upper roll, the CVC roll profile is

$$y_{u0}(x) = A_0 + A_1 x + A_2 x^2 + A_3 x^3 \qquad (1)$$

where, A_0, A_1, A_2 and A_3 are the parameters which have to be determined in the optimizing design of CVC roll profile.

According to symmetry the bottom roll profile is given by

$$y_{b0}(x) = A_0 + A_1(2L - x) + A_2(2L - x)^2 + A_3(2L - x)^3 \qquad (2)$$

where, L is the half length of roll.

If the upper and bottom rolls shift a distance (s) in opposite direction, then the roll profile of the upper and bottom rolls become respectively,

$$y_{u1}(x) = A_0 + A_1(x - s) + A_2(x - s)^2 + A_3(x - s)^3 \qquad (3)$$

$$y_{b1}(x) = A_0 + A_1(2L - x - s) + A_2(2L - x - s)^2 + A_3(2L - x - s)^3 \qquad (4)$$

The equivalent crown (C_e) of the roll gap is the gap difference between the centre and the edge. C_e can be calculated using the following equation.

$$C_e = 6A_3L^2s - (6A_3L + 2A_2)L^2 \qquad (5)$$

C_e is the linear function of shifting distance s, and the coefficients ($6A_3L^2$ and $-(6A_3L+2A_2)L^2$) are related to A_2 and A_3.

3 Optimizing design

3.1 Design of A_2 and A_3

It is assumed that the range of the equivalent crown and shifting distance are [C_1, C_2] and [S_1, S_2] respectively. These values are known for the design. According to the linear relationship of equivalent crown and shifting distance (Eqn.(5)), the equivalent crown should be C_1 for the shifting distance S_1, and the equivalent crown should be C_2 for the shifting distance S_2, namely

$$C_1 = C_e(S_1) = 6A_3L^2S_1 - (6A_3L + 2A_2)L^2 \qquad (6)$$

$$C_2 = C_e(S_2) = 6A_3L^2S_2 - (6A_3L + 2A_2)L^2 \qquad (7)$$

Solving the linear equations (6) and (7), we can obtain the following results of A_2 and A_3.

$$A_2 = -\frac{C_1}{2L^2} + 3A_3(S_1 - L) \qquad (8)$$

$$A_3 = \frac{C_1 - C_2}{6L^2(S_1 - S_2)} \qquad (9)$$

3.2 Design of A_1

The forces acted on the roll from a slab of strip are shown in Fig.4. The axial force (dF_2) exists because of CVC roll profile. The axial force is harmful to rolling process. Too large axial force can cause damage to the equipment or industrial accident. Therefore, the axial force should be minimized in the CVC system. From Fig.4 we know

$$\frac{dF_2}{dF_1} = \frac{dy}{dx} \qquad (10)$$

where, dF_1 is the rolling force of slab dx. dF_2 is the axial force of slab dx.

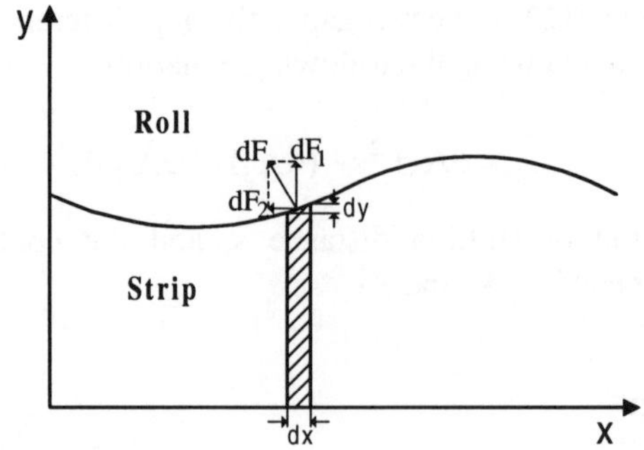

Figure 4. The schematic diagram of the forces acted on roll

It is assumed that the specific rolling pressure is constant, namely

$$\frac{dF_1}{dx} = p_0 \tag{11}$$

where, p_0 is the specific rolling pressure.

Furthermore, the total axial force (F_2) for the strip of 2b width can be obtained by integrating Eqn.(10).

$$F_2 = \int_{y_{u1}(L-b)}^{y_{u1}(L+b)} p_0 dy = p_0(y_{u1}(L+b) - y_{u1}(L-b)) \tag{12}$$

where, b is the half width of strip.

In the above equation, $y_{u1}(L+b) - y_{u1}(L-b)$ represents the effect of CVC roll profile on the axial force. Therefore, the following equation can be defined as the influence of CVC roll profile on the axial force.

$$R_t = (y_{u1}(L-b) - y_{u1}(L+b))^2 \tag{13}$$

Substitution of Eqn.(3) into Eqn. (13), we can obtain

$$R_t = 4b^2(A_1 + 2A_2(L-s) + 3A_3(L-s)^2 + A_3 b^2)^2 \tag{14}$$

Because the shifting distance s is much smaller than the length of the roll, normally s/2L<5%, the effect of roll shifting on the axial force will be neglected in the following analysis. Eqn.(14) becomes

$$R_t = 4b^2(A_1 + 2A_2 L + 3A_3 L^2 + A_3 b^2)^2 \tag{15}$$

R_t is the function of the half strip width b. The different R_t will be produced for the strip with different width as the other process values, such as the entry thickness and exit thickness are kept constant. Therefore, the maximum of R_t with respect to b exists and may be the stationary point or boundary points.

The stationary point of R_t is

$$b_1 = \sqrt{-(A_1 + 2A_2L + 3A_3L^2)/(3A_3)} \tag{16}$$

$$R_t(b_1) = -16(A_1 + 2A_2L + 3A_3L^2)^3 /(27A_3) \tag{17}$$

It is assumed that the width range of product is from B_1 to B_2. Therefore, there are two boundary points, namely B_1 and B_2 ($B_1 < B_2$).

$$R_t(B_1) = 4B_1^3(A_1 + 2A_2L + 3A_3L^2 + A_3B_1^2)^2 \tag{18}$$

$$R_t(B_2) = 4B_2^3(A_1 + 2A_2L + 3A_3L^2 + A_3B_2^2)^2 \tag{19}$$

From Eqns.(17), (18) and (19), it is known that $R_t(b_1)$, $R_t(B_1)$ and $R_t(B_2)$ are the function of parameter A_1 as A_2 and A_3 have already been designed according to Eqns.(8) and (9). The design of A_1 is to minimize the maximum of $R_t(b_1)$, $R_t(B_1)$ and $R_t(B_2)$ in order to reduce the axial force. In this paper only the following design results of A_1 are given because of the limited pages.

As $B_2 \geq 2B_1$,

$$A_1 = -2A_2L - 3A_3L^2 - A_3B_1^2 + A_3B_1B_2 - A_3B_2^2 \tag{20}$$

As $B_2 < 2B_1$,

$$A_1 = -2A_2L - 3A_3L^2 - \frac{3}{4}A_3B_2^2 \tag{21}$$

3.3 Design of A_0

Considering the gauge control, the diameter in the centre of CVC roll should be equal to the nominal diameter in the case without the roll shifting, namely

$$y_{u0}(L) = D_R / 2 \tag{22}$$

where, D_R is the nominal diameter.

Therefore, we can design A_0 using the following equation.

$$A_0 = D_R / 2 - A_1L - A_2L^2 - A_3L^3 \tag{23}$$

For the case with the roll shifting, the compensation must be added to the nominal diameter.

4 Design results

The data in Table.1 are used to design the CVC roll profile. Based on Eqns.(8), (9), (21) and (23), the optimising results are A_0=379.75, A_1=1.22*10^{-3}, A_2=-1.34*10^{-6} and A_3=4.02*10^{-10} respectively. The CVC roll profile is shown in Fig.5.

Table. 1. Process and Equipment Parameters

Parameters	Value (mm)	Parameters	Value (mm)
L	1125	S_1	100
D_R	760	S_2	-100
C_1	0.27	B_1	450
C_2	-0.34	B_2	950

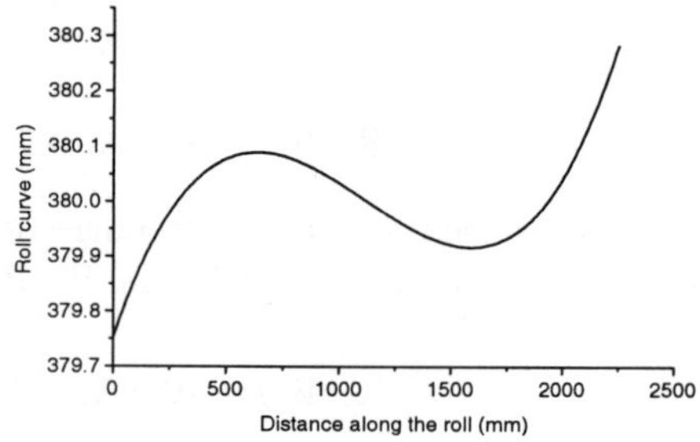

Figure.5. Design results of the CVC roll profile

5 Conclusion

CVC system is an effective method for controlling the flatness and profile of strip. In this paper a cubic curve is used as the CVC roll profile. Based on the analysis of the CVC roll profile, it is known that the equivalent crown is a linear function of the shifting distance, and the coefficients are related to A_2 and A_3. The design of A_2 and A_3 is to meet the desired adjusting range of the equivalent crown. An index (R_t) representing the effect of the CVC roll profile on the axial force acted on the rolls is obtained. A_1 can be designed to minimize the axial force.

References

1. Bald W., Beisemann G., Feldmann H., and Schultes T., Continuously variable crown (CVC) rolling. *Iron and Steel Engineer* **64** (1987) No.3 pp. 32–41.
2. Bald W. and Klamma K., CVC technology for cold rolling mills – Plant examples, *Iron and Steel Engineer.* **65** (1988) No.5 pp. 24–28.
3. Prachowsky R., Klamma K. and Setzer H., Modernization of the cold rolling mill at Krupp-VDM GmbH, *Metallurgical Plant & Technology International.* **18** (1995) No.6 pp. 72–75.
4. Espenhahn M., Friedrich K. E., Osterburg H., Tese P. and Wolpert W., Modernization of the hot strip mill of Thyssen Stahl AG, *Metallurgical Plant & Technology International.* **18** (1995) No.1 pp. 50–64.
5. Hoegberg R., Experience gained on the 5-stand cold rolling mill at SSAB with a 6-high stand using CVC technology, *Metallurgical Plant & Technology International.* **13** (1990) No.6 pp. 58–67.
6. Grell J. R., Latest cold rolling process technology illustrated by the Aluminium Norf wide strip mill with CVC stand, *Metallurgical Plant & Technology International.* **13** (1990) No.5 pp. 86–95.
7. Zhang J. and Chen X. L., Roll. *Chinese Practical Patent (89202773.8).*
8. Jiang Z. L., Wang G. D. and Zhang Q., Shifting-roll profile and control characteristics. *Journal of Materials Processing Technology.* **37** (1993) No.1 pp. 53–60.

ANALYSIS OF TRANSIENT TEMPERATURE FIELD ON THE TRANSVERSE SECTION OF HOT STRIP

W. H. SUN,* A. K. TIEU*, Z. Y. JIANG*, X. H. LIU**, G. D. WANG**

*Mechanical Engineering, Faculty of Engineering, University of Wollongong, Wollongong, NSW 2522, Australia weihua@uow.edu.au

**State Key Laboratory of Rolling Technology and automation, Northeastern University, Shenyang, 110006, P. R. China liuxh@mail.neu.edu.cn

In this paper, a finite element model was used to determine the transient temperature field of the transverse section of strip in the whole hot strip mill production line. The simulation shows that the calculated values are in good agreement with the measured values obtained from a hot strip mill. The temperature field across the thickness becomes uniform at the exit of finishing stands, but it shows that there is about 100 °C difference between the center and the edge of the strip. Before the strip enters the finishing stands, a temperature drop of 110 °C between the head and tail of strip has been found. However, the temperature drop reduces to 10 °C as the strip is processed at an accelerated speed of 0.14m/s^2 through the finishing stands.

1 Introduction

Accurate prediction and control of the temperature within slabs and strips during hot rolling is an essential requirement for efficient pass scheduling [1]. In this paper, a finite element model to determine the transient temperature field of the transverse section of strip on a hot strip mill is presented. The simulated hot strip mill consists of 3 roughing mills, seven finishing stands, a set of laminar cooling system and coilers, as shown in Figure 1. During the hot strip processing, the temperature will change continuously due to heat transfer by heat radiation and convection into the air, cooling by water such as descaling, spraying and laminar water, and conductive heat loss into the work roll. There is also a temperature rise due to plastic deformation and interface friction work in the roll bite. Calculations were carried out for the slab and strip from the reheating furnace to the strip coiler and the calculated values were compared with the measured data.

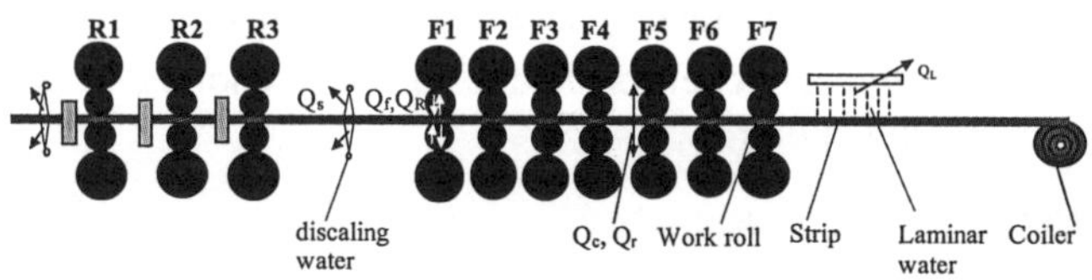

Figure 1. A typical simplified arrangement of hot strip rolling system

2. Thermal resources

2.1 Temperature change due to heat transfer to the air

The heat flux q_a (w/m^2) and heat transfer coefficient h_a (w/m^2·K) due to heat convection and radiation in air can be described as

$$q_a = q_r + q_c \quad ; \quad h_a = h_r + h_c \tag{1}$$

where a, r and c refer to air, radiation and convective respectively. For heat convection to the air, Hollander [1] considered that heat loss by convection in a strip cooling from 1200 to 900°C is only about 4-6% of the total loss and, therefore, is negligible. Other authors [2-5] use the empirical equation to calculate the effect of convection and radiation on a cooling piece of steel. In this paper, an empirical equation is applied [6].

$$h_c = 1.74 \times (T - T_\infty)^{1/3} \tag{2}$$

Heat flux by radiation can be described by

$$q_r = \varepsilon\sigma(T^4 - T_\infty^4) = \left[\varepsilon\sigma(T^2 + T_\infty^2)(T + T_\infty)\right]\{T - T_\infty\} = h_r(T - T_\infty) \tag{3}$$

where T and T_∞ are respectively the temperatures on the surface and in the surrounding media, $\sigma = 5.6699 \times 10^8 \mathrm{W/m^2K^4}$ is the Stefan-Boltzmann constant, ε the emissivity, selected as shown in Table 1.

Table 1. The emissivity of the work-piece on the production line[7]

Position on the production line	Emissivity
After exiting from the furnace	0.8
After roughing	0.6
After finishing	0.5

2.2 Temperature change due to water cooling

The heat flux during water cooling zone can be calculated as follows.

$$q = q'_r + q_w \quad ; \quad q_w = h_w(T - T_w) \tag{4}$$

which q_w, h_w and T_w are the heat flux, heat transfer coefficient to cooling water and water temperature respectively. q'_r is the heat flux of heat radiation into the air which is much smaller than that of q_w and is neglected.

Assuming water flow is C_1 ($\mathrm{m^3/minm^2}$) and the strip surface temperature before cooling is T_0(K), the heat transfer coefficient due to spray cooling between stands is described by the following equation [7].

$$h_w = 124.67 C_1^{0.167} \times 10^{-0.00147 T_0} \tag{5}$$

In the descaling zone, h_w is selected as 1163w/m^{2}K at the striking point of the descaling water, and 200 w/m^{2}K in the affected zone[8].

For laminar water cooling, h_w is described by equation (6).

$$h_w = \frac{510.78 \times 10^5 \times C_2^{0.353}}{(T - T_w)} \left\{ \frac{12.5 - 1.15 \log T_w \times D}{P_l P_c} \right\}^{0.645} \tag{6}$$

where C_2, T_w, P_l, P_c, D are the water flow($\mathrm{m^3/minm^2}$), temperature of water, space between the pipes along the rolling direction and width, and the diameter of the pipe respectively [9].

2.3 Temperature change in the roll bite

Three kinds of heat source influence the temperature change in the roll bite, namely heat generation due to plastic deformation work and friction, and the heat conduction between the interface of the rolls and the strip.

The amount of heat generation $Q_p(J)$ by plastic deformation and friction work power $\dot{q}_f$ (w/m^2) can be calculated as follows.

$$Q_p = \eta_p p_m \ln(h_1 / h_2) \quad ; \quad \dot{q}_f = \eta_f \mu p_m v_r \tag{7}$$

where p_m (N/m^2) the average pressure the work roll acts on the strip, h_1, h_2 the strip inlet and exit thickness, μ the coefficient of friction, v_r the relative slippage speed between the work roll and the strip, η_p the heat proportion of the plastic deformation work and η_f the proportion of friction works dissipated as heat. η_p and η_f are selected as 0.9 and 2/3 respectively[10-12]. Both of heat generations by plastic deformation and by friction work are taken as internal heat resources.

Conductive heat flux from strip to the roll bite is calculated by equation[7,10]

$$q_R = \frac{2\lambda_s (T_{s0} - T_m)}{(\alpha_s \pi t)^{1/2}} \tag{8}$$

where, $T_m = (\lambda_s T_{s0} \alpha_s^{1/2} + \lambda_R T_{R0} \alpha_R^{1/2}) / (\lambda_s \alpha_s^{1/2} + \lambda_R T_{R0} \alpha_R^{1/2})$, the equivalent roll temperature, λ , T , α , t represent the thermal conductivity of the material (w/mK), temperature(K), the thermal diffusivity (m$^2/s$) and the contact time(s) respectively. The subscription s, R represent the strip and roll respectively, $_0$ indicates the state before inlet. The heat transfer coefficient then can be assumed as

$$h_R = q_R / (T_{s0} - T_{R0}) \tag{9}$$

3 Finite Element Model for Analyzing the Transverse Temperature Field of Hot Strip

Without considering heat conduction parallel to the rolling direction, the heat generation due to plastic deformation and friction works are taken as internal thermal resources. By assuming the width direction as x axis and thickness as y axis, the transverse temperature field can be determined by equation (10).

$$\begin{cases} \dfrac{\partial^2 T}{\partial x^2} + \dfrac{\partial^2 T}{\partial y^2} + \dfrac{\dot{q}}{k} = \dfrac{1}{\alpha} \dfrac{\partial T}{\partial t} \\ -k \left(\dfrac{\partial T}{\partial x} l_x + \dfrac{\partial T}{\partial y} l_y \right) + h(T - T_\infty) = 0 \\ T(x, y, t_0) = T_0 \end{cases} \tag{10}$$

where $k, \alpha, q, l_x, l_y, T_\infty$ are the thermal conductivity, thermal diffusivity, internal heat generation rate, outer normal cosine of the boundary surface and the surrounding temperature respectively, and h can be one of the follows: h_r, h_c, h_R, h_w etc. This boundary problem can be expressed as the minimization of the following functional with Euler-Lagrange equation[11].

$$I = \frac{1}{2} \iint_s \left\{ \left[\left(\frac{\partial T}{\partial x} \right)^2 + \left(\frac{\partial T}{\partial y} \right)^2 \right] - \alpha \left(\dot{q} - \rho \frac{\partial T}{\partial t} \right) T \right\} ds + \frac{1}{2} \int_{s3} h(T - T_\infty)^2 dl \qquad (11)$$

which ρ, c are density and specific heat of the material respectively, s the transverse area of the strip and s_3 the perimeter of the transverse section where there is heat transfer to the surrounding.

Based on the finite element theory, the minimization of the functional I of equation (11) results in the following equations:

$$[K_T]\{T\}^t + [K_3]\left\{\frac{\partial T}{\partial t}\right\} = \{P\} \qquad (12)$$

$\partial T/\partial t$ can be treated with the backwards differential method, then

$$([K_T] + \frac{1}{\Delta t}[K_3])\{T\}_t = \frac{1}{\Delta t}[K_3]\{T\}_{t-\Delta t} + \{T\}\{P\} \qquad (13)$$

The transient transversal temperature field of the hot strip can be obtained from equation (13).

4 Calculation and discussion

A FORTRAN program has been developed according to the analysis above and the calculation was carried out based on a hot strip. To start with, one fourth of the strip section is divided into 2000 meshes, in which there 20 meshes through thickness and 100 along the width, assuming uniform temperature within the slab at the exit of the furnace and symmetrical cooling on the top and bottom surfaces and along the width. Moreover, the number of meshes is changed to 4000 while the strip entered the last two finishing stands, where the number is fourfold along the width and halved through the thickness. The properties of the steel are considered to be temperature dependent [9,13]. Appropriate boundary conditions are applied corresponding to the cooling and processing characteristics of the hot strip mill.

Figures 2 to 4 describe the calculation results for a 6.06 mm thick strip. The temperature evolution at the surface and the centre as well as the average one of the transverse section on top end of the strip are illustrated in Figure 2. The error between the measured value and the calculated one is within ± 15°C. After the roughing rolling stage, the center temperature of the strip increases due to plastic deformation, while the surface temperatures reduce distinctively due to the heat conduction into the roll.

Figure 3 compares the average temperature evolution on the two ends. The temperature difference is only about 30 °C after roughing, while it reaches 110°C before entering the finishing stands. Accelerated finishing is useful to eliminate the temperature drop, which increases the heat generation due to plastic deformation. At the exit of the finishing stands the temperature difference reduces to about 10°C at an accelerated finishing speed of 0.14m/s^2.

Figure 4 shows the transient temperature fields on the transverse section of the strip at the typical stages during processing. Obvious temperature gradient appears in the roughing stage, and there are about 50 to 100°C of temperature gradient through the thickness in the middle of the strip. With thickness reduction, it reduces while the strip is in the finishing stands. It is about 4°C at the exit of the finishing stands and less than 3°C at the inlet of the coiler. However, at the exit of the finishing mill, the temperature gradients have been found along the width. There is only about 4/5 area along the strip width in which the uniform distribution of temperature can be obtained. It is about 100°C temperature difference along the width.

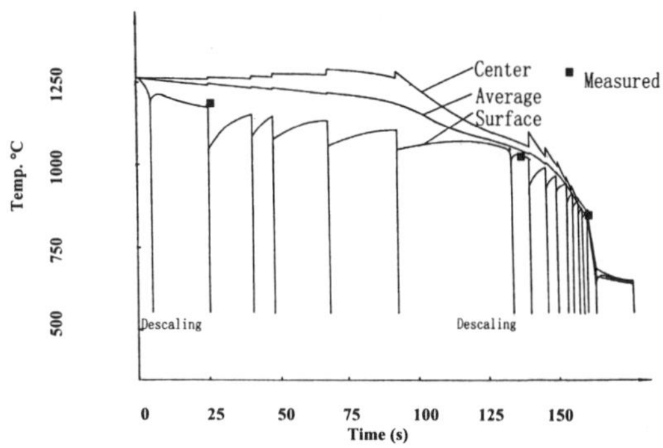

Figure 2. Temperature evolution on the transverse section of strip

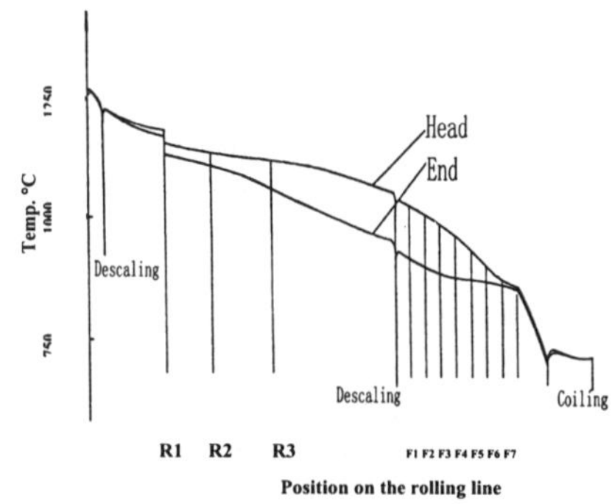

Figure 3. The average temperature evolution on the two end of strip

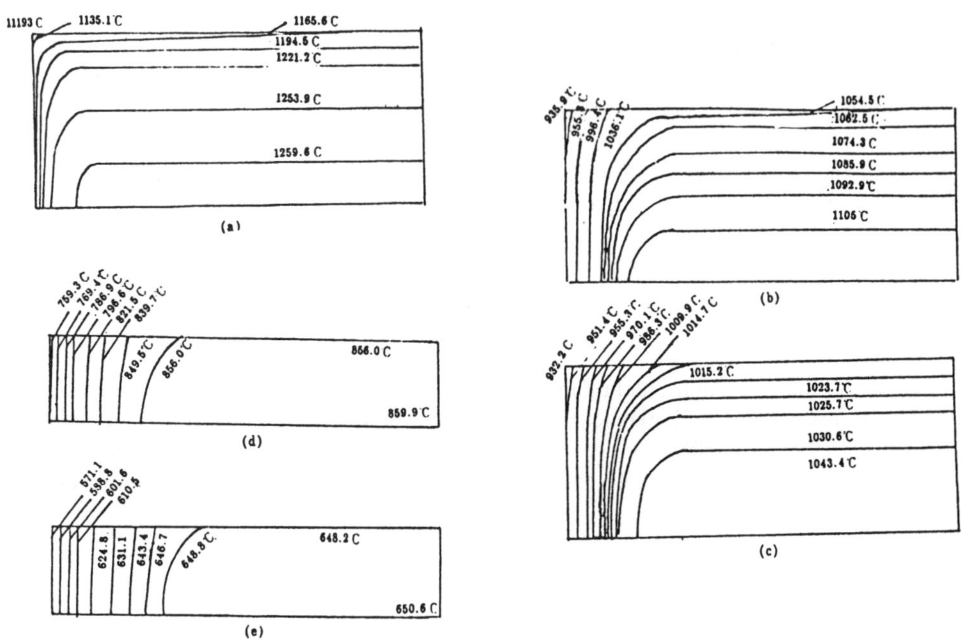

Figure 4. Transient temperature fields on the transverse section of strip

a-head at the entrance of roughing b-tail at the exit of roughing c-head at the entrance of finishing d-head at the exit of finishing e-head at the entrance of coiler

5 Conclusion

A simulation has been conducted for the transient transverse temperature fields of a strip from the start of the roughing mill to the coiler in hot strip mill. Thermal transfer model, including heat radiation, convection, cooling water and thermal exchange during rolling, was analyzed according to the practical production. Heat generation due to the strip deformation and friction between the strip and the work rolls has also been taken into account. The calculated values are in good agreement with the measured values obtained on a hot strip mill. An optimum acceleration on the finishing mills has been obtained. It can be concluded the variation of the flow of the laminar cooling water along the strip width should be implemented to obtain a uniform property.

Acknowledgment

This work was supported by the Australian Research Council (ARC) IREX Award No. X00106473 which is gratefully acknowledged.

References

1. Hollander, F., A model to calculate the complete temperature distribution in steel during hot rolling. Mathematical Models in Metallurgical Process Development, London ISJ Conference, Feb. 1969, pp.46-79
2. R. Harding, Ph.D. Thesis, University of Shefield, UK, 1976.
3. L.A. Leduc, Ph.D. Thesis, University of Shefield, UK, 1980.
4. C.M. Sellars, Mater. Sci. Technol. 1 (1985) 325.
5. R. Colás, Mater. Sci. Technol. 14 (1988) 388.
6. Okatokatsu et al, Iron and Steel of Japan. 10(1977): 28
7. Iron and Steel Society of Japan, Handbook of Iron and Steel: Steel Rolling. (1984): 17-20
8. Youno Masakiyo et al, Kobe Steel Technique Report. 3(1986): 47-49
9. Iron and Steel Society of Japan, Theory and Practice on Hot Strip Rolling. 1984:181-184
10. Yanagi Kanisa, Mitsubitsu Technique Report. 1(1974): 56-59
11. Siziki Kaiitsi, Iron and Steel of Japan. 8(1976): 49-52
12. X. H. Liu, Rigid and plastic finite element method and its application in metal rolling. 1994: 300-305
13. J. L. Ma et al, Practical hand book on thermal engineering, Hydro-electric Publication, 1988.

INFLUENCE OF THE SPINDLE ANGLES AND FRICTION ON THE STRESS STATES OF UNIVERSAL JOINT

W. H. SUN*, J. M. LI**, A. K. TIEU*, Z. Y. JIANG*

Mechanical Engineering, Faculty of Engineering, University of Wollongong, Northfields Avenue, Wollongong, NSW 2522, Australia

weihua@uow.edu.au

***Technical Center, Jinan Iron and Steel Group Co., Jinan, 250101, P.R.China*

jianminli@21cn.com

The safety of the universal joint plays a key role in a plate rolling mill. The spindle angles and friction have significant effects on the stress states of jaw and jaw coupling of the universal joint. In this paper, the authors considered the additional bending moment due to the spindle angles of the joint and the bending moment due to the friction between the bearing and the jaw and its couple respectively in addition to the torque effect from the main drive. The equivalent stress and the Tresca stress have been analyzed with the ALGOR Finite Element Analysis System. The results show that the higher is the friction coefficient, the greater the influence of the friction moment plays on the jaw couple, while the smaller the friction coefficient, the greater the influence of the additional bending moment on the jaw couple. The friction moment has been found to have a greater influence on the stress state of the jaw couple than the jaw. These were verified by the failure of the work roll at the coupling end.

1 Introduction

Universal joints (U-joint), also referred to as Cardan joints, have been used in the mechanical components of metallurgical equipment. They are used to connect misaligned shafts that are intersecting, transmitting rotational motion from one shaft to another. An exploded view of a typical "bearing block" U-joint applied on heavy plate mill is shown in Fig. 1. The joint consists of input and output yokes and a cross trunnion, which are named as jaw couple, jaw and bearing block respectively.

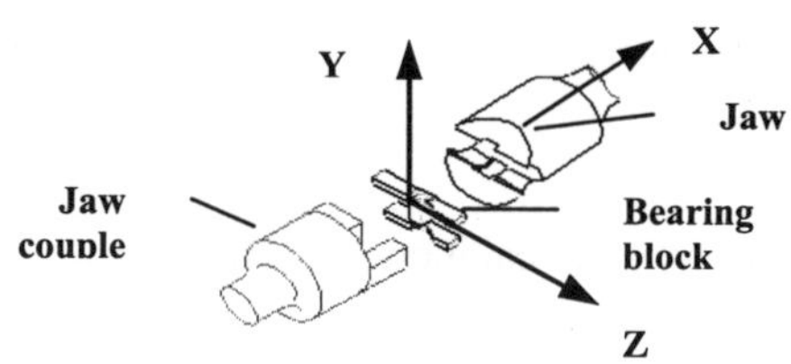

Fig. 1 Exploded view of the bearing block type of U-joint

The difference between the U-joint of a plate mill and the others is the joint angle, which is generally less than 12°. The U-joint used in the plate mill is able to withstand high input and output torque, and it can cause catastrophes if operated improperly. It has been shown that the extremely high internal forces will occur if the contact occurs between the various parts of the universal joint[1]. Wagner and Cooney[2],which gave a comprehensive guide to the design of the universal joint, analyzes the kinematics and strength of an ideal universal joint without considering interference, assuming small joint angles. General guidelines to the design and dynamics of U-joint have been shown by Shigley and Mischke[3], Lee[4], Machine Design[5] and Lingaiah[6]. These articles mainly deal with the kinematics and performance of an ideal U-joint.

The objective of this paper is to derive the relationships necessary to optimize the operation of the U-joint and to check geometric interference between its components. In addition to the torque effect from the main drive, an additional bending moment(ABM) due to the spindle angles of the joint and the bending moment due to the friction(FBM) between the bearing and the jaw and its couple are also considered in this analysis.

2 Static mechanic analysis of the universal joint

The operating mechanism of the bearing block type of universal joint is the same as the Cardan joint, and the bearing block is similar to that of the later. The center line of the semi-circle shape of bearing block turns 90° to that of the inter-shaft, the same as the cross trunnion on that of the Cardan joint. Therefore the mechanic analysis of the bearing block type of the U-joint is carried out on the Cardan joint.

Fig. 2 illustrates the operating mechanism of the universal joint, where α is the spindle angle, $\overrightarrow{M_2}, \overrightarrow{M_3}$ the torsion moments on the driving and driven shafts respectively, and φ_2 and φ_3 are the rotating angles corresponding to the main drive and the work roll of the mill respectively.

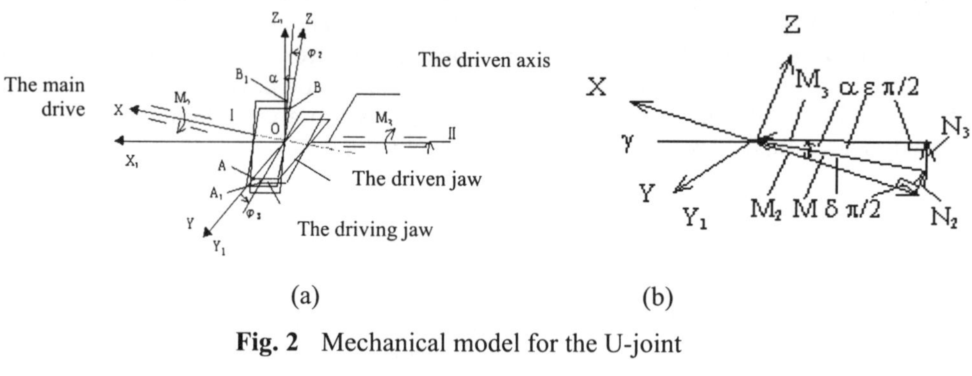

Fig. 2 Mechanical model for the U-joint

$\overrightarrow{N_2}, \overrightarrow{N_3}$ are ABMs on the U-joint and the work-roll induced by the force transmitted to cross trunion from the U-joint (Fig. 2), which are rectangular to the center-lines of the U-joint and the work-roll respectively. The coupled system of $\overrightarrow{M_2}, \overrightarrow{M_3}, \overrightarrow{N_2}$ and $\overrightarrow{N_3}$ satisfy the static equilibrium of the U-point, as shown in Fig. 2(b). The total moment $\vec{M} = \overrightarrow{M_2} + \overrightarrow{N_2}$ or $-\vec{M} = \overrightarrow{M_3} + \overrightarrow{N_3}$, is rectangular to the cross trunnion plane.

As shown in Fig. 2, when the U-joint rotates an angle φ_2, the ABMs on the U-joint and the work-roll can be calculated as

$$\begin{cases} N_2 = M_1 \sin\alpha \sin\varphi_2 (1+\sin\varphi_1 \tan^2\alpha) \\ N_3 = M_1 \cos\alpha \sin\alpha \cos(a\tan\dfrac{\tan\varphi_3}{\cos\alpha}) \end{cases} \tag{1}$$

With the rotation of the U-joint, the position of the normal pressure P', which is acted on the jaw from the bearing, will changes continuously, as shown in Fig. 3. The friction moment acted on the jaw couple from the bearing block can be described as

$$N_3' = M_3 \cdot \mu / \sin\varphi_{43} \tag{2}$$

where μ the friction coefficient of the two contacts, φ_{43} the angle between the X-Z axes which can be calculated with the following equation.

$$\varphi_{43} = 90° + \alpha\cos(a\tan\frac{\tan_2}{\cos\alpha}) \tag{3}$$

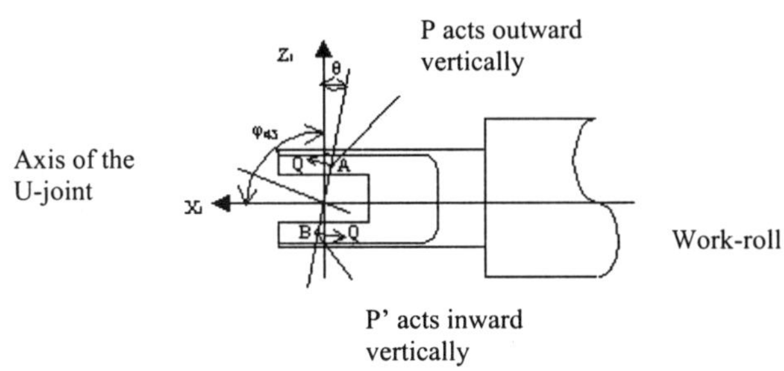

Fig. 3 Mechanical model for the jaw couple on the work-roll of an universal joint

3 Modeling and results

3.1 Finite element model

ALGOR Finite Element Analysis System was used in this investigation. Hexahedron with eight-points and triangular prism with six-points type of element meshes is used. The lubrication holes and the oil cavities are neglected in order to simplify the calculation. Rigid boundary elements are introduced in the area on the jaw couple holding the force from the bearing block. For the jaw, there are 3,268 elements and 4,498 nodes, where 144 elements are the boundary ones. There are 1,888 elements and 2,797nodes for the jaw couple, where 180 elements are the boundary ones.

3.2 *Effects of spindle angles on the stress state of the jaw and its couple*

Based on the above discussed condition, calculation has been carried out for the stress of the jaw and its couple, and the effects of the U-joint spindle angle and friction coefficient on the stress state are investigated. An index k is introduced to describe the change of the stress as follows

$$k_j = (\sigma_j - \sigma_{0j}) / \sigma_{0j} \tag{7}$$

where σ_0 is the stress without consideration of ABM and FBM, j (1,2,3 and 4) refers to von Mises, Bi-Tresca, the maximum and the minimum respectively.

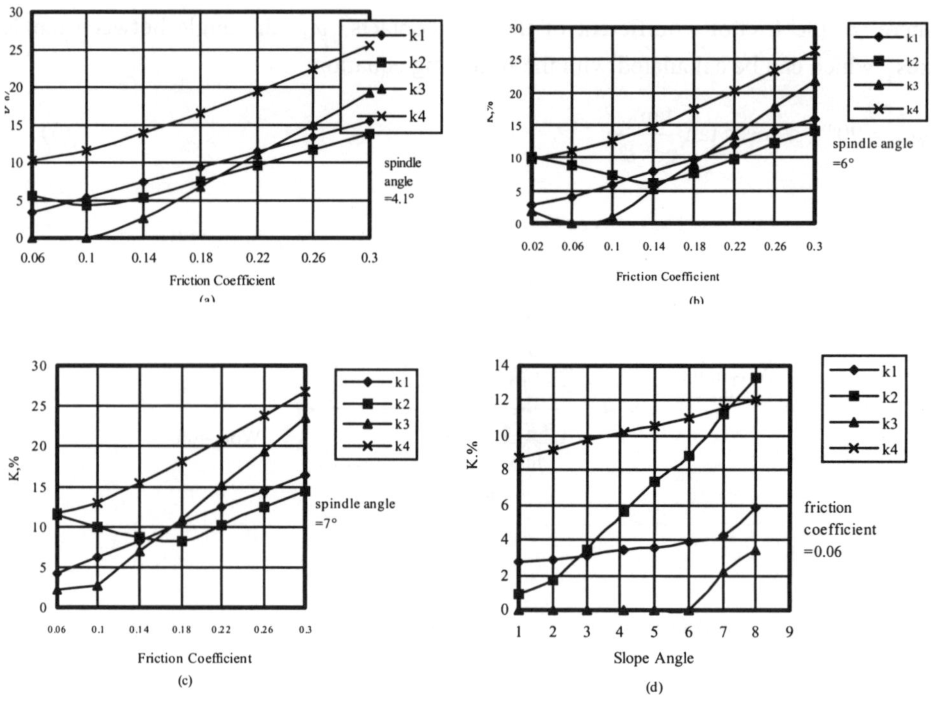

Fig. 4 The effects of friction coefficient and spindle angle on the jaw stress index

The influence of the friction coefficient on the stress state of the jaw for the given angles of 4.1°, 6° and 7° was investigated, and the results show that the von Mises stress and the absolute value of the minimum stress increase linearly with the angle, while the maximum stress and the Bi-Tresca stress reduces first and then increases due to the different effects of the ABM and FBM. The relationship between k and the friction coefficient for the given angles of 4.1°, 6° and 7° is illustrated in Fig. 4. For the given friction coefficient of 0.06, the four characteristic stresses do not change significantly with the angle. Comparing the stress state at the pure torque moment condition, the von Mises, Bi-Tresca, the minimum stresses increase 3.9, 8.8 and 11% respectively, but the maximum stress is constant when the spindle angle is 6°. When the friction coefficient increases, the characteristic stresses increase significantly. For example, if $\mu = 0.3$ and $\alpha = 6°$, the characteristic stresses will increase 15.9, 14.2, 21.7 and 26.2% respectively.

For the jaw couple, the von Mises stress, Bi-Tresca stress and the maximum stress increase linearly with the spindle angle while the minimum stress decrease for friction coefficient, of 0.06 and 0.3, as it has been shown in Fig. 6(a) and (b). When the friction coefficient is 0.06, except for the minimum stress, the von Mises stress, Bi-Tresca stress and the maximum stress increase by 6.9, 3.1 and 14.1% respectively, compared with the ones under pure torque moment state. When the friction coefficient changes to 0.3, the increases become 25.6, 18.9 and 49.3% respectively, while the minimum stress also has changed considerably by 30.6%. At the fixed angle of 6°, the characteristic stresses increase sharply with the friction coefficient, as shown in Fig. 5(c).

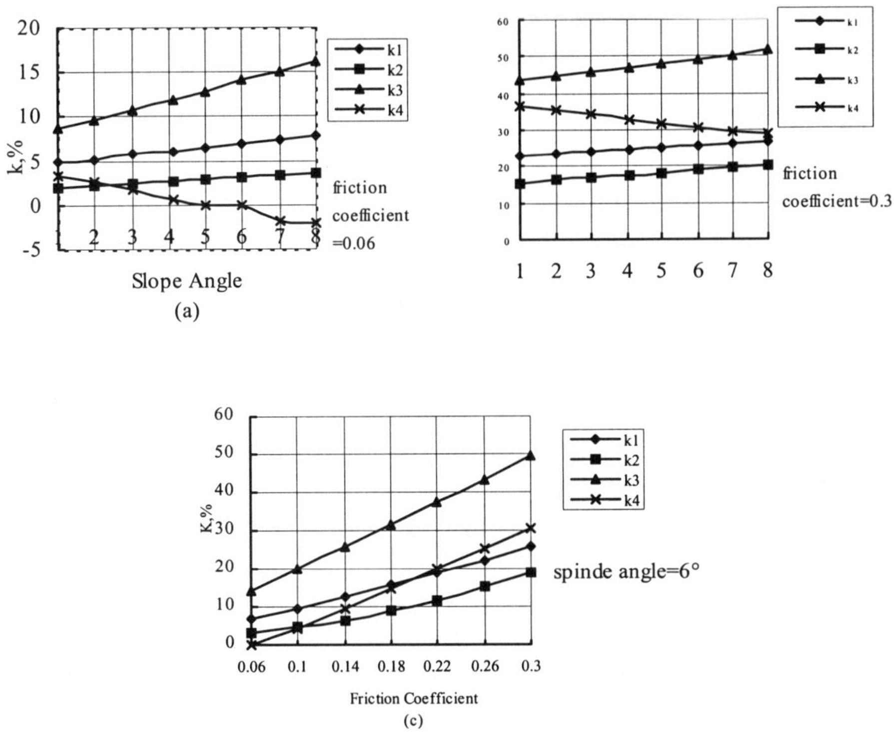

Fig. 5 The effects of friction coefficient and spindle angle on the jaw

3.3 Practical failure of the U-joint

Once there was a failure of the bearing block type of U-joint on the heavy plate mill of Jinan Iron and Steel Company, P.R. China in 1998, in which crack occurred at the coupling end on the work roll. Severe wearing and deformation on the bearing was found. Fatigue cracks were discovered on the ruptured section of the coupling. It was found that wear in the bearing blocks increase the impact load on the coupling, and when coupled with stress concentration will lead to fatigue stresses and ultimate failure. In fact, it is difficult to lubricate the U-joint in practice and dry friction often occurs between the jaw and its couple. Therefore, the effects of friction should be taken into account while designing the U-joint.

4 Conclusion

A stress analysis of a universal joint has been conducted. The model considers the additional bending moment due to the spindle angles of the joint, the bending moment due to the friction between the bearing and the jaw and its couple, as well as the torque effect from the main drive. The results show that the higher is the friction coefficient, the greater the influence of the friction moment on the jaw couple, while the smaller the friction coefficient, the greater the influence of the additional bending moment on the jaw couple. Friction moment has been found to have a more significant influence on the stress state of the jaw couple than that of the jaw.

Acknowledgment

This work was supported by the Australian Research Council (ARC) IREX Award No. X00106473 which is gratefully acknowledged.

References

1. Fischer, I. S., Internal force and torque transmission in a Cardan joint with manufacturing tolerance. Engineering Sci. D. dissertation, Columbia University, New York, 1985.
2. Wagner, E. R. and Cooney, C. E., Universal Joint and Driveshaft Design Manual, Advances in Engineering Series No. 7. Society of Automotive Engineers, Warrendale, PA, 1979.
3. Shigley, J. E. and Mischke, C. R.. Standard Handbook of Machine Design. McGraw-Hill, New York, 1986. pp. 19-28.
4. Lee, D. A., *Machine Design*, 1965. 37, 151-153.
5. Anonymous, *Machine Design*, 1939. 11, 47-48.
6. Lingaiah, K., in *Machine Design Data Handbook*. Mcgraw-Hill, New York, 1994, pp.25-90.

A SIMPLIFICATION OF NON-LINEAR RIGID PLASTIC FEM FOR ROLLING PROCESS

CHENG LU, A KIET TIEU AND ZHENGYI JIANG

University of Wollongong, Wollongong
NSW 2522, Australia
E-mail: chenglu@uow.edu.au

In this paper a new computing algorithm for the rigid-plastic finite element method is presented. The iteration method is used to solve the non-linear stiffness equations, and the error generated from the linearization of Newton Raphson Method is removed. Numerical tests show that the computing algorithm has a good convergence, and the same convergence results can be obtained for different initial guesses. The computing algorithm has shown that it is suitable for the simulation of rolling process

1 Introduction

The rigid-plastic finite element method (FEM) with the Lagrange multiplier was developed by Lee and Kobayashi [1], and it was extended to rigid-visco-plastic material by Oh et al. [2]. Zienkiewicz and Godbole [3] considered the metal forming process as a flow of plastic solid and developed a rigid plastic finite element technique with the penalty function constraints. Mori and Osakada [4] presented the rigid plastic finite element method for a slightly compressible material. In recent years the rigid plastic finite element method has been used in the simulation of different rolling process and has proved to be a useful and powerful tool [5-6].

In general, the Newton Raphson Method has been used to solve the non-linear stiffness equations of the rigid-plastic finite element method. However, the initial guess should be selected close enough to the actual solution in the simulation of rolling process. Some studies on the selection of the initial value have been carried out [2,4]. In this paper a new computing algorithm used in the rigid plastic finite element method is presented. The iteration method is used to solve the non-linear stiffness equations. The numerical tests show that this method has a good convergence and the same convergence results can be obtained for different initial guesses.

2 Theories

Consider a rigid plastic body V with tensile stress applied at the front and tail ends (S_F), and the frictional stress acted on the surface S_c between the workpiece and roll. The body, V, is divided into M elements interconnected at N nodal points. The functional is given as follows for the m-th element of the plastically deforming body.

$$\phi^{(m)} = \iiint_{V^{(m)}} \bar{\sigma}\dot{\bar{\varepsilon}}^{(m)} dV - \iint_{S_c^{(m)}} \left(\int_0^{\Delta u_r} f_s(\Delta u_r) d\Delta u_r \right) dS - \iint_{S_F^{(m)}} \mathbf{u}^T \mathbf{T} dS + \iiint_{V^{(m)}} \lambda^{(m)} \dot{\boldsymbol{\varepsilon}}^{(m)T} \mathbf{C} dV \tag{1}$$

where, $\bar{\sigma}$ and $\dot{\bar{\varepsilon}}$ are the effective stress and strain-rate respectively, f_s frictional stress, Δu_r relative velocity at the surface S_c, **T** tensile stress vector, **u** velocity vector, λ Lagrangian multiplier, $\dot{\boldsymbol{\varepsilon}}$ strain-rate vector. **C** is the proper matrix vector notation of the Kronecker symbol.

According to the variational principle it can be shown, for the actual solution, the functional becomes stationary with respect to a kinematically complete velocity fields, which satisfy the compatibility condition and velocity boundary conditions. The requirement of the functional ϕ to be stationary results in the following non-linear stiffness equations.

$$\sum_{m=1}^{M} \frac{\partial \phi^{(m)}}{\partial \mathbf{u}^{(m)}} = 0 \qquad \sum_{m=1}^{M} \frac{\partial \phi^{(m)}}{\partial \lambda^{(m)}} = 0 \tag{2}$$

where, $\mathbf{u}^{(m)}$ is the velocity vector of m-th element. $\lambda^{(m)}$ is the Lagrangian multiplier of *m*-th element.

Based on the standard finite element procedure:

$$\mathbf{u} = \mathbf{N}\mathbf{u}^{(m)} \qquad \dot{\boldsymbol{\varepsilon}} = \mathbf{B}\mathbf{u}^{(m)} \qquad \mathbf{Q} = \iiint_{V^{(m)}} \mathbf{B}^{T}\mathbf{C}\,dV \tag{3}$$

where, **N** is the shape function. **B** is the strain-rate matrix.

Substitution of Eqn. (1) into Eqn. (2) results in

$$\begin{aligned}\frac{\partial \phi^{(m)}}{\partial \mathbf{u}^{(m)}} = & \iiint_{V^{(m)}} \bar{\sigma} \frac{\partial \dot{\bar{\varepsilon}}^{(m)}}{\partial \mathbf{u}^{(m)}} dV - \iint_{S_c^{(m)}} f_s(\Delta u_r) \frac{\partial \Delta u_r}{\partial \mathbf{u}^{(m)}} dS \\ & - \iint_{S_F^{(m)}} \mathbf{N}^{T}\mathbf{T}\,dS + \iiint_{V^{(m)}} \lambda^{(m)}\, \mathbf{B}^{T}\mathbf{C}\,dV\end{aligned} \tag{4}$$

$$\frac{\partial \phi^{(m)}}{\partial \lambda^{(m)}} = \mathbf{Q}^{T}\mathbf{u}^{(m)} \tag{5}$$

The actual solution (**u** and $\boldsymbol{\lambda}$) can be obtained by solving the non-linear stiffness equations (Eqn.(4)). The Newton Raphson Method is normally used to solve the non-linear stiffness equations.

The first and second term of Eqn. (4) are changed as follows.

$$\begin{aligned}\frac{\partial \phi^{(m)}}{\partial \mathbf{u}^{(m)}} = & \iiint_{V^{(m)}} \frac{\bar{\sigma}}{2\dot{\bar{\varepsilon}}} \frac{\partial \dot{\bar{\varepsilon}}^{(m)2}}{\partial \mathbf{u}^{(m)}} dV - \iint_{S_c^{(m)}} \frac{f_s(\Delta u_r)}{2\Delta u_r} \frac{\partial \Delta u_r^2}{\partial \mathbf{u}^{(m)}} dS \\ & - \iint_{S_F^{(m)}} \mathbf{N}^{T}\mathbf{T}\,dS + \iiint_{V^{(m)}} \lambda^{(m)}\, \mathbf{B}^{T}\mathbf{C}\,dV\end{aligned} \tag{6}$$

where,

$$\dot{\bar{\varepsilon}}^{2} = \frac{2}{3}\mathbf{u}^{T}\mathbf{B}^{T}\mathbf{B}\mathbf{u} \qquad \frac{\partial \dot{\bar{\varepsilon}}^{(m)2}}{\partial \mathbf{u}^{(m)}} = \frac{4}{3}\mathbf{B}^{T}\mathbf{B}\mathbf{u}^{(m)} \tag{7}$$

For a three-dimensional rolling process, Δu_r of nodal point i can be defined by the following equation.

$$\Delta u_{ri} = \sqrt{(u_{xi}\sec\theta - u_r)^2 + u_{yi}^2} \tag{8}$$

where, θ is the angular position of nodal point i. u_r is the roll speed.

Furthermore

$$\frac{\partial \Delta u_r^2}{\partial \mathbf{u}^{(m)}} = 2(\mathbf{G}\mathbf{u}^{(m)} - \mathbf{G}_r) \tag{9}$$

where, $\mathbf{G}$ and $\mathbf{G_r}$ are related to $\mathbf{N}$ and u_r.

Substitution of Eqns. (7) and (9) into Eqn (6) results in

$$\frac{\partial \phi^{(m)}}{\partial \mathbf{u}^{(m)}} = \mathbf{P}^{(m)}\mathbf{u}^{(m)} + \mathbf{Q}^{(m)}\lambda^{(m)} - \mathbf{H}^{(m)} \tag{10}$$

where

$$\mathbf{P}^{(m)} = \iiint_{V^{(m)}} \frac{2}{3}\frac{\bar{\sigma}}{\dot{\bar{\varepsilon}}}\mathbf{B}^T\mathbf{B}\,dV - \iint_{S_c^{(m)}} \frac{f_s}{\Delta u_r}\mathbf{G}^{(m)}\,dS \tag{11}$$

$$\mathbf{H}^{(m)} = \iint_{S_F^{(m)}} \mathbf{N}^T\mathbf{T}\,dS - \iint_{S_c^{(m)}} \frac{f_s}{\Delta u_r}\mathbf{G}^{(m)}\,dS \tag{12}$$

Substituting of Eqns. (10) and (5) into Eqn. (2), the elements are assembled into a global system of equations. The global equations for the n-th iteration becomes

$$\begin{pmatrix} \mathbf{P} & \mathbf{Q} \\ \mathbf{Q}^T & \mathbf{0} \end{pmatrix}_{(n-1)} \begin{pmatrix} \mathbf{u} \\ \lambda \end{pmatrix}_{(n)} = \begin{pmatrix} \mathbf{H} \\ \mathbf{0} \end{pmatrix}_{(n-1)} \tag{13}$$

where

$$\mathbf{u} = \sum_{m=1}^{M} \mathbf{u}^{(m)}, \quad \lambda = \sum_{m=1}^{M} \lambda^{(m)}, \quad \mathbf{P} = \sum_{m=1}^{M} \mathbf{P}^{(m)}, \quad \mathbf{Q} = \sum_{m=1}^{M} \mathbf{Q}^{(m)}, \quad \mathbf{H} = \sum_{m=1}^{M} \mathbf{H}^{(m)} \tag{13}$$

Firstly **P**, **Q** and **H** are calculated by an initial guess. Eqn. (12) becomes linear equations. Then the solution of linear equations is used in the next iteration. The iterative process continues until the convergence criterions are satisfied.

3 Results and discussions

The finite element computation was carried out under the condition of plane-strain rolling of rigid plastic materials. As the deformation is symmetric, one-half of the deformation body, as shown in Figure 1, was studied.

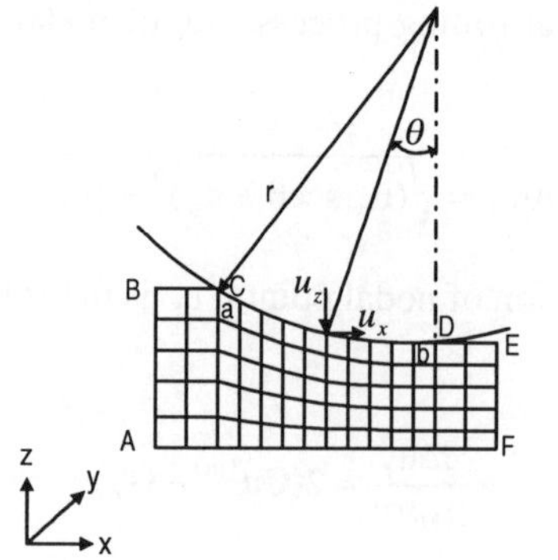

Figure 1. Schematic view of element for rolling process.

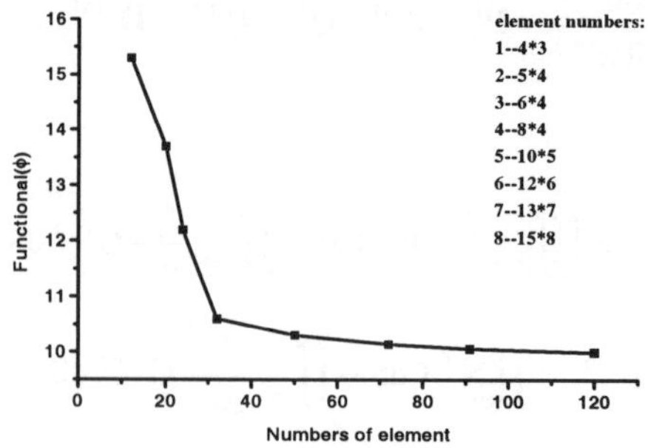

Figure 2. Effect of the numbers of element on the convergence result.

The ten cases, as shown in Table 1, are calculated in order to compare the effects of different initial velocity and Lagrangian multiplier on the convergence, the convergence results and the rate of convergence of the new computing algorithm. $\mathbf{u}_0$ and λ_0 are determined by the slab method [5]. The effect of the numbers of element on the convergence result is shown in Fig.2. Convergence of the results can be obtained for the numbers of element larger than 30. In the remaining part of this paper 10 and 5 are chosen as the numbers of elements along the x, z directions respectively.

The same convergence results of functional can be obtained using the new computing algorithm, and the rate of convergence of the new computing algorithm is rapid (Fig.3). The functional is close to the convergence value after only four iteration steps, and fifteen steps are needed at most for convergence under the convergence condition of $|\phi_n-\phi_{n-1}|/\phi_n<10^{-6}$. The rate of convergence for high initial velocity field (case 4 and 5) is faster than that for low velocity field (case 1 and 2).

Table 1. Initial Value

Case	Initial velocity field	Case	Initial Lagrangian multiplier
1	0.1 $\mathbf{u}_0$	6	0.1 λ_0
2	0.5 $\mathbf{u}_0$	7	0.2 λ_0
3	1.0 $\mathbf{u}_0$	8	1.0 λ_0
4	2.0 $\mathbf{u}_0$	9	5.0 λ_0
5	3.0 $\mathbf{u}_0$	10	10.0 λ_0

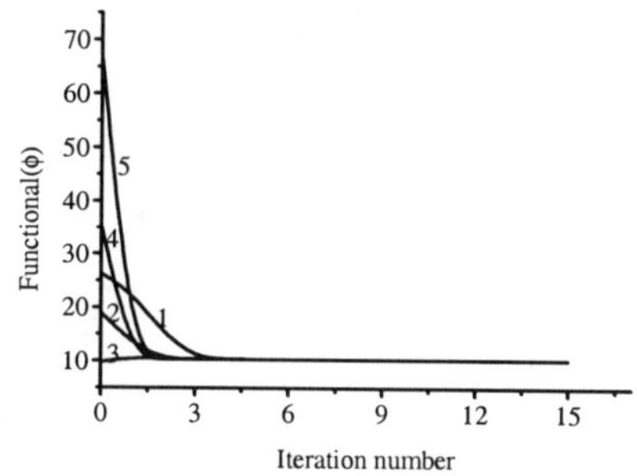

Figure 3. Variation of functional with respect to iteration number using the new algorithm

The variation of the velocity of nodal point A and F are shown in Fig. 4 and Fig.5 respectively when the new algorithm is used. Although the difference of the initial velocity is large for cases 1-5, the convergence results of the velocity of nodal point are identical. Fig.6 and Fig.7 also show that the convergence results of Lagrangian multiplier of element a and b are identical for the different initial values of the Lagrangian multiplier (cases 6-10). These indicate that the same solution can be obtained for the different initial guesses. The comparison between the measured and calculated rolling pressure is shown in Figure 8. The calculated values are in good agreement with the measured value [5].

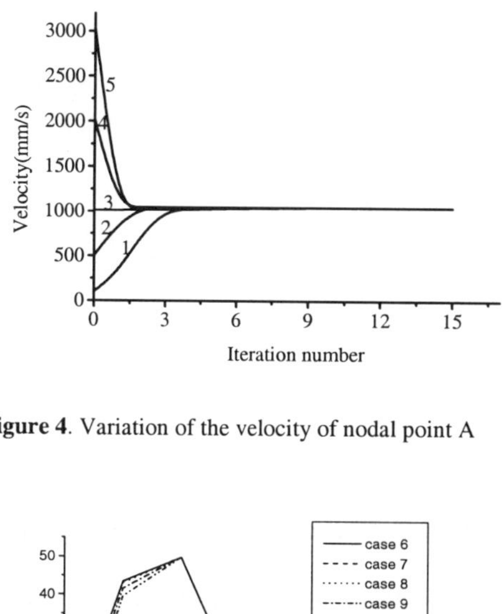

Figure 4. Variation of the velocity of nodal point A

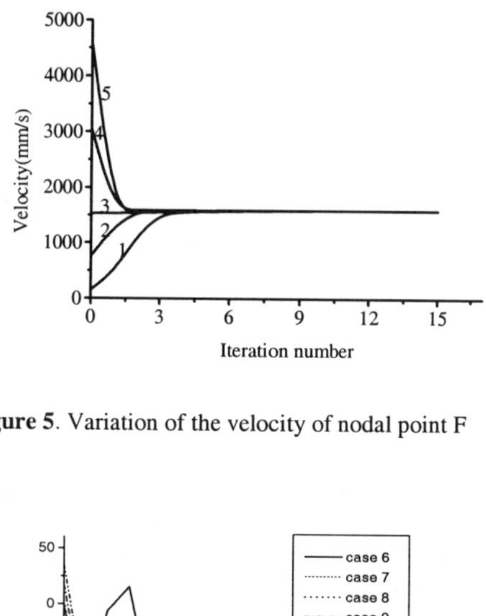

Figure 5. Variation of the velocity of nodal point F

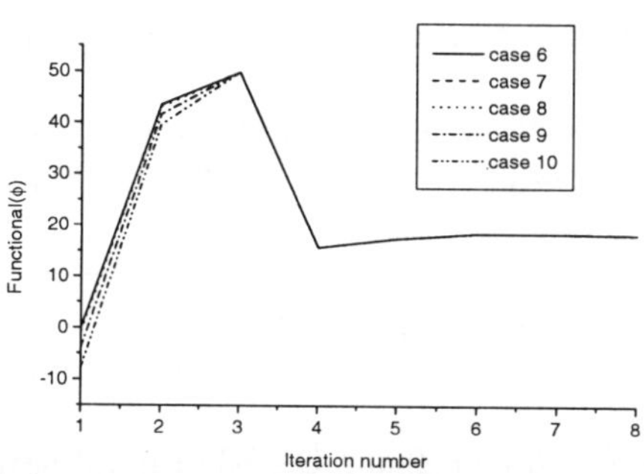

Figure 6. Variation of λ of element a.

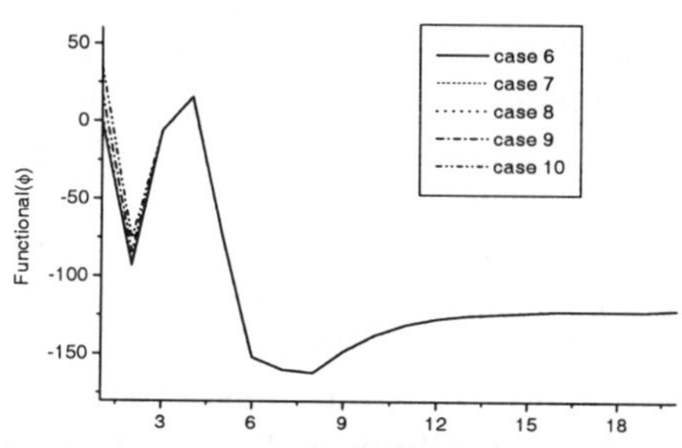

Figure 7. Variation of λ of element b.

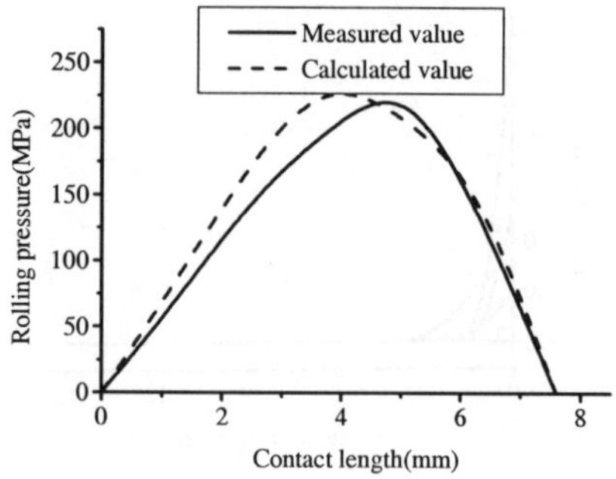

Figure 8. Comparison of the measured and calculated value

From the numerical tests it has been shown that this new computing algorithm has a rapid rate of convergence. The convergence is independent of the initial guesses and the same convergence results can be obtained for different initial guess. The algorithm is quite suitable for the simulation of the complex metal forming process because the choice of a good initial velocity field of the complex metal forming process is difficult in the traditional Newton Raphson method.

4 Conclusion

A new computing algorithm used in the rigid-plastic finite element method is presented. The iteration method is used to solve the non-linear stiffness equations. The linearization of the Newton Raphson Method is avoided, and the associated error from linearization is removed. The results of numerical tests show that the proposed algorithm has a good convergence, and the same convergence results can be obtained for different initial guesses. Therefore, the computing algorithm is suitable for the simulation of the complex rolling process.

References

1. Lee C. H., and Kobayashi S., New solutions to rigid-plastic deformation problems using a matrix method. *Journal of Engineering for Industry* **95** (1973) pp. 865–873.
2. Oh S. I., Finite element analysis of metal forming processes with arbitrarily shaped dies, *Int. J. Mech. Sci.* **24** (1982) pp. 479–493.
3. Zienkiewicz O. C. and Godbole P. N., Flow of plastic and visco-plastic solids with special reference to extrusion and forming processes, *Int. J. Numer. Methods Engrg.* **8** (1974) pp. 3–16.
4. Mori K., Osakada K and Oda T., Simulation of plane-strain rolling by the rigid-plastic finite element method, *Int. J. Mech. Sci.* **24** (1982) pp. 50–64.
5. Liu X. H., *Rigid-Plastic FEM and Its Application in Steel Rolling,* Beijing: Metallurgical Industry Press, 1994.
6. Jiang Z. Y., Xiong S. W., Liu X. H., Wang G. D. and Zhang Q., 3-D Rigid-Plastic FEM Analysis of the Rolling of a Strip with Local Residual Deformation, *J. Materials Processing Technology.* **79** (1998) pp. 109–112.

SIMULATION OF FRICTION VARIATION IN HOT STRIP ROLLING BY 3D RIGID PLASTIC FINITE ELEMENT METHOD

Z. Y. JIANG, A. K. TIEU, C. LU AND W. H. SUN

Mechanical Engineering, Faculty of Engineering, University of Wollongong, Wollongong, NSW 2522, Australia

E-mail: jiang@uow.edu.au

In both rolling theory and practice, the important factor that must be considered is friction on the surface of the roll-strip. In order to improve the accuracy of on-line controlling models for hot strip rolling, in this paper, the rigid plastic finite element method is employed, considering constant friction, friction variation and mixed friction models to simulate this strip rolling. The results of simulation, such as rolling pressure and forward slip, taking into account the friction variation with suitable models in the roll bite are in good agreement with the experimental values. The modelling of friction variation with suitable models and mixed friction in the roll bite is stable and applicable, which can improve the accuracy of on-line control models.

1 Introduction

In the hot strip rolling of steels, the temperatures are in the range of 850 – 1200 °C so that the deformation resistance is low and ductility high. Along the interface between the roll and strip, the tribology [1] occurs, that is friction, lubrication and wear, which has a significant influence on the dimensional accuracy and surface quality of the strip. One of these tribology components, friction has specially been considered in both rolling theory and practice, and it is an important factor affecting the rolling models of mechanics. The traditional approach is to assume that the frictional force in the roll bite is proportional to the normal force, with the friction coefficient remaining constant. However, previous researchers have shown that the friction varied in the roll bite [2], and it affected the rolling pressure and the accuracy of on-line control models. If the friction variation in the roll bite is not accounted for, it will incur a loss of accuracy in the roll gap model, and affect the thickness, shape and flatness of strip. In order to improve the accuracy of rolling models of strip, the friction variation between the roll and strip is investigated.

The 3D Finite Element Method (FEM), including rigid plastic/visco-plastic FEM, has been used in the analysis of strip rolling [3-4], shape rolling [5] and slab rolling [6, 7] in the past decades, and the authors have used the FEM to solve this special shape strip rolling [8, 9, 10]. However, the modelling of friction variation in hot strip rolling has not been reported until now. In this work, the authors use the rigid plastic FEM to solve the hot strip rolling taking into account the modelling of the friction variation in the deformation zone, and compare the computed rolling pressures, forward slip with the experimental values. The numerical simulation shows that it is effective and suitable for the modelling of the strip rolling with friction variation in the roll bite.

2 FEM Formulation

The slightly compressible materials model has been employed to solve many kinds of rolling problems by the 3D rigid plastic FEM [4]. According to the variational principle [9], the real velocity field must minimise the following functional:

$$\Phi = \iiint_v \bar{\sigma}\,\dot{\bar{\varepsilon}}\,dv + \iint_{S_f} \tau_f \Delta V_f\,ds + \iint_{S_K} \tau_K \Delta V_K\,ds \pm \iint_{S_v} Tvds$$
$$= \phi^p + \phi^f + \phi^j + \phi^t \tag{1}$$

where the first term on the right hand side is the work rate of plastic deformation (ϕ^p), $\bar{\sigma}$ is the equivalent stress and $\dot{\bar{\varepsilon}}$ is the equivalent strain rate. The second term on the right hand side is the work rate of friction (ϕ^f); ΔV_f is the relative slip velocity at the interface of the rolled material and the rolls where the frictional shear stress τ_f is applied. There are several kinds of frictional shear stress models in this analysis.

One is a constant friction model in the deformation zone as shown in Equation (2).

$$\tau_f = m_1 \frac{\sigma_s}{\sqrt{3}} \tag{2}$$

where m_1 is friction factor; σ_s is yield stress. The others are friction variation models. In the friction variation models, friction is a variable along the contact length of the deformation. In this study the frictional stress model [3] has been modified as follows:

$$\tau_f = K_i \frac{m_1 \sigma_s}{\sqrt{3}} \left(\frac{2}{\pi} \tan^{-1} \left\{ \frac{V_g}{V_a} \right\} \right) \tag{3}$$

where K_i is a coefficient describing the changes of frictional shear stress in the deformation zone, which has two parameters K_1 and K_2. Here K_1 is for the forward zone and K_2 for backward zone; m_1 and σ_s have the same values as in Equation (2); V_a is a small positive constant. The relative slip velocity, V_g, between the rolled material and roll is calculated by

$$V_g = \sqrt{(v_x \sec\beta - V_R)^2 + v_y^2} \tag{4}$$

where v_x and v_y are the velocity components in the x and y directions respectively, β is the angular position of the node, V_R is the tangential velocity of the roll. ΔV_f is the relative slip velocity between the rolled material and roll, and can be calculated by

$$\Delta V_f = \sqrt{(v_x \sec\beta - V_R)^2 + v_y^2 + k_2^2} \tag{5}$$

where k_2 is a small positive constant to prevent second singular point to occur [4].

Another friction variation model is as follows:

$$K_1 \frac{m_1 \sigma_s}{\sqrt{3}} \left(1 - \frac{V_g^{1.5}}{|\Delta V|_{f\max}} \right) \text{(Forward slip zone)};\ -K_2 \frac{m_1 \sigma_s}{\sqrt{3}} \left(1 - \frac{V_g^{1/3}}{|\Delta V|_{b\max}} \right) \text{(Backward slip zone)} \tag{6}$$

where $|\Delta V|_{f\max}$ is the maximum relative velocity between the roll and strip in the forward slip zone and $|\Delta V|_{b\max}$ is the maximum relative velocity between the roll and strip in the backward slip zone. When the forward slip zone is the sticking friction, the friction variation model is as given in Equation (7).

$$K_1 \frac{m_1 \sigma_s}{\sqrt{3}} \text{ (Forward slip zone)};\ -K_2 \frac{m_1 \sigma_s}{\sqrt{3}} \left(1 - \frac{V_g^{1.5}}{|\Delta V|_{b\max}} \right) \text{ (Backward slip zone)} \tag{7}$$

The distributions of these frictional shear stress models are shown in Figure 1 (k is the shear yield stress). When the sticking friction occurs in the vicinity of the neutral point, there is a mixed friction in the roll bite. From Figure 1d, it is seen that the frictional shear stress is $\pm K_i m_1 k$ in the vicinity of the neutral point, where there is a sticking friction zone (see Fig. 2), but it is the sliding friction with non-linear curve beyond this zone.

The third term on the right hand side of Equation (1) is the additional shearing work rate ($\dot{\phi}^j$) at a surface on which there is a velocity discontinuity. ΔV_K is the velocity discontinuity across the velocity discontinuity surface s_K within the volume V, τ_k is the shear yield stress, $\tau_k = k$ (k is the shear yield stress). The fourth term on the right hand side is the work rate ($\dot{\phi}^t$) of tension. T is the tension and v is the velocity of the cross section with tension. Here "-" indicates the forward tension, and "+" the backward tension. The standard method of discretization [4] is used, incorporating eight-node, isoparametric quadrilateral elements.

$$\sigma_{ij} = \frac{\overline{\sigma}}{\dot{\overline{\varepsilon}}}\left(\frac{2}{3}\dot{\varepsilon}_{ij} + \delta_{ij}\left(\frac{1}{g} - \frac{2}{9}\right)\dot{\varepsilon}_v\right) \tag{8}$$

where δ_{ij} is the Kronecker delta, $\dot{\varepsilon}_{ij}$ the strain rate tensor and $\dot{\varepsilon}_v$ volumetric strain rate.

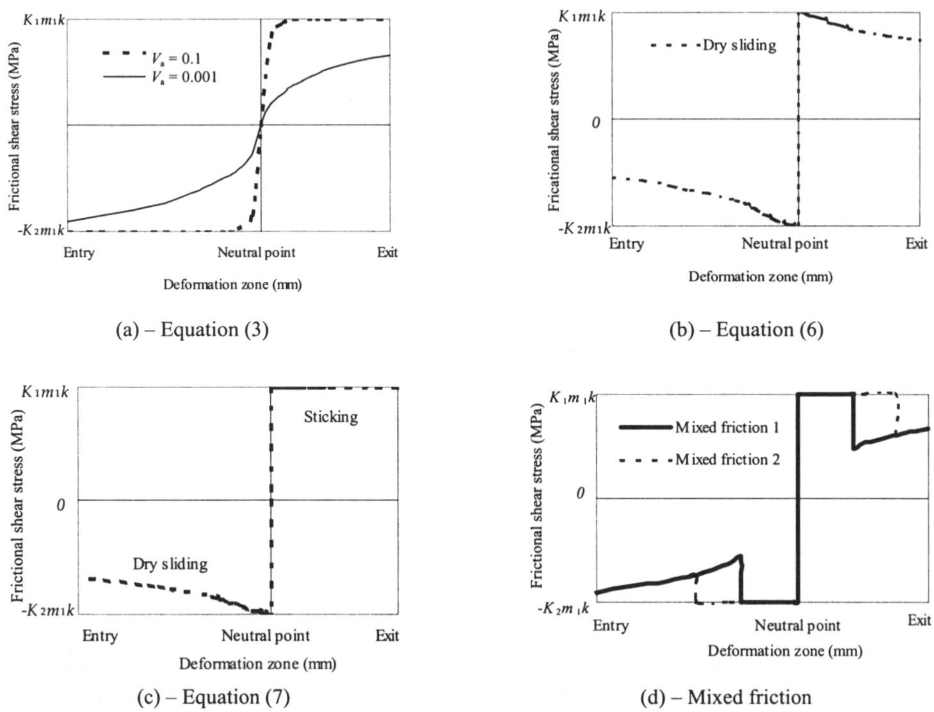

Figure 1. Frictional shear stress models with friction variation.

3 Simulation Conditions

3.1 *FE mesh*

As the deformation is symmetric about both the *y-x* and *z-x* planes, only one quarter of the deforming workpiece as shown in Figure 3, was studied. Isoparametric hexahedral elements with eight Gauss points were used throughout the strip, including the corner at entry to the roll gap. The number of elements along the *x*, *y* and *z* directions respectively, can vary according to the different conditions of simulation. The simulations were performed on a Pentium III 450 computer, and the expenditure of the computer running time will be examined.

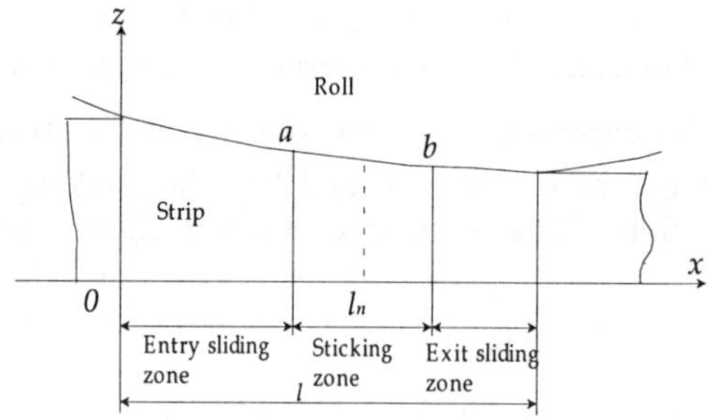

Figure 2. Distribution of friction zones.

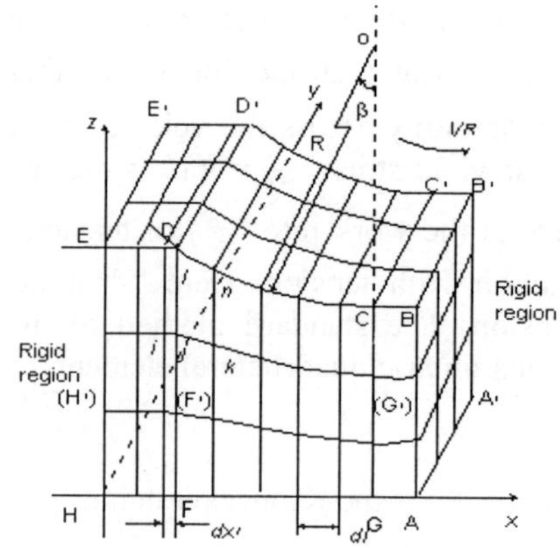

Figure 3. FE mesh used for analysis of hot strip rolling.

3.2 *Material Flow Model*

Aluminum was used to simulate the strip rolling. The constitutive model used to describe the behavior of this aluminum was

$$\sigma = 31.7 + 113.77\,\bar{\varepsilon}^{0.5539} \quad \text{(MPa)} \tag{9}$$

where $\bar{\varepsilon}$ the plastic strain. In the simulation, the following conditions were investigated: strip: initial thickness H_0= 2.84-2.92 mm, width B_0 = 50 mm; reduction: 13.07-29.79 %; roll diameter: D_e = 110 mm; roll speed is 35 rpm and there is the friction variation between the roll and strip [4].

4 Results

4.1 *Effect of Friction on the Results of Simulation*

When the friction factor m_1 = 1.0, V_a = 0.1 and $K_1 = K_2$ = 1.0 (Equation (3)), and mesh is 10×5×4 (10 rows in *x* direction, 5 rows in *y* direction and 4 rows in *z* direction, see Figure 3), the comparison of the calculated rolling pressures and forward slips for constant friction and friction variation with the measured values [4] are shown in Figures 4 and 5. It can be seen that the rolling pressures and forward slips increase with the thickness reduction, and the calculated rolling pressures and forward slips with the constant friction are close to the values obtained with the friction variation models (Equations (3) and (7)). However, there is a big difference between the calculated rolling pressure with the Equation (6) and experimental values, this means that the friction variation model described by Equation (6) is not suitable in this study. It is also seen from the Figure 4 that the calculated rolling pressure is quite close to the measured value when the reduction is 18-23%, and the difference between the calculated rolling pressure and measured values becomes larger when the reduction is less than 18% and over 23 %.

However, the time expenditure of running the program is significantly different for constant friction and friction variation models which is shown in Figure 6. It is found that the computer running time decreases for the friction variation model Equation (3) when the reduction increases, but it increases for the models of Equations (6) and (7), and it is smaller than that for the constant friction model when the reduction is 15–25 %. This result indicates that the simulation with the friction variation models can improve the convergence than that with the constant friction model.

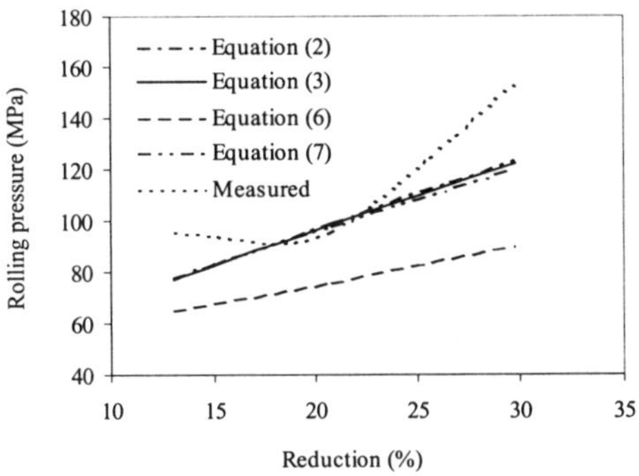

Figure 4. Effect of reduction on rolling pressure.

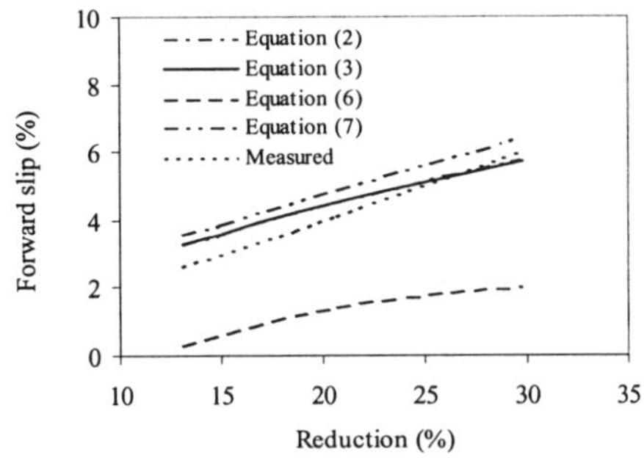

Figure 5. Effect of reduction on forward slip.

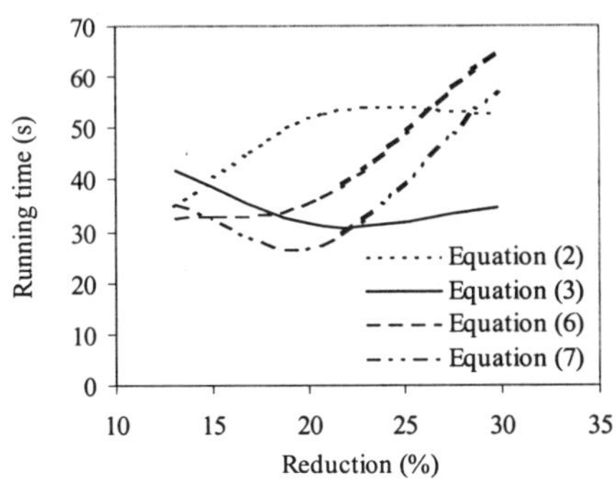

Figure 6. Relationship between reduction and the program running time.

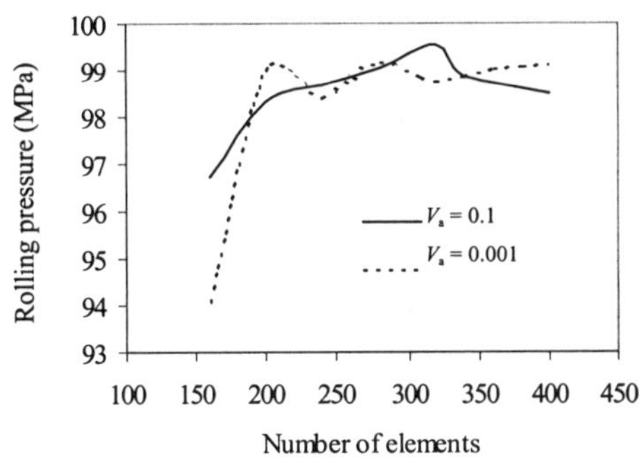

Figure 7. Effect of the number of elements on rolling pressure.

Figure 7 given the relationship between the rolling pressure and constant V_a. It is seen that rolling pressure is only slightly effected by different V_a and the value of the rolling pressure is more stable when the number of elements is over 200. The maximum relative error is 1.06 % when the number of elements is more than 200. Therefore, it can obtain a stable result when V_a changes and the number of elements is over 200.

4.2 *Effect of Mixed Friction on the Results of Simulation*

Mixed friction occurs in strip rolling, especially in hot strip rolling. When the $m_1 = 0.8$, mesh is 15×5×4, the effect of mixed friction and constant friction on the results of the simulation is shown in Table 1. It is seen that the results of the simulation with constant friction such as the rolling pressure, forward slip and the expenditure of running time are higher than that with the mixed friction variation models. And the results of mixed friction (the friction shear stress in the vicinity of neutral point is more than that in the other zone) variation model 1 (see Fig. 1d) are quite close to the measured values (rolling pressure 96.04 MPa and forward slip 4.1 %) [4], and the program running time is the smallest. Therefore, the mixed friction mode 1 is more suitable in the simulation of practical strip rolling process.

Table 1. Comparison of the results of simulation with different mixed friction models.

Friction model	Rolling pressure (MPa)	Forward slip (%)	Spread (mm)	Running time (s)
Constant friction	96.57	4.30	0.121	234.47
Mixed Friction 1	96.00	4.20	0.123	141.33
Mixed Friction 2	95.53	4.30	0.124	149.78

5 Conclusions

When the friction variation models are employed in the modelling of hot strip rolling, the calculated results such as rolling pressure, forward slip are in good agreement with the experimental values. These models can reduce the expenditure of computer running time and improve the convergence of the simulation better than the constant friction case when the reduction is 15 – 25 %. The modelling of the mixed friction variation can produce more accurate models that can improve the accuracy of the simulation results, and the computer running time is reduced significantly. This model is also applicable and effective in the modelling of hot strip rolling.

6 Acknowledgements

This work was supported by the Australian Research Council (ARC)/University of Wollongong Small Grant, Australia.

References

1. Munther P. and Lenard J.G., Tribology during hot, flat rolling of steels, *Cirp Annals* **44**(1), 1995, pp213-216.
2. Liu Y.J., Tieu A.K. Wang, D.D. and Yuen W.Y.D., Friction Measurement in Cold Rolling, *J. Mater. Proc. Technol.* **111**(1-3), 2001, pp142-145.
3. Kobayashi S., Oh S.I. and Altan T., *Metal Forming and Finite Element Method*, New York: Oxford University Press, 1989.
4. Liu X.H., *Rigid-Plastic FEM and Its Application in Steel Rolling*, Beijing: Metallurgy Industrial Press, 1994.
5. Yanagimoto J. and Kiuchi M., General Purpose FEM Simulator for the Three-dimensional Deformation Analysis of Strip, Bar/Wire and Shape Rolling Processes, *Proc. of the 7th Int. Conf. on Steel Rolling*: 1998, pp278–283. Tokyo: The Iron and Steel Institute of Japan.
6. Xiong S.W., Liu X.H., Wang G.D. and Zhang Q., Simulation of Vertical-Horizontal Rolling Process During Width Reduction by Full Three-Dimensional Rigid-Plastic Finite Element Method, *J. Mater. Eng. & Performance* **6**(6), 1997, pp757-765.
7. Jiang Z.Y., Hu W.P., Thomson P.F. and Lam Y.C., Analysis of Slab Edging by a 3-D Rigid Viscoplastic FEM with the Combination of Shifted Incomplete Cholesky Decomposition and the Conjugate Gradient Method, *J. Finite Elements in Analysis and Design* **37**(2), 2001, pp145-158.
8. Jiang Z.Y., Xiong S.W., Liu X.H., Wang G.D. and Zhang Q., 3-D Rigid-Plastic FEM Analysis of the Rolling of a Strip with Local Residual Deformation, *J. Mater. Proc. Technol.* **79**(1-3), 1998, pp109-112.
9. Jiang Z.Y., Liu X.L., Liu X.H. and Wang G.D., Analysis of Ribbed-Strip Rolling by Rigid-Viscoplastic FEM, *Int. J. Mech. Sci.* **42**(4), 2000, pp693-703.
10. Jiang Z.Y. and Tieu A.K., Modelling of Rolling of Strips with Longitudinal Ribs by 3-D Rigid Visco-Plastic Finite Element Method, *ISIJ Int.* **40**(4), 2000, pp373-379.

SEMI-CONTINUOUS EXTRUSION OF ALUMINIUM TUBES

SINNATHAMBY THIRUVARUDCHELVAN AND TAY MENG LEONG
Manufacturing Engineering Division,
School of Mechanical and Production engineering,
Nanyang Technological University,
Nanyang Avenue, Singapore 639798.
E-mail: mthiru@ntu.edu.sg

A semi-continuous extrusion technique for the manufacture of metal tubes has been developed recently: In this method a tubular billet gripped in a urethane clamp is advanced repeatedly to extrude through a die. A machine utilizing this principle has been built and tested. The extrusion cycle consists of clamping the billet; advancing it to extrude a pre-determined length; releasing the clamping force (freeing the tube to stay onto the die); and, finally retracting the clamp to its initial position. A mandrel sliding inside the billet prevents it from collapsing under high radial pressure caused by clamping. The clamped billet, gripping the mandrel, is moved toward the extrusion die. Extrusion is stopped before the front-end of the mandrel reaches the deforming metal. At the end of each extrusion cycle the mandrel is free to slide, and it is retracted by a spring attached to its back-end, the spring being stretched by the mandrel moving forward with the billet during extrusion. Results from the investigations conducted are presented in this paper.

1 Introduction

In the drawing of metal wires and tubes, the possible reduction in cross sectional area is limited, as the maximum possible draw stress is the ultimate tensile strength of the metal. To obtain larger reductions in cross sections, several drawing stages in series are used. In extrusion, however, the extrusion pressure is not limited by the strength of the metal and, substantial reduction in cross sectional area is usually achieved. During the last forty years several attempts have been made to develop continuous extrusion techniques. More attention has been directed at the easier-to-implement semi-continuous extrusion process.

In a continuous hydrostatic extrusion process [1], the extrusion pressure is generated by viscous drag along a long container using a very viscous fluid: this process is limited to soft metals. In another process [2], a large rotating wheel having a U-shaped groove around its periphery forms a continuously moving container. A stationary shoe carrying the die is pressed against the rotating wheel. This closes up the U-shaped groove and forms the extrusion container. The feedstock, when inserted between the shoe and the U-groove from the open end, is dragged automatically into the container by friction: friction from the U-shaped groove drags the feedstock into the container, while that from the shoe opposes the motion of the billet. The beneficial frictional drag from the wheel is larger than the resistance from the shoe having less contact area. The extrusion pressure thus built up by the net frictional forces in the container help extrude the rod through the die at large extrusion ratios. Tubes may also be extruded with or without a flexible mandrel inside the bore, though no research in this area has been reported in the literature.

A comprehensive review of semi-continuous extrusion processes has been given in reference [3]. Several attempts have also been made to make the hydrostatic extrusion process semi-continuous [4,5,6]. In conventional extrusion, if the container moves a little faster than the billet and in the same direction, friction will then drag the billet towards the die (building up extrusion pressure) and only a small extrusion pressure is needed at the back of the billet. Based on this principle, and making use of a urethane clamp, a machine was developed for semi-continuously extruding aluminium rods [7,8]. For rod extrusion an extrusion container was used and the friction between the billet and the container helped build-up the extrusion pressure at the die. This machine has been modified to perform the present tube extrusion process.

The present tube extrusion process is open-die extrusion without a container. The extrusion pressure and, thus the possible extrusion ratio are limited by the possibility of the billet bulging or buckling. In an earlier investigation the tube was extruded without the support of a mandrel inside [9]: The process worked well but the clamping pressure was limited to 11 MPa, but the tube collapsed inside the clamp when large clamping pressure is higher. To overcome this limitation a floating mandrel with a conical end resting against the deforming metal was tried. During the tests it was found that frictional force between the mandrel and the tube inside the clamp forced the mandrel against the deforming metal. This caused excessive stresses at the deforming zone resulting in failure of the process.

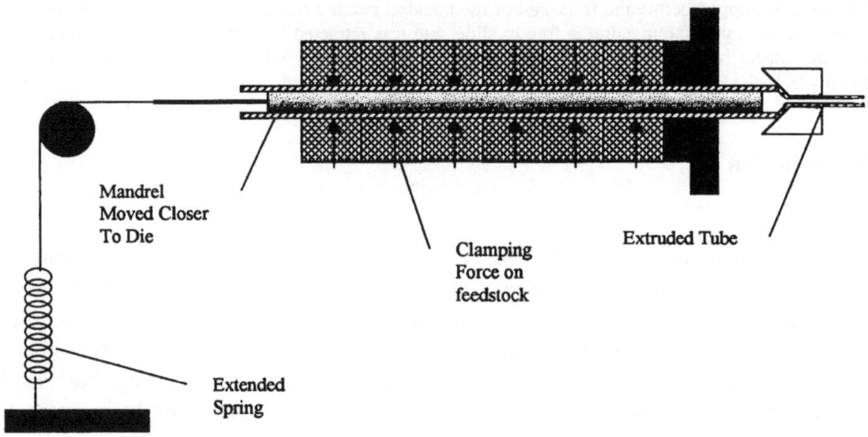

Figure 1: Schematic diagram of the semi-continuous tube extrusion process

To overcome the above problems the present technique was introduced. Here a mandrel supports the tube from collapsing with the radial clamping pressure exerted on it. However, the extrusion is stopped before the front end of the mandrel reached the deforming metal in each operating cycle, as shown in Figure 1. This process is a viable alternative to tube drawing or tube sinking. The limiting percentage reduction in open-die extrusion and drawing is more or less the same. Lengths of tube similar to those drawn on a draw bench can be extruded in several extrusion cycles using the present machine.

2 Theoretical considerations

The open-die extrusion of tube is similar to tube sinking. Any change in thickness of the tube can be taken into account. Using the analysis of Hoffman and Sachs [10], ignoring the redundant work done, the extrusion pressure in the present process can be given by,

$$\frac{p}{1.1Y}\left(\frac{r_i}{r_o}\right)^B = \frac{1+B}{B}\left[1-\left(\frac{r_1}{r_0}\right)^B\right] \tag{1}$$

Where,

B $= \mu_0 \cot \alpha$

p extrusion pressure

r_i the inner radius of the emerging tube

r_o the outer radius of the tubular billet

α the die semi-angle

μ_0 coefficient of friction at the die

Y yield strength of the tube

A more accurate expression for the extrusion pressure may be obtained from Avitzur's upper bound analysis given in reference [11]. This analysis makes use of three regions in which the velocity fields are continuous. In the entry and exit zones the velocities are uniform with axial component only. This analysis takes into account redundant work and friction. The wall thickness can vary depending on the geometry and friction. A versatile flow pattern permits wall-thickness variation, and has several variables when optimized yields a realistic solution.

The maximum axial force that can be generated by the clamp on the feedstock is [12],

$$F_{\max} = \mu \pi d_1 L_1 p_0 \qquad (2)$$

Where,

μ is the coefficient of friction between the feedstock and the urethane in the clamp,

d_1, d_2 are the outer and the inner diameters of the tubular feedstock, respectively,

L_1 is the length of the urethane in contact with the feedstock, and

p_0 is the initial axial clamping pressure exerted on the urethane.

Thus, the maximum extrusion pressure that can be generated on the feedstock is,

$$\frac{F_{\max}}{\pi\left[\dfrac{d_1^2 - d_2^2}{4}\right]} = \frac{4\mu d_1 L_1 p_0}{\left(d_1^2 - d_2^2\right)} \qquad (3)$$

3 Extrusion equipment

A photograph of the semi-continuous extrusion machine with its control unit is shown in Figure 2. At the beginning of an extrusion cycle, six pull-type hydraulic cylinders, equally spaced around the circumference of the cylindrical clamp, exert axial pressure on the annular urethane rod, causing it to grip the tube to be extruded [12]. The billet in turn grips the mandrel inside. The tube passes freely inside the urethane tube when no axial pressure is applied on the urethane. With the tube and the mandrel firmly gripped, the clamp-cylinder assembly is advanced by a hollow hydraulic cylinder (through which the billet passes) towards the extrusion die [9]. After a pre-determined length of the tube has been extruded, the gripping force is released. The clamp-cylinder-assembly is then moved back, leaving the tube to stay onto the die. The mandrel is pulled back to the initial position by a spring attached via a string to its back end. The above extrusion cycle is repeated automatically using an electro-hydraulic logic circuit.

The extrusion system consists of two sub-systems, namely the hydraulic sub-system and the electronic sub-system (PLC). In the hydraulic sub-system, there are two groups of cylinders, namely the hollow extrusion cylinder and the six clamp cylinders. An Enerpac 10000 psi (69 MPa) hydraulic pump powered all the hydraulic cylinders.

A programmable logic controller (PLC) is used to repeat the extrusion cycle automatically. It enabled the system to run in either the manual or the automatic mode. In the manual mode, the user has control over four press buttons. These buttons are: Advance extrusion press; Retract extrusion press; Advance clamp; and Retract clamp. In the auto mode, the machine can extrude in a fully automatic fashion aluminium tubes of length 1.5m or more, leaving behind 250mm of unextruded tube as discard.

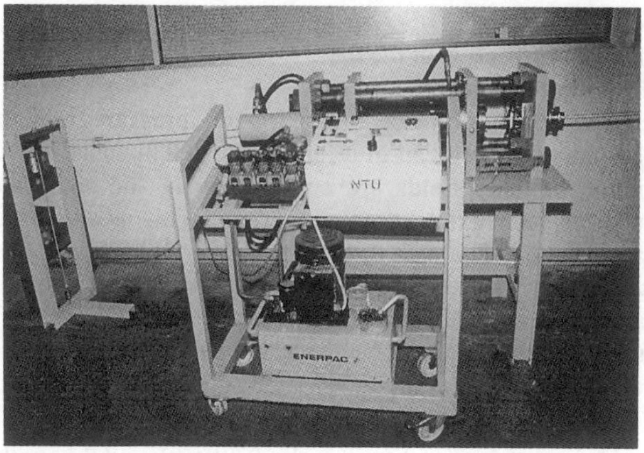

Figure 2: Photograph of the semi-continuous tube extrusion machine

The mandrel was made of high-speed tool steel heat-treated to obtain good wear resistance and toughness: It is 800mm long – more than sufficient to support the tubular feedstock within the clamp. A smaller diameter rod is screwed at the back of the mandrel to ease the insertion of the mandrel inside the feedstock. The theoretical minimum allowance of the gap needed between the mandrel and the feedstock was determined to be 0.041 mm. However, an allowance of 0.05 mm was allowed and the mandrel diameter was set at 17mm. In order to pull back the mandrel at the end of each cycle of extrusion, a mandrel retraction unit was incorporated (to the left in Figure 2). This unit must have a pulling force larger than the frictional force between the aluminium tube and the mandrel, when the clamping pressure is released. This frictional force is created by the unevenness of the bore of the tubular feedstock, as well as by any lack of straightness of it. From experiments conducted for this purpose a force of 75 N was found to be sufficient to retract the mandrel. The spring used in this unit pulls the mandrel with a force of 120 N when extended to a length of 80mm, which is less than the extrusion stroke.

In the initial trials the extruded product emerged bent in a random manner. To overcome this problem, a straightener fitted downstream of the die prevents the extruded tube from bending as it emerges from the die. The main components of this machine consist of annular urethane rods for clamping contained in a clamping unit actuated by six pull cylinders, a 30-ton (300 kN) hollow hydraulic cylinder to extrude the tube, an extrusion die and a load cell. These are mounted on the horizontal press as seen in Figure 2. The clamping unit is carried on the sliding unit, which slides on four cylindrical tie bars of the press. The load cell is mounted behind the die onto the end plate of the press through which the product emerges. The pull cylinders and the hollow hydraulic cylinder are connected to the electro-hydraulic system. More details of the various features of this machine are found in references [7,9]. At the back end of the machine the unit for retraction of the mandrel can be seen (Figure 2): a string attached to the back of the mandrel passes vertically down at a pulley attached to the retraction unit. The other end of the string is attached to a spring at the base of the retraction unit.

For extruding the tube a clamping pressure of about 2000 psi (13.8 MPa) is first applied on the urethane by the pull cylinders. The hollow cylinder is then advanced to start the extrusion. The unsupported length of billet is about 120 mm. Grease is applied to the die to lubricate the extrusion process. When a pre-determined length of the tube is extruded the main cylinder is stopped. The clamping cylinders are then allowed to unload and release the clamping of the tube. The hollow cylinder is then retracted to its starting position. The tube is held in position (without moving back), while the sliding unit carrying the clamp retracts with the hollow cylinder. The mandrel also retracts to the initial position. The machine is now ready for another cycle of extrusion.

Figure 3: Photograph of a tube that buckled during extrusion

4 Results and discussion

Experimental work was carried out with 6063T4 aluminium tubes with the machine on manual mode. Two sets of aluminium tubes were used as in the Table 1:

Table 1

Tubular billet diameter		Tubular product diameter	
outer	inner	outer	inner
19.05	16.45	14.05	10.70
19.06	17.10	14.10	11.54

The semi-angle of the die used was 15 degrees. In the tube extrusion experiments conducted the clamping pressure on the tube was 1950 psi (13.4 MPa). More details of the urethane clamp can be found in reference [10]. To ensure high coefficient of friction between the urethane clamp and the tube, rosin powder was applied on the tube before inserting into the machine. Average length of feedstock extruded per stroke was 80 mm. The thickness of the thicker tube increased from 1.30 mm to 1.69 mm, and the thinner tube thickness increased from 0.98 to 1.28 mm, respectively after extrusion. The increase in thickness is about 30% in both cases. The extrusion force was measured using a load cell installed on the machine. The extrusion force measured with the 0.98 mm thick tube was 10.12 kN, while that for the thicker tube was 13.2 kN.

The photograph in Fig. 3 shows a buckled tube. One of the main causes of this buckling is the eccentricity between the feedstock axis and the die axis. When the feedstock is not aligned properly with the die axis, the tube may buckle. With careful alignment of the die axis with the feedstock axis this problem was eliminated. High frictional resistance at the die may also cause the billet to buckle. With good lubrication of the die-metal interface the resistance is minimal and buckling avoided.

If the friction condition at the die is not steady, the extrusion force needed may be larger than the buckling load of the unsupported portion of the tube and buckling may occur, especially at the initial stages of the extrusion cycle. However, an unsupported billet length of 120 mm was found to be satisfactory under normal conditions.

Operation of the machine in the automatic mode was found to work well, thus making it feasible for industrial application, and compares favourably with tube sinking.

5 Conclusion

An automated semi-continuous extrusion process using a urethane clamp for the extrusion of aluminium tube was found to be feasible. To enhance the process friction at the urethane clamp is increased using rosin powder, while the extrusion die is lubricated with grease. Buckling of the tubular billet is prevented by limiting the unsupported length, and by minimizing the friction at the die-metal interface. Thickness of the tube increased by about 30 % after extrusion. By automatically repeating the extrusion cycle substantial lengths of the feedstock can be extruded.

References

1. Fuchs F. J., Continuous hydrostatic extrusion makes conductor wires, Wire Journal, November (1973), pp. 53-57.
2. Green, D., Continuous extrusion forming of wire sections, J. Inst. Met., **100** (1972), pp. 295-300.
3. Said, M. E., Continuous extrusion processes - A review, Proceedings of the 1st Cairo University Conference on Machine Design and Production, **1**, Pergamon Press, (1981), pp. 447-457.
4. Alexander J. M., and Lengyel, B., Semi-continuous extrusion of wire, Proc. Inst. Mech. Eng., **180**, (1966), pp. 317-328.
5. Pugh, H. Ll. D. and Donaldson, C. J. H., Hydrostatic extrusion - A review, Annals of CIRP, **21**, (1972), pp. 167-186.
6. Green D., and Slater, H. K., Augmented hydrostatic extrusion of continuous bar, Proc. Inst. Mech. Eng., **182** (1968), pp. 217-225.
7. Thiruvarudchelvan, S., Friction-aided semi-continuous extrusion of aluminium rod using a urethane clamp, J. Mater. Process. Technol., **29**, (1992), pp. 159-171.
8. Thiruvarudchelvan, S., Semi-continuous extrusion of round bars, J. Mater. Process. Technol., **31**, (1992), pp. 441-443.
9. S.Thiruvarudchelvan and M.L.Tay, Automated Semi-Continuous Open-Die Extrusion of Aluminium Tubes, Pacific Conference on Manufacturing '96, Korea, pp. 654-658.
10. Hoffman, O., and Sachs G., Introduction to the Theory of Plasticity for Engineers, McGraw-Hill, New York, (1953), pp. 163-175.
11. B. Avitzur, Handbook of Metal Forming Processes, John Wiley, N.Y., (1983), pp. 480-502.
12. Thiruvarudchelvan, S., A urethane clamp as a metal forming tool, J. Mater. Process. Technol., **38** (1993), pp. 491-500.

A GEOMETRIC MAPPING APPROACH TO STRAIN PREDICTION IN SHEET METAL FORMING

LAWRENCE TSAI

Department of Engineering , Australian National University , ACT0200, Australia
E-mail: Lawrence.Tsai@faceng.anu.edu.au

MICHAEL CARDEW-HALL

Department of Engineering , Australian National University , ACT0200, Australia
E-mail: mch@faceng.anu.edu.au

In this paper we investigate the use of Tchebychev nets to determine shear strains that occur in the mapping of flat blanks in sheet forming processes to free form surface shapes with Gaussain curvature. A draping algorithm has been adapted which deforms an inextensible net over a desired forming shape. Such a net deforms through shearing of the elements. However, there is an overall loss of area. Therefore, techniques of adjusting it to maintain continuity of area are presented. Such a modified net is able to give reasonable predictions of strain and is appropriate for assessing formability of a part early in the design process. A demonstration system based on CAD based NURBS surfaces has been generated for comparison to experimental results.

1 Introduction

Die design, development and try-out are critical elements in the lead time for new vehicles. The knowledge, tools and expertise of the die designer can make the critical difference between a good die that performs in a robust manner or one that gives continual problems in try out and throughout its working life. In this paper we investigate the application of fundamental geometry, particularly mapping techniques and differential geometry, to provide approximate and rapid tools to assist in the early tool design stages; these are based on the mathematics of surface transformation and allow the designer to explore quantitatively the intuitive concepts on which present and future die design is and will be based.

Many tool designers approach the problem by considering the blank as a square grid or mesh and imagining how this would be deformed to reach the final surface. In this paper we provide a means of achieving this transformation in a computer aided system, bearing in mind that such a deformation would not satisfy all the conditions imposed by materials, friction and other variables. Nevertheless it gives a starting point in the design process and provide information that the designer can understand in a physical and practical sense.

As a start, we adapt a mapping method [1] to map the initial square, orthogonal net or mesh on the blank onto the specified design surface as a Tchebychev net. Such a net deforms by distortion of the net elements without change in the line length between nodes [2]. Physically this is the same as a square fish net of inextensible strings. Such a net can usually be draped over a surface such that it conforms to it completely [3]. The mathematical properties of this net is well known [4] and it provides a convenient method of specifying the surface in terms of parametric points (u,v), that, in a sense, reduce the deformation problem from three to two dimensions in (u,v) space. A disadvantage of the Tchebychev net is that elements deform from a square to a rhombus with surface area decreasing with deformation. Real sheet, on the other hand, tends to deform by stretching or shear with minimal change in surface area.

To obtain more realistic strain distributions, we use geodesic properties of surface and adjust the net under a hypothesis that the elements deform keeping the surface area unchanged. As an example we apply the mapping approach to a dome hemisphere die surface.

2 Geodesic and geodesic polar coordinates

Geodesic is a fundamental concept in mathematics to generalize the idea of a "straight line" to curves on arbitrary surfaces and in general manifolds. There is a formal analogy between the properties of "straight lines" in non-Euclidean geometry, and the properties of "geodesics" on curved surfaces. In Riemannian geometry a geodesic plays the role of a straight line [5]. As a simple definition geodesic is the straightest possible path constrained to lie on a curved surface, the famous example is a great circle on a sphere. Given two points on a surface, the shortest path on the surface connecting them is a geodesic, and the path length is the distance between these two points on the surface. From a point on the surface there is a geodesic in every direction. A curve in which all points are equal distance from a center forms a geodesic circle. The geodesics from the center is called the radial geodesics (Fig.1). A surface which is a complete C^{∞} Riemannian 2-manifold has a global geodesic polar coordinate system. This means that geodesics emanating from **p** can be extended indefinitely without ever running into each other, and we can identify each point of the surface uniquely by telling how far (r) it is from **p**, and the direction (θ) you have to go to get there [6].

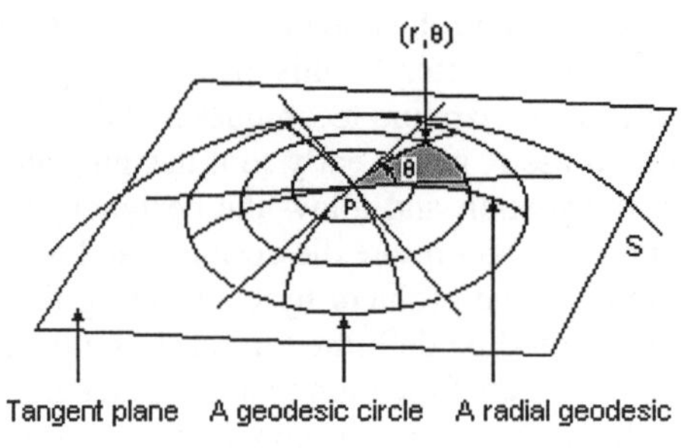

Figure 1 Geodesic polar coordinate system on a surface

Geodesics have many interesting properties:

6. Geodesic's curvature is due entirely to the surface.
7. Geodesic has no tangential curvature.
8. Geodesic curvature vector coincides with surface normal vector.
9. Geodesic transports a vector to itself.
10. Homogenous: every geodesic sub-segment is a geodesic.

3 Mapping approach

We establish a geodesic polar coordinate system on a free form surface as showed in (Fig.1). We assume that a given circle on the blank will be mapped to a geodesic circle on the surface which encloses the same area on the blank and the surface respectively. A radius of the circle on the blank will be mapped onto a radial geodesic on the surface at the same polar angle. The ratio of the length of the radial geodesic (r) to the radial length of the circle on the blank (r_b) is the extensional rate in the radial direction and can be calculated from the equal area hypothesis. Therefore, (r) is the function of (r_b). A point (r_b,θ_b) on the circle will be mapped into a point (r,θ) on the geodesic circle.

$$r = f(r_b)$$

$$\theta = \theta_b$$

Thus, we can map any pattern, for instance, a grid, on a blank onto a free form surface to simulate sheet deformation.

4 Mapping a dome

In this section we apply the mapping approach to forming a round blank (to avoid complicated boundary conditions), a disc, into a dome hemisphere (Fig.2a). Conveniently, we establish a plane polar coordinate system on the disc (Fig.2b) and a spherical polar coordinate system on the dome (Fig.2c). This is actually a geodesic polar coordinate system. From the rotationally symmetric geometry characteristics of both the blank and the dome we can imagine the blank will deform by bending and stretching the radials isometrically. A given circle on the blank will be mapped onto a circle, a geodesic circle, on the dome. Each circle encloses the same area on the blank and the dome respectively. A radius of the circle on the blank will be mapped onto a part of great circle through the pole of the dome. This is a radial geodesic. The ratio of the length of the radial on the blank (r_b) to the length of the great arc (r_d) is the extensional rate in the radial direction and can be calculated from the constant area hypothesis. A point (r_b,θ_b) on the circle of the blank will be mapped onto a point (r_d,θ_d) on the circle on the dome.

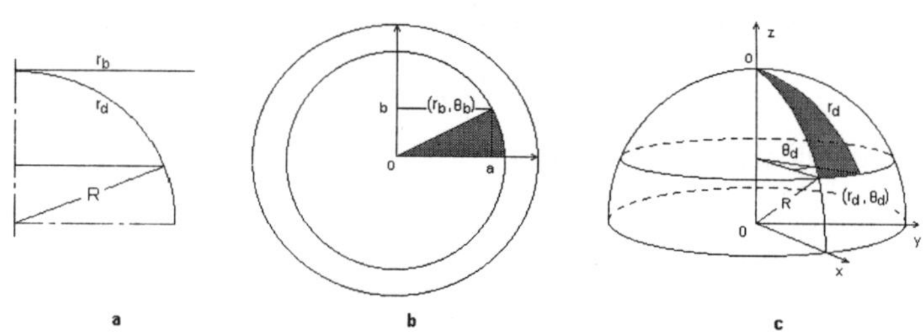

Figure 2 (a) a section. (b) plane polar coordinate system on blank. (c) spherical polar coordinate system on dome.

4.1 Transformation from plane polar coordinates to geodesic polar coordinates

The area enclosed by the circle on the blank is,

$$A_b = \pi r_b^2$$

The area enclosed by the circle on the dome is,

$$A_d = 2\pi R^2 (1 - \cos(r_d / R))$$

equating

$$2\pi R^2 (1 - \cos(r_d / R)) = \pi r_b^2$$

we obtain

$$r_d = R \arccos(1 - r_b^2 / (2R^2)) \qquad (1)$$

where R is the radius of the sphere. From the equal area hypothesis again, the point on the circle on the blank will be mapped into a point on the circle of the dome such that the enclosed triangular segment on the blank (Fig.2b) has the same area as the sphere segment (Fig.2c) on the dome. The area of the triangular segment on the blank is

$$a_b = A_b (\theta_b / (2\pi))$$

The area of the triangular segment on the dome is,

$$a_d = A_d (\theta_d / (2\pi))$$

Such that

$$\theta_d = \theta_b \qquad (2)$$

Now consider the strains occur from the mapping process. A point is under a displacement in the radial direction

$$u = r_d - r_b$$

and a displacement in the circumferential direction

$$v = \theta_b (R \sin(r_d / R) - r_b)$$

The strain with respect to the original distance from the center of the blank is

$$\varepsilon_r = \partial u / \partial r_b = (1 - (r_b / R)^2 / 4)^{-1/2} - 1 \qquad (3)$$

The strain in the circumferential direction is

$$\varepsilon_\theta = (\partial v / \partial \theta_b) / r_b = (1 - (r_b / R)^2 / 4)^{1/2} - 1 \qquad (4)$$

From rotationally symmetric geometry characteristics, shear strain

$$\gamma_{r\theta} = 0 \qquad (5)$$

This is the reason why radials of the blank are mapped on the radial geodesics on the dome. The resultant extensional strains has been illustrated in (Fig.3).

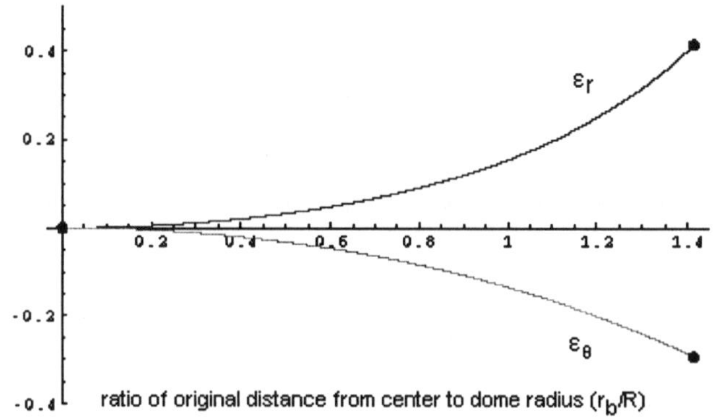

Figure 3 Extensional strains occur from mapping

4.2 Transformation from plane rectangular coordinates to 3D rectangular coordinates

Now we map a grid on the blank onto the dome hemisphere using this method. The grid on the blank is a rectangular system (a,b) (Fig.2b). There is a transformation between the rectangular and the plane polar system given below.

$$r_b = (a^2+b^2)^{1/2}$$

$$\cos(\theta_b) = a/(a^2+b^2)^{1/2}$$

$$\sin(\theta_b) = b/(a^2+b^2)^{1/2}$$

A rectangular system r(x,y,z) (Fig.2c) is used with the dome. There is a transformation between the rectangular system and the geodesic radial polar system.

$$x = R \sin(r_d / R)\cos(\theta_d)$$

$$y = R \sin(r_d / R)\sin(\theta_d)$$

$$z = R \cos(r_d / R)$$

Using the mapping relationship (1) and (2) between the plane polar system and the geodesic polar system the mapping relationship between the rectangular system on the blank and the rectangular system with the dome has been obtained.

$$x = a\,T$$

$$y = b\,T$$

$$z = R\,T^2$$

where

$$T = (1 - (a^2 + b^2) / (4R^2))^{1/2}$$

The comparison between the Tchebychev net and adjusted net on the dome is shown in (Fig.4). The latter more closely resembles the grid deformation that is observed in an actual grid etched sheet.

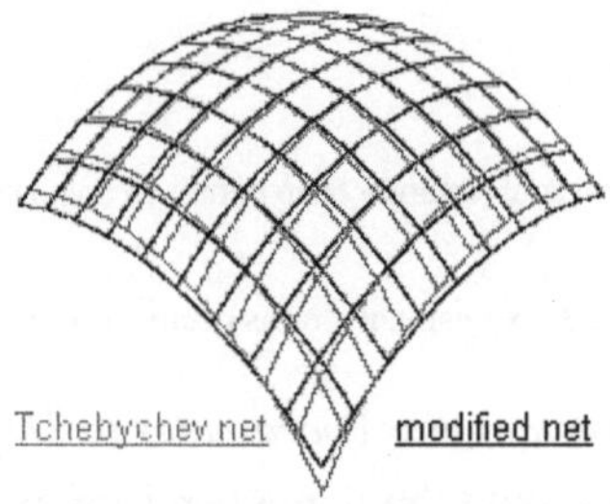

Figure 4 Comparison between the Tchebychev net and the adjusted net on a dome

Strains with respect to (a) and (b) and shear angle occur during the mapping are calculated as follow.

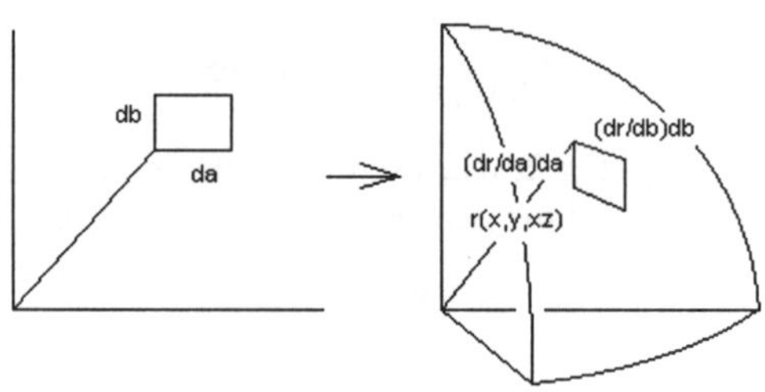

Figure 5 Element analysis

From (Fig.5) strains (ε_a) and (ε_b) respect to (a) and (b) and shear angle (γ_{ab}) occur during the mapping are calculated as follow.

$$\varepsilon_a = ((\partial \mathbf{r}/\partial a)\, da \; - da) / da = ||\partial \mathbf{r}/\partial a|| - 1 \qquad (6)$$

$$\varepsilon_b = ((\partial \mathbf{r}/\partial b)\, db - db) / db = ||\partial \mathbf{r}/\partial b|| - 1 \qquad (7)$$

$$\gamma_{ab} = \pi/2 - \arccos((\partial \mathbf{r}/\partial a)(\partial \mathbf{r}/\partial b) \; / (||\partial \mathbf{r}/\partial a|| \; ||\partial \mathbf{r}/\partial b||)) \qquad (8)$$

The distribution of the strains and shear angle are showed in (Fig.6).

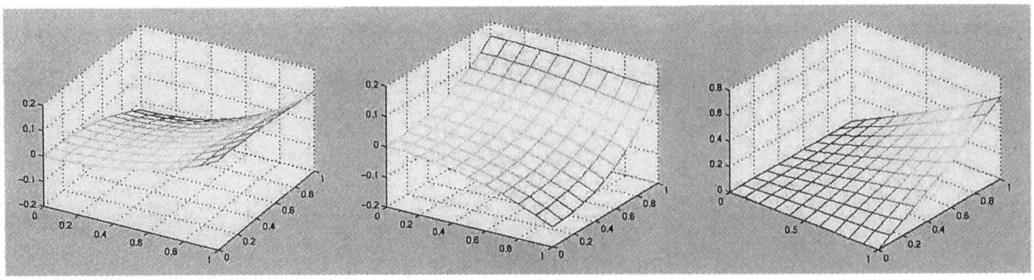

Figure 6 (a)Strains respect to a. (b)Strains respect to b. (c)Shear angle.

5 Conclusion

The mapping approach derived from the equal area hypothesis using geodesic properties provides a means to predict strains occur during sheet metal forming. Compared to Tchebychev net on a blank surface, the deformed net using the equal area geodesic criteria gives a relatively realistic prediction for the simple dome case. This needs to be confirmed experimentally and extended to more complex freeform shapes.

Acknowledgements

The authors would like to acknowledge Professor John Duncan for suggesting problem in the first instance.

References

1. Aono M, Denti P, Breen D E and Wozny M J, Fitting a woven cloth model to a curved surface: Mapping algorithm. *Computer Aided Design.* Vol 26, No 4, (1994) pp.278.
2. Pipkin A C, Equilibrium of Tchebychev nets. *Arch Rational Mech.* 85 (1984) pp.81.
3. S Samelson and W P Dayawana, On The Existence of Global Tchebychev nets. *Trans. Amer.Math.Soc.* 347 (1995) pp.651-661.
4. Caladine C R, *Theory of shell structures* (Cambridge University Press, Cambridge, 1983).
5. M P do Carmo, *Riemannian GeoMetry* (Birkhauser, 1992).
6. M P do Carmo, *Differential Geometry of Curves and Surfaces* (Prentice Hall, 1976)

Appendix - The strains and shear angle calculation

$$\partial \mathbf{r} / \partial a = (\partial x / \partial a)\, \mathbf{i} + (\partial x / \partial a)\, \mathbf{j} + (\partial x / \partial a)\, \mathbf{k}$$

$$\partial \mathbf{r} / \partial b = (\partial x / \partial b)\, \mathbf{i} + (\partial x / \partial b)\, \mathbf{j} + (\partial x / \partial b)\, \mathbf{k}$$

$$\partial x / \partial a = T - a^2 / (4R^2T)$$

$$\partial y / \partial a = - ab / (4R^2T)$$

$$\partial z / \partial a = - a / R$$

$$\partial x / \partial b = - ab / (4R^2T)$$

$$\partial y / \partial b = T - b^2 / (4R^2T)$$

$$\partial z / \partial b = - b / R$$

$$||\partial \mathbf{r}/\partial a|| = ((\partial x / \partial a)^2 + (\partial y / \partial a)^2 + (\partial z / \partial a)^2)^{1/2}$$

EFFECT OF PARTICLE MATERIAL ON THE ELASTIC PROPERTIES OF POWDERS DURING COMPACTION

M. L. HENTSCHEL AND N. W. PAGE

Department of Mechanical Engineering, The University of Newcastle. Newcastle, N.S.W., 2308

Shear and longitudinal mode ultrasonic wave speeds have been measured in powders of copper, aluminium and stainless steel during compaction. At the outset, all powders were nominally spherical, providing similar pore shapes. The wave speed measurements were used to calculate effective elastic moduli (Young's modulus, Poisson's ratio) for the powders, and monitor their evolution with compact density. Effective Young's modulus at a given density was found to be different for each material, though all increased monotonically with density in a qualitatively similar fashion. Normalisation of the measured Young's modulus by the respective solid phase value for each particle material accounted for the quantitative differences in the density dependence. Effective Poisson's ratio had a concave-shaped dependence on density for all powders, which was quantitatively similar over the range of densities studied. These results indicate that for compacts with equivalent pore shapes, the dependence of Young's modulus on density scales with the value of the solid phase Young's modulus. In contrast, the effective Poisson's ratio of the compact showed no material dependence, indicating it may be due primarily to the shape of pores in the compact.

1 Introduction

A condensed phase material will deform elastically at stress states that do not exceed its yield strength. The magnitude of the deformation produced is determined by the elastic moduli of the material. As such, elastic moduli are important physical properties, having wide application for engineering materials in general. In the specific case of granular materials, knowledge of the elastic properties has application in crushing operations, component manufacture by powder metallurgy, tableting of pharmaceutical powders, size enlargement (granulation) processes, foundations engineering, seismological events and, discharge of bulk solids from storage facilities. Following processes of powder densification, the elastic moduli of the component (tablet) on release of the compacting pressure will determine the internal stresses and dimensional changes (springback) developed during die ejection of the compact. In addition to these practical applications, interest in the elastic properties of granular materials also arises at a fundamental level. The elastic behaviour of the most common states of matter (solid, liquid, gas), is well understood. This is not the case for granular materials. The effect of microstructural variables describing the granular state on the elastic behaviour is not well known, even though the properties of the constituent phases are well characterised.

In loosely consolidated granular materials, it is difficult to observe purely elastic behaviour of the bulk due to their low strength. A powder that has been poured into a container can be densified (plastically deformed) simply by tapping the container. Further densification requires the application of pressure and leads to an increase in yield strength. This does not necessarily mean that stresses below the maximum compaction pressure will produce exclusively elastic deformation: granular materials can yield during unloading. For this reason, Hardin and Blandford [1] argue that methods for the measurement of elastic properties that involve subjecting the sample to significant macroscopic stresses will not be valid, as the measured deformation will contain both elastic and plastic components. They suggest that the measurement of elastic moduli requires infinitesimal changes in the stress state, a situation that can be achieved by propagation of low amplitude (elastic) waves.

This paper describes a study of the elastic properties of powders determined by shear and longitudinal ultrasonic wave speed measurements as the powder undergoes compaction. Along with compact density, these two wave speeds can be used to calculate the two independent elastic moduli required to fully characterise the elastic behaviour of an isotropic, homogeneous, linearly elastic medium. Application of relations based on linear elasticity is not meant to imply that the powder compact behaves in a linear elastic manner over a large range of applied stresses, as is the case for many solid materials. Rather, as the powder densifies it is assumed to evolve through a series of metastable states, with the onset of plastic deformation from each state preceded by a region in which the stress-strain response is linear elastic. Physically, this requires that the passage of the elastic waves does not induce slip at particle contacts. This is a situation that should be well fulfilled by ultrasound, which is estimated to induce deformations of the order of 10^{-10} m [2]. With this approach, it is possible to characterise the elastic properties over a very wide range of bulk densities. However, these values will only be valid locally, depending on the bulk density and microstructure of the medium.

The work undertaken in this paper is part of a more extensive body of work investigating the effect of sample microstructure on the elastic properties of powders during compaction. Microstructural variables of importance are: particle material, density (porosity), amount of solid phase bonding (sintering) and, pore character including pore shape, pore size, distributions of pore shape and size, pore homogeneity, pore anisotropy. In this study attention is focussed on the effects of particle material. Three different ductile metal powders with approximately spherical particle shape were studied. Particle size for all powders was standardised by selecting the sieve fraction +75-106 μm. By using similar shaped particles, the initial pore character in each material was also similar. Given that all materials yielded by plastic flow, the evolution of pore character with compact density was similar. Thus, at equivalent densities, the only major differences between the compacts were the properties of the solid phase material. The effect of some aspects of pore character has been studied in a companion paper [6].

In porous or dense granular materials (i.e. touching particles) the solid phase carries most or all of the load when these materials are subjected to stress. Thus, the elastic properties of the solid phase have an important effect on the elastic properties of porous and granular materials. For this reason it is common to normalise the elastic properties of the porous or granular material against the relevant value for the solid phase (see, for example Rice [7]). Usually this is performed to illustrate the effect that different pore character has on the elastic properties. With one exception discussed below, no effort has been made to explicitly compare the effect of solid phase properties by preparing a range of samples with equivalent pore character. The purpose of this paper is to explore the universality of this approach by studying a range of powders of different composition, but otherwise similar characteristics.

Studies on the elastic properties of a powder during compaction are rare. Jones and Blessing [4, 5] used ultrasonic techniques to measure elastic moduli of alumina powder during uniaxial compaction. As alumina was the only particle material studied, these works reveal nothing of solid phase effects.

Carnavas and Page [3] performed quasi-static measurement of elastic moduli on copper and stainless steel powders of equivalent morphology (spherical) during compaction. Elastic moduli were calculated from strains measured during unloading from various density levels during compaction. As argued above, these compacts should have equivalent evolution of pore character with density. However, doubts exist over the accuracy of this testing method because purely elastic unloading could not be proven. As discussed elsewhere [1, 8], it is quite likely that this requirement is not met. Neglecting this, the results of Carnavas and Page suggested, that for the case of Young's modulus, the different dependencies of the two materials on density could be accounted for by normalising by the solid phase values of Young's modulus.

2 Experiment

2.1 Materials Tested

Three different spherical metal powders were chosen for comparison. Micrographs of the powders prior to testing are shown in Figures 1(a)-(c). When pressed to 150 MPa, the stainless steel powder did not form a cohesive compact. A micrograph of particles recovered from this compact (69% of the solid phase value) is shown in Figure 1(d). Similarly, for the maximum applied pressures used in this work, the spherical copper powder did not form a cohesive compact. Figures 1(e) and (f) show particles recovered from tests for which the maximum applied pressures were 150 and 560 MPa respectively. These pressures correspond to bulk densities of 81 and 96% of solid phase density.

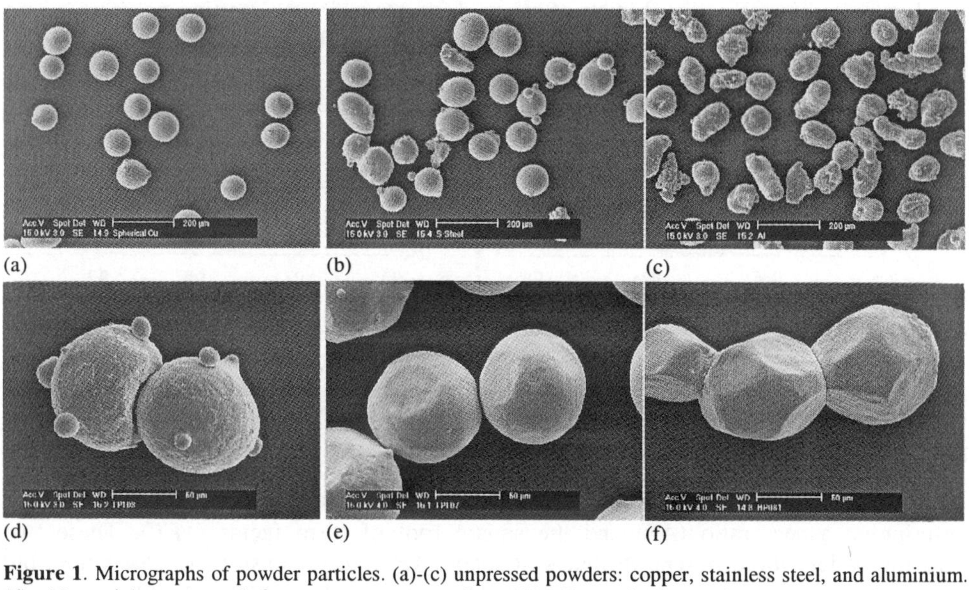

(a) (b) (c) (d) (e) (f)

Figure 1. Micrographs of powder particles. (a)-(c) unpressed powders: copper, stainless steel, and aluminium. (d), (e) particles recovered from compacts pressed to 150 MPa: stainless steel, and copper. (f) particles recovered from copper powder compressed to 560 MPa.

Relevant physical properties of the powders prior to testing are summarised in Table 1 along with quantitative shape descriptors of the powder particles obtained by image analysis.

Comparison of the micrographs for the unpressed powders (Fig. 1a-c) shows that the copper particles best approximate smooth spheres, while the aluminium particles are the most irregular, with the stainless steel powder intermediate between the two. The particles recovered from the stainless steel compact (Fig. 1d) exhibit little sign of deformation. In comparison, the copper powder pressed to 150 MPa shows modest amounts of deformation, while the particles recovered from the compact pressed to 560 MPa are heavily deformed.

Table 1. Physical and shape characteristics of powder particles.

Particle material	Apparent density	Median AR	Median $\sqrt{FF}$
Aluminium	1312 (± 2)	0.773	0.931
stainless steel (unpressed)	4250 (± 40)	0.843	0.957
stainless steel (150 MPa)	-	0.838	0.953
copper (unpressed)	5170 (± 50)	0.934	0.994
copper (150 MPa)	-	0.890	0.989
copper (560 MPa)	-	0.816	0.973

2.2 *Shape analysis*

Image analysis was used to quantify the shape of the particles shown in Figure 1. A small sample of the powder was dispersed on a microscope slide and the projected image of individual particles captured using a camera attached to a light microscope. These images were then thresholded and relevant particle dimensions measured (perimeter, area, Feret diameters, etc.), enabling calculation of two-dimensional shape descriptors. Further details of two-dimensional shape descriptors may be found in [9].

Figure 2 shows the distribution of shape descriptors for the starting powders, along with those for particles recovered from the stainless steel and copper compacts.

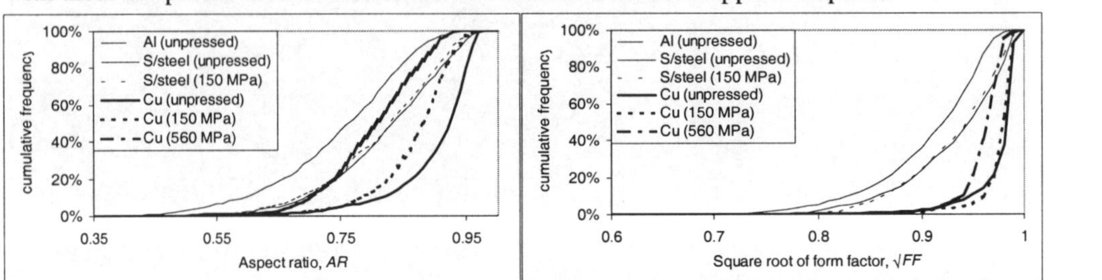

(a) Aspect ratio, AR:

$$AR = B / L \quad (1)$$

(b) Root form factor, $\sqrt{FF}$:

$$\sqrt{FF} = (2\sqrt{\pi A}) / P \quad (2)$$

B, L are minimum and maximum Feret diameters respectively, A area, and P perimeter.

Figure 2. Distributions of shape descriptors.

In keeping with the companion study [6], particle shape is quantified using two descriptors: aspect ratio (AR) and the square root of form factor ($\sqrt{FF}$). These two descriptors describe different attributes of shape. AR is most sensitive to elongation, but insensitive to boundary irregularity whereas $\sqrt{FF}$ is the converse. AR and $\sqrt{FF}$ are defined in terms of image measurements as per (1) and (2).

The range of values obtained for each powder demonstrates the distributed nature of particle shape within a sample. Although the powders are described in general as being 'spherical', all are seen to contain some elongate particles with modest amounts of boundary irregularity. In terms of both features, the aluminium powder deviates most from sphericity followed by stainless steel, and finally copper.

The particles recovered from the stainless steel compact show no change in shape for either descriptor, while the copper particles become increasingly elongate with compaction pressure.

2.3 Apparatus

Elastic moduli were calculated from measurements of shear and longitudinal ultrasonic wave speeds in the powder during compaction. Further details of the experimental method are provided elsewhere [6].

Wave speeds are related to the elastic moduli and sample density by (3) and (4).

$$E = \rho c_S^2 \frac{3c_L^2 - 4c_S^2}{c_L^2 - c_S^2} \quad (3) \qquad \nu = \frac{c_L^2 - 2c_S^2}{2(c_L^2 - c_S^2)} \quad (4)$$

where E is Young's modulus, ν is Poisson's ratio c_L, c_S are the bulk longitudinal and shear elastic wave speed and ρ the density.

3 Results

Effective elastic moduli for the powders during compaction are shown in Figure 3.

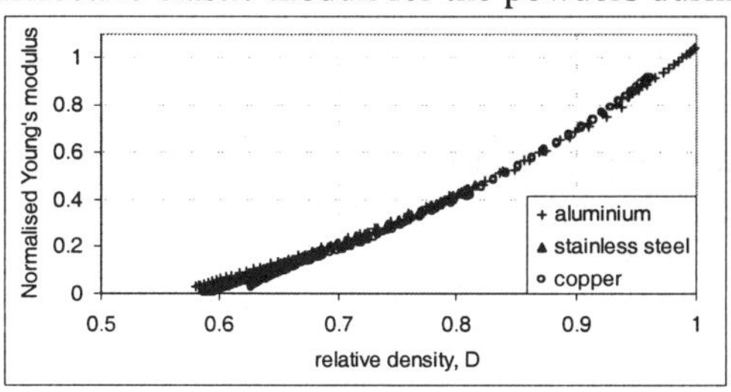

(a) Normalised Young's modulus

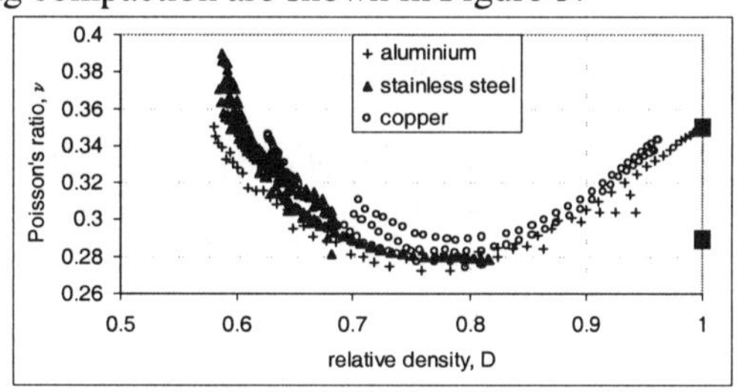

(b) Poisson's ratio

Figure 3. Evolution of effective elastic moduli of spherical metal powders during compaction. Density is normalised by the appropriate solid phase values, as is Young's modulus. Poisson's ratio is **not** normalised (the filled squares represent ν_{solid}). $\rho_{solid} = 2700, 8000, 8960$ kg m^{-3}; $E_{solid} = 68, 195, 115$ GPa; $\nu_{solid} = 0.35, 0.29, 0.35$, for aluminium, stainless steel, and copper respectively.

Normalisation of effective Young's modulus for each material by the solid phase value is seen to account for the material differences. In general, each material exhibits a steady increase of Young's modulus with density, which increases slightly as the density approaches that of the solid phase. The minor differences between the measured Young's modulus evident at low densities are likely to be due to the differences in particle shape between the materials, as discussed above.

Figure 3(b) shows effective Poisson's ratio for the different powders. All three materials are seen to have very similar dependencies over the common density range, namely an initial decrease with density, followed by a region in which Poisson's ratio is constant, and finally, for the aluminium and copper powders, an increase towards the solid phase value. In contrast to the case for Young's modulus, it is evident that normalisation by the solid phase value of Poisson's ratio will serve to separate the material types.

4 Conclusions

This work investigated the effect of particle material on the elastic properties of granular materials during compaction. To isolate the effect of solid phase properties, three different powders with equivalent particle shape were tested. Metal powders were selected so that the evolution of pore character with density was similar for each material.

Elastic properties of the powders during compaction were described by Young's modulus and Poisson's ratio. Differences in the dependence of Young's modulus on relative density for each material type could be accounted for by normalising by the solid phase value for each material. Thus, Young's modulus of compacts with equivalent microstructure depends only on the solid phase material properties. In contrast, Poisson's ratio over the common density range appears independent of the solid phase properties, indicating it may be solely dependent on pore character. In must be pointed out though, that this can not hold for the entire density range, as the density dependence of the stainless steel powder must diverge from that of the copper and aluminium as the solid phase density is approached. Thus, as the stainless steel powder is further densified, the properties of the solid phase must become increasingly important. This suggests that pore character controls the initial decrease in Poisson's ratio with density, but the eventual approach to the solid phase value is dominated by the solid phase material properties.

5 Acknowledgements

The authors gratefully acknowledge scholarship support for MLH through the ARC Small Grants Scheme. This project forms part of the research program of the ARC supported Centre for Bulk Solids and Particulate Technologies.

References

1. Hardin B.O. and Blandford G.E., Elasticity of particulate materials. *Journal of Geotechnical Engineering* **115** (1989) pp. 788-805.
2. Matikas T.E., Kapur P. and Shamasundar S., Measurement of the dynamic elastic moduli of porous titanium aluminide compacts. *Journal of Materials Science.* **32** (1997) pp. 1099-1103.
3. Carnavas P.C. and Page N.W., Elastic properties of compacted metal powders. *Journal of Materials Science.* **33** (1998) pp. 4647-4655.
4. Jones M.P. and Blessing G.V., Ultrasonic evaluation of spray-dried ceramic powders during compaction. In *International Advances in Nondestructive Testing. Volume 13*, ed. by W.J. McGonnagle. (Gordon & Breach, New York, 1988) pp. 175-191.
5. Jones M.P. and Blessing G.V., The dynamic Poisson's ratio of a ceramic powder during compaction. *IEEE 1987 Ultrasonics Symposium Proceedings.* Pt. 2, **1186**, (1987) pp. 587-590.
6. Hentschel M.L. and Page N.W., Effect of particle shape on the elastic properties of powders during compaction. To be presented at: 7th International Conference on Bulk Materials, Storage, Handling and Transportation. Newcastle. 3-5 October 2001.
7. Rice R.W. Porosity of ceramics. (Marcel Dekker, New York, 1998).
8. Straffelini G., Fontanari, V and Molinari A., True and apparent Young's modulus in ferrous porous alloys. *Materials Science and Engineering* A **260** (1999) pp. 197-202.
9. Russ J.C., Computer assisted microscopy: The measurement and analysis of images, (Plenum Press, New York, 1990).

DYNAMIC MECHANICAL PROPERTIES OF LIQUID-BOUND POWDER COMPACTS

S.M. IVESON[A,1] AND N.W. PAGE[B]

Departments of Chemical[a] and Mechanical[b] Eng, University of Newcastle, NSW 2308, Australia

The effect of particle size on the dynamic strength of liquid-bound granular materials was investigated. Four different size distributions of glass ballotini were used. Three narrowly sized fractions with mass median sizes of 28, 48 and 90 μm and one broad size distribution with a mass median size 71 μm, but a surface-mean size of only 35 μm. Pellets of length 25 mm, diameter 20 mm, porosity 35% and liquid saturation 70% were deformed at speeds varying from 0.03 to 150 mm/s. Water, glycerol and various silicone oils were used as the liquid binder. The strength of the pellets was strain rate dependent. The strength measured at low strain rates was not necessarily indicative, *even qualitatively*, of the relative strength ranking of different formulations at high strain rates. The peak flow stress (σ_{pk}) versus strain rate ($\dot{\varepsilon}$) data was found to collapse onto a single curve when plotted in terms of two dimensionless groups.

1 Introduction

In agitated wet powder granulation processes, particles are agglomerated together into larger, semi-permanent aggregates by spraying a liquid binder onto the particles as they are agitated in a tumbling drum, fluidised bed, high shear mixer or similar device (Ennis and Litster, 1997). The liquid binds the particles together by a combination of capillary pressure, surface tension and viscous forces until solid bonds are formed by subsequent drying or sintering. Some advantages of agglomerated materials include improved flow properties, reduced dustiness, increased bulk density and the co-mixing of particles which might otherwise segregate. Granulation unit operations occur in a wide range of industries including minerals processing, agricultural products, pharmaceuticals, detergents and food stuffs.

The strength of wet powder compacts helps determine their growth behaviour as they are agitated in a granulator. Many coalescence growth models include the yield strength of the colliding granules as a parameter (eg. Ouchiyama and Tanaka, 1975; Thornton and Ning, 1998; Liu *et al.*, 2000). Weak granules will form a large area of contact during collisions which helps to bind the two granules together, promoting granule coalescence growth. Granules which do not deform greatly during impact will not coalesce together until they have consolidated sufficiently to become surface wet (Iveson and Litster, 1998a).

The strength of liquid-bound granules has traditionally been measured at low and/or invariant strain rates ie. pseudo-static conditions (eg. Rumpf, 1962; Schubert *et al.*, 1975; Kristensen *et al.*, 1985). However, inside granulators, granules will experience a wide range of impact strain rates, with impact velocities as high as 10 m/s expected in some of the high-shear mixer type devices. At these higher strain rates, dynamic effects such as viscous dissipation and liquefaction may become significant (Ennis *et al.*, 1991; Wynnyckyj, 1993). For instance, Iveson and Litster (1998b) found that using glycerol (a viscous liquid) instead of water greatly reduced the amount of impact deformation experienced by granules.

[1] Corresponding author: Ph: +61 2 4921 6192; Fax: 4921 6920; E-mail: cgsmi@cc.newcastle.edu.au

Discrete element simulations of granule impacts have also predicted that viscous binders will reduce granule deformation during collisions (Lian *et al.*, 1998). In contrast, at low strain rates (5 mm/min), Mills *et al.* (2000) found that the crushing strength of granules made with 90-180 μm sized sand was unaffected by binder viscosity for four different silicone oils of viscosity ranging from 0.02 to 0.5 Pa·s.

Until recently, the only available experimental work on the strength of wet granular media at high strain rates was restricted to the soil mechanics literature. This was concerned with the effects of earthquakes to liquefy foundations and hence only considerd saturated systems with water as the liquid (a relatively non-viscous binder).

Our research group has recently begun to directly measure the dynamic strength of granular materials partially-saturated with viscous liquids. Iveson *et al.* (2000) measured the flow stress of pellets 25 mm in length, 20 mm diameter as they were uni-axially compressed at speeds varying from 0.01 to 150 mm/s. They studied pellets made from a single broad size fraction of glass ballotini and constant porosity (35%) and liquid saturation (70%). Six different liquid binders were used: water, glycerol and four different viscosity silicone oils. There were two regions of behaviour. At low strain rates, the mechanical properties were insensitive to strain rate. This suggests that viscous forces were insignificant. At high strain rates, pellet flow stress increased with increasing strain rate, suggesting that viscous forces had become significant. The strain rate of transition between these two regions depended on the viscosity and surface tension of the binder. The relative strength ranking of formulations at low strain rates was not indicative, *even qualitatively*, of their ranking at high strain rates. This is significant because it means that the traditional methods of measuring granule strength at pseudo-static conditions may be misleading when used to predict the collision deformation of granules.

In further work, Iveson *et al.* (2001) found that these results all collapsed onto a single curve when plotted in terms of two dimensionless groups:

$$\left(\frac{\sigma_{pk} d_{32}}{\gamma_{LV}}\right) = 5.2 + 310\left(\frac{\dot{\varepsilon}\mu d_{32}}{\gamma_{LV}}\right)^{0.62} \qquad (1)$$

where σ_{pk} is the peak flow stress, d_{32} is the surface-mean particle size, γ_{LV} is the liquid-vapour surface tension, $\dot{\varepsilon}$ is the strain rate and μ is the liquid viscosity. The group on the left-hand side is the ratio of peak flow stress to capillary forces. The group on the right-hand side is the so-called capillary number – the ratio of viscous to capillary forces.

However, although particle size appears as a parameter in equation (1) as a result of dimensional analysis, this parameter was not varied by Iveson *et al.* (2001) in their experiments. Hence the presence of d_{32} in equation (1) requires verification. The current paper reports the results of further work which has examined the effects of particle size on peak granule flow stress. The experimental method is reviewed briefly, followed by the results, discussion and final conclusions.

2 Experimental

2.1 Materials

Only one liquid binder was used in the current work. This was a silicone oil with nominal viscosity of 1000 mPa·s supplied by Fluka Products, Sigma-Aldrich, Australia. The measured properties of this oil were: density 0.965 g/ml, surface tension 0.026 N/m and viscosity 0.72 Pa·s. This oil is identical to the one used by Iveson *et al.* (2000, 2001).

For this work, four different size ranges of glass ballotini were used. All the ballotini was supplied by Potters Industries (Local Distributor: Burwell Abrasive Blasting Equipment, Newcastle, Australia). The glass density is 2.45 g/ml. The AH size fraction (nominal 45-90 microns) was used as supplied. This was the same fraction as used in previous work (Iveson *et al.*, 2000, 2001). The three other size fractions were prepared by hand sieving the ranges of 0-38, 45-53 and 75-90 microns. The size distributions of these four size ranges were measured using a Malvern Mastersizer E and the results are shown in Figure 1. The calculated surface mean (x_{32}), mass mean (x_{43}) and mass-median (d_{50}) size are shown in Table 1. The span is the difference between the d_{90} and d_{10} sizes, divided by the d_{50} size, and gives an indication of the breadth of the size distribution.

It is immediately apparent that the three hand-sieved fractions are quite narrowly sized. By comparison, the as-supplied AH fraction has a large span indicating a broad size distribution. In particular, it has a long tail of fines below 40 micron in size. These serve to greatly reduce its surface mean size to below that even of the 45-53 micron size fraction, whereas its mass-mean size is almost equal to that of the 75-90 size fraction.

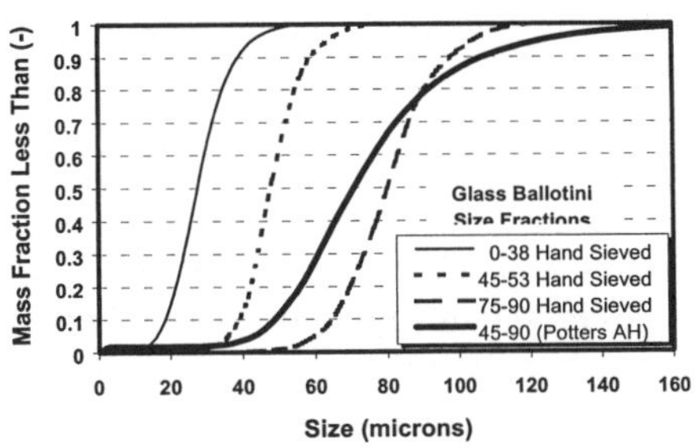

Figure 1: Cumulative mass less-than size distributions of the four different glass ballotini size fractions.

Table 1: Measured mean sizes and spread of the four ballotini size fractions.

	x_{32}	x_{43}	d_{90}	d_{50}	d_{10}	Span
0-38	26	28	39	28	19	0.74
45-53	47	49	60	48	39	0.44
75-90	78	81	97	80	64	0.41
Potters AH	35	78	108	71	49	0.83

2.2 Method

The experimental method used was identical to that reported previously (Iveson *et al.*, 2000, 2001). In brief, known weights of ballotini and binder were first hand-kneaded in a plastic bag until the binder was well-distributed. A known mass of this mixture was then weighed out and loaded into a cylindrical, uniaxial die where it was compressed by hand to form pellets of length 25 mm and diameter 20 mm. The mass of solid and liquid were chosen such that the final pellet had an overall porosity of 35% and a liquid saturation of 70%. For the current work, this meant the pellets had a mass of 14.40 g and a liquid to solid ratio of 0.148 g/g.

Each pellet was then extracted and placed between two smooth Teflon® platens in an Instron Dynamight™ load frame. This hydraulic load frame could generate cross-head speeds of up to 150 mm/s and was fitted with a 100 N load cell capable of accurately measuring loads down to the order of 1 N. The pellet was deformed at constant speeds ranging from 0.01 to 150 mm/s and the load-history recorded on a computer data-logger. Enough mixture was made to test 7 pellets across this range of speeds.

2.3 Data Analysis
As described previously, the measured load was converted into a stress by dividing by the pellet cross-sectional area. This was estimated by assuming the pellet deformed by uniform cylindrical barrelling. The natural axial strain of the pellet was calculated as the negative of the logarithm of the current pellet length divided by its original length ie. $\varepsilon = -\ln(h/h_0)$ where h is the current length and h_0 = 25 mm is the initial length. The strain rate $\dot{\varepsilon}$ was a function of the extent of deformation. The value used was the strain rate at which the peak flow stress occurred.

3 Results and Discussion

Figure 2 shows the stress versus strain results for the 75-90 micron ballotini at different deformation speeds. This behaviour was typical of all the size fractions tested and was also observed in previous work (Iveson *et al.*, 2000). The flow stress initially increases with increasing strain until it reaches a peak. This is not an elastic region – releasing the stress does not result in any springback of the pellet. After the peak, which always occurs between 5 and 10% natural strain, the flow stress drops and appears to approach a steady value. The mechanisms responsible for this characteristic stress-strain profile behaviour are as yet unknown, but presumably have something to do with the re-arrangement of the particles within the pellet as the material flows.

The deformation speed clearly effects the pellet flow stress. Between 0.01 and 1 mm/s there is little change in the magnitude of the flow stress, although there is considerably more noise in the results at the lowest speed (possibly because the hydraulic load frame found it difficult to maintain a constant speed at such low speeds). However, above 1 mm/s, the flow stress increases with increasing strain rate. This suggests that viscous effects start to become significant at higher speeds.

Figure 3 shows the flow stresses of the four different ballotini size fractions plotted in terms of two dimensionless groups – the dimensionless peak flow stress (*Str**) and the capillary number (*Ca*) – which have been evaluated using the surface-mean particle size. Also shown in Figure 3 are the results from the previous study for the as-received Potters AH ballotini which were performed by a different operator to the current work (Iveson *et al.*, 2000).

Three things are immediately obvious from Figure 3. Firstly, the results for the three narrow, hand-sieved size fractions all collapse on top of one another. Secondly, the duplicate experiments using AH ballotini from the current work agree well with the previous study. Thirdly, the results for the three narrow size fractions do not lie on top of the results for the broadly sized AH ballotini.

The excellent agreement between the duplicate experiments from this and the previous study show that the experimental technique is operator-independent. The fact that the results for the three narrow ballotini size fractions all collapse together confirms the functionality of the dependence on particle size found in the dimensionless peak flow stress and the capillary number, as predicted by dimensional analysis.

However, the fact that the results for the narrow and broadly-sized ballotini fractions do not collapse together indicates that the *shape* of the particle size distribution also has an effect on the material's flow stress. This is not surprising. Theoretical considerations suggests that the strength of a granular materials depends on the particle coordination number – the average number of inter-particle contacts. This is determined by how the particles are packed together, which is strongly influenced by the shape of the size distribution. Surface-mean size is usually used as the measure of particle size during calculations of granule strength because it is believed that this average best correlates with the number of inter-particle contacts per unit volume of material (Rumpf, 1962).

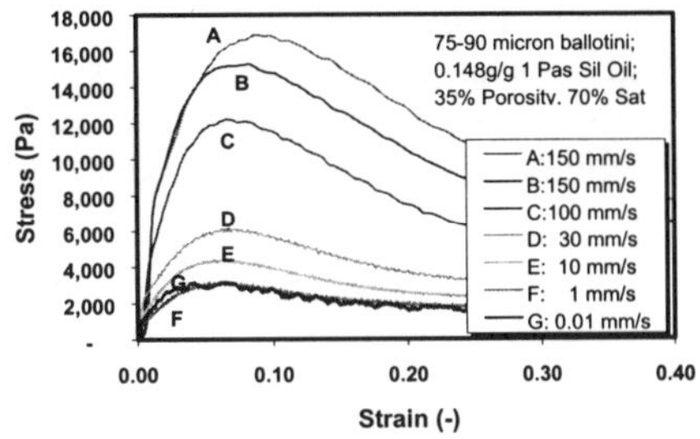

Figure 2: Stress vs strain for pellets made from 75-90 micron ballotini across a range of deformation speeds.

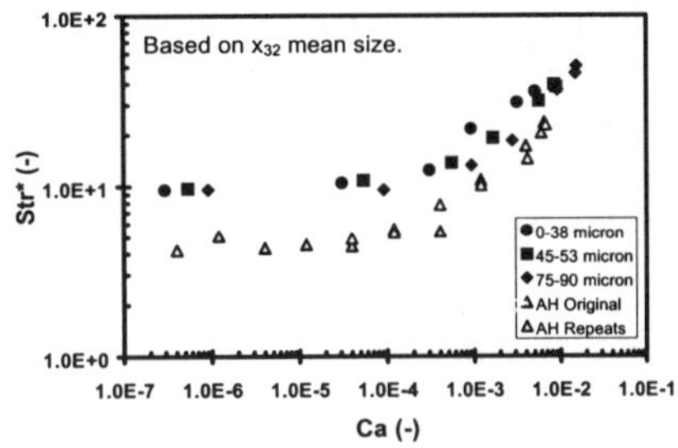

Figure 3: *Str** vs *Ca* for the four different ballotini size fractions. *Str** and *Ca* calculated using the surface-mean size, x_{32}.

However, in this case, surface-mean size is inadequate. The long tail of fines in the AH material gives a surface mean size of 35 microns (Table 1), which is smaller than that of the 45-53 micron size fraction, even though the majority of the material is actually much larger than this size fraction. Therefore, a descriptor of particle size distribution shape needs to be included as well as the average particle size itself.

Figure 4 plots the dimensionless peak flow stress versus capillary number for the four different size fractions of ballotini, but this time the measure of average size used is the mass-median size, d_{50}. In this case the results all collapse together quite nicely. Figure 5 shows the results of the current work together with those of the previous work using different liquid binders.

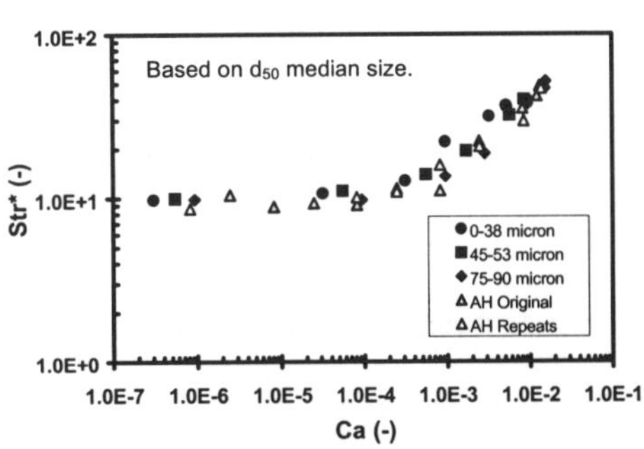

Figure 4: *Str** vs *Ca* for the four different ballotini size fractions. *Str** and *Ca* calculated using the mass-median size, d_{50}.

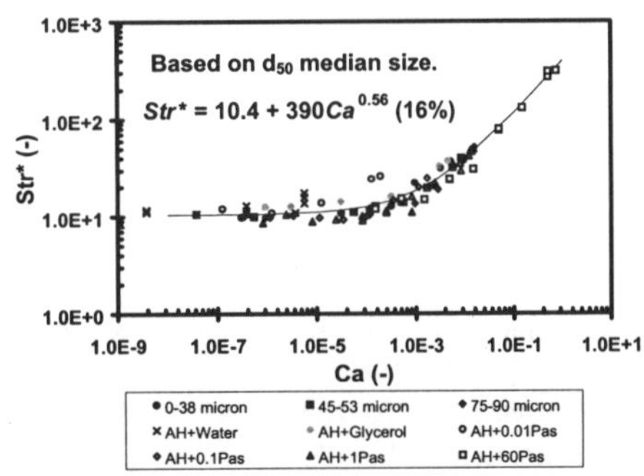

Figure 5: *Str** vs *Ca* calculated using the mass-median size, d_{50}, for the current work together with the previous results of Iveson *et al.*, (2000).

The combined results are well described by the equation:

$$\left(\frac{\sigma_{pk} d_{50}}{\gamma_{LV}}\right) = 10.4 + 390\left(\frac{\dot{\varepsilon}\mu d_{50}}{\gamma_{LV}}\right)^{0.56} \tag{2}$$

written in terms of the mass-median size. There are a total of 89 points plotted in Figure 4, and the average deviation of the experimental data from equation (2) is only ±16%. This is a very promising result, especially when it is considered that the combined correlation covers particle sizes from 28 to 80 microns, surface tensions from 0.02 to 0.07

N/m, viscosities from 0.001 to 41 Pa·s and axial strain rates from 0.001 to 10 s^{-1}. However, the fact that in this case the results all collapse together when plotted in terms of d_{50} rather than x_{32} is fortuitous. There is no reason to expect this relationship to hold for all possible shapes of the particle size distribution eg. bimodal. Describing the influence of the shape of the particle size distribution will require additional parameters.

There are of course many other parameters held constant in the current work which will also have a strong influence on pellet flow stress. These include particle shape and surface roughness, solid-liquid contact angle, pore saturation and pellet porosity. These variables are being studied in on-going work.

4 Conclusions

The effects of particle size, strain rate, binder viscosity and binder surface tension on the dynamic flow stress of liquid-bound granular materials have been shown to be well accounted for by the dimensionless empirical correlation given in equation (2). There are two dimensionless groups in this equation: *Str** is the ratio of peak flow stress to capillary forces and *Ca* is the ratio of viscous to capillary forces. At low *Ca* ($< 10^{-4}$) the dimensionless peak flow stress is independent of *Ca*,. Therefore viscous forces are insignificant in this region. At high *Ca* ($> 10^{-4}$), *Str** increases with increasing *Ca*, indicating that viscous forces have become significant.

Acknowledgments

This work has been funded by an Australian Research Council Large Grant (A89918062) and Australian Postdoctoral Fellowship (F89918124).

References

1. Ennis, B.J., G.I. Tardos and R. Pfeffer (1991), *Powder Technol.*, **65:** 257-272
2. Ennis, B.J. and J.D. Litster (1997), Eds. Perry, R. and Green, D., 7th Ed., pp. 20-56 to 20-89. McGraw-Hill.
3. Iveson, S.M. and J.D. Litster (1998a), *AIChE J.*, **44**, 1510-1518
4. Iveson, S.M. and J.D. Litster (1998b), *Powder Technol.*, **99**, 234-242
5. Iveson, S.M., J.A. Beathe and N.W. Page (2000), 1st Asia Particle Technology Forum (APT2000), Bangkok, Thailand, 13-15 Dec.
6. Iveson, S.M., J.A. Beathe and N.W. Page (2001), in preparation for submission to *Powder Technol.*
7. Kristensen, H.G., Holm, P. and Schæfer, T., (1985a), *Powder Technol.*, 44: 227-238
8. Lian, G., C. Thornton, M.J. Adams (1998), *Chem. Eng. Sci.*, **53**, 3381-3391
9. Liu, L.X., J. D. Litster, S.M. Iveson and B.J. Ennis (2000), *AIChE Journal*, **46**, 529-539.
10. Mills, P.J.T., J.P.K. Seville, P.C. Knight and M.J. Adams (2000), *Powder Technol.*, **113**, 140-147
11. Ouchiyama, N. and Tanaka, T. (1975), *Industrial & Engineering Chemistry, Process Design and Development*, **14**, 286-289.
12. Rumpf, H., (1962), in Knepper W.A. (ed.), AIME, Agglomeration, Interscience, New York 379-418
13. Schubert, H., Herrmann, W. and Rumpf, H., (1975), *Powder Technol.*, **11**: 121-131
14. Thornton, C. and Z. Ning (1998), *Powder Technology*, **99**, 154-162.
15. Wynnyckyj, J.R., (1993), *6th Int. Symp. Agglomeration*, Nov 15-17 1993, Nagoya, Japan, 143-159.

LARGE DEFORMATION COSSERAT CONTINUUM MODELLING OF GRANULATE MATERIALS

E. PASTERNAK[a] AND H.-B. MÜHLHAUS[b]

[a]*Department of Civil and Resource Engineering, University of Western Australia, Crawley, WA 6009 Australia. E-mail: elena@civil.uwa.edu.au;* [b]*CSIRO Division of Exploration and Mining, Nedlands, WA 6009 Australia. E-mail: hans@ned.dem.csiro.au*

The equations of motion for finite macrocsopic deformations of an idealised, monodisperse granular material are obtained by considering a statistically representative particle. The discrete equations of motion of the reference particle are homogenized by differential expansion of the difference expressions of the discrete model. We derive objective constitutive relationships relating the Jaumann rates of stress and moment stress to the relative strain- and curvature rates respectively. The elastic moduli are expressed in terms of the contact stiffnesses, solid volume fraction, coordination number and diameter of the sphere.

1 Introduction

Continuum modelling of materials with microstructure is based on the introduction of the continuous quantities and the determination of corresponding balance and constitutive equations. In many cases the phenomenological determination of constitutive equations with parameters derived from sample testing meets considerable difficulties, e.g. in the case of history dependence or when direct testing is impractically complicated or simply if the spatial and temporal scales are very large. In such situations the constitutive equations maybe derived from a conceptual model for the microstructure of the material in combination with a suitable homogenization procedure.

In the case of granulate materials continuum models can be derived by direct homogenisation of the discrete equations of motion of the particles based on Taylor expansions as suggested in [1]. This technique was used in [2, 3] taking into account the resistance to relative rotations at particle contacts and a 3D Cosserat continuum for infinitesimal deformations was also obtained. The deformations are finite in many practically relevant cases however. In this paper we generalize our previous results to large deformations. Problems with finite deformations are solved incrementally such that within each increment the deformations are infinitesimal. This leads to the formulation of the problem in terms of stress, stress rates and deformation rates. As usual in rate formulation the introduction of the objective deformation measures is required.

2 Large deformations in Cosserat continuum theory

2.1 Kinematics

We describe the motion of the material points in terms of the Cartesian coordinates of their current and initial (reference) position: x_k (current) and X_k (reference) respectively; we assume $x_k(t=0)=X_k$ (k=1,2,3) and $x_i=x_i(X_j,t)$. The deformation gradient tensor and the rate of the deformation gradient tensor are $F_{iJ}=\partial x_i/\partial X_J$, $\dot{F}_{iJ}=\partial \dot{x}_i/\partial X_J=\partial v_i/\partial X_J=v_{i,J}$, ($i, J$ = 1,2,3), where $v_i=\dot{x}_i$ are the components of the velocity vector and ($\dot{\ }$) means material derivative. We introduce the velocity gradient

$\left(\dot{\mathbf{F}}\mathbf{F}^{-1}\right)_{ik} = v_{i,k}$, the symmetric part $v_{(i,k)}$ of which is the so called stretching and anti-symmetric part $v_{[i,k]}$ is the spin. The rates of the Cosserat deformation measures, strains and curvature twists maybe introduced as follows

$$\gamma_{ji} = v_{i,j} - W_{ij}^c,\ W_{ij}^c = \dot{R}_{ik}^c R_{jk}^c,\ \kappa_{ji} = \partial\varphi_i^c / \partial x_j, \tag{1}$$

where φ^c is the Cosserat rotation rate, $\boldsymbol{W}^c$ is the Cosserat spin.

2.2 Objectivity of the deformation measures and rates

Two motions $\boldsymbol{x}^*$ and $\boldsymbol{x}$ are called objectively equivalent if $\boldsymbol{x}^*(\boldsymbol{X},t') = \boldsymbol{Q}(t)\,\boldsymbol{x}(\boldsymbol{X},t) + \boldsymbol{c}(t)$, $t' = t - a$, where $\boldsymbol{Q}$ is an orthogonal (rotation) matrix, $\boldsymbol{QQ}^T = \mathbf{I}$, $\det \boldsymbol{Q} = 1$ (in rectangular coordinates). $\boldsymbol{Q}$ and $\boldsymbol{c}$ represent, respectively, the rotation and translation of one frame with respect to the other or, in other words, the orthogonal transformation of the current configuration. $\boldsymbol{Q}$ is a function of time t only since $\boldsymbol{Q}$ represents the orthogonal transformation of the current configuration (rigid body rotation).

Any tensorial quantity is objective if in any two equivalent motions it obeys the appropriate tensor transformation law at all times. It means that a second order tensor $\boldsymbol{A}$, a vector $\boldsymbol{a}$ and a scalar α are objective if in any two equivalent motions their components transform according to the rule $\boldsymbol{A}^* = \boldsymbol{QAQ}^T$, $\boldsymbol{a}^* = \boldsymbol{Qa}$, $\alpha^* = \alpha$, where $\boldsymbol{Q} = \boldsymbol{Q}(t)$.

It can be shown that the gradient deformation tensor and its inverse transforms according to the following rule $\boldsymbol{F}^* = \boldsymbol{QF}$, $(\boldsymbol{F}^{-1})^* = \boldsymbol{F}^{-1}\boldsymbol{Q}^T$. Then the components of the velocity gradient, the Cosserat rotation and the Cosserat spin $\boldsymbol{W}^c$ transform as follows:

$$\mathbf{v}^* = \dot{\mathbf{Q}}\mathbf{Q}^T + \mathbf{QvQ}^T,\ \mathbf{v} = [v_{i,j}],\ \mathbf{R}^{c^*} = \mathbf{QR}^c,\ \mathbf{W}^{c^*} = \dot{\mathbf{Q}}\mathbf{Q}^T + \mathbf{QW}^c\mathbf{Q}^T \tag{2}$$

It follows from (1), (2) that the Cosserat strain rate $\boldsymbol{\gamma}$ is objective. It can be shown that the curvature rate $\boldsymbol{\kappa}$ (1) is also objective.

2.3 Force equilibrium equation

Neglecting body forces for simplicity the balance of momentum reads

$$\int_V \rho a_i dV = \int_S T_i dS,\ a_i = \dot{v}_i. \tag{3}$$

After substituting the stress vector (traction) $T_i = \sigma_{ji}\lambda_j$ on the right hand side of (3), applying Gauss' theorem and taking into account the arbitrariness of the elementary volume V, one finds:

$$\rho a_i = \sigma_{ji,j}. \tag{4}$$

The material time derivative of (3) is obtained as

$$\rho \dot{a}_i = (\overset{\nabla}{\sigma}_{ji})_{,j} + (\sigma_{jk}W_{ik})_{,j} + (W_{jk}\sigma_{ki})_{,j} + (\sigma_{ji}v_{k,k})_{,j} - (\sigma_{ki}v_{j,k})_{,j}, \tag{5}$$

where the objective Jaumann- or co-rotational rate $\overset{\nabla}{\boldsymbol{\sigma}}$ and the material derivative $\dot{\boldsymbol{\sigma}}$ are related through

$$\dot{\sigma}_{ji} = \overset{\nabla}{\sigma}_{ji} + \sigma_{jk}W_{ik} + W_{jk}\sigma_{ki}. \tag{6}$$

2.4 Moment equilibrium equation

In a Cosserat continuum the angular momentum balance of the standard continuum is extended in two respects. Firstly, a surface moment **M** is introduced in addition to the moment of the stress tractions (again, volume forces and volume moments are neglected for simplicity); secondly we have to include the angular momentum of the particles $\int_V \mathbf{J}\varphi^c dV$, where **J** is the moment of inertia of the microstructure. For spherical grains we have, $J_{ij}=(2mR^2N/5V)\delta_{ij}=\rho(D^2/10)\delta_{ij}$.

The angular momentum balance reads:

$$\frac{D}{Dt}\int_V (\mathbf{x}\times\rho\mathbf{v}+\mathbf{J}\varphi^c)dV = \int_S \mathbf{x}\times\mathbf{T}dS + \int_S \mathbf{M}dS. \tag{7}$$

With $T_i=\sigma_{ji}\lambda_j$ and $M_i=\mu_{ji}\lambda_j$ and applying Gauss' theorem we obtain

$$\int_V \frac{D}{Dt}(J_{ij}\varphi_j^c)dV = \int_V \varepsilon_{ikj}\sigma_{kj}dV + \int_V \mu_{ji,j}dV. \tag{8}$$

Because of the arbitrariness of the elementary volume, the integrand of (8) must be zero, which assuming spherical granules, gives the following moment equilibrium equation for every point of the elementary volume:

$$\rho(D^2/10)\psi_i = \varepsilon_{ikj}\sigma_{kj} + \mu_{ji,j},\ \psi_i = \dot{\varphi}_i^c. \tag{9}$$

The material derivative of (7) is obtained as:

$$\rho(D^2/10)\dot{\psi}_i = \varepsilon_{ikj}\overset{\nabla}{\sigma}_{kj} + \varepsilon_{ikj}\sigma_{kp}W_{jp} + \varepsilon_{ikj}W_{kp}\sigma_{pj} + \varepsilon_{ikj}\sigma_{kj}v_{q,q} +$$

$$+(\overset{\nabla}{\mu}_{ji})_{,j} + (\mu_{jk}W_{ik})_{,j} + (W_{jk}\mu_{ki})_{,j} + (\mu_{ji}v_{k,k})_{,j} - (\mu_{ki}v_{j,k})_{,j}. \tag{10}$$

3 Constitutive equations

We consider a three-dimensional assembly of identical spherical grains. The diameter D of the grain is assumed as much smaller than typical dimensions of the problem under consideration. The particle arrangements is assumed to be statistically homogeneous, ie for the sphere positioned at $\boldsymbol{x}$ there will be always a sphere coming into a contact. In our idealised model, every point of the equivalent continuum corresponds to the centroid of the reference sphere in the discrete material.

The starting point for our derivation are the equations of motion of the discrete system. The force and moment equilibrium equations for our representative sphere read $m\dot{\mathbf{v}}_{sp} = \mathbf{F}$ and $J_{sp}\dot{\varphi}_{sp} = \mathbf{l}\times\mathbf{F}+\mathbf{M}$, where m is the mass of the sphere, $\boldsymbol{v}_{sp}$ is the velocity vector, $\boldsymbol{\varphi}_{sp}$ is the spin vector, J_{sp} is the rotational inertia of the sphere, $J_{sp}=2mR^2/5$, **F** and **M** are the force and moment resultants exerted by the neighbouring spheres on the representative sphere, **l** is the lever-arm of the force (vector between the sphere center and the point of contact).

Consider a representative volume V containing N spheres. The equations of force and moment equilibrium may be written as $\rho\dot{\mathbf{v}}_{sp} = (6\nu_s/\pi D^3)\mathbf{F}$, $\rho(D^2/10)\dot{\varphi}_{sp} = (6\nu_s/\pi D^3)(\mathbf{l}\times\mathbf{F}+\mathbf{M})$, where the volume fraction $\nu_s = V_s/V = N\pi D^3/(6V)$, V_s is the volume of all N spheres in the volume V they occupy, ρ is density.

Following [1] the summation over the particle contacts in the discrete system is replaced by an equivalent integral representation in which the integration is performed over the sphere's surface and a distribution function $A(\boldsymbol{x},\boldsymbol{n})$ is introduced to account for the orientational distribution of the contacts:

$$\sum_{contacts}(\cdot) \mapsto \int_{\alpha} dnA(\mathbf{x},\mathbf{n})(\cdot),$$

$$\mathbf{n} = \{\sin\theta\cos\phi,\ \sin\theta\sin\phi,\ \cos\theta\},\ \int_{\alpha} dn = \int_0^{2\pi}\int_0^{\pi} \sin\theta d\theta d\phi, \tag{11}$$

$\boldsymbol{n}$ is the unit vector from the centre of a sphere to a contact on its surface, θ and ϕ are coordinate angles of a spherical coordinate system with the origin at the sphere (grain) centre, $\boldsymbol{x}$ is the position vector with respect to a spatially fixed frame of reference. The function $A(\boldsymbol{n})$ is such that $A(\boldsymbol{x},\boldsymbol{n})dn$ is a probable number of contacts in the element dn centred at $\boldsymbol{n}(\theta,\phi)$. For distributions which are isotropic and independent of position we have $A(\boldsymbol{x},\boldsymbol{n})=k/4\pi$, where $k(t)$ is the coordination number (the average number of contacts per grain). We note that generally k maybe time dependent, however we will constraint ourselves to the time independent distribution. (If particles are in rolling contact then momentum is only transferred in the normal direction. A simple way to consider (not done here) rolling contacts is to distinguish between normal and shear contacts, ie introduce separately coordination numbers, normal k^n and shear k^s, where $k^s k^n$ because rolling contacts do not contribute to momentum transfer.)

Applying the integral operation (11) to the force and moment equations one has: $\mathbf{F} = \int_{\alpha} A\mathbf{F}^n dn$, $\mathbf{l}\times\mathbf{F}+\mathbf{M} = (D/2)\int_{\alpha} A\mathbf{n}\times\mathbf{F}^n dn + \int_{\alpha} A\mathbf{M}^n dn$, where $\mathbf{F}^n$ and $\mathbf{M}^n$ are the force and moment acting on the sphere at the point of contact at $\boldsymbol{x}+(D/2)\boldsymbol{n}$, and $\boldsymbol{l}=(D/2)\boldsymbol{n}$ is the lever-arm of the force $\mathbf{F}^n$ (Fig. 1). The averaged equations of motion are obtained as:

$$\rho\mathbf{a}_{sp} = \frac{6\nu_s}{\pi D^3}\int_{\alpha} A\mathbf{F}^n dn,\ \rho\frac{D^2}{10}\psi_{sp} = \frac{3\nu_s}{\pi D^2}\int_{\alpha} A\mathbf{n}\times\mathbf{F}^n dn + \frac{6\nu_s}{\pi D^3}\int_{\alpha} A\mathbf{M}^n dn,$$

$$a_{i\,sp} = \dot{v}_{i\,sp},\ \psi_{i\,sp} = \dot{\varphi}_{i\,sp} \tag{12}$$

where $\boldsymbol{v}_{sp}=\boldsymbol{v}_{sp}(\boldsymbol{x},t)$ and $\varphi_{sp}=\varphi_{sp}(\boldsymbol{x},t)$ are the expected values of the velocities and spins, ie the averages over all positions of the particles in which the reference sphere centre is originally at $\boldsymbol{x}$.

The equations of motion (12) are written for each sphere of the discrete system. When we consider the continuum description, each point of the continuum is thought as being associated with the centroid of the reference sphere. Velocities $\boldsymbol{v}_{sp}$ and spins φ_{sp} of the sphere are understood in the continuum description as translational and rotational degrees of freedom of each point of the continuum.

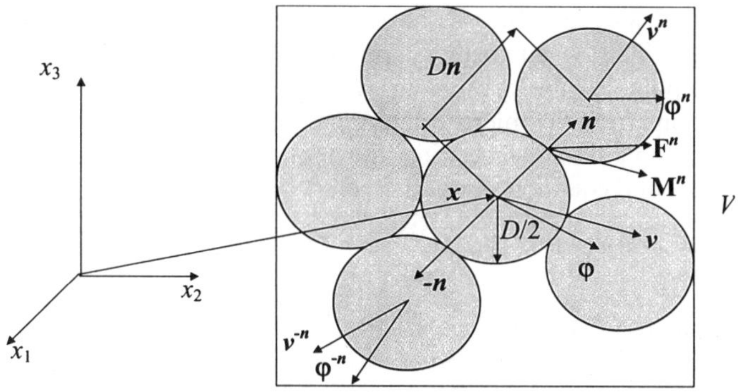

Figure 1. Reference sphere and contacting spheres in a volume element.

Next we suppose that the ergodicity (as a property that every sequence or sizable sample is equally representative of the whole) is satisfied, ie the average over realisations, that is over all positions of the reference sphere, is equal to the average over large volume element. The average over realizations is given by averaging of the force equilibrium equation over all positions of contacts at the representative reference sphere:

$$\left\langle \rho a_{i\,sp} \right\rangle_{realizations} = \rho a_{i\,sp} = \frac{6\nu_s}{\pi D^3} \int_\alpha A F_i^{\,n} dn . \tag{13}$$

According to the ergodicity hypothesis

$$\left\langle \rho a_{i\,sp} \right\rangle_{realizations} = \left\langle \rho a_{i\,sp} \right\rangle_{\substack{volume \quad in \\ discrete \quad medium}} \equiv \frac{1}{N} \sum_{k=1}^{N} \rho a_{i\,sp}^{\;(k)} , \tag{14}$$

which is the average over large volume element V containing N particles, $a_{i\,sp}^{\;(k)} = \dot{v}_i^{\,(k)}$ is the acceleration of the kth sphere in the volume V. The summation in (14) is conducted over all particles, which is a discrete equivalent to the volume integration. We keep in mind that we aim to construct an equivalent continuum representation of the discrete material such that their macroscopic (averaged over large volume elements) behaviours are equivalent. That is why the volume average, the right hand part of (14), must be equal to the volume average in the equivalent continuum, ie

$$\left\langle \rho a_{i\,sp} \right\rangle_{\substack{volume \quad in \\ discrete \quad medium}} = \left\langle \rho a_{i\,sp} \right\rangle_{\substack{volume \quad in \\ continuum}} = V^{-1} \int_V \rho a_i dV , \tag{15}$$

where in the right-hand side of (15) a_i is the acceleration of a point of the continuum. Equation (4) is the equation of motion for the continuous medium. Since all macroscopic field variables are assumed to be smooth and change little within the volume element (in the sense that the characteristic lengths of the variations must be much greater than the size of the volume element), one can write $V^{-1} \int_V \rho a_i dV \approx \rho a_i$. The insertion of (4) into this formula together with (13)-(15) gives

$$\frac{6\nu_s}{\pi D^3} \int_\alpha A F_i^{\,n} dn = \sigma_{ji,j} . \tag{16}$$

In the left-hand side of (16) we have $\rho a_{i\,sp}$, where $a_{i\,sp}$ is the acceleration of a reference sphere, which is equated by ρa_i in the right-hand side, where a_i is the acceleration of a point of the continuum. We are now in the position to deduce the constitutive equations from the microstructure.

Following the same line of reasoning, the application of the ergodicity hypothesis to the moment equilibrium equation gives

$$\frac{3\nu_s}{\pi D^2}\int_\alpha A[\mathbf{n}\times\mathbf{F}^n]_i\,dn+\frac{6\nu_s}{\pi D^3}\int_\alpha AM_i^{\,n}dn=\varepsilon_{ijk}\sigma_{jk}+\mu_{ji,j}\,. \qquad (17)$$

In the left-hand side of (17) we have $\rho(D^2/10)\psi_{i\,sp}$, where $\psi_{i\,sp}$ is the angular acceleration of a reference sphere, which is equated by $\rho(D^2/10)\psi_i$ in the right-hand side, where ψ_i is the angular acceleration of a point of the continuum.

One can show that the stress and moment stress tensors can be defined as (the definition of stress tensor consistent with the one suggested in [4]):

$$\sigma_{ji}=\frac{3\nu_s}{\pi D^2}\int_\alpha AF_i^{\,n}n_j\,dn,\ \ \mu_{ji}=\frac{3\nu_s}{\pi D^2}\int_\alpha AM_i^{\,n}n_j\,dn\,. \qquad (18)$$

Insertion of (18_1) into (5) and $(18_{1,2})$ into (10) respectively gives:

$$\rho\dot{a}_i=\left(\overset{\nabla}{\overline{\frac{3\nu_s}{\pi D^2}\int_\alpha AF_i^{\,n}n_j\,dn}}\right)_{,j}+(\sigma_{jk}W_{ik})_{,j}+(W_{jk}\sigma_{ki})_{,j}+(\sigma_{ji}v_{k,k})_{,j}-(\sigma_{ki}v_{j,k})_{,j}\,, \qquad (19)$$

$$\rho\frac{D^2}{10}\dot{\psi}_i=\varepsilon_{ikj}\left(\overset{\nabla}{\overline{\frac{3\nu_s}{\pi D^2}\int_\alpha AF_j^{\,n}n_k\,dn}}\right)+\left(\overset{\nabla}{\overline{\frac{3\nu_s}{\pi D^2}\int_\alpha AM_i^{\,n}n_j\,dn}}\right)+\varepsilon_{ikj}\sigma_{kp}W_{jp}+\varepsilon_{ikj}W_{kp}\sigma_{pj}{}_{,j}+$$

$$+\varepsilon_{ikj}\sigma_{kj}v_{q,q}+(\mu_{jk}W_{ik})_{,j}+(W_{jk}\mu_{ki})_{,j}+(\mu_{ji}v_{k,k})_{,j}-(\mu_{ki}v_{j,k})_{,j}\,. \qquad (20)$$

We also assume (Mühlhaus, [5]) that the analogue of linear response to force and moment [3] holds for the Jaumann derivative of force and the velocity and for the Jaumann derivative of moment and the Cosserat rotation rate respectively, that is

$$\overset{\nabla}{\mathbf{F}^n}=K\Delta\mathbf{v}^n,\ \overset{\nabla}{\mathbf{M}^n}=L\Delta\varphi^{c^n}, \qquad (21)$$

$$\Delta\mathbf{v}^n=\mathbf{v}^n-\mathbf{v}+D\mathbf{n}\times\varphi^c+\frac{D}{2}\mathbf{n}\times\left(\varphi^{c^n}-\varphi^c\right),\ \Delta\varphi^{c^n}=\varphi^{c^n}-\varphi^c, \qquad (22)$$

$$K_{ij}=(k_n-k_s)n_in_j+k_s\delta_{ij},\ L_{ij}=(k_{\varphi n}-k_{\varphi s})n_in_j+k_{\varphi s}\delta_{ij}, \qquad (23)$$

K and L are the matrixes of the translational and rotational spring stiffnesses, k_n, k_s and $k_{\varphi n}$, $k_{\varphi s}$ are the normal and shear (tangential) contact stiffnesses of the translational and rotational springs and the indices refer to a spatially fixed Cartesian coordinate system.

We assume that the distribution function A (particle arrangement/packing) and area of integration dn do not change (the distribution of normal vectors over the particle surface is independent of t). We replace the integration over the spherical angle with the integration over half spherical angle and take into account the symmetry of the distribution of the contacts function A in integrals in (19), (20). The corresponding integrands contain finite differences (velocities and Cosserat rotation rates for diametrically opposite contacts, ie at $\boldsymbol{n}$ and $\boldsymbol{-n}$). Replacing these v- and φ^c- differences by the corresponding Taylor expansions, keeping the leading the terms in D, and extracting the strain rates and curvature rates (1) one arrives to the following motion equations:

$$\rho\dot{a}_i = \frac{6\nu_s}{\pi D}\left[(k_n - k_s)A_{ijlm}\gamma_{lm} + k_s A_{lj}\gamma_{li}\right]_{,j} + (\sigma_{jk}W_{ik} + W_{jk}\sigma_{ki} + \sigma_{ji}v_{k,k} - \sigma_{ki}v_{j,k})_{,j}, \quad (24)$$

$$\rho\frac{D^2}{10}\dot{\psi}_i = \varepsilon_{ijk}\frac{6\nu_s}{\pi D}\left[(k_n - k_s)A_{kjlm}\gamma_{lm} + k_s A_{lj}\gamma_{lk}\right] + \frac{6\nu_s}{\pi D}\left[\left(k_{\varphi_n} - k_{\varphi_s}\right)A_{ijlm}\kappa_{lm} + k_{\varphi_s}A_{lj}\kappa_{li}\right]_{,j} +$$

$$+\varepsilon_{ikj}\sigma_{kp}W_{jp} + \varepsilon_{ikj}W_{kp}\sigma_{pj} + \varepsilon_{ikj}\sigma_{kj}v_{q,q} + (\mu_{jk}W_{ik})_{,j} + (W_{jk}\mu_{ki})_{,j} + (\mu_{ji}v_{k,k})_{,j} - (\mu_{ki}v_{j,k})_{,j}. \quad (25)$$

Comparing (24) and (25) with (5) and (10) respectively we obtain the following objective constitutive relationships

$$\overset{\nabla}{\sigma}_{ji} = \frac{6\nu_s}{\pi D}\left[(k_n - k_s)A_{ijlm}\gamma_{lm} + k_s A_{lj}\gamma_{li}\right],$$

$$\overset{\nabla}{\mu}_{ji} = \frac{6\nu_s}{\pi D}\left[\left(k_{\varphi_n} - k_{\varphi_s}\right)A_{ijlm}\kappa_{lm} + k_{\varphi_s}A_{lj}\kappa_{li}\right]. \quad (26)$$

The constitutive equations (26) provide the frame invariant constitutive relationships. They are consistent with the principle of the material objectivity since the Jaumann derivatives are objective, so are the rates of the Cosserat strain and curvatures in the right hand side. The beauty of these constitutive relationships is that the non-standard elastic moduli C_{ijlm}, C_{lj}, D_{ijlm}, D_{lj}

$$C_{ijlm} = \frac{6\nu_s}{\pi D}(k_n - k_s)A_{ijlm}, \quad C_{lj} = \frac{6\nu_s}{\pi D}k_s A_{lj}, \quad D_{ijlm} = \frac{6\nu_s}{\pi D}\left(k_{\varphi_n} - k_{\varphi_s}\right)A_{ijlm},$$

$$D_{lj} = \frac{6\nu_s}{\pi D}k_{\varphi_s}A_{lj}, \quad (27)$$

$$A_{lj} = \int_{\alpha/2} A n_l n_j dn = \frac{k}{6}\delta_{lj},$$

$$A_{ijlm} = \int_{\alpha/2} A n_i n_j n_l n_m dn = \frac{k}{30}\{\delta_{ij}\delta_{lm} + \delta_{il}\delta_{jm} + \delta_{im}\delta_{jl}\} \quad (28)$$

are the same as in the constitutive relationships for infinitesimal deformations [2, 3]. They are obtained from the microstructure (grain and contact distribution) and thus do not need to be calibrated from the experiments like in the phenomenological modelling.

4 Conclusion

Frame invariant constitutive relationships for the components of the stress and moment stress rate tensors are obtained for the isotropic and independent of the position distribution of contacts. The elastic moduli are expressed in terms of the contact stiffnesses, solid volume fraction, coordination number and diameter of the sphere. These constitutive equations hold for other statistically homogeneous particle arrangements; only components A_{ijlm} and A_{ij} in (26-28) need to be reevaluated according to the specified packing architecture, ie to the specified distribution function A. We have not attempted to include every possible micro structural feature of a real granular material into our model. Instead we have concentrated on a few typical properties such as coordination, discreteness of contacts, the six kinematically independent degrees of freedom of the individual grains. These form the basic ingredients for a macroscopic theory inspired by a selection of micro-mechanical features. The model may provide order of magnitude estimates if experiments are not available or provide guidance to the experimentalist in the selection of quantities to measure or observe.

Acknowledgment

The authors acknowledge the support of the UWA Research Grant 2001. The first author acknowledges the support of the F S Shaw Memorial Postgraduate Scholarship, UWA (January-July 2001).

References

1. Mühlhaus, H-B. and Oka, F. Dispersion and wave propagation in discrete and continuous models for granular materials. *Int. J. Solids Structures*, **33**, (1996) pp. 2841-2858.
2. Pasternak, E. and Mühlhaus, H.-B. Non-classical continua for modelling of granulate materials. In: *The 2001 ANZIAM Applied Mathematics Conference Abstracts* (Denier, J.P. and Tuck E.O., eds), Barossa Valley, South Australia, February 3-7, 2001, p. 64.
3. Pasternak, E. and Mühlhaus, H.-B. Cosserat continuum modelling of granulate materials, *APCOM01 The 1st Asian-Pacific Congress on Computational Mechanics*, Sydney, November 20-23, 2001, Elsevier, (accepted).
4. Jenkins, J.T. Volume change in small strain axisymmetric deformations of a granular material, In: *Micromechanics of Granular Materials*, (Satake, M. and Jenkins, J.T., Eds.), Elsevier Science Publishers B. V., Amsterdam, (1988) pp. 245-252.
5. Mühlhaus, H.-B., Moresi, L. and Sakaguchi, H. Discrete and continuum modelling of granular materials, In: *Constitutive Modelling of Granular Materials*, (Kolymbas, D., Ed.), Springer, Berlin, (2000) pp. 209-224.

GRANULAR MECHANICS OF SLUGS IN PNEUMATIC CONVEYING

R. PAN AND M.G. JONES
Department of Mechanical Engineering, The University of Newcastle, Australia
E-mail: merp@engmail.newcastle.edu.au

In this paper, an experimental method is presented to measure the properties of slug (an aerated material bed). Based on the measured properties, pressure drop in low-velocity slug-flow pneumatic conveying can be predicted accurately. The method can be applied to bulk solid materials with regular, irregular and/or unusual physical properties (e.g. different shape, density, size and size distribution), as long as they are good candidates for this mode of pneumatic conveying.

1 Introduction

In industry, a large amount of bulk solid materials are stored and transported in the form of an aerated packed-bed. A typical example is low-velocity slug-flow pneumatic conveying. Due to its low power consumption, product damage and pipeline wear, low-velocity slug-flow has been used widely in industry, especially in food industry. In such flow mode, bulk solids are transported in slugs and there is no relative motion between the bulk solids in the slug [1-3].

When slugs move along a horizontal pipeline, the slug picks up the particles from the stationary bed in front of it and leaves the same quantity of particles behind it. During the collection of particles from the stationary bed, the slug is subjected to an axial stress, which results in a radial stress. Therefore, the friction between the slug and the pipeline wall is produced due to the radial stress and the slug weight. The pressure drop required to carry the slugs along the horizontal pipeline mainly is to overcome such friction.

Based on the mechanism of friction produced between the slug and the pipeline wall, a simple and specific test chamber was designed and built up. The bulk solids were filled into the test chamber and held by two porous plates to form a slug. The slug in the chamber could be oriented either horizontally or vertically. When located vertically, different weights were applied on the top of the slug. The slug with weights was aerated and moved up by increasing air mass flow rate. Therefore, the relationship between the pressure gradient and the axial stress was developed. When located horizontally, the pressure gradient to overcome the friction due to slug weight was determined.

White plastic pellets were used as test material. After the properties of the slug (i.e. the relationship between the pressure gradient and the axial stress and the pressure gradient to overcome the friction due to slug weight) were determined, the pressure drop was predicted in a full-scale installation where the white plastic pellets were transported in slugs. By comparing the predicted pressure drop to the experimental values, a good agreement was achieved.

2 Theory

When material is transported in slugs, there is no relative motion between bulk solids in the slug [1-3]. Figure 1 shows a particle slug element which is subjected to air pressure and stresses in a horizontal pipe [2].

If the material obeys the Coulomb failure criterion, then [1]:

$$\tau_w = \mu_w \sigma_w + c_w = \mu_w (k_w \sigma_x + k_h \rho_b g) + c_w \tag{1}$$

where k_w is the force (stress) transmission coefficient or the ratio of radial to axial stress and k_h is the slug weight transmission coefficient. Usually, k_w is constant and k_h varies along the circumference of the pipeline.

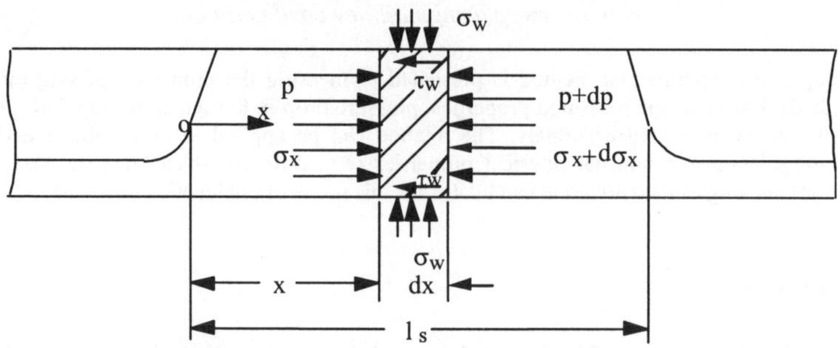

Figure 1. Air pressure and stresses acting on a horizontal particle slug [2].

When a moving slug reaches steady state, there is an equilibrium between the driving force and the resistance force. Assuming the axial stress and its radial transmission stress are functions of x only, the balance of the forces acting on the element of length dx for cohesionless materials (i.e. $c_w = 0$) results in:

$$\frac{dp}{dx} + \frac{d\sigma_x}{dx} + \frac{\int_0^{2\pi} \tau_w R d\theta}{A} = 0 \tag{2}$$

Substituting Equation (1) into Equation (2) gives:

$$\frac{dp}{dx} + \frac{d\sigma_x}{dx} + \frac{4\mu_w k_w}{D}\sigma_x + \frac{2\rho_b g \mu_w k_{hh}}{\pi D} = 0 \tag{3}$$

where: $k_{hh} = \int_0^{2\pi} k_h d\theta$

For a given bulk solid material and pipeline wall, it is believed that μ_w, k_w, ρ_b and k_{hh} in Equation (3) are constant. Also, the pressure gradient ($\delta p = dp/dx$) can be assumed constant as long as the mean conveying conditions (based on average air density) are used. After integrating Equation (3), an equation for pressure gradient in a single horizontal slug can be determined:

$$\delta p_m = \frac{4\mu_w k_w}{D}\sigma_f + \frac{2\rho_b g \mu_w k_{hh}}{\pi D} \tag{4}$$

Equation (4) clearly shows that the pressure gradient is the function of the friction due to the axial stress, σ_f and the slug weight. Assuming $B1 = 4\mu_w k_w$ and $B2 = 2\rho_b g \mu_w k_{hh}/\pi D$, therefore:

$$\delta p_m = B1\frac{\sigma_f}{D} + B2 \tag{5}$$

For a given bulk solid material and pipeline wall, B1 and B2 also are constant. The stress σ_f in Equation (4) or (5) is caused by the "collection" of particles from the stationary bed. An estimation of its value can be determined from a momentum balance:

$$\sigma_f = \rho_b (U_p - U_s) U_s \tag{6}$$

Note that U_p is the slug velocity and U_s is the average particle velocity in the slug. By using a gas/liquid analogy [4], the relationship between U_p and U_s is obtained as:

$$U_p = U_s + 0.542\sqrt{gD} \tag{7}$$

Substituting Equation (7) into (6), and then into Equation (5) gives:

$$\delta p_m = 0.542\rho_b B1\sqrt{\frac{g}{D}}U_p - 0.294gB1\rho_b + B2 \tag{8}$$

Equation (8) contains two unknown parameters, δp_m and U_p. To determine these two parameters, another equation is required. The following packed-bed model [5] is used quite often for this purpose:

$$\delta p_m = aU_{sp} + b\rho_{fm}U_{sp}^2 \tag{9}$$

where:

$$U_{sp} = U_a - U_p \tag{10}$$

Substituting Equation (10) into Equation (9) results in:

$$\delta p_m = b\rho_{fm}U_{sp}^2 - (a + 2b\rho_{fm}U_a)U_p + aU_a + b\rho_{fm}U_a^2 \tag{11}$$

Hence, using Equations (8) and (11), δp_m and U_p can be determined. That is, for a cohesionless material:

$$A1U_p^2 + A2U_p + A3 = 0$$

where

$$A1 = b\rho_{fm}$$

$$A2 = -\left(a + 2b\rho_{fm}U_a + 0.542\rho_b B1\sqrt{\frac{g}{D}}\right)$$

$$A3 = aU_a + b\rho_{fm}U_a^2 + 0.294gB1\rho_b - B2$$

It is expected that U_a - U_p is greater than zero. Therefore:

$$U_p = \frac{-A2 - \sqrt{A2^2 - 4A1A3}}{2A1} \tag{12}$$

Substituting Equation (12) into Equation (8) or Equation (11) provides the pressure gradient, δp_m. The pressure drop across the entire slug can be obtained if the length of the slug is known.

The total length of slugs in the pipeline can be calculated by the following correlation [1]:

$$L_s = \frac{m_s L}{A(1-\alpha)\rho_b U_p}$$

where:

$$\alpha = \frac{A_{st}}{A} = \frac{0.542\sqrt{gD}}{U_p}$$

Therefore, the total pressure drop is:

$$\Delta p = \delta p_m L_s \tag{13}$$

3 Test Program

3.1 Test rig

Equation (5) indicates clearly that the pressure gradient is to overcome the friction due to the axial stress and the slug weight. B1 represents the relationship between the pressure gradient and the friction due to the axial stress and B2 is the pressure gradient for the friction due to the slug weight. Also from Equations (8) and (11), it is clear that the pressure drop and slug velocity can be predicted after B1, B2, a and b have been determined.

Based on Equations (5) and (9), a simple and specific test chamber was designed and built, as shown in Figure 2.

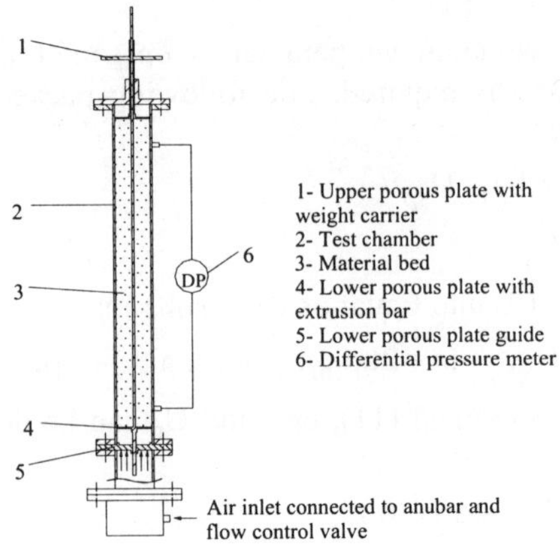

Figure 2. Schematic layout of test chamber.

The internal diameter of the chamber was 105 mm ID and the length of the test section was 1 m. The slug was formed in-between two porous plates which were linked firmly by the extrusion bar and could be moved together. Both porous plates had guides to refrain them from touching the side wall of the chamber whilst in motion. All guides and bushes were made from Teflon to reduce any unnecessary friction.

The top porous plate also was fitted with a carry plate. When the chamber was oriented vertically, different weights were placed on the top of the slug to simulate different values of σ_f. The pressure gradient was obtained after the pressure drop across a given distance of the slug (650 mm) was measured with a Differential Pressure (DP) meter. Also a pressure transmitter was fitted underneath the lower porous plate for determining the mean air density in the material bed. The air mass flow rate was monitored with an annubar.

When the chamber was vertical, a predetermined weight was placed on the weight carrier in each experiment. Air flow was increased using a manually operated globe valve until the material bed could be seen to start moving in an upward motion. Once this movement was detected, the air flow was left constant for the material bed to travel about 40 mm up. At the completion of the test, the air was shut off and porous plates returned to starting position. When the chamber was horizontal, the same procedure was applied with zero weight added on the weight carrier.

3.2 Test materials

White plastic pellets, which are good candidates for low-velocity slug-flow pneumatic conveying, were employed as test material. The physical properties of white plastic pellets were ρ_s=834 kg m^{-3}, ρ_b=578 kg m^{-3}, d_p=4100 μm (equivalent volume diameter), $\phi=44^0$ and $\phi_w=12^0$.

3.3 Test results and analysis

After the white plastic pellets were filled into the test chamber, a slug with a length of 930 mm was formed. The tests were run when the slug in the chamber was oriented vertically and horizontally. Table 1 lists the test results for white plastic pellets when the chamber was located vertically.

Table 1. Test results for white plastic pellets when the chamber was vertical.

Exp No	m_f (kg s^{-1})	p_l (kPag)	δp_m (kPa m^{-1})	w (N)	δp_m-$\rho_b g$ (kPa m^{-1})	σ_f (kPa)	U_a (m s^{-1})
1	0.0128	12.40	13.06	7.10	7.40	0.82	1.162
2	0.0199	26.45	26.20	26.72	20.54	3.09	1.682
3	0.0266	40.02	40.64	56.15	34.98	6.48	2.117
4	0.0369	66.52	67.97	85.58	62.31	9.88	2,638
5	0.0396	74.56	75.29	105.20	69.63	12.15	2,738
6	0.0445	91.67	88.16	154.24	82.50	17.81	2.850

To determine B1 for horizontal slug-flow, δp_m - $\rho_b g$ is plotted against σ_f/A. The determined value for B1 is 0.696.

The test chamber also was oriented horizontally. The air flow was increased until the slug started moving. The measured pressure gradient was 4638 Pa m^{-1} (i.e. B2 = 4638 Pa m^{-1}). Rewriting Equation (9) gives:

$$\frac{\delta p_m}{U_{sp}} = a + b\frac{m_f}{A} \tag{15}$$

Since the material bed in the test chamber was stationary, $U_{sp} = U_a$. Using the results in Table 1, the superficial air velocity, U_a (based on mean air density) can be determined, see Table 1. Therefore, the plot of $\delta p_m/U_a$ against m_f/A is obtained to determine the values for a and b. The determined values are a = 1382 Pa s m^{-2} and b = 6495 m^{-1}.

4 Pressure Drop Prediction

After B1, B2, a and b have been determined, the equations in Section 2 can be used to predict the pressure drop. Table 2 lists the predicted and experimental values of pressure drop for white plastic pellets in a full-scale conveying pipeline (L=21,7 m and D=60.14 mm ID).

Table 2. Predicted and experimental pressure drop.

m_f (kg s^{-1})	m_s (kg s^{-1})	Δp_e (kPa)	Δp_c (kPa)	m_f (kg s^{-1})	m_s (kg s^{-1})	Δp_e (kPa)	Δp_c (kPa)
0.0278	0.396	17.96	18.02	0.0307	0.726	29.98	32.70
0.0251	0.395	16.24	18.50	0.0252	0.689	30.20	32.76
0.0226	0.389	15.90	18.84	0.0260	0.656	30.26	30.86
0.0201	0.399	22.30	20.19	0.0256	0.526	25.23	24.66
0.0199	0.399	23.45	20.24	0.0266	0.512	21.85	23.73
0.0309	0.750	30.13	33.75	0.0253	0.494	22.94	23.21
0.0340	0.710	26.57	31.16	0.0206	0.366	20.77	18.28

Table 2 clearly demonstrates that, based on the determined properties of the slug in Section 3, the pressure drop can be predicted accurately. Since the slug in the test chamber was formed between two porous plates and its properties (i.e. the relationship between the pressure gradient and the axial stress and the pressure gradient to overcome the friction due to slug weight) were measured directly, the test material could have regular or irregular particles.

5 Conclusions

In low-velocity slug-flow, the bulk solids are transported in slugs and there is no relative motion between the bulk solids in slug. Along a horizontal pipeline, the pressure drop is to overcome the friction due to the axial stress and the slug weight.

Based on the mechanism of the friction produced between the slug and the pipeline wall, a simple and specific test chamber was designed and built. The test chamber could be located either vertically or horizontally. When located vertically, the relationship between the pressure gradient and the axial stress can be obtained. When located horizontally, the pressure gradient to overcome the friction due to the slug weight is measured.

The properties of the slug measured from the test chamber can be used directly to predict the pressure drop in low-velocity slug-flow pneumatic conveying. A good accuracy can be achieved.

The method developed in this paper is very simple and accurate and also can be applied to bulk solid materials with regular, irregular and/or unusual physical properties (e.g. different shape, density, size and size distribution), as long as they are good candidates for this mode of pneumatic conveying.

References

1. Konrad, K., Harrison, D., Nedderman, R.M. and Davidson, J.F., Prediction of the Pressure Drop for Horizontal Dense Phase Pneumatic Conveying of Particles, Pneumotransport 5, Fifth Int. Conf. on the Pneumatic Transport of Solids in Pipes, London, UK, April, 1980.
2. Mi, B. and Wypych, P.W., Pressure Drop Prediction in Low-Velocity Pneumatic Conveying, Powder Technology, Vol. 81, 1994, pp. 125-137.
3. Pan, R. and Wypych, P.W., Pressure Drop and Slug Velocity in Low-Velocity Pneumatic Conveying of Bulk Solids, Vol. 94, 1997, pp. 123-132.
4. Davidson, J.F. and Harrison, D., Fluidised Particles, Cambridge University Press, 1963.
5. Ergun, S., Fluid Flow through Packed Columns, Chemical Engineering Progress, Vol. 48, 1952, pp. 89-94.

Nomenclature

A	Cross-sectional area of pipe, m^2
A_{st}	Cross-sectional area of stationary bed, m^2
c_w	Particle cohesion at wall
D	Internal diameter of pipe, m
g	Acceleration due to gravity, m s^{-2}
l_s	Length of single slug, m
L	Total horizontal pipeline length, m
m_f	Air mass flow rate, kg s^{-1}
m_s	Product mass flow rate, kg s^{-1}
p	Interstitial air pressure, Pag
p_l	Pressure underneath the slug in the test chamber, Pag
U_a	Superficial air velocity, m s^{-1}
U_{sp}	Slip velocity, $U_{sp} = U_a - U_p$, m s^{-1}
w	Weight placed on the top of the slug, N
x	Horizontal coordinate
α	Cross-sectional area ratio of stationary bed to pipe
ϕ	Internal friction angle, Degree
ϕ_w	Wall friction angle, Degree
μ_w	Wall friction coefficient, $\mu_w = \tan\phi_w$
ρ_b	Loose-poured bulk density, kg m^{-3}
ρ_f	Air density, kg m^{-3}
ρ_s	Particle density, kg m^{-3}
σ_w	Wall pressure, ie total normal wall stress, Pa
τ_w	Shear stress at wall, Pa

Subscript

c	Calculated or predicted
m	Mean conveying condition (based on average air density)
e	Experimental

DEM – A VIRTUAL LABORATORY FOR STUDYING COMPLEX PARTICULATE FLOW PROBLEMS

J.G. LOUGHRAN
School of Engineering, James Cook University, Australia

E-mail: Jeffrey.loughran@jcu.edu.au

S.I. ANDERSON
Rockfield Technologies Australia Pty. Ltd.

Email: scott.Anderson@rocktech.com.au

This paper presents research into the development of a computational system for modelling of complex particulate flow problems using the discrete element method (DEM). Important features of the system are: 1) particle generation with user-specified particle angularity and distribution in two and three dimensions; 2) a range of adhesion laws to enable clumping of particles and subsequent fracture / floculation based on a user-defined tensile force; 3) specification of interparticle and particle / wall friction and cohesion; 4) general prescription of drag kinematics; 6) solution efficiency; 7) post processing involving user defined cutting planes and animation.
With increasing computational power DEM is becoming a tool which can be used to explore complex modelling processes in three dimensions on a desktop computer. Results are given for two and three dimensional modelling of industrial flows: dragline bucket filling; flow from bottom dump rail wagons and silo flow.

1 Introduction

Many problems in process and manufacturing industries involve the flow of discontinuous and general discontinua such as granular materials and rock assemblies. Historically, these problems have been either idealized and modelled with continuum mechanics or dealt with empirically. With industrial problems, however, there is often a need to define flow behaviour around critical geometry. Furthermore, there may well be a marked interaction effect between movement of the granular material and the adjacent flow boundary, for example, elastoplastic deformation of the walls or even fracture and fragmentation. The discrete element method [9] when considered in association with nonlinear finite element mechanics [8, 12] represents a powerful tool for investigating such phenomena on a virtual platform.

This paper reports on the development of a computational system for modelling of such processes and gives results for particular industrial applications.

1.1 Background

From a continuum perspective soil and granular materials are usually modelled with Mohr-Coulomb or critical state models [4]. These models perform well under small to medium levels of strain but are inadequate for industrial problems involving general flow behaviour. Engineers traditionally employ bulk materials handling principles [11] to predict the flow behaviour for problems involving discharge. The flow behaviour is dependent primarily on particle size, particle cohesion, particle to wall adhesion and the geometry of discharge sections. For simple geometry and well-behaved material the flow behaviour can be predicted using empirical rules. For complex geometry and fine cohesive material, the performance of existing models is less satisfactory. The same comment can be made for the flow of rock assemblies.

Over recent years there has been a marked advance in both contact mechanics and computational power to the extent that problems involving a large number of contacts between essentially solid entities can now be modelled. The underpinning technique is the Discrete (or Distinct) Element Method which was first implemented by Cundall and Strack [3] in 1979 where it was applied to the stability of fractured rock slopes. Over recent times the method has been applied to a wide range of industrial applications [2, 5, 6, 7, 10].

2 The Mechanics of DEM

DEM is a numerical technique that treats granular material as an assembly of 2D or 3D particles (rigid or deformable) with an appropriate rheological prescription for interparticle and particle to facet contact. All researchers recognize the immense computations required when simulating industrial problems and generally utilize rigid particles with an appropriate penalty based interaction law for both particle to particle and particle to facet contact. Assuming particles are allowed a certain amount of penetration the penalty based contact models impose the *impenetrability condition* in a weak form [5]. The interparticle collision force between two particles (F_n) is determined from the magnitude of the penetration (u_n) and the normal and tangential relative velocities (v_n),

$$F_n = P_n(u_n) + C_n v_n \tag{1}$$

P_n and C_n represent the linear or nonlinear elastic spring force and the viscous damping coefficient respectively. The latter is specified as a percentage of critical damping giving

$$C_n = \xi_n \sqrt{\frac{M_1 M_2}{M_1 + M_2} K_n} \tag{2}$$

where the normal contact stiffness is $K_n = dP_n / du_n$ and M_i represents the masses of the contacting discrete elements. A similar set of expressions can be written for the tangential component of contact force [6]. Energy dissipation is often critical to discrete element problems and it is important for the contact models adopted to be able to represent this feature of energy loss in the simulation. Figure 1 illustrates 2D DEM and shows a spring-dashpot model in the normal and tangential directions for particle-particle contact.

If the particle to facet contact deformation is small the interaction laws (linear Hooke, Hertz, Winkler) can be written in power form

$$F_n = \alpha K_n \delta^m \tag{3}$$

where $F_n = \|F_n\|$ and the different values of α and m define the contact models. In practice (3) can be scaled for partial contact [5]. Friction also plays an important role with DEM and a modified classical Coulomb friction model is used [5].

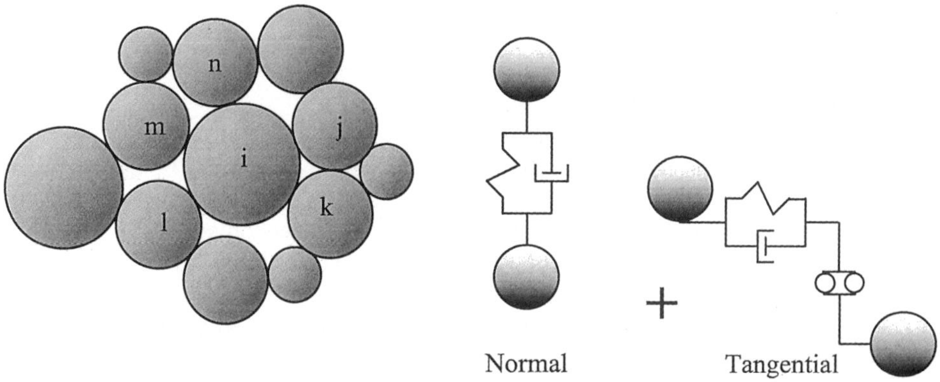

Figure 1. Schematic showing local equilibrium scheme for the 2D DEM with utilization of a spring-dashpot model for normal and tangential contact.

High moisture content materials exhibit cohesive properties in practice. Assuming that such behaviour results from interparticle attraction associated with surface tension, this can be approximately modelled by activating a cut-off force when a positive gap is experienced between particles [6].

A significant challenge to all DEM codes when modelling industrial processes is the need to have highly efficient contact detection algorithms given the potentially large number of particles involved. Here we adopt the alternating digital tree algorithm [1].

The local equilibrium scheme for the discrete element particles is given by Newton's second law, i.e.

$$M_p \ddot{u}_p = F_n \tag{4}$$

where M_p is the diagonal mass matrix and u_p and $\ddot{u}_p$ are respectively the particle displacement and acceleration vectors. Finite element discretization of the faceted structure, with appropriate boundary and initial conditions, gives

$$M_s \ddot{u}_s + C_s \dot{u}_s + F_s^{in} = F_s^{ex} \tag{5}$$

where M_s and C_s are respectively the mass and damping matrices of the boundary structure, u_s, $\dot{u}_s$, and $\ddot{u}_s$ are respectively the displacement, velocity and acceleration vectors, and F_s^{in} and F_s^{ex} the internal and external forces respectively. The impact forces between particles and between particles and the faceted structure are governed by contact interaction laws and are often both nonlinear functions. This effectively couples the relations (4) and (5).

The coupled system of equations are solved explicitly by the central difference algorithm with varying time steps.

Box 1 summarizes key DEM features available in the general-purpose finite element / discrete element code ELFEN.

- CAD interface for development of complex 2 D and 3D geometric models
- Particle assemblage generation
    - Regular-polygonal region
    - Circular region
    - User specified limits with a statistical population
    - Loose / tight packing
- Discs and ellipses in 2D
- Spheres in 3D
- Particle to particle contact with:
    - Normal and tangential stiffness
    - Normal and tangential damping
    - Rotational viscous damping
- 3D double-sided shell elements with particle to facet contact
- Artificial angularity for rolling resistance – 2D discs (only)
- 3D shell to rigid body FLAG
- Clumping spheres model with user defined groups (1,5,8) and bond strength
- Adhesion laws to model inter-particle surface tension effects
- Explicit joint elements for chain and rigging simulation
- COG (centre of gravity) feature specific to dragline bucket filling[1]
- Full geometric and material nonlinear boundary modeller with wide range of elasto-plastic constitutive models including continuum to discrete fracture
- Post processing features
    - Standard finite element output (displacement, velocity, reaction force, normal pressure etc.)
    - User defined cutting planes
    - Animation
    - Volume estimators for number of particles retained in user-specified region

Box 1. Summary of key DEM features in ELFEN

2.1 *Tuning Models for Application*

From a theoretical standpoint it is possible to determine inter-particle micro properties such as normal and tangential damping. However, in an industrial setting this may not be practical. A possible way forward is to tune the DEM parameters such that the global response of the DEM solution agrees with experimental observations, for example flow rates, drag forces and wall pressures. A post sensitivity analysis then allows both material and geometry parameters to be perturbed, thus giving insight into the flow mechanics.

3 Application

In this paper we give typical results for three industrial applications of DEM: the filling of dragline buckets; the flow of coal from bottom dump rail wagons and modelling of the

[1] The COG feature takes inertia data from a solid CAD model and imposes this at the centre of gravity of a rigid finite element shell model of the bucket. The kinematics of an empty bucket is then completely defined through the COG feature.

flow of coal from a hopper via a conveyer system. The first two applications involve 3D modelling and the latter 2D. Details of the dragline modelling are reported elsewhere [6]. Particular results are given here (Figure 2a) to show the 3D spatial response of a bucket as it cuts a path through the discrete element muck-pile. Note the captured material in the bucket during the lift stage. The steep front on the retained material is indicative of moist particulates and is achieved here by activating the adhesion function in ELFEN.

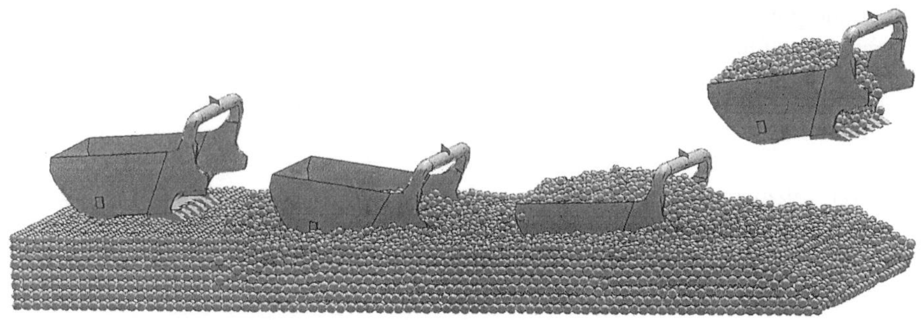

(a)

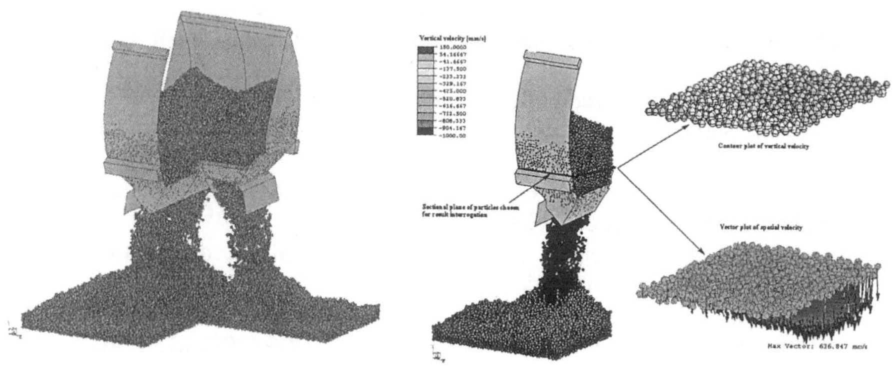

(b)

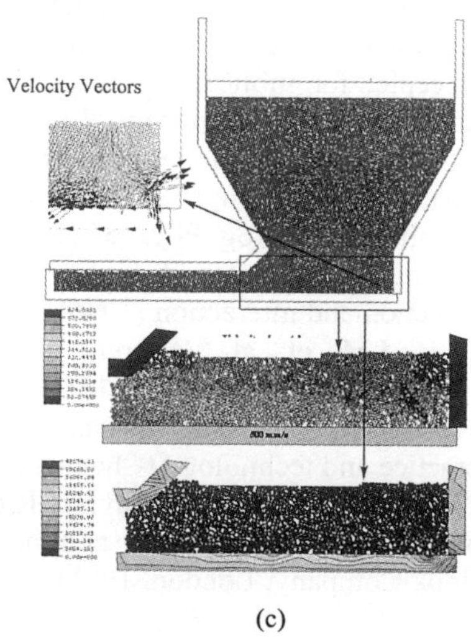

(c)

Figure 2. Examples of DEM of industrial applications: (a) Dragline bucket filling; (b) Investigation of the flow behaviour of coal from bottom dump coal wagons; (c) Investigation of the flow field around the exit region of a coal-hopper: *Centre response shows a velocity contour while the lower response shows vertical reaction force.*

The wagon results (Figure 2b) show the power of post processing DEM output. As material flows out of the wagon under gravity loading a contour plot of Y-velocity is reported. This is further interrogated by applying iso-surface cuts to the 3D solution. The end result is a single horizontal layer of particles at specified location with Y-velocity contours and vectors superimposed. Note the hold-up of material adjacent to a wall caused by particle to wall friction. Figure 2(c) shows the flow of coal on a conveyer belt below a hopper. The contour plots (from the bottom) show vertical reaction force on the conveyer and particle velocity respectively. The results were used to give insight into the flow behaviour of this industrial system and in particular wear of the conveyer belt.

4 Conclusions

The discrete element method when considered as an extension of conventional finite element technology represents a powerful tool for investigating industrial flow problems in industry. The paper describes the background mechanics and the key DEM features required to model complex flow problems.

References

1. Bonet, J. and Peraire, J., An alternating digital tree (ADT) algorithm for 3D geometric searching and intersection problems. Int. J. Num. Meth. Engng. **31** (1991) 1-17.
2. Cleary, P.W. (1998b). Predicting charge motion, power draw, segregation, wear and particle breakage in ball mills using discrete element methods. Minerals Eng. **11** (1998b) 1061-1080.
3. Cundall, P.A. and Strack, O.D.L., A discrete numerical model for granular assemblies. Geotechnique **29** (1979) 47-65.
4. Drucker, D.C. and Prager, W., Soil mechanics and plastic analysis or limit design. Quartely of Applied Mathematics **10** (1952) 157-165.
5. Han, K. et al, A combined finite /discrete element simulation of shot peening process, part I: Studies on 2D interaction laws, Eng. Comp. **17:5** (2000) 1593-619.
6. Loughran J.G. et al, Development of computational system for modelling of ground engaging processes with application to mining. APCOM01, Sydney. (2001) (accepted for publication).
7. Mishra, B.K. and Rajamani, R.K., The discrete element method for simulation of ball mills, App. Math. Model **16** (1994) 598-604.
8. Peric, D. and Owen, D.R.J.O., Finite element applications to the nonlinear mechanics of solids, Rep. Prog. Phys. **61** (1998) 1495-1474
9. Petrinic, N., Aspects of discrete element modelling involving facet to facet contact detection and interaction (1996), Dept. Civil Eng, Univ. of Wales PhD Thesis.
10. Ting, J.M. et. al., Discrete numerical models for soil mechanics, Journal of Geotechnical Engineering **115:3** (1988a) 379-397.
11. Woodcock, C.R. and Mason, J.S., Bulk solids handling: an introduction to the practice and technology (Chapman and Hall, New York, 1987)
12. Zienkiewicz, O.C. and Taylor, R.L., The finite element method: Solid and fluid mechanics, dynamics and nonlinearity (Fourth Edition), volume 2 (McGraw-Hill Book Company, London, 1991)

PEELING EFFECTS ON PERFORMANCE OF PIEZOELECTRIC ACTUATORS AND SENSORS IN SMART BEAMS

QUANTIAN LUO AND LIYONG TONG

School of Aerospace, Mechanical and Mechatronic Engineering, University of Sydney, NSW 2006

This paper demonstrates peeling effects on performances of piezoelectric actuators and sensors in a smart beam with symmetrically bonded piezoelectric patches. The numerical results including the peel stress are presented and compared with those of the shear stress model. Interacting shear and peel stresses induced by the applied voltage, the actuated forces and displacements in the host beam are calculated. The effects of the actuated peel stress on the performance of the host beam are studied. The transferring shear and peel stresses, and sensing electric charges caused by a bending moment are also calculated and compared with those of the shear stress model, and then the effects of the sensed peel stress and the sensing charge are investigated. The results show that the peeling effects on the smart beams with piezoelectric patches are significant.

Key words: Smart structure, Piezoelectric actuator and sensor, Peel stress

1 Introduction

Crawley and de Luis (1987) developed an analytical model for smart beams with symmetrically bonded piezoelectric (PZT) patches. Their derivation was based on the following assumptions: (a) only extensional deformation was permitted in the PZT patch; (b) there was only shear stress in the adhesive layer and the shear stress was constant along the thickness direction; (c) the host beam is deformed in extension or bending. As the shear stress transferring energy in the interface between PZTs and the host structure is a key to monitor and control the structure by itself, we refer this smart beam to as the shear stress model.

In the present model, referring to the exact solutions, it has been assumed that: (a) there were extensional and bending deformations in the PZT patch; (b) there existed the shear and peel stresses in the adhesive layer and the stresses were constants along the thickness direction; (c) the host beam could be deformed in both extension and bending. The present model can be applied to smart beams under the actions of any applied voltages and forces.

It is well known that beam bending deformation is one important aspect in structural analysis and design, and the flexible structures are widely used in aerospace engineering. Therefore, in this paper, we will mainly consider the host beam deformed in bending, and the smart beam with smaller thickness ratios of the host beam to the PZT patches.

We have demonstrated that, the thinner the adhesive layer, the larger the shear stress and the peel stress. However, the actuated forces or the sensing charges are almost the same for different thickness ratios of the PZT to the adhesive layer if this thickness ratio is satisfies the Goland and Reissner's assumption (1944), and thus the numerical results for different thickness ratios of the PZT patch to the adhesive layer will not be included in this paper.

2 Differential equations for the host beam deformed in pure bending

Referring to Figure 1, we can obtain the *equilibrium equations* and *constitutive relations*:

$$\left.\begin{aligned}&\frac{dN_a}{d\xi}+\tau_a=0,\ \frac{dQ_a}{d\xi}+\sigma_a=0,\ \frac{dM_a}{d\xi}+\frac{1}{2}r_1\tau_a-Q_a=0\\&\frac{dQ_{nh}}{d\xi}-2\sigma_a=0,\ \frac{dM_{nh}}{d\xi}+r_h\,\tau_a-Q_{nh}=0\end{aligned}\right\}\tag{1}$$

$$\left.\begin{aligned}&N_a=E_{n1}\,r_1\frac{du_a}{d\xi}+\frac{e_{31}V_a}{G_aL},\ M_a=-\frac{1}{12}E_{n1}\,r_1^3\frac{d^2w_a}{d\xi^2}\ \ M_{nh}=-\frac{1}{12}E_{nh}\,r_h^3\frac{d^2w_{nh}}{d\xi^2}\\&\tau_a=[-r_a\,u_a+\frac{R_{ta}}{2}(R_{ht}\frac{dw_{nh}}{d\xi}+\frac{dw_a}{d\xi})],\ \sigma_a=r_{av}\,(w_{nh}-w_a)\end{aligned}\right\}\tag{2}$$

where,

$$\left.\begin{aligned}&2N_a=N_{n1}-N_{n2},\ 2Q_a=Q_{n1}+Q_{n2},\ 2M_a=M_{n1}+M_{n2}\\&2u_a=u_{n1}-u_{n2},\ \ 2w_a=w_{n1}+w_{n2}\\&2\tau_a=\tau_{n1}+\tau_{n2},\ \ 2\sigma_a=\sigma_{n1}-\sigma_{n2},\ 2V_a=V_1-V_2\end{aligned}\right\}\tag{3}$$

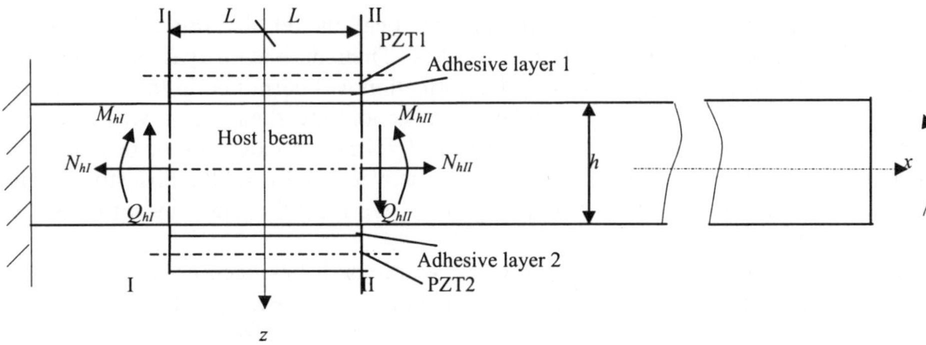

Figure 1 a host beam with the symmetrically bonded PZTs

The solutions for the host beam in pure bending are (refer to the exact solutions):

$$\tau_a=A_1\sinh\beta_1\xi\ +A_4\cosh\beta_2\xi\sin\beta_3\xi+A_5\sinh\beta_2\xi\cos\beta_3\xi\tag{4}$$
$$\sigma_a=B_2\cosh\beta_1\xi\ +B_3\sinh\beta_2\xi\sin\beta_3\xi+B_6\cosh\beta_2\xi\cos\beta_3\xi\tag{5}$$

Non-dimensional parameters are:

$$\left\{\begin{aligned}&N_{ni}=\frac{N_i}{G_aL},\ Q_{ni}=\frac{Q_i}{G_aL},\ M_{ni}=\frac{M_i}{G_aL^2},\ E_{ni}=\frac{E_i}{G_a},\ \tau_n=\frac{\tau}{G_a},\ \sigma_n=\frac{\sigma}{G_a},\ u_{ni}=\frac{u_i}{L},\ w_{ni}=\frac{w_i}{L}\\&r_i=\frac{t_i}{L},\ \xi=\frac{x}{L},\ r_a=\frac{L}{t_a},\ r_{av}=\frac{E_a}{G_a}r_a,\ R_{ta}=\frac{t_1}{t_a},\ R_{ht}=\frac{h}{t_1}\qquad(i=1,2,h)\end{aligned}\right.\tag{6}$$

In the above equations, E_i, and t_i ($t_i = t_1$, h) are elastic moduli and the thickness of the substructures; u_i and w_i (i =1, 2, h) are the displacements; N_{ni} (i = 1, 2, h) are the internal forces; e_{31} and V_i (i =1, 2) are piezoelectric permittivities of the PZTs and the applied voltages; E_a, and G_a are the elastic and shear modulus of the adhesive layer. A_i, B_i and β_i are the integration constants and the eigenvalues.

For the shear stress model, the equations and solution are:

$$\frac{dN_a}{d\xi} - \tau_a = 0, \quad \frac{dM_{nh}}{d\xi} + r_h\,\tau_a = 0, \quad \tau_a = r_a\,u_a + \frac{R_{ta}R_{ht}}{2}\frac{dw_{nh}}{d\xi} \tag{7}$$

$$\tau_c = A_{c1}\sinh\beta_c\xi + A_{c2}\cosh\beta_c\xi, \text{ where, } \beta_c = \sqrt{\frac{R_{ta}}{r_1^2}\left(\frac{1}{E_{n1}} + \frac{6}{E_{nh}R_{ht}}\right)} \tag{8}$$

On the basis of solutions to the stresses, the equilibrium and the sensing equation (Lee, 1992), we can calculate, the actuated forces and displacements in the host beam, and the sensing charges in the PZTs. The numerical results are calculated using the data: $E_1 = 7\times10^{10}$ N/m^2, $e_{31} = -5.2$ m/v, $t_1 = 0.001$ m. $L = 0.01$ m, $V_a = -100$ v, $E_a = 3\times10^9$ N/m^2, $\nu = 0.4$, $R_{ta} = 10$, $E_h = 7\times10^{10}$ N/m^2, $R_{ht} = (1\sim30)$. These data are equivalent to the properties of PZT4, epoxy resin and aluminum (Robbins and Reddy, 1991).

3 The actuated performances

The maximum stresses are plotted in Figure 2, which shows that the shear stresses by Crawley's model are larger than those of the present one. To compare stresses in both models, the Von Mises stress is defined as: $\sigma_{von} = \sqrt{\sigma^2 + 3\tau^2}$, and the absolute relative difference between the Von Mises stresses predicted is also plotted in the Figure.

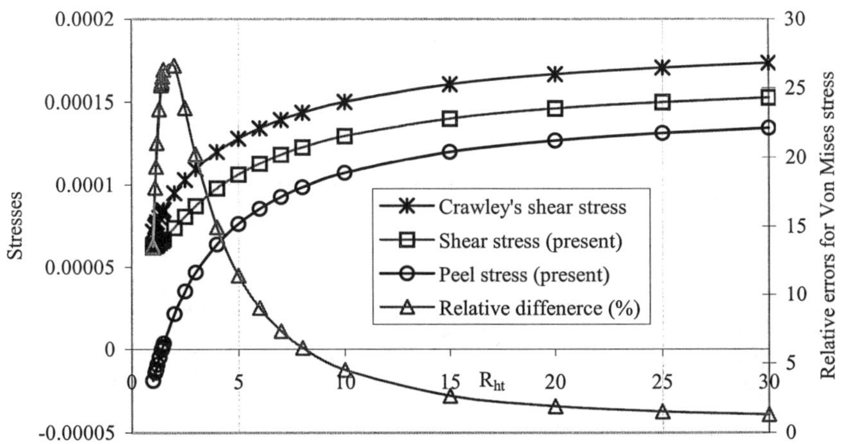

Figure 2 The actuated stress at the edge Vs. R_{ht} and the comparison

For the given example, the relative errors (%) for Von Mises stresses are 4.5, 6.1, 8.9, 11 and 20 when R_{ht} are equal to 10, 8, 6, 4 and 3. That is, if the shear stress model is employed, the relative error is larger than 5% when R_{ht} is less than 10 and the relative error is larger than 10% when R_{ht} is less than 5.

The actuated bending moment is plotted in Figure 3, in which, $M^* = M_{nh}/R_{ht}$ and $M_c^* = M_{ch}/R_{ht}$. Subscript c all stands for the results of using the Crawley's model in this paper. From Figure 2, we can see that, the direction of bending moment produced by the peel stress is opposite to that by the shear stress if $E_h R_{ht}^2$ is less than $2E_1$, and the direction of bending moment by the peel stress is the same as that by the shear stress if $E_h R_{ht}^2$ is larger than $2E_1$. Figure 3 shows that, with the increase of R_{ht}, the actuated moment of the present model is less than, equal to and then larger than that of the shear stress model.

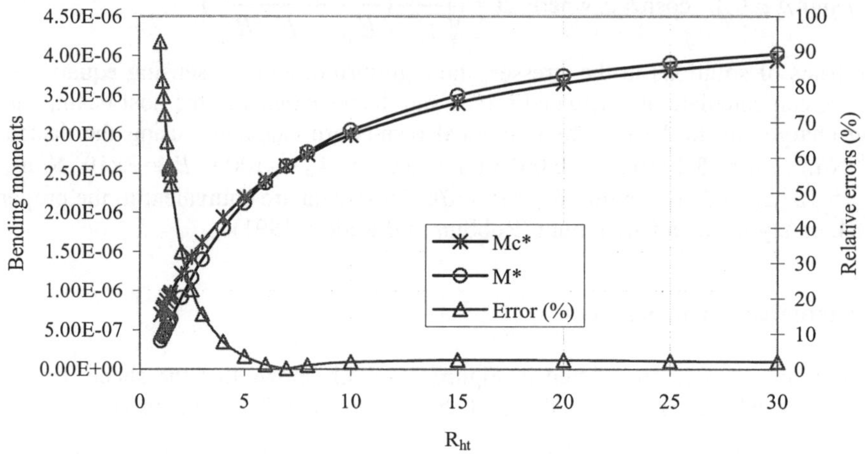

Figure 3 The actuated bending moment Vs. R_{ht} and the comparison

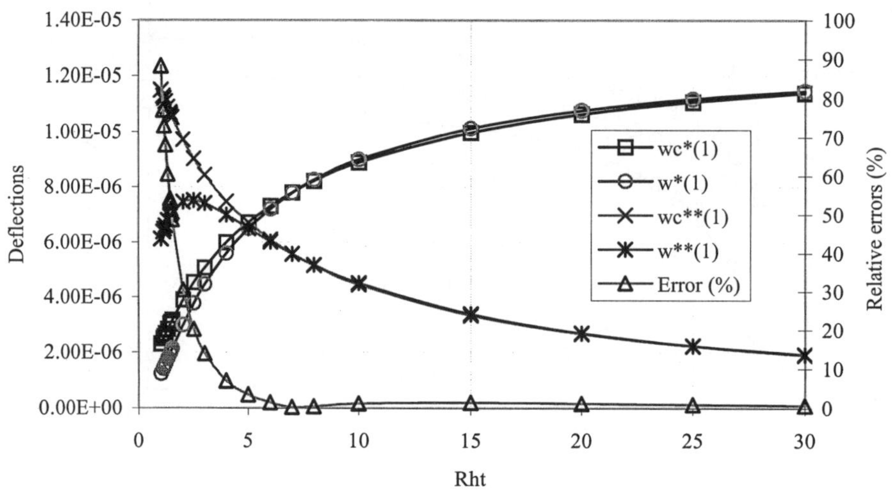

Figure 4 The actuated deflections Vs. R_{ht} and comparisons

The actuated deflection at $\xi=1$ is shown in Figure 4. To compare the curves clearly, we have let: $w^* = w_{nh} R_{ht}^2$, $w_c^* = w_{ch} R_{ht}^2$, $w^{**} = w_{nh} R_{ht}$, $w_c^{**} = w_{nh} R_{ht}$. Figures 3 and 4 indicate that, the absolute relative errors (%) for the bending moment (deflection) are 92 (88), 33 (30), 16 (14), and 7.7 (6.9) when R_{ht} are equal to 1, 2, 3, and 4 respectively.

4 The sensing performances

Extensional strain ε_m (= ± 7.42857) produced by the applied bending moment is assumed to apply to the top and bottom surfaces of the host beam respectively. The maximum stresses are plotted in Figure 5, and the Von Mises stress is also used for the purpose of comparison. Figure 6 indicates the sensing charge q_n (= q/L, q is the electric charge).

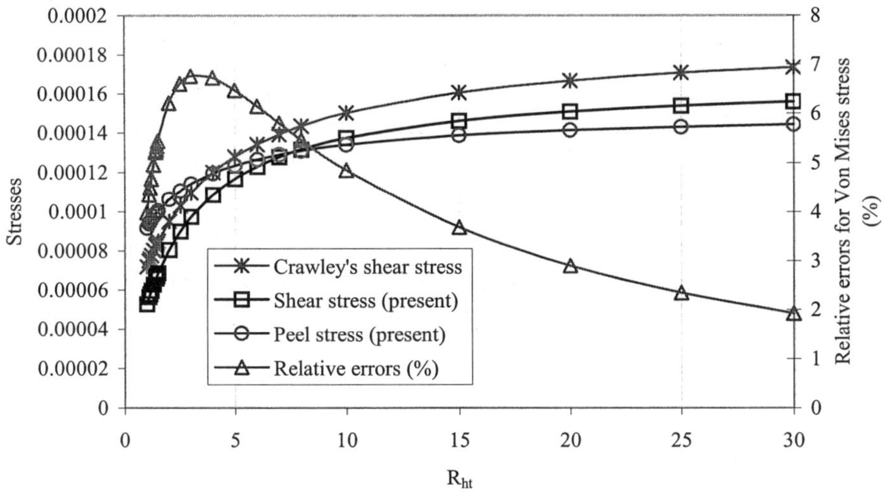

Figure 5 The stresses for the applied moment Vs. R_{ht} and comparisons

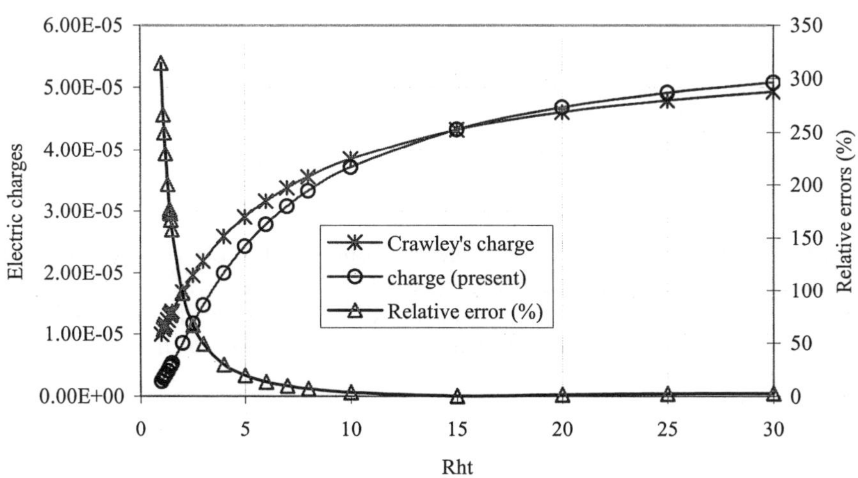

Figure 6 The sensing charge Vs. R_{ht} and comparisons

For the specific example, the absolute relative errors (%) for Von Mises stresses are 3.6, 7.0, 14, 30 and 49 when R_{ht} are equal to 10, 8, 6, 4 and 3 respectively. Namely, if the shear stress model is employed, the error will be larger than 13 % when R_{ht} is less than 6. It can be seen by Figure 5 that, the peel stress becomes larger as compared with the shear stress when R_{ht} tends to 1. In Figure 6, the relative error in electric charge is also shown.

It is noted that the relative error becomes quite noticeable when R_{ht} is less than 10, and it is as large as 300% when R_{ht} is equal to 1.

5 Discussion and Conclusion

The differences of the numerical results between the present model and the shear stress model are also related to the elastic modulus of the host beam and PZT. The actuated peel stress will be zero if $E_h R_{ht}^2 = 2E_1$, but the peel stress caused by the applied bending moment will not be equal to zero for any ratio of $E_h R_{ht}^2$ to $2E_1$.

As the adhesive layer become thinner, the increase of the peel stress will be more than that of the shear stress. For the very thin adhesive and smaller elastic modulus of the PZT, the peel stress may be twice as large as the shear stress. In this case, effects of the peel stress on the actuation and sensing functions of the PZT patches will also become more important.

For the bending actuation, the significant differences between the shear stress model and the present model were observed. In this case, the actuated strains in PZT patches are not uniform and thus the different deflections between the host beam and PZT causes peel stresses. Therefore, if the host beam is mainly deformed in bending, the peel effects on performance of piezoelectric actuators and sensors in smart beams can be significant.

Acknowledgement

The authors are grateful to the support of the Australian Research Council through its Large Grant Scheme (Grant No. A10009074).

References

1. Crawley E. F. and de Luis J., "Use of Piezoelectric Actuators as Elements of Intelligent Structures", *AIAA Journal*, **25** No.10, 1987, pp.1373-1385.
2. Goland M. and Reissner E., "The Stresses in Cemented Joints", *Journal of Applied Mechanics*, March 1944, A-17 - A-27.
3. Lee C-K, "Piezoelectric Laminates: Theory and Experiments for Distributed Sensors and Actuators", *Intelligent Structural Systems*, edited by Tzou H. S. and Anderson G. L., Kluwer Academic Publishes, 1992, pp.75-135.
4. Robbins D. H. and Reddy J. N., "Analysis of Piezoelectrically Actuated Beams Using A Layer-Wise Displacement Theory", *Computers and Structures*, **41**, No.2, 1991, pp.265-279.

VALIDATED SIMULATION TOOLS FOR STRUCTURAL WAVE ACOUSTO ULTRASONICS BASED ON PIEZOELECTRIC TRANSDUCERS

MARTIN VEIDT

Department of Mechanical Engineering, University of Queensland, Brisbane, Australia
E-mail: veidt@mech.uq.edu.au

TIERANG LIU

Department of Civil Engineering, University of Queensland, Brisbane, Australia
E-mail: t.liu@mailbox.uq.edu.au

SRITAWAT KITIPORNCHAI

Department of Building and Construction, City University of Hong Kong, Hong Kong
E-mail: bcskit@cityu.edu.hk

This paper reports on the application of multiple integral simulation tools to determine the characteristics of structural wave ultrasonic non-destructive evaluation systems for isotropic and anisotropic materials using adhesively bonded piezoelectric transducers. The electro-mechanical transduction behaviour of circular, rectangular and interdigital piezoelectric transmitter and receiver elements is discussed. The results demonstrate the importance of the application of advanced simulation tools to gain a better understanding of the complex electro-mechanical transduction processes and transducer-structure interactions, which is essential to enable the development of reliable quantitative non-destructive evaluation and health monitoring systems. In addition, it is shown that a hybrid numerical-experimental methodology is an efficient approach to optimise the performance of the electro-mechanical components of an acousto ultrasonic measurement system for a specific application and an important tool in the validation process of improved models.

1 Introduction

Due to the general pressure for increased efficiency of technical systems and the increased application of advanced materials in primary structural components, the development of in-situ, condition monitoring systems has attracted considerable attention throughout the entire engineering community. An evaluation methodology that has

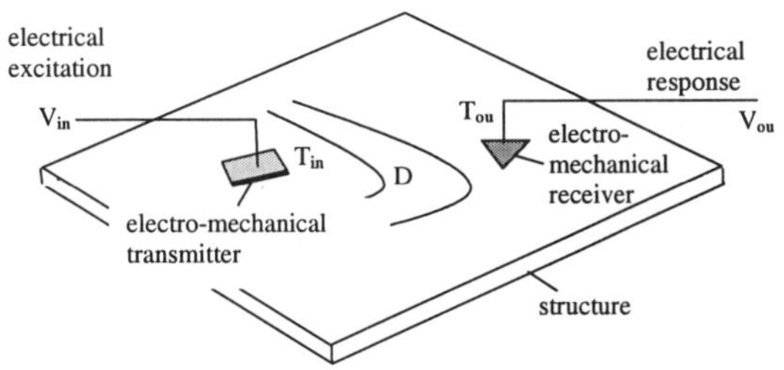

Figure 1. Schematic of AU measurement system.

proven to be very attractive in this context is the application of structural wave acousto ultrasonics using piezoelectric transducers, e.g. [1-3], which enables the inspection of large areas compared to conventional ultrasonic methods. Fig. 1 shows a general schematic of an acousto-ultrasonic measurement system.

The design of optimised measurement systems relies on the ability to quantify the three fundamental processes of wave generation, propagation and detection. Assuming linearity of the individual, physical processes the input-output characteristic of the system can be written as

$$V_{ou}(\omega) = T_{ou}(\omega) D(\omega) T_{in}(\omega) V_{in}(\omega) \quad (1)$$

where ω is the circular frequency, V_{in} and V_{ou} are the frequency spectra of the applied and measured voltage signals, respectively, T_{in} and T_{ou} are the transfer functions of the transmitter and receiver elements, respectively and D describes the propagation behaviour of the (dispersive) stress waves in the structure.

In this paper the electro-mechanical transduction function T_{in} of adhesively bonded piezoelectric transmitters are determined using a hybrid experimental-analytical methodology. In addition, the characteristics of piezoceramic receiver elements for flexural wave detection and interdigital transducers are discussed based on transducer models, which include dynamic piezoelectric effects and transducer-to-specimen interactions.

2 Transfer Characteristics of Piezoelectric Transmitters and Receivers

2.1 Calibration Methodology for Piezoceramic Transmitter Elements

In the measurement system used to determine the transfer functions of a piezoelectric transmitter element the transducer is excited using high voltage narrow band burst or broad band sweep pulses. The generated stress waves are measured using a heterodyne Doppler laser vibrometer as a non-contact sensor. The transfer function T_{in} is computed by rearranging (1) and using the transfer function of the laser vibrometer for T_{ou}, which has a constant value of one over the entire frequency range. As mentioned above, D(ω) describes the propagation characteristics of the travelling stress waves in the structure. For the present study a multi-layer spectral method, e.g. [4,5], is used to accurately model the influence of guided wave aspects.

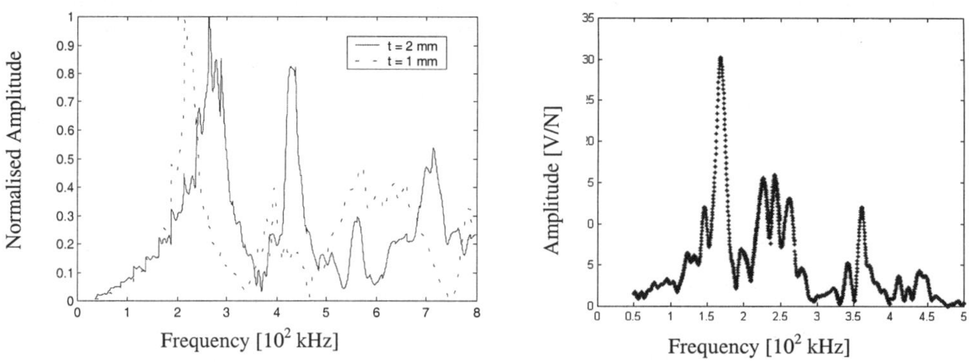

Figure 2. Transduction funtions of piezoceramic transducers a) 5-mm diameter, 2-mm thick disc on 1-mm and 2-mm thick aluminium plate; b) 10-mm diameter, 2-mm thick disc on 1.8-mm thick glass fibre-reinforced laminate

In Fig. 2a the magnitudes of the transfer functions of identical transmitters bonded to two aluminium plates with different thicknesses are compared. The scale factors used for normalising the amplitudes are 0.0035 for thickness 1-mm and 0.0245 for thickness 2-mm. Fig. 2b shows the magnitude of the transfer function of a 10mm diameter, 2mm thick piezoceramic disc attached to a 1.8mm thick, unidirectional glass fibre-reinforced laminate. The response signals used to calculate the transfer function were measured at 100mm from the centre of the transducer in the fibre direction.

The results presented in Figs. 2a and 2b are characteristic for all different transducer and transducer-to-specimen configurations investigated, e.g. [6,7]. The major findings can be summarised in the following way:

- The transducer-to-specimen transfer functions are very complex with frequency regimes of high and low amplitude electro-mechanical transduction characteristics whose magnitudes and locations in frequency space are different for individual transmitter-to-specimen configurations. Thus, it is generally not possible to predicting the behaviour of a piezoelectric transmitter bonded to one substrate based on its measured or calculated transfer characteristics for another transducer-to-specimen configuration.
- The knowledge of the transfer characteristic is essential to enable the efficient development of optimised acousto-ultrasonic measurement systems.
- The development of quantitative models accurately describing the transducer-to-specimen interaction is a challenging task.
- The presented measurement technique based on the use of the heterodyne Doppler laser vibrometer as a non-contact receiver device and a theoretical model to represent the propagation behaviour of the stress waves in the structure has great potential to be used as a generic calibration methodology to determine the transfer characteristics of piezoelectric transmitter elements.

2.2 *Sensitivity of Piezoelectric Sensors to Flexural Wave Detection in Composite Plates*

Fig. 5 shows a schematic of the problem under investigation. A piezoelectric crystal disc is attached to an orthotropic composite plate. It is used to detect incoming flexural waves generated by a single force at the origin of the cartesian coordinate system as shown. Investigating this configuration is important to enable the development of: 1) optimised acousto-ultrasonic measurement systems, 2) in-situ foreign object impacts detection systems and 3) sensor components in structural vibration control systems.

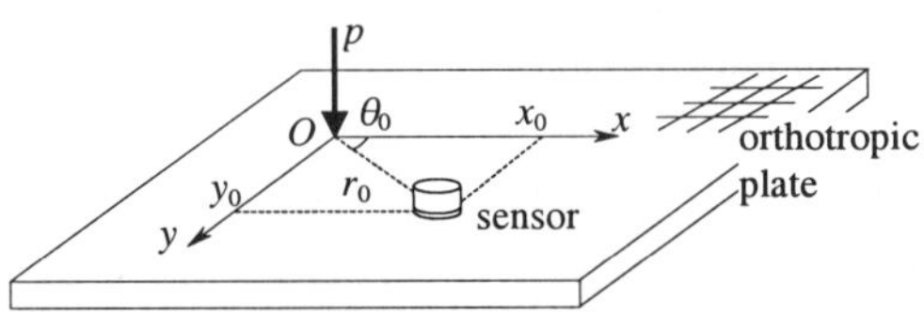

Figure 3. Piezoceramic sensor to detect flexural waves in an orthotropic plate

Using one-dimensional dynamic theory of piezoelectricity, the corresponding longitudinal wave theory in backing materials, which might be attached to the piezoelectric element, and thin plate wave propagation theory for flexural waves in an orthotropic plate, the governing equations of the problem are developed, [8].

The solution is calculated formulating the boundary and continuity equations for the out-of-plane normal stress and deflection on the surface and at the interfaces, respectively. The essential quantities to be calculated are the contact stress and the out-of-plane displacement under the transducer. They are determined by the superposition of the incoming and scattered wave fields, which are both calculated using the analytical solution of the Green's function of an orthotropic plate based on a multiple integral transform method, [5].

Figs. 4 and 5 show the influence of the thickness of the laminate and the thickness of some backing material on the sensitivity of the sensor system. The material parameters of the plate are ρ=1800kg/m^3, E_x=38.6.7GPa, E_y=8.27GPa, G_{xy}=4.14GPa and ν_{xy}=0.26; the piezoelectric material characteristics are s_{33}^E=18.5×10^{-12}m^2/N, d_{33}=374×10^{-12}m/V, ε_{33}^T=15045×10^{-12}F/m and ρ_1=7700kg/m^3; and the backing material is copper with ρ_2=8900kg/m^3 and E_2=117.2GPa.

Fig. 4 shows the influence of the thickness h of the orthotropic plate on the sensing characteristics of a 3mm diameter (a=3mm), 2mm thick (h_1=2mm) piezoelectric sensor, which is located along the principal direction x (θ_0=0°), 200mm away (r_0=200mm) from the excitation point. The admittance of the attached electric circuit is Y=10^{-7} 1/Ω.

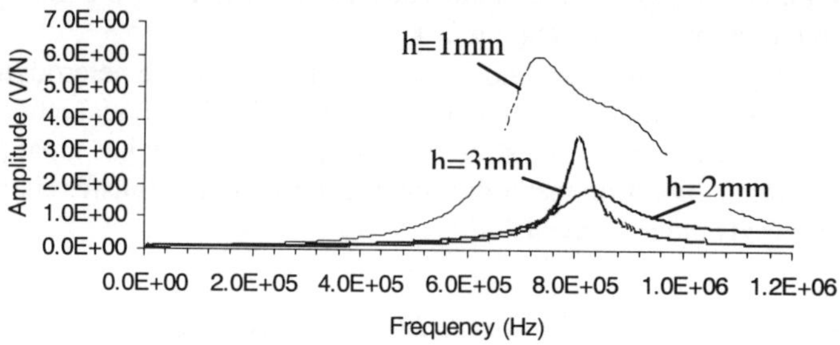

Figure 4. Sensor sensitivity for different laminate thicknesses (r_0=0.2m, θ_0=0°, a=3mm, h_1=2mm, Y=10^{-7} 1/Ω).

It is observed that completely different voltage output sensitivities result, especially in the high frequency regime. The magnitude as well as the bandwidth are changing. This implies that for a given structure, it is possible to design optimised sensors with maximum sensitivity in a specific frequency regime.

As expected, the characteristics of a sensor completely changes by attaching copper backings of various thicknesses, which is demonstrated in Fig. 5 for a 3mm diameter, 2mm thick sensor on a 2-mm thick laminate.

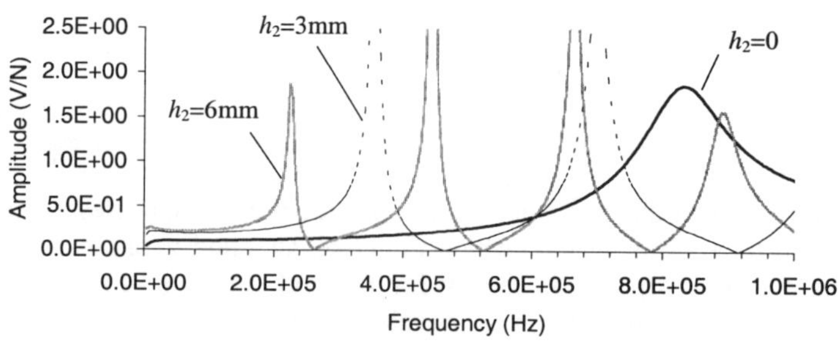

Figure 5. Sensor sensitivity for different copper backings (r_0=0.2m, θ_0=0°, a=3mm, h_1=2mm, Y=10^{-7} 1/Ω).

It is shown that, as expected, a thicker backing shifts the first resonance to lower frequencies, however makes the overall sensing behaviour of the transducer more complicated. But by carefully choosing the backing material for a sensor attached to a specific structure, it is possible to optimise the performance of the measurement system.

In summary, the results of the numerical simulation of the sensitivity of piezoelectric sensors with and without backing materials attached to orthotropic laminates show that:

- the sensitivity of a piezoelectric sensor attached to a structure depends not only on the properties of the sensor itself, but also on the characteristics of the tested structure and the admittance of the attached electric circuit. Hence, to achieve maximum sensitivity in a specific frequency regime, sensors must be designed according to the characteristics of the structure to be monitored and the electric characteristics of the attached measurement device.
- adding a backing material to a piezoelectric sensing element can improve the sensor's sensitivity, but the thickness of the backing needs to be designed carefully.

2.3 Interdigital Transducers (IDT)

IDT have been successfully used as single mode, structural wave transmitter and receiver elements, e.g. [9]. The multi-layer spectral method introduced in [5] has been modified to enable the detailed investigation of their transfer characteristics [10].

Fig. 6 shows an example of mode isolation using an IDT. The waves in a 2mm thick aluminium plate are considered. The waves are created by a 10 cycles, Hanning windowed excitation pulse with a centre frequency of f_0=1.25MHz and measured 0.3m away from the transmitter. According to the group velocity-frequency graph shown in Fig. 7, three (non-dispersive) wave modes are created using a conventional transducer with full top and bottom electrodes without mode isolation capability. This is illustrated in Fig. 6a. The application of an IDT enables the isolation of an individual wave mode. This is demonstrated in Fig. 6b for the s_0 mode using a 20 finger, 1.6mm finger spacing IDT.

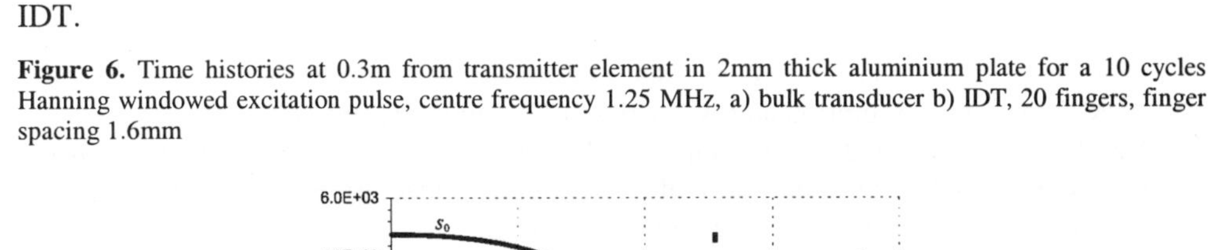

Figure 6. Time histories at 0.3m from transmitter element in 2mm thick aluminium plate for a 10 cycles Hanning windowed excitation pulse, centre frequency 1.25 MHz, a) bulk transducer b) IDT, 20 fingers, finger spacing 1.6mm

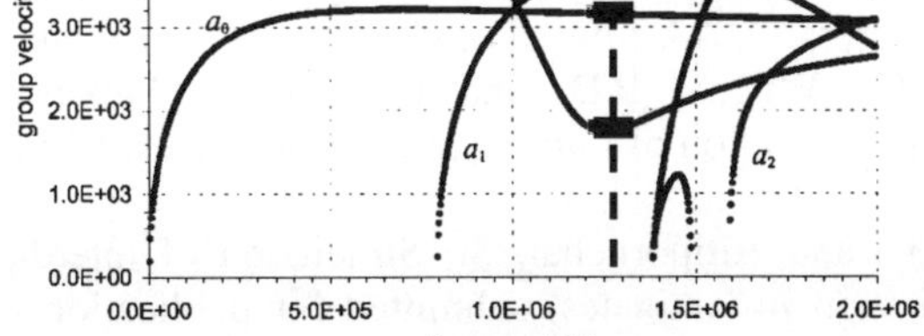

Figure 7. Group velocity-frequency graph for 2mm thick aluminium plate.

In summary, the results illustrate that IDT can be used for single mode, non-dispersive Lamb wave excitation, but careful design is necessary to avoid any unexpected effects.

3 Conclusion

It has been demonstrated that the electro-mechanical transfer characteristics of piezoelectric transmitter and receiver elements depend not only on the transducer's inherent features such as piezoelectric properties and geometry but also on the transducer-to-specimen interaction and the characteristics of possible backing materials. Hence, the optimisation of the input-output behaviour of an acousto-ultrasonic measurement system is a challenging task. Thus, the presented experimental-numerical calibration methodology for transmitter elements and the analytical models derived for a piezoelectric sensor attached to an orthotropic laminate and interdigital transmitters for single mode excitation are essential tools in the development of optimised health monitoring systems and smart structures.

4 Acknowledgment

This work has been supported by the Australian Research Council under Grant No. A89937136.

References

1. Chimenti, D.E., Guided waves in plates and their use in materials characterization, Applied Mechanics Review, 1997, 50, pp. 247-284.
2. Rose, J.L., Recent advances in guided wave NDE, IEEE Ultrasonic Symposium, 1995, pp. 761-770.
3. Alleyne D.N. and Cawley, P., Optimization of Lamb wave inspection techniques, NDT&E International, 1992, 25, pp. 11-22.
4. Liu, T., Kitipornchai, S. and Liew, K. M., Analysis of acousto-ultrasonic input-output characteristics for transmitting and receiving contact-type transducer coupled to a Timoshenko beam, AIAA Journal, 36, 1998, pp. 644-653.
5. Liu, T., Kitipornchai, S., Liew, K. M. and Wang, G., Analysis of acoustic-ultrasonic characteristics for contact-type transducers coupled to an orthotropic composite plate, ASME Journal of Vibration and Acoustics, 121, 1999, pp. 460-467.
6. Veidt, M., Liu, T. and Kitipornchai, S., Experimental investigation of the acousto-ultrasonic transfer characteristics of adhesively bonded piezoceramic transducers, Smart Materials and Structures, 9, 2000, pp. 19-23.
7. Veidt, M., Liu, T. and Kitipornchai, S., Flexural waves transmitted by rectangular piezoceramic transducers, in press: Smart Materials and Structures.
8. Veidt, M., Liu, T. and Kitipornchai, S., Analysis of piezoelectric sensor to detect flexural waves, in press: AIAA Journal of Guidance, Control and Dynamics.
9. Monkhouse R.S.C., Wilcox, P.D. and Cawley, P., Flexible interdigital PVDF transducers for the generation of Lamb waves in structures, Ultrasonics, 1997, 45, pp. 1184-1194.
10. Liu, T., Veidt, M. and Kitipornchai, S., Simulation of interdigital transducers for Lamb wave excitation in laminates, submitted for publication in: Journal of Sound and Vibrations.

SHAPE CONTROL OF SMART COMPOSITE PLATE STRUCTURES WITH NON-RECTANGULAR SHAPED PZT ACTUATORS

QUAN NGUYEN AND LIYONG TONG

School of Aerospace, Mechanical & Mechatronic Engineering, The University of Sydney, NSW, 2006 Australia

Shape control of a smart structure is deemed as achieving the desired shape of the structure through optimizing the electric fields applied to and the loci and sizes of the piezoelectric actuators attached to the structure. In this paper, a finite element analysis software has been developed for analyzing static deformation of smart composite plate structures with non-rectangular shaped PZT patches as actuators. The mechanical deformation is modeled using a 3rd order plate theory, while a layer wise theory is used to model the electric field. Numerical examples are obtained for the optimum values of electric fields in actuators to achieve the desired shapes using recently developed Linear Least Square.

Key Words: Shape control; Piezoelectric actuators/sensors; Composite plates; Finite element analysis

1 Introduction

The focus of many previous researches on shape control was to find the optimum values of electric fields in rectangular-shaped actuators for achieving the designed shape. In this case, the electric fields in the actuators are design variables and their optimum values have been sought to minimize the difference between the actuated and the desired shapes. Koconis et al[1] developed analytical methods for determining the optimal values of voltages applied to fixed rectangular-shaped actuators for achieving the specified shapes for sandwich plates and shells. Chee et al [2-3] employed the 3rd order plate theory[4] for mechanical deformation and the layer-wise theory[5] for modelling the electric field in the finite element formulation for shape control of smart composite plate structures with rectangular actuators. They developed the Buildup Voltage Distribution, Linear Least Square and Simplex algorithms for finding the optimal values of voltages in all actuators to achieve the desired shapes, and also introduced a new and challenging specified shape: twisting shape of a cantilever plate. In most research on shape control, the shape and location of an actuator are not treated as design variables, which might lead to a high-energy consumption because high voltages may be required. To overcome this issue, one possible approach is to optimise the distribution pattern of actuators, such as location, shape and size of an actuator, in addition to the values of the applied electric fields. This paper is presented to investigate the effect of actuator pattern change on optimum values of applied electric fields in shape control of smart plate structures.

2 Finite Element Formulation

In the finite element formulation, the 3rd order plate theory[3] and layer-wise theory[5] are used for mechanical deformations and electric fields respectively for developing 8-node non-rectangular plate element using typical shape functions for deformations[6].

The mechanical deformations are modelled using the 3rd order plate theory proposed by Lo et al [3] as

$$\begin{aligned} U(x,y,z) &= u_0(x,y) + z\psi_x(x,y) + z^2\zeta_x(x,y) + z^3\varphi_x(x,y) \\ V(x,y,z) &= v_0(x,y) + z\psi_y(x,y) + z^2\zeta_y(x,y) + z^3\varphi_y(x,y) \\ W(x,y,z) &= w_0(x,y) + z\psi_z(x,y) + z^2\zeta_z(x,y) \end{aligned} \tag{1}$$

where $U(x, y, z)$, $V(x, y, z)$ and $W(x, y, z)$ are total displacement function in x, y, z plane. They are composed in plane sub-functions as $u_o(x, y)$, $v_o(x, y)$, etc., which are separated from the thickness in z-dimension. The real displacements can be interpolated using shape functions and nodal variable for the 8-node isoparametric element as follows:

$$[U(x,y,z) \quad V(x,y,z) \quad W(x,y,z)]_e^T = [A(z)][N_u(\xi,\eta)][u_e] \tag{2}$$

where

$$[A(z)] = \begin{bmatrix} 1 & 0 & 0 & z & 0 & 0 & z^2 & 0 & 0 & z^3 & 0 \\ 0 & 1 & 0 & 0 & z & 0 & 0 & z^2 & 0 & 0 & z^3 \\ 0 & 0 & 1 & 0 & 0 & z & 0 & 0 & z^2 & 0 & 0 \end{bmatrix}$$

$$[N_u(\xi,\eta)] = \begin{bmatrix} N_1 & . & . & N_2 & . & . & . & . & N_8 & . & . \\ . & * & . & . & * & . & . & . & . & * & . \\ . & . & N_1 & . & . & N_2 & . & . & . & . & N_8 \end{bmatrix}$$

$$u_e = [u_1 \quad v_1 \quad w_1 \quad \ldots \quad \phi_{x1} \quad \phi_{y1} \quad \ldots \quad \ldots \quad u_8 \quad \ldots \quad \phi_{y8}]^T$$

and $[N_u(\xi,\eta)]$ is an interpolation function matrix of [11x88]. Star(*) represents the shape functions N_i (i = 1,…,8), which are the typical shape functions used for 8-node isoparametric element[6]. Dot (.) in the above matrix represents zeros.

The linear layer-wise formulation[5] has been considered as follows:

$$\Phi(x,y,z) = \sum_{j=1}^{n\_layer+1} L_j(z)\varphi_j(x,y) \tag{3}$$

The electric potential within the jth layer in term of nodal variables can be written as

$$\Phi(x,y,z) = [N_\phi]\times[\phi_{11} \quad \ldots \quad \phi_{(n+1)1} \quad \phi_{12} \quad \ldots \quad \phi_{(n+1)8}] \tag{4}$$

where: $N_\phi = [\ldots \quad L_j^d(z) \quad L_j^u(z) \quad \ldots]\times[N_u(\xi,\eta)]$; $L_j^d(z) = \dfrac{z - z_{j+1}}{z_j - z_{j+1}}$; $L_j^u(z) = \dfrac{z - z_j}{z_{j+1} - z_j}$

L_j is transverse layer-wise function. φ_j is electric field planar. L^d and L^u is top and bottom interface of layer respectively. z_{j+1} & z_j is top and bottom coordinates of L_j respectively.

For quasi-static shape control of a smart structure, the Hamilton's variational principle can be employed to develop the finite element formulation. The final system of equations, in terms of the nodal variables including both mechanical and electric quantities, is given by

$$\begin{bmatrix} K_{uu} & K_{u\varphi} \\ K_{\varphi u} & K_{\varphi\varphi} \end{bmatrix} \begin{bmatrix} u_g \\ \varphi_g \end{bmatrix} = \begin{bmatrix} F_g \\ -Q_g \end{bmatrix} \tag{5}$$

where u_g and φ_g are vectors of global nodal displacements and voltages respectively. F_g and Q_g are vectors of the applied mechanical loading and electrical charges respectively. K_{uu}, $K_{u\varphi}$, $K_{\varphi u}$ and $K_{\varphi\varphi}$ are matrices representing the total strain potential contribution of element irrespective of whether it is piezoelectric or otherwise, and they are given as follows:

$$
\begin{aligned}
K^e_{uu} &= \sum_{k=1}^{nlayer} \int_{\xi=-1}^{1} \int_{\eta=-1}^{1} \int_{z=z_k}^{z_{k+1}} (B^T_{bu} c_b B_{bu} + B^T_{su} c_s B_{su}) \left| \frac{\partial(x,y)}{\partial(\xi,\eta)} \right| dz\, d\eta\, d\xi \\
K^e_{u\phi} &= \left[K^e_{\phi u} \right]^T = \sum_{k=1}^{nlayer} \int_{\xi=-1}^{1} \int_{\eta=-1}^{1} \int_{z=z_k}^{z_{k+1}} (B^T_{bu} e_b B_{bu} + B^T_{su} e_s B_{\phi i}) \left| \frac{\partial(x,y)}{\partial(\xi,\eta)} \right| dz\, d\eta\, d\xi \\
K^e_{\phi\phi} &= \sum_{k=1}^{nlayer} \int_{\xi=-1}^{1} \int_{\eta=-1}^{1} \int_{z=z_k}^{z_{k+1}} -(B^T_{\phi i} \chi_i B_{\phi i} + B^T_{\phi o} \chi_o B_{\phi o}) \left| \frac{\partial(x,y)}{\partial(\xi,\eta)} \right| dz\, d\eta\, d\xi
\end{aligned}
\tag{6}
$$

where: $c_{b,s}$, $e_{b,s}$, $\chi_{o,i}$ are elastic, piezoelectric strain constants, electric permittivity matrix respectively and

$$
B_{bu} = [D_b][N_u] \qquad B_{su} = [D_s][N_u]
$$

$$
B_{\phi i} = [\frac{\partial}{\partial x} \quad \frac{\partial}{\partial y}][N_\phi] \qquad B_{\phi o} = [\frac{\partial}{\partial z}][N_\phi]
$$

where:
$$
D_b = \begin{bmatrix}
\frac{\partial}{\partial x} & 0 & 0 & z\frac{\partial}{\partial x} & 0 & 0 & z^2\frac{\partial}{\partial x} & 0 & 0 & z^3\frac{\partial}{\partial x} & 0 \\
0 & \frac{\partial}{\partial y} & 0 & 0 & z\frac{\partial}{\partial y} & 0 & 0 & z^2\frac{\partial}{\partial y} & 0 & 0 & z^3\frac{\partial}{\partial y} \\
0 & 0 & 0 & 0 & 0 & 1 & 0 & 0 & 2z & 0 & 0 \\
\frac{\partial}{\partial y} & \frac{\partial}{\partial x} & 0 & z\frac{\partial}{\partial y} & z\frac{\partial}{\partial x} & 0 & z^2\frac{\partial}{\partial y} & z^2\frac{\partial}{\partial x} & 0 & z^3\frac{\partial}{\partial y} & z^3\frac{\partial}{\partial x}
\end{bmatrix}
$$

$$
D_s = \begin{bmatrix}
0 & 0 & \frac{\partial}{\partial y} & 0 & 1 & z\frac{\partial}{\partial y} & 0 & 2z & z^2\frac{\partial}{\partial y} & 0 & 3z^2 \\
0 & 0 & \frac{\partial}{\partial x} & 1 & 0 & z\frac{\partial}{\partial x} & 2z & 0 & z^2\frac{\partial}{\partial x} & 3z^2 & 0
\end{bmatrix}
$$

Chee et al[3] proposed the Buildup Voltage Distribution (BVD), Linear Least Square (LLS) and Simplex algorithms for finding optimum values of voltages applied to actuators in shape control of plate structures. All three methods have been used to investigate the effect of non-rectangular shape on the optimum values of voltages and match to the desired shape. However, due to space limit, only the results obtained using the LLS method will be presented for two typical examples.

3 Numerical Examples

3.1 Example 1: Bending Test

Consider a long thick cantilever plate clamped at left end and subject to applied mechanical load, which has a length of 0.15 m, width of 0.06m and consists of 2 layers of thickness of 0.01m. The desired shape is described by equation $w^d(x,y)=10^{-5}\text{Cos}(10x-1)$. The material of the plate has the following stiffness constants: $c_{11}=c_{22}=c_{33}=82.68$GPa,

$c_{12}=c_{13}=c_{23}$=27.56GPa, $c_{44}=c_{55}=c_{66}$=26.5GPa. Three active patches are bonded to plate, which are modelled using six finite elements. The patch #1 and #3 are bonded at both ends and patch #2 is bonded at middle of the plate as numbered in Figure 1. The properties of patches include the stiffness constant: $c_{11}=c_{33}$=84.8GPa, c_{22}=29.68Gpa, $c_{12}=c_{13}=c_{23}$=36.35GPa, $c_{44}=c_{55}=c_{66}$=24.2Gpa; the electric permittivity: $\chi_{11}=\chi_{22}=15.3\times10^{-9}$F/M, $\chi_{33}=15.0\times10^{-9}$F/M; and piezoelectric strain constants d_{31}=254pm/V, d_{32}=-204pm/V, d_{33}=374pm/V, d_{24}=484pm/V, d_{15}=584pm/V.

The FE model of this structure composes of 10 elements. The model A has rectangular patches. In Model B, the patches are shifted backward of 14 degrees. The top and bottom element of the patches in model C & D has been shifted back and forward of 14 degrees respectively as shown in Figure 1. The total area of the patches is kept as constant.

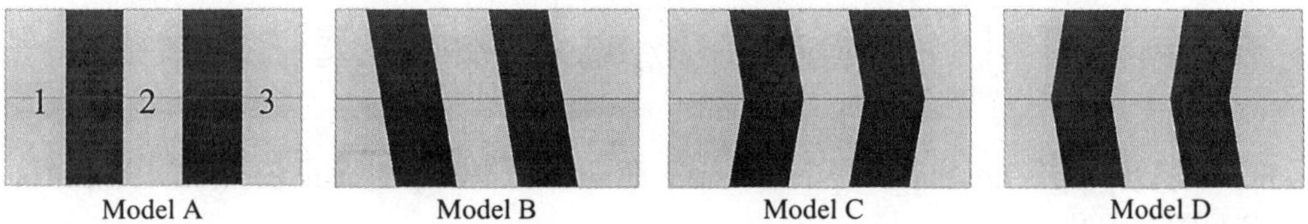

Figure 1: Cantilever plate with different patch pattern

3.2 Performance of bending shape control of cantilever plate with different patch pattern

The approach of this test is to investigate the effect of different patch pattern on the bending shape control of the cantilever plate. The voltages required for each patch to achieve the desired bending shape obtained the LLS algorithm[3] are listed in Table 1.

Comparing model B, C and D to model A. it shows that the model B has greatest achievement of desired shape, i.e., shape match factor Lnm[3] is reduced 16%, but the applied voltage on patches is high, patch #1, #2, #3 is increased 13%, 4% and 16% respectively. In model C, the applied voltage on patch #1, #2, #3 are decreased 4%, 2%, and 13% respectively with a slight increase of 1.5% in the shape factor.

Table 1:Voltage Distribution values for bending control cantilever plate with different patches pattern

Model	A	B	C	D
Patch/Voltage				
1	393.22	444.2	377.64	414.3
2	477.66	495.7	466.81	487.3
3	135.68	157.6	117.87	147.9
Lnm ($1e^{-5}$)	2.1172	1.7778	2.1497	1.9551

3.3 Example 2: Twisting test

Consider a cantilevered plate that has a length of 0.292 m, width of 0.152 m and with a thickness of 0.83 mm. There is no mechanical load in structure. The plate has the following material constant: $c_{11}=c_{22}=c_{33}$=82.68GPa, $c_{12}=c_{13}=c_{23}$=27.56GPa, $c_{44}=c_{55}=c_{66}$=27.6GPa. Fifteen piezoelectric ceramic patches of 0.25 mm thick are attached to top and bottom of the structure and has been indicated as shown in Figure 2 (model F). Patch #1, #6 and #11, each consist 2 elements. The properties of the patches include the elastic constants: $c_{11}=c_{22}=c_{33}$=84.8GPa, $c_{12}=c_{13}=c_{23}$=36.35GPa, $c_{44}=c_{55}=c_{66}$=24.2Gpa;

electric permittivity: $\chi_{11}=\chi_{22}=15.3\text{x}10^{-9}$F/M, $\chi_{33}=15.0\text{x}10^{-9}$F/M; piezoelectric strain constant d_{31}=-254pm/V, d_{32}=-204pm/V, d_{33}=374pm/V, d_{15}=584pm/V, d_{24}=484pm/V. The piezoelectric patch #1, #2, #3, #4, #5, #11, #12, #13, #14 and #15 have an area of 1950.75mm^2, patch #6 has area of 637.5mm^2, and patch #7, #8, #9 and #10 have area of 1275mm^2. The finite element mesh has 84 elements and total of 3 layers. This example focuses on the effect of the patch pattern for twisted problem. Consider 3 rows of patches along the plate, the middle row will be unchanged, while the patches on top and bottom row can be rotate. For model G, H and I, the top and bottom row patches can be rotated forward and backward of 14 degrees as shown in Figure 2. The left end of all models has been clamped. The desired shape is described by equation $w^d(x,y) = 10^{-3}(cosh(x)-1)sin(y)$. By using the LLS method[3], the optimum values of voltages applied to each patch are obtained and listed in Table 2.

Model F Model G Model H Model I

Figure 2: Twisted Cantilever plate with different patch pattern

3.4 Performance of displacement based twisted shape control with different patch pattern

Model F, G, H and I has highest magnitude of voltage is 4.23V, 5.51V, 6.52V, 3.15V, total of magnitude of voltage is 21.10V, 21.29V, 30.09V, 18.88 V and shape match factor Lnm is $2.744\text{x}10^{-3}$, $4.388\text{x}10^{-3}$, $2.285\text{x}10^{-3}$, $2.089\text{x}10^{-3}$ respectively. Hence the comparison of model I to model F show that the maximum magnitude of voltage, total voltage and shape match factor is reduced 25.5%, 10.5% and 23.9% respectively. The results show that there exist three major advantages in twisted shape control with different patch patterns: the reduction in magnitude of the maximum voltage applied to a single actuator patch, reduction in total voltage applied to all actuators and achievement to desired shape.

Table 2: Voltage distribution for cantilever plates with different patch patte

Model	F	G	H	I
Patch/Volt				
1	***4.23278***	***5.47422***	*2.91268*	*1.77098*
2	*1.76554*	*0.847548*	*1.5226*	***3.1535***
3	*2.02429*	*2.42914*	*2.19954*	*1.64567*
4	*0.962311*	*0.786611*	*0.997747*	*1.06248*
5	*-1.56294*	*-0.8426*	*0.395244*	*-1.78146*
6	*-1.88E-09*	*0.218038*	***6.5203***	*0.011625*
7	*-1.52E-09*	*-0.15879*	*-0.3752*	*-0.04749*
8	*-2.36E-09*	*0.100437*	*0.665053*	*0.012646*
9	*-2.78E-10*	*-0.03346*	*-1.74582*	*-0.02522*
10	*-1.84E-09*	*0.010638*	*-1.10428*	*-0.00242*
11	***-4.23278***	***-5.50992***	*-3.54658*	*-1.77719*
12	*-1.76554*	*-0.78919*	*-3.1073*	***-3.11253***
13	*-2.02429*	*-2.44551*	*-1.70254*	*-1.6255*
14	*-0.96231*	*-0.7926*	*-0.97724*	*-1.0575*
15	*1.56294*	*0.846482*	*2.31381*	*1.79104*
Total Voltage	**21.09572**	**21.28518**	**30.08593**	**18.87725**
Lnm	**0.002744**	**0.004388**	**0.002285**	**0.002089**

Figure 3: Desired Shape

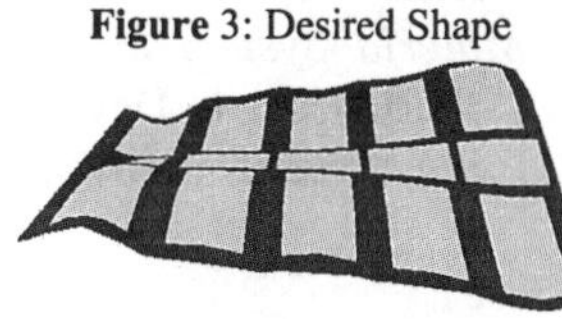

Figure 4: Calculated Shape of model F

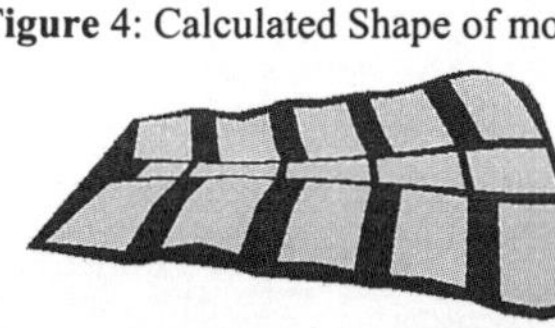

Figure 5: Calculated Shape of model I

Table 3: Comparison for cantilever plates with different patch pattern

Compare to Model F			
Error	G	H	I
Voltage (%)	0.898097	42.61627	-10.51621
Lnm (%)	59.91093	-16.72504	-23.86662

4 Concluding Remarks

This study presents a preliminary investigation into the effect of actuator pattern on the optimum values of the applied voltages and the shape match factor in shape control of smart plate structures. Numerical results show that actuator patterns may have an important influence on the values of the optimum voltages applied to each individual actuator and the final shape match factor. It is shown that by adequately designing the actuator pattern, the total voltage and the maximum voltage in all actuators may be reduced while also achieving an improved desired shape match factor. It is worth pointing out that research is in progress in developing optimisation algorithms for selecting voltages, location, shape and size of attached actuators in shape control of plate structures.

REFERENCES

1. Koconis DB, Kollar LP & Springer GS, "Shape Control of Composite Plates and Shells with Embedded Actuators. II Desired Shape Specified", J. Composite Materials, 28 (3) (1994) pp. 262-285
2. Chee, C., Tong, L. and Steven, G., "A Mixed Model for Adaptive Composite Plates with Piezoelectric for Anisotropic Actuation", *Computers ad Structure,* Vol.77, 2000, pp. 253-268.
3. Chee, C., Tong, L. and Steven, G., "A Buildup Voltage Distribution (BVD) algorithm for shape control of smart plate structures", *computational mechanics*, Vol. 26, 2000, pp 115-128.
4. Lo, K.H., Christensen, R.M. and Wu, E.M., "A High Order Theory of Plate Deformation –Part 1 & 2: Homogenous Plates", *Journal of Applied Mechanics*, Vol.44, 1977, pp 663-676.
5. Saravanos. D.A. and Heyliger, P.R., "Coupled Layerwise Analysis of Composite Beams with Embedded Piezoelectric Sensors and Actuators", *Journal of Intelligent Material System and Structures,* Vol.6, 1995, pp 350-363.
6. Robert D.Cook, Davis S.Malkus and Micheal E.Plestia, *Concepts and application of finite elements analysis*, chapter 6 p-163-176.

FREQUENCY RESPONSE OF BEAMS WITH DEBONDED PIEZOELECTRIC ACTUATOR LAYER

DONGCHANG SUN AND LIYONG TONG

School of Aerospace, Mechanical and Mechatronic Engineering, The University of Sydney, NSW 2006, Australia. Email: Dsun@aeromech.usyd.edu.au, Ltong@aeromech.usyed.edu.au

The debonding between the piezoelectric actuator/sensor layers and the host structures can result in significant changes to both open loop and closed loop properties. To investigate the effects of the debonding of the piezoelectric actuator layer from the host beam on its dynamic behavior, an analytical model of a beam with debonded piezoelectric layer is presented. In addition to the host beam and the piezoelectric layer, the adhesive layer is also considered. In the perfect bonding region, the adhesive layer is modeled to carry uniform peel and shear stresses throughout its thickness, while in the debonding area, it does not transfer any peel and shear stresses between the host beam and the actuator layer. Both displacement and force continuity are imposed at the interfaces between the debonded and perfectly bonded regions. Based on this model, a semi-analytical solution is given to obtain the frequency response of the beam with debonded piezoelectric actuator layer by employing integration transformation, and the debonding effects on the frequency response of the beam can be investigated.

1 Introduction

Vibration control of flexible structures using distributed piezoelectric sensors and actuators has been extensively studied in the past decade [1-5]. Recently, an increasing attention has been paid to the effects of the imperfect bonding of the piezoelectric transducers on the open loop and closed-loop systems [6-10]. When a debonding between the piezoelectric actuator/sensor layers and the host structures occurs, it will result in significant changes to both open loop and closed-loop properties, and may even destabilize the closed-loop control. To investigate the effects of the debonding of the piezoelectric actuator layer from the host beam on the dynamic behavior of the composite structures, an analytical model of a beam with debonded piezoelectric layer is presented. In this model, the host beam and the piezoelectric layer as well as the adhesive layer are taken into account. The adhesive layer is modeled to carry both peel and shear stresses along its thickness in the perfect bonding region. In the debonding area, it is assumed that there is no peel and shear stresses between the host beam and the actuator layer. Both displacement and force continuity conditions are considered at the interfaces between the debonded and perfectly bonded regions. To obtain the frequency response of the composite beam to a voltage applied on the actuator, a semi-analytical solution of the presented model of the smart beam with debonded piezoelectric actuator layer is given.

2 Actuator model with debonding

Consider a slender composite beam with length L, on which a piezoelectric actuator layer is bonded onto its upper surface, as shown in Figure 1. A debonding region between the actuator and the host beam is introduced at the right end of the beam. It is assumed that the debonding occurs throughout the width of the beam and the debonding front lines are straight and perpendicular to the x-axis. The adhesive layer is assumed to only carry constant transverse shear and peel strains. In the debonded region, it is assumed that there is no stress transferring between the host beam and piezoelectric actuator layer. In addition, contact and friction between the two debonded surfaces are not considered for simplicity.

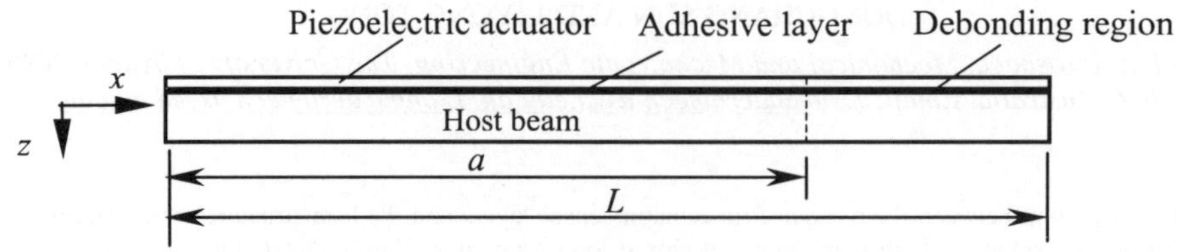

Figure 1. A beam with debonded piezoelectric actuator layer

Figure 2 depicts the free-body diagram of the composite beam including the host beam, the actuator layer and the adhesive layer, and from this figure the following equations of motion can be obtained.

$$\rho_1 b h_1 \ddot{u}_1 = T_{1,x} + b\tau \quad \text{(1a)}$$

$$\rho_1 b h_1 \ddot{w}_1 = Q_{1,x} + b\sigma \quad \text{(1b)}$$

$$M_{1,x} + bh_1\tau/2 - Q_1 = 0 \quad \text{(1c)}$$

$$\rho_2 b h_2 \ddot{u}_2 = T_{2,x} - b\tau + f_l(x,t) \quad \text{(2a)}$$

$$\rho_2 b h_2 \ddot{w}_2 = Q_{2,x} - b\sigma + f_t(x,t) \quad \text{(2b)}$$

$$M_{2,x} + bh_2\tau/2 - Q_2 = 0 \quad \text{(2c)}$$

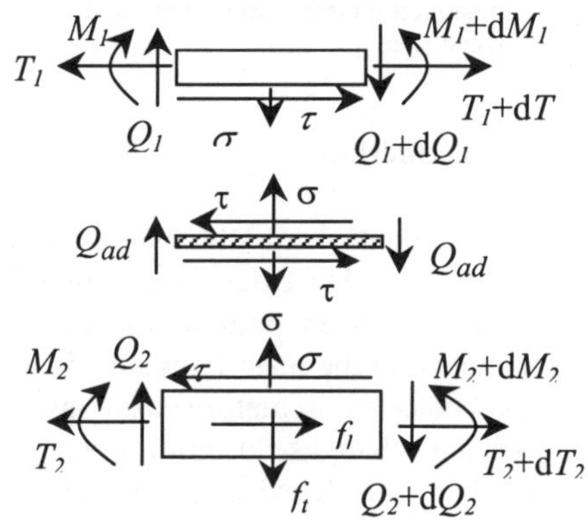

Figure 2. Free body diagrams of the infinitesimal element

where the subscripts 1 and 2 represent the actuator layer and the host beam, respectively, and u is the longitudinal displacement in mid-plane, w is the transverse displacement, h denotes the thickness, b is the width of the composite beam, τ and σ are shear and peeling stress of the adhesive layer, T, Q and M are the axial force, transverse shear force and bending moment respectively, $f_l(x,t)$ and $f_t(x,t)$ are the axial and transverse loads per unit length, ρ is the equivalent mass densities, in which the mass of the adhesive layer is considered.

The axial stress resultants and the bending moments for the actuator and the host beam can be written as follows

$$T_1 = E_1 b h_1 u_{1,x} - b e_{31} V(t), \qquad M_1 = -\frac{E_1 b h_1^3}{12} w_{1,xx}$$
$$T_2 = E_2 b h_2 u_{2,x} \qquad , \qquad M_2 = -\frac{E_2 b h_2^3}{12} w_{2,xx} \quad (3)$$

where $V(t)$ is the uniform voltage applied on the actuator along its thickness direction, E is the Young's modulus, e_{31} is the piezoelectric stress constant of the actuator layer.

Using the constant shear and peel strain assumption, the shear and peel stress in the adhesive layer are given by

$$\tau = k_b \frac{Y_{ad}}{2(1+\nu_{ad})h_{ad}}[(h_1 w_{1,x} + h_2 w_{2,x})/2 + (u_2 - u_1)] \tag{4}$$

$$\sigma = \frac{k_b E_{ad}(1-\nu_{ad})}{(1-2\nu_{ad})(1+\nu_{ad})h_{ad}}(w_2 - w_1) \tag{5}$$

where h_{ad} is the thickness of the adhesive layer, Y_{ad} and ν_{ad} are the Young's modulus and Poisson's ratio of the adhesive layer respectively, and k_b is a parameter characterizing the bonding condition

$$k_b = \begin{cases} 0 & \text{debonding} \\ 1 & \text{perfect bonding} \end{cases} \tag{6}$$

By introducing the following nondimensional notations

$$\begin{aligned}
&\eta_i = \frac{\rho_i L^2}{Y_2}, \alpha_2 = \frac{h_2}{L}, \beta_i = \frac{E_i}{E_2}, \varphi_i = \frac{h_i}{h_2}, \ \varphi_v = \frac{h_{ad}}{h_2}, \ \beta_v = \frac{E_{ad}}{E_2} \\
&\xi = \frac{x}{L}, u_{ni} = \frac{u_i}{h_2}, w_{ni} = \frac{w_i}{h_2}, T_{ni} = \frac{T_i}{E_2 b h_2}, \ Q_{ni} = \frac{Q_i}{E_2 b h_2}, \quad i = 1,2 \\
&M_{ni} = \frac{M_i}{E_2 b h_2^2/12}, \ t_n = \frac{t}{L^2}\sqrt{\frac{E_2 h_2^2}{12\rho_2}}, \ V_n(t_n) = \frac{E_1 d_{31} V(t_n)}{E_2 h2}
\end{aligned} \tag{7}$$

where d_{31} is the piezoelectric strain constant of the actuator layer, Eq.(1)-(5) can be nondimensionalized as

$$\begin{aligned}
&m_1 k_t^2 \ddot{u}_{n1} = T_{n1,\xi} + \tau_n \\
&m_1 k_t^2 \ddot{w}_{n1} = Q_{n1,\xi} + \sigma_n \\
&M_{n1,\xi} + 6\varphi_1 \tau_n - \frac{12}{\alpha_2} Q_{n1} = 0 \\
&m_2 k_t^2 \ddot{u}_{n2} = T_{n2,\xi} - \tau_n + f_{nl}(\xi, t_n) \\
&m_2 k_t^2 \ddot{w}_{n2} = Q_{n2,\xi} - \sigma_n + f_{nt}(\xi, t_n) \\
&M_{n2,\xi} + 6\varphi_2 \tau_n - \frac{12}{\alpha_2} Q_{n2} = 0 \\
&T_{n1} = \beta_1 \alpha_1 u_{n1,\xi} - V_n \\
&M_{n1} = -\beta_1 \varphi_1 \alpha_1^2 w_{n1,\xi\xi} \\
&T_{n2} = \beta_2 \alpha_2 u_{n2,\xi} \\
&M_{n2} = -\beta_2 \varphi_2 \alpha_2^2 w_{n2,\xi\xi} \\
&\tau_n = k_b r_a \left[u_{n2} - u_{n1} + \frac{1}{2}(\alpha_1 w_{n1,\xi} + \alpha_2 w_{n2,\xi}) \right] \\
&\sigma_n = k_b r_a^e (w_{n2} - w_{n1})
\end{aligned} \tag{8}$$

where the double dot represents the second derivation with respect to t_n, and the parameters are defined as

$$\alpha_i = \varphi_i \alpha_2,\ m_i = \alpha_i \eta_i\ (\ i=1,2),\quad \alpha_v = \varphi_v \alpha_2$$
$$k_v = \frac{1-\nu}{(1-2\nu)(1+\nu)},\ k_g = \frac{1}{2(1+\nu)},\ r_a = \frac{k_g \beta_v}{\alpha_v},\ r_a^e = \frac{k_v \beta_v}{\alpha_v} \tag{9}$$
$$f_{nl} = \frac{Lf_l}{E_2 b h_2},\ f_{nt} = \frac{Lf_t}{E_2 b h_2},\ k_t = \frac{1}{L^2}\sqrt{\frac{E_2 h_2^2}{12\rho_2}} = \sqrt{\frac{\alpha_2^3}{12 m_2}},\ V_{n2} = 0$$

Introducing $\mathbf{Y}(x) = (u_1, T_1, w_1, w_{1,x}, Q_1, M_1, u_2, T_2, w_2, w_{2,x}, Q_2, M_2)^T$ as the state vector and only the applied voltage being considered (i.e. $f_{nl} = f_{nt} = 0$) , Eq. (8) can be rearranged in the following state form

$$\mathbf{Y}_{,\xi} = \mathbf{M}\ddot{\mathbf{Y}} + \widetilde{\mathbf{A}}\mathbf{Y} + \mathbf{B}V,\qquad 0 < \xi < 1 \tag{10}$$

where $\mathbf{M} \in R^{12\times12}$ is the mass matrix, $\widetilde{\mathbf{A}} \in R^{12\times12}$ is the state matrix, and $\mathbf{B} \in R^{12}$ is the vector related to the control voltage on the actuator layer.

At the interface between the perfectly bonded and debonded regions, the continuity conditions of all displacements and stress resultants are imposed, i. e.

$$\mathbf{Y}^b = \mathbf{Y}^d \tag{11}$$

where the superscripts *b* and *d* represent the bonded and debonded regions respectively. In general, the boundary conditions of the beam can be expressed as

$$\mathbf{D}_0\mathbf{Y}(0) + \mathbf{D}_L\mathbf{Y}(1) = \mathbf{D},\quad \text{for } \forall t_n \tag{12}$$

where $\mathbf{D}_0 \in R^{12\times12}, \mathbf{D}_L \in R^{12\times12}$ and $\mathbf{D} \in R^{12}$ are known matrices and vector corresponding to the given boundary conditions.

3 Solutions

To obtain the frequency response of the beam to the applied voltage from equations (8) or (10), taking Fourier transformation on both sides of Eq.(10), we have

$$\overline{\mathbf{Y}}_{,\xi} = \mathbf{A}\overline{\mathbf{Y}} + \mathbf{B}\overline{V},\ 0 < \xi < 1 \tag{13}$$

where the over bar stands for the Fourier transformation, $\mathbf{A} = \widetilde{\mathbf{A}} - \Omega^2\mathbf{M}$ and Ω is the nondimensional frequency. After taking Fourier transformation, equation (13) becomes ordinary differential equations with parameter Ω. For the beam with perfectly bonded actuator $(k_b = 1)$ and debonded actuator $(k_b = 0)$ respectively, the state equation (13) has different form, i.e.

$$\begin{aligned} \overline{\mathbf{Y}}_{,\xi} &= \mathbf{A}_1\overline{\mathbf{Y}} + \mathbf{B}_1\overline{V},\ 0 < \xi \le \xi_a = a/L \\ \overline{\mathbf{Y}}_{,\xi} &= \mathbf{A}_2\overline{\mathbf{Y}} + \mathbf{B}_2\overline{V},\ \xi_a < \xi < 1 \end{aligned} \tag{14}$$

The solutions of Eq. (14) can be expressed as

$$\begin{aligned} \overline{\mathbf{Y}}(\xi) &= \mathbf{\Phi}_1(\xi)\overline{\mathbf{Y}}(0) + \mathbf{\Psi}_1(\xi)\overline{V}(\omega) & 0 < \xi \le \xi_a \\ \overline{\mathbf{Y}}(\xi) &= \mathbf{\Phi}_2(\xi - \xi_a)\overline{\mathbf{Y}}(\xi_a) + \mathbf{\Psi}_2(\xi)\overline{V}(\omega) & \xi_a < \xi \le 1 \end{aligned} \tag{15}$$

where the matrices $\mathbf{\Psi}_1 \in R^{12\times12}$ and $\mathbf{\Psi}_2 \in R^{12\times12}$ are given by

$$\mathbf{\Psi}_1(\xi) = \int_0^{\xi} \mathbf{\Phi}_1(\xi - x)\mathrm{d}x\mathbf{B}_1, \quad \mathbf{\Psi}_2(\xi) = \int_{\xi_a}^{\xi} \mathbf{\Phi}_2(\xi - x)\mathrm{d}x\mathbf{B}_2 \tag{16}$$

and $\mathbf{\Phi}_i(\xi) = e^{\mathbf{A}_i\xi} = \sum_{k=0}^{\infty} \frac{\mathbf{A}_i^k \xi^k}{k!}$ is the transition matrix, which can also be obtained by employing Laplace transformation as follow

$$\mathbf{\Phi}_i = ☹^{-1}\{[s\mathbf{I} - \mathbf{A}_i]^{-1}\} \tag{17}$$

where $☹^{-1}$ represents the inverse Laplace transformation with respect to complex variable s.

The matrix $[s\mathbf{I} - \mathbf{A}_i]^{-1}$ can be changed into the following form

$$[s\mathbf{I} - \mathbf{A}_i]^{-1} = \frac{\mathbf{P}_i(s)}{\mathrm{Det}[s\mathbf{I} - \mathbf{A}_i]} \tag{18}$$

where $\mathbf{P}_i(s)$ is the adjoint matrix of $[s\mathbf{I} - \mathbf{A}_i]$. Assume that the polynomial $q_i(s) = \mathrm{Det}[\mathrm{s}\mathbf{I} - \mathbf{A}_\mathrm{i}]$ has distinct complex roots α_{ik} $(k = 1,2,\cdots,12)$, it can be expressed as

$$q_i(s) = \mathrm{Det}[s\mathbf{I} - \mathbf{A}_i] = (s - \alpha_{i1})(s - \alpha_{i2})\cdots(s - \alpha_{i12}) \tag{19}$$

Substituting Eq.(18) and (19) into Eq. (17), and employing Heaviside's expansion theorem [11], we have

$$\mathbf{\Phi}_i(\xi) = \sum_{k=1}^{12} \frac{\mathbf{P}_i(\alpha_{ik})}{q_i'(\alpha_{ik})} e^{\xi\alpha_{ik}}, \quad i = 1,2 \tag{20}$$

In this case, when the roots of the polynomial $q_i(s)$ are found, the transition matrices can be explicitly obtained from Eq. (20).

From Eq. (15), we have

$$\overline{\mathbf{Y}}(\xi_a) = \mathbf{\Phi}_1(\xi_a)\overline{\mathbf{Y}}(0) + \mathbf{\Psi}_1(\xi_a)\overline{V}(\omega) \tag{21}$$

Substituting Eq. (21) into the second equation in Eq.(15), yields

$$\begin{aligned} &\overline{\mathbf{Y}}(\xi) = \mathbf{\Phi}_1(\xi)\overline{\mathbf{Y}}(0) + \mathbf{\Psi}_1(\xi)\overline{V}(\omega) && 0 < \xi \le \xi_a \\ &\overline{\mathbf{Y}}(\xi) = \mathbf{\Phi}_2(\xi - \xi_a)\mathbf{\Phi}_1(\xi_a)\overline{\mathbf{Y}}(0) + [\mathbf{\Phi}_2(\xi - \xi_a)\mathbf{\Psi}_1(\xi_a) + \mathbf{\Psi}_2(\xi)]\overline{V}(\omega) && \xi_a < \xi \le 1 \end{aligned} \tag{22}$$

Eq. (22) gives the solution of the state equation (14) in terms of $\overline{\mathbf{Y}}(0)$, which can be obtained by the following procedure. The state at the right end of the beam can be expressed by Eq.(22) as

$$\overline{\mathbf{Y}}(1) = \mathbf{\Phi}_2(1 - \xi_a)\mathbf{\Phi}_1(\xi_a)\overline{\mathbf{Y}}(0) + [\mathbf{\Phi}_2(1 - \xi_a)\mathbf{\Psi}_1(\xi_a) + \mathbf{\Psi}_2(1)]\overline{V}(\omega) \tag{23}$$

Noting the boundary conditions of the beam in Eq.(12) together with Eq. (23), the state vector at the two ends of the beam can be solved by

$$\begin{bmatrix} \mathbf{\Phi}_2(1 - \xi_a)\mathbf{\Phi}_1(\xi_a) & -\mathbf{I} \\ \mathbf{D}_0 & \mathbf{D}_L \end{bmatrix} \begin{Bmatrix} \overline{\mathbf{Y}}(0) \\ \overline{\mathbf{Y}}(1) \end{Bmatrix} = \begin{Bmatrix} -[\mathbf{\Phi}_2(1 - \xi_a)\mathbf{\Psi}_1(\xi_a) + \mathbf{\Psi}_2(1)]\overline{V}(\omega) \\ \overline{\mathbf{D}} \end{Bmatrix} \tag{24}$$

Once the state vector $\overline{\mathbf{Y}}(0)$ at the left end of the beam is solved from Eq.(24), the state vector at any point of the beam, which includes the displacements and stress resultants in frequency domain , can be obtained from Eq. (22).

4 Conclusion

In this paper, a mathematic model of a beam with partly debonded actuator layer is presented. To obtain the frequency response of the smart beam to the excitation voltage applied on the actuator, a semi-analytical solution of the presented model is given based on the Fourier transformation. Due to the space limitation, the example and its analysis will be given in the full paper.

Acknowledgement

The authors are grateful to the support of the Australia Research Council through a Large Grant (Grant No. A10009074).

References

1 Im S. and Atluri S. N., Effect of a piezo-actuator on a finitely deformed beam subject to general loading, *AIAA Journal.* **27** (1989) pp.1801-1807.
2 Tzou H. –S. and Anderson G. L., *Intelligent Structural Systems.* (Kluwer Academic Publishers, Dordrecht/Boston/London, 1992)
3 Crawley E. F., Intelligent structures for aerospace: A technology overview and assessment, *AIAA Journal.* **32** (1994) pp.1689-1699.
4 Chee C., Tong L. and Steven G. P., A review on the modeling of piezoelectric sensors and actuators incorporated in intelligent structures, *Journal of Intelligent Material Systems and Structures*, **9** (1998) pp. 3-19.
5 Sun D. C. and Tong L., Modal control of smart shells by optimized discretely distributed piezoelectric transducers, *International Journal of Solids and Structures.* **38** (2001) pp. 3281-3299.
6 Seeley C. E. and Chattopadhyay A., Modeling of adaptive composites including debonding. *International Journal of Solids and Structures*, **36** (1999) pp. 1823-1843.
7 Seeley C. E. and Chattopadhyay A., Experimental investigation of composite beams with piezoelectric actuation and debonding, *Smart Materials and Structures*, **7** (1998) pp. 502-511.
8 Wang X. D. and Meguid S. A., On the electroelastic behavior of a thin piezoelectric actuator attached to an infinite host structures, *International Journal of Solids and Structures*, **37** (2000) pp. 3231-3251.
9 Tylikowski A., Effects of piezoactuator delamination on the transfer functions of vibration control systems, *International Journal of Solids and Structures*, **38** (2001) pp. 2189-2202.
10 Tong L., Sun D. C. and Atluri S. N., Sensing and actuating behaviors of piezoelectric layers with debonding in smart beams, *Smart Materials and Structures*, in press
11 Debnath L., *Integral transforms and their applications*, CRC Press, Boca Raton/New York/London/Tokyo, 1995.

THE APPLICATION OF PIEZOCERAMIC ELEMENTS TO THE DETECTION OF DISBOND GROWTH IN A BONDED COMPOSITE REPAIR PATCH

Y.L. KOH AND W.K. CHIU

DSTO-Centre of Expertise for Structural Mechanics, Department of Mechanical Engineering, Monash University, Clayton, 3800, Victoria.

N. RAJIC AND S.C. GALEA

Defence Science and Technology Organisation, Aeronautical and Maritime Research Laboratory, 506, Lorimer Street, Fishermans Bend, 3207, Victoria.

This paper reports on a preliminary experimental study in which an array of surface mounted lead zirconate titanate elements (PZT) are used for the in-situ detection of disbond growth in a bonded composite repair patch. The electromechanical impedance and acoustic stress wave methods are used to track the evolution of disbond growth. Both techniques were found to provide a reliable and robust basis for the detection of disbond growth.

1 Introduction

Bonded composite repair patch technology has provided a relatively cost-effective means of extending the operational life of ageing aircraft. However, because these repairs are themselves exposed to in-service loads and impact events, there is a need to provide ongoing assurance of their structural integrity. This is usually achieved through prescribed periodic non-destructive inspection, a relatively expensive exercise.

In-situ heath monitoring principles employing piezoelectric materials in acoustic sensing and actuation roles offer an encouraging basis for the development of relatively low-cost methods for providing structural integrity assurance for composite patch repairs. Recent work has underscored the practical versatility of these materials with elements applied to in-situ health monitoring in the following configurations, as: surface mounted devices [1], embedded devices [2] and combinations of embedded and surface mounted devices [3]. The primary practical advantage in using piezoelectric materials like polymer film and piezoceramic wafers over strain gauges is that they do not require an external power source and they generate signals far stronger than those of traditional amplified strain gauges (see Penn et al [2]).

This paper will provide demonstration of the efficacy of detecting disbond growth in a bonded composite repair patch based on measurements of the electrical impedance and acoustic stress-wave response of surface mounted piezoceramic wafer elements.

2 Impedance Method

The impedance-based method exploits the electromechanical coupling described by the piezoelectric effect. This effect has both a direct and a converse sense as described by the following relations:

$$S_i = s_{ij}^E T_j + d_{mi} E_m$$

$$D_m = d_{mi} T_i + \varepsilon_{mk}^T E_k$$

The mechanical variables in the above equations are the stress **T** and the strain **S**, and the electrical variables are the electric field **E** and the electric displacement **D**. The other variables are **s** the mechanical compliance of the PZT, **d** the piezoelectric constant and ε the dielectric constant. The impedance method utilises a single piezoelectric wafer element in both the direct and converse sense. It relies on the relationship between the mechanical impedance, defined as the ratio of the applied force and the resulting velocity. Excitation of a surface mounted piezoceramic element induces a vibrational response in the host that serves to modulate the current flowing through the PZT. Consequently, the measured electrical impedance of the PZT conveys information relating to the structural characteristics of the host, including any structural change arising from the formation and/or growth of damage.

Variations in the impedance amplitude over the course of fatigue testing was used as the basis for assessing structural change in the composite patched specimen to be described shortly. Specifically, the impedance spectrum is condensed into a single measure, defined as,

$$IM = \int_0^{3500} Z(f)df$$

where **f** is the frequency, **Z** is the impedance and the upper frequency limit of 3.5 kHz is chosen to exclude the first structural resonance. In the results presented in the following sections, the **IM**-parameter was normalised with respect to the value obtained prior to the application of cyclic loading.

3 Test Specimen

Figure 1 shows a schematic drawing of the specimen considered in this experimental study. The specimen, given the designation "DD2", comprises an 11-ply unidirectional Boron/Epoxy patch bonded onto an aluminium plate to simulate a bonded composite repair. A dynamic loading regime was applied to the specimen consisting of a simple constant-amplitude sinusoidal waveform with a fixed frequency of 5 Hz. The minimum load on the specimen was kept at 0.1kN and the maximum load started at 16kN. The peak load was increased in 1kN steps every 50,000 cycles until damage was observed, after which the loading level was kept constant. During the experiments the disbond growth on Side P of the adhesive layer was monitored using the impedance method and strain gauge. The disbond on Side S was monitored using the stress wave technique and strain gauge. On Side P, impedance measurements were taken only for element PP2 ($4mm \times 20mm$ in size) on step 1 and comparison is to be made with strain gauge P3 on step 2. While on Side S, stress wave results are to be compared with strain gauge S3 on step 2.

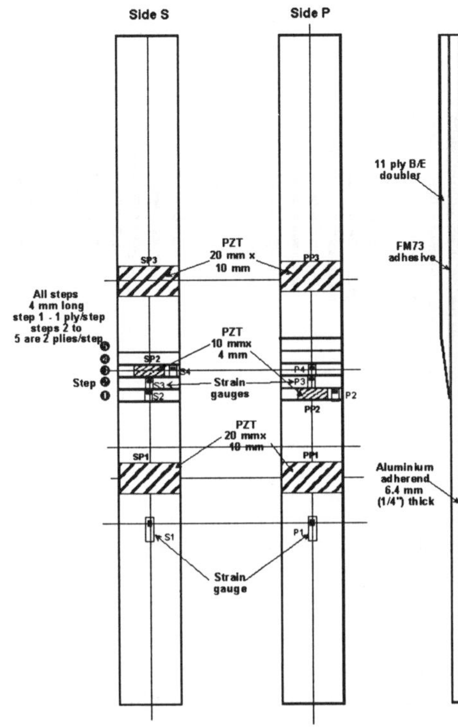

Figure 1: Specimen (DD2 Used for the Verification of Methods Impedance Method

Figure 2 suggests that disbond growth commenced relatively soon after increasing the load amplitude to 20kN after exposure to some 200,000 loading cycles at lesser amplitudes. The decline in the IM-parameter is observed to continue until the disbond grows past the PZT element, at which stage further disbond growth appears to have no effect on the element impedance.

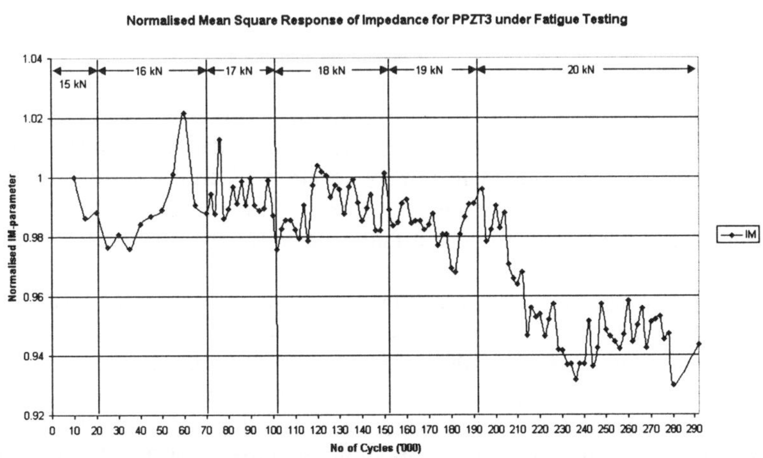

Figure 2: Impedance Readings from Sensor on Edge of Repair Patch Taper (PPZT3)

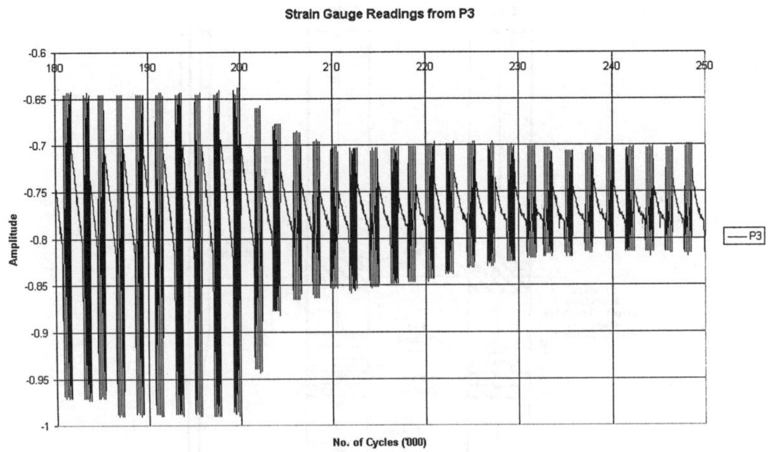

Figure 3: Strain Gauge Reading taken from Edge of Repair Patch Taper, S3

The impedance response of the PZT element is broadly consistent with the behaviour predicted in a complimentary numerical investigation of this specimen configuration reported earlier (Koh et al [4]). There, it was found that mechanical impedance was relatively insensitive to further growth of the disbond when past the sensor element. Strain gauge readings taken at P3 (second step) show a reduction in strain after 200,000 cycles, which indicate that the disbond had indeed reached the second step. This result is broadly consistent with that suggested by the impedance measurements. Initial growth of a disbond appears to commence at around 205,000 cycles, rather than 190,000 as indicated by the impedance measurements. The difference is due to the strain gauge (located on the second step) being located above the PZT sensor (located on the first step) therefore detecting the disbond growth at a later stage, since damage initiated at the edge of the patch.

4 Stress-Wave Technique

The excitation of elastic stress waves is relatively straightforward given the broadband frequency response characteristic of piezoceramic materials. One of the inherent advantages of the stress-wave arrangement over the impedance approach is that it does not require a priori knowledge of the likely location of damage initiation and should therefore allow for a lower sensor density where large structures are considered. The broader sensitivity range also provides a better basis for the quantitative assessment of disbond growth.

In this study, the approach was to excite elastic stress waves in the host by applying a short voltage pulse to a PZT element surface mounted to the metal substrate (SP1) some distance from the patch edge as shown in Figure 1. Part of the elastic energy is transmitted through the bond-line into the patch and received at the two piezosensor locations (SP2, located on step 3 of the patch, and SP3, located on the far-field region of the patch). The expectation was that continuous monitoring of the elastic energy at this sensor location would provide a robust basis for the assessment of disbond growth from the patch edge.

Computations of the power of the received signal were made at frequent intervals in the course of fatigue testing. The signal power at both sensor stations is plotted as a function of the number of loading cycles in Figure 4.

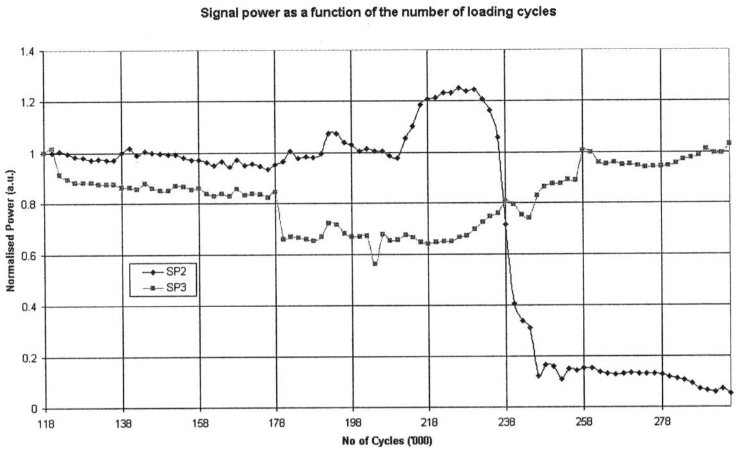

Figure 4: Signal power versus number of cycles for piezosensor SP2 and SP3

Several interesting observations can be made here. Firstly, the response at sensor SP2 reflects a higher level of sensitivity to disbond growth than that recorded at sensor SP3. This is an intuitive result since the former element is located directly above the region where disbond growth occurred. Another interesting aspect is the markedly superior noise-effective sensitivity relative to that of the strain and impedance measurements. In the case of SP2 for example, disbond growth through the sensor footprint results in a systematic drop in signal of some 90%, compared to 60% for the strain measurements and 4% for the impedance measurements. One potentially concerning aspect of these measurements is the apparent complexity of the relationship between disbond growth and the evolution of signal power. In the case of SP3 for example this is demonstrated by the 20% increase in power immediately preceding the more substantial long-term decline in signal.

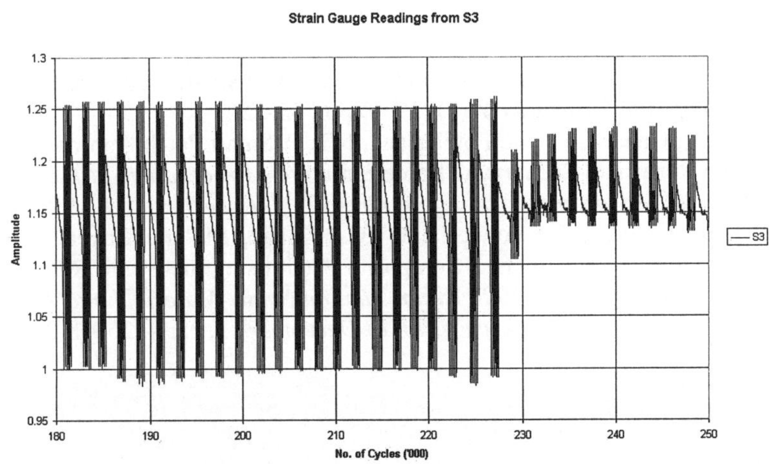

Figure 5: Strain Gauge Reading taken from Edge of Repair Patch Taper, S3

Such complexities are likely to arise from wave diffraction behaviour. Work is presently underway to try to understand such behaviour with a view to exploiting it for the quantitative assessment of disbond growth. Results from strain gauge S3 are shown in Figure 5. Note that in general the strain on step 2 should drop to zero as the damage progresses past this step [5]. The fact that the strains do not has been addressed in a previous study [6] and is due to the complex interaction between the disbonded patch and some adhesive overflow on the aluminium after the patch has been bonded to the specimen. Both the impedance and the stress wave technique were robust enough not to be affected by this phenomenon.

5 Conclusion

The experimental results described above show that surface-mounted PZT elements provide an efficacious means for the detection of disbond growth in bonded composite patch repairs. While the stress wave technique was found to have a substantially better sensitivity to disbond growth compared to the impedance approach, the latter method requires only a single surface-mounted element as opposed to the pair of elements required by the former method. That said the feasibility of the impedance method is highly reliant on the judicious placement of the element within close proximity of the likely region of disbond formation. The stress wave technique is likely to be more robust in this regard.

References

1. Penn, L.S., Jump, J.R., Greenfield, M.J., Blandford, G.E., Use of the Free Vibration Spectrum to Detect Delamination in Thick Composites, *Journal of Composite Materials* **33** (1999) pp.54–72.
2. Chaudry, Z., Lalande, F., Ganino, A., Rogers, C. and Chung, J., Monitoring the Integrity of Composite Patch Structural Repair via Piezoelectric Actuators/Sensors. *AIAA/ASME/ASCE/AHS/ASC 36th SDM Conference New Orleans LA* **4** (1995) pp.2243–2248.
3. Islam A.S., Craig, K.C., Damage Detection in Composite Structures using Piezoelectric Materials, *Smart Materials and Structures* **3** (1994) pp.318–326.
4. Koh, Y.L., Rajic, N., Chiu, W.K. and Galea, S.C., Smart Structure for Composite Repair. *Composite Structures* **47(1)** (1999) pp.745–752.
5. Baker, A. A., Galea, S C. and Powlesland, I G., A Smart Patch Approach for Bonded Composite Repairs to Primary Airframe Structures, *Proceedings of the Second Joint FAA/DOD/NASA Conference on Aging Aircraft, Williamsberg VA*, 1998.
6. Moss, S.D., Galea, S.C., Powlesland, I.G., Konak, M. and Baker, A.A., In-Situ Health Monitoring of a Bonded Composite Patch using the Strain Ratio Technique. *Proceedings of the SPIE's 2000 Symposium on Smart Materials and MEMS, Smart Structures and Devices Conference, 13-15 December*, Paper 4235-41, 2000.

CRAÇK MONITORING OF A STRUCTURAL COMPONENT

N. WIBOWO AND W.K. CHIU
Department of Mechanical Engineering, Monash University, Wellington Road, Clayton 3168, Victoria, Australia
E-mail: w.chiu@eng.monash.edu.au

This paper addresses the use of a transfer function technique for crack detection in a metallic plate. It will be shown that an array of sensor/actuator is best suited for this application. This paper will also present a simple technique of analyzing the data obtained from this array of sensor/actuator to approximately locate the position of the crack. Some experimental results will be presented.

1 Introduction

The use of smart structures concept for structural health monitoring has attracted significant interest. In some applications, an actuator excites a point-location of the host structure and the response from another selected point location is received by a sensor. Any change in the condition of the host structure also results in a change in the response. However, successful detection relies on the method that is used to analyse the data received from the sensor. A reliable monitoring method will enable an engineer to make decisions to prevent total failure of the structure.

This paper presents a transfer function (TF) method to detect and simultaneously locate the general area of damage i.e. a crack in a structure at an early stage. This TF method works by acquiring the vibration signature in a certain range of broad band frequency from the piezoceramic (PZT) patches attached on the surface of the structure. In this method, PZT patches are distributed as arrays of actuators and sensors where they generate and measure the vibration signature. By taking the ratio of the vibration spectrum in a particular position of a sensor to an actuator one can produce the TF of this actuator-sensor pair. The arrays of the PZT patches can also be used to locate the general area of the damage in the structure and/or degradation of a particular sensor- actuator pair. If the TF of a particular sensor-actuator pair shows an anomaly, the region of the damage or degradation can be easily identified.

2 Methods

When the mounted PZT patch is actuated using an electrical field, it produces strain. This strain is also experienced by the host structure. The in-plane actuation or sensing strain ε of the PZT is proportional to the mechanical and electrical coupling coefficient d_{31} of the ceramic and an applied electric field V, from Crawley et al (1991) it can be expressed as:

$$\varepsilon = d_{31}V \tag{1}$$

If the bonding layers are thin and stiff, the strain of the PZT patch acting as an actuator is equal to the strain of the corresponding area of the plate below. This is also true for the patch that acts as a sensor. Crawley et al (1987) referred to this situation as perfectly bonded actuator or sensor . The strain equation can then be written as:

$$\varepsilon_{pzt} = \varepsilon_{structure} \tag{2}$$

On any actuator-sensor pair, a broadband input voltage is applied to the actuator and the sensor detects the structural response to this input. The Transfer Function (TF) between the pair is then computed using equation (3):

$$TF = \frac{V_{ks}}{V_{kact}} \tag{3}$$

where V_{ks} is a vibration spectrum from the sensor and V_{kact} is a vibration spectrum from the actuator

The vibration spectrum V is defined as:

$$V_k(\omega) = A_k(\omega) e^{i\theta_k(\omega)} \tag{4}$$

where ω is a frequency, θ_k is a phase angle and A_k is an amplitude

In the experimental investigation, this vibration spectrum is measured with a frequency analyser. The main purpose of this study is to determine if the TF obtained by Equation (3) can be used to monitor the development of a crack on a structure.

In this investigation, a thin aluminium plate, clamped on the shorter edges, was selected. The plate, shown in Figure 1a, has a dimension of 200 x 400 x 2 mm^3. The arrays of actuators and sensors, denoted as rows of A, B and C, were bonded on the surface of the plate. The detailed configuration of the actuator-sensor pairs is shown in Figure 1b. The Transfer Function was measured from each actuator-sensor pair. Actuator-sensor pairs were combined in two ways to produce the transfer function: A TF measured between sensor in row A and actuators in row B (i.e. AnBn) , and between sensor in row B and actuators in row C (i.e. BnCn). Three-damage conditions in the structure, damage1 (with a 10mm crack), damage2 (with a 20mm crack) and damage3 (with a 30mm crack) were used in order to identify changes in the TF for each damage state. A no-damage (no-crack) case was also considered.

The placement of distributed PZT patches in the configuration shown in Figure 1a and 1b was designed to answer the following questions:

- Can the crack be predicted when it is located between rows A and B?
- How sensitive are the rows B and C to the occurrence of a crack in between rows A and B?
- Can the crack tip location be approximately determined?

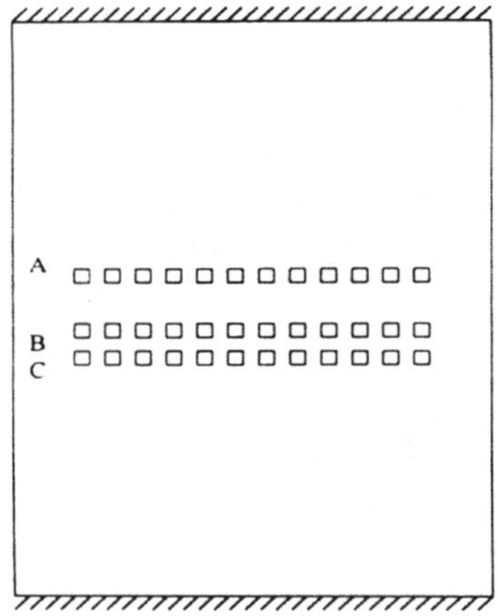

Figure 1a. A thin aluminium plate clamped on the 2 shorter edges with distributed PZT actuator and sensors

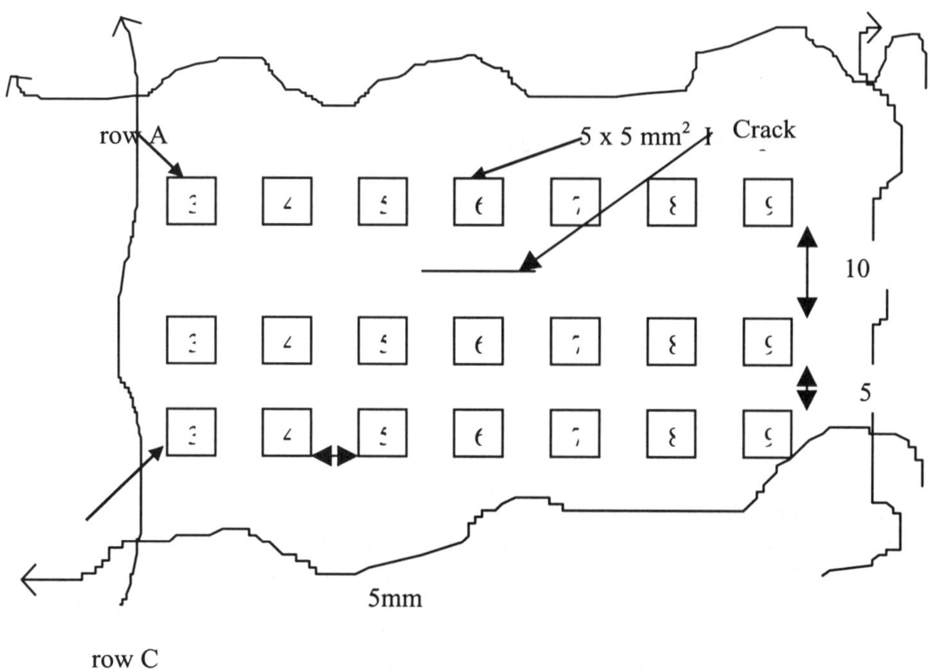

Figure 1b. Detailed Configuration of the Actuator-Sensor Pairs

3 Experimental results

In this experimental work, the test equipment consisted of a signal generator (SG) Hewlett-Packard 33120A, a dual channel oscilloscope HAMEG HM 305, a frequency analyser AND AD-3525 and a plate specimen to which the PZT patches were attached. Three rows A, B, C of 5x5 mm^2 of PZT patch were mounted on the plate (shown in Figures 1a and 1b). The distance between rows A and B was 10 mm with the crack laid exactly in the middle of the rows. No crack was made between rows B and C where the distance between them was 5 mm. Multiple switches were installed on the test plate to help the connections between sensors and actuators to the frequency analyser.

A 10V peak to peak broadband Gaussian white noise in a range of 250 to 100000Hz was supplied by the signal generator to excite the PZT actuator. Before the test was conducted all PZT actuators-sensors were tested by measuring their capacitances and their response to a given input voltage.

The frequency analyser calculates the transfer function of each actuator-sensor. When calculating the transfer function, a total of 30 spectral averages were taken. Hanning window was used.

The R-curves were derived from the transfer function obtained. Here the R-value as a function of frequency is defined as:

$$R(f_n) = \int_{f_0}^{f_n} A_{TF}\, df \qquad [5]$$

where A_{TF} is the amplitude of the TF frequency spectrum

This R-Curve method (Equation 5) will be applied to the experimental investigation in this study. The results are in the form of a comparison of the area under the TF frequency spectrum for all the damage conditions, no damage, and 10mm, 20mm, 30mm cracks.

4 Analysis of the R-Curve of AnBn Pairs

Figure 2 shows the R- value for the pair A1B1 for every damage condition. The results showed that there are no significant changes in the R-value for the various damage conditions. This is because the crack is located far away from these pairs. In Figure 3, the R-values for A5B5 pair of sensor-actuator for the 10mm, 20mm, and 30mm crack cases are lower than the no-damage case.

The R-values at the upper limit of the spectrum obtained were normalised with the no-damage cases. The percentage reduction of the R-values compared to the undamaged case were calculated. The results for the AnBn combination of actuator and sensors are shown in Figure 4(a). This figure suggests that the development of the crack can be tracked by analysing the array of AnBn actuator/sensor combination. This figure shows that progressive development of a crack in the plate gave rise to a more significant reduction of the R-value.

To get a better representation of the extent of the crack in the plate, similar analysis were performed using the sensor/actuator combination BnCn. The results are shown in Figure 4(b). Figure 4(b) shows that the change in the R-value was confined to the B5C5 combination (i.e 5th row) of the sensor/actuator when the crack was 10 mm. When the crack in the plate was 20 mm and 30 mm long, Figure 4(b) showed that these cracks were confined within the region occupied by B4C4 (i.e. 4th row of the sensor and actuator combination). Figure 4(a) and (b) show to obtain a clear overall picture of the crack development, an array of sensor and actuator is required.

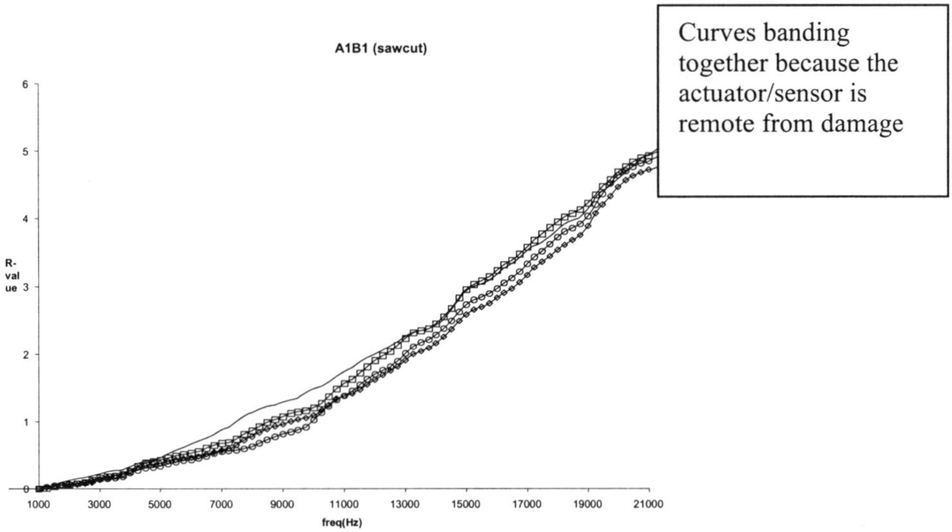

Figure 2: R-curve for the A1B1 actuator/sensor combination

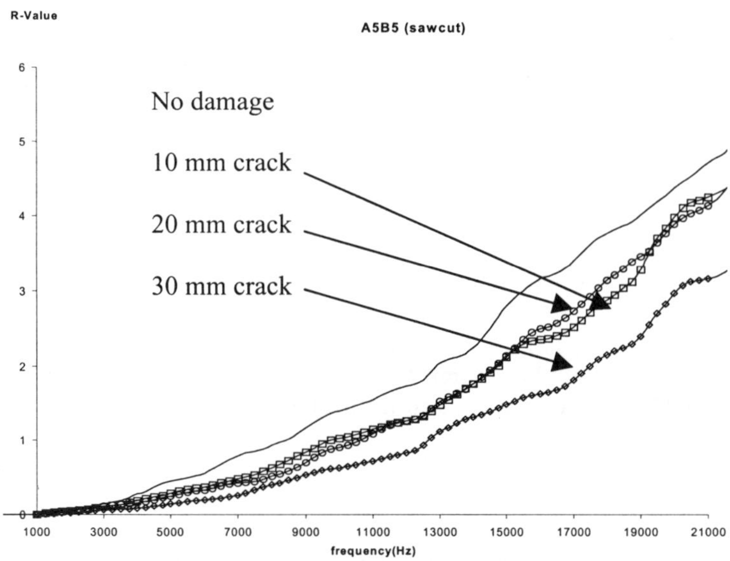

Figure 3: R-curve for the A5B5 actuator/sensor combination

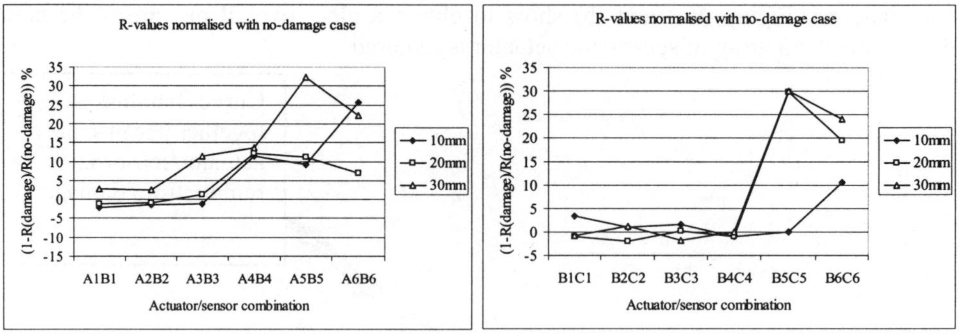

(a) (b)

Figure 4: Distirbution of R-parameter normalised with the no-damage case
(a) for AnBn actuator/sensor combination
(b) for BnCn actuator/sensor combination

The main findings described above consolidates the following inferences drawn in the experimental investigation are

- The pair of sensor-actuator, which encompasses the crack (i.e. AnBn), will show a reduction in the R-value prior to the BnCn pairs. This can be used to indicate across which row of sensor-actuator the crack appears (i.e. between A and B or B and C).
- It is important also to note that the BnCn are also an important part of the monitoring procedure. The R-values of the BnCn pairs can be used to indicate the extent of the crack. This is because a large R-value reduction is obtained when the crack face spans across a BnCn pair.

5 CONCLUSIONS

The experimental work described above show that the placement/distribution of PZT patches in the configuration shown in Figure 1a can be used for the detection of a crack. Thus, the TF results from the arrays of the PZT patches can be used to locate the general area of the damage in the structure. In general, this TF method is shown to offer great potential for detecting and locating damage in a structure at an early stage.

REFERENCES

Crawley E.F., and Lazarus K.B.," Induced Strain Actuation of Isotropic and Anisotropic Plates", AIAA Journal, Vol. 29, No.6, 1991, pp. 944-951

Crawley E.F., and de Luis J., " Use of Piezoelectric Actuator as Elements of Intelligent Structures ", AIAA Journal, Vol. 25, No.10, 1987, pp. 1373-1385

DEVELOPMENTS IN SHAPE OPTIMISATION OF BONDED REPAIRS

R. KAYE AND M. HELLER
Airframes and Engines Division, DSTO, PO Box 4331, Melbourne VIC 3001, Australia
E-mail: manfred.heller@dsto.defence.gov.au

Initially an automated sensitivity-based shape optimisation procedure is developed for the optimal design of free-form bonded repairs and lap joints, with the aim of achieving reduced adhesive stresses. The finite element approach is demonstrated through application to both a double-sided lap joint configuration, as well as for the reinforcement to an F/A-18 bulkhead. Here both the shapes of the adhesive layer and the outer adherend are allowed to vary. Significant improvements over conventional designs are obtained, as assessed by the reduction in peak adhesive stresses. This is followed by analysis to investigate stress magnification effects due to in-plane load attraction arising from bonded doublers of various shapes in remotely loaded plates. The optimality of the candidate shapes is then discussed.

1 Introduction

Over the last two decades, the use of bonded repairs (i.e. doublers) has been demonstrated to be an effective procedure for the life extension of cracked and uncracked metallic airframe components. Typically for practical applications, the design of bonded repairs has been undertaken through trial and error finite element analyses, or experiment. However, some recent experience indicates that for more complex practical applications, unacceptably high stresses can occur in the adhesive layer, or in the parent material surrounding the repair. There are essentially two load transfer mechanisms contributing to the magnification of these stresses. Firstly, there is through-thickness load attraction, where load is transferred via the adhesive layer from the parent material to the doubler. In this case, high adhesive stresses can occur at the two extremities of the load transfer regions, i.e. at: (i) the end of the bonded repair and (ii) over the crack. Secondly, there is in-plane load attraction, where a doubler bonded to a large plate attracts load in excess of the remote stress due to the local stiffness increase. This causes increased stresses in the adhesive layer and the parent material near the end of the doubler.

Automated shape optimisation offers the potential to improve the design of bonded repairs to minimise the stress magnifications discussed above. However, only limited work has been undertaken to date, and this has focussed on doubler through-thickness tapering. For example, an analytical treatment of the tapering at the ends of an isotropic doubler of a uniaxialy loaded lap joint is given by Ojalvo [1]. In other work the same problem is considered by using a 2D finite element gradientless optimisation method [2] or design sensitivity methods [3]. These analyses do not consider variation in adhesive thickness. Also no work appears to have been undertaken to minimise the adhesive stresses over the crack or to optimise the in-plane shape of the patch. Hence in this paper, the shapes of the adhesive layer and the outer adherend are first optimised to minimise adhesive stresses in both the taper and crack regions. This is followed by analyses to investigate stress magnification effects due to in-plane load attraction for doublers of various shapes.

2 Through-thickness shape optimisation of bonded crack repair/lap joint

2.1 Lap-joint loading and initial configuration

The configuration under study is a bonded double lap joint as defined in Fig. 1. The geometry is also representative of the two dimensional idealisation of a bonded repair to a cracked plate [4]. In this case the two outer adherends represent the doublers, while the two inner adherends represent the two sections of a cracked plate. Consequently, the terms joint and crack are both used to describe the centreline discontinuity of the base structure. The inner adherends have a thickness of 4mm and are subjected to a remote uniaxial stress of 100MPa. The outer adherends are bonded to the inner adherends by an adhesive layer having a nominal thickness of η=0.15 mm. The outer adherends are both 120mm in length and have a thickness of 2mm and an initial linear taper with a 1:10 slope at the ends to reduce the magnitude of the local adhesive stress concentration. For crack repair representations, the geometry is considered to be a worst case 2D slice of unit thickness through the repaired plate, where the slice is (i) aligned with the remote applied stress direction, (ii) perpendicular to the crack, and (iii) is located in the centre of the plate. The configuration has been divided into six load transfer regions, symmetric about the line x=60 mm, each having an arbitrary length of 20mm. In the taper and joint/crack regions most of the load transfer takes place between the doublers and the plate, and this is where the shape changes for optimisation were undertaken. It is common practice to have a region in between where there is no load transfer, termed here as the separation region.

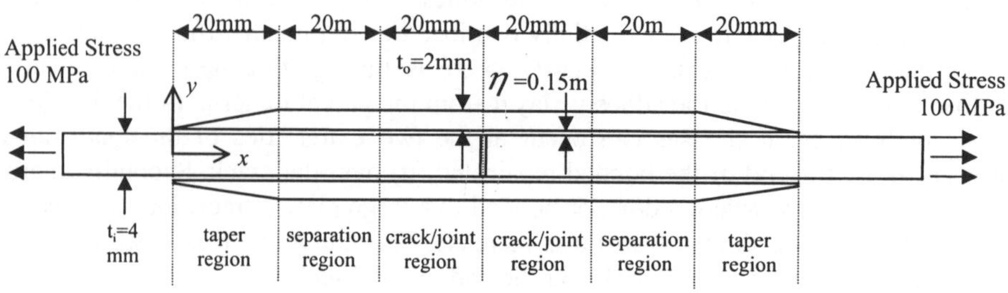

Figure 1. Dimensions and loading arrangement for double lap-joint.

The finite element mesh used quarter symmetry and plane strain conditions were assumed, where both the inner and outer adherends were aluminium alloy with Young's modulus E=70,000MPa, and Poisson's ratio ν=0.35. The stresses obtained from this analysis are presented in Fig. 4. The opening displacement of the centreline node, half way through the plate thickness in the x direction was 0.0086 mm. For 2D crack patching, the stress intensity factor in the inner adherend is directly related to the opening displacement by the equation from [4]:

$$K_1 = \sqrt{\tfrac{1}{2} E_p \phi \sigma \Delta} \qquad (1)$$

where K_1 is an upper bound stress intensity factor for a long crack under a patch, E_p is the elastic modulus of patch, ϕ is the stress reduction in the uncracked plate due to the patch at the prospective crack location, σ is the remote applied stress, and Δ is the crack opening.

2.2 Optimisation method

The aim was to optimise the repair shape to minimise peak von Mises adhesive stresses, such that the repair effectiveness (ie opening) was unchanged as compared to the above nominal geometry. Shape optimisation was achieved using the general sensitivity-based optimisation technique available in the MSC/NASTRAN code [5]. In broad terms, the approach involves changing the position of nodes defining a boundary shape, based on the computation of rates of change of nodal stresses with respect to shape changes. The general problem definition for shape optimisation can be stated in three parts as; (i) find values of thedesign variables, (ii) to minimise/maximise objective function, and (iii) subject to constraint functions. The objective function used here has been termed the least squares objective function in prior AMRL work [3,6]. It seeks to minimise the deviations from the average von Mises adhesive stress value in both the joint/crack region and the taper region, and is given in equation (2) as:

$$\text{minimise: } f_{obj} = \sum_{i=n}^{i=1}\left(\sigma_{avt} - \sigma_i\right)^2 + \sum_{j=k}^{j=1}\left(\sigma_{avj} - \sigma_j\right)^2 \qquad (2)$$

where i and j are the number of an individual adhesive element in the taper and joint regions respectively, n and k are the total number of adhesive elements in the taper and joint regions respectively, and σ_{avt} and σ_{avj} are the average adhesive element von Mises stresses in the taper and joint region respectively. The adhesive was modelled using three rows of elements and the stress values are taken at the centroids of centre row adhesive elements. This objective function takes advantage of the constant stress characteristic of optimal geometries where achievement of uniform boundary stress is known to be synonymous with minimising the peak stress value [2]. A key feature of the optimisation process is the way that shape changes are defined as design variables. Here a set of displacement fields was generated by using an auxiliary model with dummy loads applied normal to the movable boundary. These displacement fields are used to link a number of nodal movements to a single design variable. The user must choose the number and nature of displacement fields so as to give the optimiser enough shape variability to achieve the optimal solution. The auxiliary model must have the same geometry and node numbers as the primary model, however the material and element properties can be varied so as to give suitable shapes. It is run once only and the deformation output is stored and used by the optimisation process as required. It should be noted that each load is applied individually as a separate load case.

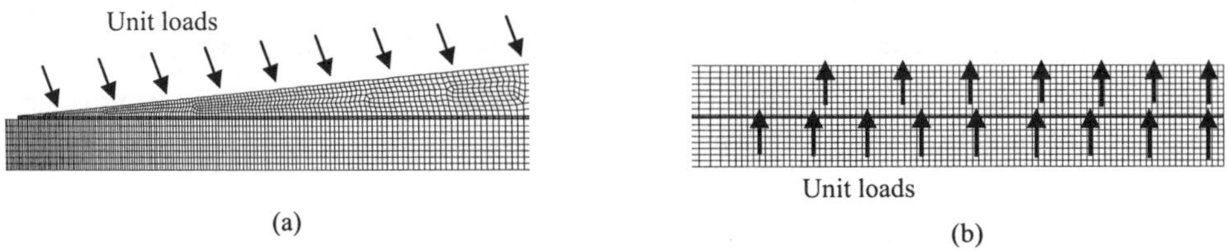

Figure 2. Finite element mesh and loading arrangement for generating basis shape vectors in auxiliary model: (a) taper region, and (b) crack/joint region.

The resultant displacement fields are called basis shape vectors in NASTRAN and can be considered as a set of vectors T_j as follows:

$$\mathbf{T}_j=\begin{bmatrix}\Delta G_{1x},\Delta G_{1y}\\ \cdots\cdots\cdots\\ \cdots\cdots\cdots\\ \Delta G_{Kx},\Delta G_{Ky}\end{bmatrix} \quad (3)$$

where j is the basis shape vector number, ΔG_x and ΔG_y are the nodal displacement in the x and y direction respectively, and K is the total number of nodes (same in both models). Hence 25 loadcases were used to generate 25 displacement fields and consequently 25 design variables were defined in the optimisation data file. These load application points, along the movable part of the boundary or interface, are shown in Fig. 2. At each iteration i+1, the new shape comes from taking the nodal positions at iteration i, and adding the displacements from all of the basis shape vectors multiplied by their scaling factors, as given by equation (4) below:

$$\begin{bmatrix}G_{1x}^{i+1},G_{1y}^{i+1}\\ \cdots\cdots\\ \cdots\cdots\\ G_{Kx}^{i+1},G_{Ky}^{i+1}\end{bmatrix}=\begin{bmatrix}G_{1x}^{i},G_{1y}^{i}\\ \cdots\cdots\\ \cdots\cdots\\ G_{Kx}^{i},G_{Ky}^{i}\end{bmatrix}+\sum_{j=J}^{j=1}X_j*T_j\;,\text{ where }\begin{bmatrix}X_1^{i+1}\\ \cdots\\ \cdots\\ X_J^{i+1}\end{bmatrix}=g\left(\begin{bmatrix}\delta f/\delta X_1^{i}\\ \cdots\\ \cdots\\ \delta f/\delta X_J^{i}\end{bmatrix}\right) \quad (4)$$

Here G_x and G_y are the x and y coordinates respectively of nodes k=1 to K (all nodes in model), i is the iteration number, j is the design variable number in the range j=1 to J, X is a design variable (i.e. multiplier applied to $\boldsymbol{T}$), $\delta f/\delta X$ is the sensitivity of the objective function with respect. to a design variable, and g is a function to represent the search algorithm described by Moore [5]. The scaling factors X_j, (one per basis shape) are the unknown design variables to be found by the optimisation process. Circular beam elements have been added along the movable boundaries of the auxiliary model to avoid shape discontinuities caused by the point loads. The circular beam elements were of 25mm radius and 0.25mm length, and were used in conjunction with plate elements of 0.5 mm thickness, noting that both had the same elastic modulus. For the taper region the doubler was allowed to vary in thickness, while the adhesive thickness was kept constant at 0.15 mm. However in the crack region, both the doubler and the adhesive thickness were allowed to vary. A constraint on crack opening was applied at the centreline node, half way through the plate thickness, so that $|\Delta/2|$ < 0.0086 mm, where $|\Delta/2|$ is the displacement in the x direction at the node described above. The optimisation regions and the constraint conditions are given schematically in Fig. 3(a).

2.3 *Optimisation results*

The solution shapes obtained after 7 iterations are shown in Figures 3(b) and 3(c) for the taper and crack/joint regions respectively, while the adhesive stresses are given in Fig. 4. For the taper region the shape obtained is very similar to the 2D FE solutions given in other investigations [2]. Here there is a rapid reduction in doubler thickness at the start of the taper leading to a very fine tip at the end of the doubler. A good result has been achieved in the taper region showing relatively uniform stresses, with a reduction in the peak of about 50%. A good result has also been achieved in the crack/joint region, with a near constant von Mises stress in the adhesive. A very large stress reduction of about 70% is evident in this linear elastic analysis. For the joint region it can be seen there is an increase both in the adhesive and the doubler thickness. The growth in doubler thickness is a less obvious result as it has occurred not merely to help satisfy the crack opening constraint but to reduce adhesive stresses.

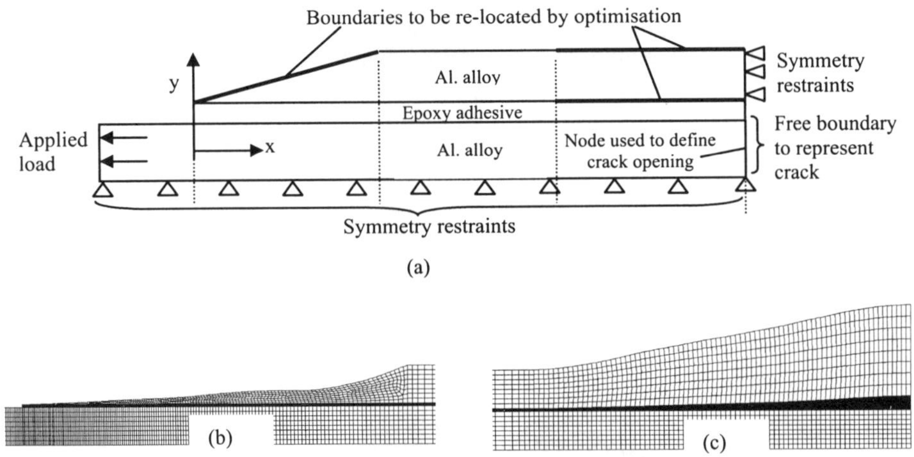

Figure 3. Configuration and finite element mesh for symmetric crack repair with aluminium doublers; (a) configuration and initial shape, (b) optimised shape in taper region, and (c) optimised shape in joint/crack region.

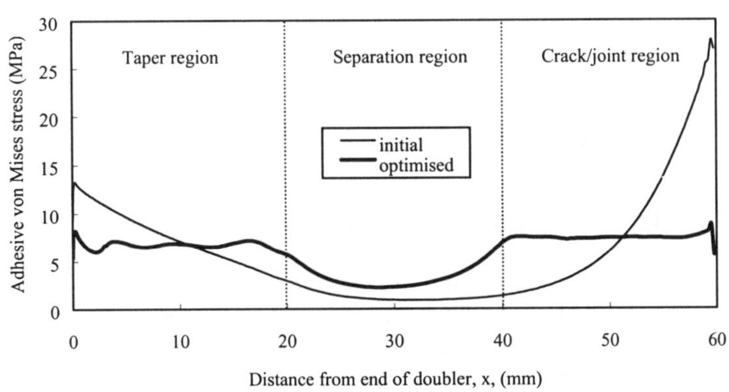

Figure 4. Adhesive stresses for symmetric crack repair with aluminium doublers.

3 Shape optimisation of bonded reinforcement to an F/A-18 FS 470 Bulkhead

The 2D finite element mesh shown in Fig. 5 indicates the in-plane geometry and dominant loading of the F/A-18 470 Bulkhead, which is prone to cracking. It is constructed as an aluminium alloy frame made up of flanges and webs of thicknesses ranging from 6 – 81 mm. The initial MSC/NASTRAN model is semi-symmetric about the *y*-axis (i.e. vertical centreline) and consists of three-noded and four-noded plane stress elements. For the bulkhead, isotropic material behaviour was assumed with Young's modulus E=72,000MPa, and Poisson's ratio ν=0.35. Also, the isotropic epoxy adhesive had Young's Modulus E=840 MPa and Poisson's ratio ν=0.3, while the unidirectional boron/epoxy reinforcement was taken as 2D orthotropic with E_{11}=207,000 MPa, E_{22}=25400 MPa, ν_{12}=0.18 and G_{12}=7200 MPa where the subscript 1 refers to the direction of the fibres and 2 and 3 are the transverse fibre directions. Initially, a standard finite element analysis was undertaken with a nominal 18-layer reinforcement, which has a uniform thickness of 2.34 mm and a uniform thin adhesive layer of 0.13 mm thickness. It provides a reduction in peak boundary stress of 25% compared to the un-reinforced case.

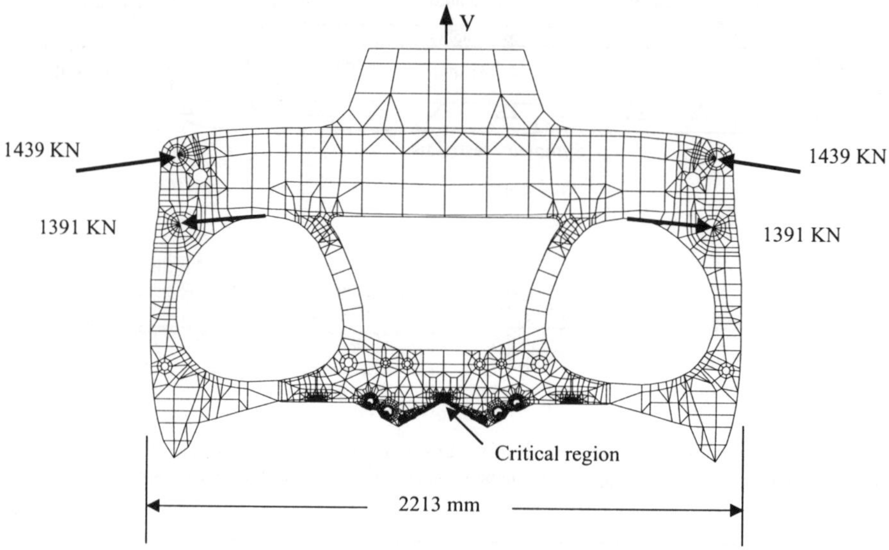

Figure 5. F/A-18 FS 470 bulkhead finite element mesh and dominant loading.

The aim in the optimisation analyses is to achieve the same reinforcement effectiveness as the nominal case, with a reduction in the peak adhesive shear stresses. There are consequently three boundaries under consideration with the following features (i) the interface between the bulkhead and the adhesive is fixed in location and is where the effectiveness of the reinforcement is determined by the bulkhead nodal stress values, (ii) the interface between the adhesive and the reinforcement is a boundary that has been made variable in position and has ultimately been located by the optimisation process, and (iii) the outer free boundary of the reinforcement has also been made variable and has been located by the optimisation process. Three rows of elements are used across the initial adhesive thickness of 0.13mm. The method as described in section 2, is again used, and the objective function is

$$f = \sum_{i=n}^{i=1} (\sigma_i - \sigma_{av})^2 \tag{5}$$

where σ_i is the stress at element *i* (i.e. structural response), σ_{av} is the arithmetic average of element stress values over a selected bond line segment (i.e. reference stress), and *n* is the number of elements included in objective function evaluation. Here the stresses used in the above equation are the mid-layer element centroid stresses. Reducing the number of stress points in the objective function by including only every fifth element enhanced numerical performance.

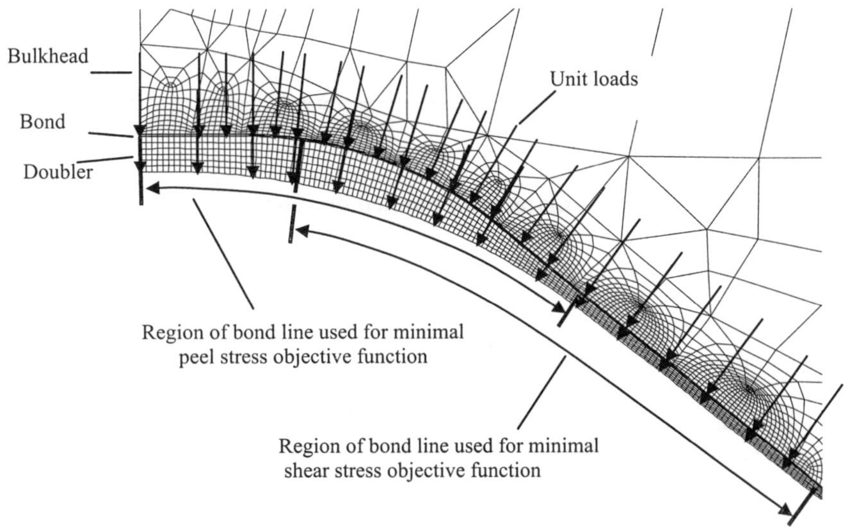

Figure 6. Finite element mesh and loading arrangement for generating basis shape vectors in auxiliary model.

The regions of bond line that were included for the minimum shear stress analysis are given in Fig. 6. Also shown is the auxiliary model loading arrangement for generating the 23 basis shapes. Here beam elements of 25 mm diameter were applied along both edges of the reinforcement to smooth the deflections caused by the point loads. All materials in the auxiliary model were changed to aluminium alloy and all 2D elements were reduced in thickness to 0.5mm. The results for the minimum shear stress solution obtained after seven iterations are shown in Figures 7 and 8. It can be seen that the shear stress has been made more uniform by the increase in adhesive thickness at the high stress region near the curved part of the boundary. Figure 8 shows that a 20% reduction in shear stress has been achieved compared to the nominal case, without an increase in adhesive peel stress or the hoop stress in the parent material of the bulkhead. It can be seen that the adhesive has been made thicker in the regions of high initial shear stress near the curved parts of the boundary. A minimum peel stress solution was also obtained, (not shown), [6].

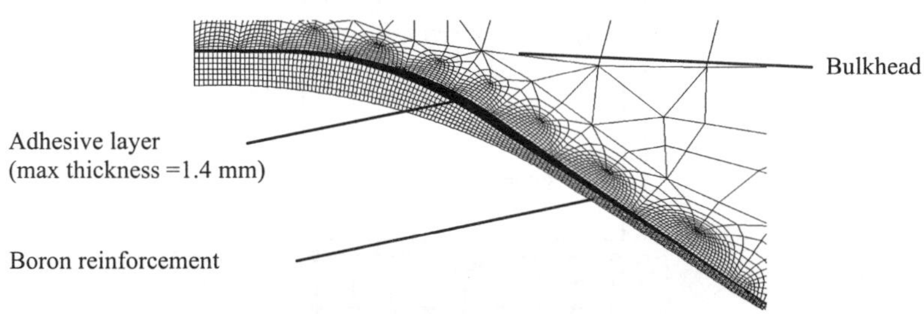

Figure 7. von Mises stress distribution in adhesive for symmetric crack repair with aluminium doublers.

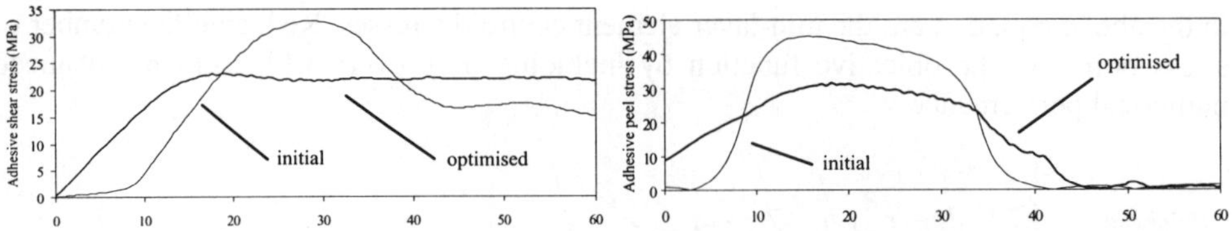

Figure 8. Stress solution for F/A-18 FS 470 bulkhead reinforcement with minimal adhesive shear stress.

4 In-plane shape effects for flat plates with bonded doublers

When a doubler is applied to a large flat surface such as a wing or fuselage skin, the in-plane load attraction creates a region of higher stress immediately outside the ends of the doubler. Historically, rectangular or elliptical doublers have been the most common for practical applications. In this section, the aim is to investigate the extent to which alternate shapes may give reduced stress concentrations, and to provide candidate configurations to be used as initial estimates for possible subsequent shape optimisation. Here numerical analyses are undertaken where the in-plane load attraction effect of a doubler is approximated as a 2D inclusion in a relatively large plate. The 2D geometry and loading under consideration is shown in Fig. 9, and consists of a flat square plate of 3mm thickness with 400mm sides. The plate was modelled using quarter symmetry and assuming 2D plane stress conditions. The bounding doubler design space was 80mm by 80mm at the centre of the plate. In all cases the doubler was represented as an inclusion, by a region of elements with an increased thickness of 3.381 mm. This thickness gives an increase in stiffness approximately equivalent to one layer of boron/epoxy. A remote unidirectional stress of 100 MPa was applied in the y direction along the edges y=+/- 200 mm. Isotropic aluminium alloy properties of E=71000 MPa and ν=0.3 were used. To investigate shape effects alone, only doublers with the same overall aspect ratio were compared. The aspect ratio is defined as the maximum length of the doubler in the y direction divided by its maximum width in the x direction. Numerical values of element centroid major principle stresses were extracted, and the peak stress concentration K_t is expressed in percentage terms as compared to the remote stress.

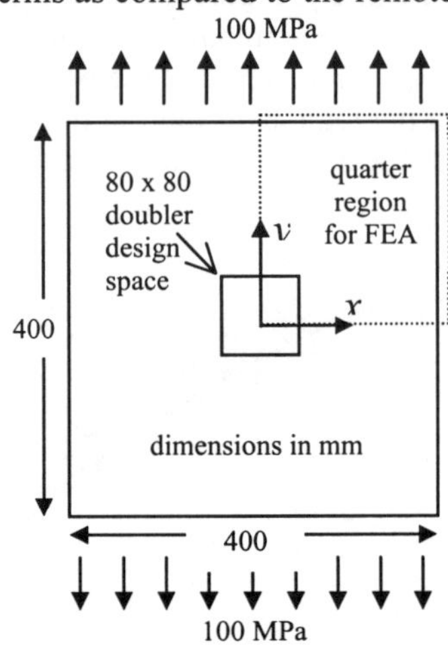

Figure 9. Geometry and loading for in-plane analysis of doubler on plate using inclusion analogy.

Results obtained for the 2:1 aspect ratio are summarised in Fig. 10. It can be seen in Fig. 10(a) that the raised stress region is distributed evenly across the end of the rectangular doubler. This is an indication of optimality. Small perturbations in the slope of the end of the doubler were then made, which gave worse K_t values and made the stresses less uniform as shown in Figures 10(b) & (c). Some other shapes such as diamonds and ellipses were also considered (not shown). From these results it is clear that the rectangular shape is very close to being fully optimal for this aspect ratio.

Rectangular and elliptical shapes were then also analysed at other aspect ratios, and the K_t values are plotted in Fig. 11. It is seen that rectangular doublers perform better than ellipses at aspect ratios larger than 3:4. It was also found that for the 1:1 and 1:2 cases the raised stress region is not evenly spread at the ends, suggesting that some improvement could be made through an optimisation process. The wider ellipses give further reductions in K_t and longer boundaries (not shown) with even stress distribution. Because of the need for long load transfer lengths and space limitations most real doubler applications have to be at least twice as long as their width. Hence this work suggests that the commonly used rectangular shapes are good if not optimal (for aspect ratios of 2:1 and higher). However if for a practical application the opportunity arose to implement a 1:1 or wider doubler, some benefit could be gained by using a fully optimised shape, or in some cases an ellipse. The results stated here are for one layer only and the actual benefit would be significantly larger for a typical multi-layer doubler.

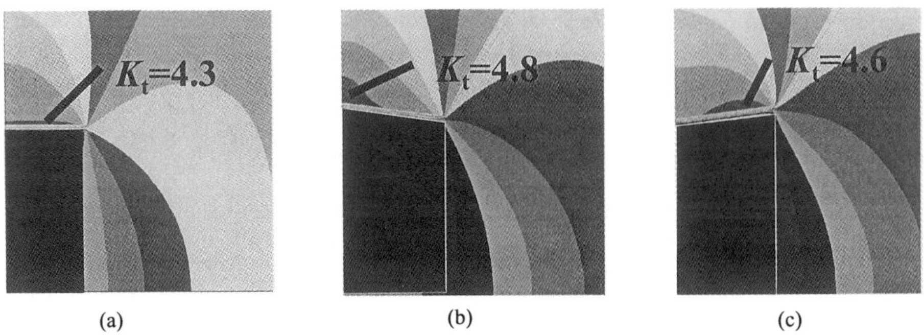

Figure 10. Stress results for doublers with aspect ratio 2:1; (a) rectangular, (b) sloped end, and (c) reverse sloped end.

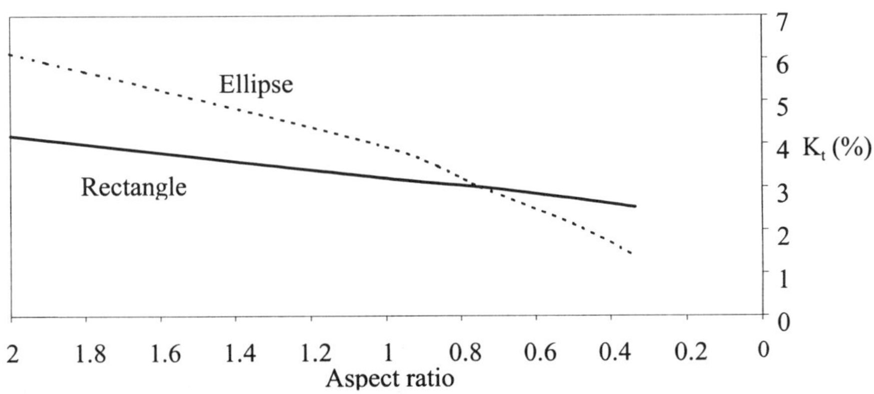

Figure 11. Kt results for rectangular and elliptical doublers as a function of aspect ratio.

5 Conclusion

Through-thickness shape optimisation provides significant scope for minimisation of peak adhesive stresses for bonded joints and repairs to ageing aircraft components. It is expected such designs would contribute to improved repair durability, and hence further experimental validation is recommended. It has also been found that the optimality of in-plane reinforcement shapes, in terms of the plate stress concentration near the doubler ends, depends on the aspect ratio of the reinforcement. For uniaxially loaded plates, rectangular doubler shapes are essentially optimal for aspect ratios of 2:1 and higher. For lower aspect ratios, neither rectangular or elliptical shapes are optimal, and hence it is recommended in future work to determine the optimal shapes through automated procedures.

References

1. Ojalvo I. U., Optimisation of bonded joints, *AIAA Journal,* **23**(10) (1985) pp.1578-1582.
2. Kaye R. and Heller M., *Structural shape optimisation by iterative finite element solution*, DSTO-RR-0105, (1997) AMRL, Melbourne, Australia.
3. Kaye R. and Heller M., Through thickness shape optimisation of bonded repairs and lap-joints, *International Journal of Adhesion and Adhesives, accepted,* (2001).
4. Rose L.F.R., *Theoretical analysis of crack patching- chapter 5*, in Bonded Repair of Aircraft Structures (Baker, A.A. and Jones, R. editors), (1988) Matrinus Nijhoff, Dordrecht.
5. Moore G. J., Design sensitivity and optimisation, the MacNeal Schwendler Organisation, (1994).
6. Kaye R. and Heller M., Investigation of shape optimisation for the design of life extension options for an F/A-18 airframe FS 470 bulkhead, *Journal of Strain Analysis,* **35**(6) (2000) pp.204-212.

SHAPE OPTIMIZATION OF A BONDED PATCH TO A CYLINDRICAL SHELL

LIYONG TONG AND XIANNIAN SUN

School of Aerospace, Mechanical and Mechatronic Engineering, The University of Sydney, New South Wales 2006, Australia.

A novel shape optimization scheme is presented for minimizing the adhesive stresses in bonded patch to curved thin-walled structures. Sequential linear programming is employed to search the optimum shape of the bonded patch. In each iteration, finite element method is employed to conduct stress analysis and sensitivity analysis using a novel adhesive element. An automated mesh generation algorithm is developed to adapt the change of geometry of the patch. Numerical examples show that the peak adhesive stresses can be reduced remarkably through optimizing the shape of bonded patch.

1 Introduction

Adhesive bonding is one of the primary joining techniques of connecting metallic and/or composite structural components, particularly in aircraft and automobile structures [1]. It has also been used to repair damaged thin-walled flat structures with bonded patch to their surfaces [2]. While there exists a large number of research focusing on boned repairs to flat surfaces, there is only very limited knowledge available on bonded patch repair to curved surfaces. Recently, Tong and Sun [3,4] investigated the effect of curvature on stresses in bonded repair to curved surfaces. Similar to the case of bonded repair to flat surface, numerical results show that high stresses in the adhesive occur along the periphery of the bonded patch, where high gradient of load transfer takes place. In addition, stresses concentrations can also happen near a crack in a parent structure. When bonding a patch to a surface, one would need to design the patch so that it would not come apart from the parent structure under the designated operational environment, including mechanical and thermal loading. This becomes even more critical when bonding a patch to a curved surface due to the effect of curvature [3,4]. Thus, one of the most practical tasks is to find an optimum shape of bonded patch to minimize the stresses in the adhesive layer so that the patch would not disintegrated from the parent curved structure.

There exists a limited number of research effort placed on shape optimization of bonded patch or even bonded joints. Optimum profile of adherend was pursued by several researchers [5-8] to minimize the peak stresses in the adhesive layer. Rispler et al [9] optimized the spew fillet of adhesive in a bonded double lap joint. Although there exist a large amount of research effort on structural optimization focusing on development of efficient searching algorithms and their applications to specific problems [10], little, if not no, effort has been devoted to optimize the geometrical configuration of a bonded patch to a flat surface. To our best knowledge, there exists no work conducting shape optimization of bonded patch to a curved surface via minimizing the adhesive stresses. In this paper, shape optimization of a bonded patch to a cylindrical shell for minimizing the stresses in the adhesive layer is conducted by using the sequential linear programming (SLP), the new adhesive element developed by Tong and Sun [3] and a novel automated mesh generation scheme.

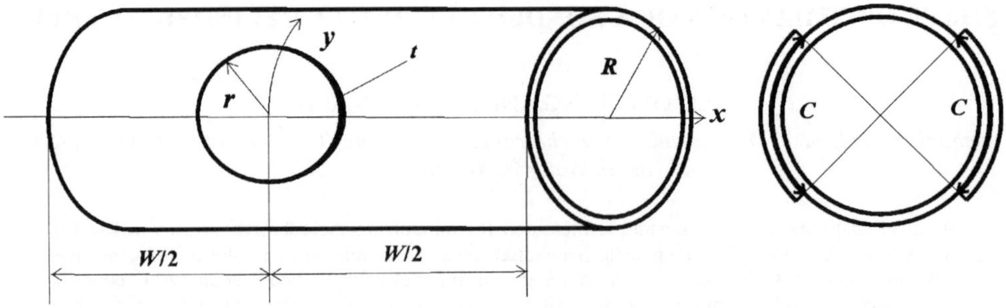

Figure 1 A cylindrical shell bonded two circle patches on its external surface

2 Stress Analysis

A cylindrical shell bonded with two circular patches on its external surface is shown in Fig.1. The patch area can be modeled using the adhesive element developed by Tong and Sun [3] to determine stresses in adhesive layer. The 16-node adhesive element, shown in Fig.2, is sandwiched by two 8-node Serendipity degenerated shell elements based on the first-order shear deformation theory. As shown in Fig. 2, the adhesive element shares the same nodes as the corresponding shell elements. Some basic assumptions made for the adhesive element are:

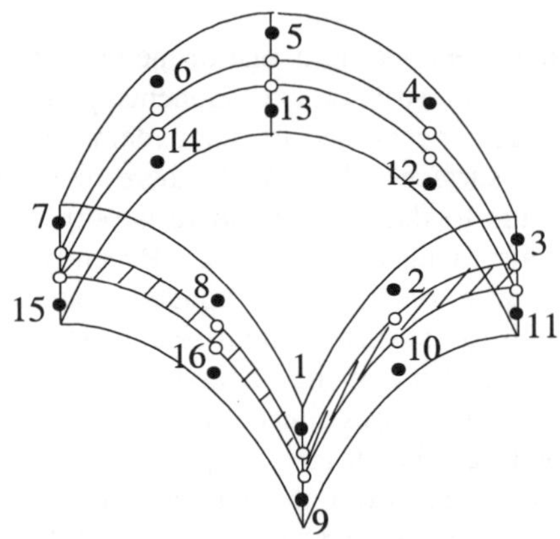

Figure 2 A 16-node adhesive element

a. The patch is perfectly bonded to the cylindrical shell with uniform thickness and no debonding occurs in the whole adhesive layer.
b. The thickness of adhesive layer is very thin comparing to that of patch and cylindrical shell. Then the strains in the adhesive layer are assumed to be uniform across the thickness of the adhesive layer.
c. Only three out-of-plane stresses σ_z, γ_{yz}, γ_{xz} are considered and other three in-plane stresses σ_x, σ_y, γ_{xy} are assumed to be equal to zero.

The detailed formulation and accuracy analysis for the adhesive element can be found in reference [3].

3 Shape Optimization Method

3.1 *SLP formulation* [6]

Our purpose is to optimize the geometrical shape with minimized the maximum peel and shear stresses in the adhesive layer. Hence, a nominal stress parameter σ_a that determines the strength of the adhesive layer is chosen to be the objective function. This stress parameter can be the maximum peel, shear or Von-Mises stress in the adhesive layer. Then the shape optimization can be expressed in a standard form by linearizing the objective function and constraints as follows:

$$\text{Minimize } \sum_{j=1}^{n} \frac{\partial \sigma_a}{\partial X_j}(X - X_j)$$

$$\text{Subject to } \begin{cases} \sigma_i^{(k)} + (\nabla \sigma_{ij}^{(k)})^T (X - X^{(k)}) \le \sigma_a & (i = 1, \ldots, m) \\ \max(X_j - d) \le X_j \le (X_j + d) & (j = 1, \ldots, n) \end{cases} \quad (1)$$

Where $\sigma_i^{(k)}$ is the nominal stress at the *i*th node in the *k*th iteration step, $\nabla \sigma_{ij}^{(k)}$ is the sensitivity vector of all design variables at the *i*th node in the *k*th iteration step, *X* is the design variable vector in the current iteration, and $X^{(k)}$ the design variable vector in the last iteration. *d* is the move limit to ensure the linear approximation. *m* is the number of nodes that control the maximum nominal stress in the adhesive layer. *n* is the number of nodes, which are allowed to move as design variables.

It is worthy noting that some limitations should be imposed to the area of the bonded patch because the weight is an important concern for most structures. On the other hand, the shape change in current analysis is arbitrary. To avoid that the shape of the bonded patch reduce infinitely in some direction to become a narrow strip, it is necessary to define a basic domain that the shape change must take place out of this domain. All these limitations can be inserted to equation (1) as more constraints.

The convergence of the optimization process is checked by the following criterion:

$$|(\sigma_{\max} - \sigma_{\max}^k)/\sigma_{\max}^k| < \varepsilon \quad (2)$$

where ε is the convergence tolerance.

3.2 *Shape design variables*

For the sake of convenience, we map the curved patch and the cylindrical shell into a flat plane. As shown in Fig.3, one quarter of one curved patch, after being mapped into *x*-*y* plane, is plotted with a basic domain defined. Some points along the periphery of the bonded patch are selected to define the boundary shape of the patch. The movements of these points are assumed to be in both the *x*- and *y*- direction. At the middle point, where *x*=0 or *y*=0, the movement is limited in one direction, i.e. *y*-direction or *x*-direction. Beside that, the movement would be neglected if the shape change with this movement were very small (refer to Fig.3).

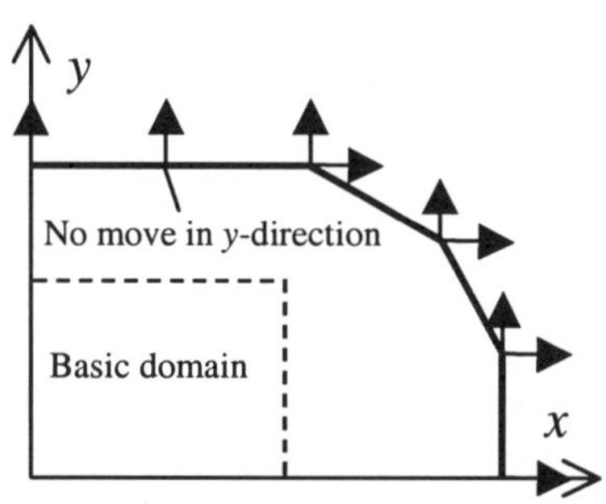

Figure 3 Shape design variables

4 Adaptive mesh generation

It is inevitable that mesh topology varies during the optimization process due to a large shape change of the bonded patch. To avoid excessive element distortions and unreasonable large aspect ratios, even unreal element, an adaptive mesh generation algorithm is employed to generate quadrilateral element [11]. The mesh is first automatically generated in a flat plane. Then the mesh topology is mapped into a curved surface for both the bonded patch and cylindrical shell. Two layers of elements by offsetting the boundary are generated first to ensure the elements along the periphery of the bonded patch, where stress concentration occur, as close to uniform square element as possible. A typical mesh occurring in the analysis is show in Fig. 4, which has 3650nodes, 1118 shell elements, and 189 adhesive elements.

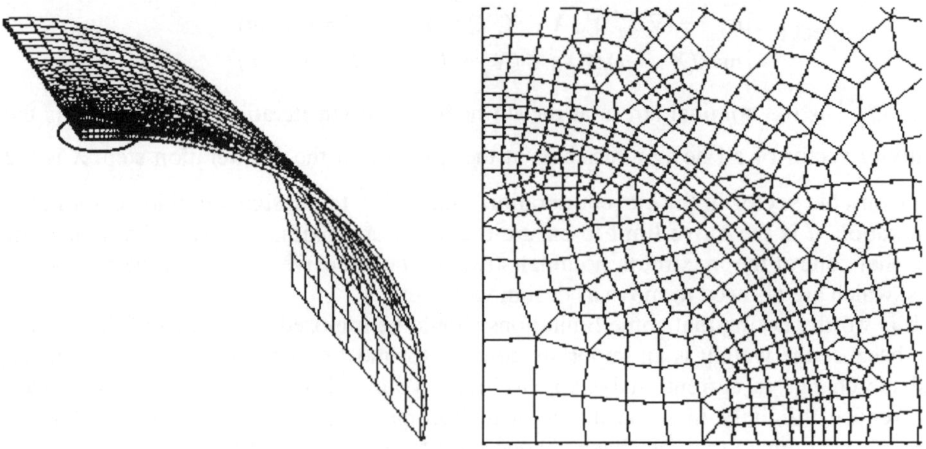

Figure 4 Typical mesh used in current analysis

5 Numerical Example

As shown in Fig.1, the two circular patches of radius r=15mm are externally bonded to a cylindrical shell. These two patches are located at the middle of the cylinder and only 1/8 of structure is modeled due to its structural symmetry. The overall length of the shell is W=300mm. The radius of the shell is R=150mm. The thickness of the cylindrical shell and the patches are all 2.0mm. The thickness of adhesive layer is t=0.15mm. The shell and the patch are assumed to be metal with a Young's modulus of E=70GPa and Poisson's ratio of v=0.3. The adhesive used has a Young's modulus of E_c=2.4GPa and a Poisson's ratio of v_c=0.33. Both ends of the cylindrical shells are supported by rigid diaphragms, which allow displacements only in the axial direction of the shell. The applied internal pressure is p=1.0MPa.

The area of the bonded patch is limited between $3.0r^2$ and $6.0r^2$. A 1.2r×1.2r rectangle domain in the center of patch is defined as the basic domain (see Fig. 3). The move limit in equation (1) is d=0.375mm.

Three nominal stress parameters are used in this optimization analysis, and they are the maximum peel stress, the maximum shear stress, and the maximum Von-Mises stress in the adhesive layer, respectively. They are defined as: peel stress= σ_z , shear stress= $\sqrt{\tau_{yz}^2+\tau_{zx}^2}$ and Von-Minses stress= $\sqrt{\sigma_z^2+3(\tau_{yz}^2+\tau_{zx}^2)}$. All the stresses used here are the nodal stresses at the same node.

Several shape evaluations for minimizing the maximum Von-Mises stress are presented in Fig. 5. It should be noted that only a few iteration steps are given in Fig. 5 to clearly show the evolved curves. Moreover, three optimum shapes for different objective nominal stresses are plotted in Fig. 6 and the stresses for the initial and optimum shape of the bonded patch are given in Table 1. In this analysis, the optimization process starts from the circular patch when the maximum Von Mises stress is selected to be objective function. For the other two cases, the optimum shape developed from minimizing the maximum Von-Mises stress is used as the initial shape to save computing time. It should be noted that the peak values of three nominal stresses might be located on different nodes for the same patch shape.

From Fig.5, it can be clearly seen that the shape optimization scheme proposed here can successfully trace into the optimum shape undergoing large movement. It can more adaptively move the shape to an optimum one based on different constraints because it can be moved in both *x*- and *y*-direction at the same time. By comparing the shapes at step 2, 3 and 4, it can be found that the algorithm can automatically move back from some sharp points according to the employed constraints.

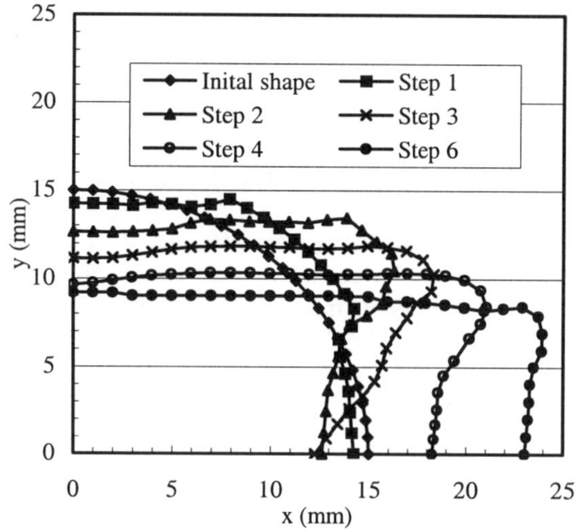

Figure 5 Shape evaluation in the optimization process (The objective function is Von-Mises stress)

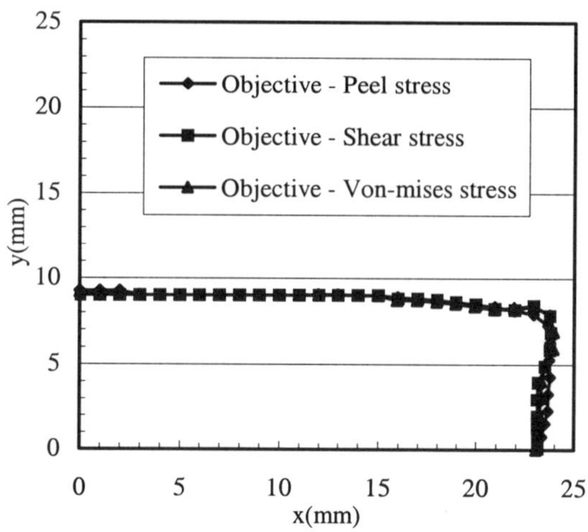

Figure 6 Optimum shape for different objective function

Table 1 Initial stresses and optimum stresses in the adhesive layer

Maximum stress		Peel	Shear	Von-Mises	$(\sigma_0 - \sigma_a)/\sigma_0$
Initial shape		16.38	19.66	21.82	
Objective function	Peel	6.75	15.40	15.74	58.8%
	Shear	6.90	14.88	15.58	24.3%
	Von-Mises	7.81	14.99	15.49	29.0%

From Table 1, it is clear that the peak stresses in the adhesive layer can be reduced dramatically by optimizing the geometrical shape of the bonded patch. Especially for the peel stress, it can be reduced by an amount up to 58.8% in this example. In fact, the peel stress is closely related to the circular length of the bonded patch. For the limitation of the basic domain, the circular length of the patch is stopped at the basic domain. The peel stress should be reduced further if a smaller basic domain is given. The shear stress and Von Mises stress are reduced by 24.3% and 29.0%, respectively. From Fig.6, we can see that the optimum shapes for different objective functions are very similar except for a little change at the corner of the patch. Obviously, cutting off the corner can slightly reduce the peak peel stress, but it can increase the peak shear stress at the same time. The optimum criterion should be selected based on the possible failure mechanism.

6 Conclusions

A novel shape optimization algorithm for minimizing the maximum stress in the adhesive layer is presented in this paper. The numerical examples show that the optimization scheme can adaptively trace to the optimum shape. The peak stresses in the adhesive layer can be reduced dramatically by optimizing the geometrical shape of the bonded patch.

Acknowledgement

The authors are grateful to the support of AOARD/AFOSR, and also would like to acknowledge technical discussion with Dr. Greg Schoeppner of the AFRL.

References

1. L. Tong and G. P. Steven, *Analysis and Design of Structural Bonded Joints*, Kluwer Academic Publishers, Boston, USA, Apr. 1999
2. A.A. Baker and R. Jones, *Bonded repair of aircraft structures*. Martinus Nijhoff: Dordrecht; 1988
3. L. Tong & X. Sun, 'Stresses in bonded repair/joints to curved structures', US AFOSR/AOARD Contract Report FEARC-00-001, Jun 2000
4. L. Tong & X. Sun, Stress analysis of bonded and curved lap joints, Proceedings of American Society Composites 15th Technical Conference Sep 24-27, 2000, Edited by O. O. Ochoa, Texas A&M University, USA, pp. 719-728
5. B.W. Cherry and N.L. Harrison, The optimal profile for lap joint. J Adhesion Vol. 2 (1970) 125-128
6. I.U. Ojalvo, Optimization of bonded joints. AIAA Journal, Vol. 23, No. 10 (1985) 1578-1582
7. H.L. Groth and P. Nordlund, " Shape optimization of bonded joints ", Int. J. Adhesion and Adhesives Vol. 11, NO. 4 October 1991
8. N. Hu, B. Wang, H. Sekine, Z. Yao, Guowen Tan, Shape-optimum design of a bi-material single-lap joint, *Composite Structures*, 1998, Vol.41, pp.315-330
9. A. R. Rispler, L. Tong, G.P. Steven, M. Wisnom, Shape optimization of adhesive fillets, *Int. J. of Adhesion & Adhesives*, 2000, Vol. 20, No.3, pp.221-231
10. R.H. Gallagher, O.C. Zienkiewicz, *Optimum Structural Design Theory and Applications*, 1973
11. J.A. Talbert, A.R. Parkinson, Development of an automatic, two-dimensional finite element mesh generator using quadrilateral elements and Bezier curve boundary definition, *Int. J. for Numerical Methods in Engineering*, 1990, Vol.29, pp.1551-1567

STRAIN MEASUREMENT AND PREDICTION IN A COMPOSITE DOUBLE-LAP SHEAR BONDED JOINT

DAVID H. MOLLENHAUER, GREGORY A. SCHOEPPNER
Air Force Research Laboratory, AFRL/MLBC, Wright-Patterson AFB, OH 45433, USA

and
ENDEL IARVE
University of Dayton Research Institute, Dayton, OH 45469, USA

Adhesive bonding has been used to join or repair metallic and/or composite structural components to achieve or restore their designed structural stiffness and strengths. It has been demonstrated in numerous programs over many years that the critical interlaminar adhesive stresses govern the lifetime performance and durability of adhesively bonded joints and repairs. Current design tools for composite bonded joints and repairs typically use laminate level analyzes in which the properties of the laminate are homogenized and the residual ply level stresses are neglected. This study uses the BSAM spline variational three dimensional ply level stress analysis method to model a double-lap shear specimen. Moiré interferometry was used to measure the surface strains on the edge of the specimen in the bondline end region. A comparison of the predicted and experimentally measured strains shows excellent agreement.

1 Introduction

The long-term goal of the current in-house effort is to establish a baseline understanding of the service life behavior of composite to composite adhesively bonded joints and repairs. Lifetime performance prediction of composite bonded joints requires an accurate description of the loading and environment history as well as their effect on the properties of the adherends and adhesives. As an example, some adhesives have been shown to be susceptible to hydrolytic degradation that can affect the constitutive and failure behavior of the bonded joints.

The preferred mode of failure for a composite joint is cohesive failure, as opposed to adhesive failure that may indicate poor surface preparation or poor bonding process control. Whether failure initiates in the adhesive or the adherends, the origin of failure occurs at the micromechanical level. For instance, Roy et al. (1-2) and others have reported that composite adherends fail on planes that are only a few fiber diameters from the adhesive/adherend interface. Examination of the failure surfaces indicates that fracture occurs along fiber/matrix interface boundaries. It is unlikely that an explicit full field micromechanical analysis of the joint will be required to accurately model the service performance and failure of composite bonded joints. However, it will require an explicit representation of the behavior of the composite joint constituents and their interactions. The behavior of the adhesive, the composite matrix material, and the fiber each must be accurately represented and characterized for the given service loading conditions

The majority of current analysis tools for composite bonded joints model the composite adherends as homogenized orthotropic bodies and neglect the residual curing stresses. However, the analysis of failure initiation and progression is unlikely to be successful if the composite is modeled as a homogeneous body and is likely to require modeling of the adherend at the ply level or lower. Furthermore, neglecting the residual curing stresses, particularly in free edge regions, will lead to an incorrect load prediction for failure initiation (3).

The United States Air Force Research Laboratory is developing an analysis package called BSAM (B-Spline Analysis Methods) that is the foundation on which our service life prediction capability will be built. The analysis package uses independent spline approximations of the displacement components *u*, *v*, and *w* for each ply of a laminate to provide accurate solutions for stresses and displacements. BSAM maintains continuity of strains and stresses throughout a homogenous ply while allowing strain discontinuity at ply interfaces in order to achieve interlaminar traction continuity, not achievable using conventional finite elements.

Results from the BSAM program have been shown to be in excellent agreement with moiré interferometry ply level results for open-hole composite laminate coupons (4). Future plans for the analysis package include extending its capability to model scarf and step lap composite joints and repairs. This paper describes the *preliminary* effort of analyzing a composite double-lap shear joint in which the adherends are modeled at the ply level and the adhesive is modeled as a linear elastic material. The analysis is compared to moiré interferometry measurements of the strain field on the edge of a double-lap shear specimen in the bondline end region.

2 Bsam - spline variational analysis method

Applications of BSAM have focused on modeling the elastic and failure behavior of bolted and bonded joints. B-spline approximations offer the most efficient solution when compared with other approximations of similar order. Finite element approximations using higher order –p elements have been implemented into a number of commercial finite element packages and shown to be beneficial in the areas of high stress concentrations. B-spline approximations offer an additional benefit of being able to prescribe the order of continuity between the finite elements. The C^0 continuity, that is typical in finite element approximations, requires excessive number of degrees of freedom within homogeneous domains. The B-spline approximation basis functions can be used to build a variable continuity approximation. Under given nodal continuity condition and given power of polynomial approximation, B-splines provide the least populated stiffness matrix. Complete details of the method can be found in Iarve (5).

Asymptotic solution corrections for 3–D free edge effects have been integrated into the solution method for composite bolted joints and are planned for implementation in the more general bonded joint code. The BSAM code has also been used to successfully model composites with curvilinear fibers and discontinuous reinforcement. The current capabilities of BSAM include the ability to model 3-D geometries, using *p* and *h* type refinements (as used in finite elements), model generally anisotropic materials (21 elastic constants), and calculate fracture mechanics parameters such as J-integrals.

3 Moiré interferometry experiments

Moiré interferometry was employed in this work to determine x- and z-displacements on the edge of the double lap shear specimen. The specimen was prepared by first applying a diffraction grating, necessary for the technique, to the edge of the specimen. The lap shear specimen was loaded with a screw-driven load frame. It was examined at three load levels (1,300 N, 2,700 N, and 5,300 N) in a moiré interferometer. The results from the highest load level are presented in this work.

Moiré interferometry is an optical interferometry technique capable of measuring in-plane displacements with high sensitivity. A rigorous explanation of the physics and principles of moiré interferometry can be found in the literature Post et al. (6). The technique yields a displacement fringe pattern that is essentially a contour map of in-plane displacements in a given direction on the face of the specimen. A phase-shifting technique of the type described in Lassahn et al. (7) was used in this research to enhance the basic sensitivity of the moiré interferometry technique.

4 Specimen fabrication, modeling, and testing

The material system used for the double-lap shear adherends was IM7/3501-6 graphite/epoxy supplied in prepreg form. Two panels with a 24 ply $[0/-45/+45/90]_{S3}$ stacking sequence were fabricated in an autoclave according to the manufacturers recommended cure cycle. The adherend panels were C-scanned after curing and found to be free of defects before being trimmed. The FM300-2M film adhesive used in this research was mistakenly cured at 177°C instead of the 121°C specified. Properties of the properly cured adhesive, however, were used in the BSAM model After the adherends where bonded together, the adhesive spew (see Fig. 1) were carefully removed using an end mill. The removal of the filets resulted in bondline end regions that were easily modeled with BSAM. The double-lap shear panel was then cut into nominal 25.4 mm wide specimens. The dimensions of specimen are shown in Fig. 2.

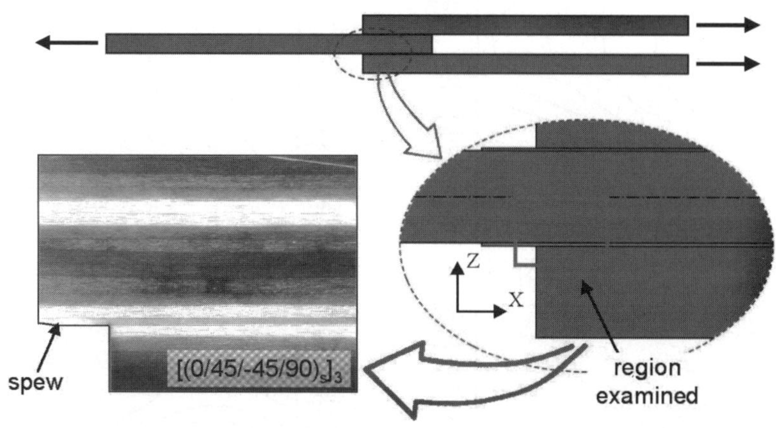

Figure 1: Specimen Preparation and Moiré Test Region

A side edge of one of the double-lap shear specimens was polished and a diffraction grating was applied to the specimen near the bondline end region. Moiré interferometry measures the *change* in the surface strains. The changes in the strains due to mechanical loading are independent of the residual strains, assuming that the specimen behaves linear elastically. Therefore, to model the change in strain due to mechanical loading, we neglect the residual curing stresses in the adherends and in the bondline in the analysis for this comparison.

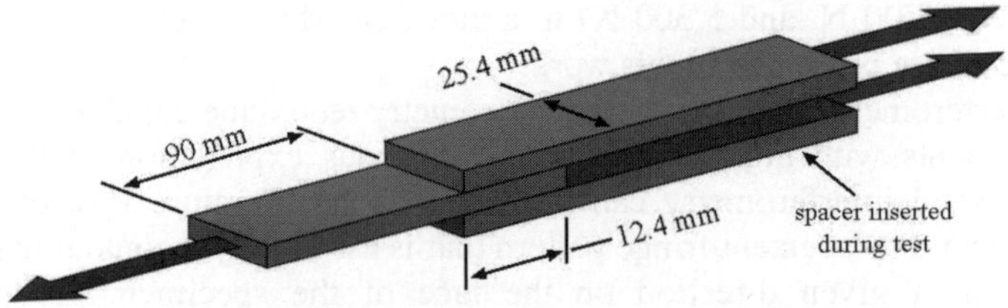

Figure 2: Double-Lap Shear Specimen Dimensions

5 Analytical/experimental comparison

For comparing the analytical and experimental surface strains, all of the analytical strain components were scaled with respect to the far field axial strain on the outer adherends. A comparison of the experimentally measured surface strain in the bondline end region to the predicted surface strain shows favorable agreement. Figure 3 displays contours of the experimental and predicted axial displacements and the transverse displacements, respectively. It can be seen that the region of the adhesive layer shows close contour lines indicating large strain gradients, as is expected. The variation of the axial and transverse displacements through the thickness of the adherends in both the moiré contour plot and the experimental contour plot support the argument that a ply level analysis is required. Figures 4-6 show a comparison of the axial, transverse and shear strains respectively, in the region of the bondline interfaces.

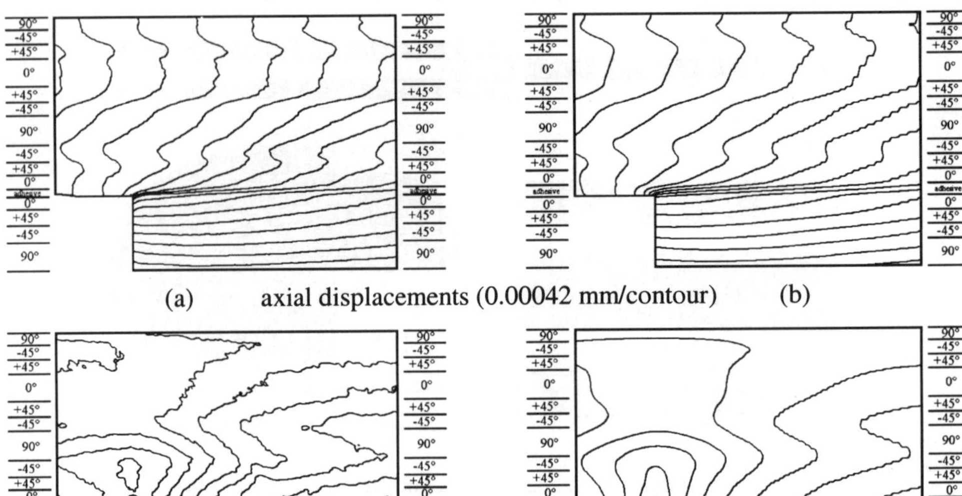

(a) axial displacements (0.00042 mm/contour) (b)

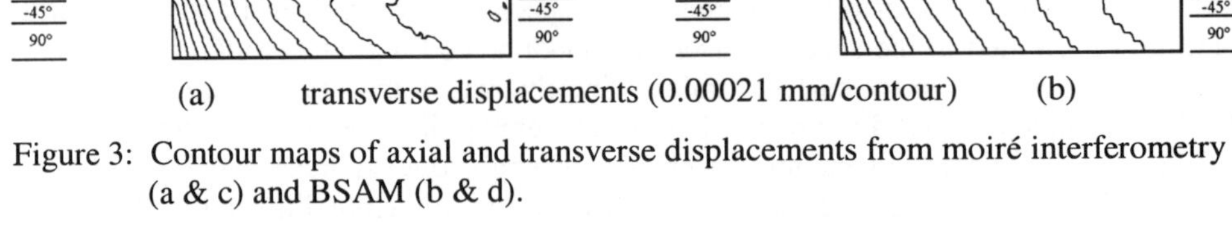

(a) transverse displacements (0.00021 mm/contour) (b)

Figure 3: Contour maps of axial and transverse displacements from moiré interferometry (a & c) and BSAM (b & d).

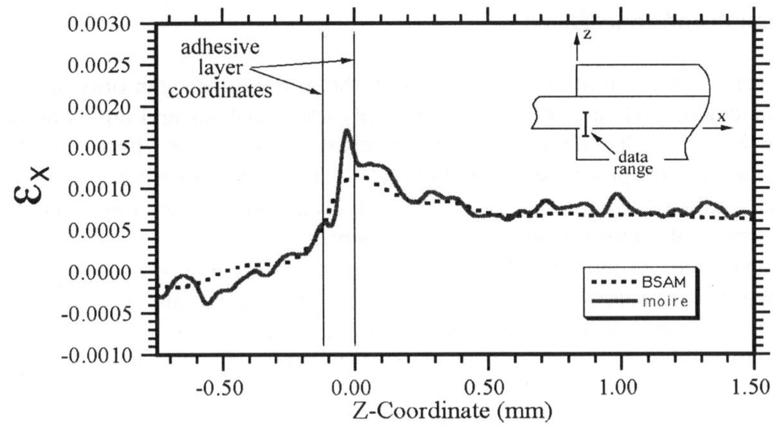

Figure 4. Axial strain variation along a line x=0.3 mm.

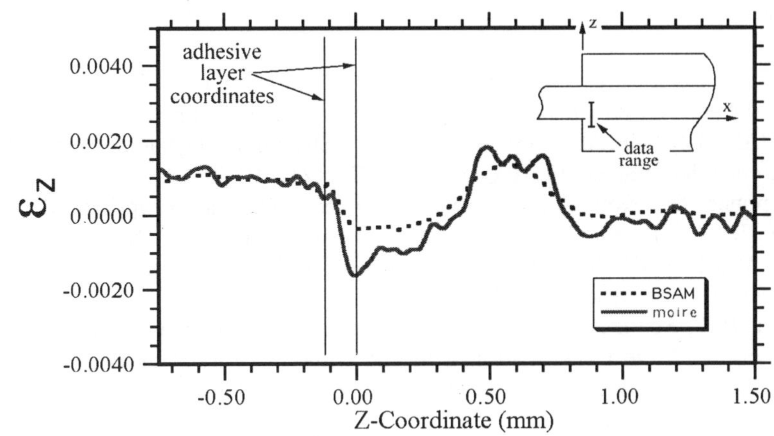

Figure 5. Transverse strain variation along a line x=0.3 mm.

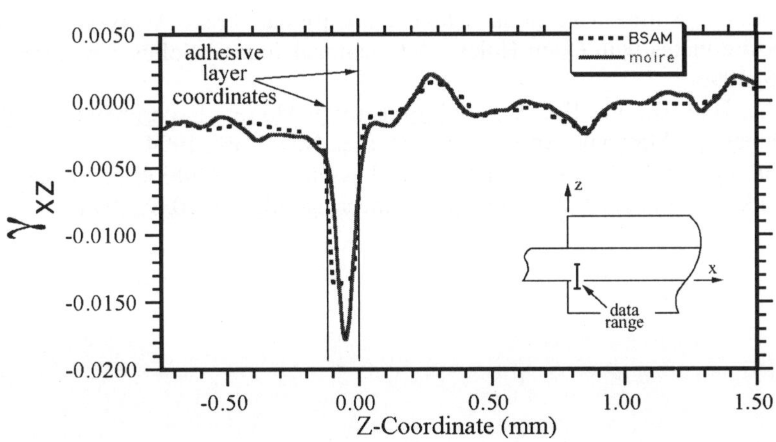

Figure 6. Shear strain variation along a line x=0.3 mm.

6 Discussion and conclusions

The ply level variation in the strains as seen in the moiré results can only be predicted if the composite adherends are modeled using a ply level analysis and not as homogeneous orthotropic bodies. The details of the ply level variations are important for failure initiation and failure progression predictions. The comparison of the BSAM three dimensional spline variational ply level analysis with the experimental moiré interferometry results showed very close agreement.

The capabilities of the BSAM code are currently being extended to include material and geometric nonlinearity which, as discussed previously, are necessary to model hygrothermal behavior in highly loaded bonded joints. Additionally, discrete damage modeling capabilities, including ply level fracture mechanics, has been incorporated in the BSAM code. Validation of the energy release rate calculations is progressing. Future extensions will include the ability to model flush bonded joints including step-lap and scarf joints.

References

1. Roy, A.K., Donaldson, S.L., and Schoeppner, G.A.,"Bonded Joints of Unidirectional and Cross-ply Laminates: An Experimental Study," Proceedings of the 38th AIAA/ASME/ASCE/AHS/ASC Structures, Structural Dynamics and Materials Conference, Kissimmee, FL (Paper AIAA-97-1121), 1997.
2. Schoeppner, G., Roy, A., and Donaldson, S.," Damage Progression in Bonded Composite Double-Lap Joints," proceedings of the 39th AIAA/ASME/ASCE/AHS/ASC Structures, Structural Dynamics and Materials Conference, Long Beach, CA (Paper AIAA-98-1885), 1998.
3. Sandhu, R.S., Sendeckyj, G.P., Schoeppner, G.A., and Pappas, J.E., "Initiation and Prevention of Edge Delamination with and without Residual Stresses," Proceedings of the 74th meeting of the Structures and Materials Panel AGARD, AGARD-CP-530, Patras, Greece, 1992
4. Mollenhauer, D. and Reifsnider, K., "Measurements of Interlaminar Deformation along the Cylindrical Surface of a Hole in Laminated Composite Materials by Moiré Interferometry," Composites Science and Technology, **60**(12-13), p 2375, 2000.
5. Iarve, E.V., "Spline Variational Three Dimensional Stress Analysis of Laminated Composite Plates with Open Holes," International Journal Solids Structures, **33**(14), p 2095, 1996.
6. Post, D., Han, B., and Ifju, P., High Sensitivity Moiré: Experimental Analysis for Mechanics and Materials, Springer-Verlag Inc., New York, 1994.
7. Lassahn, G., Taylor, P., Deason, V., and Lassahn, J., "Multiphase Fringe Analysis with Unknown Phase Shifts," Optical Engineering, **33**(6), p 2039, 1994.

TRANSVERSE DAMAGE INDUCED STRESS REDISTRIBUTION IN OPEN HOLE COMPOSITE LAMINATES

E.V. IARVE, [1] D. MOLLENHAUER[2] AND R. KIM[1]

[1] *University of Dayton Research Institute, Dayton, OH 45469, USA*

[2] *Air Force Research Laboratory, AFRL/MLBC, Wright-Patterson AFB, OH 45433, USA*

The spline approximation approach for three-dimensional stress analysis in laminates containing open holes has been extended to account for matrix cracking in each ply. Crack surface where the displacement jump takes place is defined by using the domain Heaviside function approximated with higher order polynomial B-splines. It is shown that the spline approximation of the Heaviside function maintains the integral properties of the step function and that the gradient of the approximation maintains the integral properties of Dirac's delta function for any order of approximation. An advantage of the proposed method is that its implementation only involves integration of the products of original shape functions and their derivatives and does not require modification of the integration domains. Uniaxial tension of a unidirectional composite with open hole is considered. Fiber direction stress relaxation due to longitudinal splitting is successfully modeled by the method proposed. Observed by using incremental x-ray technique split origination and extension loads correlate well with predictions.

1 Introduction

Development of a high fidelity strength prediction tool for laminated composites with stress concentrations is an important problem for the aerospace industry as well as for less traditional areas of composite applications, such as rail-transportation and sports equipment industries. High fidelity modeling involves simulation of actual damage mechanisms such as matrix cracking, delaminations and their combined influence on the redistribution of fiber stresses, which determines the strength. Discrete modeling of transverse cracking represents a formidable problem. Normally the mesh configuration is dictated by the boundaries of the specimen, such as the presence of a hole. Transverse cracking directions in the plies are defined by fiber orientations and change from ply to ply. Higher order shape functions will be employed below to construct an efficient method for mesh independent discontinuity modeling.

The concept of partition of unity approximation by Babuska and Melenk [1], which encompasses the moving least square approximation approach, provides a powerful tool for building higher order approximations for mesh independent field discontinuity modeling. Recently a novel method based on finite element approximation as a partition of unity basis was proposed for modeling crack propagation by Belytschko and Black [2], Moes, et al. [3], and Sukumar, et al. [4]. In this method the Heaviside step function is used for local enrichment. Practical implementation of this method involves several steps such as element partitioning and calculation of the cross integrals of the shape functions in the partitioned elements. Care also has to be taken to avoid crack trajectories generating near-zero volume partitions.

The method of mesh independent crack modeling by using higher order shape functions proposed below is based on replacing the Heaviside step function with its B-spline approximation. While an arbitrary crack face can be modeled in the limit of increased power of approximation, the implementation involves only standard Gaussian integration over original mesh cells (finite elements). Application of a three-dimensional B-spline approximation to stress analysis in laminated composites, including boundary conditions, implementation issues and asymptotic enrichments in the vicinity of stress singularities, are discussed in Iarve [5-7] and Iarve and Pagano [8].

2 Variational formulation

Consider an elastic volume V and a set of piecewise polynomial three-dimensional functions $X_i(\mathbf{x})$, which provide a partition of unity-type basis function for displacement approximation, so that

$$\mathbf{u}(\mathbf{x}) = \sum_{i \in \Omega} X_i(\mathbf{x})\mathbf{U}_i, \tag{1}$$

where $\mathbf{U}_i$ are displacement approximation coefficients, not necessarily associated with nodal displacements except for the boundary ∂V where boundary conditions are imposed. All values of index i in equation (1) compose set Ω. Functions $X_i(\mathbf{x})$ are constructed from one-dimensional sets of B-spline basis functions of maximum order n and nodal defect k as standard tensor products. Nodal defect of spline designates maximum number of discontinuous derivatives in this node. If k=1 then $u_i(\mathbf{x}) \in C^{(n-1)}$; and in the case of $k=n$-1, one obtains C^0 continuous $p=n$ approximation. In our present work a procedure based on polynomial representation (Iarve [5]) was used. Each function $X_i(\mathbf{x})$ possesses local support of no more than $(n+1)^3$ mesh cells and

$$\sum_{i \in \Omega} X_i(\mathbf{x}) \equiv 1, \quad \mathbf{x} \in V.$$

The subdivision into mesh cells or finite elements in Figure 1 is essential for implementation because we assume that there is an integration procedure associated with these cells, which allows us to calculate volume integrals of shape functions, and their cross products and the products of their derivatives. Our goal is to model a displacement field discontinuous along a surface Γ_α for a given mesh configuration and approximation (1).

The step function enrichment of approximation (1) proposed in Belytscko and Black [2] for modeling the desired discontinuity for a bisected domain V can be written as

$$\begin{gathered}\mathbf{u}(\mathbf{x}) = \tilde{H}(\mathbf{x})\mathbf{u}^{(1)}(\mathbf{x}) + (1 - \tilde{H}(\mathbf{x}))\mathbf{u}^{(2)}(\mathbf{x}) \\ \tilde{H}(\mathbf{x}) = H(f_\alpha(\mathbf{x}))\end{gathered}, \tag{2}$$

where $\mathbf{u}^{(1)}$ and $\mathbf{u}^{(2)}$ are approximated by shape functions (1), $H(x)$ is the Heaviside step function, and $f_\alpha(\mathbf{x})$ is the signed distance function for the surface Γ_α. To extend equation (2) for a crack ending inside volume V in point B, we shall define the extension of the signed distance function for a given point, in the case when the point of the crack surface nearest to it is not the orthogonal projection of that point to the crack surface. The signed distance function, Belytschko and Black [2], is defined as

$$f_\alpha(\mathbf{x}) = sign(\mathbf{n}(\overline{\mathbf{x}})(\mathbf{x}-\overline{\mathbf{x}}))\min_{\overline{\mathbf{x}}\in\Gamma_\alpha}\|\mathbf{x}-\overline{\mathbf{x}}\|, \tag{3}$$

where Γ_α is the crack surface. In the case of the bisecting crack point, $\overline{\mathbf{x}}$ is the orthogonal projection of $\mathbf{x}$ on surface Γ_α. However, in the case when the crack ends in point B inside the region, we will find such $\mathbf{x}$ that $\overline{\mathbf{x}} = B$ according to (3), and $\mathbf{x}$- $\overline{\mathbf{x}}$ is not orthogonal to Γ_α. It is in this context that the signed distance function will be used below.
Let us denote the set of all index values in (1) for which the shape functions are nonzero at the crack surface by Ω_α, so that

$$\exists \mathbf{x}\in\Gamma_\alpha : X_i(\mathbf{x}) \neq 0 \Leftrightarrow i\in\Omega_\alpha .$$

The enriched approximation for the domain V and arbitrary crack is defined in the following form

$$\mathbf{u} = \tilde{H}\mathbf{u}^{(1)} + (1-\tilde{H})\mathbf{u}^{(2)} + \mathbf{u}^{(3)}, \tag{4}$$

$$\mathbf{u}^{(1)} = \sum_{i\in\Omega_\alpha} X_i \mathbf{U}_i^{(1)}, \quad \mathbf{u}^{(2)} = \sum_{i\in\Omega_\alpha} X_i \mathbf{U}_i^{(2)}, \tag{5}$$

and

$$\mathbf{u}^{(3)} = \sum_{i\in\Omega/\Omega_\alpha} X_i \mathbf{U}_i^{(3)} \tag{6}$$

We have also omitted the spatial argument for conciseness. Strictly speaking, the strains are assumed independently, as

$$\boldsymbol{\varepsilon} = \tilde{H}\boldsymbol{\varepsilon}^{(1)} + (1-\tilde{H})\boldsymbol{\varepsilon}^{(2)} + \boldsymbol{\varepsilon}^{(3)}, \tag{7}$$

where the strain tensors $\boldsymbol{\varepsilon}^{(1)}$, $\boldsymbol{\varepsilon}^{(2)}$ and $\boldsymbol{\varepsilon}^{(3)}$ are expressed through $\mathbf{u}^{(1)}, \mathbf{u}^{(2)}$ and $\mathbf{u}^{(3)}$ by using Konhsi equations, and the step function is not differentiated. Considering the elastic stress-strain relationship:

$$\boldsymbol{\sigma} = \mathbf{C}\boldsymbol{\varepsilon},$$

one can write the strain energy in the volume V as

$$W = \int_V \left\{ \frac{1}{2}\tilde{H}\left[\boldsymbol{\varepsilon}^{(1)}\right]^{\mathrm{T}}\mathbf{C}\boldsymbol{\varepsilon}^{(1)} + \frac{1}{2}(1-\tilde{H})\left[\boldsymbol{\varepsilon}^{(2)}\right]^{\mathrm{T}}\mathbf{C}\boldsymbol{\varepsilon}^{(2)} + \frac{1}{2}\left[\boldsymbol{\varepsilon}^{(3)}\right]^{\mathrm{T}}\mathbf{C}\boldsymbol{\varepsilon}^{(3)} + \right.$$
$$\left. + \tilde{H}\left[\boldsymbol{\varepsilon}^{(1)}\right]^{\mathrm{T}}\mathbf{C}\boldsymbol{\varepsilon}^{(3)} + (1-\tilde{H})\left[\boldsymbol{\varepsilon}^{(2)}\right]^{\mathrm{T}}\mathbf{C}\boldsymbol{\varepsilon}^{(3)} \right\} dV \tag{8}$$

The work of external traction $\mathbf{T}$ applied at the portion S_T of the boundary ∂V is expressed as

$$A = \int_{S_T} \mathbf{Tu} dS\,, \tag{9}$$

where the displacement is approximated by using equation (4). The minimum potential energy principle to obtain the equations for determining unknown displacement approximation coefficients can be expressed as:

$$\delta(W - A) = 0\,. \tag{10}$$

The idea of the method proposed in this paper is to replace the step function in equations (4)-(10) by an approximation based on shape functions (1), so that

$$\tilde{H}(\mathbf{x}) = \sum_{i \in \Omega} X_i(\mathbf{x}) h_i \tag{11}$$

In this case all integrals appearing in the system of equations resulting from (10) will contain products of unaltered original shape functions and their derivatives. The orders of polynomials to be integrated in the mesh cells are higher but always the same, thus a standard Gaussian quadrature can be employed. Equations (4)-(11) completely determine the displacement approximation coefficients in equations (5) and (6) provided that the coefficients h_i are defined. It is worth noting that extending the crack, i.e. changing the set Ω_α, the total number of degrees of freedom in equation (4) also changes, since duplicate degrees of freedom are required for each function with the index belonging to Ω_α .

3 Results and discussion

Cracks in a $[0°]_8$ IM7/5250-4 composite with a *0.25"* hole were analytically examined using the mesh independent damage modeling technique proposed above. Axial tensile loading was applied through displacement boundary conditions $u_x(0,y,z)=-u_x(L,y,z)=u_0$. Other displacement components were zero at $x=0,L$. The crack orientation was parallel to fiber direction. The coefficients of the step function approximation (11) were determined for each crack according to the following rule: $h_i=1$ if

$$\int_V X_i(\mathbf{x}) dV > \int_V X_i^+(\mathbf{x}) dV$$

and $h_i=0$ otherwise, where

$$X_i^+(\mathbf{x}) = \begin{cases} X_i(\mathbf{x}), f_\alpha(\mathbf{x}) > 0 \\ 0, f_\alpha(\mathbf{x}) \le 0 \end{cases}$$

In this initial study assumptions regarding the crack length increment were made for ease of modeling. The specimen plane was subdivided into 20 regions, as shown by bold lines in Figure 1.a, and the cracks were incremented so that on each step they will bisect a new region. Maximum stress failure criterion was used to predict failure initiation and failure in the fiber direction. Initial crack length was defined by the near-hole region size shown in Figure 1.b, which is a close-up of the central region of the specimen. Crack growth was predicted by comparing the J-integral calculated around the crack tip to the Mode II critical energy release rate value G_{IIC}. Besides that the maximum stress criteria everywhere else in the laminate was checked to account for additional damage origination or fiber breakage. In the present case crack growth resulted in complete cracking of the specimen into four separate sections. After that the fiber failure in the ligaments was predicted.

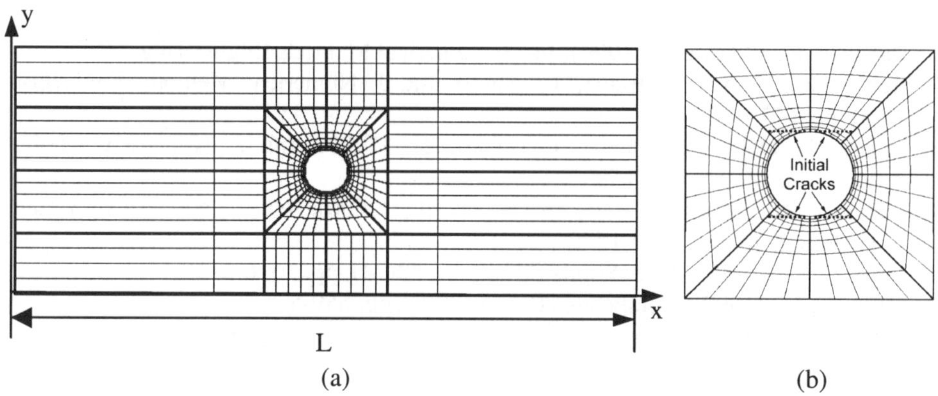

Figure 1: Specimen mesh and crack locations.

Axial stress results from this analysis are shown in Figure 2. These figures contain axial (fiber direction) stress distributions for the undamaged specimen (Figure 2.a) and that for two crack lengths for a close-up of the central region of the specimen. Of special interest is the reduction of the stress concentration at the hole edge as the cracks lengthen. When the cracks have extended to the ends of the specimen (Figure 2.c), the concentration is completely eliminated. This result is important in the sense that this effect can not be readily modeled by using finite element property degradation-based techniques. Extension of the splits eliminates the stress concentration as the physical reality of the situation demands.

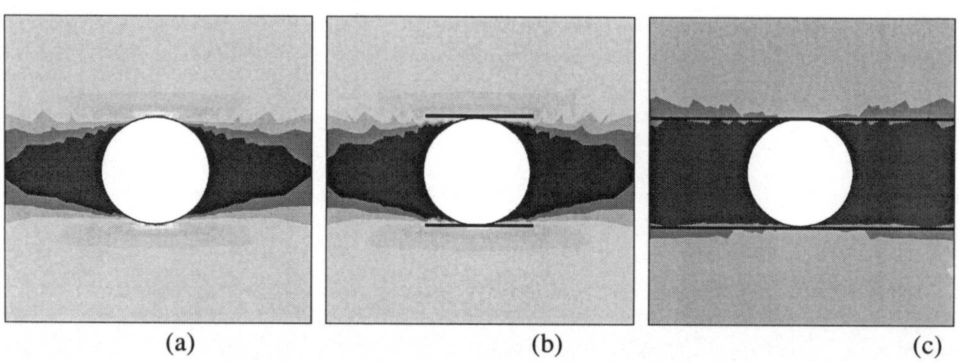

Figure 2: Axial stress distribution in a region close to the hole. (a) no crack; (b) initial crack, (c) cracks split the specimen.

The proposed mesh independent crack modeling technique was integrated into progressive failure analysis framework. The analytically generated stress strain curve for unidirectional composites as well as the loading range in which the splits were originated and propagated through the specimen was in good agreement with experimental observations .

References

1. Babuska, I. and Melenk, (1996), Int. J. Numerical Methods Engineering,
2. Belytschko, T. Y. and Black, T, (1999), Int. J. Numerical Methods Engineering, 45(5), pp. 601-620.
3. Moes, N., Dolbow, J., and Belytschko, T., (1999), Int. J. Numerical Methods Engineering, 46, pp. 131-150.
4. Sukumar, N., Moes, N., Moran, B., et al., (2000), Int. J. Numerical Methods Engineering, 48, pp. 1549-1570.
5. Iarve, E. V., (1996), Int. J. Solids Structures, 33(14), pp. 2096-2118.
6. Iarve , E. V., (1997), Composites Part A, 28(6), pp. 559-571.
7. Iarve, E. V., (2000), Composites Science and Technology, 60, pp. 2365-2374.
8. Iarve, E. V. and Pagano, N. J., (2001), Int. J. Solids Structures, 38(1), pp. 1-28.

RESIDUAL STRESS PREDICTION AND MEASUREMENT IN COMPOSITE BONDED JOINT ADHESIVE

G.A. SCHOEPPNER, D.H. MOLLENHAUER, K.B. BOWMAN
United States Air Force Research Laboratory, AFRL/MLBC, 2941 P Street Room 243, Wright-Patterson AFB, OH 45433-7750 USA
E.V. IARVE,
University of Dayton Research Institute, 300 College Park Ave., Dayton, OH 45469-0168

Residual strain predictions for an adhesive layer between composite adherends are performed using piece-wise linear elasticity. The analysis was performed using an in-house stress analysis program based on a higher order displacement spline approximation method. The residual strain measurements were conducted by using the moiré interferometry technique. A diffraction grading was installed at room temperature through the entire thickness of a polished edge of the fully cured adhesive joint. The joint was cut through the thickness along a vertical cross-section in the middle of the diffraction grating area. A full field deformation pattern was obtained in the grating area by analyzing the recorded fringe patterns. The deformation field induced by the cut is caused by the residual stress redistribution and can be predicted by the piece-wise linear elasticity analysis outlined above. A good agreement between the analysis and the experiment was observed. It is noted that the temperature dependent stress relaxation in the AF-191 adhesive is critically important for accurate residual stress prediction. Linear analysis based on room temperature thermo-mechanical properties leads to overestimation of the residual stress values.

1 Introduction

The emergence of composite materials in unitized highly loaded primary aerospace structures creates new challenges for adhesive bond joints and adhesively bonded repairs. The renewed interest in composite-to-composite bonded joints comes with awareness that the ability to predict long term behavior under service environments is lacking. The long-range objective of the current program is to develop analysis methods that accurately predict the failure of composite bonded joints and to validate the developed methods with detailed experiments. The majority of current analysis tools for composite bonded joints model the composite adherends as homogenized orthotropic bodies and neglect the residual curing stresses. It is unlikely that analysis procedures that neglect the residual curing stresses in the adherends and in the adhesive, particularly in free edge regions, will be successful in accurately predicting joint failure initiation loads, damage propagation and joint strengths (1). The objective of the present study is to experimentally measure and analytically predict the residual strains in composite-to-composite adhesively bonded joints.

The initial form of damage in composite laminates manufactured from unidirectional plies is believed to be micro-damage in the form of fiber/matrix debonds and matrix micro-cracks. The initial ply-level failure is typically transverse cracking and delamination resulting from the coalescence of micro-damage. The efficiency of the designs of composite structures is highly dependent on our ability to accurately predict damage initiation and subsequent damage growth. The development of design tools to predict the performance of composite structures will require analysis methods that accurately describe ply-level, and possibly micro-level, stress and strain fields to allow for the prediction of damage formation. For instance Roy et al. (2,3) and others have reported that composite adherends fail on planes that are only a few fiber diameters from the adhesive/adherend interface. Methods that model the adherends as homogenized laminates can not capture this type of behavior.

The U.S. Air Force Research Laboratory is developing a three dimensional ply-level analysis package called BSAM (B-Spline Analysis Methods) that is being used to predict the behavior of composite bolted and bonded joints. The ply-level analysis in BSAM maintains continuity of strains and stresses throughout a homogenous *ply* while allowing strain discontinuity at ply interfaces in order to achieve interlaminar traction continuity, not achievable using conventional finite elements. Results from the BSAM program have been shown to be in excellent agreement with moiré interferometry ply-level results for open-hole composite laminate coupons (4-5). This paper describes an effort of analyzing a composite-to-composite bonded specimen in which the adherends are modeled at the ply-level and the adhesive is modeled as a piece-wise linear elastic material. The analysis is compared to moiré interferometry measurements of the strain field on the edge of the specimen in the bondline region.

2 BSAM - Spline Variational Analysis Method

The BSAM computer code (formally referred to as SVELT) uses B-spline approximation basis functions to build a variable continuity approximation for laminated composite structures. The displacement components in the x, y and z coordinate directions are approximated by using polynomial splines as follows

$$\begin{aligned} u_x &= \mathbf{C}_1^{(S)}\mathbf{X}^{(S)}\mathbf{U}_S^* + \mathbf{X}^{(S)}\mathbf{E}_0^* \cdot u_0 + \mathbf{X}^{(S)}\mathbf{E}_L^* \cdot u_L, \\ u_y &= \mathbf{C}_2^{(S)}\mathbf{X}^{(S)}\mathbf{V}_S^*, \\ u_z &= \mathbf{C}_3^{(S)}\mathbf{X}^{(S)}\mathbf{W}_S^*. \end{aligned} \tag{1}$$

where vectors $\mathbf{U_S}, \mathbf{V_S}, \mathbf{W_S}$ contain unknown displacement spline approximation coefficients as detailed in Iarve (6). See the coordinate system in Fig 1. Bold type is used to designate matrices and vectors and the superscript star indicates the transpose operation. The same spline approximation functions are used for interlaminar traction approximation at the lower $z = z^{(S)}$ and upper $z = z^{(S+1)}$ surfaces of the s-th ply of a laminate.

$$p_i^{(s_1)}(x,y) = \mathbf{C}_i^{(s)}\mathbf{X}(x,y,z^{(s_1)})\mathbf{P}_i^{(s_1)}, i = 1,2,3, s_1 = s-1, s.$$

vectors $\mathbf{P}_i^{(s)}$ contain the unknown spline approximation coefficients of the interlaminar traction components and x, y and z correspond to i=1,2,3. The vector of the three dimensional spline approximation functions is defined as:

$$\left\{\mathbf{X}^{(s)}\right\}_q = R_i(x)\Phi_j(y)Z_l^{(s)}(z),$$

$$q = l + (j-1)(n_s+3) + (i-1)(n_s+3)k, l = 1,\ldots,n_s+3, j = 1,\ldots,k, i = 1,\ldots,m+3.$$

where sets of basis spline functions $\{R_i(x)\}_{i=1}^{m+3}$, $\{\Phi_i(y)\}_{i=1}^{k+3}$, $\{Z^{(s)}{}_i(z)\}_{i=1}^{n_s+3}$ along each coordinate are built upon subdivisions: $0 = x_0 < x_1 < \ldots < x_m = 1$, $0 = y_0 < y_1 < \ldots < y_k = A$, $z^{(s-1)} = z_0 < z_1 < \ldots < z_{n_s} = z^{(s)}$ so that the s-th ply occupies a region $z^{(s-1)} \le z \le z^{(s)}$, and n_S is the number of sublayers in each ply. The subdivisions of the x and y coordinates are in general nonuniform. Numbers of intervals of subdivision m, k, n_S in each direction along with the mesh nonuniformity characteristics such as the consecutive interval ratio, determine the accuracy of the solution and the size of the problem. Nonsquare boundary matrices $\mathbf{C}_1^{(s)}, \mathbf{C}_2^{(s)}, \mathbf{C}_3^{(s)}$ and constant vectors $\mathbf{E}_0$, $\mathbf{E}_L$ are defined so that the approximation (1) provides a kinematically admissible displacement field for any coefficients $\mathbf{U_S}, \mathbf{V_S}, \mathbf{W_S}$ i.e. satisfying prescribed boundary conditions. The components of vectors $\mathbf{E}_0$, $\mathbf{E}_L$ are equal to 1 if the same components of $\mathbf{X}^{(s)}$ are nonzero at edges $x = 0$ and $x = L$, respectively. All other components of the vectors $\mathbf{E}_0$, $\mathbf{E}_L$ are equal to zero. The boundary matrices are obtained by deleting a number of rows from the unit matrix. The rows deleted are the ones having a nonzero scalar product with $\mathbf{E}_0$ or $\mathbf{E}_L$.

The strain energy density can be written as

$$U(\varepsilon_{ij}) = \frac{1}{2}\sigma_{ij}(\varepsilon_{ij} - t_{ij}^{(s)}\Delta T), \tag{2}$$

where ΔT is uniform temperature change and $t_{kl}^{(s)}$ are thermal expansion coefficients of the s-th ply. All components of equation (2) are expressed through the displacement spline approximation coefficients. To calculate the strain components ε_{ij}, the equations (1) are differentiated symbolically. Stress components are calculated at each point through Hooke's law

$$\sigma_{ij} = C_{ijkl}^{(s)}(\varepsilon_{kl} - t_{kl}^{(s)}\Delta T)$$

where $C_{ijkl}^{(s)}$ are the stiffness coefficients of the s-th ply. No approximation error in displacement-strain or stress-strain relationship is permitted at any point. Thus all components of the potential energy functional for the composite plate have been constructed. The current capabilities of BSAM include the ability to model 3-D geometries, using *p* and *h* type refinements (as used in finite elements), model generally anisotropic materials (21 elastic constants), and calculate fracture mechanics parameters such as J-integrals. Additional details of the formulation can be found in Iarve (6). In the present study, piece-wise linear elasticity was used to model the adhesive layer and the laminated adherends.

3 Moiré Interferometry Experiments

Moiré interferometry was employed in this work to determine x- and z-displacements on the edge of the bonded specimen. Moiré interferometry is an optical interferometry technique capable of measuring in-plane displacements with high sensitivity (7). A rigorous explanation of the physics and principles of moiré interferometry can be found in the literature (7,8). The technique yields a displacement fringe pattern that is essentially a contour map of in-plane displacements in a given direction on the face of the specimen. Typically, fringe patterns are obtained in two orthogonal directions. In this research, x- and z-displacement fringe patterns were obtained during the experiment. A measure of the *change* in displacement field on the edge of the bonded specimen was obtained by subtracting the null or initial displacement field from the modified displacement field. Displacements are related to the x- and z-displacement fringe patterns by the expressions shown in Equation 3.

$$U_X(x,z) = \frac{1}{f}N_X(x,z), \qquad U_Z(x,z) = \frac{1}{f}N_Z(x,z) \tag{3}$$

where U_X and U_Z are the displacements in the x- and z- directions, N_X and N_Z are the fringe orders from the x- and z- displacement fringe patterns, and f is two times the frequency of the diffraction grating on the specimen (1200 lines/mm in this research). The *change* in strain on the specimen edge corresponding to the change in displacement, is obtained by applying the linear elastic strain-displacement relationships.

4 Fabrication of the Bonded Specimen

The material system used for the adherends was IM7/5250-5 graphite/bismaleimide supplied in prepreg form. A 24 ply panel with a $[0/\pm 45/90]_{S3}$ stacking sequence was fabricated in an autoclave according to the manufacturers recommended cure cycle. The adherend panels were C-scanned after curing and found to be free of defects. Two 10.2 cm by 10.2 cm adherend plates were cut from the panel and bonded together using 3 layers of AF-191-U adhesive film (thickness of film .20 mm). AF-191-U is a unsupported structural epoxy film adhesive manufactured by 3M. The adherends were bonded together such that the plate had an unsymmetric stacking sequence given by $\left[(0/\pm 45/90)_{S3}/Adh/(0/\mp 45/90)_{S3}\right]$. The adhesive was cured in the autoclave at 177°C. After curing, the 10.2 × 10.2 cm plate was C-scan and found to have isolated minor defects in the adhesive bondline. Two 50.8 × 25.4 mm specimens were cut from the 10.2 × 10.2 cm plate in regions where no defects were located. One of the 50.8 mm sides of each of the specimens was polished and a diffraction grating, necessary for the moiré technique, was applied to the edge of the specimen.

After taking null displacement field measurements on the moiré interferometer, a slit was cut through the upper adherend, adhesive, and part way through the bottom adhesive as shown in Fig 1. The top 10 plies of the bottom adherend were cut, leaving an unsymmetric un-cut region of the bottom adherend. The slit serves to release the residual curing stresses in the upper adherend, the adhesive, and the first few plies of the lower adherend in the region of the slit. A slit rather than a through cut was used so that the specimen could be placed back into the moiré interferometer holding fixture used to measure the null displacement field.

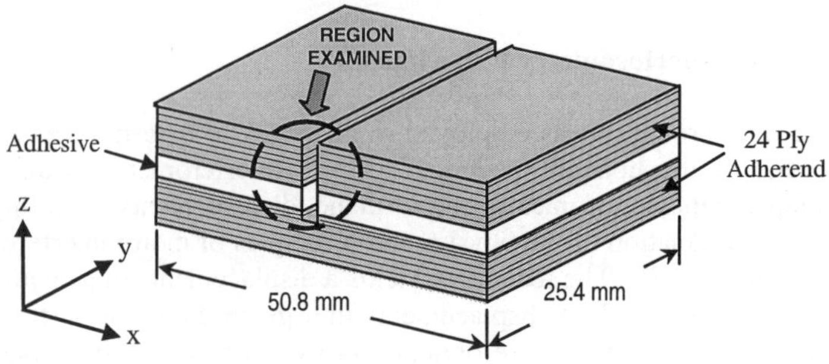

Figure 1: Bonded Specimen

5 BSAM Model and Moiré Testing

All of the elastic properties of the IM7/5250-4 graphite/bismaleimide were determined using ASTM D3039 and D3518 standard tests except ν_{23}. This Poisson's ratio was determined from a $[90_8]$ coupon under tension in which axial strain and transverse strain in the z-direction were measured using small strain gages (0.81 mm long). The transverse gage was mounted on the free-edge and two axial strain gages, one on the flat surface and one on the free-edge were mounted for comparison. Both axial strains gages gave nearly identical readings. The transverse ν_{23} was calculated by taking the negative of the transverse strain divided by the axial strain. Thermal expansion coefficients α_1 and α_2 were measured using strain gages in conjunction with a computer-controlled temperature chamber. The measured material properties for the adherends used for the linear elastic analysis of the adherends are:

E_1=165.475 GPa, E_2=E_3=10.342 GPa, G_{12}=G_{13}=5.7922 GPa,

G_{23}=3.3095 GPa, ν_{12}=ν_{13}=0.31, ν_{23}=0.56

$$\alpha_1=0.45\times10^{-6}/°C, \qquad \alpha_2=\alpha_3=24.66\times10^{-6}/°C$$

The study includes an investigation of the stress-free temperature for the adherends through the use of a $[0_2/90_2]$ laminate. The stress free temperature was measured by heating the unsymmetric laminate until it flattened. Additionally, 5 cm by 5 cm cured specimens were used to measure the thermal expansion coefficients over a temperature ranging from 20°C to the measured stress free temperature.

The thermomechanical properties of the AF-191-U adhesive material were measured over a temperature range from room temperature to 177°C. Temperature dependent stiffness properties of neat adhesive AF-191 were characterized by using uniaxial static tension tests at 20^0C increments. The tests were conducted after 30 minute temperature stabilization periods. The Poisson's ratio and thermal expansion coefficients were also measured at the various temperatures and found to be nearly constant over the considered temperature range. The details of the testing and nonlinear adhesive properties are given in Bowman (9). The thickness of the 24 ply adherends was 3.287 mm and the thickness of the adhesive was 0.592 mm giving a total specimen thickness of 7.166 mm.

Each ply of the unsymmetric specimen was modeled as a separate homogeneous orthotropic three-dimensional body. Quadratic spline interpolation functions for the three direction displacement approximations are used, resulting in a quadratic distribution of the through thickness displacement for each ply of the adherends. The interpolation is refined in the thickness direction for the adherend plies adjacent to the adhesive and for the adhesive. Additionally, refinement near the free edges of the x-y plane, where the experimental strain measurements were recorded, is required.

6 Analytical/Experimental Comparison

A comparison of the experimentally measured and predicted surface strains in the adherends and in the bondline near the slit that releases the residual curing stresses, shows favorable agreement. Figs 2 and 3 are contours of the experimental surface displacements for the x- and z-directions, respectively. It can be seen that in the region of the adhesive layer (the shaded area) and regions within the adherends the close contour lines indicating large strain gradients. The large strain gradients are attributed to the redistribution of the residual stress due to cutting the slit in the specimen. The variation of the axial and transverse displacments through the thickness of the adherends in the experimental contour plot support the argument that a ply-level analysis is required. In particular, a homogenized analytical representation of the composite adherends would not capture any of the detailed residual strain release shown in the figures. It is noted that symmetry of the displacement and strain fields about the plane of the slit is not expected due to the fact that the adherends have a quasi-isotropic stacking sequence. The BSAM analysis shows favorable agreement for both the left and right side regions of the specimen.

7 Discussion and Conclusions

The residual curing strains for composite-to-composite bonded joints is investigated with a novel experimental technique and a 3-dimensional analysis. The results illustrate the importance of using a ply-level analysis to describe the residual strain and stress fields. The details of the ply-level variations are important for failure initiation and failure progression predictions. The capabilities of the BSAM code are currently being extended to include material and geometric non-linearity which are necessary to model hygrothermal behavior in highly loaded bonded joints.

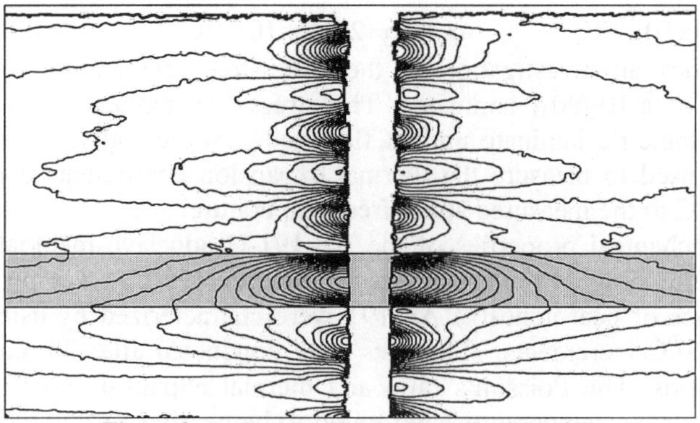

Figure 2: Inplane (x-) displacement in slit region

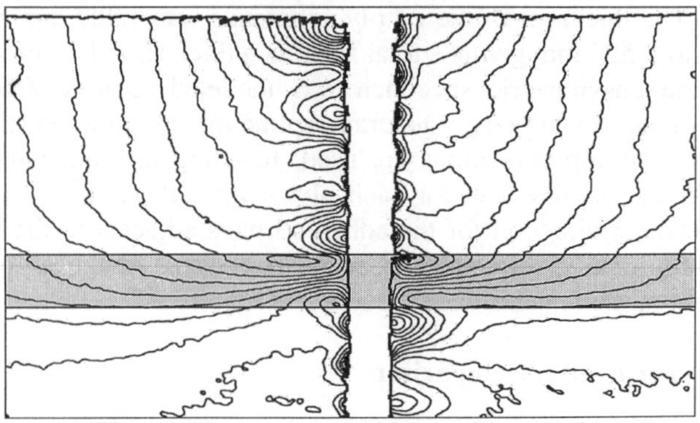

Figure 3: Transverse (z-) displacement in slit region

8 References

1. Sandhu, R.S., Sendeckyj, G.P., Schoeppner, G.A. and Pappas, J.E.,"Initiation and Prevention of Edge Delamination with and without Residual Stresses," Proceedings of the 74th meeting of the Structures and Materials Panel AGARD, AGARD-CP-530, Patras, Greece, 1992
2. Roy, A.K., Donaldson, S.L. and Schoeppner, G.A.,"Bonded Joints of Unidirectional and Cross-ply Laminates: An Experimental Study," Proceedings of the 38th AIAA/ASME/ASCE/AHS/ASC Structures, Structural Dynamics and Materials Conference, 1997, Kissimmee, FL (Paper AIAA-97-1121)
3. Schoeppner, G., Roy, A., and Donaldson, S.," Damage Progression in Bonded Composite Double-Lap Joints," proceedings of the 39th AIAA/ASME/ASCE/AHS/ASC Structures, Structural Dynamics and Materials Conference, 1998, Long Beach, CA (Paper AIAA-98-1885
4. Mollenhauer, D. and Reifsnider, K.,"Measurements of Interlaminar Deformation along the Cylindrical Surface of a Hole in Laminated Composite Materials by Moiré Interferometry," Composites Science and Technology, **60**(12-13), 2375, 2000
5. Mollenhauer, D.,"Interlaminar Deformation at a Hole in Laminated Composites: A Detailed Experimental Investigation Using Moiré Interferometry," Ph.D. Dissertation, Virginia Polytechnic Institute and State Univeristy, 1997
6. Iarve, E.V.,"Spline Variational Three Dimensional Stress Analysis of Laminated Composite Plates with Open Holes," Int. J. Solids Structures, **33**(14), 2095 (1996)
7. Post, D., B. Han, and P. Ifju, 1994, High Sensitivity Moiré: Experimental Analysis for Mechanics and Materials, Springer-Verlag Inc., New York.
8. Dai, F., J. McKelvie, and D. Post, 1990, "An Interpretation of Moiré Interferometry from Wavefront Interface Theory," *Optics and Lasers in Engineering*, Vol. 12, pp. 1-18.
9. Bowman, K.B.,, Development of an Experimental Technique to Evaluate the Residual Stresses on the Free Edges of Layered Material Systems, Ph.D. Dissertation, University of Dayton, Dayton, Ohio, 2001

DETERMINATION OF RESIDUAL STRESSES ON FREE EDGES OF LAYERED MATERIALS USING THE MATERIAL FOR FREE EDGE EVALUATION (MRFEE) METHOD

KEITH B. BOWMAN AND DAVID H. MOLLENHAUER
Air Force Research Laboratory, Materials and Manufacturing Directorate
Wright-Patterson Air Force Base Ohio 45433-7750, USA
Email: keith.bowman@wpafb.af.mil

As two dissimilar materials are bonded or co-cured at elevated temperatures, residual stresses result upon cooling the layered material system to room temperature. The residual stresses result from the thermal expansion mismatch between the materials. Current closed form solutions and experimental methods used to determine the stresses yield average stresses in the center of the material, away from edges or discontinuities. It is well known that the free edges of composite laminates possess interlaminar shearing stresses experienced during applied mechanical loading or from processing of the material. These shearing stresses are significant and may lead to failure. Until now, no experimental technique was available to assess the stress-state on the free edge. This paper describes the development of a novel experimental technique using moiré interferometry that is able to determine the residual stresses resulting from elevated temperature processing at the free edges of layered materials. The novel experimental method, referred to as Material Removal for Free Edge Evaluation (MRFEE), combines the versatility of moiré interferometry with a destructive material removal procedure to assess the free edge stress distribution.

1 Introduction

The subject of bonding dissimilar materials or applying coatings to high temperature substrates has resulted in years of intense research trying to quantify and qualify the issues leading to durable designs. Materials with different properties joined together at elevated temperatures will experience residual stresses as the combination is brought to a lower temperature. Accurate prediction of the residual stresses is essential before the problem can be understood and compensated for in the design. Lightweight organic composite airframes benefit from secondary bonding, co-curing, patching/repair, and even the use of environmental coatings. In all of these examples, the materials that are joined possess different structural characteristics, including coefficients of thermal expansion, that lead to residual stresses. Analytical approaches to characterize the problem require experimental validation to insure confidence in the model. Experimental techniques, such as layer removal, x-ray diffraction, hole-drilling and strip curvature methods, have been used extensively in industry to quantify the residual stress-state in layered materials, but these methods are valid at points away from edges and discontinuities. From the research performed by Puppo and Evensen [7] and Pipes and Pagano [5] thirty years ago, it is commonly acknowledged for layered materials under both mechanical and thermal loads that the free edge is characterized by a three-dimensional stress-state that includes interlaminar shear. Other forms of stress concentrations, such fastener holes or defects in the substrate, can have the same effect on layered materials systems. Unfortunately for most applications, edges and other discontinuities are a fact of life and unavoidable. Therefore, an experimental method is needed that can accurately determine the residual stress behavior at edges where the failure initiates, as well as validate analytical models used to predict the free edge state.

2 Experimental Method Development

The object of this research is a unique method to assess the stresses on the free edge of a layered material system. This method, referred to as the Material Removal for Free Edge Evaluation (MRFEE) method, is based on the same principle as the layer removal [8] and its modified derivative [2]. The premise is that as the material from one side of the layered system is removed, the strain energy is relieved resulting in relaxation of the specimen's curvature. Strain gages placed on the opposite face of the specimen (away from the edges) record the relaxation, and the strain values are used to determine the corresponding residual stress in the material removed. With the MRFEE technique, a whole-field technique such as moiré interferometry is used to detect the change in displacement along the free edge. The following paragraphs describe the methodology behind the MRFEE method. A detailed explanation of moiré interferometry can be found Post et al [6].

Figure 1 shows an illustration defining the material removal steps of the MRFEE method.

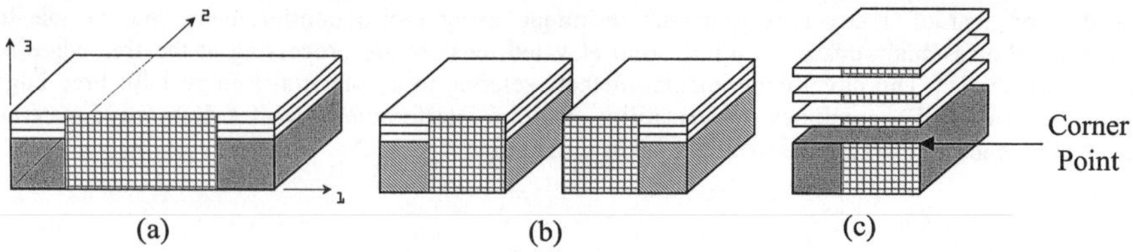

Figure 1. Schematic of the MRFEE Sectioning Process (a) the specimen with grating, (b) after sectioning, and (c) left half of the sample as the top surface is removed in layers (corner point marked).

The general constitutive relationship at any point in the material system is given as

$$\begin{Bmatrix} \varepsilon_1 \\ \varepsilon_2 \\ \varepsilon_3 \\ \gamma_{23} \\ \gamma_{13} \\ \gamma_{12} \end{Bmatrix} = \begin{bmatrix} S_{11} & S_{12} & S_{13} & S_{14} & S_{15} & S_{16} \\ & S_{22} & S_{23} & S_{24} & S_{25} & S_{26} \\ & & S_{33} & S_{34} & S_{35} & S_{36} \\ & & & S_{44} & S_{45} & S_{46} \\ & symm & & & S_{55} & S_{56} \\ & & & & & S_{66} \end{bmatrix} \begin{Bmatrix} \sigma_1 \\ \sigma_2 \\ \sigma_3 \\ \tau_{23} \\ \tau_{13} \\ \tau_{12} \end{Bmatrix} + \begin{Bmatrix} \int \alpha_1 dT \\ \int \alpha_2 dT \\ \int \alpha_3 dT \\ 0 \\ 0 \\ 0 \end{Bmatrix} \quad (1)$$

The cross pattern in Figure 1a on the front surface represents the diffraction grating where the edge displacements and strains will be determined. On that surface due to the boundary conditions, the stresses σ_2, τ_{23}, and τ_{12} are zero. The first step in the material removal occurs when the layered sample is cut through the thickness as shown in Figure 1b. The focus of the experiment is now only on one-half of the original specimen. The sectioning of the sample relieves strain along the cut edge with the stresses σ_1 and τ_{13} being eliminated from the boundary. The constitutive relationship reduces to the following where $\{\varepsilon\}$ contains the strains that we are trying to determine, and $\{\Delta\varepsilon'\}$ represents the strain relief measured along the edge using moiré interferometry.

$$\begin{Bmatrix} \varepsilon_1 \\ \varepsilon_2 \\ \varepsilon_3 \\ \gamma_{23} \\ \gamma_{13} \\ \gamma_{12} \end{Bmatrix} + \begin{Bmatrix} \Delta\varepsilon_1' \\ - \\ \Delta\varepsilon_3' \\ - \\ \Delta\gamma_{13}' \\ - \end{Bmatrix} = \begin{bmatrix} S_{11} & S_{12} & S_{13} & S_{14} & S_{15} & S_{16} \\ & S_{22} & S_{23} & S_{24} & S_{25} & S_{26} \\ & & S_{33} & S_{34} & S_{35} & S_{36} \\ & & & S_{44} & S_{45} & S_{46} \\ & symm & & & S_{55} & S_{56} \\ & & & & & S_{66} \end{bmatrix} \begin{Bmatrix} 0 \\ 0 \\ \sigma_3' \\ 0 \\ 0 \\ 0 \end{Bmatrix} + \begin{Bmatrix} \int \alpha_1 dT \\ \int \alpha_2 dT \\ \int \alpha_3 dT \\ 0 \\ 0 \\ 0 \end{Bmatrix} \quad (2)$$

The sectioned sample is then machined through the thickness (Figure 1c) in increments between 76μm and 127μm to eliminate the unknown stress σ'_3 that resulted from the prior removal method. Equation 2 can now be rewritten as

$$\begin{Bmatrix}\varepsilon_1\\ \varepsilon_2\\ \varepsilon_3\\ \gamma_{23}\\ \gamma_{13}\\ \gamma_{12}\end{Bmatrix}+\begin{Bmatrix}\Delta\varepsilon_1'\\ -\\ \Delta\varepsilon_3'\\ -\\ \Delta\gamma_{13}'\\ -\end{Bmatrix}+\begin{Bmatrix}\Delta\varepsilon_1''\\ -\\ \Delta\varepsilon_3''\\ -\\ \Delta\gamma_{13}''\\ -\end{Bmatrix}=\begin{bmatrix}S_{11} & S_{12} & S_{13} & S_{14} & S_{15} & S_{16}\\ & S_{22} & S_{23} & S_{24} & S_{25} & S_{26}\\ & & S_{33} & S_{34} & S_{35} & S_{36}\\ & & & S_{44} & S_{45} & S_{46}\\ & symm & & & S_{55} & S_{56}\\ & & & & & S_{66}\end{bmatrix}\begin{Bmatrix}0\\0\\0\\0\\0\\0\end{Bmatrix}+\begin{Bmatrix}\int\alpha_1 dT\\ \int\alpha_2 dT\\ \int\alpha_3 dT\\ 0\\ 0\\ 0\end{Bmatrix} \tag{3}$$

The strains {Δε"} are determined at the corner of the specimen where the removal occurred, and are the sum of the incremental strain changes experienced at a location as the material above that point is removed in layers. With moiré interferometry, the {Δε'} and {Δε"} can be obtained with only one measurement as long as the interferometer, initially set with null field at the beginning of the test, is not altered. This is comparable to not "zeroing" a strain gage between measurements, and allowing the cumulative change of strains to be summed to a final value. Replacing {Δε'} and {Δε"} with the one measurable strain-change quantity {Δε}, and combining Equation 3 with Equation 1 (with the appropriate free edge boundary conditions), the following expression is derived relating the residual stresses to the experimentally obtained strains. Equation 4 is applied for each layer removed.

$$\begin{aligned}-\Delta\varepsilon_1 &= S_{11}\sigma_1 + S_{13}\sigma_3 + S_{15}\tau_{13}\\ -\Delta\varepsilon_3 &= S_{13}\sigma_1 + S_{33}\sigma_3 + S_{35}\tau_{13}\\ -\Delta\gamma_{13} &= S_{15}\sigma_1 + S_{35}\sigma_3 + S_{55}\tau_{13}\end{aligned} \tag{4}$$

3 Material Model

A layered material system consisting of an AF-191-U unsupported structural epoxy film adhesive cured onto a 2024-T3 aluminum was utilized to demonstrate the MRFEE method. AF-191-U is manufactured by 3M and has a cure and use temperature of 177°C. The thickness of the aluminum was 3.175-mm and the thickness of the cured AF-191-U was 1.70-mm (nine layers). The decision to use these two materials was based on their structural characteristics, machinability, and adherence to the polymeric diffraction grating required with the moiré interferometry set-up. The AF-191-U film adhesive was characterized from room temperature to the cure temperature of 177°C. The modulus is nonlinear with regards to temperature and ranges from 2.86GPa at room temperature to 0.61GPa at 177°C. The Poisson's ratio was determined to be 0.42. The value used for the coefficient of thermal expansion (CTE) was a compilation of the material's actual CTE and the polymer cure shrinkage and stress relaxation experienced during the cure cycle. The "effective" coefficient of thermal expansion was set as 68ppm/°C.

4 Analytical Model Development

The analysis package used to validate the MRFEE method was the B-Spline Analysis Method (BSAM) computer program developed utilizing the methodology described in Iarve et al. [3,9]. The BSAM code was essential in the validation of the MRFEE method, but the model input had to be verified first against an accepted experimental technique. The modified layer removal method (MLRM) developed by Greving et al [2] was used as the baseline for the analytical model. The MLRM evaluates the stress distribution away from the edges, eliminating any uncertainty in the analysis due to edge effects. Details of the BSAM validation can be found in Bowman and Mollenhauer [1].

5 Results and Discussion

The sample that was used to demonstrate the MRFEE technique was 5.08-cm long and 2.54-cm wide. The polymer and aluminum were 1.70-mm and 3.18-mm thick, respectively. A crossed-line diffraction grating was replicated on the edge of the layered sample using EPON 828 epoxy with a Jeffamine® D-230 hardener. A diagram of the specimen and how the layer removal process progresses is given in Figure 2. A detailed description of the moiré interferometer used to record fringe patterns for the components of x-displacement (U_x) and y-displacement (U_y) in available in Mollenhauer et al. [4].

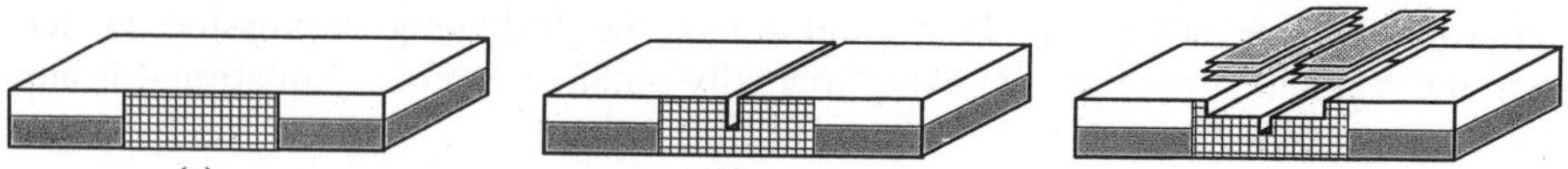

Figure 2. Schematic Showing the (a) uncut specimen, (b) after sectioning, and (c) during the milling procedure perfromed in the MRFEE Method.

Referring to Figure 2b, the initial cut through the thickness of the specimen was performed using a 3.17-mm end mill cutter. The cutter depth is set so that the cut penetrates through the polymer and approximately 0.89mm into the aluminum, maintaining the overall original geometry while effectively isolating the two halves of the specimen. This is necessary for ensuring consistent placement of the sample in the milling jig throughout the milling process. Subsequent to the end mill cut, the displacement change along the face is recorded and the change in strains $\{\Delta\varepsilon'\}$ along the cut edge are determined. The end mill cutter is replaced with a 7.62-cm diameter cutter, and the polymer is milled away in 0.127-mm thick layers as shown in Figure 2c. After each layer is removed, the sample is removed from the jig and displacement changes along the face are recorded with the interferometer. The total change in strains $\{\Delta\varepsilon\}$ at the newly formed corner are derived from the displacement fields. Figure 3 shows the displacement contours that were obtained as the polymer was removed. The contours on the left are the horizontal (U field) and on the right the vertical (V field) directions. The displacement fields are noticeably different as the milling progresses through the thickness of the sample. The corner point of interest is in the upper right-hand corner of each contour image. In total, fourteen layers were removed before the aluminum substrate was reached and the milling operation terminated. It was not necessary to proceed into the aluminum since the polymer material that was responsible for inducing the stresses was removed. The $\Delta\varepsilon_1$, $\Delta\varepsilon_3$, and $\Delta\gamma_{13}$ values obtained at the corner after each layer removal step were compared to the analytical predictions, which were determined using BSAM derived stress distributions along the center of the free-face together with Equation 4 to "reverse" engineer the expected experimental strains. The actual experimental strains were far less in magnitude than the BSAM-derived values, indicating that the cut edge and milled surface were displacing less than expected. A comparison of the experimental strain field (comparable in size to the displacement field in Figure 3) to the analytical results revealed that the diffraction grating's stiffness and thickness affected significantly the experimental data for a distance up to 250μm from the edge.

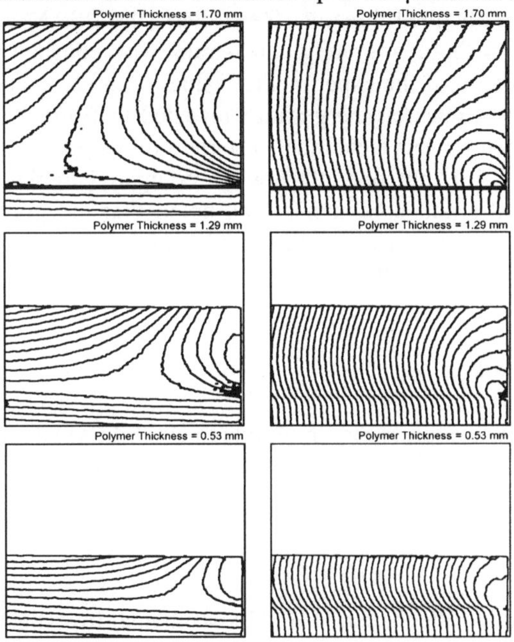

Figure 3. Displacement Contours on the Polymer Free Edge in the (a) Horizontal and (b) Vertical Fields. Note: the aluminum-polymer interface is clearly marked in the top two images.

The grating's stiffness was comparable to the underlying AF-191-U's modulus, and since the grating was applied after the specimen was cured, it effectively reinforced the polymer layer when the layer was sectioned. In addition, the 51-μm thickness of the grating resulted in shear lag, thereby, reducing the displacements observed on the grating surface. Analytical runs were made of the layered sample with and without the polymeric grating included in the model. A comparison of the two runs revealed the grating effect was negligible outside of the affected region. It was also noted that the strain was relatively constant through the region when the grating was absent. With this finding, the BSAM-derived strains were compared to the experimental strains outside of the edge region. For both $\Delta\varepsilon_1$ and $\Delta\varepsilon_3$, the comparisons agreed within a few percent, away from the polymer-aluminum interface. Near the interface, the experimental values failed to capture the rapid strain change (seen with the BSAM-derived strains) indicating that perhaps material conditions such as plasticity or stress relaxation were present. The shear experimental strain was nearly zero for the entire thickness until the interface was approached at which point it increased in magnitude. The overall trend of the experimental shear agreed favorably with the BSAM-derived value, but was nearly a magnitude lower in magnitude. This discrepancy could be attributed to the fact that the shear component is determined from displacement points entirely located in the affected edge region, with no points available outside of the region as with the other two strain components to extrapolate the strain.

In conclusion, it can be seen from the above discussion, that the MRFEE method is a valuable tool in determining the stresses along a free-edge. In this effort, it was shown that the strains (and hence the stresses) at the corner point compared well to the analytical predictions. Had it not been for the polymeric diffraction grating, the strains at the corner point could have been obtained directly. Regardless, the BSAM analysis demonstrated that for a specimen without a grating, the strain was constant through the region. Therefore, the comparison of the experimental values with the analytical BSAM-derived was valid. Sensitivity studies on the gratings revealed that thinner and more compliant gratings are better suited for this technique. The ratio of the grating stiffness to that of the AF-191-U was nearly 1.0, but only 0.04 when compared to the aluminum substrate. From the experimental strain contours, it was noted that the grating applied over the aluminum surface showed no evidence of the edge effect, supporting the hypothesis on the importance of the grating's stiffness.

6 Acknowledgements

The authors would like to express their gratitude to Drs Ed Rybicki and John Shadley from the University of Tulsa in Oklahoma for providing valuable assistance in using the modified layer removal technique.

References

1. Bowman, K.B. and Mollenhauer, D.H., "Experimental Investigation of Free Edge Residual Stresses in Layered Materials," Proceedings of the 42nd AIAA/ASME/ASCE/AHS/ASC Structures, Structural Dynamics, and Materials Conference, 16-19 April 2001, Seattle WA.
2. Greving, D.J., Rybicki, E.F., and Shadley, J.R., "Through-Thickness Residual Stress Evaluations for Several Industrial Thermal Spray Coatings Using Modified Layer-Removal Method," Journal of Thermal Spray Technology, Vol. 3, No. 4, December 1994, pp. 379-388.
3. Iarve, E.V., "Spline Variational Three Dimensional Stress Analysis of Laminated Composite Plates with Open Holes," Int. J. Solids Structures, Vol. 33, No. 14, 1996, pp. 2095.
4. Mollenhauer, D., Ifju, P., and Han, B., "A Compact, Robust, and Versatile Moiré Interferometer," Optics and Lasers in Engineering, Vol. 22, pp. 29-40, 1994.
5. Pipes, R.B. and Pagano, N.J., "Interlaminar Stresses in Composite Laminates Under Uniform Axial Extension," Journal of Composite Materials, Vol. 4, October 1970, pp. 538-548.
6. Post, D., "Moiré Interferometry," Handbook of Experimental Mechanics, Chapter 7, A.S. Kobayashi, ed., Prentice-Hall, Englewood Cliff NJ, 1987.
7. Puppo, A.H. and Evensen, H.A., "Interlaminar Shear in Laminated Composites Under Generalized Plane Stress," Journal of Composite Materials, Vol. 4, 1970, pp. 204-212.
8. SAE J936, "Methods of Residual Stress Measurement," Handbook Supplement J936, Society of Automotive Engineers, December 1965.
9. Schoeppner, G.A., Mollenhauer, D.H., and Iarve, E., "3D Spline Variational Analysis of a Composite Double-Lap Shear-Joint," Proceedings of the 46th International SAMPE Symposium, Long Beach CA, May 2001, pp. 72-84.

A NEW CLASS OF COMPOSITE MATERIALS BASED ON TOPOLOGICAL INTERLOCKING

AV DYSKIN

Department of Civil and Resource Engineering, The University of Western Australia, 35 Stirling Highway Crawley WA 6009, Australia, E-mail: adyskin@cyllene.uwa.edu.au

Y ESTRIN

Institut für Werkstoffkunde und Werkstofftechnik, Technische Universität Clausthal, D-38678 Clausthal-Zellerfeld, Germany, E-mail: juri.estrin@tu-clausthal.de

A J KANEL-BELOV

Youth Centre of Science and Technology, Moscow, Russia, E-mail: kanel@mccme.ru

E PASTERNAK

Department of Civil and Resource Engineering, The University of Western Australia, 35 Stirling Highway Crawley WA 6009, Australia, E-mail: elena@civil.uwa.edu.au

Assemblies of topologically interlocked tetrahedron-shaped elements form layers in which each individual block is held in place by neighbouring blocks. Under lateral constraint, such assemblies keep their integrity without any binder phase or adhesive. Failure of a single block cannot cause failure of the assembly, as with a single missing or damaged block interlocking within the assembly is still retained. Weak adhesion between the blocks (or total absence of adhesion) arrests propagating cracks, induced e.g. by impact, and prevents them from spreading into neighbouring blocks. For block fracturing occurring at random and in an uncorrelated way, the total assembly disintegration will require fracturing of as many as about 59% of all blocks. This suggests the suitability of the proposed layers of interlocking elements as shields against impacts. The large fraction of 'vacancies' the system can tolerate makes it possible to fill them with blocks of different material that does not have to possess good mechanical properties and can be used to fulfil other functional requirements (heat insulation, sound absorption, etc.)

1 Introduction

It was recently found that regular assemblies of identical topologically interlocked elements are possible [1]. An example of such a building block is an equilateral tetrahedron [1]. The topology of packing of the elements prevents each individual 'building block' from breaking out. Under lateral constraint, such assemblies keep their integrity without any connectors, binder phase or adhesive. Since the properties of the assemblies are determined by their topology, rather than by the size of the 'building blocks', the proposed principle of topologically motivated materials design can be used in large scale structures, as well. In the present paper, we investigate some mechanical properties of materials designed using the topological interlocking principle. On this basis, it is suggested that assemblies of topologically interlocked elements can be used as skeleton structures for novel composite materials.

2 Construction of self-locking assembly of tetrahedrons

The material architecture considered is based on the fact [1] that equilateral tetrahedrons can form regular interlocked assemblies. In essence, the interlocking shapes are such that while some parts of the blocks prohibit the displacements of an individual block in one direction, other parts prevent it from moving in the opposite direction, Fig. 1. When placed in a confinement at its periphery, the assembly will preserve its structural integrity without any adhesive or connectors.

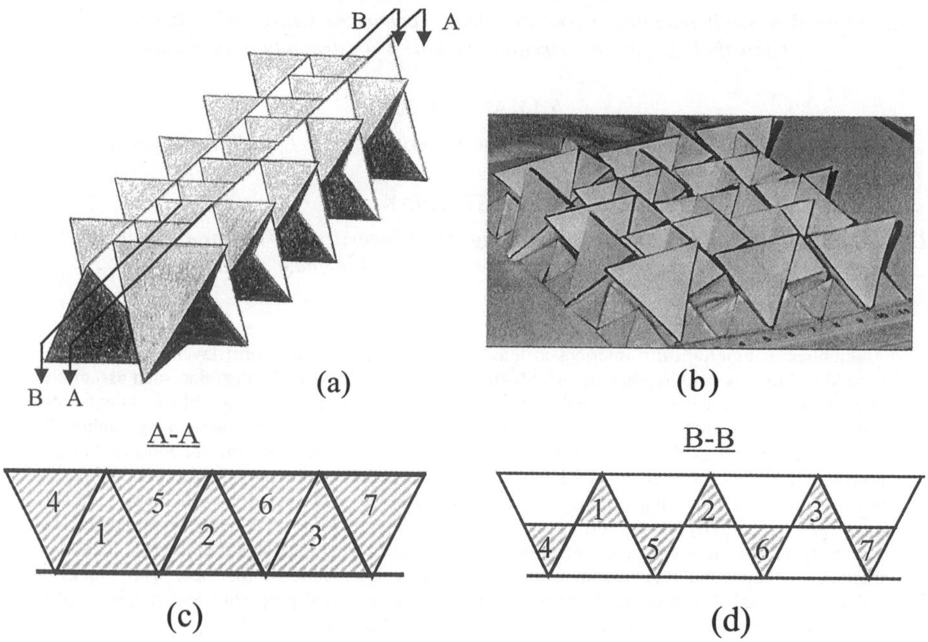

Figure 1. Assembly of interlocked tetrahedrons: Schematics (a), a geometrical model (b), and two parallel cross-sections (c, d). In the central cross-section, A-A (c), any of the numbered elements can be removed by downward or upward movement. In the other cross-section, B-B (d), each element is locked by elements of the adjacent row of tetrahedrons. (The elements with the same number in (c) and (d) belong to the same tetrahedron.)

The assembly of identical tetrahedrons is shown in Fig. 1. A fragment of the whole assembly are sketched in Fig. 1a; the structure is visualised with the help of a simple model in Fig. 1b. Figs. 1c and 1d show two parallel cross-sections. While in the central cross-section A-A (Fig. 1c) blocks numbered 1-3 are prevented from moving upwards, in an offset cross-section, B-B (Fig. 1d) they are prevented from downward movement by elements located in the adjacent row of tetrahedrons. The arrangement obtained imposes kinematic constraints that provide self-locking of the structure. Two of the four neighbours of each individual tetrahedron preclude its removal by upward movement, while the two others lock it against removal in the opposite direction.

Additional external constraints have to be imposed to fix the elements at the periphery of a finite assembly. This can be achieved by placing it in a rigid frame, as was done in our experiments described below, cf. Fig. 2.

3 Deformation behaviour

To investigate the mechanical response of an assembly of interlocked tetrahedrons, a model assembly of 100 tetrahedrons with the edge length of $a = 1$ cm was manufactured from a commercial Al-Mg-Si alloy [1]. The assembly was placed in a rigid steel frame, Fig. 2a and tested in an Instron machine under concentrated load. The displacement of the indentor (i.e. that of the cross-head of the machine) and the applied force were recorded. In addition, the cental deflection, i.e. the displacement of the edge of the central tetrahedron at the bottom side of the layer, was measured. The loading was stopped when the tetrahedrons adjacent to the indentor had rotated towards it to the extent that they actually touched the indentor. The diameter of the cylindrical indentor (1 cm) was chosen to exactly match the size of the 'unit cell' of the assembly.

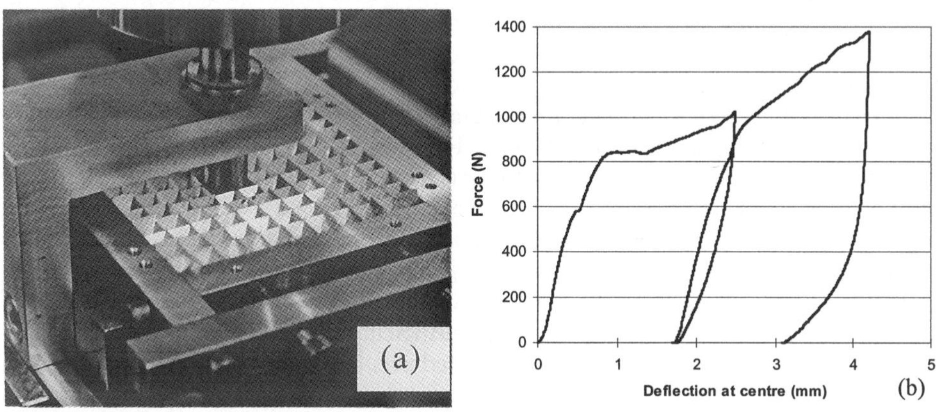

Figure 2. Concentrated loading of the self-locking assembly. Rigid frame restricting the lateral movement of tetrahedrons (a); force vs. deflection diagram for the assembled layer (b).

Fig. 2b shows the force vs. central deflection diagram for the assembly. A two-stage deformation behaviour is observed, with a nearly elastic and an apparently plastic part. As evidenced by the unloading curve, the nearly elastic component is fairly non-linear. A comparison with the results for a massive plate of the same thickness shows [1] that the bending stiffness of the assembly is as low as about one eighth of that of the plate. The residual deformation upon total unloading is very pronounced, but inspection of the individual tetrahedrons, even in the most heavily loaded central part of the structure, showed no sizeable remanent deformation of those. It is to be concluded that the remanent displacement of the tetrahedrons forming a dome-shaped indent is associated with *collective* behaviour of the assembly, rather than with plastic deformation of individual elements.

4 Failure mechanisms

The integrity of the assembly is based upon the interlocking property of its elements. Each internal block is held in place due to constraints imposed by its immediate neighbours. Therefore, provided the external lateral constraint is intact, the failure of the assembly associated with the breakage or removal of blocks will occur by reduction and eventual loss of the support for the remaining blocks. The failure mechanism based on local block fracturing will be addressed here.

Fig. 3a shows the top view of a fragment of an assembly. If just one block (e.g. the one indicated by white broken line) is removed (that is to say, broken into fragments that cannot be held in place by the adjacent blocks), it will not cause disintegration of the assembly. Indeed, support provided to the immediate neighbours of the failed block by their remaining intact neighbours is sufficient for maintaining structural integrity of the assembly. Therefore, failure of the assembly as a whole should be associated with breakage of a certain number of blocks. Two extreme cases can be considered: (a) *long distance propagation of a crack* initiated in one block and (b) distributed damage caused by an external action such that blocks fail at random and *accumulation of independently damaged blocks* can be assumed.

Long distance crack propagation requires growth of a crack generated in a particular block into neighbouring blocks. However, as the blocks are not strongly connected to each other, interfaces act to hinder crack propagation in a way similar to Cook-Gordon crack retardation [2]. The mechanism can be rationalised as follows. Consider a mode I crack in a fractured block propagating along the $\underline{x}$-axis normal to the interface. At a distance r from the crack tip the crack creates normal tensile stress $\sigma_x=K_I(2\pi r)^{-1/2}$ (cf., eg, [2]). As the crack tip approaches the interface, the magnitude of this stress increases and eventually becomes sufficient to separate the blocks along part of the interface. The interface void/crack formed acts to reduce the stress concentration at the tip of the primary crack and, by arresting it, prevents the fracture from propagating into the adjacent block. Therefore, it is the lack of adhesion, or weak adhesion in the presence of a binder phase, that isolates the broken block and preserves the integrity of the assembly at large.

Accumulation of independently damaged blocks will now be considered as a possible mechanism of failure. Since an individual missing or broken block cannot cause failure, the accumulated damage should at least form a connected chain of removed or broken blocks for the assembly to start disintegrating. If one assumes that the blocks are destroyed at random, overall failure should be attributed to the concentration of broken blocks reaching the *percolation threshold.* In the case under consideration, one can consider a square lattice with an intact block represented by a black dot (site) and a failed block by a white one, Fig. 3b. Failure is attributed to a cluster of white sites traversing the whole plane. This corresponds to the formation of an infinite (critical) cluster occurring when the concentration of the white sites (broken blocks in the assembly) reaches a critical value referred to as the percolation threshold.

For the site problem on a square lattice, the percolation threshold was calculated to be $x_c \approx 0.59$ [3]. Therefore, for the random block fracturing case, around 59% of the blocks need to be destroyed for an assembly to loose its integrity.

This consideration applies for assemblies of *infinite* size. For assemblies of finite site L, one has to deal with the percolation problem on finite lattices. In this case the percolation threshold x_{cL} becomes a random variable. This random variable is approximately normally distributed with the mean value $\langle x_{cL} \rangle = x_c$ and standard deviation δ_L that in the 2-D case considered is equal to (cf. Ref.. 3)

$$\delta_L \cong (a/L)^{1/\nu} = N^{-(1/2\nu)}, \quad \nu \approx 1.35 \tag{1}$$

where a is the lattice period, that is the block size and N is the number of elements in the assembly. Thus, the assembly exhibits a scale effect in that the scatter (uncertainty) in the failure threshold decreases with the increase in the assembly size. In this probabilistic approach, the survival probability of an assembly of interlocked elements can be assessed for a given number of elements.

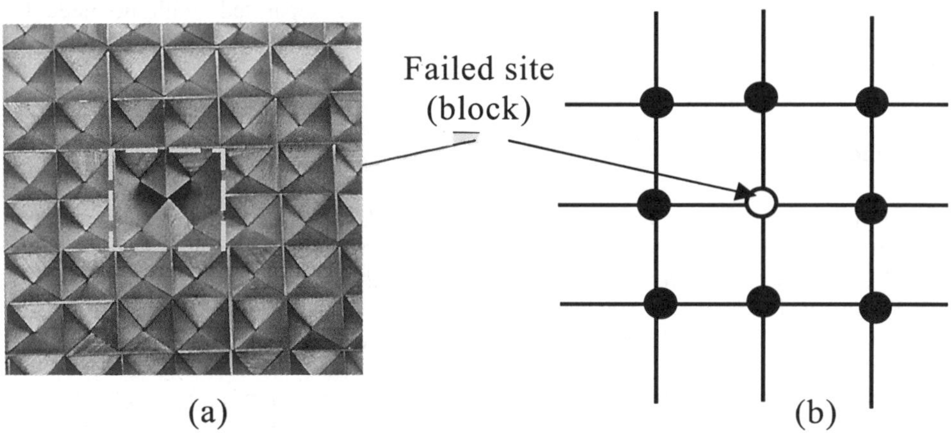

Figure 3. Percolation model of failure of an assembly due to random block destruction: (a) assembly of tetrahedrons with a missing block (indicated by white broken line. (This block was in contact with four neighbours.) (b) 2-D site percolation on square lattice.

5 Conclusions

The proposed assembly of topologically interlocked tetrahedron-shaped elements forms a layer in which each individual block is held in place by neighbouring blocks. The layer has a low bending stiffness, but can withstand considerable loads even if no binder is used to hold the elements together.

Failure of a single block cannot cause failure of the assembly, as with a single block being damaged interlocking within the assembly is still retained. Weak adhesion between the blocks (or total absence of adhesion) arrests propagating cracks, induced e.g. by impact, and prevents them from spreading into neighbouring blocks. Therefore, the only way to break an assembly is to destroy a certain number of connected blocks. For block fracturing occurring at random and in an uncorrelated way, the total assembly disintegration will require fracturing of a large fraction of the blocks given by the percolation limit. In the limit of an infinitely large assembly, the percolation limit is close to 59%. For finite size assemblies, the percolation limit is distributed probabilistically, the standard deviation from the value of 59% decreasing with the number of elements according to a power-law. The large value of the percolation limit suggests that the proposed layers of interlocking elements can be effective as shields against impacts.

An interesting avenue is also opening up for the development of composite materials in which topologically interlocked elements from a mechanically strong and tough material provide a structural 'skeleton', while elements from a second material are used to fulfil functions other than structural. Indeed, with an extremely high tolerance of an assembly to vacant sites discussed above, a large proportion of sites (theoretically up to 59%!) can be filled with tetrahedrons made from a material with e.g. good thermal insulation properties, large sound absorption capacity, etc. In such a composite, the functions of the two materials can be separated, with no need for stringent requirements on the mechanical properties of the minor phase (or multiple phases chosen to optimise the functional profile of the composite).

References

1. Dyskin A. V., Estrin Y., Kanel-Belov A J., and Pasternak E., A new concept in design of materials and structures: Assemblies of interlocked tetrahedron-shaped elements. *Scripta Materialia* 44, (2001) pp.2689-2694.
2. Parton, V. Z., *Fracture Mechanics. From Theory to Practice.* (Gordon and Breach Science Publishers, Philadelphia, Reading, Paris, Montreux, Tokyo, Melbourne, 1992).
3. Shklovskii B. I. and Efros A L., *Electronic Properties of Doped Semiconductors.* (Springer Series in Solid-State Sciences, No. 45. Springer-Verlag, Berlin,1984).

A CRITICAL ASSESSMENT OF THE APPLICABILITY OF ESHELBY'S INCLUSION ANALOGY TO A REINFORCED HOLE

C. PICKTHALL, L. R. F. ROSE

Aeronautical and Maritime Research Laboratory,
Defence Science and Technology Organisation, Fishermans Bend, Australia

A plate containing a circular hole, and reinforced by a patch may as a first approximation, be analysed by treating the reinforced structure as a plate containing an inclusion. This leads to a state of uniform stress in the inclusion. In this paper, an exact solution is obtained using a layered-plate theory where a general remote biaxial loading is split into a combination of hydrostatic tension and pure-shear. The focus is on the pure-shear component where it is found that the stress in the reinforcement over the hole is not uniform, contrary to the expectation from the two-dimensional inclusion analogy.

1 Introduction

An effective means of restoring the strength of damaged thin cross-section structures is to reinforce the afflicted area with a composite patch. For simplicity, the efficiency of such bonded repairs can be analysed by treating the reinforced structure with a part-through hole representing the damage, as a plate containing an inclusion, reducing the problem to a two-dimensional elasticity problem. However, such an approach implicitly assumes that the stresses in the equivalent inclusion are uniform, according to Eshelby's inclusion analogy [2].

To assess the validity of the inclusion analogy, a comparison is made between the inclusion-analogy solution and the exact solution of an infinite plate with a circular hole reinforced by a patch. The exact solution is obtained using a layered-plate theory which accounts for the load transfer between the plate and the patch via shear deformation in the adhesive layer. The general remote biaxial loading can be split into a combination of hydrostatic tension and pure-shear. This paper will focus on the case of pure-shear as the solution for the hydrostatic tension case is the same as that obtained using Eshelby's inclusion analogy.

2 Formulation of the Equations of Equilibrium

A remote pure-shear with $\sigma_{xx}=-\sigma_{yy}$, $\sigma_{xy}=0$ is considered to be acting on the plate and reinforcement shown in Figure 1 with zero relative displacement at infinity. In this two-dimensional analysis, all displacements are confined to the x-y plane, and the adhesive is assumed to transfer shear stress only. The analysis proceeds by splitting the problem into a superposition of three simpler parts:

(1) plate and reinforcement loaded at infinity with the hole through both,
(2) the same configuration with the reinforcement loaded on the hole surface only,
(3) a disk of reinforcement which neatly fills the hole, loaded on its boundary.

The tractions along the hole surface are determined by ensuring the continuity of appropriate stresses and displacements in the reinforcement across the hole boundary. The final solution is basically a superposition of the solutions corresponding to the above three cases.

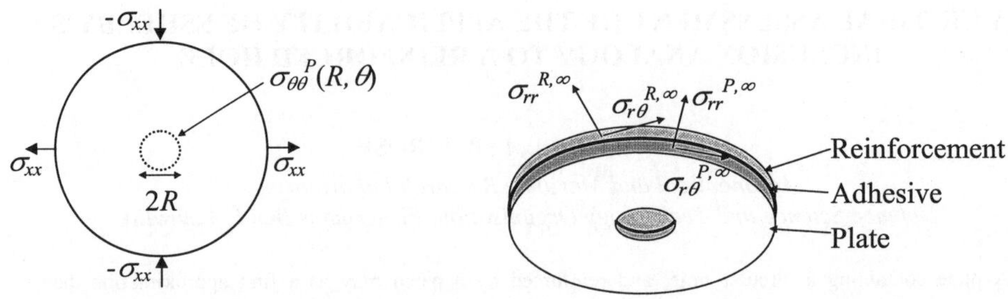

Figure 1 (a) A large plate with a circular hole, reinforced by a patch covering the hole is loaded in pure shear. The (hoop) stress of interest is shown. (b) Detail showing the hole and far-field stresses in polar coordinates.

In two dimensions and polar coordinates, the stress-strain relations under plane stress conditions can be written as

$$\sigma_{rr} = \frac{E}{1-\nu^2}\left(e_{rr} + \nu e_{\theta\theta}\right) \qquad \sigma_{\theta\theta} = \frac{E}{1-\nu^2}\left(\nu e_{rr} + e_{\theta\theta}\right) \qquad \sigma_{r\theta} = 2\mu e_{r\theta} \quad E = 2\mu(1+\nu). \qquad (1)$$

The shear modulus is μ, Young's modulus E and Poisson's ratio ν. The strains in turn involve derivatives of displacement (radial u_r, tangential u_θ) as

$$e_{rr} = \frac{\partial u_r}{\partial r} \qquad e_{\theta\theta} = \frac{1}{r}\frac{\partial u_\theta}{\partial \theta} + \frac{1}{r}u_r \qquad 2e_{r\theta} = \frac{\partial u_\theta}{\partial r} + \frac{1}{r}\frac{\partial u_r}{\partial \theta} - \frac{1}{r}u_\theta. \qquad (2)$$

Substituting the above formulae into the equations of equilibrium produces the general equations for the displacements in the plate and reinforcement. For the reinforcement,

Radial :

$$\left[2\left(\frac{\partial^2}{\partial r^2} + \frac{1}{r}\frac{\partial}{\partial r} - \frac{1}{r^2}\right) + \frac{(1-\nu_R)}{r^2}\frac{\partial^2}{\partial \theta^2}\right]u_r^R + \left[\frac{(1+\nu_R)}{r}\frac{\partial^2}{\partial r \partial \theta} + \frac{(-3+\nu_R)}{r^2}\frac{\partial}{\partial \theta}\right]u_\theta^R = \frac{\mu_A}{t_A}\frac{(1-\nu_R)}{t_R \mu_R}\left(u_r^R - u_r^P\right) \qquad (3)$$

Tangential :

$$\left[(1-\nu_R)\left(\frac{\partial^2}{\partial r^2} + \frac{1}{r}\frac{\partial}{\partial r} - \frac{1}{r^2}\right) + \frac{2}{r^2}\frac{\partial^2}{\partial \theta^2}\right]u_\theta^R + \left[\frac{(1+\nu_R)}{r}\frac{\partial^2}{\partial r \partial \theta} + \frac{(3-\nu_R)}{r^2}\frac{\partial}{\partial \theta}\right]u_r^R = \frac{\mu_A}{t_A}\frac{(1-\nu_R)}{t_R \mu_R}\left(u_\theta^R - u_\theta^P\right)$$

The equations for the plate are obtained by interchanging "R" and "P", which also takes care of the reversal of the force exerted by the adhesive. In these formulae, the arguments of the displacements, (r, θ), have been omitted for clarity, and the t are thicknesses.

3 Single plate assumption

Setting the adhesive force term (μ_A/t_A in equations 3) to zero, and then only taking the equations for the reinforcement, gives the single layer problem. This applies to the reinforcement over the hole. Dropping the "R" superscript, (3) becomes:

Radial :

$$\left[2\left(\frac{\partial^2}{\partial r^2} + \frac{1}{r}\frac{\partial}{\partial r} - \frac{1}{r^2}\right) + \frac{(1-\nu_R)}{r^2}\frac{\partial^2}{\partial \theta^2}\right]u_r^R + \left[\frac{(1+\nu_R)}{r}\frac{\partial^2}{\partial r \partial \theta} + \frac{(-3+\nu_R)}{r^2}\frac{\partial}{\partial \theta}\right]u_\theta^R = 0 \qquad (4)$$

Tangential :

$$\left[\frac{(1+\nu_R)}{r}\frac{\partial^2}{\partial r \partial \theta} + \frac{(3-\nu_R)}{r^2}\frac{\partial}{\partial \theta}\right]u_r^R + \left[(1-\nu_R)\left(\frac{\partial^2}{\partial r^2} + \frac{1}{r}\frac{\partial}{\partial r} - \frac{1}{r^2}\right) + \frac{2}{r^2}\frac{\partial^2}{\partial \theta^2}\right]u_\theta^R = 0$$

In the angularly dependent solution, the displacements (anticipating a pure shear type loading as in Figure 1) must be of the forms:

$$u_r(r,\theta) = \sum_m F_m r^m \cos(n\theta) \qquad u_\theta(r,\theta) = \sum_m G_m r^m \sin(n\theta), \tag{5}$$

n being an even integer equal to 2 for pure shear. The radial parts of each function have been taken as power series in r because all terms in (4) lead to the same power m-2 for a given power m. Solution of the resulting eigenvalue – eigenvector problem leads to the displacements

$$\begin{aligned} u_r(r,\theta) &= \left[F_{-3}r^{-3} + F_{-1}r^{-1} + F_1 r + F_3 r^3\right]\cos(2\theta) \\ u_\theta(r,\theta) &= \left[F_{-3}r^{-3} - \frac{1-\nu}{2}F_{-1}r^{-1} - F_1 r - \frac{3+\nu}{2\nu}F_3 r^3\right]\sin(2\theta) \end{aligned} . \tag{6}$$

The F_{-3} and F_{-1} coefficients are zero for the reinforcement disk, as the displacements vanish towards the centre $r \to 0$. Using equations (2), the stresses can be evaluated. Two cases are identified: a pure shear and an "extra mode".

The pure shear case results from a pure shear stress of magnitude σ_d acting on the boundary r=R. The stresses are independent of radius, in agreement with the inclusion analogy.

$$\begin{aligned} u_r(r,\theta) &= \frac{\sigma_d R}{2\mu}\left[\frac{r}{R}\right]\cos(2\theta) \\ u_\theta(r,\theta) &= \frac{\sigma_d R}{2\mu}\left[-\frac{r}{R}\right]\sin(2\theta) \end{aligned}, \qquad \begin{aligned} \sigma_{rr}(r,\theta) &= \sigma_d\cos(2\theta) \\ \sigma_{r\theta}(r,\theta) &= -\sigma_d\sin(2\theta) \\ \sigma_{\theta\theta}(r,\theta) &= -\sigma_d\cos(2\theta) \end{aligned} . \tag{7}$$

For the extra mode, the radial and tangential boundary stresses at r=R have the same sign coefficients σ_e leading to

$$\begin{aligned} u_r(r,\theta) &= \frac{\sigma_e R}{2\mu}\left[\frac{r}{R} - \frac{4\nu}{3(1+\nu)}\left(\frac{r}{R}\right)^3\right]\cos(2\theta) \\ u_\theta(r,\theta) &= \frac{\sigma_e R}{2\mu}\left[-\frac{r}{R} + \frac{2(3+\nu)}{3(1+\nu)}\left(\frac{r}{R}\right)^3\right]\sin(2\theta) \end{aligned}, \qquad \begin{aligned} \sigma_{rr}(r,\theta) &= \sigma_e\cos(2\theta) \\ \sigma_{r\theta}(r,\theta) &= \sigma_e\left[-1 + 2\left(\frac{r}{R}\right)^2\right]\sin(2\theta) \\ \sigma_{\theta\theta}(r,\theta) &= \sigma_e\left[-1 + 4\left(\frac{r}{R}\right)^2\right]\cos(2\theta) \end{aligned} . \tag{8}$$

All three stresses in this case depend on radius. This behaviour is inconsistent with the inclusion analogy discussed later.

4 Two bonded plates

One method to solve equations (3) for the displacements is to introduce two potential functions and let the displacements be defined in terms of these. This method has been presented in [1], and will not be detailed below. In two dimensions, the displacements in the reinforcement are expressed in terms of the potentials P and Q by:

$$u_r^R(r,\theta) = \frac{\partial}{\partial r}P^R + \frac{1}{r}\frac{\partial}{\partial\theta}Q^R \qquad u_\theta^R(r,\theta) = \frac{1}{r}\frac{\partial}{\partial\theta}P^R - \frac{\partial}{\partial r}Q^R . \tag{9}$$

Similar expressions apply in the plate. After a considerable amount of algebra, equations (3) lead to separated equations for the potentials in the plate and reinforcement and their differences ΔP and ΔQ. Although not necessary to solve the equations, equal Poisson's ratios are assumed: $\nu_P = \nu_R$. This simplifies the equations and the stress states in the plate and reinforcement. Assuming angular dependences for ΔP and ΔQ similar to those given by (5), the potentials are all found to have forms similar to

$$P^R(r,\theta) = \left[G_{nP}^R r^n + G_{-nP}^R r^{-n} + G_{n+2P}r^{n+2} + G_{-n+2P}r^{-n+2} + \gamma_P^R F_{KnP}K_n(\beta_P r)\right]\cos(n\theta) \tag{10}$$

where the G and F are constants determined by the boundary conditions, and K_n is the modified Bessel function which is non-singular at infinity. The Q potentials have a $\sin(n\theta)$ angular dependence. These potentials are then substituted back into (9) to obtain the displacements and thence using (2), the stresses.

5 Superposition to solve the reinforced hole problem

Once the component cases have been solved, the principle of superposition is invoked to combine these using appropriate scaling factors to solve the problem of Figure 1. This approach takes case (1) with the desired far-field loading, and adds undetermined multiples of cases (2) and (2e), corresponding to applying pure shear and "extra mode" stresses of magnitudes σ_D and σ_E respectively on the reinforcement hole surface. These same stresses are then applied to the outer edge of a disk of reinforcement, using equations (7) and (8), re-named as cases (3) and (3e) for the pure shear and extra modes respectively. The resulting displacements of the disk edge and reinforcement hole are then equated. To perform this analysis, the displacements in the reinforcement at R are expressed as follows. For simplicity, the "stiffness ratio" has been defined as:

$$\Sigma = \frac{t_R \mu_R}{t_P \mu_P} . \tag{11}$$

Case (1): Plate and reinforcement loaded at infinity with the hole surfaces stress-free

$$u_r^R(R,\theta) = \sigma_R^\infty g_{1c}(R) R\cos(2\theta) \qquad u_\theta^R(R,\theta) = \sigma_R^\infty g_{1s}(R) R\sin(2\theta) . \tag{12}$$

The g functions are given by

$$g_{1c}(R) = 4/(2\mu_R(1+\nu)) \qquad g_{1s}(R) = -4/(2\mu_R(1+\nu)) . \tag{13}$$

Case(2): Plate and reinforcement with a pure shear load on the reinforcement hole surface. In the interest of brevity, the g functions are not given for this case or for (2e) below.

$$u_r^R(R,\theta) = \sigma_D g_{2c}(R) R\cos(2\theta) \qquad u_\theta^R(R,\theta) = \sigma_D g_{2s}(R) R\sin(2\theta) . \tag{14}$$

Case(2e): Plate and reinforcement with the extra mode stress on the reinforcement hole surface

$$u_r^R(R,\theta) = \sigma_E g_{2c}^e(R) R\cos(2\theta) \qquad u_\theta^R(R,\theta) = \sigma_E g_{2s}^e(R) R\sin(2\theta) . \tag{15}$$

Case (3): Reinforcement disk with pure shear on the perimeter, obtained from equation (7) with r set equal to R.

$$\begin{array}{ll} u_r^R(R,\theta) = \sigma_d g_{3c}(R) R\cos(2\theta) & u_\theta^R(R,\theta) = \sigma_d g_{3s}(R) R\sin(2\theta) \\ g_{3c}(R) = 1/(2\mu_R) & g_{3s}(R) = -1/(2\mu_R) \end{array} . \tag{16}$$

Case (3e): Reinforcement disk with the extra mode applied; from equation (8).

$$\begin{array}{ll} u_r^R(R,\theta) = \sigma_e g_{3c}^e(R) R\cos(2\theta) & u_\theta^R(R,\theta) = \sigma_e g_{3s}^e(R) R\sin(2\theta) \\ g_{3c}^e(R) = (3-\nu)/(2\mu_R \times 3(1+\nu)) & g_{3s}^e(R) = (3-\nu)/(2\mu_R \times 3(1+\nu)) \end{array} . \tag{17}$$

The superposition and equal displacement requirement is now enacted by expressing the total radial and tangential displacements at R under the combination of cases (1), (2) and (2e), and equating these to the displacements in the disk due to cases (3) and (3e).

$$\begin{array}{l} \dfrac{u_{r\,TOT}^R(R,\theta)}{R\cos(2\theta)} = \sigma_R^\infty g_{1c}(R) + \sigma_D g_{2c}(R) + \sigma_E g_{2c}^e(R) \equiv \dfrac{u_{r\,Disk}^R(R,\theta)}{R\cos(2\theta)} = \sigma_d g_{3c}(R) + \sigma_e g_{3c}^e(R) \\ \dfrac{u_{\theta\,TOT}^R(R,\theta)}{R\sin(2\theta)} = \sigma_R^\infty g_{1s}(R) + \sigma_D g_{2s}(R) + \sigma_E g_{2s}^e(R) \equiv \dfrac{u_{\theta\,Disk}^R(R,\theta)}{R\sin(2\theta)} = \sigma_d g_{3s}(R) + \sigma_e g_{3s}^e(R) \end{array} \tag{18}$$

The continuity of stresses across the boundary at R requires

$$\sigma_D = \sigma_d \qquad \sigma_E = \sigma_e \,, \tag{19}$$

and thus there are two equations for the two undetermined multipliers σ_D and σ_E. This equation pair has a solution given by

$$\begin{array}{lll} \sigma_D = \sigma_R^\infty \left(g_s^e g_{1c} - g_c^e g_{1s}\right)/\left(g_s^e g_c - g_c^e g_s\right) & g_c = g_{3c}(R) - g_{2c}(R) & g_c^e = g_{3c}^e(R) - g_{2c}^e(R) \\ \sigma_E = \sigma_R^\infty \left(g_c g_{1s} - g_s g_{1c}\right)/\left(g_s^e g_c - g_c^e g_s\right) & g_s = g_{3s}(R) - g_{2s}(R) & g_s^e = g_{3s}^e(R) - g_{2s}^e(R) \end{array} . \tag{20}$$

It is convenient to normalise the g functions by multiplying by $2\mu_R(1+\Sigma)$. From (12),

$$g_{1c}' \equiv 2\mu_R(1+\Sigma)g_{1c} = \frac{4(1+\Sigma)}{(1+\nu)} \qquad g_{1s}' \equiv 2\mu_R(1+\Sigma)g_{1s} = -\frac{4(1+\Sigma)}{(1+\nu)} . \tag{21}$$

The other g functions are similarly normalised so that (20) can be rewritten as

$$\begin{array}{l} \sigma_D/\sigma_R^\infty = \dfrac{4(1+\Sigma)}{(1+\nu)}\left(g_s^{e\prime} + g_c^{e\prime}\right)/\left(g_s^{e\prime} g_c' - g_c^{e\prime} g_s'\right) \\ \sigma_E/\sigma_R^\infty = \dfrac{4(1+\Sigma)}{(1+\nu)}\left(-g_s' - g_c'\right)/\left(g_s^{e\prime} g_c' - g_c^{e\prime} g_s'\right) \end{array} . \tag{22}$$

Once these stresses have been determined, it is possible to calculate the stress distribution of most interest which is the hoop stress in the plate at the hole surface.

6 Hoop Stress in the Plate

For the plate, the hoop stress is given by summing the separate contributions from the far field case (1) loading, and both the pure shear and extra mode loadings (cases (2) and (2e) respectively), caused by the disk of reinforcement over the hole in the plate. Inserting the values of all the coefficients, and summing, the stress is

$$\frac{\sigma_{\theta\theta}^P(r>R,\theta)}{\sigma_P^\infty \cos(2\theta)} = \left[\begin{array}{l} -1 + \left[-3 + \dfrac{\Sigma}{1+\Sigma}\left(3\dfrac{\sigma_D}{\sigma_R^\infty} - \dfrac{\sigma_E}{\sigma_R^\infty}\right)\right](r/R)^{-4} + \\ \dfrac{\Sigma}{1+\Sigma}\left[\left(\dfrac{\sigma_D}{\sigma_R^\infty}\dfrac{N_P}{D} + \dfrac{\sigma_E}{\sigma_R^\infty}\dfrac{N_P^e}{D}\right)H_P(r/R) - \left(\dfrac{\sigma_D}{\sigma_R^\infty}\dfrac{N_Q}{D} + \dfrac{\sigma_E}{\sigma_R^\infty}\dfrac{N_Q^e}{D}\right)H_Q(r/R)\right] \end{array} \right] . \tag{23}$$

$$H_P(r/R) = (r/R)^{-3}\left[K_1(s_P)\big/K_1(s_{RP})\right]\left[\left(6(1-\nu) - \nu s_P^2\right)G_K(s_P) + (1-\nu)s_P^2\right]$$

$$H_Q(r/R) = (r/R)^{-3}\left[K_1(s_Q)\big/K_1(s_{RQ})\right]\left[12G_K(s_Q) + 4s_Q^2\right] \qquad G_K(s) \equiv sK_0(s)/K_1(s)$$

$$s_P = s_{RP} \times (r/R) \qquad s_Q = s_{RQ} \times (r/R) \qquad s_{RP} = \beta_P R \qquad s_{RQ} = \beta_Q R$$

This equation incorporates the plate stress at infinity $\sigma_P^\infty = \sigma_R^\infty \times \mu_P/\mu_R$. The other constants in the above equation are given by

$$\begin{array}{ll} N_P = \left(24 + s_{RQ}^2\right)G_K(s_{RQ}) + 6s_{RQ}^2 & N_P^e = s_{RQ}^2\left[G_K(s_{RQ}) - 2\right] \\ N_Q = \left(12(1-\nu) + s_{RP}^2\right)G_K(s_{RP}) + 3(1-\nu)s_{RP}^2 & N_Q^e = -s_{RP}^2\left[G_K(s_{RP}) - (1-\nu)\right] \\ 2D = s_{RP}^2\left[G_K(s_{RP}) - (1-\nu)\right]N_P + s_{RQ}^2\left[G_K(s_{RQ}) - 2\right]N_Q & \end{array} \tag{24}$$

There are two "load transfer lengths" β_P^{-1} and β_Q^{-1} in (23) above, given by

$$\beta_P^2 = \frac{\mu_A}{2t_A}\left(\frac{1-\nu_R}{t_R\mu_R} + \frac{1-\nu_P}{t_P\mu_P}\right) \qquad \beta_Q^2 = \frac{\mu_A}{t_A}\left(\frac{1}{t_R\mu_R} + \frac{1}{t_P\mu_P}\right) \tag{25}$$

7 Circular Inclusion Model

The equations for stresses in the case of a large plate of material (A) loaded at infinity in pure shear (amplitude σ_A), where the plate contains a circular region of radius R of another material (B), are obtained from the single plate equations (4), and matching tangential and radial stresses and displacements at the boundary r=R. The large plate and inclusion are assumed to have the same thickness. The applicable equations are summarised as:

$$\frac{\sigma^A_{\theta\theta}(r>R,\theta)}{\cos(2\theta)}=\frac{\sigma_A}{D^*}\left[-3(1+\nu_A)(1-\mu_n)(r/R)^{-4}-D^*\right] \qquad \frac{\sigma^B_{\theta\theta}(r<R,\theta)}{\cos(2\theta)}=\frac{\sigma_A}{D^*}\left[-4\mu_n\right]. \quad (26)$$

$$D^*=1+3\mu_n+\nu_A(1-\mu_n) \qquad \mu_n=\mu_B/\mu_A$$

It is clear that the inclusion B must have the same properties (Poisson's ratio, thickness and shear modulus) as the reinforcement R, while plate A has to be equivalent to the plate rigidly bonded to the reinforcement:

$$\mu_B=\mu_R \qquad t_B=t_A=t_R \qquad t_A\mu_A=t_R\mu_R+t_P\mu_P \qquad \sigma_At_A=\sigma^\infty_Rt_R+\sigma^\infty_Pt_P\,. \quad (27)$$

From these equations,

$$\mu_n=\Sigma/(1+\Sigma) \qquad D^*=(1+\nu_A+4\Sigma)/(1+\Sigma) \qquad \sigma_A=\sigma^\infty_R(1+\Sigma)/\Sigma\,. \quad (28)$$

Inserting these last expressions into the hoop stress equations (26), the result is

$$\frac{\sigma^A_{\theta\theta}(r>R,\theta)}{\sigma_A\cos(2\theta)}=-1-\frac{3(1+\nu_A)}{1+\nu_A+4\Sigma}(r/R)^{-4} \qquad \frac{\sigma^B_{\theta\theta}(r<R,\theta)}{\sigma_A\cos(2\theta)}=-\frac{4\Sigma}{1+\nu_A+4\Sigma}\,. \quad (29)$$

These agree with the limiting s_{RP},s_{RQ}→∞ expressions from (23), (7), and (8), where it is noted that

$$\sigma_D/\sigma^\infty_R\to\frac{4(1+\Sigma)}{(1+\nu+4\Sigma)}\{1-O(1/s_{RP})\} \qquad \sigma_E/\sigma^\infty_R\to O(1/s_{RP})\,. \quad (30)$$

and ν_A is identified as ν with the resulting exponential term in (23) neglected.

8 Conclusion

In superimposing the separate components, it was found that a stress-distribution which is not a pure-shear had to be included for components (2) and (3) above to ensure displacement continuity across the hole boundary. Thus although the remote stress is pure-shear, the stresses in the reinforcement over the hole and those in the plate and reinforcement in the vicinity of the hole are not pure-shear. This finding highlights the inadequacy of the inclusion analogy in dealing with reinforced holes. The inclusion analogy, however, is correct in the limiting case of a large hole relative to the load transfer length.

References

1. Pickthall, C and Rose, L. R. F., Size Effect for the Stress Concentration Around a Patched Hole in a Biaxially Loaded Plate, *Proceedings of the Second Australasian Congress on Applied Mechanics; ACAM99*. (Australian Defence Force Academy, 1999) paper U-079.
2. Eshelby, J. D., 1957, The determination of the elastic field of an ellipsoidal inclusion, and related problems, *Proceedings of Royal Society of London*, Vol. **A241**, 376-396.

PLASTIC DEFORMATION AT A NOTCH IN A FINITE THICKNESS PLATE

C. H. Wang
Aeronautical and Maritime Research Laboratory, Defence Science and Technology Organisation, 506 Lorimer Street, Fishermans Bend VIC 3207, Australia.
E-mail: Chun.Wang@dsto.defence.gov.au

A. Kotousov
Department of Mechanical Engineering, Monash University, P.O. Box 31, VIC 3800, Australia.
E-mail: Andrei.Kotousov@eng.monash.edu.au

An elasto-plastic analysis of the three-dimensional stress-state at blunt notch root is presented, with an emphasis on the influence of plate thickness on the stress distributions. By assuming that the through-the-thickness extensional strain is uniform in the thickness direction, explicit solutions of the three-dimensional stresses are obtained for blunt V-notch with a circular tip using the generalised plane-strain theory. Making use of the elastic solutions and the deformation theory of plasticity, an analytical method is developed to predict the elastic-plastic stress distribution ahead of a notch tip in a finite thickness plate.

1 Introduction

The cross-sectional thickness of plate-like structures has been found to strongly affect the propagation rates of fatigue cracks and hence a structure's total fatigue life [1]. This effect is especially prominent for fatigue cracks emanating from a notch plastic zone where the crack growth driving force depends markedly on the plate thickness. However, due to the complexity of three-dimensional stress-state at a notch root, conventional engineering practices are to treat the problem either as plane-stress or plane strain. Although such approaches would normally give satisfactory solutions of the in-plane stresses under elastic conditions, significant difference arise between the plane-stress and the plane-strain solutions within the notch plastic zone. This difference is primarily due to the different level of through-thickness stress. For instance, the out-of plane constraint, which is defined as the ratio of out-of-plane stress to the sum of in-plane stresses at the mid-plane,

$$C_z = \frac{\sigma_z}{\nu\left(\sigma_x + \sigma_y\right)} \tag{1}$$

can vary between 0 (plane stress) and 1 (plane strain). In the case of elastic deformation, this out-of-plane constraint has only minor effect on the in-plane stress distribution [2]. However, when plastic yielding occurs at notch roots, this three-dimensional constraint becomes a very important factor. For instance, under plane-strain conditions, when the notch plastic zone size is comparable with notch root radius the in-plane stress directly ahead of a notch root can exceed the material's yield stress by up to 70 per cent, according to Hill's slip-line field solution [3,4]. Approximate methods [5] have been proposed to account for the effect of three-dimensional constraint on the elastic-plastic deformation at notch root by introducing this constraint factor C_z into existing two-dimensional analysis. However, such an approach would require knowledge of this constraint factor for a given notch configuration, which is not yet available in the literature.

According to the Michell theory [6], in the limiting cases of very thin plate and very thick plate, corresponding respectively to plane-stress and plane strain conditions, the three-dimensional governing equations result in identical solutions for the in-plane stresses. For intermediate plate thickness, the exact solution is known to consist of an interior component and layer components [7]. While the interior solution is significant throughout the plate, the layer solution has only a localised effect in regions near plate edges. Nevertheless, it has been confirmed by several numerical computations that the corresponding plane solutions of the theory of elasticity provide a good approximation to the in-plane stresses. However, it is also well recognised that these plane solutions are not applicable when assessing the out-of-plane stress and deformation [8-11]. Only a few analytical layer solutions exist in the literature for particular geometry and boundary conditions. These particular solutions give a little help in the analysis of practical problems of finite thickness plates with cut-outs of irregular shape.

The generalised plane strain theory [12] has been found to be a promising method to derive simple yet accurate solutions for the layer components that prevail near crack-tip [13] and notch root [14]. The purpose of this paper is to present an analytical solution, based on the generalised plane-strain theory, of the three-dimensional stress state ahead of blunt V-notches in plates of arbitrary thickness. By assuming that the through-the-thickness extensional strain is uniform in the thickness direction, closed-form solutions are obtained for the stress constraint factor pertinent to notches with arbitrary opening angles. Making use of these elastic solutions and assuming that the stress-state at a notch tip is a proportional, an analytical method is developed to determine the elastic-plastic stresses at notch root. Wherever possible comparison is made between theoretical solutions and finite element results.

2 Elastic Solution

The theory of generalised plane-strain [12] assumes that the through-the-thickness extensional strain is uniform in the thickness direction. With this assumption, the three-dimensional field equations reduce to a set of two-dimensional equations [12-14] in terms of stress resultants. Solution of the out-of-plane constraint factor for a blunt V-notch with a circular tip of radius ρ and an opening angle of θ, subjected to remote tension at the infinity, as shown in Fig.1(a), is [14],

$$C_z(r,\phi=0)=\frac{2nr^n K_n(\kappa r)}{(1-\nu^2)\kappa r_0^{1+n}[K_{n-1}(\kappa r_0)+K_{n+1}(\kappa r_0)]+2n\nu^2 r^n K_n(\kappa r)} \tag{2}$$

where (r,ϕ) denote the polar coordinates whose centre being located a distance r_0 behind the notch tip. The parameter r_0 is given by [15],

$$r_0=\frac{\pi-\theta}{2\pi-\theta}\rho \tag{3}$$

The parameter n is often denoted as the order of singularity, which is equal to the eigenvalue of the following characteristic equations with the smallest real part [16],

$$\sin[(1-n)(2\pi-\theta)]+(1-n)\sin(2\pi-\theta)=0 \tag{4}$$

Plot of n as a function of the notch opening angle θ is shown in Fig.1(b).

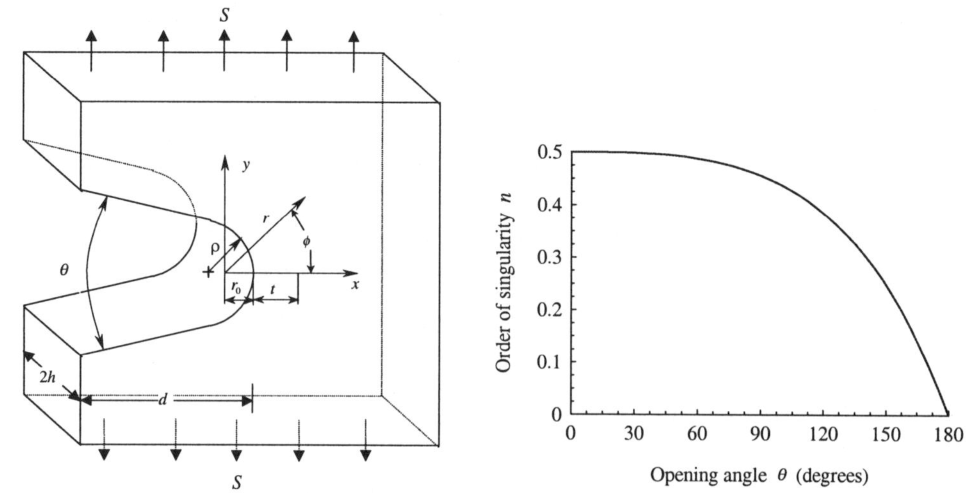

Figure1 A blunt V-notch in a finite-thickness plate (a) the configuration and the co-ordinate system, (b) the order of singularity versus notch opening angle.

In the case of U-shaped notch, which has a zero opening angle ($\theta = 0$), we have $n = \frac{1}{2}$ and $r_0 = \frac{1}{2}\rho$. The out-of-plane constraint factor can now be simplified to become,

$$C_z(t) = \frac{1}{(1-\nu^2)(1+\kappa\rho)e^{\kappa t} + \nu^2} \tag{5}$$

where $t\ (= r - \rho/2)$ denotes the distance ahead of the notch tip.

To verify the present solutions, the theoretical results of the out-of-plane constraint factor are compared with the results of a finite element analysis [11], as shown in Fig.3. It can be seen that the present theory suggests that opening angles of V-notches have only minor influence (less than 2%) on both the maximum constraint factor at notch root and the distribution of the

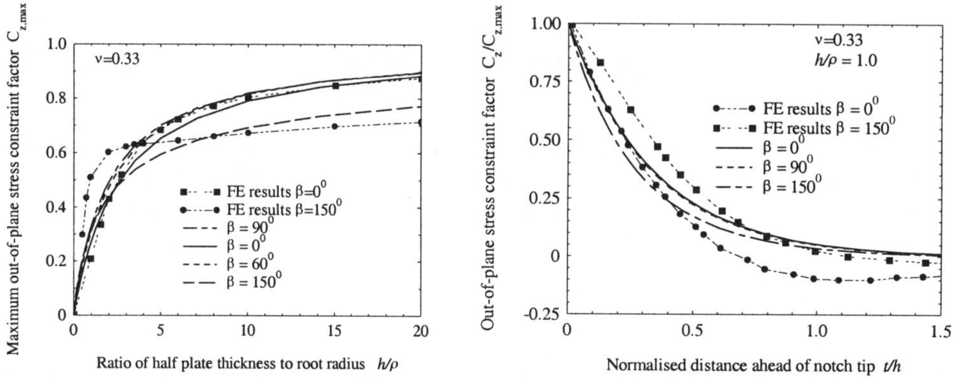

(a) (b)

Figure3 Maximum constraint factor at the roots of V-shaped notches. Symbols indicated finite element results and lines indicate present theory.

constraint factor ahead of notch tip, except when the opening angle exceeds 90 degrees. The correlation between the theoretical results and the numerical results seems to be best in the case of U-shaped notches.

3 Elastic-plastic Deformation

For a perfectly plastic material obeying the von Mises yield criterion, the equivalent stress within the notch plastic zone is equal to the material's yield stress σ_{YS}. In this case, the *y*-stress can be expressed in terms of the in-plane stress ratio $T_x = \sigma_{xx} / \sigma_{yy}$ and the out-of-plane stress constraint $T_z = \sigma_{zz} / (\sigma_{xx} + \sigma_{yy}) = \nu C_z$,

$$\sigma_{yy} = \sigma_{YS} / g(C_z, T_x) \tag{6}$$

with

$$g(T_z, T_x) = \left[(1 - T_z + T_z^2)(1 + T_x^2) - (1 + 2T_z - 2T_z^2) T_x \right]^{1/2} \tag{7}$$

A first order estimate of the plastic zone size can be made by equating the hypothetical elastic stress to the yield stress,

$$\sigma_{eq}^E(\omega_0) = \sigma_{YS} \tag{8}$$

where ω_0 denotes the first order estimate of the plastic zone size. In doing so the effect of the stress redistribution induced by the plastic deformation has been ignored, and the estimated ω_0 is expected to be smaller than the actual extent of the plastic deformation. Improvement can be achieved by using a method similar to that proposed by Irwin for sharp cracks. The basic idea is that the occurrence of plasticity makes the notch behave as if it were deeper than its physical size. The physical notch is replaced by a deeper notch of size, $d + \delta$, and the actual plastic size is $\omega = \delta + \omega_0$. In Fig.4, the elastic stress distribution at the tip of the effective notch is the same as that for the original notch except that the origin of the coordinate is shifted to o' as shown in the figure. Overall equilibrium requires that the load carried by the net section should remain the same, in other words, the area *A* should be equal to the area *B*; see Fig.4(b). This can be mathematically expressed as,

$$\int_0^{\omega} \sigma_{yy}(x) dx = \int_0^{\omega_0} \sigma_{yy}^E(x) dx \tag{9}$$

Due to the complexities of T_x, no closed-form solution of the above equation is possible. However, since the only unknown in the equation is the plastic zone size ω, it can be readily obtained numerically by means of Newton-Raphson's method using ω_0 as an initial guess.

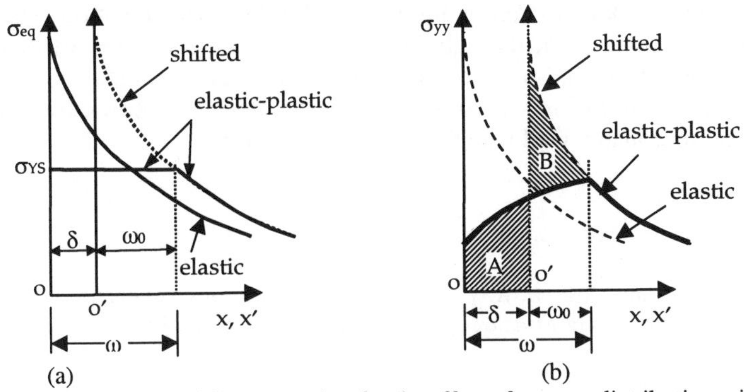

Figure 4 A method for accounting for the effect of stress redistribution within the notch plastic zone on the equivalent stress distribution.

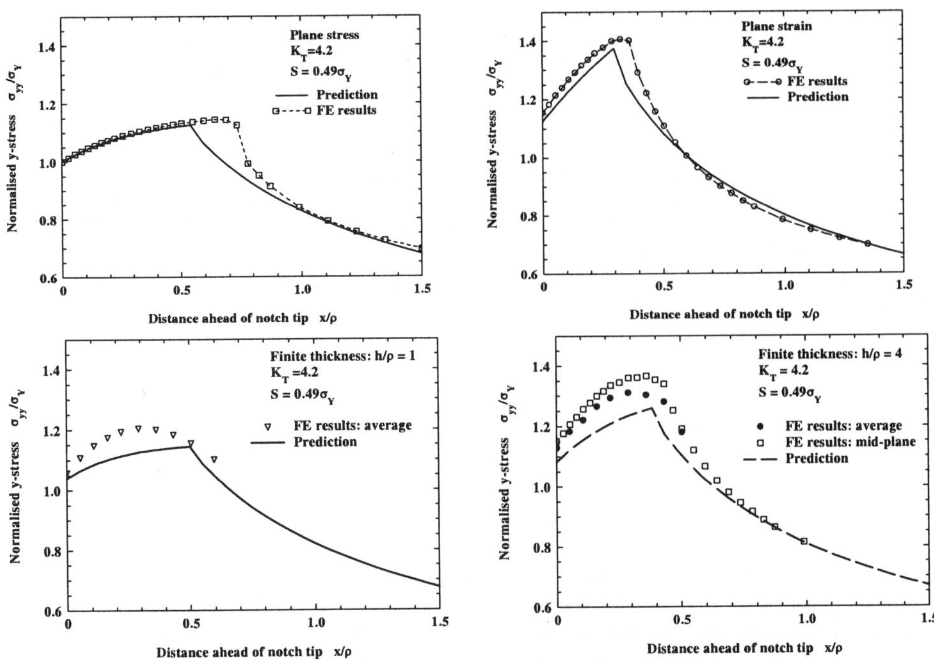

Figure 5 Comparison of theoretical predictions and finite element solutions.

4 Conclusion

An explicit solution of the three-dimensional stresses at the root of V-shaped notches in a finite thickness plate has been obtained using the generalised plane-strain theory, which assumes that the through-thickness strain is uniform in the thickness direction. The present theory, which gives the average of the through-the-thickness stress, correlates reasonably well the finite element results for the mid-plane of the plate. An analytical method for determining the elastic-plastic stress distribution within a notch plastic field has been developed, with predictions comparing reasonably well with finite element solutions.

References

1. Guo, W., Wang, C. H., and Rose, L. R. F. (1999) The influence of cross-sectional thickness on fatigue crack growth, *Fatigue and Fracture of Engineering Materials and Structures*, Vol.**22**, 437-444.
2. Sternberg, E. and Sadowsky, M.A. (1949) Three-dimensional solution for the stress concentration around a circular hole in a plate of arbitrary thickness, *J. of Applied Mechanics, Trans., ASME,* Vol.**16**, 27-38
3. Hill, R. (1949) The plastic yielding of notched bars under tension, *Q.J. Mechanics and Applied Mathematics,* Vol.**2**, p.40-50.
4. Wang, C. H., Guo, W. and Rose, L. R. F. (1999) A method for determining the elastic-plastic response ahead of a notch tip, *Journal of Engineering Materials and Technology*, Vol. **121**, 313-320.
5. Ball, D. (1999) Fatigue crack growth analysis for localised plastic yielding problems, *NASGRO Interagency Working Group (IWG) Meeting*, NASA Johnson Space Centre, Houston, TX, 14- 15 September, 1999.
6. Michell, J. H. (1899) On the direct determination of stress in an elastic solid, with application to the theory of plates, *Proc. Lond. Math. Soc.,* Vol.**31**, 100-124. (also in *The Collected Mathematical Works of J. H. and A.G.M. Michell*, edited by Niedenfuhr and Radok, Noordhoff Ltd, Groningen, the Netherlands, 1964).
7. Gregory, R. D. and Wan, F. Y. M., (1988) The interior solution for linear problems of elastic plates, *J. Applied Mechanics, Trans., ASME,* Vol.**55**, 551-0559.
8. Young, C. K. and Sternberg, E. (1966) Three-dimensional stress concentration around a cylindrical hole in a semi-infinite elastic body, *J. Applied Mechanics, Trans., ASME* Vol.**88,** 855-865.
9. Folias E.S. and Wang, J.J. (1990) On the three dimensional stress field around a circular hole in a plate of arbitrary thickness, *Computational Mechanics*, Vol.**6**, 379-391.
10. Li, Z., Guo W. and Kuang Z. (2000) Three dimensional elastic stress fields near notches in finite thick plates, *International Journal of Solids and Structures*. Vol.**37**, 7617-7631.
11. Li, Z. and Guo, W. (2001) Three dimensional elastic stress fields ahead of blunt V notches in finite thickness plate, *International Journal of Fracture*. Vol.**107**, 53-71
12. Kane, T.R. and Mindlin, R.D. (1956) High frequency extensional vibrations of plates, *Journal of Applied Mechanics, Trans., ASME,* Vol.**23**, 277-283
13. Yang, W. and Freund, L.B. (1985) Transverse shear effects for through-cracks in an elastic plate, *International Journal of Solids and Structures*, Vol. **21**, 977-994
14. Kotousov, A. and Wang, C. H. (2001) Three-dimensional stress constraint in an elastic plate with a notch, *Int. J. Solids and Structures* (in press).
15. Lazzarin, P and Tovo, R. (1996) A unified approach to the evaluation of linear elastic fields in the neighborhood of cracks and notches, *International Journal of Fracture*, Vol.**78**, 3-19
16. Williams, M.L. (1952) Stress singularities resulting from various boundary conditions in angular corners of strips in extension, *J. Applied Mechanics, Trans., ASME,* Vol.**74**, 526-528.

EXACT SOLUTIONS FOR BUCKLING OF MULTI-SPAN RECTANGULAR PLATES

Y. XIANG
School of Engineering and Industrial Design, University of Western Sydney, Locked Bag 1797, Penrith South DC NSW 1797, Australia.
E-mail: y.xiang@uws.edu.au

This paper presents exact solutions for buckling of multi-span rectangular plates. The plate is assumed to be simply supported on two opposite edges and the other two edges may take any combination of free, simply supported or clamped conditions. The Levy solution procedure in connection with a domain decomposition technique is employed to develop an analytical approach for the buckling analysis of multi-span plates. Buckling factors, most of which are first-known exact solutions, are given in tabular form for several multi-span plates subjected to uni-axial in-plane load in the x or y directions and bi-axial in-plane load. The exact buckling solutions presented in this paper are of benchmark values for such plates.

1 Introduction

Multi-span plates have important applications in many man-made structures such as aeroplane surfaces, ship decks, slabs in house construction, bridges and glass window panels. Multi-span plates may be modeled as plates with internal stiffeners or line supports. An early study on buckling of multi-span plates was conducted by Yamazaki et al. [1]. They presented buckling solutions for two-way multi-span orthotropic square plates with different combination of simply supported and clamped edges. Since then, the buckling of multi-span plates has been investigated by many researchers using the analytical [2], the finite strip [3], the p-Ritz [4], and the Galerkin [5] methods.

This paper will present an investigation on the buckling behaviour of rectangular plates having multiple internal line supports (multi-span plates). An analytical approach, based on the Levy solution method, is developed to deal with buckling of multi-span rectangular plates. It is noted that apart from the exact solutions for simply supported rectangular plates [2], there are no exact solutions for buckling of multi-span plates involving free and clamped edges. This paper aims to fill this gap by providing exact solutions for plates with free and clamped edges.

2 Problem Modeling

Figure 1 shows a flat isotropic rectangular plate with n spans in the x direction. The plate is of uniform thickness h, width L, length aL, Young's modulus E, Poisson's ratio ν and shear modulus $G = E/[2(1 + \nu)]$. The plate is subjected to uni- or bi-axial in-plane loads and has $(n - 1)$ internal line supports that impose zero displacement constraint along the line supports.

The governing differential equations for the i-th span of the plate can be derived as follows

$$D\left(\frac{\partial^4 w_i}{\partial x^4}+2\frac{\partial^4 w_i}{\partial^2 x \partial^2 y}+\frac{\partial^4 w_i}{\partial y^4}\right)+\beta N\frac{\partial^2 w_i}{\partial x^2}+\gamma N\frac{\partial^2 w_i}{\partial y^2}=0 \tag{1}$$

in which the subscript i (= 1, 2, …, n) refers to the i-th span in the plate, $w_i(x, y)$ is the transverse displacement, x and y are the Cartesian coordinates, $D = Eh^3/[12(1-\nu^2)]$ is the flexural rigidity of the plate, N is the in-plane compressive load, and β and γ are switch parameters that take the values 0 or 1 for different in-plane load combination.

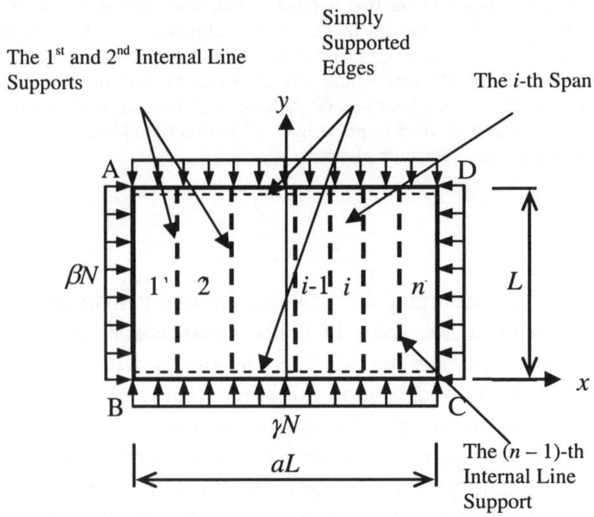

Figure 1. Geometry and coordinate system for a Levy plate with $(n-1)$ internal line supports

It is assumed that the two parallel edges (edges *AD* and *BC* in Fig. 1) are simply supported and the other remaining two edges may take any combination of free, simply supported or clamped edge conditions. The essential and natural boundary conditions for the two simply supported edges at $y = 0$ and $y = L$ in the *i*-th span are

$$w_i = 0\,,\ (M_y)_i = D\left(\frac{\partial^2 w_i}{\partial y^2}+\nu\frac{\partial^2 w_i}{\partial x^2}\right)=0. \tag{2, 3}$$

The well-known Levy type solution method is employed to derive the solutions for the *i*-th span in the plate. The displacement function for the *i*-th span can be expressed as

$$w_i(x, y) = \sin\left(\frac{m\pi}{L}y\right)X_i(x) \tag{4}$$

where *m* is the number of half waves of the buckling mode in the *y* direction and $X_i(x)$ is an unknown function to be determined. Equation (4) satisfies the boundary conditions (2) and (3) for the two simply supported edges at $y = 0$ and $y = L$.

Using the state-space technique, a homogenous differential equation system for the *i*-th span can be derived in view of (4) and (1):

$$\boldsymbol{\Psi}'_i - \mathbf{H}_i \boldsymbol{\Psi}_i = 0, \quad \boldsymbol{\Psi}_i = \begin{bmatrix} X_i & X_i' & X_i'' & X_i''' \end{bmatrix}^T \tag{5, 6}$$

where the prime denotes differentiation with respect to x and $\mathbf{H}_i$ is a 4 x 4 matrix. The non-zero elements of $\mathbf{H}_i$ are derived as

$$H_{12} = H_{23} = H_{34} = 1, \quad H_{41} = -\left(\frac{m\pi}{L}\right)^4 + \frac{\gamma N}{D}\left(\frac{m\pi}{L}\right)^2 \tag{7, 8}$$

$$H_{43} = 2\left(\frac{m\pi}{L}\right)^2 - \frac{\beta N}{D}. \tag{9}$$

The procedure for solving (5) has been detailed by Xiang et al. [6]. The solution for (5) can be expressed as

$$\boldsymbol{\Psi}_i = \mathbf{e}^{\mathbf{H}_i x}\mathbf{c}_i \tag{10}$$

in which $\mathbf{e}^{\mathbf{H}_i x}$ is a general matrix solution for (5), $\mathbf{c}_i$ is a 4 x 1 constant column matrix that will be determined by the plate boundary conditions at the two side spans and/or interface conditions between spans.

The essential and natural boundary conditions for the two edges at x = -aL/2 and x = aL/2 (see Fig. 1) are given by

if simply supported:

$$w_i = 0, \quad (M_x)_i = D\left(\frac{\partial^2 w_i}{\partial x^2} + \nu\frac{\partial^2 w_i}{\partial y^2}\right) = 0 \tag{11, 12}$$

if clamped:

$$w_i = 0, \quad \frac{\partial w_i}{\partial x} = 0 \tag{13, 14}$$

if free:

$$(M_x)_i = D\left(\frac{\partial^2 w_i}{\partial x^2} + \nu\frac{\partial^2 w_i}{\partial y^2}\right) = 0, \tag{15}$$

$$(V_x)_i = D\left(\frac{\partial^3 w_i}{\partial x^3} + (2-\nu)\frac{\partial^3 w_i}{\partial x \partial y^2}\right) + \beta N\frac{\partial w_i}{\partial x} = 0. \tag{16}$$

Along the interface between the i-th and (i + 1)-th spans the following essential and natural boundary conditions must hold to ensure the continuity of the plate and the satisfaction of the internal line support condition:

$$w_i = 0\,,\; w_{i+1} = 0\,,\; \frac{\partial w_i}{\partial x} = \frac{\partial w_{i+1}}{\partial x}\,,\; (M_x)_i = (M_x)_{i+1} \tag{17-20}$$

In view of (10), a homogeneous system of equations can be derived by implementing the boundary conditions of the plate along the two edges parallel to the y axis [(11-16)] and the interface conditions between two spans [(17-20)] when assembling the spans to the whole plate

$$\mathbf{K}[\mathbf{c}_1 \quad \mathbf{c}_2 \quad \cdots \quad \mathbf{c}_i \quad \mathbf{c}_{i+1} \quad \cdots \quad \mathbf{c}_n]^T = \{\mathbf{0}\} \tag{21}$$

where **K** is a $4n$ x $4n$ matrix. The buckling load N is evaluated by setting the determinant of **K** in (21) to be zero.

3 Results and Discussions

The proposed method is applied in this section to obtain exact buckling solutions for Levy rectangular plates having multiple spans in the x direction. The buckling load is expressed in term of a non-dimensional buckling factor $\lambda = NL^2/(\pi^2 D)$. A two-letter symbol is used to describe the plate boundary conditions on the two edges parallel to the y axis. For instance, an *SF* Levy plate has the edge *AB* simply supported and edge *DC* free (see Fig. 1). The Poisson ratio $\nu = 0.3$ is used when needed in the calculation.

Table 1 presents the exact buckling factors for *SS*, *CC* and *FF* Levy square plates having two and three uneven spans, respectively. The plates are subjected to uni-axial in-plane load in the x direction and bi-axial in-plane load. The internal line support is moving from the left edge to the centre for two-span plates. The two internal line supports for the three-span plates shift from the plate centre to the edges. The influence of the internal line supports, plate edge conditions and in-plane load conditions on the buckling factors may be observed from Table 1. It is seen that the three-span *FF* plate behaves like a simply supported plate when the two internal line supports are close to the plate edges ($b = 0.999$).

Table 2 shows the buckling factors for rectangular Levy plates having two, three and four even spans. It is observed from Table 2 that the buckling factors for the three multi-span *SS* plates are the same as for a simply supported square plate. It is because the internal line supports in the considered multi-span *SS* plates force each span in the plates to buckle in the same pattern as the one for a simply supported square plate. Table 2 reveals that when the number of spans increases in the *CC* plates, the buckling factors approach the ones for a simply supported square plate. The influence of the clamped edge constraints becomes less significant.

Table 1. Exact buckling factors $\lambda = NL^2/(\pi^2 D)$ for *SS*, *FF* and *CC* Levy square plates

Cases	*b*	Uniaxial Loading ($\beta = 1$, $\gamma = 0$)			Biaxial Loading ($\beta = 1$, $\gamma = 1$)		
		SS	*FF*	*CC*	*SS*	*FF*	*CC*
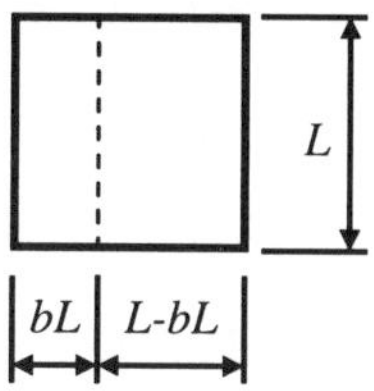	.001	4.8489	2.3660	6.7481	2.6652	1.0553	3.8352
	0.1	5.0467	2.3891	7.3051	2.9582	1.0785	4.4443
	0.2	5.3165	2.3966	8.0315	3.3631	1.1076	5.2776
	0.3	5.6652	2.2564	8.9336	3.9087	1.1417	6.3961
Two-Uneven-Span Plate	0.4	6.0482	2.0936	9.8920	4.5774	1.1746	7.7621
	0.5	6.2500	2.0429	10.386	5.0000	1.1899	8.6205
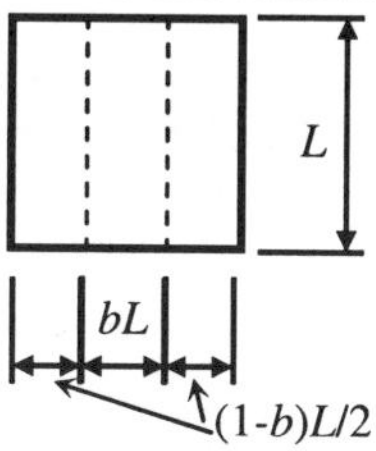	0.1	10.516	2.6213	18.135	8.8626	1.8536	15.559
	.333	11.111	2.5935	15.665	10.000	2.0218	14.225
	0.5	10.386	2.6277	12.114	8.6205	2.2040	9.9270
	0.7	8.6119	2.8740	9.0574	6.0147	2.3133	6.3532
Three-Uneven-Span Plate	0.9	7.2345	3.5149	7.3092	4.3680	2.1388	4.4374
	.999	6.7476	3.9948	6.7481	3.8346	2.0014	3.8352

Table 2. Exact buckling factors $\lambda = NL^2/(\pi^2 D)$ for Levy rectangular plates having two, three and four even spans. The plate aspect ratio is a = 2, 3 and 4 for the two-, three- and four-span plates, respectively

Cases	β	γ	*SS*	*FF*	*CC*
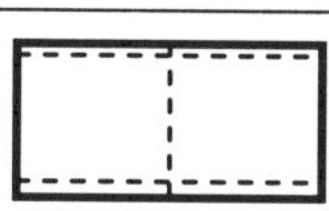	1	0	4.0000	2.3658	4.8471
	0	1	4.0000	1.4016	5.7402
Two-Even-Span Plate	1	1	2.0000	1.0551	2.6627
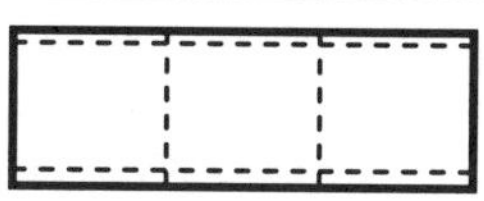	1	0	4.0000	2.3764	4.4108
	0	1	4.0000	1.4962	4.7905
Three-Even-Span Plate Spans	1	1	2.0000	1.0904	2.3083
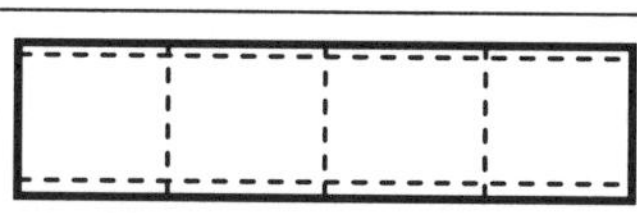	1	0	4.0000	2.3782	4.2389
	0	1	4.0000	1.5123	4.4471
Four-Even-Span Plate	1	1	2.0000	1.0971	2.1763

4 Conclusions

This paper presents exact buckling solutions for multi-span rectangular plates having two parallel edges simply supported and the remaining two edges being either free, simply supported or clamped. The exact results provided in this paper may serve as benchmark solutions for buckling of multi-span plates and are also useful to engineers who are designing plate structures.

References

1. Yamazaki, T., Hikisaka, H. and Katsuragi, K., A calculation method of the buckling load for continuous orthotropic plates. *Technical Report, Kyusyu University,* **41** (1968) pp. 61-68.
2. Sakata, T., A reduction method for vibration and buckling problems of orthotropic continuous plates. *Journal of Sound and Vibration,* **49** (1976) pp. 45-52.
3. Cheung, Y. K. and Delcourt, C., Buckling and vibration of thin flatwalled structures continuous over several spans. *Proceedings of the Institution of Civil Engineers,* **63** (Part 2) (1977) pp. 93-103.
4. Wang, C. M., Liew, K. M., Xiang, Y. and Kitipornchai, S., Buckling of rectangular Mindlin plates with internal line supports. *International Journal of Solids and Structures*, **30** (1993) pp. 1-17.
5. Saadatpour, M. M., Azhari, M. and Bradford, M. A., Buckling of arbitrary quadrilateral plates with intermediate supports using the Galerkin method. *Computational Methods for Applied Mechanics and Engineering,* **164** (1998) pp. 297-306.
6. Xiang, Y., Liew, K. M., and Kitipornchai, S., Exact buckling solutions for composite laminates: proper free edge conditions under in-plane loadings, *Acta Mechanica*, **117** (1996) pp. 115-128.

BUCKLING OF TRIANGULAR PLATES WITH ELASTIC EDGE SUPPORTS

Y. XIANG

School of Engineering and Industrial Design, University of Western Sydney, Locked Bag 1797, Penrith South DC NSW 1797, Australia.

E-mail: y.xiang@uws.edu.au

This paper is concerned with the elastic buckling of triangular plates having translational and rotational elastic edge supports. The energy functional of a general triangular plate with elastic edge supports is derived and the *p*-Ritz method is utilised to obtain the governing eigenvalue equation for the buckling problem. First-known buckling factors are presented for several selected isosceles triangular plates of various edge support conditions and subjected to isotropic in-plane compressive load.

1 Introduction

It is interesting to note that studies on buckling of triangular plates have received far less attention than its rectangular counterpart. Woinowsk-Krieger [1] was probably the first researcher to study the buckling of triangular plates in 1933. He derived the exact buckling load for a simply supported equilateral triangular plate under an isotropic in-plane compressive load. Since then, several researchers have investigated the buckling behaviour of thin [2-4] and thick [5] triangular plates under different loading and support conditions. However, to the best of the author's knowledge, there is hitherto no study available in the open literature on the buckling of triangular plates with elastic edge supports.

This paper presents the energy functional for the buckling of a general triangular plate subjected to isotropic in-plane compressive load and with both translational and rotational elastic edge supports. The *p*-Ritz method [6] is employed to derive the governing eigenvalue equation for a general triangular plate and the first-known buckling results for several selected isosceles triangular plates with elastic edge supports are presented.

2 Governing Eigenvalue Equation

Consider a general isotropic triangular plate with length a, a vertex angle $\alpha + \beta$, uniform thickness h, modulus of elasticity E, and Poisson's ratio ν , as shown in Fig. 1. The edges of the plate may be free, simply supported, clamped or elastically supported against translation and/or rotation. The problem at hand is to determine the elastic buckling load of such a triangular plate subjected to isotropic in-plane compressive load N.

Adopting the classical thin plate theory, the strain energy U due to bending is given by

$$U = \frac{D}{2}\int_A \left\{ \left(\frac{\partial^2 w}{\partial x^2} + \frac{\partial^2 w}{\partial y^2} \right)^2 - 2(1-\nu)\left[\frac{\partial^2 w}{\partial x^2}\frac{\partial^2 w}{\partial y^2} - \left(\frac{\partial^2 w}{\partial y \partial x} \right)^2 \right] \right\} dA \tag{1}$$

and the potential energy V of the isotropic in-plane load N can be expressed as

$$V = -\frac{N}{2}\int_A \left[\left(\frac{\partial w}{\partial x}\right)^2 + \left(\frac{\partial w}{\partial y}\right)^2\right] dA \tag{2}$$

in which $w(x, y)$ is the transverse displacement at the midplane of the plate, $D = Eh^3/[12(1-\nu^2)]$ is the flexural rigidity of the plate and A is the area of the plate.

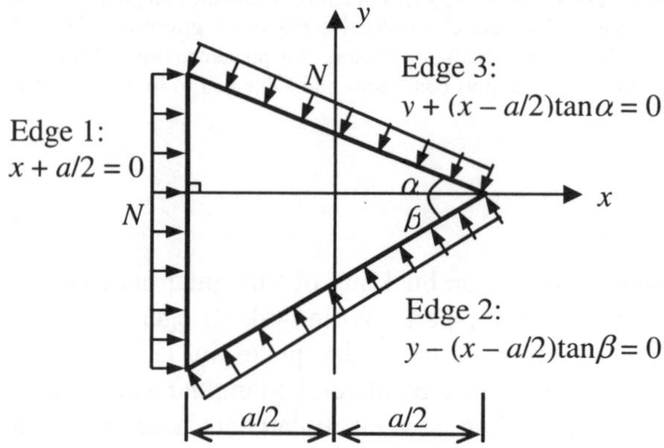

Figure 1. The geometry and coordinate system for a triangular plate subjected to isotropic in-plane load N

It is assumed that k^w is the stiffness for the transverse elastic support and k^r is the stiffness for the rotational elastic support, respectively. If an edge has a rotational elastic support, only the rotation along the edge is constrained in this study. The potential energy stored in the elastic supports in the three edges can be derived as

$$S = \frac{1}{2}\int_{-a\tan\beta}^{a\tan\alpha} \left[k_1^w w^2 + k_1^r \left(\frac{\partial w}{\partial x}\right)^2\right]_{x=-a/2} dy$$
$$+\frac{1}{2}\int_{-a/2}^{a/2} \left[k_2^w w^2 + k_2^r \left(\frac{\partial w}{\partial x}\sin\beta - \frac{\partial w}{\partial y}\cos\beta\right)^2\right]_{y=(x-a/2)\tan\beta} \frac{1}{\cos\beta}\, dx$$
$$+\frac{1}{2}\int_{-a/2}^{a/2} \left[k_3^w w^2 + k_3^r \left(\frac{\partial w}{\partial x}\sin\alpha + \frac{\partial w}{\partial y}\cos\alpha\right)^2\right]_{y=-(x-a/2)\tan\alpha} \frac{1}{\cos\alpha}\, dx \tag{3}$$

where the subscript i (i = 1, 2 and 3), refers to the i-th edge in the plate (see Fig. 1). The total potential energy functional Π for the plate system is given by

$$\Pi = U + S + V \tag{4}$$

Employing the p-Ritz method [6], the transverse displacement of the plate may be approximated by

$$w(x,y) = B(x,y)\sum_{i=1}^{m} c_i \Phi_i(x,y) = B(x,y)\{\Phi(x,y)\}_{1\times m}^{T}\{c\}_{m\times 1} \tag{5}$$

where the subscript $r \times c$ denotes a matrix with *r* rows and *c* columns; $\{\Phi(x,y)\}_{m\times 1}$ is a column matrix containing 2-D complete polynomials of *m* terms; $\{c\}_{m\times 1}$ is a column matrix containing the unknown Ritz coefficients; *m* is the number of terms in the Ritz trial function and is defined by the degree of the 2-D polynomial set *p* as $m = (p+1)(p+2)/2$; and $B(x,y)$ is the basic function formed by taking the product of the equations of the plate edges [6]. Equation (5) satisfies the geometric boundary conditions of the plate on outset [6].

By substituting (5) into (4), the total potential energy functional for the plate can be expressed as

$$\Pi = \frac{1}{2}\{c\}_{1\times m}^{T}\left([K]_{m\times m} + [L]_{m\times m} - N[G]_{m\times m}\right)\{c\}_{m\times 1} \tag{6}$$

in which [*K*] is the plate stiffness matrix, [*L*] is the stiffness matrix due to elastic edge supports and [*G*] is the plate geometric matrix. The minimization of the total potential energy functional Π in (6) with respect to the unknown Ritz coefficients leads to the governing eigenvalue equation for the plate system

$$\left([K]_{m\times m} + [L]_{m\times m} - N[G]_{m\times m}\right)\{c\}_{m\times 1} = \{0\}_{m\times 1} \tag{7}$$

where the stiffness and geometric matrices are given by

$$[K]_{m\times m} = \int_A D\left[\mathbf{\Omega}^{2020} + \mathbf{\Omega}^{0202} + \nu\left(\mathbf{\Omega}^{2002} + \mathbf{\Omega}^{0220}\right) + 2(1-\nu)\mathbf{\Omega}^{1111}\right] dA \tag{8}$$

$$\begin{aligned}
[L]_{m\times m} &= \int_{-a\tan\beta}^{a\tan\alpha} \left[k_1^w \mathbf{\Omega}^{0000} + k_1^r \mathbf{\Omega}^{1010}\right]_{x=-a/2} dy \\
&+ \int_{-a/2}^{a/2} \left\{k_2^w \mathbf{\Omega}^{0000} + k_2^r \left[\mathbf{\Omega}^{1010}\sin^2\beta - \left(\mathbf{\Omega}^{1001} + \mathbf{\Omega}^{0110}\right)\sin\beta\cos\beta \right.\right. \\
&\qquad \left.\left. + \mathbf{\Omega}^{0101}\cos^2\beta\right]\right\}_{y=(x-a/2)\tan\beta} \frac{1}{\cos\beta}\, dx \\
&+ \int_{-a/2}^{a/2} \left\{k_3^w \mathbf{\Omega}^{0000} + k_3^r \left[\mathbf{\Omega}^{1010}\sin^2\alpha + \left(\mathbf{\Omega}^{1001} + \mathbf{\Omega}^{0110}\right)\sin\alpha\cos\alpha \right.\right. \\
&\qquad \left.\left. + \mathbf{\Omega}^{0101}\cos^2\alpha\right]\right\}_{y=-(x-a/2)\tan\alpha} \frac{1}{\cos\alpha}\, dx
\end{aligned} \tag{9}$$

$$[G]_{m\times m} = \int_A \left(\mathbf{\Omega}^{1010} + \mathbf{\Omega}^{0101}\right) dA \tag{10}$$

in which

$$\mathbf{\Omega}^{pqrs} = \frac{\partial^{p+q}\left(B(x,y)\{\Phi(x,y)\}_{m\times 1}\right)}{\partial x^p \partial y^q}\frac{\partial^{r+s}\left(B(x,y)\{\Phi(x,y)\}_{1\times m}^T\right)}{\partial x^r \partial y^s} \tag{11}$$

The buckling load N can be obtained by solving the generalised eigenvalue problem defined by (7).

3 Results and Discussions

Although the proposed p-Ritz method can be used to study buckling of general triangular plates, this paper will focus on the buckling of isosceles triangular plates subjected to isotropic in-plane compressive load and with elastic edge supports. To describe the boundary conditions in a plate, the letters F, S and C are used to denote free, simply supported and clamped edges, respectively. The letters T and R are used to denote an edge having a translational and a rotational elastic edge supports, respectively.

Figure 2 shows three selected isosceles triangular plates studied in this paper. The buckling load N is expressed in terms of a non-dimensional buckling factor $\lambda = Nb^2/(\pi^2 D)$, where b is the dimension of the plates as marked in Figs. 2. The Poisson ratio $\nu = 0.3$ is used in this study. Although the model for elastic edge supports allows one to choose the translational and rotational elastic support stiffnesses independently for each edge, it is considered in this section that the stiffness factors are uniform for all edges with elastic supports, i.e. $k_i^w = k^w$ and $k_i^r = k^r$, i = 1, 2 and/or 3. The elastic support stiffnesses are expressed in terms of non-dimensional elastic stiffness parameters $S_w = k^w b^3/D$ and $S_r = k^r b/D$ for translational and rotational elastic edge supports, respectively.

As the p-Ritz method used in this study is an approximate numerical approach, convergence and comparison studies are essential to ensure the correctness and convergence of the buckling results.

Table 1 presents the buckling factors λ for a free isosceles triangular plate with both translational and rotational elastic edge supports (Case 3 in Fig. 2). The vertex angle α_o of the plate is set to be 60 degrees. The elastic stiffness parameters (S_w, S_r) have several combinations as shown in Table 1. The degree of polynomial set p used in the calculation varies from 6 to 13. It is seen that the buckling factor decreases monotonically as the polynomial degree p increases. It is observed that the buckling factors are well converged with p = 10. We notice that when (S_w, S_r) = (10^6, 0), the free plate behaves like a simply supported plate. And when (S_w, S_r) = (10^6, 10^6), the plate acts like a clamped plate. As shown in Table 1, the buckling factors from this study compare very well with the available solutions from open literature [5] for simply supported and clamped isosceles triangular plates.

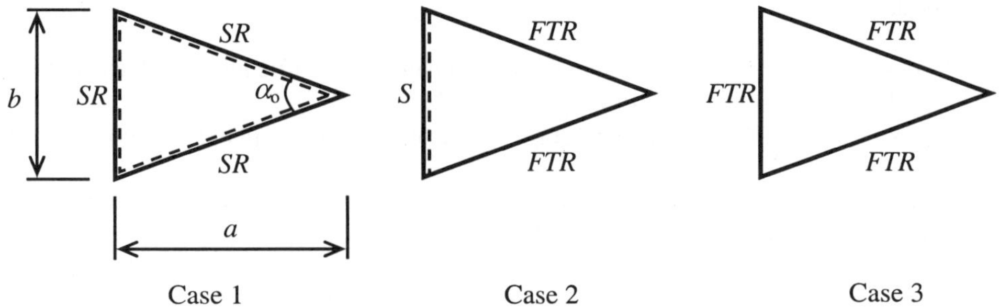

Figure 2. Geometry and boundary conditions of analysed isosceles triangular plates

Table 1. Convergence and comparison studies of buckling factors $\lambda = Nb^2/(\pi^2 D)$ for a free isosceles triangular plates with elastic edge constraints (Case 3 in Fig. 2)

α_o (Degree)	p	(S_w, S_r)			
		$(10^2, 0)$	$(10^2, 10^2)$	$(10^6, 0)$	$(10^6, 10^6)$
60	6	2.4757	6.7402	5.3694	17.824
	8	2.4755	6.7343	5.3333	15.017
	10	2.4755	6.7341	5.3332	14.843
	11	2.4755	6.7341	5.3332	14.833
	12	2.4755	6.7341	5.3332	14.832
	13	2.4755	6.7341	5.3332	14.832
	Others			5.3329 [5]	14.8430 [5]

Table 2 presents the buckling factors λ for the three selected isosceles triangular plates with various vertex angles. It is observed from Table 2 that the buckling factors λ increase as the elastic stiffness parameters increase. While an individual elastic support, *i.e.* a translational or a rotational elastic support, on a free edge has limited effect on the buckling factors of a plate, the combined elastic supports have more pronounced influence in increasing the buckling factors of the plate. It is seen that the buckling factors increase when the vertex angle of the plates varies from 30 to 120 degrees for most cases in Table 2.

Table 2. Buckling factors $\lambda = Nb^2/(\pi^2 D)$ for isosceles triangular plates with elastic edge constraints

Cases	α_o	(S_w, S_r)							
		$(0, 0)$	$(10^2, 0)$	$(0, 10^2)$	$(10^2, 10^2)$	$(10^3, 0)$	$(0, 10^3)$	$(10^3, 10^3)$	$(10^6, 10^6)$
1	30	2.8494		7.7455			8.1020		
	60	5.3333		13.890			14.733		
	90	10.000		25.850			28.014		
	120	21.880		56.354			63.527		
2	30	0.0000	1.7231	0.1477	3.5816	2.8366	0.1805	6.1175	6.4968
	60	0.0000	2.5158	0.6819	4.8869	5.2324	0.7232	9.5061	10.604
	90	0.0000	2.9318	1.8872	6.0454	9.0930	1.9882	15.101	18.664
	120	0.0000	2.1682	4.9384	7.0829	12.026	5.2915	22.439	39.837
3	30	0.0000	1.7228	0.3979	4.4905	2.7887	0.4153	7.5717	8.1445
	60	0.0000	2.4755	1.6694	6.7341	5.1469	1.7663	12.617	14.832
	90	0.0000	1.8995	1.8298	7.6357	8.9048	1.9816	19.457	28.271
	120	0.0000	1.0879	1.8645	9.2398	7.8226	2.1585	25.081	64.669

References

1. Woinowsky-Krieger, S., Berechnung der ringsum frei aufliegenden gleichseitigen Dreiecksplatte. *Ing. Arch.* **4** (1933) pp. 254-262.
2. Cox, H. L. and Klein, B., The buckling of isosceles triangular plates. *J. Aero. Sci.* **22** (1955) pp. 321-325.
3. Wakasugi, S., Buckling of a right-angled isosceles triangular plate under edge thrust. *Trans. Jpn. Soc. Mech. Eng.* **21** (1960) pp. 474-480.
4. Wang, C. M., and Liew, K. M., Buckling of triangular plates under uniform compression. *Engineering Structures*, **16** (1994) pp. 43-50.
5. Xiang, Y., Wang, C. M., Kitipornchai, S., and Liew, K. M., Buckling of triangular Mindlin plates under isotropic inplane compression. *Acta Mechanica*, **102** (1994) pp. 123-135.
6. Liew, C.M. Wang, C. M., Xiang, Y., and Kitipornchai, S., *Vibration of Mindlin plates: Programming the p-version Ritz Method*, Elsevier, Oxford (1998).

HULL SLAMMING USING NON LINEAR FINITE ELEMENT METHODS

JOHN MARCO

Defence Science and Technology Organisation, Aeronautical and Maritime Research Laboratory, PO Box 4331, Melbourne 3001, Australia

E-mail: john.marco@dsto.defence.gov.au

The phenomenon of hull slamming results when the bow of a ship emerges from the water and then submerges again with a vertical velocity greater than a certain threshold velocity. If the bow geometry of the ship is such that the angle between the bottom plates and the water (dead rise angle) is small, this action can produce large forces for short time durations. Traditionally, a statistical approach has been used whereby the maximum slamming load is applied as a static force. New numerical techniques are available which model the ship and waves as interacting dynamic components in the scenario. Multi-materials, Eulerian methods when coupled with Lagrangian procedures can be employed to simulate this fluid structure interaction. A generic example of a ship structure impacting a wave profile will be used as an example to demonstrate the application of these non-linear dynamic finite element methods.

1 Introduction

Damage to ship hulls by bottom slamming has been recognised for many years. The phenomenon of hull bottom slamming results when the ship's bow emerges from the water and subsequently submerges at an attitude such that the angle between the bottom plates and water is small. This action produces large forces for short time durations. The applied forces can cause local buckling and plastic deformations in the ship structure within the forward region of a ship. It can also cause an increase in the global responses (shear forces, bending moments and stresses) of the entire ship. Hard slamming produces large impulses that set the entire ship into vibratory motion which may persist for up to a minute [1]. Slamming can also cause damage to cargo, and loss or damage to shipboard equipment. It can also be an issue with motion induced illness on the crew. Through-life operational costs can therefore increase due to the effects of hull slamming.

Traditionally, designs were investigated through rigid body motion studies and then verified by model tests for regular and irregular sea conditions [1]. In general, both hull flexibility and the dynamic loading produced from sea loads in irregular seas were not explored. These loads were replaced by the statistical maximum slamming load as a static force. The true dynamic interaction between hull and fluid was not examined. Recently, an attempt was made to consider hull flexibility, this approach uses two dimensional beam theory and therefore is limited to the global structural response [2].

The development of new numerical features incorporating a combination of Lagrangian and Eulerian techniques has opened up the possibility of exploring this interaction. An idealized example of a ship structure impacting a wave profile will be used to demonstrate the application of these features, namely, a non linear dynamic finite element approach. The present analysis is limited to local structural response within the first 0.1 second of interaction. In order to analyse the global response of the whole structure a simulation run time of up to a minute is required and would be the subject of further studies.

2 The Scenario

The scenario explored in this paper consists of an idealized ship structure, see Figure 1 for details. The ship structure has a nominal length of 45m, with a maximum beam of 6m, hull depth of 4m, displacement 538.4 tonnes and is positioned in a sagging condition as shown in Figure 1, with the bow and stern ends on two wave crests, wavelength 40m, and peak to peak wave height of 2m.

The motion of the ship relative to the wave is as follows, sway V_x=0 m/s, heave V_y= -1 m/s, forward speed V_z=+10 m/s (approximately 20 knots) and a clockwise pitch motion of 0.1 radian/s, the rotation point is at the aft end of fluid model.

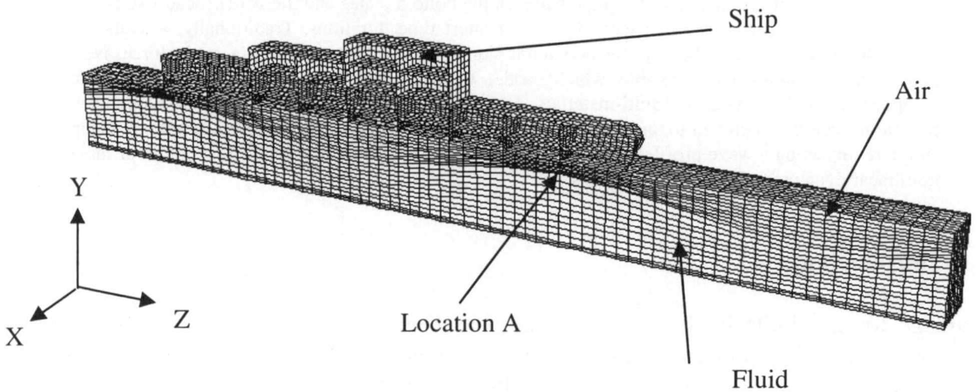

Figure 1 Ship Fluid Scenario Modeled

3 The Finite Element Code

The LSTC [3,4] finite element suite of codes were used to model, solve and process the numerical data. The solver LS-DYNA is an explicit three dimensional finite element code for analysing the large deformation dynamic response of inelastic solids. A contact impact algorithm permits gaps and sliding along material interfaces for Lagrangian solutions. These interfaces can be rigidly tied to allow variable zoning of the adjacent meshes, without the need of transition regions. Spatial discretisation is achieved by the use of eight node solid elements, two node beam elements, three and four node shell elements, truss elements, membrane elements, discrete elements, and rigid bodies. The equations of motion are integrated in time by the central difference method. LS-DYNA contains over thirty material models and ten equations of state to cover a wide range of possible material behaviour.

4 The Finite Element Model

The scenario modeled includes the metal ship structure, fluid and surrounding air, half symmetry was incorporated to reduce the size of the global numerical model. Therefore limiting the responses to symmetric loads, that is, heads seas only. The ship structure was assembled from 5500 shell elements, and point masses were used to distribute additional mass on board the structure, simulating equipment and crew, see Table 1. The sea model was modeled using 10296 solid elements, and the surrounding air model used 7722 elements.

Table 1. Finite Element Scenario Details

Part ID	Element Type	#Elements	Mass(tonnes)
Air	Solid	7722	n/a
Fluid	Solid	10296	n/a
Hull	Shell	2825	144.9
Bulkheads	Shell	1125	46.1
Deck A	Shell	450	143.5
Deck B	Shell	450	146.7
Super Structure	Shell	700	57.3
		Total Ship Mass	538.5 tonnes

5 The Numerical Technique

This scenario was modeled using a combination of Lagrangian and Eulerian methods. The air and water components were modeled using an Eulerian system (i.e. fixed in space) incorporating multi-material capabilities within each solid element (A multi-material element is when more than one material type (i.e. air and water) can coexist within the single Eulerian mathematical element). The ship structure was modeled using a Lagrangian system (i.e. deformable in space). The two coordinate systems were 'coupled' in order to transfer the kinetic energy from the ship onto the fluid (i.e. the fluid is assumed to be initially stationary). This transfer of energy is handled within the code through an automatic contacting surface algorithm which is set-up between the air/water and ship interfaces. Nodes from the Lagrangian mesh (i.e. structure) are constrained to move with the Eulerian mesh (i.e. fluid) as it comes into contact. Sizing of respective finite elements is important as 'leakage' may occur, that is, some fluid may miss seeing the structure and penetrate into the hull space.

The software was executed on a Silicon Graphics Indigo^2 workstation for a simulated run time of 100 ms, the CPU time consumed was 20.4 hrs, and the elapsed real time was 22.0 hrs.

6 Discussion

Figures 2 and 3, show the transient vertical displacement and velocity responses for location A, which is positioned on the ships bottom plating on the longitudinal centerline, directly below the first bulkhead. Over the simulated impact time of 100 ms, the response at location A shows the hull moving in a downward total displacement of about 27 cm, where as, the local velocity of that location falls to maximum value of -3 m/s. In Figure 3, two plots are shown, one referring to the point (ie location A) on the ship hull, the other referring to the fluid particle adjacent the hull point. Both curves show nominally the same velocity response over the analysed time, thereby indicating a coupled response.

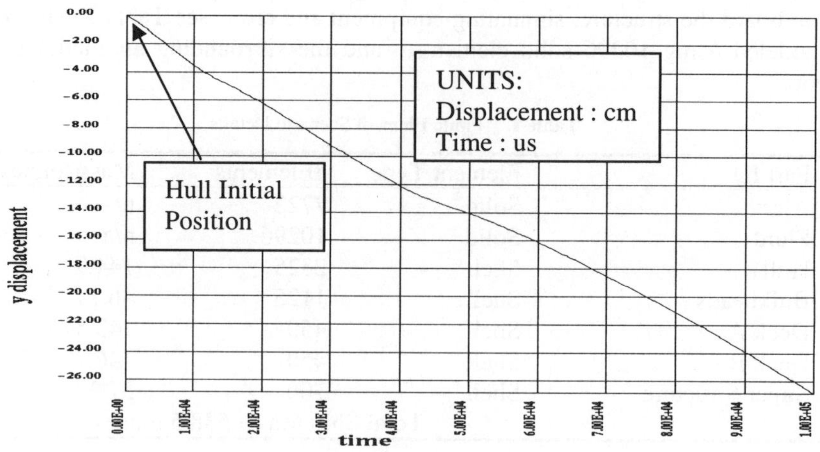

Figure 2 Vertical Transient Displacement Of Location 'A'

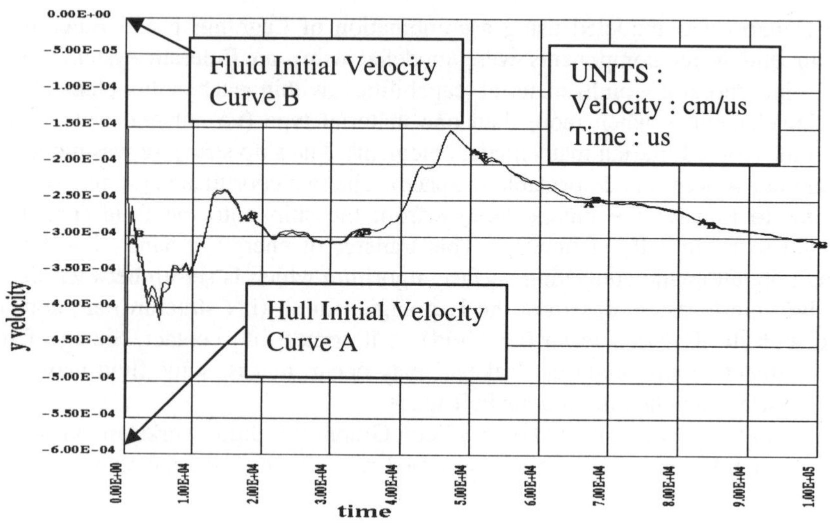

Figure 3 Vertical Transient Velocity Of Location 'A' Ship And Fluid

Figures 4 and 5 show the time sequence plots of the ship's structural deformation and the fluid's vertical velocity for the forward section of the ship and surrounding fluid. The structural deformation plots show the deformation occurring to the ship's hull structure, as well as, the oscillatory response in the decks. Figure 5 shows the vertical velocity fringe plots of the fluid below the forward section of the ship. The plots show

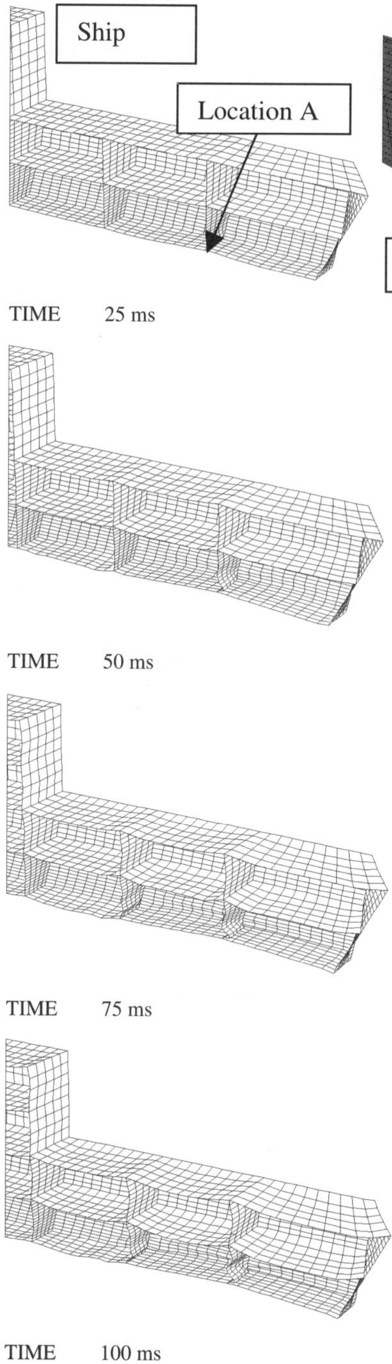

Figure 4 Structural Deformation Plots

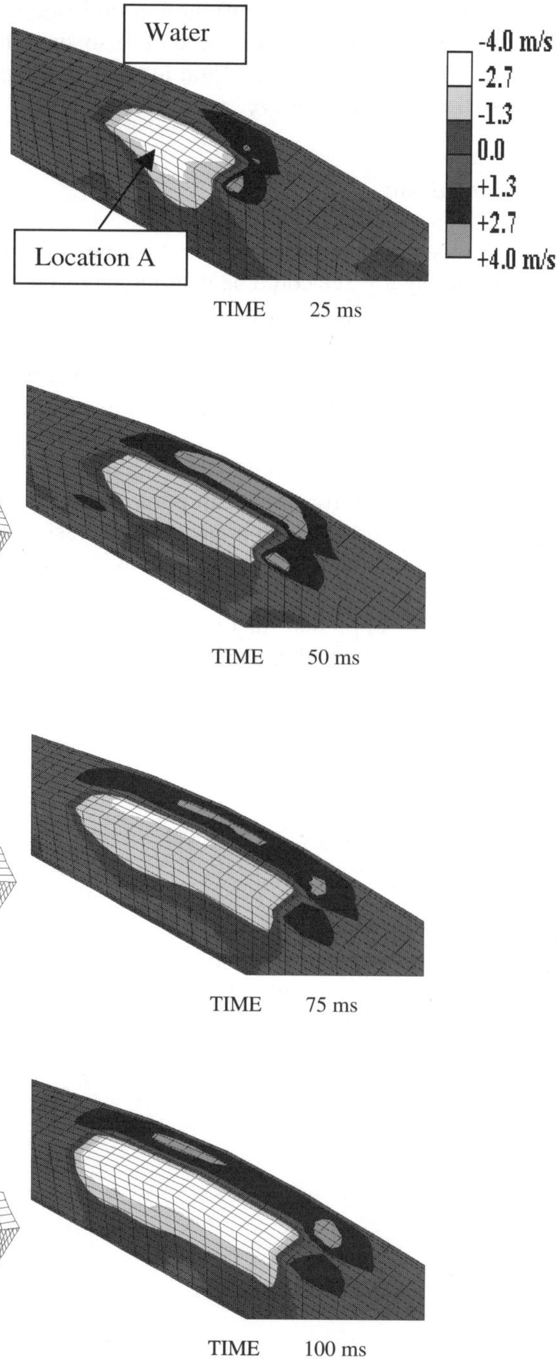

Figure 5 Water Vertical Velocity Plots

the downward (ie white region) motion (ie velocity 2.7 to 4.0 m/s) of that region of fluid directly below the keel plate. The plots also show the upward (ie dark region) motion (ie velocity 1.3 to 2.7 m/s) of the fluid located on the port side of the ship's hull as it pierces through the crest of the wave.

7 Conclusion

Using an idealized ship simulation, it has been demonstrated that dynamic response for a ship impacting a wave could be determined using a non-linear finite element technique, experimental data from actual ship wave impacts needs to be used to validate this approach.

New advances in finite element techniques have the potential to explore the phenomenon of hull slamming through the application of Eulerian Multi-material techniques coupled to Lagrangian techniques. This technique coupled with full-scale tests for validation purposes would allow the dynamic fluid structure interaction of the impact to be investigated for practical purposes. The results show the short term duration dynamic responses of a ship impacting the crest of a wave.

Further numerical studies of real ships need to be examined with trials data from actual ships to validate this approach for short and long term impact durations. If successful, this approach may clarify the extent of local and global structural responses due to hull slamming, thus enabling better design of ship structures under these types of loads.

References

1. Slamming of Ships: A Critical Review of the Current State of Knowledge, Ship Structures Committee, 1970, U.S. Coast Guard Headquarters.
2. Bishop, R.E.D and Price, W.G (1979) Hydroelasticity of Ships. Cambridge University Press
3. Livermore Software Technology Corporation (LSTC), Developer of LS-NGRID, LS-DYNA, LS-POST suite of finite element codes, Livermore, California, USA
4. LS-DYNA Keyword User's manual, May 1999 Version 950, Livermore Software Technology Corporation.

NONLINEAR THEORY OF MULTILAYERED SHELLS WITH ACTUATION STRAINS

L.H. HE[1], C.W.H. LAU[2] AND C.W. LIM[3]

[1]*Department of Modern Mechanics, University of Science and Technology of China Hefei, Anhui 230026, P. R. China. E-mail: lhhe@ustc.edu.cn*

[2]*Department of Construction, Hong Kong Institute of Vocational Education New Territories, Hong Kong. E-mail: whclau@vtc.edu.hk*

[3]*Department of Building and Construction, City University of Hong Kong Tat Chee Avenue, Kowloon, Hong Kong. E-mail: bccwlim@cityu.edu.hk*

The objective of this paper is to propose a general theory for multilayered shells with actuation strain. The constituent layers in the shell are in perfect bond, and the actuations strain is constant within every layer. A layer-wise representation of the displacement field is assumed, and the compatibility condition of traction at layer interfaces and on the major surfaces is imposed to reduce the number of unknowns. Hamilton's principle is used to derive the governing equations that involve only five unknowns, as in the first-order shear deformation theory. Dynamic effect and the Karman-type nonlinearty are incorporated. Application of the present theory in the analysis of some smart structures integrating actuation layers is discussed.

1 Introduction

Multilayered plates and shells incorporating sensors and actuators have been considered for smart structure designs which are used for active shape control, structural vibration and noise control applications [1-4]. The most commonly used actuators are piezopolymeric and piezoceramic layers. Actuators made of other materials such as shape memory alloys and electro-strictives etc. have also received great attention. The sensors sense the state of the structure by generating some signals, which are fed back to the actuators according to the control law implemented, while the actuators react to the feedback signals and generate certain actuation strains. Therefore, these structures are capable of adapting to or correcting for changing operating conditions.

There are a number of theoretical models proposed for piezoelectric laminates; for the literature one is referred to [5] and related references. In contrast, multilayered shells with integrated actuators have seen less investigation. Most of the existing models, proposed with special reference to multilayered piezoelectric shells, are formulated based on Kirchhoff-Love hypothesis [6-8] or the first-order displacement assumption [9-10]. These theories are unsatisfactory as they predict discontinuous transverse shear stress across layer interfaces. In fact, transverse shear stress may play a pronounced role, even if the structure is not thick. For instance, the results in [11] reveal that the modal amplitudes of smart beams predicted by these theories differ significantly from those based on more accurate discrete-layer displacement approximations. It is one of the central issues to accurately account for the effects of transverse shear deformations. Due to this reason, a linear theory of multilayered shells with actuation strain is proposed recently based on discrete-layer approximation [12]. The theory satisfies the stress compatibility conditions at layer interfaces and on the major surfaces. The number of unknowns are five, as in the first-order shear deformation theory. In this paper, the theory is extended to incorporate the Karman-type geometric nonlinearity.

2 Geometric Relations

The system to be considered is a multilayered shell composed of k perfectly bonded elastic layers with constant thickness. The top and bottom surfaces as well as the k-1 layer interface are denoted, in turn, by ${}_{(k)}\omega$, ${}_{(0)}\omega$, ${}_{(1)}\omega$, ..., ${}_{(k-1)}\omega$, the distance between ${}_{(m)}\omega$ and ${}_{(0)}\omega$ by ${}_{(m)}\zeta$, and the actuation strain within the mth layer by ${}_{(m)}\overset{*}{e}_{ij}$ (${}_{(m)}\overset{*}{e}_{ij} = {}_{(m)}\overset{*}{e}_{ji}$). As shown in Fig. 1, a curvilinear coordinate system $(\theta^1, \theta^2, \theta^3)$ is introduced, where the θ^3 coordinate line is perpendicular to the bottom surface of the shell which coincides with θ^3=0. In the following, the Einsteinian summation convention is adopted for repeated indices, where Latin indices range from 1 to 3 while Greek ones from 1 to 2.

The spatial and surface covariant base vectors, $\mathbf{g}_i$ and $\mathbf{a}_i$, are defined by

$$\mathbf{g}_i = \mathbf{r}_{,i}, \quad \mathbf{a}_\alpha = \mathbf{R}_{,\alpha}, \quad \mathbf{a}_3 = \mathbf{a}_1 \times \mathbf{a}_2 / |\mathbf{a}_1 \times \mathbf{a}_2|, \tag{1}$$

with $(\mathbf{g}_1 \times \mathbf{g}_2) \cdot \mathbf{g}_3 > 0$ and $(\mathbf{a}_1 \times \mathbf{a}_2) \cdot \mathbf{a}_3 > 0$. Here a comma refers to partial differentiation with respect to the suffix indices, **r** and **R** are position vectors of a point in the shell space and a point on surface ${}_{(0)}\omega$, respectively. The corresponding spatial and surface contravariant base vectors, $\mathbf{g}^i$ and $\mathbf{a}^i$, are defined through the relations

$$\mathbf{g}^\alpha \cdot \mathbf{g}_\beta = \delta^\alpha_\beta, \quad \mathbf{g}^3 = \mathbf{g}_3, \quad \mathbf{a}^\alpha \cdot \mathbf{a}_\beta = \delta^\alpha_\beta, \quad \mathbf{a}^3 = \mathbf{a}_3, \tag{2}$$

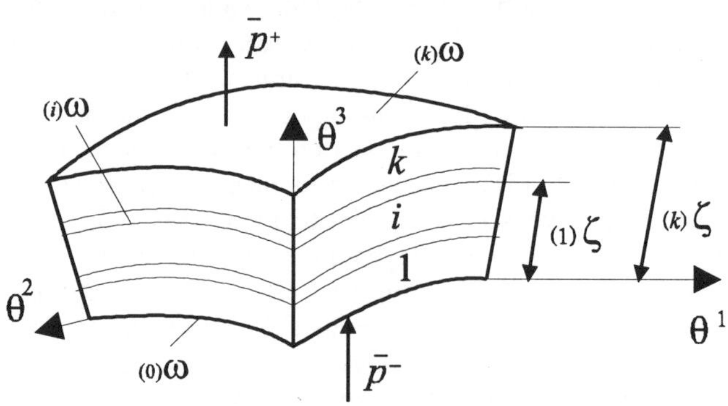

Fig. 1. Geometry of the shell

where δ^α_β is the mixed Kronecker delta. There exist the following connections between the space and surface base vectors:

$$\begin{aligned} &\mathbf{g}_\alpha = \mu^\beta_\alpha \mathbf{a}_\beta, \quad \mathbf{g}_3 = \mathbf{a}_3, \quad \mathbf{g}^\alpha = \overset{-1}{\mu}{}^\alpha_\beta \mathbf{a}^\beta, \quad \mathbf{g}^3 = \mathbf{a}^3, \\ &\mathbf{g}_\alpha = g_{\alpha\beta} \mathbf{g}^\beta, \quad \mathbf{g}^\alpha = g^{\alpha\beta} \mathbf{g}_\beta, \quad \mathbf{a}_\alpha = a_{\alpha\beta} \mathbf{a}^\beta, \quad \mathbf{a}^\alpha = a^{\alpha\beta} \mathbf{a}_\beta, \end{aligned} \tag{3}$$

in which $\mu_\alpha^\beta = \delta_\alpha^\beta - \zeta b_\alpha^\beta$ is the shifter tensor with the inverse $\overset{-1}{\mu}{}_\beta^\alpha = \varepsilon^{\beta\lambda}\varepsilon_{\alpha\nu}\mu_\lambda^\nu / \mu$, $g^{\alpha\beta} = \mathbf{g}^\alpha \cdot \mathbf{g}^\beta$ and $g_{\alpha\beta} = \mathbf{g}_\alpha \cdot \mathbf{g}_\beta$ are metric measure tensors in the shell space, $a^{\alpha\beta} = \mathbf{a}^\alpha \cdot \mathbf{a}^\beta$ and $a_{\alpha\beta} = \mathbf{a}_\alpha \cdot \mathbf{a}_\beta$ are metric measure tensors on surface ${}_{(0)}\omega$. In addition, $b_\alpha^\beta = -\mathbf{a}_{3},_\alpha \cdot \mathbf{a}^\beta$ is the curvature tensor of ${}_{(0)}\omega$, $\mu = \det(\mu_\alpha^\beta)$ the determinant of μ_α^β, $\varepsilon_{\alpha\nu} = (\mathbf{a}_\alpha \times \mathbf{a}_\nu)\cdot \mathbf{a}_3$ and $\varepsilon^{\beta\nu} = (\mathbf{a}^\beta \times \mathbf{a}^\nu)\cdot \mathbf{a}^3$ the two-dimensional permutation tensors.

Under the combined action of the actuation strains and the external load, the shell deforms. In this paper, we focus our attention on the case that the shell undergoes a Karman-type nonlinear deformation. Under this condition the strain tensor of the shell, $\mathbf{e} = e_{ij}\mathbf{g}^i \otimes \mathbf{g}^j$, relates to the displacement vector $\mathbf{v} = V_i\mathbf{g}^i = v_i\mathbf{a}^i$ by

$$e_{\alpha\beta} = \tfrac{1}{2}(V_{\alpha\|\beta} + V_{\beta\|\alpha} + V_{3\|\alpha}V_{3\|\beta}), \quad e_{\alpha 3} = \tfrac{1}{2}(V_{\alpha\|3} + V_{3\|\alpha}), \quad e_{33} = V_{3\|3}, \tag{4}$$

in which the double vertical line denotes space covariant derivative. The components of displacement and its derivatives with respect to the spatial and surface base vectors are linked by

$$V_\alpha = \mu_\alpha^\beta v_\beta, \quad V_3 = v_3, \quad V^\alpha = \overset{-1}{\mu}{}_\beta^\alpha v^\beta, \quad V^3 = v^3, \quad V_{\alpha\|\beta} = \mu_\alpha^\nu(v_{\nu|\beta} - b_{\nu\beta}v_3), \tag{5}$$
$$V_{\alpha\|3} = \mu_\alpha^\nu v_\nu,_3, \quad V_{3\|\alpha} = v_3,_\alpha + b_\alpha^\nu v_\nu, \quad V_{3\|3} = v_3,_3,$$

where the single vertical line is surface covariant derivative.

3 Basic Assumptions

Since the constituent layers of the shell are in perfect bond, the displacement and traction are continuous across the layer interfaces. As discussed in [12 -14], the displacement can be approximately expressed by

$$v_\alpha = u_\alpha + \eta_\alpha\zeta + \varphi_\alpha\zeta^2 + \psi_\alpha\zeta^3 + \sum_{m=1}^{k-1} {}_{(m)}u_{\alpha\,(m)}R(\zeta), \quad v_3 = u_3, \tag{6}$$

where u_i are displacement components of a point on ${}_{(0)}\omega$, $\eta_\alpha, \varphi_\alpha, \psi_\alpha$ and ${}_{(m)}u_\alpha$ are unknown functions of coordinates θ^β and time *t*, and ${}_{(m)}R(\zeta)$ is a ramp function which is defined in term of the Haviside step function ${}_{(m)}H(\zeta)$ that equals 1 if $\zeta \geq {}_{(m)}\zeta$ and 0 if $\zeta < {}_{(m)}\zeta$:

$${}_{(m)}R(\zeta) = (\zeta - {}_{(m)}\zeta)\,{}_{(m)}H(\zeta). \tag{7}$$

The layers in the shell are elastic. The constitutive equation can be written as

$$\sigma^{ij} = {}_{(m)}c^{ijkl}(e_{kl} - {}_{(m)}\overset{*}{e}_{kl}), \tag{8}$$

with σ^{ij} being stress components and ${}_{(m)}c^{ijkl} = {}_{(m)}c^{klij} = {}_{(m)}c^{jikl} = {}_{(m)}c^{ijlk}$ the elasticity tensor of the *m*th layer. It is assumed and assumed that the normal stress component σ^{33} can be ignored in comparison with the other stress components. In conjunction with (8) this requires

$$e_{33} - {}_{(m)}\overset{*}{e}_{33} = -{}_{(m)}c^{33\omega\rho}(e_{\omega\rho} - {}_{(m)}\overset{*}{e}_{\omega\rho}) / {}_{(m)}c^{3333}. \tag{9}$$

Thus, inserting (9) into (8) provides

$$\sigma^{\alpha\beta} = {}_{(m)}\tilde{c}^{\alpha\beta\omega\rho}(e_{\omega\rho} - {}_{(m)}\overset{*}{e}_{\omega\rho}), \quad \sigma^{\alpha 3} = 2{}_{(m)}c^{\alpha 3\nu 3}(e_{\nu 3} - {}_{(m)}\overset{*}{e}_{\nu 3}), \tag{10}$$

where

$${}_{(m)}\tilde{c}^{\alpha\beta\omega\rho} = {}_{(m)}c^{\alpha\beta\omega\rho} - {}_{(m)}c^{\alpha\beta 33}\,{}_{(m)}c^{\omega\rho 33} / {}_{(m)}c^{3333}. \tag{11}$$

This simplified constitutive relation will be used in the following.

By substituting (6) into (4), the transverse shear strain components are derived:

$$\begin{aligned} e_{\alpha 3} &= \tfrac{1}{2}(\eta_\alpha + u_{3,\alpha} + b_\alpha^\beta u_\beta) + \varphi_\alpha \zeta + \tfrac{1}{2}(3\psi_\alpha - b_\alpha^\beta \varphi_\beta)\zeta^2 \\ &\quad - b_\alpha^\beta \psi_\beta \zeta^3 + \tfrac{1}{2}\sum_{m=1}^{k-1}(\delta_\alpha^\beta - {}_{(m)}\zeta b_\alpha^\beta)_{(m)} u_{\beta(m)} H(\zeta). \end{aligned} \tag{12}$$

Note that in the derivation all derivatives with respect to ζ are stipulated as right-hand derivatives so as to maintain the uniqueness of the shear strains. The top and bottom surfaces of the shell are assumed free of tangential force. Upon (10) and (12), this requires

$$\begin{aligned} \eta_\alpha &= -u_{3,\alpha} - b_\alpha^\beta u_\beta + 2{}_{(1)}\overset{*}{e}_{\alpha 3}, \\ \psi_\alpha &= d_\alpha^\beta \varphi_\beta + \sum_{m=1}^{k-1} {}_{(m)}f_\alpha^\beta\,{}_{(m)}u_\beta - l_\alpha^\beta({}_{(k)}\overset{*}{e}_{\beta 3} - {}_{(1)}\overset{*}{e}_{\beta 3}), \end{aligned} \tag{13}$$

with

$$\begin{aligned} d_\alpha^\beta &= -\varepsilon_{\alpha\sigma}\varepsilon^{\lambda\nu}\bar{\mu}_\nu^\sigma(2\delta_\lambda^\beta - h b_\lambda^\beta)/3h\hat{\mu}, \quad l_\alpha^\beta = 2\varepsilon_{\alpha\sigma}\varepsilon^{\beta\nu}\bar{\mu}_\nu^\sigma / 3h^2\hat{\mu}, \\ {}_{(m)}f_\alpha^\beta &= -\varepsilon_{\alpha\sigma}\varepsilon^{\lambda\nu}\bar{\mu}_\nu^\sigma(\delta_\lambda^\beta - {}_{(m)}\zeta b_\lambda^\beta)/3h^2\hat{\mu}, \end{aligned} \tag{14}$$

in which h is the total thickness of the shell, $\bar{\mu}_\nu^\sigma$ and $\hat{\mu}$ are, respectively, the values of μ_ν^σ and μ at $\zeta = 2h/3$.

To ensure the continuity of transverse shear stress at layer interface, the following relation is imposed

$${}_{(i+1)}c^{\alpha 3\omega 3} e_{\omega 3}(\theta^\nu, {}_{(i)}\zeta) = \lim_{\varepsilon \to 0} {}_{(i)}c^{\alpha 3\omega 3} e_{\omega 3}(\theta^\nu, {}_{(i)}\zeta - \varepsilon). \quad (i = 1,2,\ldots,k-1) \tag{15}$$

With the help of (12), this results in

$$\begin{aligned} \sum_{m=1}^{k-1} {}_{(i)}\overset{(m)}{\Delta}{}^{\alpha\beta}\,{}_{(m)}u_\beta &= {}_{(i)}\Pi^{\alpha\beta}\varphi_\beta + {}_{(i+1)}c^{\alpha 3\omega 3}\,{}_{(i+1)}\overset{*}{e}_{\omega 3} - {}_{(i)}c^{\alpha 3\omega 3}\,{}_{(i)}\overset{*}{e}_{\omega 3} \\ &\quad + ({}_{(i+1)}c^{\alpha 3\omega 3} - {}_{(i)}c^{\alpha 3\omega 3})[(\tfrac{3}{2}\delta_\omega^\lambda - b_{\omega\,(i)}^\lambda\zeta)_{(i)}\zeta^2 l_\lambda^\nu({}_{(k)}\overset{*}{e}_{\nu 3} - {}_{(1)}\overset{*}{e}_{\nu 3}) - {}_{(1)}\overset{*}{e}_{\omega 3}], \end{aligned} \tag{16}$$

From these equations one possesses

$$ {}_{(m)}u_{\alpha} = {}_{(m)}a_{\alpha}^{\beta}\varphi_{\beta} + \sum_{i=1}^{k} {}_{(m)}^{(i)}k_{\alpha}^{\beta}\,{}_{(m)}\overset{*}{e}_{\beta 3}, \tag{17} $$

where ${}_{(m)}a_{\alpha}^{\beta}$ and ${}_{(m)}^{(i)}k_{\alpha}^{\beta}$ are quantities depending only on the curvature tensor b_{α}^{β} and the material constants of the layers. Substitution of (13) and (17) into (6) finally gives the expression of the displacement

$$ v_{\alpha} = \mu_{\alpha}^{\beta}u_{\beta} - \zeta u_{3},_{\alpha} + h_{\alpha}^{\beta}\varphi_{\beta} + \sum_{i=1}^{k} {}^{(i)}g_{\alpha}^{\beta}\,{}_{(i)}\overset{*}{e}_{\beta 3}, \quad v_{3} = u_{3}, \tag{18} $$

The corresponding strain components can be obtained by (4), (5) and (18).

4 Governing Equations

Hamilton's principle is used here to derive the governing equations and the associated boundary conditions of the shell. To this end, denote the stress acting on the lateral edge of the shell by $\overline{\sigma}^{ij}$, and the normal stresses on the top and bottom surfaces by $\overline{p}^{+}$ and $\overline{p}^{-}$, respectively. Then one has

$$ \begin{aligned} &\int_{t_0}^{t}[\int_{V}(\sigma^{\alpha\beta}\delta e_{\alpha\beta} + 2\sigma^{\alpha 3}\delta e_{\alpha 3})dV - \int_{V}\rho\dot{V}^{i}\delta\dot{V}_{i}dV \\ &\quad - \int_{V}\overline{\sigma}^{ij}n_{j}\delta V_{i}dV - \int_{A^0}\overline{p}^{-}\delta V_{3}dA - \int_{A^k}\overline{p}^{+}\delta V_{3}dA]dt = 0, \end{aligned} \tag{19} $$

where δ is the variational operator, ρ the mass density, V the volume of the shell, A^{0} and A^{k} are the areas of the bottom and top surfaces, the superposed dot denotes differentiation with respect to time, n_{i} the outward unit vector normal to the edge of the shell.

5 Conclusions

A general theory for multilayered shells with constant actuation strains has been formulated based on a novel kinematical assumption, via Hamilton's principle. The theory satisfies the continuity conditions of displacement and traction at layer interfaces, and the governing equations contain only five unknown variables. The source of the induced strains is not restricted, and, therefore, the present theory is applicable to many cases, including e.g. deformation of multilayered shells with piezoelectric layers or shape memory alloy actuators. For example, for shells composed of piezoelectric layers with poling direction along the thickness, the relevant actuation strain in the *i*th layer can be written as

$$ {}_{(i)}\overset{*}{e}_{\alpha\beta} = {}_{(i)}s_{\alpha\beta}(\Phi^{+} - \Phi^{-})/{}_{(i)}h, \quad {}_{(i)}\overset{*}{e}_{\alpha 3} = 0, \tag{20} $$

provided the so-called "strain induced method" is adopted [12]. Here ${}_{(i)}s_{\alpha\beta}$ is related material constant, ${}_{(i)}h$ the thickness of the *i*th layer, and Φ^{+}, Φ^{-} are electric potential on the top and bottom electrodes covering that layer. Accordingly, the responses of the shells can be analyzed by using the present theory. Due to the limitation of space, however, numerical example will be given elsewhere.

6 Acknowledgements

The work described in this paper was substantially supported by a grant from the Research Grants Council of the Hong Kong Special Administrative Region, China (Project No. CityU 7116/99E).

References

1. Wada, B. K., Fanson, J. L. and Crawley, E. F. (1990) Adaptive structures. J. Intelligent Mat. Syst. Struct. 1, 157-174.
2. Tzou, H. S. and Anderson, G. L. (1992) Intelligent structural system. Boston/Dordrecht: Kluwer Academic Publishers.
3. Rao, S. S. and Sunar, M.(1994) Piezoelectricity and its use in disturbance sensing and control of flexible structures: a survey. Appl. Mech. Rev. 47, 113-123.
4. Crawley, E. F.(1994) Intelligent structures for aerospace: a technology overview and assessment. AIAA 32, 1689-1698.
5. Gopinathan, S. V., Varadan, V. V. and Varadan, V. K. (2000) A review and critique of theories for piezoelectric laminates. Smart Mater. Struct. 9, 24-48.
6. Tzou, H. S. and Gadre, M. (1989) Theoretical analysis of a multi-layered thin shell coupled with piezoelectric shell actuators for distributed vibration controls. J. Sound Vib. 132, 433-450.
7. Jia, J. and Rogers, C. A. (1990): Formulation of a laminated shell theory incorporating embedded distributed actuators. J. Mech. Des. 112, 596-604.
8. Chaudhry, Z., Lalande, F. and Rogers, C. A. (1995) Modeling of induced strain actuation of shell structures. J. Acoust. Soc. Am. 97, 2872-2877.
9. Tzou, H. S. and Zhong, J. P. (1993) Electromechanics and vibrations of piezoelectric shell distributed systems. J. Dyn. Syst. Meas. Control 115, 506-517.
10. Tzou, H. S. and Bao, Y. (1995) A theory on anisotropic piezothermoelastic shell laminates with sensor/actuator applications. J. Sound Vib. 184, 453-473.
11. Robbins, D. H. and Reddy, J. N. (1991) Analysis of piezoelectrically actuated beams using a layer-wise displacement theory. Comput. Struct. 41, 265-279.
12. He, L. H. (1998) Laminated shells with actuation strains: general theory and applications. Arch. Appl. Mech. 68, 637-649.
13. He, L. H. (1994) A linear theory of laminated shells accounting for the continuity conditions of displacements and transverse shear stresses at layer interfaces. Int. J. Solids Struct. 31, 613-627.
14. He, L. H. (1995) Non-linear theory of laminated shells accounting for the continuity conditions of displacements and tractions at layer interfaces. Int. J. Mech. Sci. 37, 161-173.

EFFECTS OF VERTICAL AND ROUND BEADS ON BUCKLING BEHAVIOUR OF COMPOSITE STRUCTURES

H. HOSSEINI – TOUDESHKY

Aerospace Engineering Department – Amir-Kabir University of Technology, Iran.
E-mail: hosseini@cic.aku.ac.ir

M. SADEGHI

H.F.DORNA Aerospace Engineering Co., Iran.
E-mail: mansoor_48@yahoo.com

In this paper, effects of vertical and round beads used for shear webs and ribs in the composite structures are studied. Using these methods, we can achieve more lightness and lower manufacturing cost. In this investigation, instability (buckling) of wing nose and main ribs made of woven E-glass material and/or P.V.C foam core with and without beads are studied using finite-element analysis and experimental methods. The results show that we can easily improve shear webs stability using vertical male beads and/or round beads instead of sandwiched webs. It may cause weight reduction of about 50%, manufacturing time reduction of about 50% and increasing the buckling load. Further studies and tests on optimum ply angles on buckling and post buckling of these new structures show that by introducing 0° and 90° plies in the core region of the laminates, higher buckling load is obtained, but substantially post buckling stiffness is increased.

1 Introduction

Since wing ribs and fuselage frames weights are considerable parts of the wing and fuselage structural weight, it is interested to optimize in terms of weight, cost of fabrication and assembly. Figure 1 shows three types of shear resistant, nonbuckling webs, reinforced by vertical and round beads which frequently used in the metallic web design. One advantage of webs with holes is creation of built-in access space for the necessary linkages in airplanes.

Composites have advantages over aluminum alloy in reducing the weight of a component or structure. This is due to their properties and their ability to exactly meet the loading requirement on a section; i.e. more layers can be easily added or removed as required. This would require complex machining in conventional alloys.

Due to the simplicity of the manufacturing and easily variation of stiffness properties by either altering the skin thickness, the core density or the thickness of the core, using of sandwich designs to stabilize composite shear webs and ribs is one of the most conventional techniques in the composite structures. Using vertical male beads and round beads or combinations of them instead of sandwiched webs causes reduction in weight, manufacturing time, manufacturing cost and increasing in the buckling load. Therefore, optimization in the post-buckling range becomes an important design consideration for laminates, which may be subjected to compressive loads higher than the buckling load.

In this paper, instability (buckling) of new beaded wing nose and main ribs, shear webs and the effect of different ply angles on the buckling loads of these new structures were studied using finite-element analysis and experimental methods.

2 Finite-Element buckling analysis

The objective of the buckling analysis is to estimate the maximum load that a structure can support before it becomes elastically unstable or it collapse. The analysis here is actually a 'bifurcation buckling'. This type of bifurcation or eigenvalue buckling problem is usually limited and noncoservative in its applications. Most practical problems with buckling behavior should be analyzed as large displacement geometrically nonlinear problems.

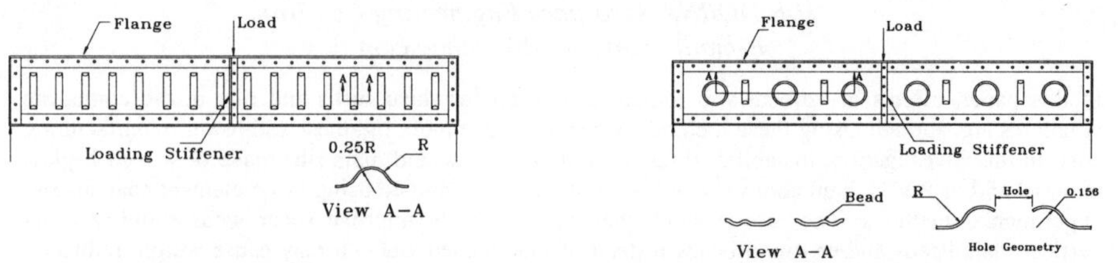

a: vertical male beads   b: Round beads and vertical male beads between holes

Figure 1: Two types of shear resistant, nonbuckling metallic webs

Buckling analysis in NISA is a two-pass analysis. The first pass is a static analysis which determines the stress for a given reference set of load. The second pass is an eigenvalue analysis which first computes the geometric stiffness matrix. The second pass results are in terms of load factors (eigenvalues) and mode shapes (eigenvectors).

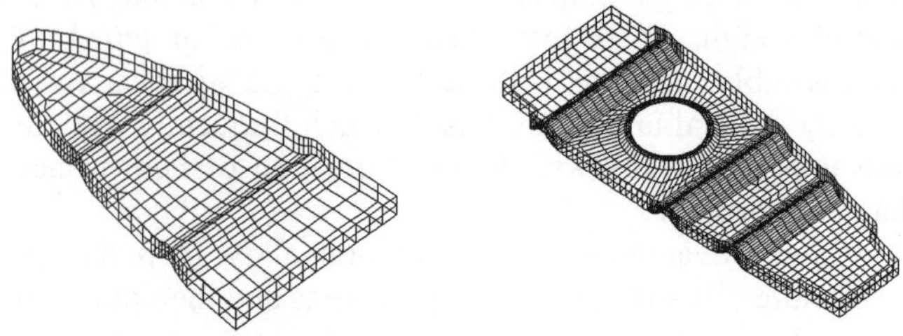

Figure 2: Finite-element mesh for specimen GE2, GE5.

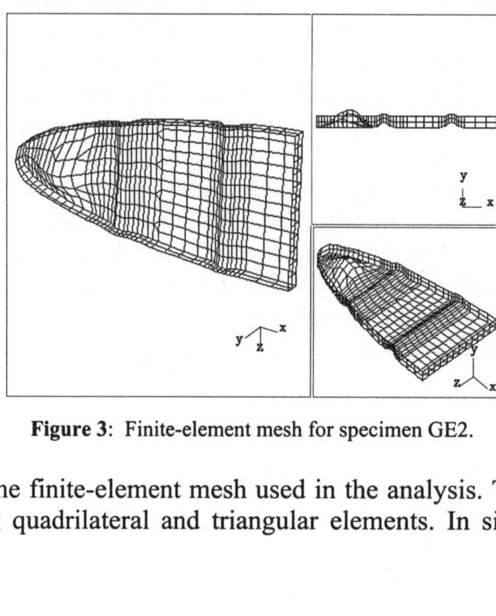

Figure 3: Finite-element mesh for specimen GE2.

Figure 2 shows the finite-element mesh used in the analysis. The specimens meshes were generated using quadrilateral and triangular elements. In simulating compressive forces, uniform pressure was applied to the one end of specimens while being reacted at the other end. This approach is more applicable in simulation of the load application of the actual tests. All specimens were analyzed using the finite-element method.

Figure 3 shows the fundamental buckling mode shape of specimen GE2, which shows a wave in the axial direction. Buckling loads obtained from the finite-element analysis are presented in Table 2.

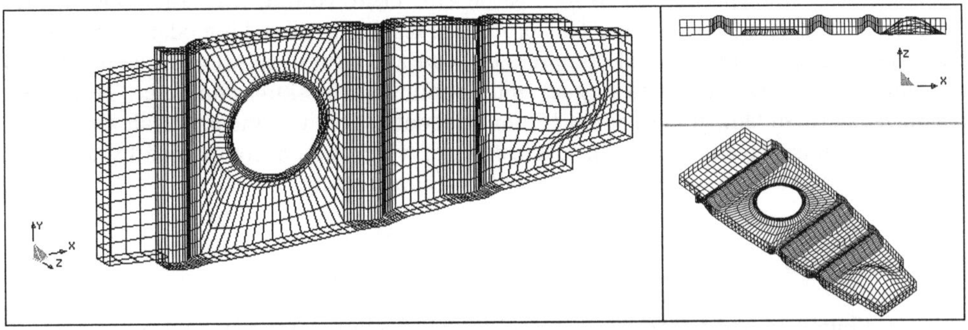

Figure 4: Fundamental buckling mode shape for specimen GE5.

3 Materials and Test Specimens

The glass-epoxy specimens were manufactured using hand lay up method with SP system RE300 and RE165 woven E-glass cloth and Airex R82.6 P.V.C foam core impregnated with SP system SP 690 epoxy resin and finally all specimens cured in room temperature and post cured in an oven after 4 hour up to 120 $^{\circ}c$ with 15 $\frac{^{\circ}c}{hr}$ ramp rate.

Because of the balanced-weave cloth, there is no distinction between properties in the warp and fill directions. Thus, the ply oriented in 0 or 90 degrees has the same contribution to laminate stiffness and strength. Five types of specimens were fabricated. Table 1 lists candidate specimens for the buckling strength tests.

Table 1. Test Specimen Configurations.

Type	Test ID	Lay up	Dimensions (mm)	Q.t.y
Nose Rib	GE1	$[\pm45_G, \pm45_g, F_8]_s$	see Fig. 5	2
Nose Rib	GE2	$[\pm45_G, \pm45_g]_s$	see Fig. 5	2
Nose Rib	GE3	$[\pm45_G, \pm90_g]_s$	see Fig. 5	2
Main Rib	GE4	$[\pm45_G, \pm45_g, F_8]_s$	see Fig. 5	2
Main Rib	GE5	$[\pm45_G, \pm45_g]_s$	see Fig. 5	2

F_8=8 mm PVC foam, G=0.3 mm woven E-glass, g=0.2 mm woven E-glass

Symmetric laminates where chosen for construction of all specimens to decouple stretching and bending deformations. If there are no imperfections in the specimens, we can assume that out-of-plane deformations will not occur under the axial compressive load until onset of buckling. Specimens were sized based on the guideline that the buckling load is no greater than the ultimate compressive strength failure load divided by a safety factor of 1.5. Therefore, it is estimated that test specimens will fail in buckling mode rather than the ultimate compressive strength.

4 Experimental Setup and Procedure

Buckling tests were performed using Instron Universal Testing Machine. Both loaded ends were fixed using special jigs. This procedure results in a fixed boundary condition at both loaded ends. Applied axial compression load and axial displacement automatically measured using a D-Link Data Logger and Pentium 133 desktop computer. Tests were conducted at room temperature and 0.15mm axial deflection increments for all specimens. All test specimens were loaded up to the failure, which was accompanied by a sudden drop of load and a loud sound associated with specimen fracture.

5 Test Results

Measured buckling and failure loads for all tested specimens are summarized in Table 2. The test data is some what scattered, particularly after buckling loads.

Table 2: Test Results

Test ID	Lay up	Buckling Load (lb.)	Failure Load (lb.)
GE1	$[\pm45_G, \pm45_g, F_8]_s$	925	1311
GE2	$[\pm45_G, \pm45_g]_s$	1300	1560
GE3	$[\pm45_G, \pm90_g]_s$	1850	2041
GE4	$[\pm45_G, \pm45_g, F_8]_s$	1014	1410
GE5	$[\pm45_G, \pm45_g]_s$	2760	2914

The obtained experimental results, applied load versus axial displacements, for specimens GE1, GE2 and GE3 are shown in Figure 6 and Figure 7. It is observed that the axial displacement is increased linearly with the load variation until the specimen begins to buckle. Upon onset of buckling, the specimen loses some stiffness, but it still able to carry additional load to the failure point.

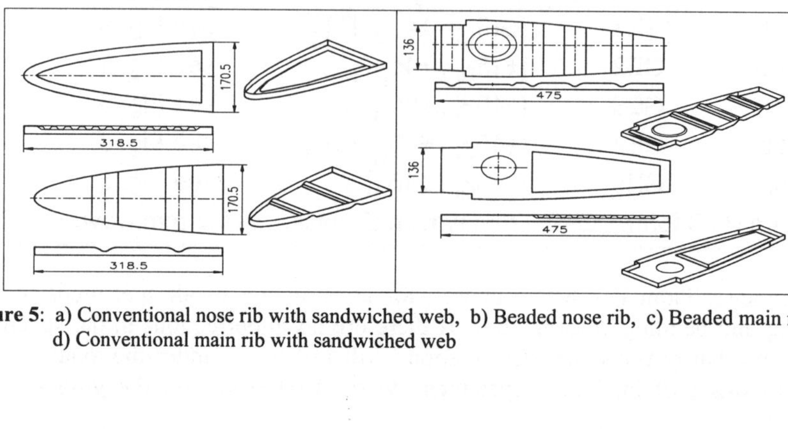

Figure 5: a) Conventional nose rib with sandwiched web, b) Beaded nose rib, c) Beaded main rib d) Conventional main rib with sandwiched web

Figure 6 shows that sandwiched specimen with 9mm total web thickness (specimen GE1) start to buckle at the load about 925 lb., but buckling load for similar beaded specimen without sandwiched web and with 1mm total web thickness (specimen GE2) is about 1300 lb. By introducing 90° plies in the core region of specimen GE2, higher buckling load of 1850 lb for specimen GE3 obtained and failure load increased to 2041 lb which shows 23.5% increasing in the buckling load with respect to the buckling load for the specimen GE2. Specimens GE4 and GE5 show similar behavior, sandwiched specimen with 9mm total web thickness (Specimens GE4) start to buckle at the load about 1014 lb., but buckling load for similar beaded specimen without sandwiched web and with 1mm total web thickness (Specimens GE5) is about 2760 lb.

6 Results and Discussion

According to the obtained results summarized in Table 2. and weight and manufacturing time compared in Table 3 it is seen that, for example, by eliminating the foam core in the GE1 specimen, buckling load increases to 28.8% for new beaded specimen GE2 with similar lay up and 50.0% for new beaded specimen GE3 with optimized lay up. Also weight reduction is about 135% and manufacturing time reduction is about 56%.

Table 3: Comparison of weight and manufacturing time.

Specimen	Lay up	Weight (lb)	Manufacturing
GE1	$[\pm45_G, \pm45_g, F_8]_s$	0.66	20.5
GE2	$[\pm45_G, \pm45_g]_s$	0.21	9.0
GE3	$[\pm45_G, \pm90_g]_s$	0.21	9.0
GE4	$[\pm45_G, \pm45_g, F_8]_s$	0.92	21.0
GE5	$[\pm45_G, \pm45_g]_s$	0.30	10.0

Experimental and finite element method calculated buckling loads are presented in Table 4. Excluding specimen GE1, finite-element analysis, in general under predict the buckling loads. This may be due to the edges of the specimens which are not ideally simply supported by the test jigs. Buckling loads predicted by the finite-element analysis are in good agreement with experimentally measured values, however, there are differences between finite-element and test results. It could arise from the initial imperfection of specimens, uneven load application and/or uncertainty of the material property data used in finite-element analysis and so on.

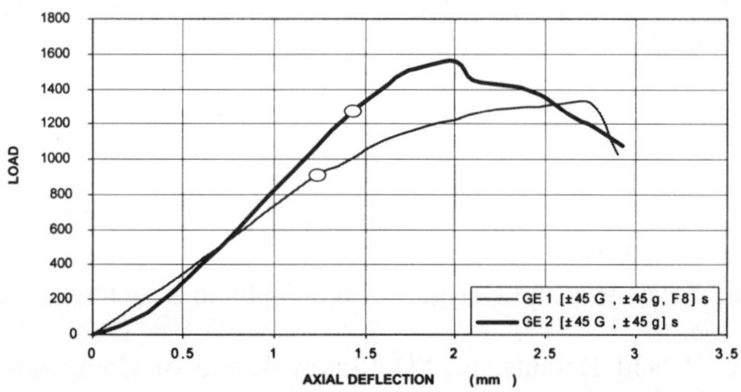

Figure 6: Comparison of the experimental results between the old shear web reinforced by foam and new shear web with the vertical male beads.

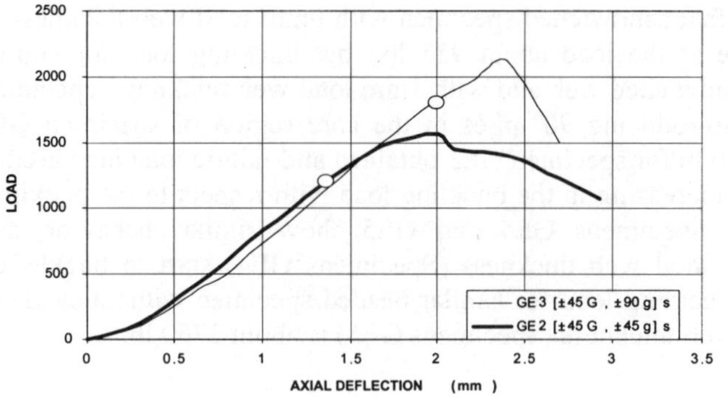

Figure 7: The experimental results for beaded shear webs with different ply angles.

Table 4: Comparison of experimental and finite element method calculated buckling loads.

Specimen	Experimental (lb)	F.E.M (lb)	Diff. (%)
GE1	925	986.2	+6.2
GE2	1300	785.5	-39.5
GE3	1850	827.6	-55.3
GE4	1014	228.4	-77.4
GE5	2760	857.5	-68.9

7 Conclusions

Numerical simulations are used in conjunction with experiments to study the buckling loads of typical wing nose and main ribs made of composite materials with and without sandwiched webs. Based on obtained results, the following conclusions can be drawn:

Test results and finite-element analysis show that we can improve composite shear webs instability using beaded webs instead of sandwiched webs. It causes weight reduction more than 50%, manufacturing time reduction more than 50% and increasing the buckling load. Tests results show that woven E-glass beaded webs supported along the edges and subjected to axial compressive loads are capable of sustaining the loads to failure considerably larger than the buckling loads. These failure loads range from 5.2% to 31.5% higher than the buckling loads. To predict the failure load, a nonlinear post buckling analysis is required. However, in view of the test results, it is believed that the post buckling strength should be investigated for the design of the composite structures to minimize the weight penalty.

References

1. Turvey, G. J. and Marshall, I. H., " Buckling and Post buckling of Composite Plates," Chapman & Hall Publishers, 1995.
2. Bruhn, E. F., " Analysis and Design of flight Vehicle Structures," S. R. Jacobs & Associates Inc.
3. Marshall, I. H. and Demuts, E., "Optimum design of Composite Structures," Elsevier Applied Science Publishers, 1990.

BUCKLING ANALYSIS OF COMPOSITE STIFFENED PLATE STRUCTURES DUE TO INTERACTION OF IN-PLANE COMPRESSION AND SHEAR LOADING, USING FINITE STRIP METHOD

HAMID REZA OVESY

Aerospace Engineering Department, Amirkabir University of Technology, Tehran, Iran
Assistant Professor
E-mail: ovesy@aku.ac.ir

HASSAN ASSAEE

Aerospace Engineering Department, Amirkabir University of Technology, Tehran, Iran
Research Assistant
E-mail: a7829552@aku.ac.ir

A semi-analytical approach called the finite strip method (FSM) is used in this paper to investigate the effects of interaction of in-plane compression and shear loading on the buckling load capacity of blade-stiffened composite panels. The panel is loaded under pure compression, pure shear and combined loading. The interaction curves are plotted for different stiffener aspect ratios. Also, the buckling mode shapes for a few selected points on the interaction curves are plotted. Based on the obtained results certain conclusions are made.

1 Introduction

Thin-walled structures have become very popular in structural engineering due to their high strength-to-weight ratio. These structures can be precisely analyzed using finite element method. However, for a certain group of problems, such as prismatic plate assemblies, the usual finite element analysis may be extravagant. Among many attempts to find an efficient analysis of such class of problems, the finite strip method is considered to be appealing. The finite strip method can be considered as a kind of finite element method in which a special element called strip is used. A great deal of research has been carried out in this area to investigate the applicability of method. Subsequently, many useful extensions have been developed.

Cheung [1], among others, may be considered as the pioneer who first presented the concept of finite strip method. Subsequently, other researchers developed different variations of the method and applied them to different problems. To name a few, Graves Smith and Sridharan [2] proposed the first FSM formulations for isotropic plate structures under edge loading. Plank & Wittrick [3] used a finite strip method to study the behavior of isotropic stiffened plates subjected to combined loading. Loughlan [4] used a FSM approach to study the interaction of in plane loading on the buckling load capacity of some composite stiffened panels. Dawe and Peshkam [5] re-examined the buckling characteristics of composite blade stiffened panels, by using strips of higher orders so called superstrips.

More recently, the authors of this paper generated a FORTRAN code on the basis of FSM approach. In the recent work [6] the related code was validated by comparing the obtained results with those already available in open literature for composite stiffened panels subjected to in-plane compression and shear loading. In the current paper, the effects of interaction of in-plane loadings on buckling properties of a two-blade stiffened panel are studied.

2 FSM Buckling Analysis

The results given in the paper are those for thin laminated panels. This allows the application of the classical laminated plate theory (CLPT) in the theoretical analysis of the paper and thus ignoring the effects of shear deformation through the thickness of laminate.

The detailed theoretical development of the FSM used in this paper is given in an early publication by the authors [6]. However, for the sake of completeness, the dominating features of the method are repeated bellow.

Figure 1. shows an example of plate structures modeled by a number of finite strips running the whole length of the structure. The finite strips are rigidly joined at their longitudinal reference lines where the compatibility conditions should be satisfied. The geometry of a multi-layered laminated composite finite strip is depicted in Figure 2., indicating the coordinate axes system pertaining to the strip displacements u, v and w.

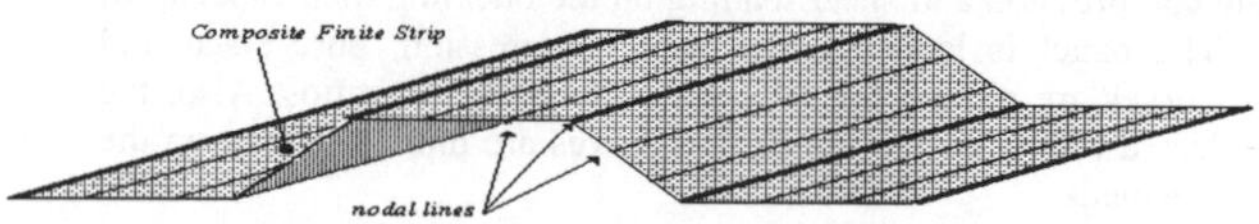

Figure 1. Plate structure discretized by finite strips

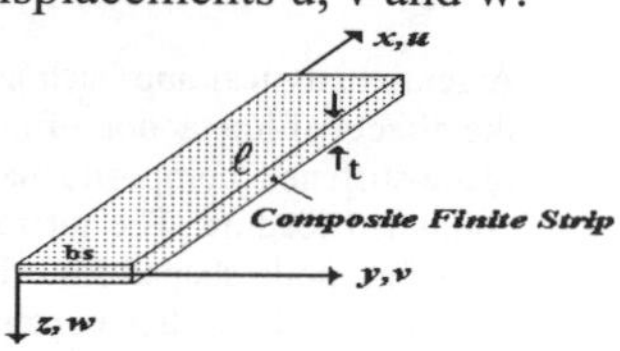

Figure2. Strip coordinate system

To determine the finite strip formulation for buckling analysis, the energy method seems to be appealing. Within the concept of CLPT, the strain energy U stored in the composite finite strip is given by equation 1. [8].

$$U=\frac{1}{2}\iint\{[\varepsilon_\ell^o]^T[A][\varepsilon_\ell^o]+2[\varepsilon_\ell^o]^T[A][\varepsilon_{n\ell}^o]+2[\varepsilon_\ell^o]^T[B][\psi]+2[\varepsilon_{n\ell}^o]^T[B][\psi] \\ +[\varepsilon_{n\ell}^o]^T[A][\varepsilon_{n\ell}^o]+[\psi]^T[D][\psi]\}\,dx\,dy \qquad (1)$$

In the above equation, $\varepsilon_L{}^o$ and $\varepsilon_{nL}{}^o$ are respectively linear and nonlinear components of the middle surface strains of the laminate ε^o. The middle surface strains are described in terms of strip displacements u, v and w as given in equation 2.

The linear bending and twisting curvatures of the middle surface of the strip, which cause additional linear strains at positions away from the middle surface through the strip thickness, are given in terms of w in equation 3.

$$\varepsilon_\ell^0=\begin{Bmatrix}u,_x\\ v,_y\\ u,_y+v,_x\end{Bmatrix}\quad \varepsilon_{n\ell}^0=\begin{Bmatrix}\frac{1}{2}(u,_x^2+v,_x^2+w,_x^2)\\ \frac{1}{2}(u,_y^2+v,_y^2+w,_y^2)\\ \frac{1}{2}(u,_x u,_y+v,_x v,_y+w,_x w,_y)\end{Bmatrix}\ (2)\qquad \psi=\begin{Bmatrix}w,_{xx}\\ w,_{yy}\\ 2w,_{xy}\end{Bmatrix}\ (3)$$

The matrices [A], [B] and [D] are the usual in-plane, coupling and flexural stiffness matrices of composite materials respectively. For a laminate with a symmetric lay-up configuration, which is the case throughout the current study, the matrix [B] is identically zero. Moreover, since this paper deals with a linear eigenvalue problem the strain energy could be written in the following form:

$$U=U_e+U_g \quad \textit{Where}$$
$$U_e=\frac{1}{2}\iint\{[\varepsilon_l^o]^T[A][\varepsilon_l^o]+[\psi]^T[D][\psi]\}\,dx\,dy$$
$$U_g=t\iint\{[\sigma]^T[\varepsilon_{nl}^o]\}\,dx\,dy \qquad (4)$$

It should be noted that the strain energy component Ue is associated with the conventional stiffness matrix of the strip and Ug is associated with the geometric stiffness matrix of the strip. The stress matrix [σ] corresponds to the stresses acting on the strip prior to the buckling, and t is the thickness of laminated plate.

It should also be noted that in CLPT analysis, the strain energy expression contains second derivatives of w and first derivative of u and v. This implies a requirement of C^1 type continuity for w and C^o type continuity for u and v. It means that Hermittian shape functions should be used for w and Lagrangian shape function should be used for u and v. Therefore, the buckling displacement functions are defined by equation 5.

$$u=\sum_{n=1}^{N}\left[(1-\eta)\cdot u_{1n}+\eta\cdot u_{2n}\right]\cos(\frac{n\pi x}{\ell}) \qquad v=\sum_{n=1}^{N}\left[(1-\eta)\cdot v_{1n}+\eta\cdot v_{2n}\right]\sin(\frac{n\pi x}{\ell})$$
$$w=\sum_{n=1}^{N}[(1-3\eta^2+2\eta^3)\cdot w_{1n}+b_s(\eta-2\eta^2+\eta^3)\cdot\theta_{1n}+(3\eta^2-2\eta^3)\cdot w_{2n}+b_s(\eta^3-\eta^2)\cdot\theta_{2n}]\sin(\frac{n\pi x}{\ell}) \quad (5)$$

Where $\eta=y/bs$ and $u_{1n},v_{1n},w_{1n},\theta_{1n}$, $u_{2n},v_{2n},w_{2n},\theta_{2n}$ are the displacement connection quantities along the edge1 and edge2 of the strip and corresponding to *n*th harmonic of displacement series along the strip length. In the present work three harmonics are postulated along the strip to determine the deflection series. By substituting u, v and w in equations 2. & 3. and then substituting the resultant linear and nonlinear strains in equation 4. and carrying out the integration the strain energy in terms of the displacement connection quantities is derived. By invoking the principle of virtual displacement in the form developed by Castigliano, the elasic and geometric stiffness matrices for a strip are developed and by assembling the related matrices for the whole structure the final set of equations for buckling stability problems are defined as equation 6.

λcr is a scalar which is related to critical level of stresses acting on the structure prior to buckling. The above equation is a kind of generalized eigenvalue problem that could be solved by appropriate solution schemes such as QZ algorithm or Cholesky QR iteration algorithm. In this paper the Cholesky QR iteration algorithm is implemented.

$$F_r=\frac{\partial U(\delta)}{\partial\delta_r} \quad\Rightarrow\quad [F]=[k_e+k_g]\cdot[\delta]$$
$$Det\left([K_e]+\lambda_{cr}[K_g]\right)=0 \quad (6)$$

3 Results and Discussion

The panel considered in this paper is depicted in Figure 3. The panel, which is subjected to combined in-plane compression and shear loadings, is stiffened by two stiffeners. The skin and the stiffeners are both laminated from orthotropic Boron-Epoxy plies. The lay up sequence, which is the same in the skin and the stiffeners, is $[0/30/60/90/-30/-60]_s$.

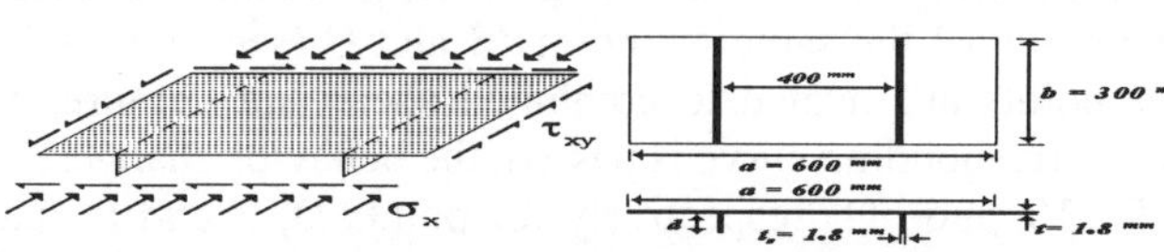

Figure 3. The description of the panel

The material properties for a single ply are:

$$E_1=206.9^{Gpa}\ E_2=20.7^{Gpa}\quad G_{12}=5.2^{Gpa}\quad \upsilon_{12}=0.31\quad t=0.15^{mm}$$

It is noted that the subject lay up configuration is of quasi-isotropic nature. Therefore, there does not exist any degree of in-plane anisotropy within the in-plane stiffness matrix associated with the skin or the stiffener. This implies that the application of pure compressional stress does not lead to any shear strains and vice versa.

A model constructed of 24 strips, as depicted in Figure 4., is used to analyze the panel.

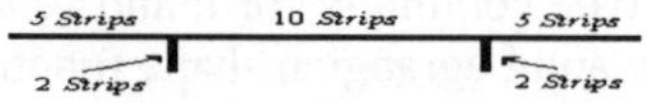

Figure 4. The FSM model for the panel

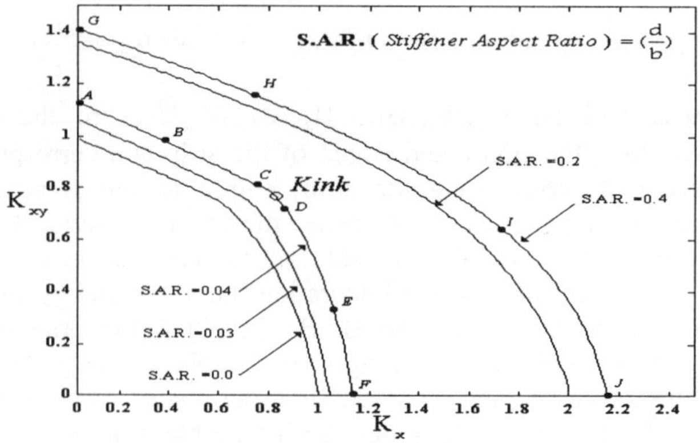

Figure 5. The buckling interaction curves for the panel

The interaction curves for different stiffener aspect ratios (d/b) are plotted in Figure 5. The value of K_x is the ratio of the critical compressive buckling stress for the stiffened panel under the combined loading to that of the unstiffened panel when the latter is subjected to pure compression. Similarly, K_{xy} is the corresponding ratio for the shear stress cases. As it can be seen, for small S.A.R (stiffener aspect ratio) values the curves have obvious kinks. However, for greater values of S.A.R the interaction curves are smooth and no such kinks are encountered. To investigate the cause of these kinks, some arbitrary points on either sides of certain kinks are choosen and the corresponding buckling mode shapes are plotted in Figures 6. to 15. It is noted that points A and G correspond to the buckling under pure shear whereas points F and J are related to the buckling under pure compression. The other points are related to the buckling under combined loads.

As far as the interaction curve for stiffener with the aspect ratio of 4% is concerned, it can easily be seen in Figure 6. that the buckling mode shape for the case of pure shear, i.e. point A, is associated with 2 half waves along the longitudinal centerline of the panel. This behavior remains almost the same for points B and C (see Figures 7. and 8.). It is noted that the latter points are related to combined shear and compression. However, when the kink on the corresponding curve is passed, the behavior changes. This can easily be seen in Figures 9., 10. and 11. respectively for points D, E and F that the buckling mode shapes are associated with only one half wave along the longitudinal centerline of the panel.

It may be concluded that for the cases with obvious kinks on the corresponding interaction curves there exists a sudden change of behavior at the kink. This happens because the buckling behavior of the panel for points between pure shear and the kink on a given interaction curve is dominated by the shear buckling mode and the behavior for points on the opposite side is dominated by the compression buckling mode. Thus, when for a given panel geometry, such as panel with S.A.R of 4%, the buckle mode shape for pure shear and that for pure compression are quite different, a drastic change of behavior at a specific point, i.e. kink, on the corresponding interaction curve is inevitable.

The above discussion is valid for the cases where no obvious kinks are experienced. This can be explained by an example such as the panel with S.A.R of 40% for which no kink can be distinguished on the corresponding interaction curve (see Figure 5.). For this panel, the mode shapes corresponding to points G, H, I and J are shown in Figures 12.,13,.14. and 15. respectively. As it can be seen in Figures 12. and 15. corresponding to pure shear and pure compression respectively, the buckling mode shapes, apart from minor differences at locations away from the longitudinal centerline of the panel, are very similar in general. To be more specific, in both cases, three main buckle half waves occur along the longitudinal centerline of the panel. Therefore, according to our previous discussion, the transition from pure shear to pure compression is expected to be fairly smooth.

For the reader's information, as it can be seen in Figures 6. to 15., the buckling modes are overall for the panel with S.A.R=4% and those are local for the panel with S.A.R.=40%.

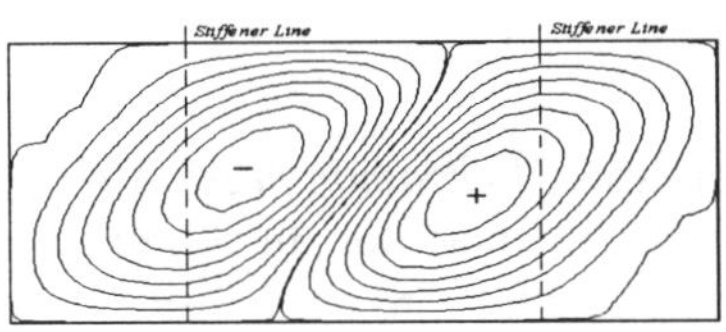

Figure 6. Buckling mode shape for point A

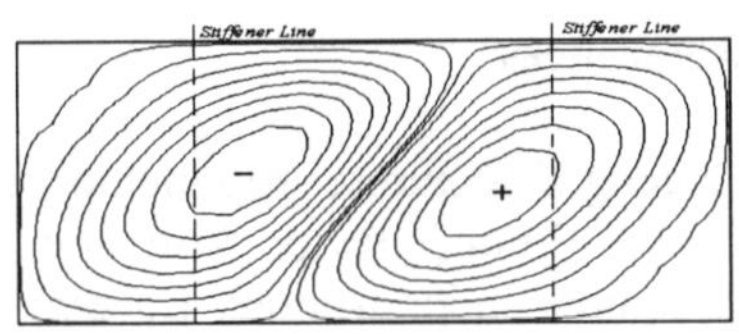

Figure 7. Buckling mode shape for point B

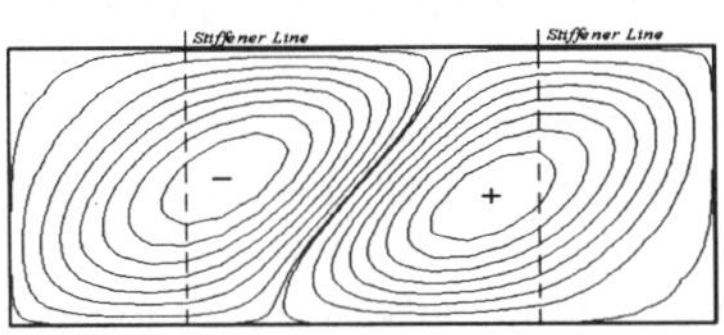

Figure 8. Buckling mode shape for point C

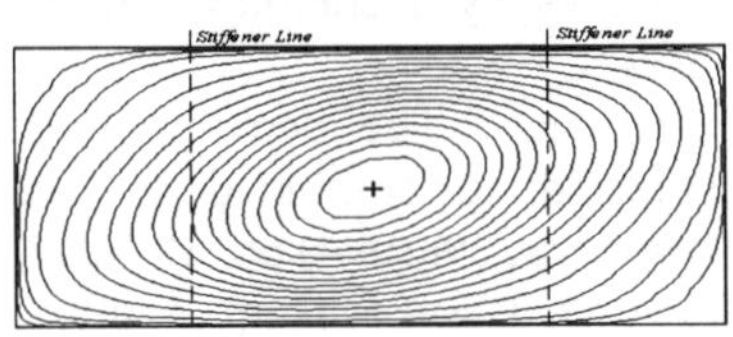

Figure 9. Buckling mode shape for point D

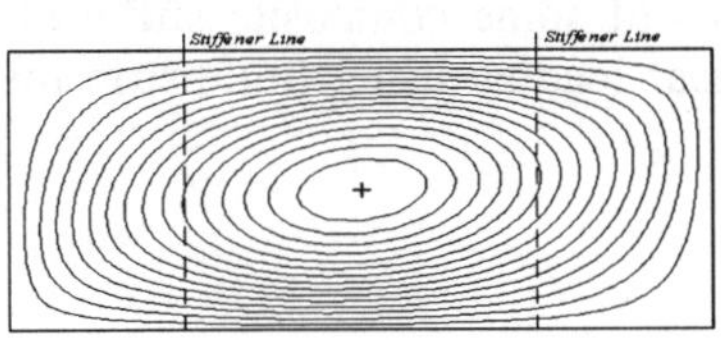

Figure 10. Buckling mode shape for point E

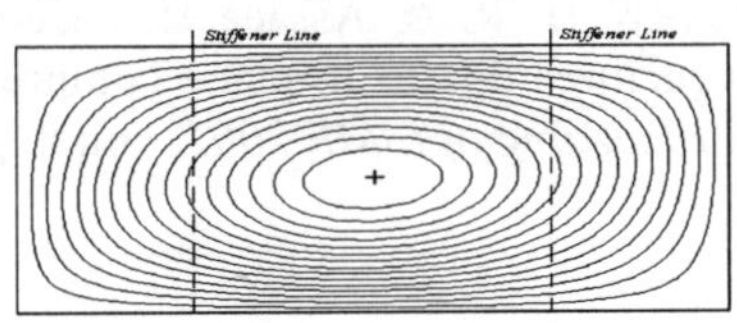

Figure 11. Buckling mode shape for point F

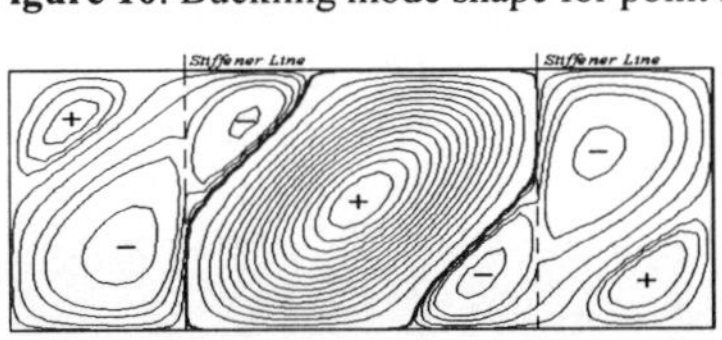

Figure 12. Buckling mode shape for point G

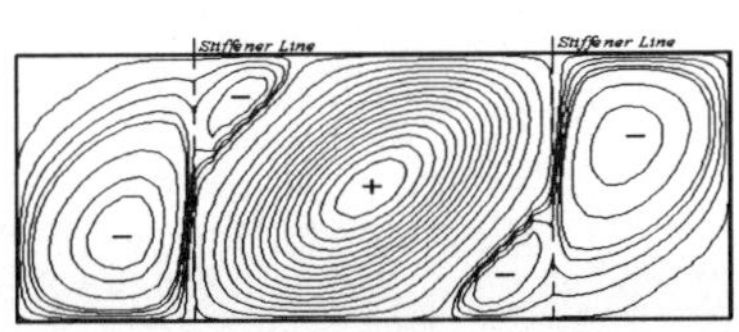

Figure 13. Buckling mode shape for point H

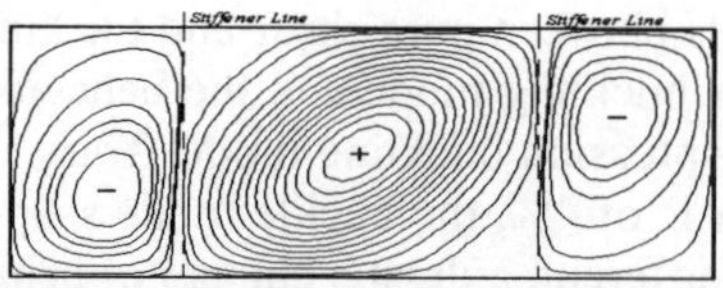

Figure 14. Buckling mode shape for point I

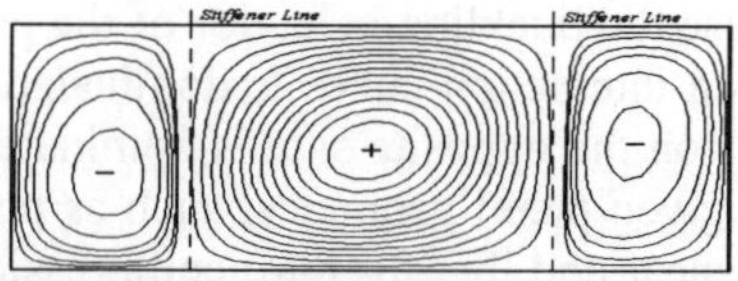

Figure 15. Buckling mode shape for point J

4 Conclusions

A multi-term finite strip formulation has been presented and used to predict the buckling load factors and the related mode shapes of stiffened plate structures due to interaction of in-plane shear and compression loading.

A two-blade stiffened composite panel is considered and the related interaction curves are plotted for different stiffener aspect ratios. The buckling mode shapes for some points on the interaction curves are plotted. The study of the results revealed the fact that if the buckle mode shape for pure shear and that for pure compression are quite different, for a given panel geometry, a drastic change of behavior at a specific point, i.e. kink, on the corresponding interaction curve is inevitable.

5 References

1. Chueng,Y.K.,Finite strip method in structural analysis , *Pergamon press*, 1976, p. 26
2. Graves-Smith,T.R. & Sridharan,S.,A finite strip method for buckling of plate structures under arbitrary Loading, *Int. J. Mech. Sci.*, **20**(1978) pp. 685-93
3. Plank R. J. & Wittrick W. H.,Buckling under combined loading of thin, flat-walled structures by a complex finite strip method , *Int. J. Num. Meth. Engng*, **8**(1974) pp.323-39
4. Loughlan, J.,The buckling performance of composite stiffened panel structures subjected to combined in plane compression and shear loading, *Composite Structures*, **29**(1994) pp. 197-212
5. Dawe D. J. & Peshkam ,A note on finite strip buckling analysis of composite plate structures, *Composite Structures*, **34** (1996) pp.163-168
6. Ovesy H. R. & Assaee H. , Buckling analysis of some composite stiffened plate structures due to in plane compression and shear loading using finite strip method, *Procceding of ISME 2001*, vol. **4**, pp. 75-82

SHEAR BAND ORIENTATION OF VERY HIGH STRENGTH (VHS) CIRCULAR STEEL TUBES

H. JIAO AND X. L. ZHAO

Department of Civil Engineering, Monash University, Clayton, Vic. 3168, Australia

This paper identifies different shear band orientations for different types of steel tubes and sheet steels. Strain gauge rosettes were used to measure the maximum shear stress orientation for very high strength (VHS) tubes. It was found for VHS tubes that a final 62.5° (rather than 45°) shear band developed during the plastic deformation stage. This paper explored the reasons behind the different orientations, including anisotropy of steels, rotation of steel crystals during plastic deformation, material properties, manufacturing process and geometric effects.

1 Introduction

The very high strength (VHS) circular steel tube is produced by OneSteel using quenching and tempering of cold-formed circular steel hollow section (called non-heated-treated (NHT) tube in this paper). VHS tubes have a yield stress of around 1350 MPa and an ultimate strength of 1500 MPa [1]. The yield stress of NHT steel was about 430 MPa. Although VHS and NHT steels have the same geometrical shape and dimensions, completely different failure modes were observed when loaded under uniaxial tension. VHS steel showed a shear band failure, whereas the NHT tubes illustrated a ductile diffuse necking (see Figure 1). The same shear band orientation was observed on full section tubes and tensile coupons for VHS steel [2]. The angle between the shear band and the loading direction was around 62.5° for VHS tubes.

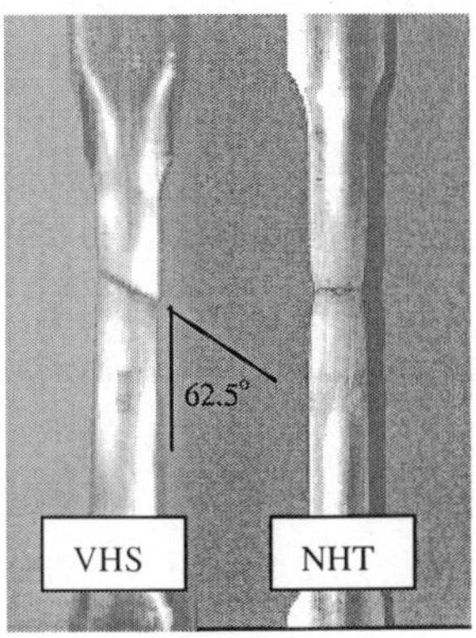

Figure 1 Failure modes of VHS and NHT

This paper identifies different shear band orientations for different types of tubes or sheet steels. It explores the reasons behind the phenomenon, such as: (1) Anisotropy of steels, (2) Rotation of steel crystals during plastic deformation, (3) Material properties, manufacturing process and (4) geometric effects.

2 Other Observations

Similar shear band orientations were observed for G550 sheet steel (62°) [3] and cold-formed C450 circular hollow sections (60°) [4]. However, different shear orientation or failure mode was found for some other thin walled steels, such as the maraging steel (38°) [5] and NHT tube and C350 tube (ductile failure, no shear band) [2, 6]. Table 1 shows the yield stress (f_y), specimen thickness (t), width (w), and the manufacturing processes of these steels. Some failure modes are shown in Figure 2.

Table 1: Steels with different shear band

Steel Name	f_y (MPa)	t	W	w/t	Orientation Angle	Manufacturing process	Reference
VHS	1350	1.6	19	12	62.5	heat treated	this paper
VHS	1350	1.6	12	8	62.5	heat treated	[1]
G550	658	0.4	13	30	62	cold reduction	[3]
C450 CHS	450	2.0	11	6	60	cold formed	[4]
Maraging steel	1750	4.0	19	5	38	hot rolled	[5]
NHT	430	1.6	12	8	0	cold formed	[2]
C350	350	1.6	11	7	0	cold formed	[6]

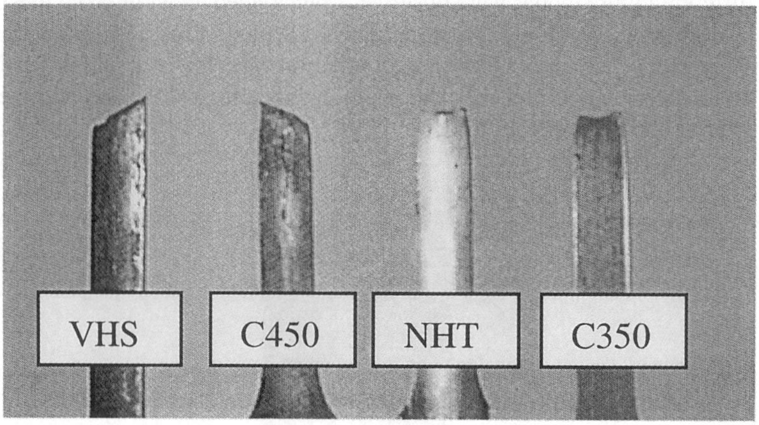

Figure 2 Failure modes of some steel tubes

3 Measurement of Maximum Shear Stress Orientation

Two specimens of full section VHS tubes with nominal diameter of 38mm and thickness of 1.8mm were tested under uniaxial tension. A 45° strain rosette (Figure 3) was mounted on the surface of each specimen. The tests were conducted in a 500 kN Baldwin universal testing machine with a strain rate of 1×10^{-4}/s.

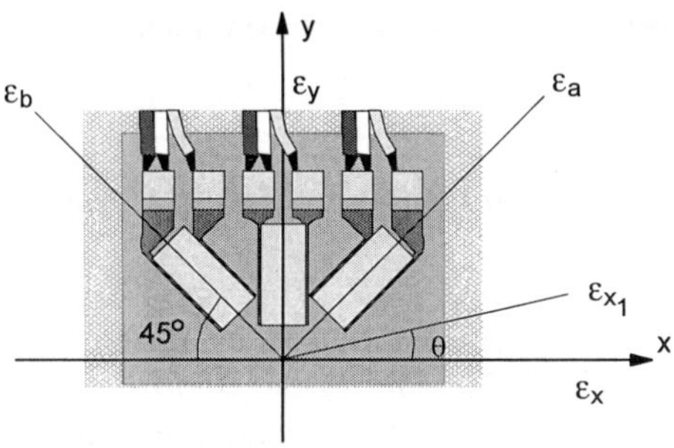

Figure 3 Strain rosette mounted on VHS tubes (y axis is the loading direction)

Since the VHS tube is thin-walled, when an infinitesimal area is selected, the stress status can be treated as plane stress. The transformation equations derived from plane strain can be used for the strains in plane stress [7]. The transformation equation for strains at any direction is expressed as:

$$\varepsilon_{x1} = \frac{\varepsilon_x + \varepsilon_y}{2} + \frac{\varepsilon_x - \varepsilon_y}{2}\cos 2\theta + \frac{\gamma_{xy}}{2}\sin 2\theta \tag{1}$$

in which, ε_{x1} is the normal strain in x_1 direction. θ is the angle between x axis and x_1 direction. According to Eq.(1), the shear strain γ_{xy} and normal strain ε_x based on the measured strains (ε_a, ε_y and ε_b) in the strain rosette of Figure 3 is expressed as:

$$\gamma_{xy} = \varepsilon_a - \varepsilon_b \tag{2}$$

$$\varepsilon_x = \varepsilon_a + \varepsilon_b - \varepsilon_y \tag{3}$$

The measured strains are shown in Figure 4 where γ_{xy} is calculated using Eq. (2).

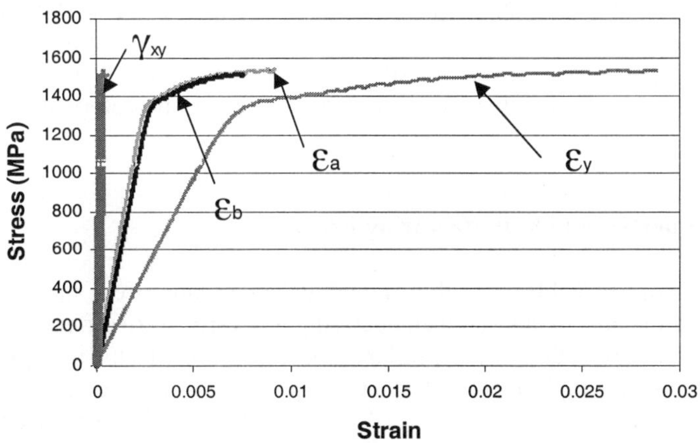

Figure 4: The measured strain elements

It can be seen from Figure 4 that there is no shear strain (γ_{xy} is almost zero) in the loading direction even when VHS steel entered into the non-linear part of the stress-strain curves, i.e., ε_x and ε_y are principle strains. The maximum shear stress orientation is 45° away from the loading direction.

4 Anisotropy of Steels and Rotation of Steel Crystals During Plastic Deformation

The anisotropy of steels may be caused by different manufacturing process, such as cold-rolled sheet metals [9, 10]. During the cold-rolling process, the randomly distributed grains in a metal may orient towards a certain direction. As a result, initially isotropic steel may become anisotropic after the plastic deformation. For example, the tensile yield stress of G550 sheet steel on transverse to the rolling direction was about 13% greater than that on parallel to the rolling direction [11]. Similar to the mechanical process, a sequence of heat treatments may also cause grains of a steel to re-crystallize to a final texture [5]. The anisotropy of steels may also form during a plastic deformation. For example, when hexagonal single crystals are stretched in tension, the base planes gradually turn towards positions parallel to the load direction. The grains of a polycrystal steel will rotate towards some limiting orientations [12]. Some of the quantitative analyses of shear band formation were based on the constitutive equations of continuum mechanics [13, 14]. The Hill's bifurcation theory [15, 16] was adopted to calculate the onset of the instability of a solid under static load [17, 18].

According to the maximum shear stress yield criterion [8], the shear orientation should be 45° to the principal stress for isotropic steel. On one hand, the final orientation of 62.5° (not 45°) might indicate that VHS tubes are not isotropic material. On the other hand, it seems that VHS tubes might be isotropic material since the measured maximum shear orientation is 45° relative to the loading direction. However, it should be noted that shear band did occur along the line on which the strain gauge rosette was located. The strain gauge rosette was damaged or peeled off when the shear band started forming. In other words, the strain gauge rosette cannot give any information on the maximum shear stress orientation during the shear band formation. It is clear that the shear band occurred during the plastic deformation of the specimen. It is very possible that the 62.5° (rather than 45°) shear band formed because of the rotation of the steel crystals during the plastic deformation as pointed out in [12].

5 Material Properties and Manufacturing Process

All of the steels shown in Table 1 are thin walled and in a plane stress status under axial tension. But the chemical contents and manufacturing processes were different, which may lead to different yield stress, ductility and failure mode. For example, the maraging steel [5] was a kind of ultrahigh strength steel that had a 0.2% proof yield stress of around 1750MPa (measured from the tensile stress-strain curve). The shear band was formed at a tensile strain of 0.034. It was manufactured by hot rolled at 1091°C and then normalized at 900°C for 30 minutes, whereas VHS steel is produced by elevating the temperature to 950°C, quenching the steel in 20°C cold water, and then tempering at 230°C for 15 minutes. Since VHS steel and the maraging steel have different chemical contents and manufactured in different way, the amount of precipitate particles in the martensite matrix, that are responsible for the ultrahigh strength property of such steels [5], are different. As a result, shear stress orientation at failure may be different (38° versus 62.5°). G550 steel is produced by hot rolled to 2.5mm in thickness, and then cold reduced to 0.42mm [19]. This cold-reduced process made it a kind of anisotropic steel [11]. Although similar shear band were observed for VHS, G550 and C450 steels, different analyses models may be adopted when conducting quantitative analyses.

It is also interesting to observe from Table 1 that materials with lower yield stress (350MPa and 430MPa) have orientation of zero, while those with higher yield stress have orientations of above 38°. This may match, to some extent, ductile necking failure for those with lower yield stress and the more brittle failure (sudden) for those with higher yield stress.

6 Geometric Effect

It was found by Okazawa and Usami [18] that failure mode of tensile coupons depends on the width-to-thickness ratio (w/t) of the specimens. Three failure modes were identified: (1) diagonal localized necking for w/t of 10, (2) concentrated necking for w/t of 4, (3) two axial symmetric localized necking for w/t of 1.0. It was pointed out by Hill [12] that "When thin strips are pulled in tension (their width being at least five times their thickness), it is observed that the neck does not form directly across the specimen, but at an oblique angle which depends on the state of anisotropy". It can be seen from Table 1 that all of the samples listed have w/t≥5, but different orientation angle are observed. This may indicate that different degrees of isotropy exist in the steels. It is also interesting to observe that the same VHS tubes with different w/t ratios (12 verus 8) have the same orientation angle. It should be noted that both w/t ratios are larger than 5. It may be concluded that the shear band orientation depends on not only the geometric w/t ratio, but also depends on the microstructure of the steel. Future research is needed to study the interaction of geometric effect and microstructure effect.

7 Conclusions

The following conclusions and observations are made.

- Different shear band orientation angles were identified for different types of steel tubes and sheet steels.
- The final shear band angle of 62.5° observed for VHS tubes formed during the plastic deformation stage.
- It was found that the shear band orientation and failure mode were influenced by the anisotropy of steels, rotation of steel crystals during plastic deformation, material properties, manufacturing process and geometric effects.

8 Acknowledgements

The authors are grateful to the Australian Research Council and OneSteel for financial support. Comments from Mr. Roger Doulis, Dr. Geoff Taplin and Mr. Mohamed Elchalakani of Monash University, and Mr. Tony Gunn and Mr. Hayden Dagg of OneSteel are appreciated.

References

1. Zhao, X.L., Section Capacity of Very High Strength (VHS) Circular Tubes under Compression. Thin-Walled Structures, 2000; 37(3), 223-240.
2. Jiao, H. and Zhao, X.L., Ductility of Very High Strength (VHS) Circular Steel Tubes in Tension. Thin-Walled Structures, 2001 (Accepted for Publication).
3. Kennedy, B., Effect of Holes on the Ductility of Condeck HP and Bondek II. 4th year student project, Monash University, 1996.
4. Zhao, X.-L., Lim, P., Joseph, P., and Pi, Y.-L., Member Capacity of Columns with Semi-Rigid End Conditions in Oktalok Space Frames. Structural Engineering & Mechanics- An International Journal, 2000; 10(1), 27-36.
5. Anand, L. and Spitzig, W.A., Initiation of Localized Shear Bands in Plain Strain. Journal of the Mechanics and Physics of Solids, 1980; 28, 113-128.
6. Mashiri, F.R., Zhao, X.L., and Grundy, P., Fatigue Tests and Design of Thin Cold-Formed Square Hollow Section-to-Plate T-Connections under Cyclic in-Plane Bending. J of Structural Engineering, ASCE, 2001(Accepted for Publication).
7. Gere, J.M. and Timoshenko, S., Mechanics of Materials. 4th SI ed. 1999, Cheltenham: Stanley Thornes. xx, 912.
8. Benham, P.P., Crawford, R.J., and Armstrong, C.G., Mechanics of Engineering Materials. 2nd ed. 1996, Harlow: Longman. xiii, 627.
9. Hatherly, M. and Malin, A.S., Shear Bands in Deformed Metals. Scripta Metallurgica, 1984; 18, 449-454.
10. Kusnierz, J., State of Stress and Strain in Cold-Rolled Metal and the Localization of Strain in Shear Bands. Scripta Metallurgica et Materialia, 1994; 30(4), 405-410.
11. Rogers, C.A. and Hancock, G.J., Ductility of G550 Sheet Steels in Tension. Journal of Structural Engineering, ASCE, 1997; 123(12), 1586-1594.
12. Hill, R., The Mathematical Theory of Plasticity. Oxford Engineering Science. 1998, Oxford, New York. 355.
13. Tvergaard, V., Studies of Elastic-Plastic Instabilities. Journal of Applied Mechanics-Transactions of the ASME, 1999; 66(1), 3-9.
14. Needleman, A. and Tvergaard, V., Analyses of Plastic Flow Localization in Metals. Applied Mechanics Reviews, 1992; 45(3), S3-S15.
15. Hill, R., A General Theory of Uniqueness and Stability in Elastic-Plastic Solids. Journal of the Mechanics and Physics of Solids, 1958; 6(236-249).
16. Hill, R. and Hutchinson, J.W., Bifurcation Phenomena in the Plane Tension Test. Journal of the Mechanics and Physics of Solids, 1975; 23, 239-264.
17. Tomita, Y., Simulations of Plastic Instabilities in Solid Mechanics. Applied Mechanics Reviews., 1994; 47(6), 171-205.
18. Okazawa, S. and Usami, T. Plastic Instability Simulation of Steel in Tension: Structural Failure and Plasticity (IMPLAST 2000). X.L. Zhao and R.H. Grzebieta, (eds). Melbourne, Australia: Elsevier Science Ltd., 2000. 775-780.
19. Rogers, C.A. and Hancock, G.J., Fracture Toughness of G550 Sheet Steels Subjected to Tension. Journal of Constructional Steel Research, 2001; 57(1), 71-89.

DETERMINATION OF PRESTRESS IN GROUT SLEEVE CONNECTIONS

YEE TECK LEE, XIAO LING ZHAO AND PAUL GRUNDY

Department of Civil Engineering, Monash University, Clayton, VIC 3168, Australia

Radial prestress in the grout is generated by adding an agent to the Portland cement so that it expands when setting. A normal constituent of cement, a complex Calcium Sulpho-Aluminate (CSA) is used. Prestress is a function of the expansion of grout which may depend on many factors such as the percentage of CSA in the mix, water cement ratio, time, surface condition of sleeve tubes, grout thickness, grouted length and ratio of tube diameter to thickness. There are insufficient data to establish a formula to predict prestress over the full range of connection parameters. The existing data only covers limited parameter ranges. A series of tests was designed to include a wide range of percentages of CSA, water cement ratio and grout thickness. Strain gauge readings were used to obtain measured prestress. A long-term data acquisition system was used to monitor the readings. The newly obtained results were used to establish a formula for confidently predicting the prestress in grouted sleeve connections and the existing data were used to verify this equation. The prestress formula will be used in predicting the capacity of the grouted sleeve connections in tension or compression.

1 Introduction

An innovative tubular steel jointing system that incorporates prestress grouted sleeve connections (see Figure 1) rather than conventional bolted or welded connections has been developed at Monash University [1]. It offers enormous benefits in fabrication and construction through minimising detailed welding and requiring low on-site technology. Research performed at Monash University in the last 10 years has investigated the behaviour of such connections under static loading and fatigue loading [2,3,4], and recently under elevated temperature [5].

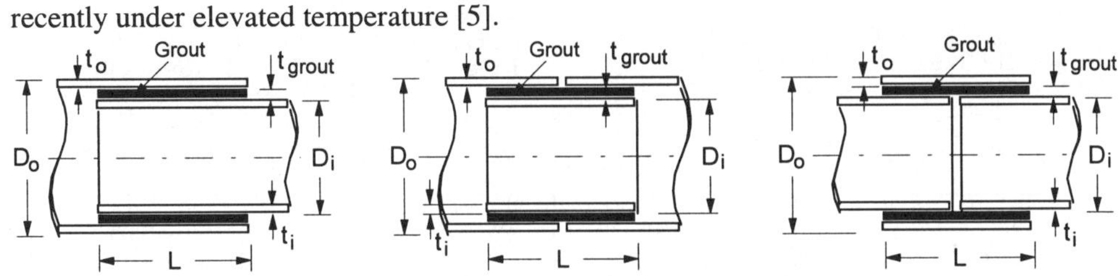

(a) half sleeve connection (b) inner sleeve connection (c) outer sleeve connection

Figure 1 Grouted sleeve connection

Expansive cement is used to fill the gap of the grouted connection, and is usually made by adding CSA expansive admixture to Normal Portland cement [6]. The factors affecting the rate and level of expansion include: (1) an increase in the ratio of expansive admixture [7,8,9,10], (2) a decrease in the C_3A content in the Portland cement [7], (3) an extension in the mixing time [11], (4) a change in the water cement ratio in the mix [11], (5) a change in curing process [9,11], (6) a change in the amount of restraint [7].

However little is known about how much and how fast the expansion in prestressed grouted connections occurs. The paper aims to quantify the expansion of the grout by testing the different proportions of CSA admixture, water cement ratio and the grout thickness in combination. Formulae for prestress are established.

2 Experimental Program

2.1 Methods

Sixteen specimens were tested. Eight specimens had 20.6 mm grout thickness while the others had 7.7 mm. For both grout thicknesses, different ratios of Denka CSA expansive agent and water cement ratios were tested. To keep the mixing time constant, only one mixing procedure was used for all the mixes. After the connection was grouted, strain readings were recorded using data loggers. There was no temperature control over the curing process so temperature compensating gauges were used. The pressure on the sleeve caused by expansion was predicted by converting the strain reading into prestress, using the equation $P = E_s \times \frac{(\varepsilon_h + v_s \times \varepsilon_l) \times t}{(1 - v_s^2) \times a}$ [7], where P is the radial prestress, t is the thickness of the steel section, a is the outer radius of the pile or the inner radius of the sleeve, v_s is the Poisson's ratio of steel, ε_h is the average of the two hoop strain reading form strain gauges and ε_l is the longitudinal strain reading form strain gauges.

2.2 Test Setup

Two different combinations of outer and inner hollow circular section (CHS 168.3 x 6.4 Grade 350 with CHS 114.3 x 6.0 Grade 350 and CHS 165.1 x 5.0 Grade 250 with CHS 139.7 x 5.0 Grade 250) were used; 20.6 mm and 7.7 mm grout thickness were used (see Figure 2). All steel surfaces were sand blasted in order to remove the mill scale on the new pipes.

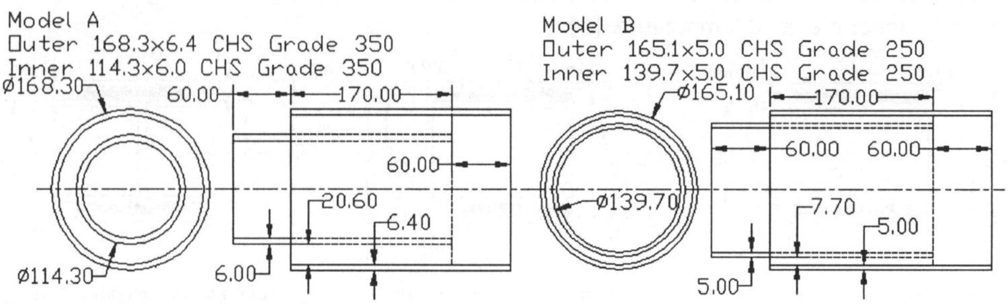

Figure 2 Detail layouts of the specimens

Three strain gauges were used to measure the strain generated on the sleeve (or outer tube) of each specimen. Two located exactly opposite each other measured the hoop strain while the third one measured the longitudinal strain.

The mixing procedure was carried out as follows. Firstly, Portland cement was placed in the bowl and blended. A given ratio of Denka CSA was added slowly. The dry material was mixed for two minutes, a given amount of water was then added at a uniform rate, with another two minutes of mixing. The mixer was then stopped, the bowl was covered and the mix was allowed to stand for 2 minutes. Finally, the mix was agitated for another 2 minutes and poured into the connection.

Different proportions of Normal Portland cement, DENKA CSA #20 expansive agent and water were used. Specimens CSA 1 to CSA 8 had grout thickness 20.6 mm and specimens CSA 9 to CSA 16 had grout thickness 7.7 mm. Both grout thicknesses had a range of CSA ratio 0.1, 0.2, 0.3, 0.4 and 0.5; water cement ratio 0.30 and 0.36.

The mix was placed by pouring it into the annular gap and a metal round bar was used to compact the grout by hand tamping. After the gap was filled silicon rubber was used to seal the exposed top end of the grout to prevent moisture loss. Two Datataker DT500 Series Data Loggers and three Datataker Channel Expansion Modules, 50 channels in all, were used to record strain readings automatically. All the specimens were placed on the table after grouting (see Figure 3).

Figure 3 Sixteen grout connection specimens

3 Test Results

The prestress generated on the sleeves was plotted on two separate graphs (see Figure 4(a), (b)).

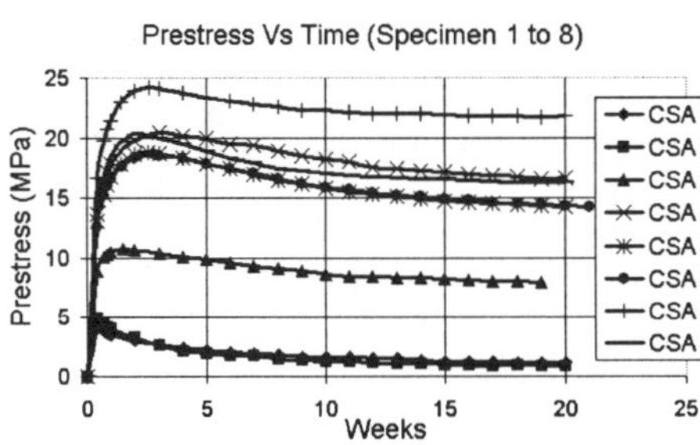

Figure 4(a) Specimens CSA 1 – 8

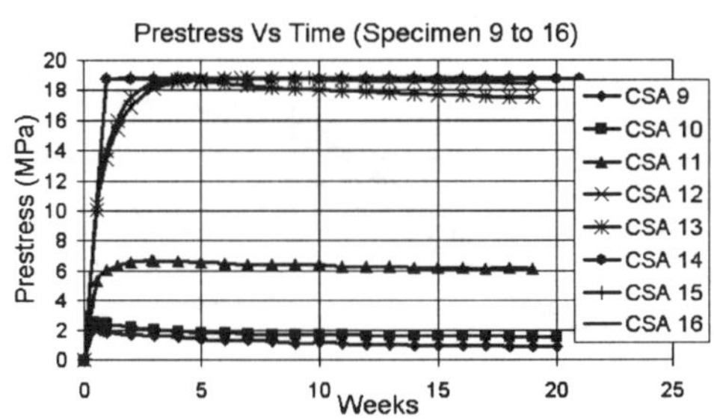

Figure 4(b) Specimens CSA 9 - 16

All the curves on the graphs had a similar shape, increasing from zero to the maximum value, and slowly reducing due to shrinkage of the grout. Prestress was limited by yield in the outer steel sleeve (see Figure 4(b), specimens CSA 14, CSA 15 and CSA 16).

Increased CSA ratio in the cement increased the expansion but not proportionally. A higher CSA ratio expanded more rapidly; when the grout expanded too fast in the initial curing state, it had insufficient initial strength and flowed in the longitudinal direction, and resulting in lower final expansion.

Increased water cement ratios in the mixture caused it to be more liquid. It took a longer time for the grout to gain initial strength; when the grout started to expand, flowed in the longitudinal direction and resulted in lower final prestress. For w/c ratio equal to 0.36, a higher prestress was generated with CSA ratio equal to 0.1, because the expansion rate was low and the expansive cement had longer to gain initial strength.

In addition, increased water cement ratio and grout thickness increased the shrinkage; shrinkage directly reduced the prestress generated (see Table 1). More reduction was observed for specimens with smaller CSA ratio. After mixing, the grout immediately started to expand and shrink. The prestress started to reduce when cement shrinkage was greater than the expansion.

Table 1 Maximum reduction in prestress over peak prestress (in percentage of peak prestress)

CSA Ratio	Grout 7.7mm W/c 0.30	Grout 7.7mm W/c 0.36	Grout 20.6 mm w/c 0.30	Grout 20.6mm w/c 0.36
0.10	59%	59%	74%	85%
0.20	9.2%	N/A	26%	N/A
0.30	2.4%	7%	20%	25%
0.40	N/A	Steel Yield	N/A	23%
0.50	Steel Yield	Steel Yield	11%	21%

4 Prestress Formula

The final prestress (measured around 20 weeks) was reached after the grout stopped shrinking (see Figure 4). The lines in Figure 5 can be separated in to two regions, from CSA ratio 0.1 to 0.3 and CSA ratio 0.3 to 0.5. For grout thickness 7.7mm the steel reached its yield stress when the CSA ratio was greater than 0.3, so these data are not included in deriving the formula.

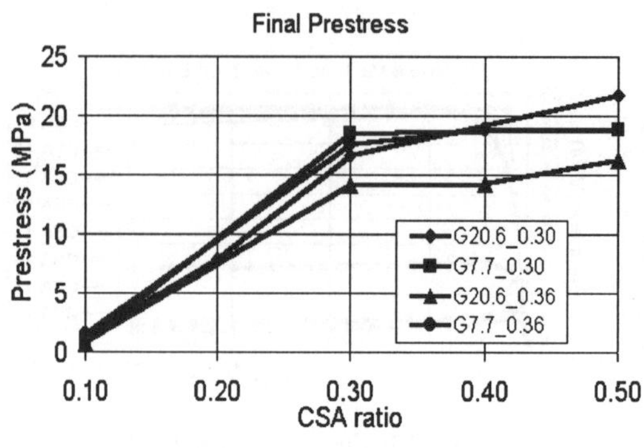

Figure 5 Final prestress

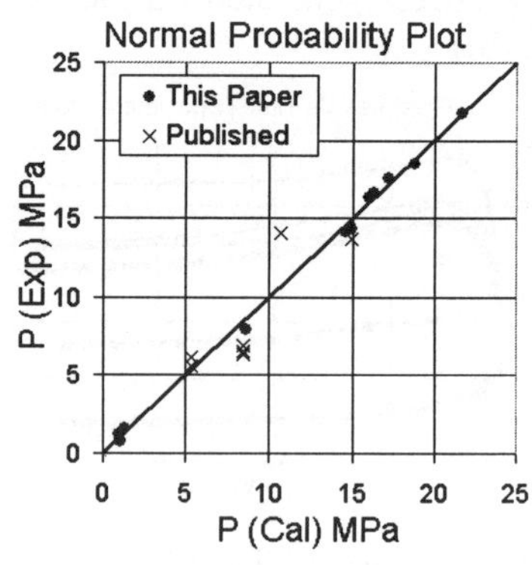

Figure 6 Comparison

The prestress formula was

Prestress (MPa) = A x CSA + B

For CSA ratio from 0.1 to 0.3,

$A = 35.1557 - 0.9088Gt - 50.0791\log(w/c)$, $B = -[3.5397 \times Gt^{-0.1465} \times (w/c)^{-0.9018}]$

For CSA ratio from 0.3 to 0.5,

$A = -251.7105(w/c) + 101.0320$, $B = 30.2867(w/c) - 0.1255$

Gt is the grout thickness in mm, w/c is the water cement ratio

From Table 2, P_{Cal}/P_{Exp} had a mean value of 1.028 and a coefficient of variation of 0.131. From Table 3, P_{Cal}/P_{Exp} had a mean value 1.075 and a coefficient of variation of 0.193. This shows that the equation fitted these and published data quite well. Figure 6 shows a comparison of predicted and measured prestress used in this paper and [6,12,13].

Specimen	Gt (mm)	W/c	CSA ratio	P_{Cal} (MPa)	P_{Exp} (MPa)	P_{Cal} / P_{Exp}
CSA 1	20.6	0.30	0.10	0.94	1.17	0.81
CSA 2	20.6	0.36	0.10	1.05	0.79	1.32
CSA 3	20.6	0.30	0.20	8.62	7.93	1.09
CSA 4	20.6	0.30	0.30	16.29	16.62	0.98
CSA 5	20.6	0.36	0.30	14.57	14.21	1.03
CSA 6	20.6	0.36	0.40	14.94	14.32	1.04
CSA 7	20.6	0.30	0.50	21.72	21.72	1.00
CSA 8	20.6	0.36	0.50	15.99	16.29	0.98
CSA 9	7.7	0.30	0.10	1.07	0.87	1.23
CSA 10	7.7	0.36	0.10	1.34	1.57	0.85
CSA 11	7.7	0.30	0.20	9.92	Instrument failure	N/A
CSA 12	7.7	0.30	0.30	18.76	18.46	1.02
CSA 13	7.7	0.36	0.30	17.20	17.53	0.98

Table 2 Comparison of the measured prestress with prestress calculated from the equation

Fitting previous data into the equation.

Specimen	Sleeve (mm)	G_t (mm)	w/c	CSA ratio	P_{Cal} (MPa)	P_{Exp} (MPa)	Days	P_{Cal} / P_{Exp}
LM5A [7]	406.4	15.90	0.36	0.20	8.44	6.30	170	1.34
LM5B [7]	406.4	15.90	0.36	0.20	8.44	6.90	170	1.22
LM5C [7]	406.4	15.90	0.36	0.20	8.44	6.50	170	1.30
A [13]	323.9	19.05	0.42	0.40	10.72	14.05	28	0.76
D [13]	323.9	19.05	0.36	0.30	14.93	13.68	28	1.09
Profile Cut [14]	165.1	7.70	0.36	0.15	5.30	6.13	42	0.87
Flattened [14]	165.1	7.70	0.36	0.15	5.30	5.63	42	0.94

Table 3 Comparison of published data with the equation

5 Conclusion

Increase of CSA content in a Portland cement mixture can increase the prestress generated, but not proportionally. Prestress increase less when CSA ratios increase from 0.3 to 0.5. A high water cement ratio combines well with a lower CSA ratio because the grout has a longer time to gain initial strength and resist the expansion in the longitudinal direction. In addition, higher water cement ratio and thicker grout increases the shrinkage of the grout, and results in lower final prestress.

The prestress formula can predict the prestress generated by the expansive cement in a grouted connection. Observations of the longitudinal expansion of the grout during setting for the higher percentages of CSA indicate that such percentages might not produce consistent and durable connections. Further work is needed to establish practical limits. Yielding of the outer sleeve affected the level of prestress achieved with higher percentages of CSA, so that the results obtained in these higher ranges are perhaps conservative and are not necessarily transferable to other connection configurations in which yielding does not occur.

6 Acknowledgements

The authors are grateful to the ARC for financial support. Thanks are given to Mr. C. Powell, Mr. G. Rundle and Mr. R. Doulis for their technical assistance.

References

1. Grundy, P., Prestressed Grouted Sleeve Tubular Joints, *Proc.,* Australasian Structural Engineering Conference, 1998, Auckland, pp 815 - 820.
2. Grundy, P. and Foo, J., Performance of Flattened Tube Connections, *Proc.,* 4th Int. Symposium on Tubular Structures, 1991, Delft, pp 251 – 258.
3. Eimanis, A. and Grundy, P., Load Capacity of Innovative Tubular X-Joints, *Proc.,* 5th Int. Symposium on Tubular Structures, 1993, Nottingham, UK, pp 485 – 494.
4. Grundy, P., Grouted Sleeve Tubular Splices, *Proc.*, Fifth Int. Society of Offshore and Polar Engineering Conference, 1995, The Hague, Vol. IV, pp 116-120.
5. Zhao, X.L., Grundy, P. and Han, L.H., Behaviour of Grouted Sleeve Connections at Elevated Temperature, 9th ISTS, April, 2001, Germany, pp 453 – 460.
6. Foo, J., Prestressed Grouted Tubular Connections, PhD Thesis. 1990, Monash Univ.
7. Polivka, M. and V. Bertero, A review of research on expansive-cement concretes Part III: Factors affecting properties of expansive-cement concrete, in The Structure of Concrete and its behaviour under load, A. E.Brooks and K. Newman, Editors. 1965, Cement and Concrete Association: London. pp 479-492.
8. Klein, A., T. Karby, and M. Polivika, Properties of an Expansive Cement for Chemical Prestressing. ACI Journal Proc. 1961. Vol. 58, pp 59-82.
9. Polivka, M., Factor Influencing Expansion of Expansive Cement Concretes. ACI, 1972 (SP-38-11).
10. Nagataki, S. and H. Gomi, Expansive admixtures (mainly ettringite). Cement and Concrete Composites, 1998(20). pp 163 -170.
11. ACI Committee 223, Standard Practice for the Use of Shrinkage-Compensating Concrete (ACI 223-90), in ACI Manual of Concrete Practice - part 1. 1993, American Concrete Institute. pp 1 – 5, 19 - 24.
12. Lim, K., Prestressed Grouted Pile/Sleeve Connections for Offshore Structures, MSc Thesis. 1992, Monash University.
13. Ure, A.J., Elastic Behaviour of Tubular Connections, PhD Thesis. 1994, Monash University.

F-111 AIRFRAME LIFE EXTENSION THROUGH REWORK SHAPE OPTIMISATION OF CRITICAL STIFFENER RUNOUT FEATURES IN THE WING PIVOT FITTING

M. MCDONALD, M. BURCHILL, M. HELLER AND K. C. WATTERS

Aeronautical and Maritime Research Laboratory,
506 Lorimer St, Fishermans Bend Vic 3207 Australia
E-mail: marcus.mcdonald@dsto.defence.gov.au

A finite-element-based *gradient-less* shape optimisation procedure has been used to develop free-form optimal rework shapes for a number of critical high stress concentrating features in the F-111 wing pivot fitting structure. A *gradient-less* method was used here, in preference to the *design sensitivity* method that is more commonly found in commercial finite element software, due to its simpler approach and compatibility with higher order *p formulation* elements. The use of higher order elements resulted in a very accurate and timely efficient stress solution as compared to standard *h formulation* elements. An added benefit of the relatively large *p formulation* elements was that significant boundary movements could be made with minimal element distortion, thus reducing the number of re-meshes required during the optimisation procedure. The resulting optimal shapes provided a predicted 30% reduction in peak elastic stresses as compared to traditional shapes consisting of circular arcs. These predictions have been confirmed by a recent full-scale static wing test. It is anticipated that these stress reductions will significantly reduce the risk of fatigue cracking, thus allowing the airframe to reach the planned withdrawal date of 2020. In addition, a lower inspection frequency is expected, which will provide large maintenance cost savings and increased aircraft availability. Fleet-wide implementation of the optimal rework program is scheduled to commence in 2001, as part of the F-111 Australian Sole Operator Program.

1 Introduction

The F-111 aircraft, the primary air strike capability of the Royal Australian Air Force, is planned to remain in operation until 2020 representing a significant extension beyond its intended design life. At present, one of the most significant potential life-limiting features of the aircraft is fatigue cracking that occurs in the wing pivot fitting at high stress-concentrator features known as stiffener runouts, in particular at locations 2U, 3U, 4U and 5U as shown in Figure 1. Current predicted crack growth rates imply a theoretical risk that the aircraft may not be able to reach the planned withdrawal date of 2020. Also, the required frequency of inspection (every 500 flying hours at runout 2U, the most severely stressed runout feature in the wing pivot fitting) imposes significant maintenance costs and reduces aircraft availability.

Shape reworking of cracked military airframe components during service life is an effective life extension practice. It is a material removal process that is traditionally used to remove existing fatigue cracks while at the same time reducing the local stress concentrations at the structural detail. Shape reworking is currently applied to the stiffener runout features in the F-111 wing pivot fitting, however the shape of the reworks are based on circular arc geometries that do not provide significant stress reductions. Therefore they do not provide an adequate life extension, or give sufficient basis to lower the inspection frequency.

Hence this paper discusses the development of precise, free-form optimal shapes to reduce the peak elastic stress at the stiffener runout features. A recently developed *gradient-less* optimisation method was used to determine the optimal profiles [1, 2]. The *gradient-less* method offers some significant advantages over the *design sensitivity* method that is commonly found in commercial finite element software including a simpler approach and it is compatible with any element type or formulation. Also, it can be implemented with any finite element code that has a programming capability.

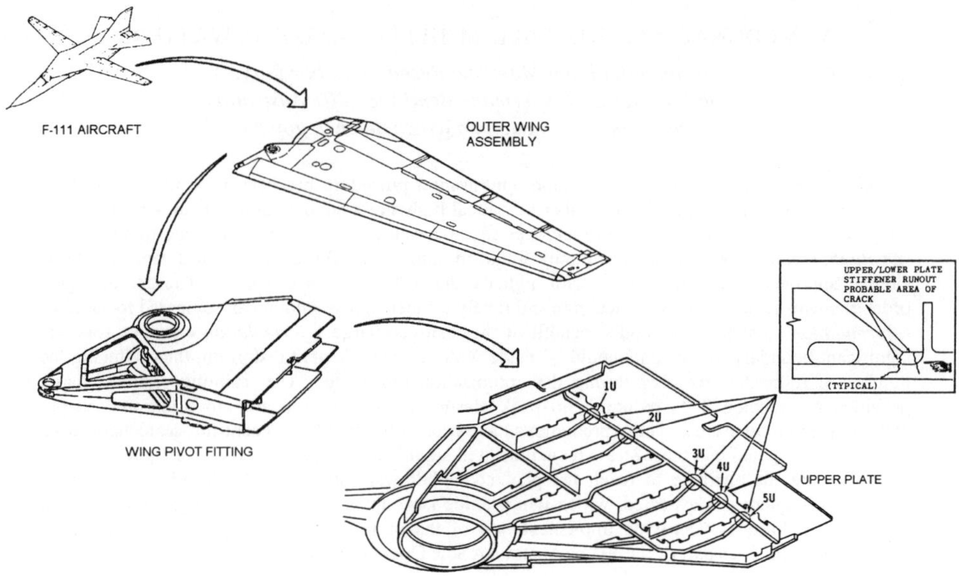

Figure 1. Location of the wing pivot fitting structure showing internal stiffener details and probable areas of cracking at the stiffener runouts.

2 The gradient-less finite element shape optimisation method

The *gradient-less* shape optimisation method involves iterative finite element solutions to alter the local geometry to achieve a constant (or near constant) boundary stress. This is consistent with the aim of minimising the peak stress. The method is based on an analogy with biological growth [3], whereby material is added to regions of high stress and removed from regions of low stress. This can in principle achieve a free-form shape that produces a constant stress on the free surface. Apart from the present approach, similar optimisation algorithms (albeit of differing effectiveness) based on biological growth analogies have been utilised experimentally by Heywood [4], and numerically by Mattheck and Burkhardt [5]. The boundary movement for any given position $\boldsymbol{i}$ on the boundary is given by

$$d_i = \left(\frac{\sigma_i - \sigma_{th}}{\sigma_{th}} \right) ls$$

where $\boldsymbol{d_i}$ is the outward normal boundary movement, σ_i is the hoop stress at position $\boldsymbol{i}$, σ_{th} is a non-zero threshold stress, $\boldsymbol{l}$ is an arbitrary characteristic length and $\boldsymbol{s}$ is an arbitrary step size scaling factor [1]. Material is added or removed at each position $\boldsymbol{i}$ depending on the chosen threshold stress. If the magnitude of the stress at position $\boldsymbol{i}$ is less than the magnitude of the threshold stress, then material is removed at this position. Otherwise, if it is greater, then material is added. The threshold stress is typically chosen equal to or less than the peak stress along the optimised region and is updated each iteration.

In this work, the gradient-less shape optimisation method was implemented with the MSC.NASTRAN finite element code, which is commonly used at the Airframes and Engines Division for 3D analyses. The use of higher order *p formulation* elements (typically of order 3 to 5) in the MSC.NASTRAN code allowed a relatively coarse mesh to be used whilst retaining the accuracy of the stress results [6]. An added benefit of the relatively large *p formulation* elements as compared to the standard *h formulation* type, was that significant boundary movements could be made with minimal element distortion. Typically though, element distortion was difficult to avoid completely, hence re-meshing was required intermittently during the optimisation. This usually resulted in a new node numbering scheme, which limited the amount of automation achieved.

3 Analysis and Results for the optimisation of the stiffener runouts

The F-111 wing pivot fitting is analysed using a detail 3D finite element model of structure as shown in Figure 2. The gradient-less procedure was implemented with this model, to develop unique optimal shapes at each runout. The initial starting shapes were developed in prior work using an automated 2D procedure [7]. For the 3D work, 5 to 20 iterations were typically required, with the product ***ls*** chosen to give the largest step size practical while maintaining convergence. Figure 3 shows a comparison of the stress contour plots from the 3D model before (i.e. with blueprint shaped runouts) and after the optimisation. The geometry and hoop stress distributions for runout 2U are shown in Figure 4, which compares the results for the blueprint, traditional circular arc rework, and the optimal rework shapes. It can be seen that the optimal reworks provide a near constant stress distribution along the optimised region which was constrained in this case to $10 < x < 55$ mm. It should be noted that the length of the optimised region could have been selected longer, resulting in even lower stresses, but this would have compromised the buckling strength of the upper plate.

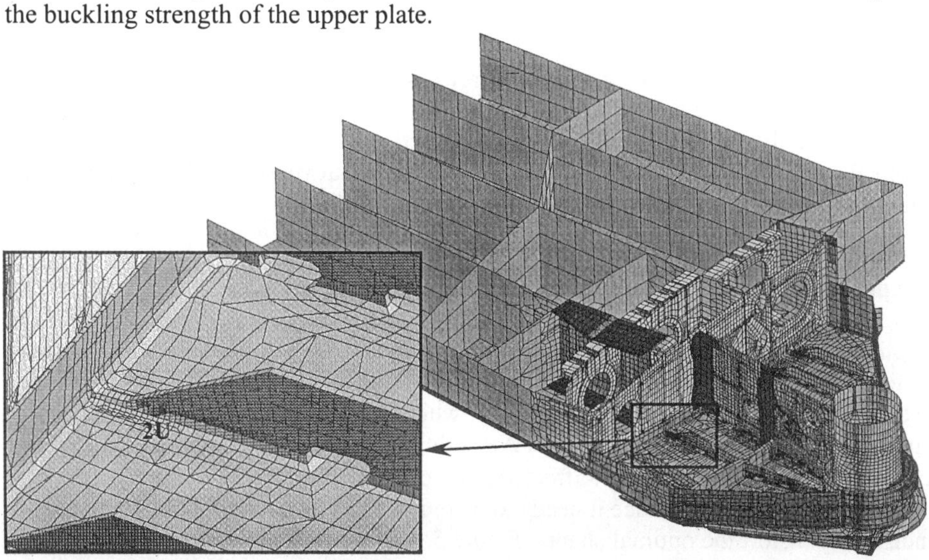

Figure 2. Section view of the 3D finite element model of the F-111 wing pivot fitting structure (shown inverted).

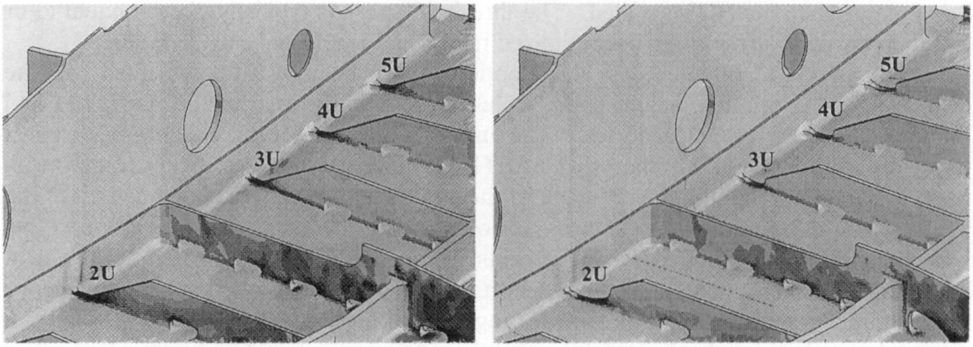

(a) (b)

Figure 3. Contour stress plots of the wing pivot fitting upper plate (shown inverted) showing the stiffener runout features with (a) original blueprint shapes, and (b) optimal rework shapes.

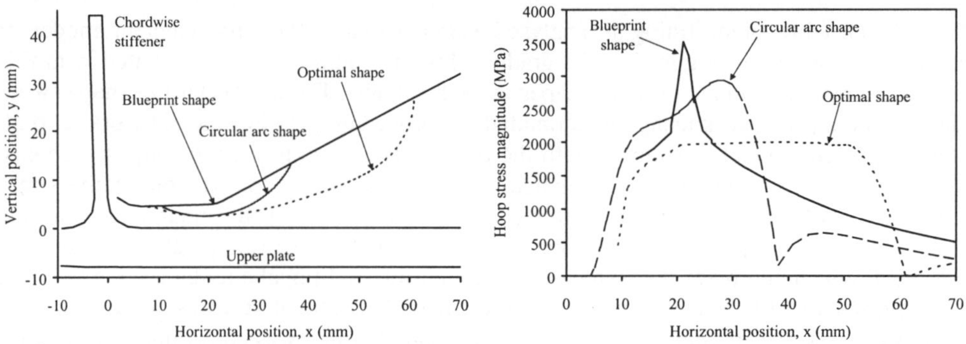

(a) (b)

Figure 4. Comparison of (a) the geometry profiles (shown inverted), and (b) scaled elastic hoop stress distributions, for runout 2U with a blueprint shape, a traditional circular arc rework shape, and the optimal rework shape.

In summary, at runout 2U the optimal rework shape provided a 43% peak elastic stress reduction compared to the blueprint shape, whereas the traditional circular arc rework shape only provided a 17% stress reduction. Similar reductions were also achieved for the other individual optimal rework shapes at runouts 3U, 4U and 5U, all resulting in final peak stresses less than that at runout 2U [7].

4 Experimental validation by full-scale wing strain survey

The optimal rework shapes were implemented at runouts 2U, 3U, 4U and 5U of a retired wing to conduct a full-scale static strain survey. An electrical discharge machining process was used to manufacture the reworks, which gave precise control over the shape and position of the optimal profiles [8]. An additional benefit of this type of machining process is that the size of the heat-affected-zone can be minimised (5 - 8 μm). Keeping this zone small is important since it needs to be removed by further polishing which tends to undesirably distort the optimal shape. Figure 5 shows an example of the electrode tool for runout 2U. Figure 6 shows runout 2U in the test wing after machining. Figure 7 shows a photo of the test wing under load.

Figure 5. Photo of the electrode tool for the final optimal shape for runout 2U.

Figure 6. Photo of runout 2U in the test wing after the electrical discharge machining process

Figure 7. Photo of the full-scale wing strain survey test under load.

The strain gauge locations along runout 2U for the different geometric profiles are shown in Figure 8(a). Figure 8(b) contains a comparison of the experimental strain results for runout 2U (the blueprint shape experimental results were obtained from a previous wing test) along with the finite element model predictions. The experimental and model results compared very well, confirming predictions of a significant reduction in strain due to the optimal shape reworking. Similar very good comparisons were also achieved at runout locations 3U, 4U and 5U [9].

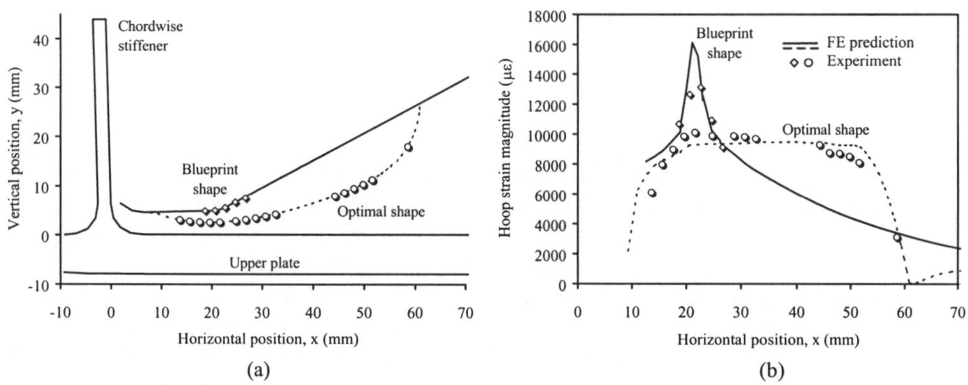

Figure 8. Comparison of (a) geometry profiles showing strain gauge locations, and (b) scaled elastic hoop strain distributions at the maximum positive wing load, for runout 2U with blueprint and optimal rework shapes.

Following the static test, crack starters were introduced at runout 2U, in the form of corrosion pits, to determine the *damage tolerance* of the wing pivot fitting with the optimal shapes. This wing will be spectrum loaded until wing failure occurs, or until enough equivalent flight hours have been applied to satisfy that cracks will not grow. The optimal shapes have also been machined into another wing for *durability* testing.

This test is scheduled to operate for up to 32000 equivalent flight hours, i.e. more than enough to cover the remaining average life of the fleet of 4000 hours. Finally, representative coupon *fatigue* testing will be undertaken to determine a new safe inspection interval for the wing pivot fitting which is expected to exceed 1000 hours, i.e. more than double the current 500 hour interval.

5 Conclusion

The peak stresses at critical fatigue locations in the F-111 wing pivot fitting have been significantly reduced by re-shaping some high stress concentrating runout features using free-form optimal profiles. The optimal shapes provide a predicted 30% reduction in peak elastic stress as compared to the traditional shapes consisting of circular arcs that have historically been used. These predictions have been confirmed by a recent full-scale static wing test, and it is anticipated that the results from further damage tolerance and durability wing tests will demonstrate that a total equivalent service life of 10000 hours can be achieved - sufficient to reach the planned withdrawal date of 2020. The stress reductions are also expected to halve the current inspection frequency for the wing pivot fitting, which will provide large fleet maintenance savings (AUD$100M) and increase the availability of F-111 aircraft. Fleet-wide implementation of the optimal rework program is scheduled to commence in 2001 as part of the F-111 Australian Sole Operator Program.

References

1. Kaye R. and Heller M., *Structural Shape Optimisation by Iterative Finite Element Solution,* DSTO-RR-0105, DSTO Aeronautical and Maritime Research Laboratory, Melbourne, Australia (1997).
2. Waldman W., Heller M. and Chen G., Optimal free-form fillet shapes for tension and bending, *International Journal of Fatigue,* **23** (2001) pp. 509-523.
3. Thompson D.W., *On growth and form,* Cambridge University Press, London (1917). [Also abridged edition. Editor Bonner J.T. (1961)]
4. Heywood R.B., Photo-Elasticity and Design Problems (Part II), *Aircraft Engineering,* **17** (1945) pp. 226–228.
5. Mattheck C. and Burkhardt S., A new method of structural shape optimisation based on biological growth, *International Journal of Fatigue,* **12**(3) (1990).
6. McDonald M., Heller M. and Wescott R.J., *Notch stress convergence studies for H and P formulation finite elements,* DSTO-TR-1151, DSTO Aeronautical and Maritime Research Laboratory, Melbourne, Australia (2001).
7. Heller M., McDonald M., Burchill M. and Watters K.C., *Shape optimisation of critical stiffener runouts in the F-111 wing pivot fitting,* DSTO-TR-1119, DSTO Aeronautical and Maritime Research Laboratory, Melbourne, Australia (2001).
8. Weller S. and Koning R., *F-111 Wing Pivot Fitting Optimisation: Summary of the Manufacturing Process,* DSTO-TN-0349, DSTO Aeronautical and Maritime Research Laboratory, Melbourne, Australia (2001).
9. McDonald M., *Validation of the F-111 wing pivot fitting optimal reworks by full-scale wing strain survey,* DSTO-TN-0392, DSTO Aeronautical and Maritime Research Laboratory, Melbourne, Australia (2001).

STRUCTURAL OPTIMISATION APPLIED TO THE DESIGN OF THE BICYCLE FRAME

OSVALDO M. QUERIN AND ROBERT NEIL BRODIE

School of Mechanical Engineering, University of Leeds, Leeds LS2 9JT, UK, E-mail: ozz@mech-eng.leeds.ac.uk

This paper looks at the design of the bicycle using the Evolutionary Structural Optimisation (ESO) method. It gives a brief overview of the history of the bicycle and a description of its anatomy. It then goes on to carefully determine equations that predict all of the riding loads experienced by the bicycle. These are applied onto a design domain and by using the ESO method various examples are produced and analysed.

1 Introduction

Over the past few centuries an apparently simple structure has been evolving. This evolution process has not been determined by nature, it has however been determined by designers, engineers and public demand. The structure is the bicycle, and more specifically the bicycle frame.

The past nine years have seen the emergence of the Evolutionary Structural Optimisation (ESO) [5,6,9,10] method as a viable practical tool for engineering design application. Using the ESO method, this research has been undertaken to determine what the shape of the bicycle frame would be if it was allowed to evolve by itself, based on structural performance criteria and with minimal human interference. Thus far the research has produced very interesting results, which have also demonstrated certain limitations in the design of current bicycles.

This paper explains in brief the design history of the bicycle. In order to successfully optimize any structure the working loads must be carefully calculated. A literature survey has been carried out to determine this. Equations that predict the loads on the pedals, handle bars and saddle at different stages of a ride will be given. Different combinations of loadings were investigated and the emerging topologies will be given and discussed.

2 Brief History of the Bicycle

The history of the bicycle is not too clear [2]. A student of Leonardo Da Vinci is widely thought to be the first person to conceive the modern bicycle. The simple sketches he had drawn in the 1490's were not discovered until 1966.

In the first half of the 19th Century, inventors produced velocipedes [2]. These were bicycles with no pedals or steering that were propelled by riders pushing with their feet while the wheels simply supported the frame. The pedals and rubber tyres were added in the 1860's, leading to the Penny-Farthing in about 1870. The later Penny-Farthings produced a major step forward being the first bike to use a gear ratio to achieve greater distance per revolution of the pedals.

The next big step in the evolution of the bicycle was the safety bicycle. Having two wheels of the same size and a chain drive with a ratio of 1:1 to the back wheel. The common diamond frame bicycle followed this in 1893. This bike had pneumatic tyres and a roller chain.

Evolution of the bicycle stopped in the first half of the 1900's, coinciding with the appearance of the automobile. Suddenly the bike looked boring and dated. However, in the 1970's, interest in the bicycle was revived, principally by the invention of changeable gear ratios, initially in-hub designs, and later the current standard derailleur multiple sprocket design.

Bicycles can now be arranged into three main groups; BMX, Racing/track and mountain bikes. Track bikes have, since 1990 been allowed to have monocoque frames. The first use of a monocoque frame at Olympic level was the 1992 Lotus Engineering bike by Chris Boardman. The Lotus bike was a groundbreaking design constructed of carbon fibre. It won the gold medal and set a New World record. The bike and the position of the rider, were the results of extensive aerodynamic research. Lotus claimed that as a result of their radical design, the turbulence behind the rider had been drastically reduced; therefore more of the riders' leg power was being used to propel the bike rather than combat the drag forces.

3 Forces Applied to the Bicycle Frame

The forces applied to the bicycle frame during cycling are very complex [1,3,4,7,8]. Forces applied at the three contact points, saddle, handlebars and bottom bracket, were investigated and related to the crank angle (θ).

3.1 Bottom Bracket Loads

$$\alpha_P = -8.44E-6\theta^3 + 0.0048\theta^2 - 0.66\theta + 39.46 \quad (1)$$

$$\alpha_X = -\alpha_{x\max} Cos(\theta) \quad (2)$$

$$H_{CRK} = CLCos(\theta) \quad (3)$$

$$L_{CRK} = CLSin(\theta) \quad (4)$$

$$F_P = \begin{bmatrix} 0 \\ F \\ 0 \end{bmatrix} \quad (5)$$

$$F_P^{'} = \begin{bmatrix} 0 \\ FCos\alpha_X \\ FSin\alpha_X \end{bmatrix} \quad (6)$$

$$F_P^{''} = \begin{bmatrix} (FCos\alpha_X)Sin\alpha_P \\ (FCos\alpha_X)Sin\alpha_P \\ FSin\alpha_X \end{bmatrix} = \begin{bmatrix} F_{PX} \\ F_{PY} \\ F_{PZ} \end{bmatrix} \quad (7)$$

Where, α_P is the angle of the pedal rotation from the horizontal axis. α_X is the angle of bicycle tilt from the vertical axis.

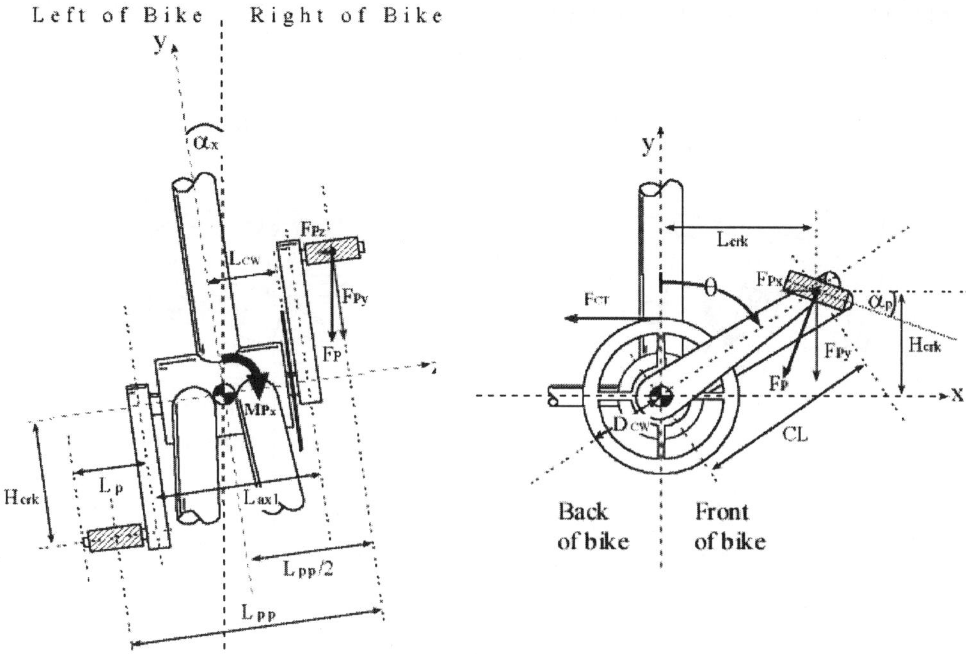

Figure 1: Bottom bracket shown in deflected position.

$$F_{CT} = \frac{L_{CRK}}{D_{CW}} \times F_{PY} \tag{8}$$

$$M_{PX} = F_{PY} \times (L_{PP}/2) - F_{PZ} \times H_{CRK} \tag{9}$$

$$M_{PY} = P_{PX} \times (L_{PP}/2) + F_{CT} \times L_{CW} - F_{PZ} \times L_{CRK} \tag{10}$$

3.2 Handlebar loads

$$F_H = \begin{bmatrix} 0 \\ F \\ 0 \end{bmatrix} \tag{11}$$

$$F_H^{'} = \begin{bmatrix} 0 \\ FCos\alpha_X \\ FSin\alpha_X \end{bmatrix} \tag{12}$$

$$F_H^{''} = \begin{bmatrix} (FCos\alpha_X)Sin45^o \\ (FCos\alpha_X)Sin45^o \\ FSin\alpha_X \end{bmatrix} = \begin{bmatrix} F_{HX} \\ F_{HY} \\ F_{HZ} \end{bmatrix} \tag{13}$$

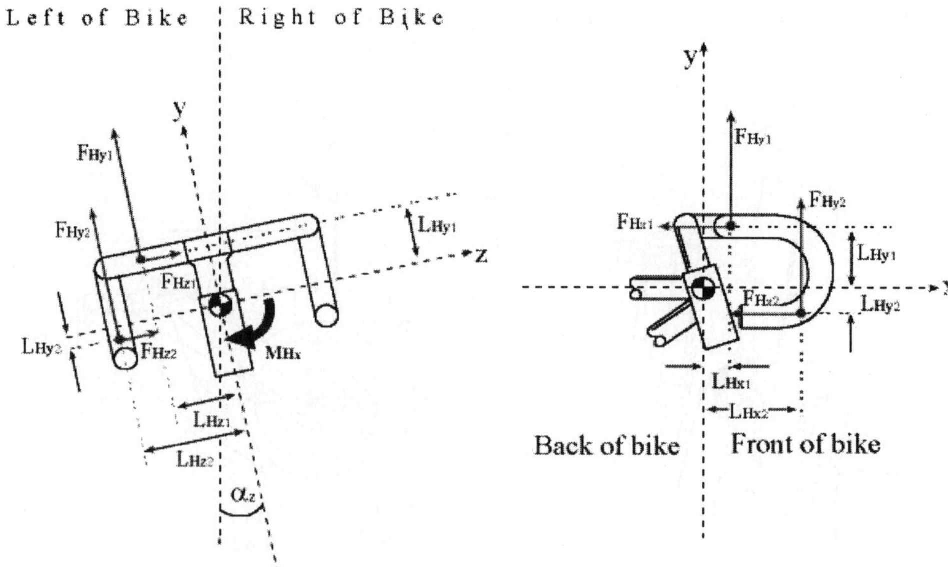

Figure 2: Handlebars in deflected state.

$$M_{HX} = F_{HY} \times L_{HZ(1,2)} + F_{HZ} L_{HY(1,2)} \tag{14}$$

$$M_{HY} = F_{HZ} \times L_{HX(1,2)} - F_{HX} \times L_{HZ(1,2)} \tag{15}$$

$$M_{HZ} = -F_{HX} \times L_{HY(1,2)} - F_{HY} \times L_{HX(1,2)} \tag{16}$$

3.3 Quantifying the loads

The maximum Normal force on the pedal, F_{Pmax}, can be estimated for each rider from the results found by Soden and Adeyefa [10] (17).

$$F_{P_{Max}} = K * Mg \tag{17}$$

Where: F_{Pmax} is the maximum normal force on the pedal in N. K is body weight multiplier (3 for accelerating / starting / climbing or 1 for speeding.). M is the rider body mass in kg. g is the gravitational acceleration in ms^{-1}.The Normal force on the pedal, F_P, is given in (18).

$$F_P = F_{P\max} \times (-8.8 \times 10^{-5} \theta^2 + 15.9 \times 10^{-3} \theta + 0.24) \tag{18}$$

Where : F_{Pmax} is the maximum normal force on the pedal in N. θ is the crank angle of the bike in degrees.

4 Evolutionary Structural Optimisation (ESO)

The method consists of removing inefficiently utilised material from a design domain in an iterative fashion. At each iteration, an FEA was carried out to evaluate the effect of the element removal process. This cycle continues until an optimum is reached. The method works as follows:

```
Define volume domain for ground structure;
Discretise volume using regular sized and shaped finite elements.
do while (optimum not reached)
        Carry out FEA;
        Calculate criteria for each finite element of ground structure;
        if (Element Criteria < Threshold Value)
                Remove these elements;
        end if,
end do,
```

Figure 3: ESO Algorithm

5 Example

The ground mesh consisted of a region discretised by 57x30 linear FE plate elements. The material properties were linear static with a Poisson's Ratio of 0.3. In this case moments about the x and y-axis are considered. The loads for this case were generated using the constants defined in table 1.

Table 1: Constants used to generate loads for case 6

α_{xmax} =	10 °	Body mass =	85kg	CL =	180mm
L_{pp} =	270mm	L_{CW} =	90mm	D_{CW} =	85mm
L_{HY1} =	115mm	L_{HX1} =	55mm	L_{HZ1} =	100mm

Table 2: Forces applied to the Handlebars

FHx	FHy	FHz	MHx	MHy	MHz
-100	95	110	19.9	-5.9	0

Table 3: Forces applied to the Bottom Bracket

FPx	FPy	FPz	MPx	MPy	MPz
-576	-1888	347	254	127	0

Table 4: Forces applied to the Saddle

FSx	FSy	FSz	MSx	MSy	MSz
0	0	0	0	0	0

The optimal frame is given in Fig. 4 and the computer-enhanced image is Fig. 5.

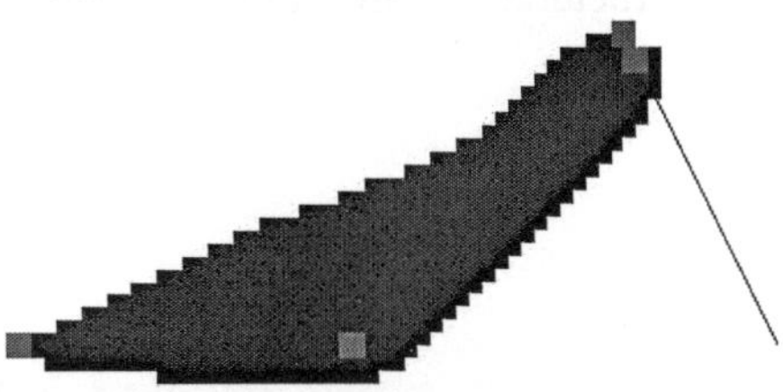

Figure 4: Selected optimal frame

Figure 5: Conception image of optimised bicycle.

6 Conclusions

In order to be ale to optimise a bicycle frame, the loads applied to it must be determined. The research presented has serviced equations to convert the riders body weight into maximum force generated on the pedals and handle, and then convert these into forces and moments acting at the contact points, Bottom Bracket, Handlebar and Saddle. These were also related to crank angle, thus allowing theoretical values for loading conditions to be estimated from riders weight and riding style.

Although not presented as part of this paper, an Excel workbook was designed and implemented that is used to generate the loads applied at the contact points. An example was shown to demonstrate the application of these loads on the final topology of the bicycle.

Work is continuing to increase the set of applied loads and on the workbook to automate the process.

References

1. Bolourchi, F. and Hull, M.L, 1985: Measurement of Rider Induced Loads During Simulated Bicycling. International Journal of Sports Biomechanics, 1, 308 -329.
2. Jones, B.J., 1999: Pedalling through the Past. http://www.aol.com/jones/Bicycle history.html
3. Kennelly, S., 1996: Structural Optimisation of the Bicycle for the Individual. Department of Aeronautical Engineering University of Sydney.
4. Neptune, R.R and Kautz,S.A and Zajac, F.E, 2000: Muscle Contributions to Specific Biomechanical Functions do not Change in Forward versus Backward Pedaling. Journal of Biomechanics, 33, 155 -164.
5. Querin, O.M. 1997: Evolutionary structural optimisation: stress based formulation and implementation. Ph.D. Thesis, University of Sydney
6. Querin, O.M.; Steven, G.P.; Xie, Y.M. 1998: Evolutionary structural optimisation (ESO) using bi-directional algorithm. Eng. Comp. 15, 1031-1048
7. Soden, P.D and Adeyefa, B.A., 1978: Forces Applied to a Bicycle during Normal Cycling. Journal of Biomechanics, 12, 527-541.
8. Stone, C. and Hull, M.L, 1993: Rider/Bicycle Interaction Loads During Standing Treadmill Cycling. Journal of Applied Biomechanics, 9, 202-218.
9. Xie, Y.M:; Steven, G.P. 1992: Shape and layout optimisation via an evolutionary procedure. Proc. Int. Conf. Comp. Engrg. (held in Hong Kong), p. 471. Hong Kong University
10. Xie, Y.M.; Steven, G.P. 1997: Evolutionary Structural Optimisation. London: Springer-Verlag.

EARTHQUAKE DESIGN OF FRAME STRUCTURES BY COMPUTER

K. J. R RASMUSSEN AND J. P. PAPANGELIS

Department of Civil Engineering, University of Sydney, Sydney, NSW 2006 Australia
E-mail: k.rasmussen@civil.usyd.edu.au

The paper first summarises the linear theory for determining displacements and internal forces developing from earthquake motion. The analysis can be a transient dynamic analysis where a ground acceleration history is used for calculating displacements and corresponding internal forces. Alternatively, the analysis can be a spectral analysis in which an equivalent external static force vector is calculated which can be used as loading in a linear static analysis. The two types of analysis constitute the "dynamic" analyses specified in the Australian earthquake standard AS1170.4. The paper describes the implementation of dynamic earthquake analyses in the general purpose program PRFSA and includes an example of a spectral analysis of a steel frame.

1 Introduction

Structural Dynamics is an advanced elective course offered to final year students in the Department of Civil Engineering at the University of Sydney. The course covers the natural and forced vibration of structures, the transient response of structures to impulsive loading including earthquake, and wind excitation.

A computer program has recently been prepared to assist the teaching of transient response and earthquake loading in the course. It has been incorporated in the Plane Rigid Frame Structural Analysis package (PRFSA) developed in the Department of Civil Engineering [1]. The new earthquake analysis module of PFRSA will be described in the paper, including the graphical visualisation of the results.

The calculation of earthquake loads for structural design is specified in AS1170.4 [2]. Depending on the structural regularity in terms of geometry, stiffness and mass, the earthquake design loads may be calculated by use of a simple equivalent static method or a dynamic method. The dynamic analysis can be a time history analysis, where site-specific earthquake records are used to determine internal actions, or a response spectrum analysis, where spectral accelerations are specified for determining an equivalent static force vector which can be applied as loading to the structure. The earthquake analysis module recently implemented in PRFSA can be used for both types of dynamic analysis.

2 Theoretical Basis

2.1 Transient Dynamic Analysis

To analyse the effect of an earthquake on a building, the governing first order dynamic equation [3],

$$\mathbf{m}\ddot{\mathbf{x}}^{t} + \mathbf{c}\dot{\mathbf{x}} + \mathbf{k}\mathbf{x} = \mathbf{0} \tag{1}$$

is combined with the relative motion equation $\mathbf{x}^{t}=\mathbf{x}^{g}+\mathbf{x}$ to produce,

$$\mathbf{m}\ddot{\mathbf{x}} + \mathbf{c}\dot{\mathbf{x}} + \mathbf{k}\mathbf{x} = -\mathbf{m}\ddot{\mathbf{x}}^{g} \tag{2}$$

where **m**, **c** and **k** are the mass, damping and stiffness matrices respectively, the superscripts *t* and *g* refer to "total" and "ground" respectively, and the ground acceleration ($\ddot{\mathbf{x}}^g$) is assumed to be known. The ground acceleration can be expressed as,

$$\ddot{\mathbf{x}}^g = \mathbf{1}_h a_h^g + \mathbf{1}_v a_v^g \tag{3}$$

where $\mathbf{1}_h$ and $\mathbf{1}_v$ are vectors describing unit translations in the horizontal and vertical directions respectively, and a_h^g and a_v^g are corresponding horizontal and vertical ground accelerations. The solution to Eqn. 2 may be obtained by modal expansion,

$$\mathbf{x} = \sum_j^N q_j(t)\,\mathbf{A}_j \tag{4}$$

where N is the number of DOF, $\mathbf{A}_j$ is the jth eigenvector and $q_j(t)$ is the time function for the jth mode. The modes of free vibration,

$$\mathbf{x}_j = \mathbf{A}_j \sin(\omega_j t) \tag{5}$$

are the solutions to the linear eigenvalue problem $\mathbf{m}\ddot{\mathbf{x}} + \mathbf{k}\mathbf{x} = \mathbf{0}$ and so satisfy,

$$(\mathbf{m} - \frac{1}{\omega_j^2}\mathbf{k})\mathbf{A}_j = \mathbf{0} \tag{6}$$

The eigenvectors are orthogonal to the mass (**m**) and stiffness (**k**) matrices, and are assumed to also be orthogonal to the damping matrix (**c**),

$$\mathbf{A}_i^T \mathbf{c}\,\mathbf{A}_j = \mathbf{0} \text{ for i} \neq \text{j} \quad \text{and} \quad \mathbf{A}_i^T \mathbf{c}\,\mathbf{A}_j \neq \mathbf{0} \text{ for i} = \text{j} \tag{7}$$

By substituting Eqns 3 and 4 into Eqn. 2, ignoring the inconsequential minus sign on the right-hand side of Eqn. 2, and utilizing the orthogonality properties of the mass, damping and stiffness matrices, such as Eqn. 7, the following *uncoupled* equation is obtained for the ith mode,

$$M_i \ddot{q}_i + C_i \dot{q}_i + K_i q_i = f_{hi} a_h^g + f_{vi} a_v^g \tag{8}$$

where M_i, C_i and K_i are the generalized mass, damping and stiffness of the ith mode, respectively,

$$M_i = \mathbf{A}_i^T \mathbf{m}\,\mathbf{A}_i \quad (9) \qquad C_i = \mathbf{A}_i^T \mathbf{c}\,\mathbf{A}_i \quad (10) \qquad K_i = \mathbf{A}_i^T \mathbf{k}\,\mathbf{A}_i \quad (11)$$

and f_{hi} and f_{vi} are participation factors for the ith mode defined as,

$$f_{hi} = \mathbf{A}_i^T \mathbf{m}\mathbf{1}_h \quad (12) \qquad f_{vi} = \mathbf{A}_i^T \mathbf{m}\mathbf{1}_v \quad (13)$$

The solution of Eqn. 8 is given by the convolution integral [3], and may be expressed as,

$$q_i(t) = \frac{f_{hi}}{M_i \omega_{Di}} V_{hi}(t) + \frac{f_{vi}}{M_i \omega_{Di}} V_{vi}(t) \tag{14}$$

where

$$V_{(h\,\mathrm{or}\,v)i}(t) = \int_0^t a^g_{(h\,\mathrm{or}\,v)}(\tau)\exp[-\eta_i\omega_i(t-\tau)]\sin[\omega_{Di}(t-\tau)]\mathrm{d}\tau \qquad (15)$$

In Eqns 14-15, ωi, ωDi, and η are the angular frequency, damped angular frequency and damping ratio for the ith mode, respectively, defined as,

$$\omega_i = \sqrt{\frac{K_i}{M_i}} \quad (16) \qquad \omega_{D_i} = \sqrt{1-\eta_i^2} \quad (17) \qquad \eta_i = \sqrt{\frac{C_i}{2M_i\omega_i}} \quad (18)$$

Once the time functions (qi(t), i=1...N) (or a truncated subset thereof) have been calculated, the relative displacement vector (x(t)) can be calculated at the time t using Eqn. 4. Subsequently, the internal forces can be found using the stiffness equation at local level.

2.2 Equivalent Static Force Vector

In designing for dynamic loading, an alternative approach to determining internal forces by a transient dynamic analysis, as described in Section 2.1, is to replace the dynamic effects by equivalent static external forces defined such that they produce the same internal forces as the dynamic loading. Recalling that the internal forces depend on the relative displacements, this definition leads to,

$$\mathbf{F}_S = \mathbf{k}\mathbf{x}_D(t) \qquad (19)$$

where the subscripts S and D refer to "static" and "dynamic" respectively. Assuming the displacement ($\mathbf{x}_D$) has been obtained from a modal analysis, as described in Section 2.1, the static equivalent force can be written as,

$$\mathbf{F}_S = \sum_i \mathbf{F}_{Si} \quad (20) \qquad \text{where} \qquad \mathbf{F}_{Si} = \omega_i^2 q_i(t)\mathbf{m}\mathbf{A}_i \quad (21)$$

To obtain Eqn 21, use has been made of Eqns 4, 6 and 19. We can substitute Eqn. 14 into Eqn. 21 to derive,

$$\mathbf{F}_{Si} = [\frac{f_{hi}\omega_i^2}{M_i\omega_{D_i}}V_{hi}(t) + \frac{f_{vi}\omega_i^2}{M_i\omega_{D_i}}V_{vi}(t)]\mathbf{m}\mathbf{A}_i \qquad (22)$$

Most structural buildings have low ratios of damping and there is no need to distinguish between the natural and damped natural frequencies, i.e. $\omega_{Di} \approx \omega_i$ thus simplifying Eqn. 22.

2.3 Spectral Analysis

We are particularly interested in the maximum values of the $V_{hi}(t)$ and $V_{vi}(t)$ functions, since these, by Eqn. 22, produce the largest statically equivalent external forces ($\mathbf{F}_{si}$) and hence the largest internal forces which must be designed for. The maximum absolute value of $V_i(t)$, which here may refer to the horizontal or vertical, is related to the spectral total acceleration (S_{ai}) as follows

$$[V_i(t)]_{\max} \approx \frac{1}{\omega_i}\mathrm{S}_{a_i}(\eta, \omega_i) \qquad (23)$$

The spectral total acceleration (S_{ai}) is specified in earthquake loading standards. For instance, the Australian standard AS1170.4 provides charts of the spectral acceleration ($s(T_i)$) as a function of the structural period $T_i=2\pi/\omega_i$ so that S_{ai} can be calculated as,

$$S_{a_i}(\omega_i) = \frac{s(T_i)gaI}{R_f} \tag{24}$$

where the acceleration coefficient (a), importance factor (I) and structural response factor (R_f) are also specified in the standard.

The statically equivalent forces to be used for design are obtained by combining Eqns 22 and 23,

$$[\mathbf{F}_{Si}]_{\max} = [\frac{f_{hi}}{M_i} S_{a_{hi}} + \frac{f_{vi}}{M_i} S_{a_{vi}}]\mathbf{m}\mathbf{A}_i \tag{25}$$

3 Implementation in PRFSA

3.1 Methods of Analysis

The transient dynamic and spectral analyses described in Sections 2.1-2.3 above have been implemented in PRFSA. They provide two distinct methods for calculating internal design forces corresponding to the "dynamic" methods of analysis detailed in Section 7.2 of AS1170.4 [2]. In the transient dynamic analysis, the ground accelerations (a_h^g, a_v^g) are input at nominated time points (t). The output consists of tables and/or diagrams of the relative displacement (x) and internal forces calculated at specified time points. In the spectral analysis, the horizontal and vertical spectral total accelerations (S_{ahi}, S_{avi}) are specified for selected periods of the structure. The output consists of the maximum equivalent static external force vector, determined using Eqn. 25, which in a subsequent analysis can used as loading in the first order elastic static module of PRFSA. The same module determines the resulting internal design actions. When several modes of vibration (i=1,2…,n) are considered, they may be combined to a resultant force vector ($\mathbf{F}_{S_{\max}}$) using the "square root of sum of squares" method (SRSS) [3].

3.2 Example

The frame structure shown in Fig. 1 has been analysed using the spectral analysis module of PRFSA. First, a natural vibration analysis is performed to obtain the vibration modes ($\mathbf{A}_i$) and corresponding periods (T_i). The first two modes are shown in Figs 2a and 2b, having periods of T_1=0.465s and T_2=0.287s.

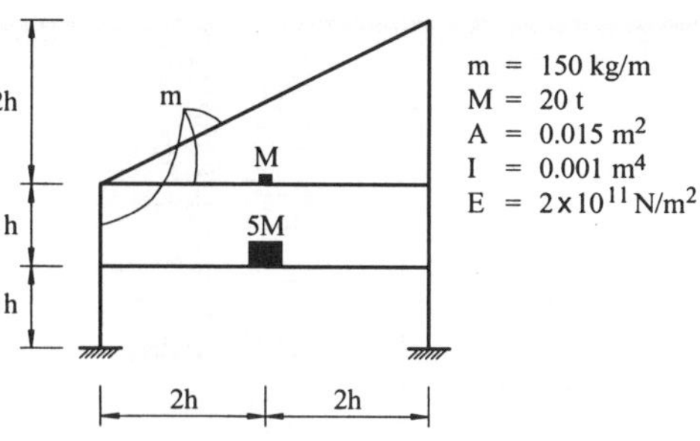

Figure 1. Frame geometry and properties.

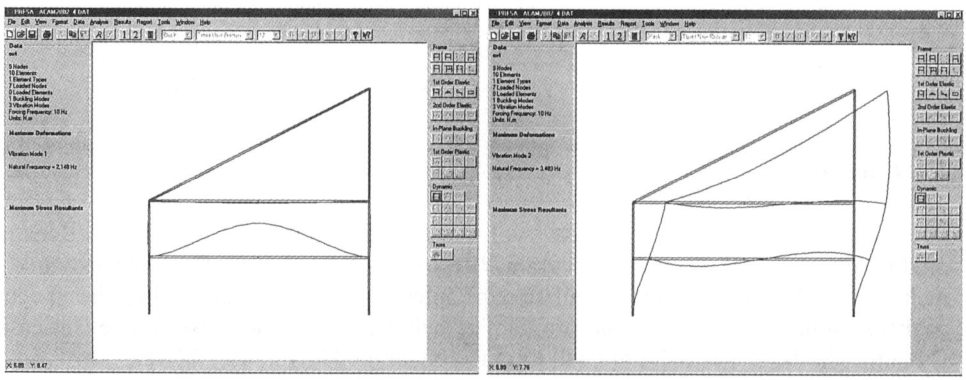

(a) (b)

Figure 2. a) First mode of vibration, b) Second mode of vibration.

Using these periods and assuming a site factor (S) of unity, the spectral acceleration can be read off the chart in Fig. 7.2 of AS1170.4 as: $s(T_1)=2$ and $s(T_2)=2.5$. Assuming typical values of $a=0.2$, $I=1$ and $R_f=4.5$, the spectral total accelerations (S_{ai}) can be obtained using Eqn. 24 as: $S_{ah1}=0.872\text{m/s}^2$, $S_{av1}=0.436\text{m/s}^2$, $S_{ah2}=1.09\text{m/s}^2$ and $S_{av2}=0.545\text{m/s}^2$, where the vertical spectral acceleration has been taken as half of the horizontal spectral acceleration in accordance with Section 6.8 of AS1170.4. The spectral analysis of PRFSA can now be executed to produce the static equivalent external force vector ($\mathbf{F}_{si}$). PRFSA has a facility to update the loading to be this force vector, and by performing a 1st order elastic analysis with this force vector, the deflections and internal design actions are achieved. The equivalent external forces and resulting deflections are shown in Fig. 3a. Figure 3b shows the internal bending moment distribution resulting from the earthquake loading. The structure is to be designed for these moments.

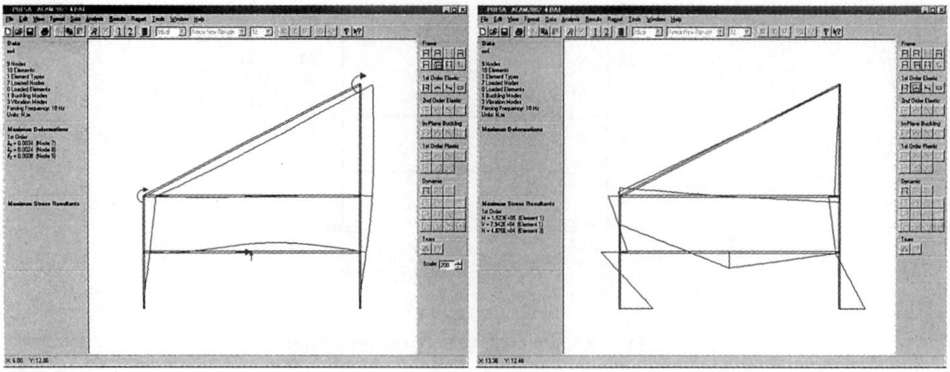

(a) (b)

Figure 3. a) Equivalent static external forces and deflections, b) Bending moment distribution.

4 Conclusions

The theoretical basis of earthquake analysis and the implementation in the dynamic module of the general purpose program PRFSA has been summarised. An example is included to demonstrate the calculation of internal design actions using the spectral analysis facility of PRFSA. The program is being used to teach the course Structural Dynamics in the Department of Civil Engineering at the University of Sydney.

References

1. PRFSA – A Computer Program for Plane Rigid Frame Structural Analysis, Department of Civil Engineering, University of Sydney, 2001.
2. AS1170 *Minimum Design Loads on Structures*, Part 4: Earthquake Loads, Standards Australia, 1993.
3. Clough R. W. and Penzien J, *Dynamics of Structures*, 2nd edition, McGraw Hill, 1993.

OPTIMUM DESIGN OF REINFORCED CONCRETE STRUCTURES USING NEURAL NETWORKS

MUHAMMAD N.S. HADI

Faculty of Engineering, University of Wollongong, NSW 2522, Australia
E-mail: mhadi@uow.edu.au

This paper presents and discusses the applications of neural networks in optimum design of concrete structures. It aims to introduce some of the neural networks applications, especially structural design. Backpropagation networks are chosen for the proposed network, which is written in the programming package MATLAB. The overall results are compared and observed for the performance of the networks. Based on the application it is found that neural networks are comparatively effective for a number of reasons, which include the amount of computer memory consumed by neural networks is less than that consumed by conventional methods and their ease of use and implementation. Neural networks provide both the users and the developers more flexibility to cope with different kinds of problems. One of the significant premises to this reason is that the modification of neural networks is much easier than the modification of most conventional methods. When many different problems are to be solved, the equations used by certain networks do not change. This results in the reduction of implementation time. Consequently, the cost of their implementation is relatively low.

1 Introduction

The design of structures is an iterative process where the designer assumes a design and then goes through the analysis process. Next the designer needs to use the design rules of the adopted design standard to design the structure. This step is followed by comparing the assumed and the calculated designs. Ideally the two designs should be the same. In reality they are seldom the same, hence the designer needs to take the new design and go through the process of analysis and design time and again. The iterations should continue until the difference between the assumed design at the beginning of the step and the output from the same step is negligible. In order to obtain economical designs quickly, optimisation techniques have been developed. They have been implemented successfully in many structural systems. These optimisation techniques are usually lengthy and complicated to implement, hence putting more burden on the designer.

One of the techniques to reduce the resources and time required for the design process is to store many optimum designs and train a neural network for the design. Thus, the neural network will come up with a design based on its training rather than conducting a full design from scratch. In this paper, the optimum design of simply supported reinforced concrete beams is introduced in simple equations form by using Lagrange multipliers. Next, these equations are used to calculate the optimum design of beams under different loadings and configurations. These optimum designs are used to train a neural network, which is established specifically for the optimum design of beams.

2 Optimum Design of Simply Supported Concrete Beams

The cost function of a beam typically includes the cost of concrete, the cost of reinforcement and the cost of formworks as illustrated in Figure 1. For simply supported beams it is sufficient to base the design only on the section at mid-span, which produces maximum bending moment, and near a support because most designs determine the maximum bending moment and maximum shear. The amount of longitudinal reinforcement governed by mid-span flexural considerations may be reduced towards the supports by curtailing some of the reinforcing bars as the bending moment envelope may allow. On the other hand, the spacing of stirrups may be adjusted to the requirements of shear force envelope. Finally, the reinforcement details will be subject to the requirement of design code AS3600 [1].

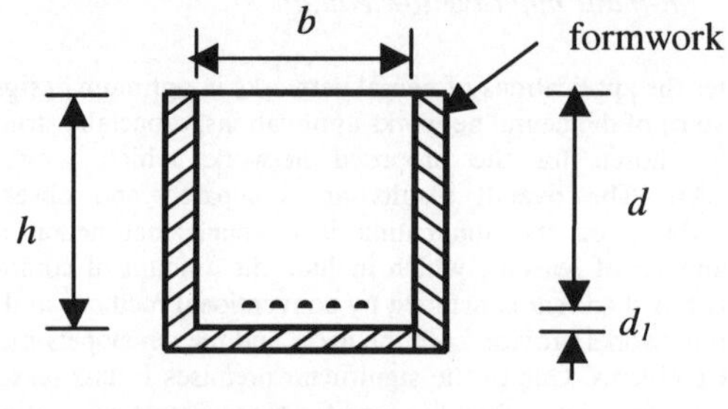

Figure 1. Beam Cross Section

In this problem, the objective function is the cost unit length of a reinforced concrete beam, which is given in the following relationship:

$$Cost = [b(d + d_1) - A_s]C_c + A_s C_s + [2(d + d_1) + b]C_f \tag{1}$$

Simplifying, we get:

$$Cost = bC_c\left(1 + 2\frac{C_f}{bC_c}\right)d + C_c\left(\frac{C_s}{C_c} - 1\right)A_s + bC_c\left(d_1 + (2d_1 + b)\frac{C_f}{bC_c}\right)$$

where C_c is the cost of concrete per unit volume; C_s is the cost of steel per unit volume; C_f is the cost of formwork per unit peripheral area; *b* is the width of the beam; *d* is the effective depth of the beam; d_1 is cover to the centroid of tensile reinforcement and A_s is the cross-sectional area of tensile reinforcement.

The constraints for the optimisation problem are the flexural strength, which can be represented as:

$$M_u \leq \phi\left[pbd^2 f_y\left((1 - 0.59p\frac{f_y}{f'_c})\right)\right] \tag{2}$$

where *p* is the steel ratio, given by $p = \frac{A_s}{bd}$; ϕ is the strength reduction factor, AS3600 [1] recommends ϕ = 0.8 for flexural, M_u is required moment capacity, f'_c is the compressive strength of concrete and f_y is the tensile strength of the reinforcement. The above equation can be rewritten as:

$$0.59\phi\frac{f_y^{\,2}}{bf'_c}A_s^{\,2} - \phi f_y A_s d + M_u \leq 0$$

The constraints on the lower and upper limits of steel ratio can be written as:

$$A_s - p_{\max} bd \leq 0 \tag{3}$$

$$p_{\min} bd - A_s \leq 0 \tag{4}$$

where $p_{\max}$ and $p_{\min}$ are respectively the maximum and minimum allowable reinforcement in a cross section. The side constraint that is incorporated with the means of limiting the depth of the beam, h to the maximum depth h_m of the beam is written as

$$d - h_m \leq 0 \tag{5}$$

The optimisation problem is to minimize the cost (equation 1) subject to constraints in the equations 2 to 5. The problem can be solved by the Lagrange multipliers technique [2]. The solution is defined in terms of optimum value of p, d and A_s. Using Lagrange multipliers technique the following expressions are obtained:

$$p_{optimum} = \frac{1 + \dfrac{2C_f}{bC_c}}{\dfrac{C_s}{C_c} + 2.356(\dfrac{2C_f f_{sy}}{cC_c f'_c}) + 1.178\dfrac{f_{sy}}{f'_c} - 1} \tag{6}$$

$$d_{optimum} = \sqrt{\frac{M_u\left(\dfrac{C_s}{C_c} + 2.356\dfrac{C_f f_{sy}}{bC_c f'_c} + 1.178\dfrac{f_{sy}}{f'_c} - 1\right)^2}{0.9 f_y b\left(\dfrac{C_s}{C_c} + 1.178\dfrac{C_f f_{sy}}{bC_c f'_c} + 0.59\dfrac{f_{sy}}{f'_c} - 1\right)\left(1 + 2\dfrac{C_f}{bC_c}\right)}} \tag{7}$$

$$A_{s(optimum)} = \sqrt{\frac{bM_u(1 + 2\dfrac{f_{sy}}{f'_c})}{0.9 f_{sy}\left(\dfrac{C_s}{C_s} + 1.178\dfrac{C_f f_{sy}}{bC_c f'_c} + 0.59\dfrac{f_{sy}}{f'_c} - 1\right)}} \tag{8}$$

The above equations are used to build a database for the different designs for the beams and then used to train the neural network.

3 Neural Network

Neural Networks are systems loosely modelled on the human brain. Like existing conventional computations, neural networks get inputs and perform some mathematical operations and produce outputs. However, unlike existing conventional computational methods, the operations in neural networks that produce outputs are defined as learning rather than executing. Neural network architectural components consist of processing elements, sometimes known as artificial neurons and connections. Each neuron is linked to certain of its neighbours with varying coefficients of connectivity that represent the strengths of these connections so-called weights. Each network has its own different architecture, however all networks share the same common features. Neurons in a particular network are divided into subgroups, and elements in each subgroup are placed in a row or column form. Each group is referred to as a layer. In any network, layers are divided into three sets: input layer, hidden layer and output layer. A network can have only one input layer and one output layer but can have more than one or none of hidden layers. A neuron in input layer collects the values from all inputs to perform particular predefined mathematical operations for producing a single value output. The outputs from preceding layers are then inputs for other neurons in a subsequent layer. Similar operations go on until the final output is produced at output layer. With training mechanisms, this process repeats until all outputs produced at layer are satisfied.

There are a number of backpropagation techniques: pure backpropagation, backpropagation with momentum and with adaptive learning, backpropagation with Levenberg-Marquardt updating rule and fast learning backpropagation [3]. All these techniques were used for solving the problem of finding the optimum design of beams.

4 Application of Neural Network for Beam Optimum Design

The developed database for the optimum design of beams, which are based on the equations above are used to train the neural network. The design input to the problem includes: applied moment, $M*$; concrete strength, f'_c; yield strength of steel reinforcement, f_y; beam width, b; maximum depth of beam, h_m; unit cost of concrete, C_c; unit cost of steel reinforcement, C_s; and unit cost of formwork, C_f. The design output includes: optimum steel ratio $p_{optimum}$; optimum area of reinforcement, $A_{s(optimum)}$; optimum effective depth of beams, $d_{optimum}$; and optimum unit cost of beams $C_{optimum}$.

Two alterative networks were considered. In the first alternative, the cost of concrete, cost of steel and cost of formwork are not part of inputs for a network. They are taken as fixed values, which are specified by the user. The network parameters are: number of training samples is 550, number of input layer neurons is 5, number of hidden layer neurons is 10, number of output layer neurons is 4, type of backpropagation is Levenberg-Marquardt Backpropagation, activation function is Sigmoidal function, learning rate is 0.01, number of epochs is 3874 and sum square error achieved is 0.0001. The network was tested with 50 samples. It has an average error of 2.71 percent. The actual outputs and the outputs from the networks of each sample are graphically represented in Figure 2.

In the second alternative, the cost of concrete, cost of steel and cost of formwork are added to the input list. Unit cost of concrete and the unit cost of steel reinforcement is $\$129/m^3$ and $\$1075/t$ [4] or $\$8730/m^3$, respectively. Unit cost of formwork is $\$68/m^2$. Because these costs are used for input, this work has adjusted the range of – 10 % and +10 % of the real cost. Network parameters used are as follows. The number of training samples is 550; number of input layer neurons is 8; number of hidden layer neurons is 10; number of output layer neurons is 4; type of backpropagation is Levenberg Marquardt Backpropagation; Activation function is Sigmoidal function; learning rate; 0.01; number of epochs is 3000; Sum Square Error achieved is 0.08. The network had been tested with 50 samples and yielded an average error of 6.1 percent. The results are presented in Figure 3.

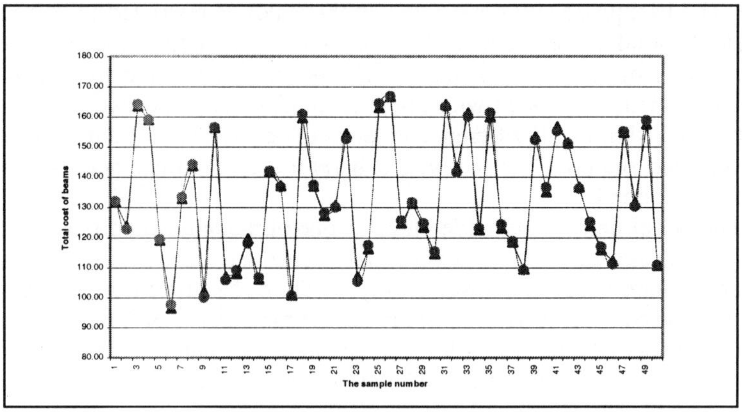

Figure 2. Total Cost of Beams Alternative 1

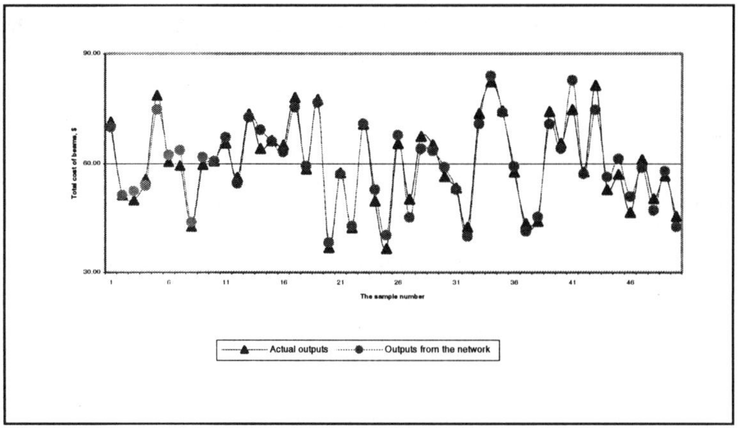

Figure 3. Total Cost of Beams Alternative 2

5 Discussion

Each of the backprogation techniques listed above were used in the current problem. Pure backpropagation is found to be extremely slow in learning. Backpropagation with momentum and with adaptive learning can be said to be the solution to a slow network. However, pure backpropagation with small learning rate shows a low possibility of oscillation. Backpropagation with momentum and with adaptive learning tends to oscillate after a large learning is used. Backpropagation with Levenberg-Marquardt updating rule and fast learning backpropagation were found to have capability to solve all the problems. The distinction of backpropagation with Levenberg-Marquardt updating rule to other algorithms is that backpropagation with Levenberg-Marquardt updating rule consumes significantly large amount of memory. In most cases, it was found that backpropagation Levenberg-Marquardt updating rule spent fewer epochs and less time to converge. The backpropagation simulator in this study is restricted to two hidden layers, which yield the total of four layers. Therefore, the observation of the effect of the number of hidden layers of networks on their performance is restricted to 2 layers. From the study, each problem was tested with both one hidden layer and two hidden layers and it did not show any significant difference in terms of accuracy but it did show the difference in terms of time required for learning. Therefore, all optimal networks in this study consist of one hidden layer.

The number of neurons of hidden layer is one of the most important considerations of selecting a network because it governs the networks' characteristics. The fewer the number of neurons used, the lower ability to learn. Meanwhile, the use of too many neurons will lead to overfitting. This study has set the number of input layer neurons to be equal to the number of input variables and the number of output layer neurons to be equal to the number of outputs parameters. Therefore the observation would focus on the effect of the number of hidden layer neurons alone. For each application the input layer neurons of 1 up to 30 were tested. According to Kasabove [5], with careful adjustment, the number of hidden layers of a typical network can be up to 32 to produce expected outputs, however some applications have as many as one thousand or so of hidden layer neurons. From this observation, there is no conclusion made from this study on the number of hidden layer neurons. However, it was found that if the difference between the number of hidden layers and the number of input layer neurons were too great, a network would not provide accurate results.

6 Conclusions

This paper presents a new form of computing, neural networks, which are increasingly used in many various areas including science and technology, education, business and military. In civil engineering fields alone, neural networks are already being extensively investigated because they show some special features to extract significant information from a massive set of data and ability to cope with ill-defined problems, which are common problems in civil engineering applications.

Beam structures are chosen for the study and the area involved are concrete beam design and cost optimisation. The beam design is to estimate the best dimensions, depth and width of its cross-section, and the cross-sectional area of reinforcement based on Australian Standard AS3600 [1]. Based on this application, it can be concluded that neural networks are easily applied for the design of concrete structures and yield accurate results.

References

1. Australian Standard for Concrete Structures. AS3600, Sydney, Australia, (1994).
2 Al-Salloum, Y.A. and Siddiqi, G.H., Cost-optimum design of reinforced concrete beams, *ACI Structural Journal* **91**(6), (1994) pp. 647-655.
3. Demuth, H. and Beale M., Neural Network Toolbox For use with MATLAB, *The Mathworks*, Inc., MA. (1995).
4. Rawlinsons , Australian Construction Handbook, Rawhouse Pub. Ltd, Perth., (1999).
5. Kasabove, N., K., Foundations of Neural Networks, Fuzzy system, and Knowledge Engineering, *A Bradford Book*, Cambridge (1996).

DISCRETE OPTIMIZATION OF TRUSS STRUCTURES WITH GEOMETRICALLY NON-LINEAR BEHAVIOUR USING GENETIC ALGORITHMS

KHALIL S. ALVANI AND MUHAMMAD N.S HADI
Faculty of Engineering
University of Wollongong, NSW 2522. Australia
E-mail: mhadi@uow.edu.au

This paper presents a formulation of Genetic Algorithm based optimization for trusses. The minimum weight of trusses with geometrically non-linear behaviour is sought. Constraints are imposed on member stresses and nodal displacements when the member cross-sectional areas are available from a discrete set of a given catalogue. The GA is based on a roulette-wheel reproduction scheme, a one-point crossover, and a standard mutation scheme. An elitist strategy is also used that passes the best designs of a generation to the next generation. GAs are well suited for unconstrained optimization problems, a transformation based on the violation of normalized constraints is used to transform the constrained problem into an unconstrained one. An incremental load approach with a Newton-Raphson type of iteration is used in the geometrically non-linear analysis procedure and in developing the optimization algorithm. Numerical results are presented that show the efficiency of the optimization procedure.

1 Introduction

There are several optimization techniques that are used to obtain optimum designs of structural system. Khot and Kamat [1] reported on the use of optimality criteria technique for optimum design of geometrically non-linear truss structures with specified constraints on displacements, stresses and system non-linear stability. Saka and Ulker [2] presented the optimal design of geometrically non-linear space truss structures with the same optimization method. Gutkowski and Zawidzka [3] used sequential and enumeration methods for discrete minimum weight of truss structures subject to member stresses, displacements, and cross-sectional area constraints.

GAs is one of the optimization techniques, which have used successfully in truss structures design problems. GAs are stochastic search techniques based on the principles of survival of the fittest and adaptation. A detailed theory and formal proofs for this optimization technique can be found in Ref. [4]. Many research works have been recently reported the solution of truss structure optimization problems via GAs [5-6]. These studies are mainly focused on the design of trusses with linear behaviour. In structural optimization there are cases, such as long-span and slender structures such as suspension bridges, where to determine the behaviour of these types of structures under the external loading non-linear analysis is needed. None of the previous work is based on non-linear analysis with GA as the discrete optimization tool.

In the present study an optimum design algorithm is developed for geometrically non-linear plane trusses composed of elements that are chosen from a given set of cross-sections. In optimization process the non-linear analysis of trusses are solved by an incremental strategy; non-linear equation are linearized and iteratively solved by the Newton-Raphson method.

2 Genetic Algorithms

GAs are stochastic search techniques based on the principles of survival of the fittest and adaptation. According to Goldberg [4] the application of genetic algorithms for adaptive systems was first proposed by J. H. Holland in 1962. GAs does not require gradient or derivative information. For this reason alone, it has been suitable to be used for hard and complex optimization problems. GA differs from traditional gradient-based optimization techniques in terms of the following: They work with a coding of the design variables rather than the variables themselves. They search from a population of possible solutions instead of a single point. GAs use stochastic reproduction instead of deterministic rules

3 Optimum design problem

The optimum design problem of a geometrically non-linear truss structure with n elements, m degrees of freedom and discrete member size can be expressed as follows:

Minimize the weight
$$W(x) = \sum_{i=1}^{n} A_i L_i \rho \tag{1}$$

subject to
$$|\sigma_i| \leq |\sigma_i^a| \tag{2}$$

$$|d_j| \leq |d_j^a|, \quad j = 1.....m \tag{3}$$

$$A_i^L \leq A_i \leq A_i^U, \quad A_i \in S \tag{4}$$

Where $\mathbf{A_i}$ and $\mathbf{L_i}$ are respectively the area and length of the **i**th member, ρ is the weight density of material and; $\boldsymbol{\sigma_i}$ is the stress in the **i**th member and $\boldsymbol{\sigma}_i^a$ is allowable stress, $\mathbf{d_j}$ is the displacement of the **j**th degree of freedom, $\mathbf{d}_j^a$ is the allowable displacement, **S** is a set of available cross-sections area, L and U are the lower and upper bound values, respectively.

Since GAs can be used directly for unconstrained optimization problems only, we transform the constrained problem into the unconstrained one by using a penalty function as suggested in Ref. [7]. The unconstrained fitness function **U(x)** is then written as:

$$U(x) = W(x) \times P \tag{5}$$

Where **P** is a penalty value, if no violation is found, then all penalty value is 1; that is no penalty is imposed on the objective function. If a constraint is violated, then the penalty is defined as:

$$P = (1 + dc \times (Cs + Cd)) \tag{6}$$

where
$$Cs = \sum_{s=1}^{n} gs \quad \text{and} \quad Cd = \sum_{d=1}^{m} gd \tag{7}$$

$$gs = |\sigma_i| / |\sigma_i^a| - 1 \quad \text{if} \quad |\sigma_i| / |\sigma_i^a| - 1 > 0 \text{ or} \tag{8}$$

$$gd = | d_j | / | d_j^a | -1 \quad \text{if} \quad | d_j | / | d_j^a | -1>0$$

where **dc** is termed as a violation parameter, and is selected depending on the required influence of a violated individual in the next generation, a value of 30 was found suitable for all the design examples presented in the study, **Cs** and **Cd** are stress and displacement violation coefficient respectively, and the number of constraints is (n+m). The minimum of the unconstrained function **U(x)** is obtained by GA. The algorithm requires a condition to perform the selection between the individuals. This is done in such a way, that the best individual has the maximum fitness. An expression for the fitness can be selected as:

$$F_i = [\, U(x)_{max} + U(x)_{min}] - U(x) \qquad (9)$$

where $\mathbf{F_i}$ is the fitness of the **i**th individual; $\mathbf{U(x)_{max}}$ and $\mathbf{U(x)_{min}}$ are the maximum and minimum values of unconstrained function of Eq. (5).

The fitness factor **Fi / FF** for each individual is calculated, where **FF** is average fitness of the population. According to the fitness factor individuals get a number of copies in the mating pool.

4 Optimum design algorithm

The algorithm begins by initialization a set of binary strings (population) and the penalty function. The population presents design variables in the initialization phase. Then geometrically non-linear analysis procedure is performed to evaluate the fitness function value for each individual and based on the fitness factor individuals are selected and placed into the mating pool. New individuals are created through crossover and mutation process.

The values of 0.8 and 0.002 are used for the probability of crossover and mutation respectively, which remain valid in all the examples. In each optimization cycle the structure is analyzed using the geometrically nonlinear procedure. This process is continued until the optimum solution (minimum total weight of truss) is obtained. The optimum design formulation presented here was writhed in Matlab and a number of examples solved

5 Design examples

To investigate the feasibility of the developed algorithm, a 10-bar truss (Fig. 1), and a 20-bar truss (Fig. 3) are solved and results are presented.

5. 1 Design of 10-bar truss

The 10-bar truss is taken from Ref. [3]. Dimensions and loads are presented in the (Fig. 1). The initial, unfeasible structure shown in (Fig. 2) is obtained by assigning the lowest cross-sectional area from the catalogue of cross-section area to all members.

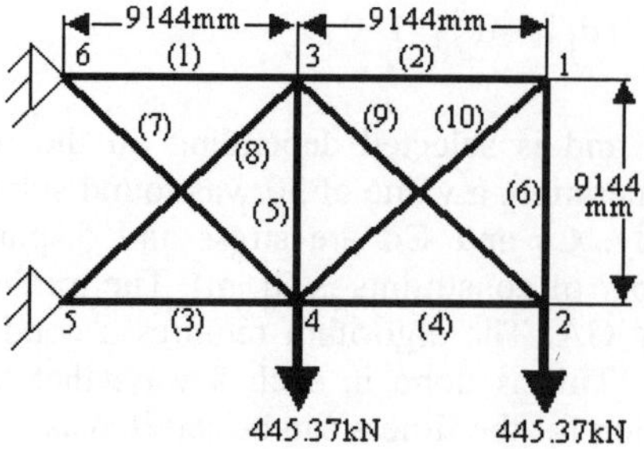

Figure 1. Geometry and loading of 10-bar truss

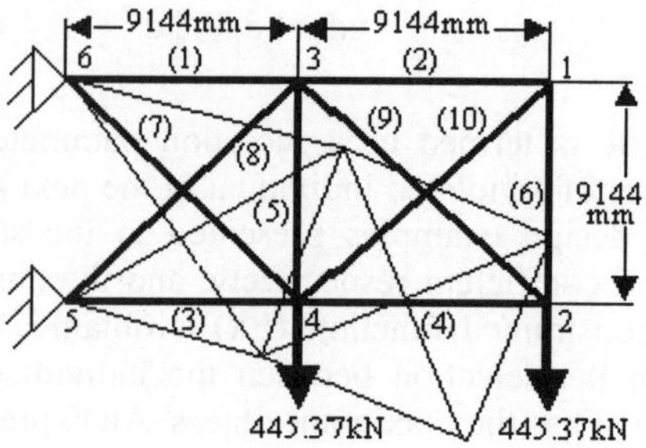

Figure 2. Deformed Shape of 10-bar truss

Table 1. The comparison table for 10-bar truss

Design Variables	Cross-section areas (mm^2)				
	Solution from Gutkowski-Zawidzka [3]			Proposed method	
	Continuous solution	Sequential algorithm	Enumeration algorithm	Continuous solution	Discrete solution
Case 1.					
A_1	19375	23226	23226	19652	17419
A_2	65	326	323	65	65
A_3	15015	17419	17419	13329	17419
A_4	9862	12258	12258	11690	12258
A_5	65	323	323	65	65
A_6	365	323	1290	65	65
A_7	4818	4516	4516	5050	4516
A_8	13676	17419	12258	13643	12258
A_9	13947	12258	12258	13458	17419
A_{10}	66	323	65	65	65
Weight (kg)	2295.9	2484.7	2429.5	2288.7	2423.2
Case 2.					
A_1	19375	19355	19355	19652	19355
A_2	65	65	65	65	65
A_3	15016	19355	16129	13329	16129
A_4	9862	9677	7742	11690	7742
A_5	65	65	65	65	65
A_6	365	65	645	65	65
A_7	4818	5161	5161	5050	5161
A_8	13676	12903	12903	13643	16129
A_9	13947	12903	16129	13458	12903
A_{10}	65	65	65	65	65
Weight (kg)	2295.9	2340.4	2339.9	2288.7	2325.2

The problem was solved for the following values of structural parameters: the modulus of elasticity is 6.89×10^4 MPa, and the material density 2.770×10^3 kg/m^3. The allowable displacement is limited to 50.8 mm and stresses to $\pm$ 172.25 MPa. The problem has 10 design variables. In the optimization process the design variables are selected from the following set of discrete values of two catalogues: the first catalog is: (65, 323, 645, 1290, 2581, 4516, 7742, 12258, 17419, 23226) (mm^2), and the second catalogue is: (65, 645, 1290, 3226, 5161, 7742, 9677, 11613, 12903, 16129, 19355, 22581, 29032) (mm^2).

The corresponding optimized truss obtained using population size of 30 and final results of the optimization procedure and a comparison optimal design in the literature are listed in Table 1. The result of optimization using the first catalogue is presented as case 1 and case 2 show the result of optimization using the second catalogue.

5. 2 Design of 20-bar truss

The geometry and nodal coordinates of this example shown in (Fig. 3). The modulus of elasticity is 200×10^3 MPa, material density is 7.85×10^3 kg/m^3, the allowable stresses are $\pm$ 350 MPa, and the allowable displacement is limited to 50 mm.

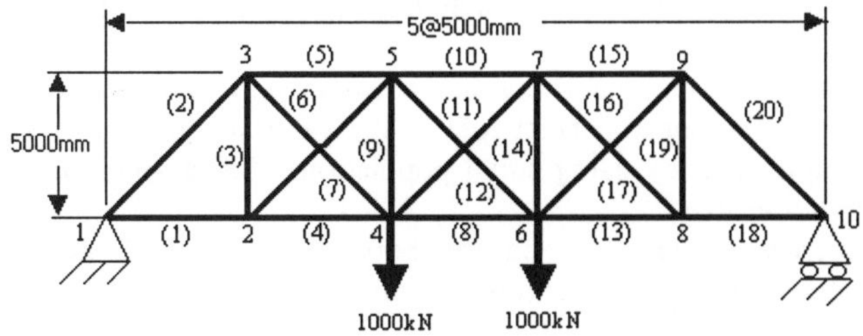

Figure 3. Geometry and loading of 20-bar truss

Table 2. Member grouping detail and result of optimization for 20-bar truss

Member grouping Group	Member	Results of 20-bar truss Cross-section area (mm^2)
A_1	1, 4, 8, 13, 18	6380.0
A_2	3, 9, 14, 19	182.0
A3	5, 10,15	9060.0
A_4	2, 20	8040.0
A_5	6	8230.0
A_6	7	325.0
A_7	11	523.0
A_8	12	108.0
A_9	16	1530.0
A_{10}	17	5360.0
Total weight (Kg)		4132.4

In the optimization process the design variables are selected from the following set of discrete values: (=(80.9, 96.6, 108, 121, 130, 153, 156, 178, 182, 199, 254, 325, 332, 414, 419, 453, 523, 533, 705, 733, 809, 862, 989, 1110, 1120, 1230, 1240, 1250, 1270, 1290, 1400, 1440, 1500, 1530, 1550, 1650, 1660, 1780, 1830, 1860, 1910, 2040, 2280, 2230, 2470, 2700, 2760, 3230, 3540, 3600, 4020, 4050, 4280, 4420, 4670, 5360, 5430, 5440, 6380, 7020, 7710, 8040, 8230, 9060, 9380) (mm^2). In total there are 65 different values. This list of discrete values is taken from Ref. [8]. To supply of the members and simplify the fabrication of the structures, it was decided to categorize the members in groups. The member grouping and optimization results are shown in Table 2.

The above final solution of a weight of 4132.4 kg has been obtained using population size of 60 and the seven iterations required for convergence.

6 Conclusions

A GA based optimization method is presented in this paper for optimal design of plane trusses with geometrically non-linear behaviour. Through several examples, we found that GA is effective and reliable to solve discrete optimization problems and can be applied to any discrete set of sections produced according to the different standards. Furthermore it is possible to obtain realistic solutions by including the geometric non-linear analysis in the design of trusses.

GA parameters such as probabilities of crossover and mutation, population size, and number of generation play an important role in the value of optimum weights. It has been proved in the examples that by using small population a better solution can be obtained in contrast large population size.

Computing increases depending on the population size and the number of design variables, large population size implies longer waiting times for convergence because in each cycle of optimization process, the non-linear analysis requires 3-15 or more iterations.

References

1. Khot N. S. and Kamat M. P., Minimum weight design of structures with geometric non- linear behaviour AIAA/ASME/ASCE/AHS, *24th Structures, Structural Dynamics and Materials Conference,* Lake Tahoe, Nevada, (1983).
2. Saka M. P. and. Ulker M, Optimum Design of Geometrically Non-linear Space Trusses, *J. of Computers & Structures* **42**, No. 3, (1992) pp. 289-299.
3. Gutkowski, W. and Zawidzka, J., A sequential algorithm for discrete minimum weight design of structures, *Second congress of structural and multidisciplinary optimization*, Zakopane Poland May 26-30, (1997).
4. Goldberg, D.E. Genetic algorithms in search, optimization and machine learning. Reading, MA: Addison-Wesley, (1989).
5. Rajeev S. and Krishnamoorthy C.S., Genetic algorithm-based methodologies for design optimization of trusses. *J. of Structural Engineering,* **123**, (1997) pp. 359-358.
6. Rajan S.D., Sizing, Shape and topology design optimization of trusses using genetic algorithm. *J. of Structural Engineering*, **121**, (1995) pp. 1480-1487.
7. Rajeev S. and Krishnamoorthy C.S., Discrete optimization of structures using genetic algorithms. *J. Struct. Engrg., ASCE*, **118**, (1992) pp. 1233-1250.
8. BHP, Structural cold formed hollow sections and profiles. BHP Structural & Pipeline Products April (1999).

MOLECULAR DYNAMICS ANALYSIS OF NANO-WEAR OF MONOCRYSTALLINE SILICON UNDER MULTI-ASPERITY SLIDING

WUN CHET DAVY CHEONG AND LIANGCHI ZHANG

School of Aerospace, Mechanical & Mechatronics Engineering, The University of Sydney, Australia

With the aid of molecular dynamics analysis, this study uses a three-asperity model to investigate the effects of the relative orientation and position of the asperities on the nano-wear mechanism of silicon. It was found that when the first asperity has created a damaged layer, the material will deform differently under subsequent sliding. A thin amorphous layer always remains and there is also an absence of dislocations when the depth of asperity penetration is small. On the other hand, the forces experienced by the asperities differ and are dependent on their relative positions. The results suggest that microstructural changes in silicon due to the initial sliding affect the subsequent deformation over a damaged region.

1 Introduction

Previous studies of a single sliding asperity using molecular dynamics simulation has provided us with important knowledge on the deformation mechanisms of friction and wear of monocrystalline materials [1-3]. Results obtained, however, are not sufficient to provide a complete understanding of the mechanism of deformation in many processes, because in a nano-system, a surface subjected to contact sliding is actually under multi-asperity interactions. When the first interaction has created a damaged zone, the material may deform differently in subsequent operations. The present study tends to take the previous research a step further by considering not only the effects of sliding over a damaged region but also the extent of the effects due to different configurations of the sliding asperities.

2 Modelling

The present mechanics model consists of three spherical diamond asperities, A, B and C, sliding on an atomically smooth silicon surface, as illustrated in Figure 1. Their relative positions and orientations are defined by their distances, L_{AB} and L_{AC}, and angles with respect to the sliding direction, α and θ. Three cases are of our particular interest. These are (I) $\alpha = \theta = 0^{\circ}$ with $L_{AB} < L_{AC}$, representing a repeated single-asperity sliding so that the effect of residual subsurface damage can be understood, (II) $\alpha = 0^{\circ}$ and $\theta = 90^{\circ}$ with $L_{AB} = L_{AC}$, standing for the interaction of two parallel asperities, and (III) $\alpha = 45^{\circ}$ and $\theta = 45^{\circ}$, indicating the case with parallel sliding asperities coupled with an interaction from a third asperity. Since diamond is much harder than silicon, the asperities are modelled as rigid spheres. These spheres slide across the silicon surface at a specified velocity 40m/s. The maximum depth of cut is 1.0nm.

The dimensions of the moving control volume of the silicon specimen are chosen respectively as 24.435nm × 9.231nm × 4.34nm containing 60,446 silicon atoms, 13.575nm × 19.791nm × 4.34nm containing 57,007 silicon atoms and 19.005nm × 14.661nm × 4.34nm containing 63,433 silicon atoms. These control volume sizes are large enough to eliminate the boundary effect. Boundary and thermostat atoms are used to cancel rigid body motion of the silicon specimen and ensure reasonable heat conduction from the moving control volume respectively. Layers of boundary atoms and thermostat atoms are arranged to surround the Newtonian atoms of silicon in the control volume except its top surface that is subjected to asperity sliding.

For covalent systems such as silicon, the directionality of bonding is important. Tersoff [4,5] proposed a simple pair-like potential where the bond order of the atoms is affected by its local environment. This replaces the two and three-body potential conventionally employed when directionality of bonding is a concern. Based on empirical data, Tersoff [5] also verified that the Tersoff potential is capable of predicting stable phases of diamond cubic silicon and body-centred tetragonal β-silicon. Hence Tersoff potential is used in the present simulation to dictate the interaction between the silicon atoms. Morse potential is employed to describe the interactions between silicon and diamond atoms [1,6]. In addition, based on careful considerations, it was found that the most appropriate equations for temperature conversion from the kinetic energy of an atom are the Debye equation for silicon and Einstein equation for diamond [7]. These equations are therefore used correspondingly in the present temperature conversion. As in previous simulations of silicon workpiece with carbon tool or asperity [1,6,8] the time step used for the molecular dynamics simulation is 2.5fs.

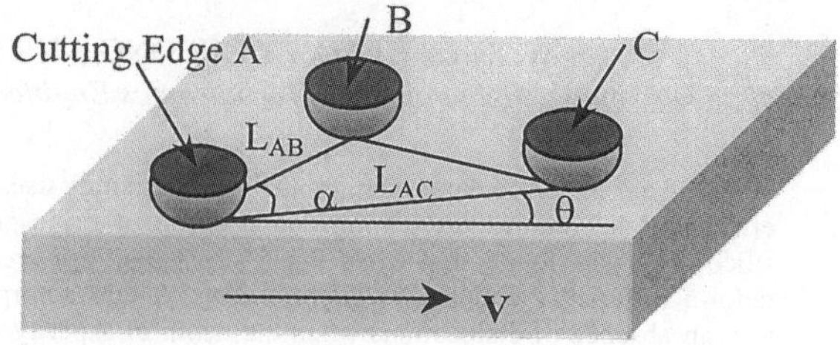

Figure 1. General mechanics model for multi-asperity cutting.

3 Results and Discussion

For each configuration of the asperities mentioned in the previous section, the following factors will be examined and discussed: (i) Mechanism of wear of silicon workpiece, (ii) Phase transformation

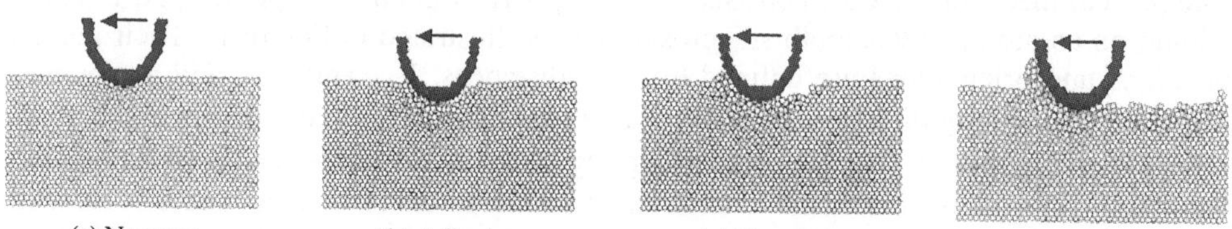

(a) No wear (b) Adhering (c) Ploughing (d) Cutting

Figure 2. General wear characteristics of diamond asperity sliding on (100) silicon surface.

3.1 Mechanism of wear

The general mechanisms of wear of silicon specimen under all the asperity configurations are found to be the same as those described by Zhang and Tananka [1,2] when only a single asperity was involved. That is, as the penetration depth of asperity increases, the mechanism of wear progresses from no-wear to adhering, ploughing and finally cutting. Deformation without wear happens at an extremely small penetration depth when the atomic lattice of silicon deforms purely elastically. By increasing the penetration depth, adhering takes place. Some surface atoms stick to the asperity surface and move together with it to cause wear. When the penetration depth is further increased, ploughing, characterised by an atomic cluster being pushed to move with the asperity, is observed. Finally cutting occurs upon further increment in the penetration depth, characterized by chip formation. Figure 2 shows these general wear characteristics obtained from present simulation.

3.1.1 Configuration II and III

In the cases of configurations II and III, the asperities do not retrace the damaged zones. At depth of cut of 1.0nm, the wear mechanism observed is that of cutting. The plastic deformation due to the sliding asperities is very localized. Figure 3(a) shows a cross-section of the silicon workpiece through the center of the asperities. It can be seen that there is almost no subsurface damage to the silicon workpiece between the two asperities. Figure 3(b) shows a cross-section of the damaged zone of the silicon workpiece behind the asperities. It is obvious that plastic defomation on the nanoscale is very localized and occurs beneath the asperities. There is little plastic deformation to the sides of the asperities and therefore little interaction among asperities A, B and C in terms the mechanism of wear.

As in nano-indentation [8], the sliding asperities A, B and C also create trails of subsurface amorphous layer in the damaged zones, see Figure 4. Phase transformation of silicon occurs at the leading edge beneath the sliding asperity resulting in the formation of amorphous chips and an amorphous trail along the path traversed by the asperities. Dislocations are absent at this particular depth of cut. This suggests that the plastic deformation is solely due to phase transformation that will be discussed later.

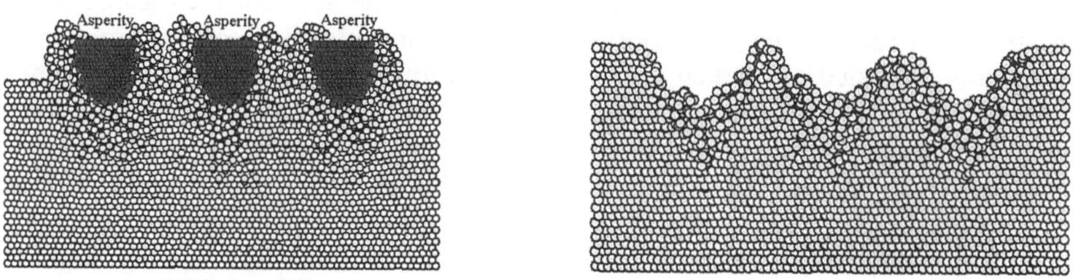

(a) during sliding (b) residual deformation

Figure 3. Cross-sectional view of plastic deformation in silicon during and after sliding loading.

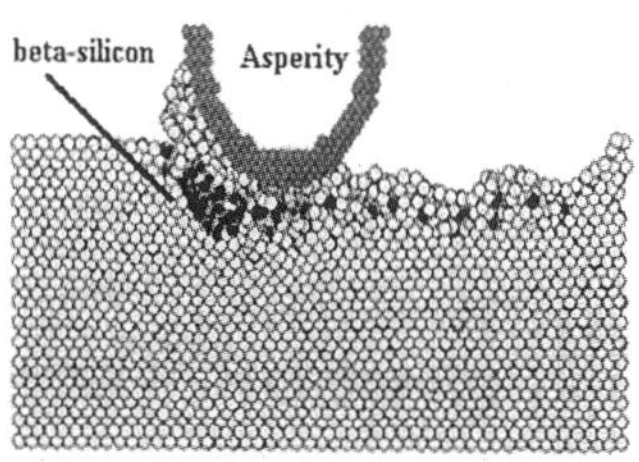

Figure 4. Cross-sectional view of a typical asperity in configuration II. The light colored circles represent atoms with 4 coordinations and the dark ones indicate β-silicon atoms with 6 coordinations.

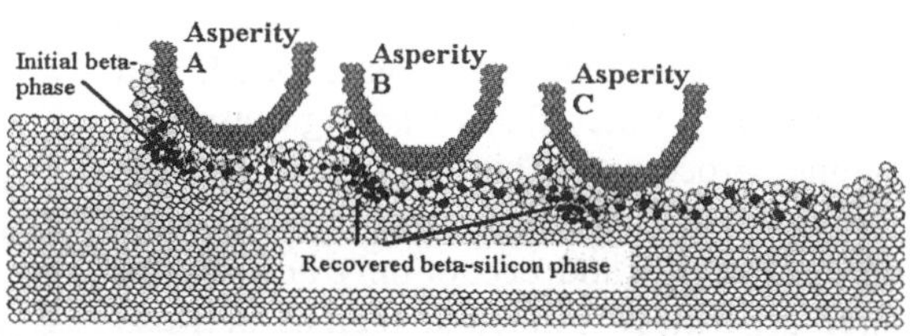

Figure 5. Cross-sectional view of asperities A, B and C in configuration I. The light colored circles represent atoms with 4 coordinations and the dark ones indicate β-silicon atoms with 6 coordinations.

3.1.2 Configuration I

In this case, the second and third asperities B and C retrace the damaged path caused by asperity A (Figure 5). Therefore, the cutting mechanism involved in the first and the following two asperities are very different. Asperity A cuts the silicon workpiece in the same fashion as the case of a single sliding asperity, causing phase transformation of the original diamond cubic silicon. Asperity B, however, ploughs through the residual amorphous layer in the wake of asperity A. Instead of the phase transformation of diamond cubic silicon, no further phase transformation occurs but the amorphous silicon atoms are pushed away as the asperity ploughs through. It should also be noted that the depth of cuts of asperities A, B and C are maintained as a basis for comparison. The effects of this differing sliding mechanism will be discussed in the following sections.

3.2 Phase Transformation

As mentioned in the previous section, something apparent in the silicon workpiece in all the three configurations is the formation of an amorphous subsurface layer in regions traversed by the diamond asperities. This is analogous to the amorphous damaged zone due to nano-indentation. Hence it is worthwhile to draw comparisons and to predict the process of phase transformation based on results obtained from nano-indentation [8].

3.2.1 Configurations II and III

Wear of silicon via cutting is achieved by chipping the amorphous silicon. By considering the exact coordinates of the silicon atoms, it is found that the transformation mechanism is similar to that of nano-indentation [8-10]. Diamond cubic silicon first transforms into its β-silicon and then upon the removal of stresses, the β-silicon transforms into an amorphous phase. This explains the trail of subsurface amorphous silicon in the damaged zone behind each sliding asperity. The mechanism of phase transformation is reflected in the coordination numbers of atoms near the regions beneath the asperities. It can be seen that a cluster of six-coordinated body-centered tetragonal β-silicon atoms indicated by darken circles (Fig. 4) forms near the leading edge beneath each asperity. Four-coordinated diamond cubic silicon transforms to its six-coordinated β-silicon form due to the stresses induced by the asperity. When the stresses are removed as the asperity slides past, the atoms transforms into an energetically more favorable amorphous form. Therefore, as the asperity slides across the silicon workpiece, diamond cubic silicon continuously transforms into β-silicon beneath the asperity and then transforms into amorphous silicon when the asperity passes, leaving a layer of subsurface amorphous silicon in its wake.

Figure 6 shows the number of β-silicon atoms formed during the cutting process with asperity configurations I, II and III. The number of β-silicon in configurations II and III are three times that of a single asperity cutting process. This indicates that the formation of β-silicon is highly stress state dependent [12] and occurs only beneath the three asperities. Although regions between the asperities are compressed due to the proximity of the asperities, it is likely that β-silicon do not form because of the absence of the required stress states and that is why there is no amorphous silicon in the zone between asperities.

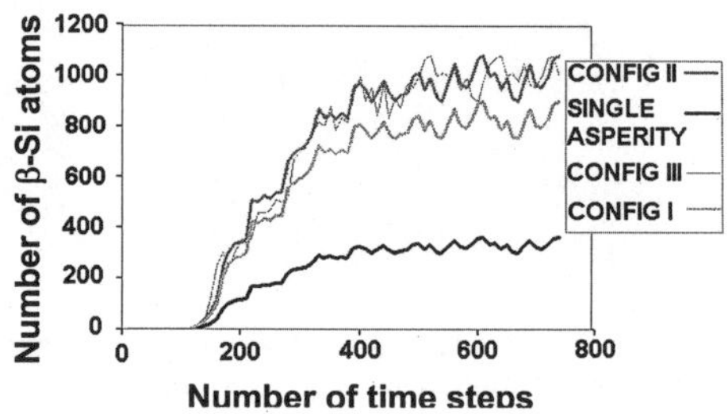

Figure 6. The number of 6-coordinated β-silicon atoms in Configurations I, II and III compared to that with a single asperity.

3.2.2 Configuration I

Asperity A in this configuration cuts the silicon workpiece in very much the same way as in the other two configurations described above. The asperity cuts through the diamond cubic silicon leaving behind a trail of subsurface amorphous silicon. However there is a vast difference in the silicon phase tranformation involved due to asperities B and C, which represents subsequent cuts into the amorphous zone. In the present case, the six-coordinated atoms also form beneath asperities B and C (dark circles in Figure 5). This implies that the β-silicon phase is recoverable from the amorphous phase provided that the required stress field is achieved. It must be noted that only some of the β-silicon is recovered as asperity B and C retraces the amorphous damaged zone. That is why the number of β-silicon atoms in Configuration I is slightly less than that of Configuration II and III (Figure 6). This agrees with the authors' previous findings on nano-indentation [8,11,12].

3.3 Cutting Force

3.3.1 Configurations II and III

In configuration II and III, there is no interaction between the cutting zones around the asperities in the direction of cut. This is because the mechanism of wear on the nanoscale is very localized as explained before. As a result, the cutting force experienced by the asperities A, B and C is similar to that of a single sliding asperity, averaged at around 150nN (Figure 7(a)).

Similarly, in configuration III, all asperities behave like that of single sliding asperity as there is no interaction between the cutting zones (Figure 7(b)).

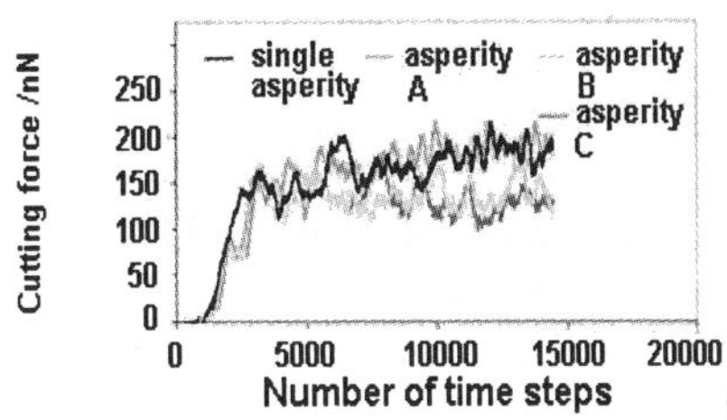

(a)

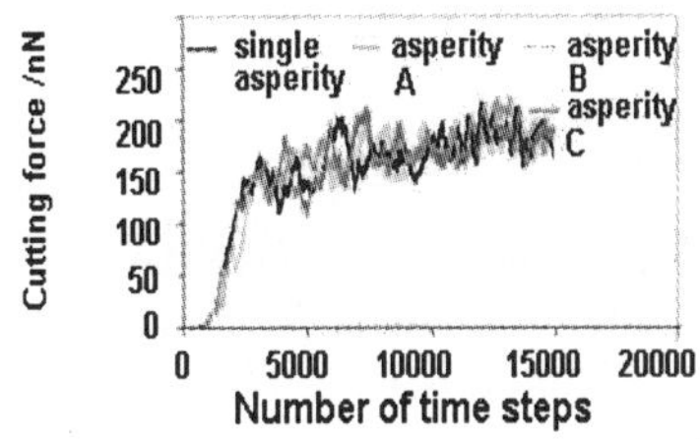

(b)

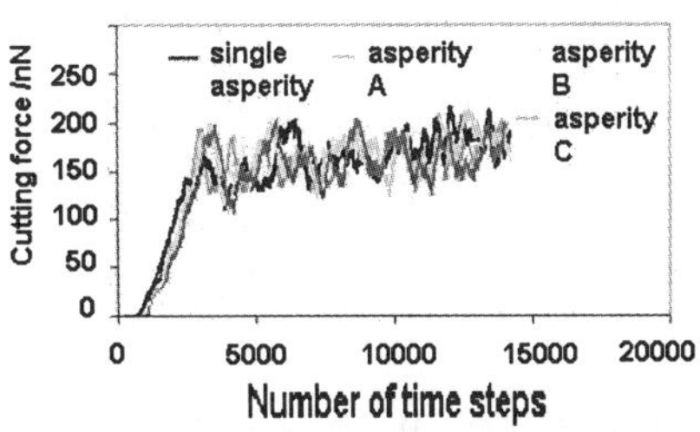

(c)

Figure 7. Variation of cutting forces on individual asperities versus cutting time:

(a) Configuration I
(b) Configuration II
(c) Configuration III

3.2.2 Configuration I

In this case, asperities B and C experience a lower cutting force around 125nN compared to that of asperity A, which is around 150nN. This shows that the asperity cutting through amorphous silicon experience less resistance compared to the cutting through diamond cubic silicon. It is interesting to note that in the case of repeated nano-indentations, residual stresses within the amorphous zone after the first indentation act to resist subsequent indentation [11]. Nevertheless, there is little difference in the cutting forces experienced by asperities B and C which represent the second and third cuts respectively (Figure 7(c)).

4 Conclusions

The above discussion leads to the following new understanding of the wear of monocrystalline silicon under multi-asperity sliding on the nanometer scale:

(a) The wear mechanism proceeds from no-wear to adhering, ploughing and finally cutting as the depth of asperity penetration increases.

(b) The effect of asperity-workpiece interaction is very localised at the asperity penetration depth of 0.6nm and hence the configurations of the asperities will not affect the mechanism of wear. The only situation where the asperities affect each other is when they retrace a damaged zone traversed by a previous asperity.

(c) Plastic deformation of silicon under multi-asperity sliding is a result of phase transformation. When an asperity cuts through, stress-induced β-silicon phase occurs, which in turn transforms to amorphous silicon upon the removal of stresses at the trailing edge of the asperity.

(d) Asperities cutting through the amorphous layer in a damaged zone experience less resistance.

Acknowledgments

The authors appreciate the continuous financial support of the Australian Research Council.

References

1. L. C. ZHANG and H. TANAKA, *Tribology International*, **31**, (1998) 425-433.
2. L. C. ZHANG and H. TANAKA, *Wear*, **211**, (1997) 44-53.
3. L.C. ZHANG, K.L. JOHNSON and W.C.D CHEONG, *Tribology Letters,* **10**, (2001) 23-28.
4. J. TERSOFF, *Phys. Rev. B* **39** (8), (1989) 5566-5568.
5. J. TERSOFF, *Phys. Rev. Lett.* **56** (6), (1986) 632-635.
6. L. C. ZHANG and H. TANAKA, *JSME International* A 42 (4), (1999) 546-559.
7. M.J. SINNOTT, *The Solid State Physics for Engineers*, (John Wiley & Sons, Inc., New York), (1958) 291.
8. W C D CHEONG and L. C. ZHANG, *Nanotechnology* **11** (3), (2000) 173-180.
9. DANIEL L. CALLAHAN and JONANATHAN C. MORRIS, *J. Mater. Res.* **7**, (1992) 1614-1617.
10. D. R. CLARKE, M. C. KROLL, P. D. KIRCHNER, R. F. COOK, and B.J. HOCKEY, *Phys. Rev. Lett.* **21**, (1988) 2156-2159.
11. W C D CHEONG and L. C. ZHANG, *Journal of Materials Science Letters* **19**, (2000) 439-442.
12. S. MINOMURA, *Localization and Metal-Insulator Transitions*, edited by H. Fritzsche and D Adler, 1985, New York:Plenum, p 63.

RHEOLOGY APPLIED TO OPEN GEAR LUBRICATION

W SCOTT[1] AND M KEENTOK[2]

[1] *Queensland University of Technology,* [2] *School of Aerospace, Mechanical and Mechatronic Engineering, University of Sydney.*

Elastohydrodynamic Lubrication (EHL) theory is frequently used to determine the most suitable lubricant for gear drives. Although it is known that the visco-elastic properties of the lubricant will affect the results of computation, a Newtonian model is chosen for simplicity. In particular, the lubricants used for open type gearing are of a sprayable, adherent type and are usually significantly non-Newtonian. Attempts to correlate the performance of open gears in the field with EHL theory based on Newtonian behaviour have been unsuccessful. This is due mainly to non-Newtonian effects. This paper reports on the rheological properties of some open gear lubricants and highlights the vast differences between the products which are supplied to perform the same task. The complexities which arise when using a grease, or grease-like, lubricant are gone into in some detail. It is concluded that, for thick film lubrication, the Maxwell relaxation time may be more important and the base oil viscosity less important than hitherto assigned.

1 Introduction

Many large machines such as draglines, crushers, ball mills and rod mills utilize gear transmissions which are of an open configuration i.e. the gears are not enclosed in a sealed gearcase. Traditionally, this type of gearing was confined to relatively low speed duties and the performance of the lubricant was not a critical factor. However, increased power, in the form of higher speed and torque, for the same material weight has resulted in a greater performance expectancy from open gear lubricants.

Opinion seems to be divided on the mode of lubrication pertaining to open gears. Many people are of the opinion that a viscous, bitumastic type lubricant is utilized to maintain a thick lubricating film between the gear teeth and not because of the adhesive qualities. Others believe that elements such as open gears, lubricated on a total loss basis must operate in the boundary lubrication mode and in which case a viscous, adhesive lubricant is superior.

Repeated attempts to evaluate open gear lubricants based on elastohydrodynamic (EHD) theory have proved to be unsatisfactory. The reasons for the lack of correlation are due mainly to the fact that almost all open gear lubricants are non-Newtonian and the simple EHD formulae do not consider the rheological properties of the lubricant. Non-Newtonian fluids may behave as an elastic solid, a liquid or a plastic solid depending on the conditions at the gear tooth contacts.

This does not mean that the application of elastohydrodynamic lubrication (EHL) is not informative, but it has to be put into context with the known (or more likely unknown) properties of the lubricant. Since it is much easier to measure the value of apparent viscosity than to determine its change under varying conditions, the former tends to be overemphasised and the latter overlooked.

It has recently become more fashionable to emphasis the base oil viscosity when specifying open gear lubricants, without recourse to the influence of pressure, temperature and shear rate on this property. This was not such a great problem in years gone by when the type of lubricant employed was based on bitumen and there was similarity between different supplier products. However, many products available today have different bases and the additive packages are more diverse. To complicate matters further, some suppliers refer to polymer thickened oil as the "base" oil.

2 Concepts: steady & oscillatory shear flow

We consider steady shear flow first. If we consider a cube of liquid and apply a shearing strain to it, we achieve a shear flow. In this shear flow, the shear stress, τ is defined as the tangential force, F, divided by the area, A. The velocity, V and the distance, d determine the velocity gradient or shear rate, γ, which is given by V/d. The viscosity, η is simply the shear stress divided by the shear rate, τ/γ. It is also possible to measure normal stresses, as well as shear stresses, in this flow and these give a measure of the elasticity of the liquid.

In steady shear flow, the shear rate is assumed constant, ignoring the effects of structural breakdown or transients. In oscillatory shear flow, the shear rate varies sinusoidally, written as $\gamma = \gamma_0 \sin \omega t$ and the shear rate has maximum and minimum values of γ_0 and $-\gamma_0$ respectively. In oscillatory shear flow, we measure the complex viscosity η^*, given by τ/γ, and this is decomposed into the storage modulus G' and the loss modulus G'', given by $\eta^* = \eta' + i\eta'' = G'/\omega + iG''/\omega$,. The loss modulus G'' is the viscous component, corresponding to the shear stress in steady shear flow, and the storage modulus G' is the elastic component, corresponding to the normal stress measured in steady shear flow.

Viscous models There are many mathematical models for viscous flow with or without a yield stress. The simplest viscous model without a yield stress is the Newtonian liquid, in which the shear stress is proportional to the viscosity at all shear rates, written as $\tau = \eta_0 \gamma$ and hence the viscosity $\eta = \eta_0$, and the viscosity shows no variation with shear rate (i.e. is a constant). This is to be distinguished from the power law model where the viscosity varies with shear rate, $\eta = k\gamma^n$ and n usually falls in the range 0 and 1, where n = 1 corresponds to the Newtonian liquid. Values greater than 1 are possible with the so-called shear thickening liquids. The simplest viscous models incorporating a yield stress are the Bingham model and the Herschel-Bulkley model.

In the Bingham model, when the shear stress is less than or equal to the yield stress, the liquid does not flow and the shear rate is zero, shown mathematically by $\gamma = 0$, for $\tau \leq \tau_y$. When the shear stress exceeds the yield stress, the Bingham liquid flows like a Newtonian liquid with a viscosity η_p, giving $\tau - \tau_y = \eta_p \gamma$ for $\tau > \tau_y$.

In the Herschel-Bulkley model, when the shear stress is less than or equal to the yield stress, the liquid does not flow and the shear rate is zero, as for the Bingham model ($\gamma = 0$, for $\tau \leq \tau_y$) When the shear stress exceeds the yield stress, the Herschel-Bulkley liquid flows like a power law liquid, giving $\tau - \tau_y = k\gamma^{n-1}$ for $\tau > \tau_y$.

Viscoelastic models Already mentioned is the existence of viscous and elastic components in the rheological properties of liquids, whether these are measured in steady shear or oscillatory shear flows. Mention is now made of the existence of viscoelastic models, some of which may be applicable for lubricating greases. This is a very large area which is out of the scope of this paper. The simplest of the viscoelastic models is the Maxwell model and the Kelvin-Meyer-Voigt model. These models represent the viscous components as idealised dashpots and the elastic components as idealised, linear springs.

The Maxwell model is mathematically equivalent to a spring and dashpot in series. When the system is compressed for a very short time, the spring becomes compressed without the piston being displaced. If the pressure is released the spring regains its original shape. In this case the system is said to be "relaxed". If the system is compressed for a relatively long time, the spring is compressed first and then the piston is slowly displaced. When the force is removed the spring regains its original shape but the piston remains in its new position. Thus the system is subjected to a certain deformation. In the case of the Maxwell fluid, the piston represents viscosity and the spring the rigidity modulus. The relaxation time is given by the ratio of the viscosity to the rigidity modulus.

The relaxation time can be evaluated by different methods. All of these methods are based on the same principles i.e. the fluid is subjected to a cyclic deformation and this is analysed as a function of frequency.

Relaxation times Measurement of relaxation times for multi-grade engine oils shows that the relaxation time increases with increasing concentration of additive.

The relaxation time varies as follows:

- for straight mineral oils between 10^{-9} and 10^{-12} seconds
- for multi-grade oils between 10^{-4} and 10^{-6} seconds
- for greases between 10^{-1} and 10^{+1} seconds

Compare these with the transit times of the lubricant in the conjunction between mating gear teeth of 10^{-4} and 10^{-6} seconds.

The Kelvin-Meyer-Voigt model has a spring in parallel to a dashpot. For rheologists dealing with computational models of viscoelastic liquids, these two models are quite crude for prediction of the behaviour of viscoelastic liquids in shear flows. It is well to bear in mind that many engineering calculations have been made using simple viscous models which neglect altogether the elasticity of grease.

3 Grease as the lubricant

A conventional grease is an oil thickened by a soap and therefore exhibits non-Newtonian behaviour. Greases are commonly used as open gear lubricants, and have been for many years. Considerable work has been carried out to determine the relative differences in EHL film thicknesses obtained with grease in comparison to oil. It would appear that the issue is complicated by the tendency of the grease to cause starvation of lubricant to the contact. Consensus is that grease will provide thicker films (by 2 ½ times according to one reference) provided that starvation does not occur. It has also been found that higher yield values give larger film thicknesses. Because of the many problems with EHL computations with grease as a lubricant it has been customary to use the base oil viscosity as the effective viscosity. This was seen as a worst case scenario. However, this cannot be done if the "base oil" is polymer thickened. It has been said that grease is an embarrassment to EHL theoreticians!

4 Measurements on a conventional Lithium Grease

Measurements were undertaken on a Reologica Stresstech, a rotational rheometer with parallel plate and cone-plate attachments in the Rheology Laboratory, Department of Mechanical and Mechatronic Engineering, University of Sydney. The sample was a conventional Lithium grease of NLGI #2 grade.

Steady shear test The steady shear test was undertaken on a Reologica StressTech rheometer, with parallel plates (roughened with garnet paper) having 25 mm diameter, with a 2.0 mm gap. The temperature was 25.0 ± 0.2 °C.

The flow curve of shear stress against shear rate under steady shear is shown in fig. 1a and 1b. The flow curve, plotted as linear shear stress against linear shear rate, is non-linear (Fig. 1a) suggestive of a Herschel-Bulkley liquid. As the curve is non linear, it is difficult to define a Bingham yield stress although it is possible to attempt to extrapolate linear portions to give yield stresses around 431 Pa and 560 Pa. The Bingham yield stress has been taken as 431 Pa (at $\gamma<1\ s^{-1}$) and the yield stress from curve fitting was 223 Pa (at $\gamma = 0.05\ s^{-1}$). When the flow curve is plotted as log shear rate against linear shear stress, it is clearer that there is a yield stress (Fig. 1b), the 'true' yield stress of 350 Pa at 0.1 s^{-1}.

Most of the yield stress values quoted here are on the low side for an NLGI #2 grease (compare NLGI#2 grease data in Keentok, 1982). The average viscosity was 19 Pa.s at a shear rate of 49 s^{-1}. The yield stress and viscosity values are relevant to the pumpability of the grease as well as the service properties.

Oscillatory shear test The oscillatory shear test was undertaken on a Reologica StressTech rheometer with parallel plates (roughened with garnet paper) having 25 mm diameter, with a 2.0 mm gap. The temperature was 25.0 ± 0.2 °C. The strain was set to be low (~0.03) so as to be as close as possible to the linear viscoelastic region.

The flow curves G' and G'' obtained with oscillatory shear are shown in fig 2. The oscillatory shear flow curves for greases show considerable variation with the formulation of the grease and the current curves have few points of similiarity with curves obtained previously for a calcium based grease (Keentok, 1997). This grease has G' > G'' over the whole range, indicating that the elastic properties are stronger than the viscous properties. This suggests that for this grease, rheological measurements of only viscous properties could be deceptive. Prater et al (1998) have reported G' > G'' for some polytetrafluoroethyelene (PTFE) thickened greases. As elasticity of oils and greases may be beneficial for load bearing capacity, this grease may have an enhanced load bearing capacity. It is possible that structural information could be derived from this rheological data, as shown by the work of Forster and Kolfenbach (1959). Following the approach of Forster and Kolfenbach (1959), the rheological curves suggest that this grease has an entangled structure with long fibres. The flow curve also shows slight evidence of an entanglement plateau on the G' curve, often seen in polymeric solutions and also greases. Using the G'' and G' curves at the limit of low frequency, one can estimate roughly that the yield stress is of order 50 Pa. It is not possible to obtain other than rough estimates of yield stress from oscillatory shear measurements (Nguyen and Boger, 1992). Further interpretation of oscillatory shear curves for greases is difficult with current levels of knowledge.

Stability tests Each of the five open gear products from the market-place were subjected to rheometer tests using the cone and plate rotary viscometer. Each product was tested to identify yield stress and stability of the structure. Procedure A consisted of ramping the frequency of oscillation first up, then down, to form a thixotropic loop and recording the stresses.

Procedure B, repeated Procedure A but after the product had been subjected to the maximum oscillation frequency for a period of five minutes. Test temperature was 60° C. The results are given in Table 1. The apparent viscosities for Sample 1 at a shear rate of 1 s^{-1} were outside the range of the instrument.

5 Lubricant properties under pressure

The measurement of viscosity under pressure has a century-long history. Among the viscometer types that have been pressurised are the capillary, Couette, rolling ball, vane, acoustic and falling-body. For high-pressure measurement of the limiting low shear viscosity, the falling-body viscometer is particularly convenient.

Table 1: Results of rheometer test

SAMPLE	Yield stress (Pa)			Viscosity (Pa.s) at 1 s^{-1}		Viscosity (Pa.s) at 100 s^{-1}	
	Proc.A	Proc.B	% change	Proc.A	Proc.B	Proc.A	Proc.B
1	1080	480	-56	large	large	20	12
2	483	400	-17	500	350	10	8
3	115	86	-25	100	80	8	4
4	47	82	74	50	70	12	14
5	436	210	-52	190	150	6	4

Although the means for direct measurement of viscosity to high pressures has been available throughout the study of concentrated contact lubrication, tribologists have recently used the contact itself for indirect measurement of pressure-viscosity behaviour. Although many workers using the contact as a viscometer claim broad agreement with direct measurements and overlapping data exist, there has been little, if any, overlapping data presented. Bair (2000) has shown that this traction-based method of measurement understates the pressure-viscosity dependence at high pressures (> 0.8 GPa).

Ohno and Hirana (2001) have conducted high-pressure rheology analysis of traction oils based on free volume measurements. They have proposed a density-pressure-temperature relationship on the basis of the free volume equation of Doolittle (1951). These relationships allow the prediction of viscosity from density measurements. The trick now is to see if these methods can be applied to greases and grease-like substances.

Limiting shear stress For Newtonian fluids, the shear stress developed in the fluid is directly proportional to the shear strain rate. If the shear rate becomes infinitely high, the shear stress also becomes infinitely high. For liquid lubricants the behaviour may be approximated by a three component model. Initially, at low shear rates, the behaviour is Newtonian but at higher shear rates a maximum shear stress is developed which can not be exceeded however much the shear rate is increased. A slip plane develops in the material instead. The change from the proportional to the limiting shear stress is progressive and thus the three distinctly different flow properties are viscous, visco-elastic and plastic. Transition from one state to another is temperature, pressure and shear rate dependent and is different for different materials.

6 DISCUSSION

It can be seen from Figs 1a and 1b that the lithium base grease appeared to follow a Herschel-Bulkley model when recorded on a linear scaled graph but that a clearer picture of the presence of a Bingham yield stress is given when recorded on a log-linear graph. This demonstrates the difficulty of characterising grease samples and emphasises the need for the rheologist to show expertise in the choice and control of the rheometer.

Fig. 2 shows that the elastic contribution G' to complex viscosity is greater than the viscous component G'' especially at the higher frequencies (and hence shear rates). These are low compared with those found in the conjunction between mating gear teeth where shear rates are in the region of $10^5 - 10^6$ s^{-1}. It would appear, therefore, that the visco-elastic behaviour is the major contributor to the formation of separating films between gear teeth, certainly where a significantly non-Newtonian lubricant (such as grease) is utilised.

Table 1 illustrates the vast differences in rheological properties of open gear lubricants from several suppliers. All of these are being used in the field (presumably successfully) by different machine operators. Either the lubricant is not critical to the performance of the gears or the influencing properties are not being measured.

7 CONCLUSIONS

The lack of correlation between the results of elastohydrodynamic lubrication (EHL) computations and the performance of grease or grease-like lubricants for open gear drives suggests that there are additional physical properties of the lubricant, other than those considered by the theory, which are influential in the lubrication mechanism.

Rheological properties such as yield stress and Maxwell relaxation time would appear to be important. However, the measurement of these properties, especially under realistic conditions of pressure and shear rate, are extremely difficult.

The 'high-viscosity-is-best' syndrome may have been taken far and impeded true progress in arriving at optimum lubrication. Certainly, problems with pumping and starvation at the contacts are more acute with high viscosity lubricants.

We are pursuing means to identify and measure these other influential properties of the lubricant. The flow imposed on the lubricant in gear drives is a complex one and since there may be a compressive/tensile component it would be profitable to measure viscosity under extensional flow, as well as under shear flow. Thus it is suggested in further work that extensional viscosity should be measured, in addition to a full characterisation under steady and oscillatory shear flow.

8 References

1. Bair, S (1994), The High Pressure Rheology of a Soap-Thickened Grease, Tribology Transactions, v.37, pp 646-650.
2. Doolittle, A.K (1951), Studies in Newtonian flow II. The dependence of the viscosity of liquids on free space, Journal of Applied Physics, v 22, pp 1471-1475.
3. Forster, E.O, Kolfenbach, J.J (1959), The viscoelastic behaviour of greases, ASLE Transactions, v. 2, pp 13-24.
4. Keentok, M, (1982), The measurement of the yield stress of liquids, Rheologica Acta, v. 21, pp 325-332.
5. Keentok, M (1997), Edge fracture in rheometry, Ph.D Thesis, University of Sydney, Sydney.
6. Nguyen and Boger (1992), Measuring the flow properties of yield stress fluids, Annual Review of Fluid Mechanics, v 24, pp 47-88.
7. Ohno, N, Hirano, F (2001), High Pressure rheology analysis of traction oils based on free volume measurements, Journal of STLE, Lubrication Engineering, v 57, pp 16-22.
8. Prater, W, Burns, J, Stone, G and Ting, T (1998), Evaluation of Perfluorinated polyalkylether grease rheology on disk drive bearing performance, Journal of Tribology, v. 120, pp 21-27.

EVALUATION OF LOW SULPHUR DIESEL FUEL WITH LUBRICITY ADDITIVE USING A HIGH-FREQUENCY RECIPROCATING RIG

R. PAI AND D.J. HARGREAVES

Tribology Research Concentration, Queensland University of Technology, Brisbane 4001
E-mail: d.hargreaves@qut.edu.au

Environmental concerns have resulted in a need for cleaner fuels. Current requirement limits the sulphur content of diesel to 500 parts per million. It is expected that the limit will be set at a low 50 parts per million by 2006, although it is expected that the industry will move to the 'ultra low sulphur' diesel before then. Diesel with low sulphur levels produces less black smoke emissions and less of the smaller particulate matter, which has adverse health effects. Removing sulphur from diesel results in the reduction of some aromatics in diesel during the distillation. This has an adverse effect on the lubricating properties of diesel fuel. The shortened life of engine components such as the fuel injection pump and injectors has been attributed to the lack of lubricity of diesel fuel. Experiments are conducted using a High-Frequency reciprocating rig built as per ASTM D-6079-99 [1] on 5 samples of diesel with sulphur content in the range of 245 to 480 parts per million. Experiments are also conducted when a readily biodegradable lubricity additive is used in each sample. The wear scar area and the coefficient of friction are measured in each case. The effect of low sulphur levels and additive on wear and friction coefficient are studied and compared. Typical results obtained are presented.

1 Introduction

Performance studies on Tribosystems are generally done to evaluate their friction and wear behaviour. Friction and wear are not intrinsic material properties, but are related to the system of interacting components. The need to simulate various interactions in Engineering systems has led to the development of a number of test configurations and rigs. The development of a successful accelerated wear test rig culminates in creating the same essential parameters in the model that gives useful indicators to the system performance required. An accelerated process gives fast and easily measurable results. The discrepancies created by the modelling assumptions are ultimately shown when materials or lubricants are put to use in actual practice or where correlation of test methods is warranted. The purpose of accelerated , laboratory scale friction and wear tests is to assess performance under controlled conditions that can be easily evaluated. In many instances, tribotesting is used to rank tribological parameters such as materials and lubricants within the system rather than fully simulate a particular situation.

Sliding, reciprocating contact is very useful with respect to the wear modes that can be simulated and the amount of information that can be extrapolated from the measured variables. Alteration of stroke length, affects the wear mode in the system. For adhesive/abrasive wear, the motion should be such that although the sliding motion repeatedly traverses the same track, one of the contact surfaces is exposed, most of the time. Klaffke [3] studied the influence of some test parameters on the results of oscillating sliding tests

In this paper, the results of a series of tests performed using the high-frequency reciprocating rig to evaluate low sulphur diesel with and without a lubricity additive, are presented. Coefficient of friction in each case is measured using a force transducer. The wear scars are studied under an optical microscope to measure the wear scar diameter. A comparative assessment of the test results reveals that the lubricity additive is essential for improving the lubricity of low sulphur diesel.

2 Description of Test Rig

The test rig (Fig.1) [4,5] comprises of - the loading system, the specimen support and Instrumentation.

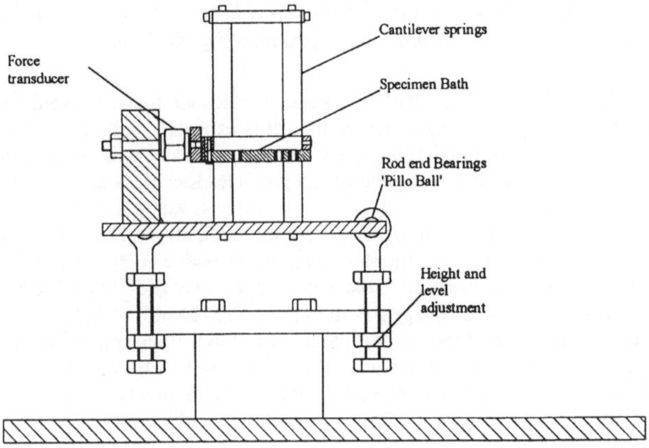

Figure 1. Bath / Table assembly

2.1 Loading system

A pneumatic chamber applies the required force to one end of the torque arm, the other end of which carries the moving specimen. A threaded section at this end allows a variety of clamping devices to be used, to hold various specimen types. The load system consists of a pressure chamber made of rubber hose clamped at both ends. The rubber hose elongates under pressure and applies a normal load on the torque arm. The hose ends are clamped using Loctite RTV Silicone sealant and plastic straps, to provide optimal clamping and no leakage. This results in a low-stiffness, leak proof and frictionless pneumatic system that is capable of providing a constant normal load over the entire reciprocating cycle, which is important in stick-slip studies. To calibrate the pressure chamber, a strain gauge is used on the torque arm. Using the strain gauge readings and the calibration chart for the strain gauge a known normal load can be applied on the torque arm. The reciprocating motion is obtained with the help of an electromagnetic oscillator, which is attached to the torque arm.

2.2 The specimen support

The purpose of the specimen support is to hold the stationary specimen in place under the applied normal load. The support used must provide minimum resistance in the tangential direction to provide the greatest accuracy whilst measuring the friction force. This is achieved by the use of a cantilever spring arrangement. A bath is provided for the specimen, so that provision can be made for lubrication and heating. The specimen support, in turn, is mounted on fine pitch threaded fasteners for precise height and level adjustment in all directions for alignment

2.3 Instrumentation

An electromagnetic oscillator controlled by a Playmaster Proseries 3 Power Amplifier and BW 141 Function generator provides the tangential drive for the load system. An accelerometer (B & K Type 4333) attached to the connector plate accurately measures the sinusoidal motion of the torque arm. By looking at the accelerometer readings, the motion can be kept constant and at the desired frequency. A force transducer Type PCB-208C01 Piezoelectric load cell, attached to the bath assembly measures the friction force. The output of the force transducer and accelerometer is observed on a Tektronix TDS 210 Oscilloscope.

3 Test Procedure

The development of the test procedure is based on the guidelines given in [1]. The ball and flat specimen shall be in a condition (composition and surface finish) consistent with the required test procedure. Clean the specimen as described in [1]. Load arm must be dismounted to get the ball specimen in the holder. Care must be taken so as not to damage the ball holder. Screw the flat specimen into its fixture and put the load arm back into its place. Do not let the ball specimen hit the flat specimen during mounting. The height of the specimen or mount may require adjustment to ensure that the reciprocating load fixture is horizontal and parallel to the surface of the flat specimen.

Charge the pneumatic system and apply the desired test load by noting the strain gauge reading. Take the load off and balance the strain gauge readings to zero levels. Repeat this again to ensure that the strain gauge is balanced and the friction in the load system will not affect the results. Do not proceed until force transducer readings have decayed to zero levels. Add lubricant to the bath until the surface of the flat specimen is covered (a small quantity is applied at the contact if it is grease). When tests are conducted at elevated temperatures, the lubricant and specimen are heated to the selected temperature and allowed to reach equilibrium conditions for not less than five minutes before commencing the test. Bath temperatures will be controlled to within ± 2° C from the desired temperature using a thermocouple as close as possible to the upper surface of the specimen. A fresh supply of lubricant will be used for each test.

On the function generator the frequency of the sinusoidal input is set to 50 Hz, and amplitude zero volts. The amplifier is switched on. The amplitude of the sinusoidal input is set to a resulting peak-to-peak value of the accelerometer signal of 5 volts. This should remain constant during the entire test. It takes about two minutes to generate a stable situation. The oscillator is set to the desired frequency and voltage and the friction force noted. The accelerometer and force transducer readings are noted every five minutes during the test duration of 75 minutes. After the prescribed duration, remove the normal force and recheck the zero point on the friction force recording system. Do not touch the test rig during a test, it might affect the test results.

Allow specimens to cool (if heated) and remove them. To measure wear, it is necessary to clean the specimen in such a way that the surface features are not altered. Ultrasonic cleaning in a suitable solvent is necessary. Specimens are thoroughly dried. Examine tip of ball to ensure that no slipping or rolling has occurred. Any ball movement within the holder during the test invalidates the test results. In addition, any slippage of the flat specimen in its fixture invalidates test results.

4 Results and Discussion

Table 1. compares the lubricity of diesel fuel having different levels of sulphur, in terms of the wear scar diameter. The comparisons are done with a readily biodegradable lubricity additive added to diesel fuel and without the addition of the additive. Without the additive, the wear scar diameter increases as the sulphur content in diesel reduces. This indicates more wear with less and less sulphur in diesel. But with less sulphur, diesel fuel produces lower atmospheric pollutants during the combustion process.

In order to reduce the pollutants during the combustion process and at the same time reduce wear of fuel injection equipment it is necessary to introduce a lubricity additive to enhance the lubricity of diesel having reduced sulphur. The test results are affected by the amount of additive and the method by which it is mixed with diesel fuel. 2 % additive was mixed with additive free diesel for the tests. The quantity involved was small and it was difficult to get a uniform dispersion of the additive in diesel fuel. Excess additive resulted in sticking of the additive to the surface of the flat specimen and caused a change in the motion of the sliding member. With a lesser amount of additive friction increased, producing a larger wear scar. A typical plot of the variation of coefficient of friction [2] with time is shown in Figure 2.

Table 1. Comparison of Wear Scar Diameter.

		Without lubricity additive			With lubricity additive		
	Sulphur Content (microns)	Major Axis mm	Minor Axis mm	Wear Scar Diameter mm	Major Axis mm	Minor Axis mm	Wear Scar Diameter mm
1	480	0.213	0.115	0.1813	0.125	0.08	0.1025
2	445	0.25	0.2	0.225	0.143	0.112	0.1275
3	410	0.263	0.238	0.25	0.159	0.146	0.1525
4	395	0.288	0.213	0.2505	0.165	0.152	0.1585
5	245	0.294	0.281	0.288	0.185	0.175	0.18

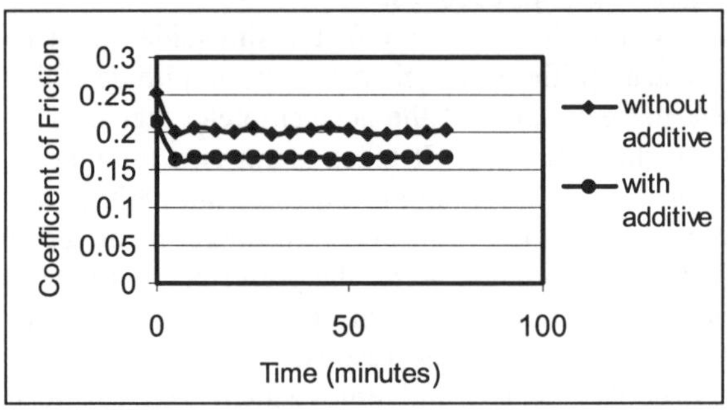

Figure 2. Plot of Coefficient of Friction versus Time for sulphur content of 480 microns.

5 Conclusions

The results of the test on diesel without the lubricity additive indicate that with less and less sulphur in diesel, the wear of components with which it is in contact increases. However, lesser the sulphur content in diesel, better it is for the environment, as less pollutants are produced. Addition of a lubricity additive to diesel will improve the friction behaviour and reduce wear. It is seen from the results of the test that there is on an average a 40 % reduction in wear with the addition of a lubricity additive.

6 Acknowledgements

The authors thank Mr. Murray Gray of BP, Brisbane, for providing the samples of low sulphur diesel fuel and the lubricity additive.

References

1. ASTM D6079, Standard test method for evaluating lubricity of diesel fuels by the high-frequency reciprocating rig, (1999).
2. ASTM G115, Guide for measuring and reporting friction coefficients, (1993).
3. Klaffke D., On the influence of some test parameters on the results of oscillating sliding tests, *Tribo-Test* **6(1)** (1999) pp 25-39.
4. Singerling, B., Redesign and testing of a reciprocating lubricant test rig, Report, Queensland University of Technology, Australia, (2000).
5. White E., Development of a reciprocating lubricant test rig, Report, Queensland University of Technology, Australia, (1999).

A NOVEL APPROACH TO DETERMINE PERMEABILITIES FOR FLOW PROCESSES IN COMPOSITE MANUFACTURING

G. BECHTOLD, L. YE

School of Aerospace, Mechanical and Mechatronic Engineering, The University of Sydney, Sydney NSW 2006, Australia

E-mail:bechtold@aeromech.eng.usyd.edu.au, ye@aeromech.usyd.edu.au

K. FRIEDRICH

Institut für Verbundwerkstoffe (IVW) GmbH, University of Kaiserslautern, 67663 Kaiserslautern, Germany

E-mail: friedrich@ivw.uni-kl.de

In most of the manufacturing processes for continuous fibre reinforced plastics, the dry fibres must be wetted by a liquid matrix, such as thermosetting resins or thermoplastic polymer melt or solution. After the solidification process, quality requirements demand a solid composite part without voids. Economical constraints are making this task to a compromise between a fast process and a good quality of the part. In order to reduce experimental efforts, computational flow simulations are very common in optimisation of the manufacturing processes. For the mathematical compilation of the problem, Darcy's law is used most of the times. The method requires a factor, the permeability. This fibre bed related constant must often be determined experimentally or is roughly estimated. In this work, a new approach for determining permeabilities, taking into consideration of the stochastic characteristics of the fibre bed, is presented.

1 Introduction

High performance polymer matrix composite materials, especially those based on thermosetting resins, have gained substantial importance in the past three decades due to the high specific stiffness and high specific strength as well as other excellent properties. They have commonly been used in the aerospace industry and in sporting goods [1]. However, due to their expensive manufacturing processes, they were not yet able to win wide recognition in automotive industry. Therefore, cost-reduction via mass production and rapid processing is the critical challenge for polymer composites in practical applications. Flow simulation of fibre reinforcement impregnation therefore is very important to save time-consuming and cost-intensive trials. Most of the simulations are based on Darcy's law, describing the dependency of flow rate on the pressure gradient in the flow field by a constant containing the fluid's viscosity and the fibre architecture's resistance to the permeating fluid, called permeability:

$$\dot{q}_z = -\frac{K_P}{\eta}\frac{dp}{dz}$$

$\dot{q}_z$:	Penetration rate	p:	Pressure
K_P:	Permeability	η:	Viscosity
z:	Length co-ordinate		

Most of the times, the permeability has to be determined experimentally for every new fibre bed, although some efforts have been made to express this factor in a more general way; an overview can be found in [2]. Very few took the influence of the stochastic fibre bed characteristics into consideration [3], although it is well known that simple models with a regular quadratic or hexagonal fibre arrangement lead to unrealistic results [3,4]. Characterisations of stochastic fibre arrangements on the other hand do often have the drawback to require special structural parameters which are not easy to determine [2]. Characterising a stochastic distribution with only a few parameters is a highly nontrivial task [5].

In this study, using a film stacking technique some glass fibre (GF) / Polyamide 6.6 (PA) composite specimens were produced with relatively low temperatures, pressures and short holding times. The cross sections of the only partly impregnated GF bundles were examined. A special algorithm was applied to characterise the stochastic fibre arrangement by only a few parameters. The dependence of those parameters on the compression pressure was examined. A finite element model was developed to simulate the impregnation process and to verify the permeability in a small region with a characteristic fibre distribution. The results correspond very well to the data in literature.

2 Specimen preparation

By the aid of a heating press at a temperature of 275°C, some composite specimens were produced by stacking PA6.6 films (thickness 50 μm, Zytel 101, DuPont) and GF mats (7581/A1100, Hexcel) with varied holding times and pressures. Specimens were 100 mm x 80 mm x 4 mm in size. They were cut, grinded, polished and examined under a light microscope.

3 Fibre distribution characterisation

The Morishita square method is often used in the fields of ecology, e.g. for determining the distribution of bacteria cultures in a dish [6]. By this method, the area under examination (marked in Fig. 1) is divided into different subsections with equal sizes. The number of fibres in each subsection is computed separately. The Morishita index I_δ plays an important role on the characterisation of the fibre distribution:

$$I_\delta = q\delta$$

$$\delta = \sum_{i=1}^{q} n_i(n_i - 1) / N(N-1)$$

N:	Total number of fibres	q:	Number of subsections
n_i:	Number of fibres in subsection i		

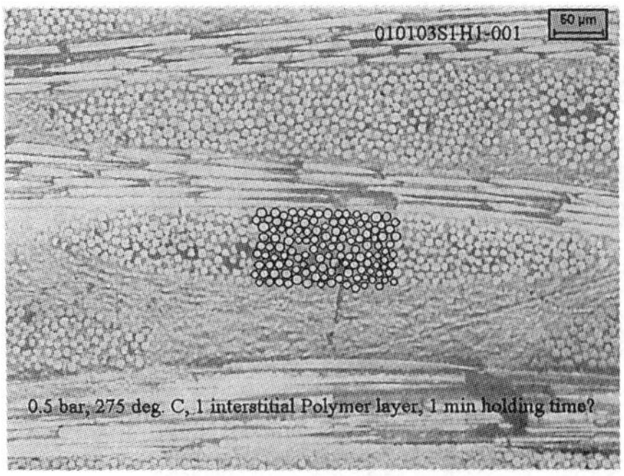

Figure 1. Example for a fibre arrangement inside a fibre tow, evaluated fibre region marked.

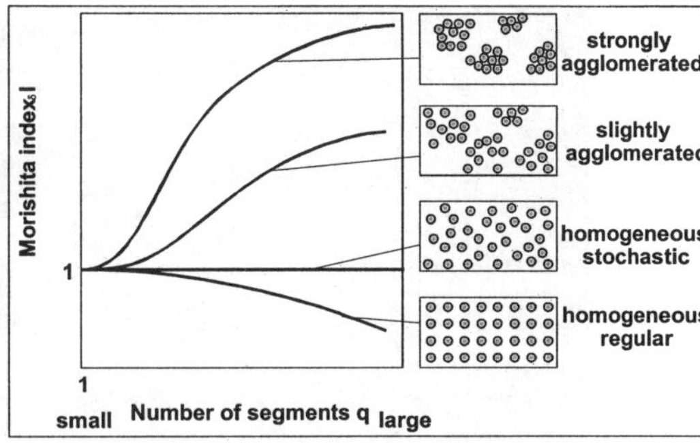

Figure 2. Morishita distribution.

The relationship of I_δ and q shows a curve characterising the distribution of the fibres, Figure 2. Characterising properties result from the shape of the curve I_δ over q, approaching more or less fast asymptotically a constant value.

From the fibre tow shown in Figure 1, a section containing about 100 fibres was chosen for a further evaluation by the Morishita algorithm. At q = 100, an I_δ of 0.44 was obtained. This means that the characteristic of the fibre arrangement is not purely stochastic, but possesses a certain regularity.

Two neighbouring 100-fibre regions of the same fibre bundle were evaluated, leading to values of 0.56 and 0.53. The fibres at the sides of the elliptical fibre bundles seem to be less regular arranged and more agglomerated compared to those at the centre area. Two other fibre bundle ellipses of the same specimen led to values of 0.49 and 0.41 for the centre areas, which confirms the certain regularity of the fibre distribution in the bundles.

Furthermore, in order to get an impression of the influence of the pressure on the Morishita distribution, two other specimens, produced with pressures of 0.42 MPa and 1.41 MPa respectively (instead of 0.71 MPa for the specimen in Figure 1), were examined. The mean I_δ of three fibre tow ellipses in the specimens was 0.47 at 0.71 MPa and 0.44 at 0.42 MPa.

Table 1 summarises the dependency of the Morishita number on the pressure. It concludes, that a higher pressure leads to a less regular fibre arrangement due to more significant squeeze flow under a high pressure.

Table 1. Morishita number dependent on pressure.

Pressure [MPa]	0.42	0.71	1.41
Morishita number I_δ	0.44	0.45	0.47

4 Flow simulation

The Morishita number of the fibre tows was used as a basis to create a finite element mesh containing 25 fibres with the same Morishita number. The mesh for I_δ = 0.45, shown in Fig. 3, was realised by PATRAN.

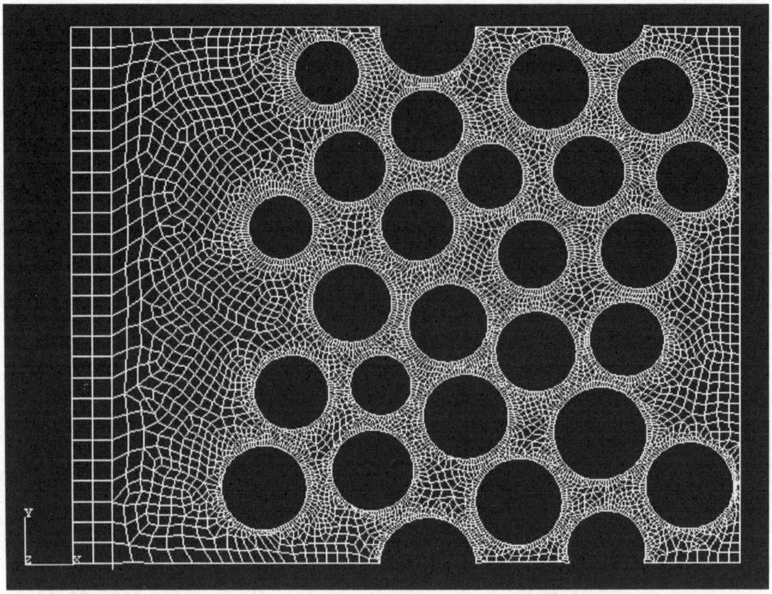

Figure 3. FE mesh of 25 fibre region.

For the flow simulation, the CFD code FIDAP (Fluent Inc.) was used. The boundary conditions and parameters were as follows for the matrix, viscosity is 65 Pas and density is 1.14 x 10^3 kg/m^3, it flows from the left boundary into the model with a constant speed of 0.1 μm/s. At the top and bottom boundaries, continuous boundary conditions were applied. On each fibre surface, no-slip boundary conditions were used. More information concerning the FE model can be found elsewhere [2].

The speed vectors after a filling time of 33.1 s are shown in Figure 4. It can be seen that about half of the region is filled with matrix.

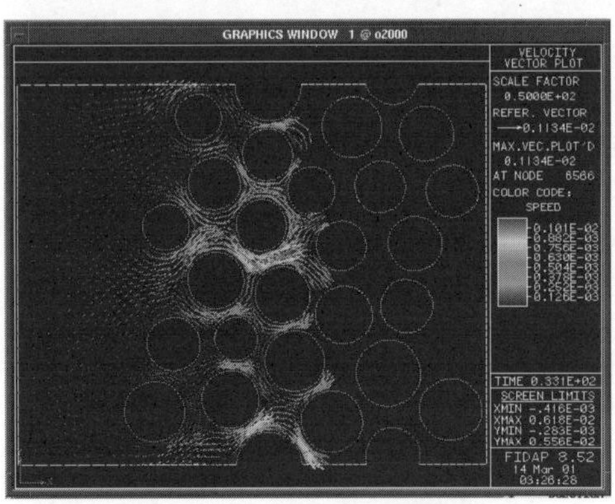

Figure 4. Filling of 25 fibre region after 33.1 s.

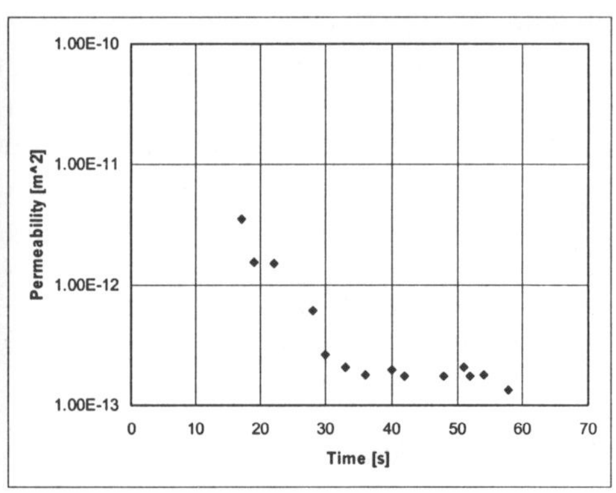

Figure 5. Permeabilities resulting from the model

The results of flow rate and pressure can now be fit into Dary's law. After a rearangement, the permeability can easliliy obtained. The results for a couple of time steps are presented in Figure 5.

It was found that a stable value for the permeability is obtained after a flow time of about 30 s. This means that a model with only half of the length used here would be sufficient for a reliable result for microgeometrical transverse permeabilities. In the stable region, the mean value for the permeability is 1.8×10^{-13} m^2, corresponding very well with results found in [7] and [8].

5 Conclusions

The Morishita algorithm offers a simple method to characterise fibre distributions with only a few parameters. The pressure applied on a fibre bed has an influence on the characteristic of the distribution of the fibres, and a higher pressure leads to a certain loss of regularity of the distribution due to squeeze flow.

The realised computational flow simulation of the polymer flow through a region of 25 fibres with a Morishita number corresponding to characteristics of the actual fibre distribution leads to realistic permeabilities.

6 Future work

In the further work, the minimum model size to obtain realistic permeabilities will be worked out in a series of simulations. Furthermore, the correlation of Morishita number and permeability will be established.

Acknowledgements

Georg Bechtold would like to express his thanks to the Alexander-von-Humboldt Foundation, Bonn, Germany, for the Feodor-Lynen scholarship, enabling him to conduct the described research work at the University of Sydney.

References

1. Colucci F, Lighter-Than-Air Materials. *Aerospace Materials* **5** (1993) pp. 9-12.
2. Bechtold G., Ye L. and Friedrich K., Influence of Stochastic Fibre Distribution on Permeabilities of Fibre Bundles. Submitted to "The Sixth Flow Process in Composite Materials", Auckland, New Zealand, 4-5 February 2002.
3. Ghaddar C. K., On the permeability of unidirectional fibrous media: A parallel computational approach *Phys. Fluids* **7** (1995) pp. 2563 – 2586.
4. Sangani A. S. and Acrivos A., Slow flow through a peroidic array of spheres. *Int. J. Multiphase Flow* **8** (1982) pp. 343 – 360.
5. Provatas N., Haataja M., Asikainen J., Majaniemi S., Alava M. and Ala-Nissila T., Fiber deposition models in two and three spatial dimensions *Colloids and Surfaces A* **165** (2000) pp. 209 – 229.
6. Nakao Y., Yabuta M. and Tomianga A., The Relation Between the Dispersion State of Polymer Microspheres in Coatings and Film Properties *Progress in Organic Coatings* **20** (1992) S. 369-381.

7. Phelan F. R. and Spaid M. A. A., Modeling of unsaturated flow dynamics in composites processing. Proceedings "5th International Conference on Automated Composites", Glasgow Moat House, Glasgow, UK (1997), pp. 33 – 41.
8. de Parseval Y., Roy, V. and Advani S. G., Effect of local variations of preform permeability on the average permeability during resin transfer molding of composites. Proceedings "ANTEC '95", Soc of Plastics Engineers (Ed.) (1995), pp. 3040 – 3044.

FINITE ELEMENT SIMULATION OF LASER SHOCK PEENING ON AN ALUMINUM ALLOY

KAN DING[1], LIN YE[1*], QIAN-CHU LIU[2]

[1]*Centre for Advanced Materials Technology (CAMT)*
School of Aerospace, Mechanical and Mechatronic Engineering, The University of Sydney
NSW 2006, Australia

[2]*Airframes and Engines Division, Aeronautical and Maritime Research Laboratory*
Fishermens Bend
Vic 3207, Australia

A novel method for generating compressive residual stresses on surfaces of metallic alloys is laser shock peening (LSP). The LSP process uses a high intensity, short duration laser pulse to create shock waves in the materials. LSP was performed on an aluminum alloy with a Nd-glass laser beam of 8 mm in diameter, operating with a 50 J per pulse and 25 ns in duration. Shock waves were generated by volume expansion of plasma when the material was laser-irradiated. The specimen was coated by the energy-absorbing black paint, and water as the plasma-confining layer was used to obtain the maximum pressure without melting the specimen surface. The effect of laser shock peening on the residual stresses of the aluminum alloy was studied and simulated using the finite element method. The finite element codes (the ABAQUS standard and ABAQUS explicit) were used to simulate the two-sided peening mechanisms on the specimens.

1 Introduction

Laser shock peening (LSP) was developed in the period between 1968 and 1981 in the USA [1]. It was particularly applied to enhance fatigue properties of fastened joints for aeronautical applications. LSP is a novel treatment technique similar to traditional shot peening to generate compressive residual stresses on surfaces of targets. The appreciable improvements of surface properties after LSP have already been demonstrated as compared with traditional shot peening because LSP is known to generate the same amplitude of compressive residual stresses but with much lower work-hardening [2]. Since 1986, many studies have been concerned with the confined interaction mode and of its limiting factors, as well as analytical modeling of the processes. The published literatures show that LSP by using a high energy pulsed laser produces residual stresses and strain hardening into the surface of targets, and it appears to achieve a much greater depth of penetration than shot peening and significantly improves the fatigue life of a number of metals such as stainless steels, titanium alloys and aluminium alloys [1]-[7].

Up to now, most researchers have investigated mechanical properties of targets using experimental approaches, but a few published papers have been presented in the development of analytical techniques to predict the residual stresses from LSP. Consequently, this paper has focused on simulating residual stresses generated by two-sided LSP on the surface of an aluminum alloy (7050) using the finite element analysis (FEA) method.

* Corresponding author

2 The Fundamentals of LSP

LSP processing was performed on aluminum alloy (7050) specimens. The specimen has a diameter of 10 mm at the center and a diameter of 30 mm at both ends. The Q-switched Nd-glass laser equipment produces a laser beam with a spot size of 8 mm, operating with 50 J per pulse and 25 ns in duration.

The area treated by LSP was locally covered with two types of overlays such as confined overlay (water) and thermo-absorptive overlay (black paint). When the laser beam is directed onto the surface of the specimen, it passes through the water layer and strikes the black paint layer. As shown in Figure 1, the source laser beam is split into two laser beams, which simultaneously focus on the surface of the black paint layer. They immediately result in vaporising the black paint. This vapor has to absorb the incoming laser energy, rapidly heating and expanding against the surfaces of the specimen and the confined water layer. The interaction between the specimen and the water layer induces a high-pressure plasma which produces high amplitude shock waves, consequently causes plastic deformation and in turn residual stresses under the surface of the specimen.

The confined ablation mode was well described in the published literature [6]. It contains a three-step process for the pressure environment generation such as heating, adiabatic cooling and final expansion of the plasma. Using this model, and considering the plasma to be a perfect gas, the scaling law of the pressure generation can be estimated by the following relationship [7]

$$\mathrm{P(GPa)} = 0.01\sqrt{\frac{\alpha}{2\alpha+3}}\sqrt{Z(\mathrm{gcm^{-2}s^{-1}})}\sqrt{I_0(\mathrm{GWcm^{-2}})} \tag{1}$$

where I_0 is the incident laser power density, P is the pressure, Z is the reduced shock impedance between the target and the confining medium (water or glass) and α is the efficiency of the interaction (α=0.1-0.2). The parameter Z is given by $Z=2Z_1Z_2/(Z_1+Z_2)$. Z_1 and Z_2 are the shock impedances of the confining medium and the target, respectively [5]. $Z_1=0.165\mathrm{x}10^6$ g cm^{-2} for the water confining layer and $Z_2=1.5\mathrm{x}10^6$ g cm^{-2} for the aluminum alloy (7050), in the current case, the pressure reaches a limit over a laser power density of 4 GWcm^{-2} with a gaussian pulse and its peak of 2 GPa can be approximately obtained from Equation (1).

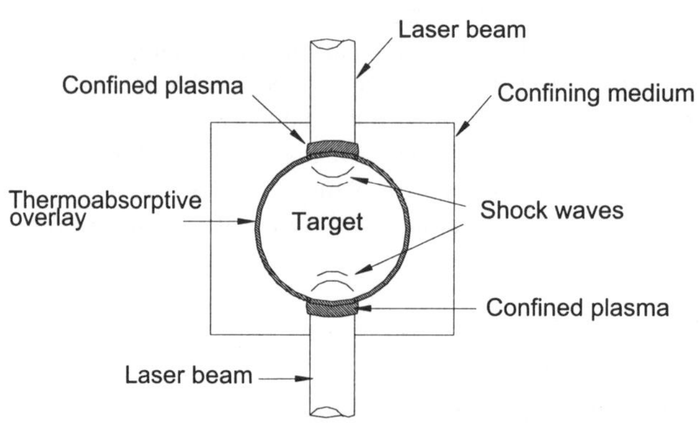

Figure 1. The two-sided LSP in confined plasma configuration

LSP can be described as a compressive stress wave creating plastic strains in materials, provided the stress amplitude exceeds the elastic limit stress defined as Hugoniot Elastic Limit (HEL) under the unaxial strain condition. When the peak pressure is below HEL, no plastic strain occurs. When the peak pressure is between 1 HEL and 2 HEL, the plastic strain occurs with a purely elastic reverse strain. When the peak pressure is above 2 HEL, the elastic reverse strain gets saturated to 2 HEL and the plastic strain occurs. The HEL value depends on mechanical properties according to [4]

$$HEL = (1-\nu)/(1-2\nu)\sigma_y^{dyn} \tag{2}$$

Where ν is the anisotropy coefficient and σ_y^{dyn} is the compressive yield strength at a high strain rate (normally exceeding 10^6 s^{-1}). The specimen (7050) has HEL of 1.1 GPa and the anisotropy coefficient (ν) of 0.33. Hence, the dynamic yield strength at the high strain rate is approximately 0.558 GPa.

3 Finite Element Analysis

The elastic-plastic finite element (FE) model was applied in current studies to predict the amount of plastic strain and residual stresses induced by LSP. The 2D-dimensional FE model was used to calculate the elastic-perfectly plastic response of the material subjected to a radial shock wave.

A symmetric two-sided peening model is presented in Figure 2. Two-sided peening was used to simultaneously impact both sides of the target. The pressure pulses on the opposite surfaces of the target balance one another and prohibit excessive deformation of the target. The shock waves induced by LSP from each surface will interact through the radial direction of the specimen resulting in a more complex residual stress field.

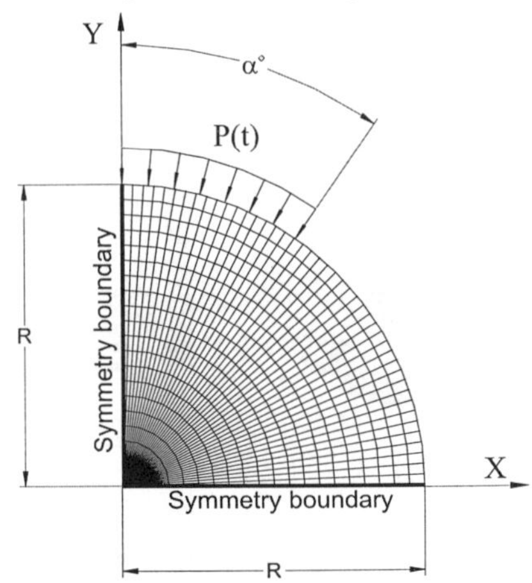

Figure 2. The two-dimensional finite element model

For simulating the two-sided peening effects, some essential assumptions have to be used for the FE simulation: (1) the model is considered to be in the plain strain condition ($\varepsilon_z = 0$); (2) the pressure pulse on the surface is uniform in space; and (3) the specimen material obeys a Von Mises plasticity criterion.

The finite element ABAQUS package was used in the simulation, including two commercial finite element codes, the ABAQUS/Standard and the ABAQUS/Explicit. The ABAQUS/Standard is a nonlinear elastic-plastic implicit time integration finite element code used primarily for static or natural frequency calculations. The ABAQUS/Explicit is a nonlinear elastic-plastic explicit time integration finite element code specifically designed for short duration transient analyses, which was used as the main tool in the present study. However, when using it to determine the final residual stress state after LSP, the stress state is extremely slow to reach the convergence solution. The method adopted by Braisted *et al.* [3] was used to solve this problem using the ABAQUS/Standard. It was noted that two finite element codes were used together to provide the most effective and efficient numerical solution. But it is quite unstable to determine the running time of the analysis to insure all plasticity has occurred in the material due to reflection and interaction of various shock waves.

4 Results and Discussion

The residual stresses predicted by the FE analysis for the two-sided LSP are plotted in Figure 3. The analytical results indicate that the tensile residual stresses have occurred at the center of the cross-section of the specimen. The field of tensile residual stresses is circled about 1 mm from the center of cross-section due to focusing of the various stress waves. The compressive residual stresses are distributed along the perimeter of the cross-section of the specimen. During LSP, the x component of stresses, comparing with the y component of stresses, has played an important role in the LSP simulation because it ultimately becomes the compressive residual stress for this particular geometry of the specimen. The equivalent plastic strains (Figure 4) have penetrated into the depth of up to 1 mm from the treated surface. In addition, the peak plastic strain is found at the center of the cross-section of the specimen.

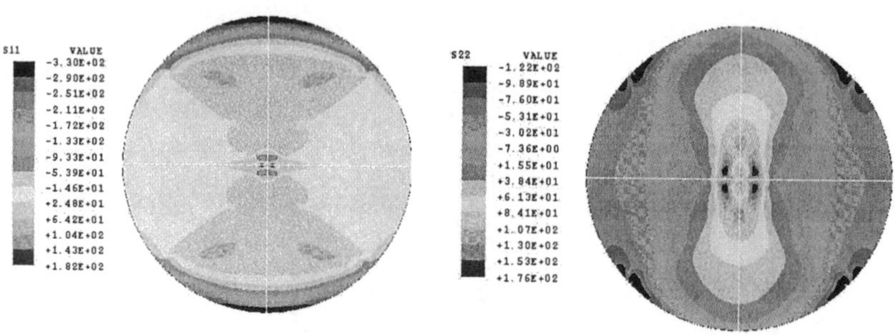

(σ_x) (σ_y)

Figure 3. Residual stresses

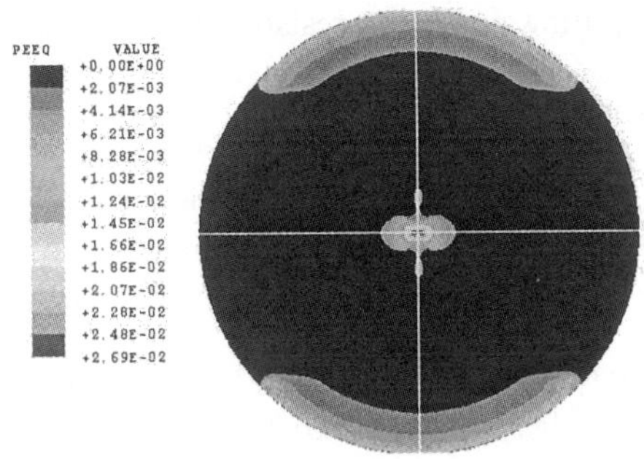

Figure 4. Equivalent plastic strains

Under the same LSP conditions, the FE analytical results indicate that the tensile residual stress field at the center area has significantly reduced when increasing the diameter from 10 mm to 12 mm. In addition, the compressive residual stresses at both the center area and on the treated surface have slightly increased, as shown in Figure 5.

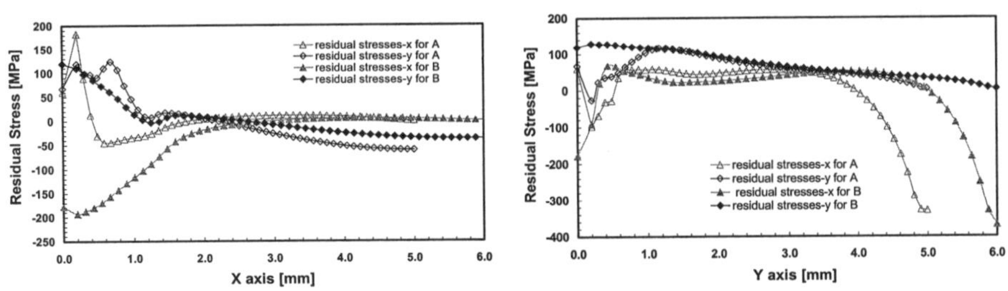

Figure 5. Residual stresses for A (D=10mm) and B (D=12mm)

5 Conclusions

The peak tensile residual stress predicted by the FE analysis was found at the center of cross section with a diameter of 10 mm. This peak tensile residual stress may cause material failure or cracking during LSP. To avoid having those tensile residual stresses at the center, it is necessary to increase the diameter of specimens or reduce the laser spot size or the laser power density. The predicted compressive residual stresses during LSP can be driven into the depth of up to 1 mm from the treated surface, distributing along the perimeter of the cross-section of the specimen. These compressive residual stresses in depth of the specimens are expected to greatly enhance fatigue behavior of the specimens.

References

1. Peyre P. and Fabbro R., Electromagnetic gauge study of laser-induced shock waves in aluminum alloys, *Journal of Physics III* France 5 (1995) 1953-1964.
2. Peyre P., Scherpereel X., Berthe L., Carboni C., Fabbro R., Beranger G. and Lemaitre C., Surface modifications induced in 316L steel by laser peening and shot-

peening. Influence on pitting corrosion resistance, *Materials Science & Engineering* A280 (2000) 294-302.

3. Braisted W. and Brockman R., Finite element simulation of laser shock peening, *International Journal of Fatigue*, 21 (1999) 719-724.
4. Peyre P., Berthe L. Scherpereel X. and Fabbro R., Laser-shock processing of aluminum-coated 55C1 steel in water-confinement regime, characterization and application to high-cycle fatigue behaviour, *Journal of Materials Science*, 33 (1998) 1421-1429.
5. Berthe L., Fabbro R., Peyre P., Tollier L. and Bartnicki, Shock waves from a water-confined laser-generated plasma, *American Institute of Physics*, S0021-8979 (1997) 03618-9.
6. Fabbro R., Fournier J. Ballard P., Devaux D. and Virmont J., Special study of laser-produced plasma in confined geometry, *Journal of Apply Physics*, 68(1990) 775.
7. Peyre P., Fabbro R., Merrien P. and Lieurade H.P., Laser shock processing of aluminum alloys, Application to high cycle fatigue behaviour, *Journal of Materials Science & Engineering*, A210 (1996) 102-113.
8. ABAQUS *user's manual version 5.8*, Pawtucket (RI): Hibbitt, Karlsson and Sorensen Inc, 1998.

EFFECT OF SHIELDING GAS COMPOSITION ON WELD BEAD GEOMETRY, MICROSTRUCTURE AND HARDNESS OF WELDMENTS USING FCAW

T. SHEHATA*, R.N. IBRAHIM*, P.T HOMSON** AND A. SADEK*

**Department of Mechanical Engineering, Monash University,*
P.O.Box 197, Caulfield East, Victoria 3145, Australia
Email: Raafat.Ibrahim@eng.monash.edu.au
***Department of Material Engineering, Monash University*
Email: P. Thomson@eng.monash.edu.au

There are several industry-attractive aspects of flux cored arc welding (FCAW) process that needs further investigation. Also, the welding parameters that can provide the required microstructure, hardness and minimum residual stresses in thick walled weldments need to be established to validate the application of flux cored arc welding process.

In the present work, the effect of welding parameters and type of shielding gas on the weld bead geometry, microstructure, hardness, spatter rates and depth of penetration have been studied and reported. The results shows that the spatter rates increase as the CO_2 content of $Ar+CO_2$ shielding gas mixtures increases. The hardness of all the weld metals is affected by the amount of pro-eutectoid ferrite and the size of acicular bainitic structure of the columnar grains of deposited metal. On the other hand, the changes in hardness values at the heat-affected zone (HAZ) were found to be a function in the changes of the heat input. In this paper, values of welding parameter were obtained to give best quality of welding using FCAW process.

1 Introduction

There are several industry-attractive aspects of FCAW that need further investigation. One of these is the application of FCAW process in repair welding of thick wall steel constructions, by using temper bead welding (TBW) technique. There has been limited research into the effectiveness of this process [1-3] but a preliminary study [4-6] indicates that FCAW might have a number of advantages over the commonly used solid wires such as MMAW or GMAW process. The welding parameters that can provide the required microstructure and minimum residual stresses in thick walled weldments need to be established to validate the application of flux cored arc welding (FCAW). The development of better welding methods, especially in the field of automatic welding, requires an accurate knowledge of the relationship between various welding parameters. To achieve this, an improved physical understanding of the welding process is needed. Most process modifications result in some kind of change in the processes within the arc and the melt. The nature of these processes depends on the characteristics of the arc, such as current, temperature, and the resulting heat input, which in turn affect the cooling rate of the weldment.

In the present work the effect of welding parameters and type of shielding gas on the weld bead geometry, hardness, spatter rates and depth of penetration have been studied and reported. The spatter rates increase as the CO_2 content of $Ar+CO_2$ shielding gas mixtures increases. Penetration is one of the most important factors controlling hot cracking susceptibility especially in the case of welding thick sections. By increasing the penetration depth the susceptibility to hot cracking is increased especially in the case of welding thick sections.

The cooling rate after welding has a great influence on the microstructure and in turn on the mechanical properties and performance of the weldments. Very limited experimental data regarding cooling rate after welding of plates by FCAW process is available in the literature. In addition, it is often necessary to place a ceiling on the heat input used to weld high strength low alloy steels. Too slow cooling rate can lead to the formation of a mixture of proeutectoid ferrite and high carbon martensite, which is also deleterious to the mechanical properties. In this paper, the cooling rate measured experimentally, as a function of welding parameters, is presented and its effects on HAZ hardness are discussed.

2 Experimental work

2.1 *Material*

The base metal used in this experimental work was 12 mm plates of commercially manufactured C-Mn steel. Typical chemical composition of the material used in this experimental work is given in Table 1. Basic flux-cored wire, AWS A5.20, E71T-5 H4 was used through out the experimental work, with 1.2 mm. diameter.

Table 1 Chemical composition of base metal

C	Si	Mn	P	S	Cu	Ni	Cr	Mo
0.22	0.55	1.7	0.04	0.03	0.4	0.5	0.3	0.1

2.2 Welding Procedure

An automatic FCAW machine using constant voltage direct current electrode positive (DCEP) was adopted in this study of bead on plates welding. The welding parameters are shown in Table 2. The base metals were prepared as strips with the dimension of 400 x 100 x 12 mm to study the effect of welding parameters on temperature distribution and weld bead geometry.

Pure Argon and mixture of Ar+CO_2 (82% Ar, 18% CO_2) were used as a shielding gas with flow rate of 7-8 l/min in case of group A and group C. However, in case of group B, CO_2 gas was used with flow rate of 10-12 l/min. Table 2 shows the welding parameters, (gas type, welding current, arc voltage, welding travel speed and tip to work piece distance for groups), which had been studied in this paper to investigate its effects on the cooling time after welding. For each group, the gas type, the arc voltage the welding travel speed and tip to work piece distance are constant while the change only was in the welding current. The welding currents were the same for groups D, E, F, G and H.

A special welding jig (rigid fixture) was designed and fabricated to hold parts in alignment and to ensure accurate fit-up with no need for tack welding. Meanwhile, by using this jig, which restrain the test strips, the hot cracking tendency can be examined. A short arc and flat bead was maintained during welding. The joint was kept fixed until it cooled down. After slag removal, the samples were released from jig.

2.3 Metallographic Investigation, Hardness Testing and Cooling Rate

All samples were cut transversely for macrostructure investigation and etched in 5% Nital (95% Ethanol and 5% Nitric acid). Specimen were photographed and enlarged to measure the weld area and HAZ width. Hardness was measured across the weldments using a Vickers diamond indenture with a load of 10 Kg. Temperature and cooling rate were measured by inserting, W, W-Rh thermocouple in the weld pool. The measured value were recorded and analyzed instantaneously by the means of Data Taker connected to PC computer.

Table 2 Welding parameters

Group	Gas Type	Current A	Voltage V	Travel speed Mm/min	Tip to work distance mm
A	Ar	350	29-30	620	23 25 29
B	CO_2	380	29-30	620	23 25 29
C	Ar+CO_2	365	29-30	620	23 25 29
D		150 170 190 210	22	703	25
E		150 170 190 210	23	638	25
F	Ar	150 170 190 210	24	510	25
G		150 170 190 210	23	380	25
H		150 170 190 210	24	100	25

3 Results and discussions

3.1 Effect of Cooling Time on Hardness of HAZ

The measured cooling time curves are shown in Fig. 1 as a function of different welding current at constant welding voltage and traveling speed, while the results shown in Fig.2 are measured at different welding speeds at constant welding current and voltage. Special consideration was given to the cooling rate from 800 to 500 °C, during solidification, and the measured cooling rate is also shown. The hardness of HAZ as a function of cooling time is shown in Fig.3.

From these results the following points were concluded:

1. The effect of welding current on the hardness of HAZ is almost small when other welding variables are fixed.
2. The welding speed has the biggest effect on the cooling time as well as the hardness value of HAZ. At lower welding speed, the amount of adjacent material affected by the heat of the arc is increased, resulting in wider HAZ, as shown in Fig.3, slower cooling rate, and lower hardness value. However, the amount of residual stresses will be increased resulting in increasing the susceptibility of hot cracking. This is comply with the previous results of hot cracking. However, at higher welding speed the amount of adjacent material affected by the heat of the arc is reduced, resulting in the decreasing of residual stresses and increasing of HAZ hardness of this type of steel

under investigation. Noting that for this steel type, the Ac_1-temperature is about 708 °C, with the Ac_3-temperature of 849 °C and Ms Of 370 °C.

3. At higher levels of welding current and voltage the effect of welding speed and cooling time on the hardness values is relatively remarkable, as shown in Fig.2 (C-1&C-3), but still higher than the acceptable range stated in ASME section VIII.
4. The effect of welding voltage on the hardness of HAZ is also small when other welding variables are fixed.

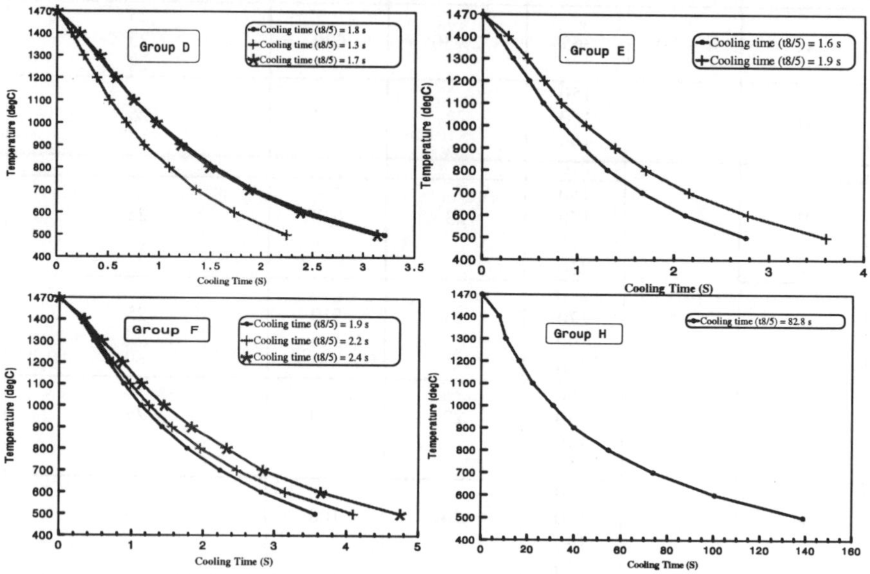

Figure 1 Cooling time as function of welding current at constant voltage and welding speed

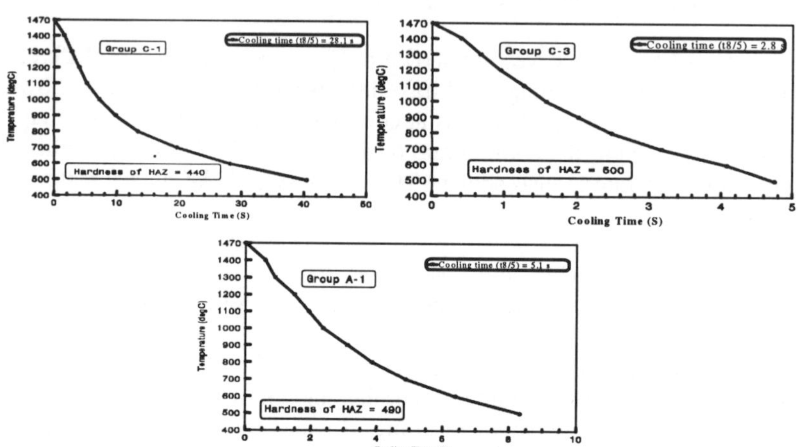

Figure 2 Cooling time as function of welding speed at constant welding current and voltage

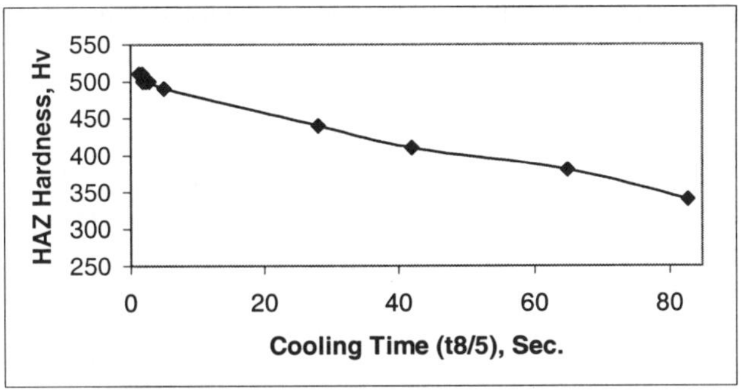

Figure 3 Effect of cooling time on the hardness of heat affected zone

3.2 The effect of welding parameters on the welding microstructure

In this part the effect of the shield gases, the welding traverse speed and welding voltages on the weld bead and HAZ microstructure have been investigated. Two types of shield gases (100 % Ar. and mixed 82 % Ar and 18 CO_2), welding traverse speed (230 and 380 mm/min), and two welding voltages (25 and 32 volts) are used in this study.

3.2.1 The effect of 100% Ar and mixed (82% Ar and 18% CO_2) gases on the microstructure

The two gases used in this investigation gave columnar structure (J) with bands of pro-eutectoid ferrite in the weld bead area, but in the case of the 100 %Ar, (fig 4), the ferrite band is wider. In the case of mixed gases (co_2 and Ar.), (fig 5) ,there is only columnar structure with bands of pro-eutectoid ferrite. While in the case of 100% Ar. free pro-eutectoid microstructure (K) appeared. In coarse grains heat affected zone, the two gases have lower bainite (L) surrounded by bands of pro-eutectoid ferrite and upper bainite microstructure (M). But in the case of mixed gases (co_2 and Ar), the pro-eutectoid ferrite is narrower and the amount of upper bainite is smaller. In the fine grains heat affected zone, the amount of the pro-eutectoid ferrite in (the case of 100%Ar) is lager than in the case of mixed gases which showed some pro-eutectoid ferrite on the boundaries of the lower bainite structure. Also the graphite in 100% Ar is spherical (N) while in the other case is laminar (O).

3.2.2 The effect of welding speed on the microstructure

In the weld bead area, higher welding speed, as shown in fig 6, gave more homogenise columnar structure (J) with narrow bands of pro-eutectoid ferrite, while in the case of lower welding speed free pro-eutectoid structure (K) started to appear, as shown in fig 4. In coarse heat affected zone, the higher welding speed still gave more homogenise structure of lower bainite (L) surrounded by upper bainite (M), but in the case of the lower speed the pro-eutectoid ferrite is narrower. In fine heat affected zone, finer structure is achieved when higher welding speed was used. The graphite in both cases is spherical (N) but it is less in the case of higher welding speed.

3.2.3 The effect of the welding voltage on the micro structure

The change in the welding voltage doesn't affect the microstructure of the heat affected zone and weld bead, as shown in fig 4 and fig 7. In weld bead, the structure is columnar (J) with bands of pro-eutectoid ferrite and free pro-eutectoid ferrite (K). In coarse heat affected zone, the microstructure is lower bainite (L) surrounded by bands of pro-eutectoid ferrite and upper bainite (M). In the fine heat affected zone, there is no noticable changes, but the amount of free pro-eutectoid ferrite is more in the case of the higher voltage in all studied areas.

3.3 Hot Cracking Test

Low alloy steels are prone to hot cracking and particularly to solidification cracking. Examination of the restraint test samples by dye pentrant test and optical microscopy showed that solidification cracks occasionally occurred as in case of specimens A-1, C-1, C-2 and H-4 while no cracks were found when using the other weld conditions. The reasons are partly due to the metallurgical microstructures, heat input and the amount of residual stresses. Accordingly, at lower heat input the amount of segregation produced, such as low melting elements like P and S or low melting carbide phases, is smaller than at higher heat input. Therefore, cracking response in the latter case will be higher due to the changes of weld pool shape and to the increase in heat input. High heat input lead to lower cooling rate, wider liquidus-solidus range, coarser cell size and high amount of segregation, as well as the change in stress distribution along the weldment as a function of temperature gradient around the weld pool

4 Conclusions

The following specific conclusions can be drawn from the results of this study:

1. With increasing tip to work distance, the bead width increases and the reinforcement and penetration decrease.
2. The measured values of weld bead dimensions are generally higher for the case of Ar gas shielding than in the case of CO_2 gas shielding.
3. The width of heat affected zone decreases with an increase in tip to work distance.
4. With increasing welding travelling speed, weld metal cross sectional area and HAZ width will be reduced. On the other hand, the hardness of HAZ will be increased.
5. The hardness of the weld bead and the heat affected zone is increased due to using CO_2 in the mixed gases or increasing the welding speed

This work will be extended to get an appropriate hardness prediction formula and another empirical formula based on base metal chemical composition and weld cooling time. Based on this formula, methods will be developed for determining the welding procedure needed to achieve a specific maximum HAZ hardness or estimating the HAZ hardness of existing welds. This will be extended to temper bead welding for repair welding.

References

1. C. D. Lundin, *WRC Bulletin* 412, 1996, p. 16 - 26.
2. T. W. Lau, M. L. Lau, et al. *WRC Bulletin* 412, 199, p. 35 - 42.
3. J. S. Edgley and M. Pitrun, *Failure, repair and life assessment of pressure vessels and pipe work II.* Institute of Metals and Materials, IEAust Materials Society, ed JWH Price, Melbourne, 1993.
4. R. N. Ibrahim, T.Shehata, *ICM8, 8th International Conference on the Mechanical Behaviour of Materials,* Canada, British Colombia, 1999, p. 771
5. R. N. Ibrahim, T. Shehata, *9th International Conference on Pressure Vessels Technology*, Australia, Sydney, April 9 - 14, 2000.
6. M. Sato, K. Suda, et al, *Journal of Japan Welding Society*, 65, 1996, p. 13
7. J.F. Lancaster *The physics of welding,* IIW Pergamon Press, Oxford 1984

COMPUTER MODELLING OF THE MESHING PROCESS OF THE ARTHUR BISHOP'S RACK AND PINION MECHANISM

V.KOSSE
School of Mechanical, Manufacturing and Medical Engineering,
Queensland University of Technology, Gardens Point Campus,
GPO Box 2434, Brisbane, Qld. 4001, Australia

Most modern cars have steering system utilising Arthur Bishop's rack and pinion mechanism. When it wears out replacement is costly. Attempts to re-manufacture them were unsuccessful because teeth have variable module and height. As a result interference takes place on re-manufactured mechanisms when the rack travels from mid-position. This may lock the steering system with catastrophic consequences. To investigate the meshing of virgin and remanufactured rack and pinion "Mechanica" software was used ("Motion" module). The clash points were detected and recommendations for profile modification were developed to prevent interference.

1 Introduction

A scheme of a typical rack and pinion mechanism of a car steering system is shown in Fig.1 below. It was invented by the Australian inventor Arthur Bishop in mid-seventies and now is standard on most small, medium and family cars. In the initial state when the rack and pinion mechanism is assembled, the screw supporting the rack is adjusted to provide zero backlash in the mid-position and locked by the locknut. It is assumed that the piston axis lies in the rack's pitch plane, and the pinion middle tooth axis of symmetry is perpendicular to the rack's pitch plane. To find the pinion's proper position, serrated spline on the output shaft has four splines removed (at 90°) to provide four groves one of which has to coincide with the mark on the casing.

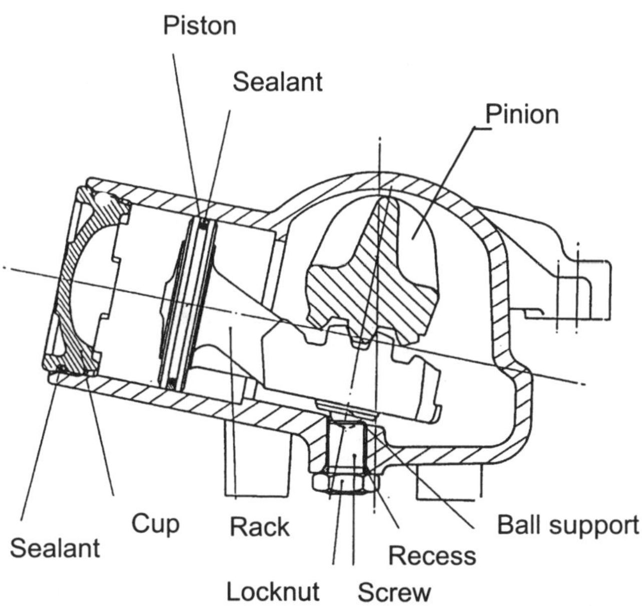

Fig.1. Scheme of the rack and pinion mechanism

Analysis of the teeth profile on the basis of the measurements taken and laser scanning showed that the profile resembles the convolute profile with substantial profile modification. The pinion has three working teeth with variable pitch circle (bigger for the mid-tooth and smaller for side teeth. The rack has two working teeth with almost flat root surface. The tip surfaces are composed of planes - one forms the tips and is parallel to the pitch line, and another is at an angle of 24° to the pitch line. The main aim of the profile modification is to provide variable backlash, which is zero when the rack is in the mid-position, and takes its maximum value at the end of the piston rack travel, as well as to provide variable turning speed of wheels when the steering wheel is turned uniformly.

New (virgin) mechanisms work perfectly and are widely used on small and medium cars. Millions of worn out mechanisms are replaced every year. Their recycling is costly. Attempts to re-manufacture them retaining the virgin profile had been unsuccessful because the meshing process on re-manufactured profiles was different.

When 1 mm allowance was machined of by grinding on worn out rack and pinion and the unit was assembled, it was detected that somewhere between the mid-position of the rack and its far most position the mechanism gets locked. To investigate the teeth meshing both rack and pinion profiles were scanned on a CNC laser scanner and then profiles were reproduced in a computer model. "Mechanica" CAE (Computer-Aided Engineering) software was used for computer modelling, in particular "Mechanica-Motion" module.

2 Description of Computer Models

For modelling of the meshing process in the rack and pinion mechanism a number of computer models were created using "Mechanica" software.

Model 1. A simplified model for modelling of the meshing process in a new (virgin) rack and pinion mechanism is shown in Fig. 2 below.

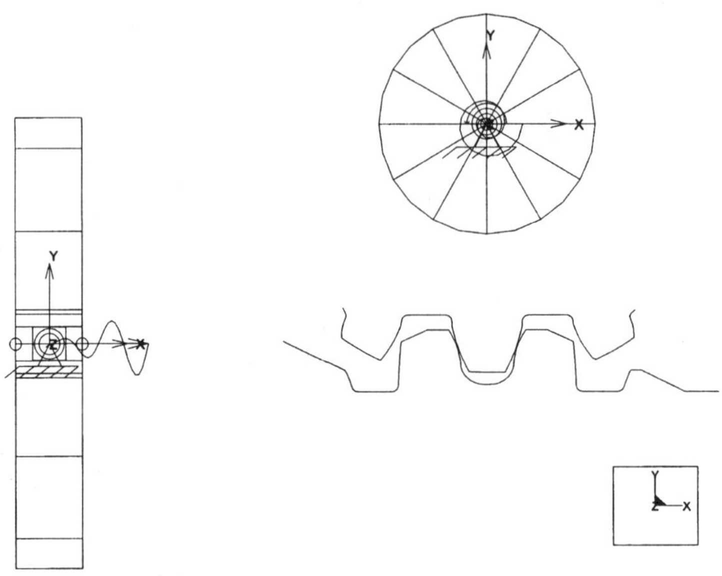

Figure2 Simplified model of a new rack and pinion mechanism

To model the rack motion it was connected with the slider joint to the "ground point" which represents the casing (here and further below specific terminology of the "Mechanica" software package is used). The pinion was connected by a pin joint to another "ground point". Spirals in Fig.2 denote "drivers", which allow to specify the displacement, velocity or acceleration of moving parts during computer modelling. Fixed or variable values of parameters can be specified. It was assumed that the rack moves in the horizontal plane, which coincides with the rack's pitch plane.

The modelling time was set to 23 seconds with the time increment of 1 second. This produced 23 frames during animation. These frames could be reproduced in motion, "frozen" individually or in groups and printed out.

During different runs the velocity of the rack was adjusted to have particular flank surfaces of the mating teeth in touch - contact.

The piston moving out of the cylinder was assumed mowing forward and the backward motion when the piston is retracted. To model the backward motion the drivers velocity was given negative value.

Model 2. The advanced model enables the modelling of the meshing process in the rack and pinion mechanism varying the angular position of the rack in space (see Fig. 3 below). To achieve this the piston was connected to a "ground point" by a combination of a slider and a pin joints, to enable the combined linear motion of the rack and its rotation about the centre point of the piston. A similar combination of two joints was created at the point where the adjusting screw supports the rack. All other features of the model were similar to the first model.

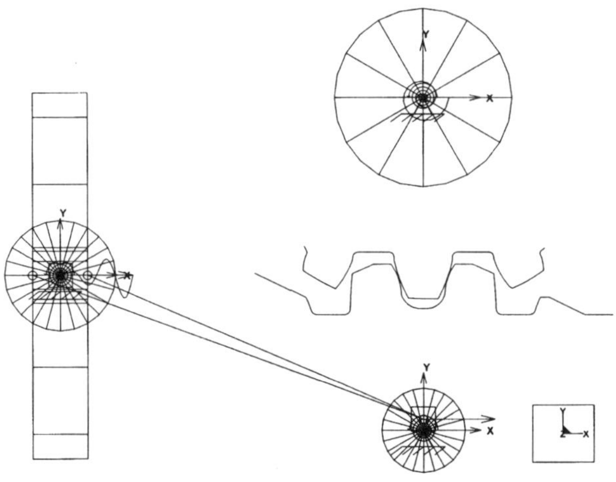

Fig.3. Advanced model of a new rack and pinion mechanism

To model the meshing process of a remanufactured pinion with a new rack, the contour of the pinion was shifted upwards 1 mm (to imitate the allowance of 1 mm removed), and the contour of the rack was also shifted by rotating it about the centre of the piston to provide the zero initial backlash.

To model the meshing process of a remanufactured pinion and a remanufactured rack, the contour of the pinion was shifted upwards 1 mm, and the contour of the rack was shifted downwards 1 mm, and also rotated to adjust its position and provide zero initial backlash in the mid-position as it takes place in a virgin mechanism.

In both cases models were run in forward and backward directions to investigate meshing process, to establish whether interference takes place and to determine the values of backlash at different points along the travel as well as to verify whether actual values of the backlash comply with the specification.

3 Discussion of Results of Computer Modelling

When a new rack and pinion mechanism is modelled using a simplified model, there is a backlash of 0.2 mm both on forward and backwards runs (see Fig 4 and Fig. 5) which is close to the value of the backlash 0.21 mm specified by the documentation.

Figure 4. Meshing of a new mechanism on a forward run

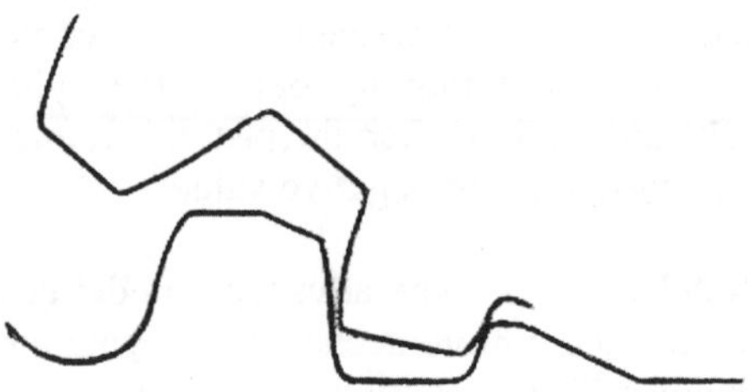

Figure 5. Meshing of a new mechanism on a backward run

When a new rack and pinion mechanism is modelled using an advanced model, the backlash is also 0.2 mm on forward run and 0.21 mm on backward run. This is not surprising, because in a new mechanism there is no shift in teeth contour position, and the advanced model resembles the simplified model. When a re-manufactured pinion is engaged with a new rack, an interference of 0.26 mm takes place at the end of the rack travel on forward run (Fig.6), and the backlash of 0.42 mm on the backward run (Fig.7).

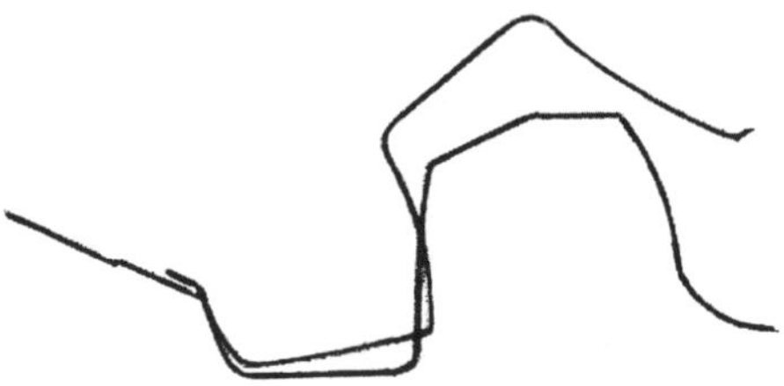

Figure 6. Meshing of a new rack with re-manufactured pinion (forward run)

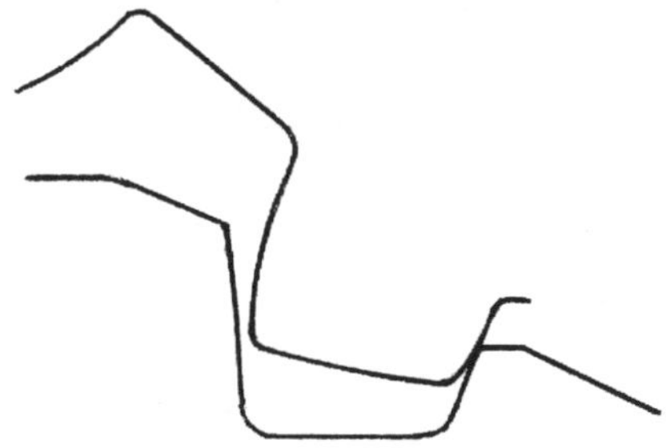

Figure 7. Meshing of a new rack with re-manufactured pinion (backward run)

When a re-manufactured pinion is engaged with a remanufactured rack, on the forward run the interference is likely to happen both between the flank surfaces and tip and root surfaces (see Fig.8). On the backward run the backlash of 0.75 mm is expected at the end of the rack travel (see Fig.9), and the flank surface of the pinion tooth interacts with fillet surface of the rack tooth causing substantial contact stresses and extensive wear.

Figure 8. Meshing of a re-manufactured rack and pinion (forward run)

Figure 9. Meshing of a re-manufactured rack and pinion (backward run)

When the rack's profile is reground at an angle, and the supporting surface is reground at an angle as well, being engaged with the remanufactured pinion it gives zero backlash both in the mid - position and at the end of the rack travel both in forward and backward directions. This happens because when the pinion is remanufactured, its pitch diameter becomes smaller, and the difference between pitch radii of the middle tooth and the side tooth becomes greater than on a new pinion. This results in decreased value of the backlash (in this case zero value) which is contrary to specifications.

Since on re-manufactured rack and pinion mechanism (when 1 mm is ground off), the pattern of teeth meshing is significantly changed (the interference takes place on the forward run and increased clearance and contact stresses on the backward run) the teeth profile correction is required.

4 Correction of the Teeth Profile

Computer modelling showed that in the case of removal of 1 mm allowance from the pinion profile (in the direction normal to the rack) the new pinion profile becomes superimposed on the old (worn out) profile. External flank surface of the side tooth is not completely machined. To prevent this the tooth has to be rotated about the origin closer to the mid-tooth. In this case the inner flank surface becomes superimposed on the old profile. As a result some dents are left on the flank surface close to the root (see Fig.10 below), and the interference takes place with the rack profile.

The only way to eliminate the problem is to reduce the tooth thickness by shifting the inner flank surface of the side tooth (see Fig.11).

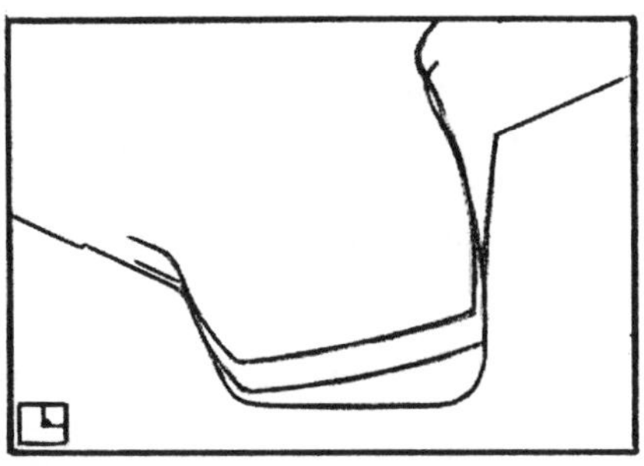

Figure 10. Superimposed initial and re-manufactured profiles of the side pinion teeth

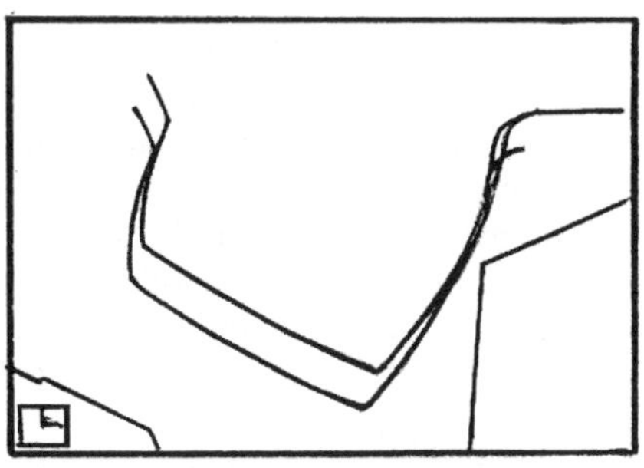

Figure 11. Modified pinion profile superimposed on the initial and re-manufactured profiles

To provide the proper meshing of the pinion and the rack, and equal backlash at the end of the travel in both directions, two conditions have to be satisfied:

1. The pitch line of the rack has to be perpendicular to the pitch circle diameter of the pinion sector mid-tooth when it is in the initial position.
2. The rack has to travel in the direction parallel to the rack's pitch line.

When 1 mm allowance is removed from the pinion sector, and the rack is shifted towards the pinion sector to provide engagement, the rack takes a tilt position and swings during travel, which causes the teeth interference and other problems discussed above. To prevent this the rack's back surface has to be re-ground at an angle to make it perpendicular to the pinion's pitch circle diameter when the pinion is in its initial position. The rack's profile has to be re-machined as well to make it parallel to the back surface.

The tilt angle has to be calculated for particular steering mechanism depending on the dimensions of mating rack and pinion.

It is necessary to make sure that the adjusting screw is long enough to bring re-manufactured rack and pinion in touch-contact when re-manufactured mechanism is assembled and tuned.

5 Development of Modified Rack and Pinion Profiles

On the basis of computer modelling modified profiles of rack and pinion were developed. New series of animations had been carried out, which proved that no interference takes place between profile elements. The backlash at the end of the travel was 0.21 mm, which is close to that for a new mechanism predetermined by specifications. More over, approximately at the mid-distance of the rack travel, two teeth are engaged at a time, which increases loading capacity of the re-manufactured mechanism.

These modified rack and pinion profiles were "exported" from "Mechanica" software as MNF and IGES files and used to program a CNC grinding machine.

6 Conclusions

A computer modelling was conducted using Mechanica – Motion software package. Simplified and advanced models of the rack and pinion mechanism of a car power steering system were developed. Modelling was carried out for the following cases: a new mechanism; a remanufactured pinion engaged with a new rack; a remanufactured pinion engaged with a remanufactured rack.

It was shown that for a new mechanism the value of the backlash at the end of the piston rack travel is close to the value of the backlash specified by the documentation, which proves the accuracy of the models.

A re-manufacturing of the mechanism results in significant change in the teeth meshing pattern, and the complication of this is the interference on the forward run, and increased backlash and contact stresses on the backwards run.

In the case of remanufacturing of the rack and pinion mechanism the teeth profile correction is required.

New profiles were developed for re-manufacturing of the rack and pinion mechanism.

Computer modelling of the mashing process showed that parameters of the remanufactured mechanism are within specifications for a new mechanism.

MATHEMATICAL AND EXPERIMENTAL MODELLING OF A TWO-ROLLER CONVEYOR FOR SEMI-SELF-FLOWING TRANSPORTATION OF PARTS

V.KOSSE
School of Mechanical, Manufacturing and Medical Engineering, Queensland University of Technology, Gardens Point Campus, GPO Box 2434, Brisbane, Qld. 4001, Australia

In conveyors utilising the principle of semi-self-flowing transportation parts are put on the surface moving in the lateral direction by means of vibration or on two rollers rotating in opposite direction. Advantages of the latter one are that parts can be supplied on demand, and very small angle of inclination is required (less than 3°). Formulas found in scientific literature for the velocity of transportation as a function of the roller velocity of rotation give large discrepancy with experimental results. Extensive experimental modelling allowed to establish the dependence of the frictional coefficient on the shape of parts and the contact angle. A new formula was derived which is in good agreement with experimental data.

1 Introduction

Self-flowing and semi-self-flowing transportation of workpieces are widely used in automated manufacturing systems. To provide self-flowing transportation of workpieces it is necessary to tilt the chute at an angle greater than the angle of repose (8° to 17° for metals). Main shortcomings of self-flowing transportation are as follows:

- Sliding velocity of workpieces cannot be controlled and for a given tilt angle is entirely determined by the coefficient of friction and the gravity forces.
- Sliding workpieces may collide and damage each other.
- Relatively large tilt angle (greater than 17°) requires large overfall (the difference in height) between the upper and the lower ends of the chute, which makes it unacceptable if workpieces are to be transported in the horizontal plane.

Semi-self-flowing transportation is used for small tilt angles (usually under 3°). For such a small tilt angles there are only two ways to make workpieces slide by themselves:

- By reducing the coefficient of kinetic friction significantly (for example, by employing an aircushion).
- By making the supporting surface of the chute move in the lateral direction.

The second method can be implemented by applying lateral vibrations to the chute or by designing a chute consisting of two parallel rollers rotating in opposite directions.

In the case of a vibrating chute sliding velocity of workpieces cannot be easily controlled. Then vibrating appliances have well-known shortcomings – unpleasant noise and vibrations transmitted to the surroundings.

A two-roller conveyor for semi-self-flowing transportation (see Fig.1) is quite simple. Longitudinal velocity of workpieces can be easily controlled by varying the angular velocity of the rollers. But the problem is that currently there is no a comprehensive theory that explains the mechanics of semi-self-flowing transportation and gives mathematical apparatus for analytical determination of the conveying velocity. For example, formula (1) suggested in [1] gives controversial results. In some cases coincidence with experimental results is excellent, in other cases the difference is by one order.

$$\mathbf{V_L} = \boldsymbol{\omega}\cdot\mathbf{R}\cdot(\tan\boldsymbol{\alpha} / \boldsymbol{\mu}) \tag{1}$$

Where: ω = angular velocity of rollers, (rad/s); μ = coefficient of kinetic friction;
$\mathbf{R}$ = radius of rollers, (m); α = tilt angle (Deg.).

In this work an attempt is made to understand the mechanics of semi-self-flowing transportation on two-roller conveyors and by means of theoretical and experimental investigation develop a more accurate formula describing longitudinal velocity of transportation.

2 Analysis of preliminary experimental data and formulation of initial assumptions

Two series of experiments have been conducted to verify known formula and collect data for further investigation.

A scheme of the experimental test rig of the two-roller conveyor is shown in Fig.1 below. Two rollers 1 are made of steel and supported by bearings 2 and 3 and synchronised by gears built in the support 3. Rollers are driven by an electric motor 5 through a variator 4, which provides variable speed of roller rotation. The whole structure is mounted on the board 6, which rests on two supports 7 and a jack 9, which allows the angle of inclination to be changed. The inclination angle can be measured (or preset) by means of a gauge 8 with bubble level. The variator allows the angular velocity of rollers to be changed gradually within the range of 8 to 330 rpm.

Major parameters of the test rig are as follows:

- Centre to centre distance of rollers: $\mathbf{A} = 0.058$ m.
- Roller radius: $\mathbf{R} = 0.023$ m.
- Angular velocity of the roller: $\omega = 0.911$ rad/sec.
- Peripheral velocity of the roller: $\mathbf{V_p} = 2.095 \cdot 10^{-2}$ m/sec.

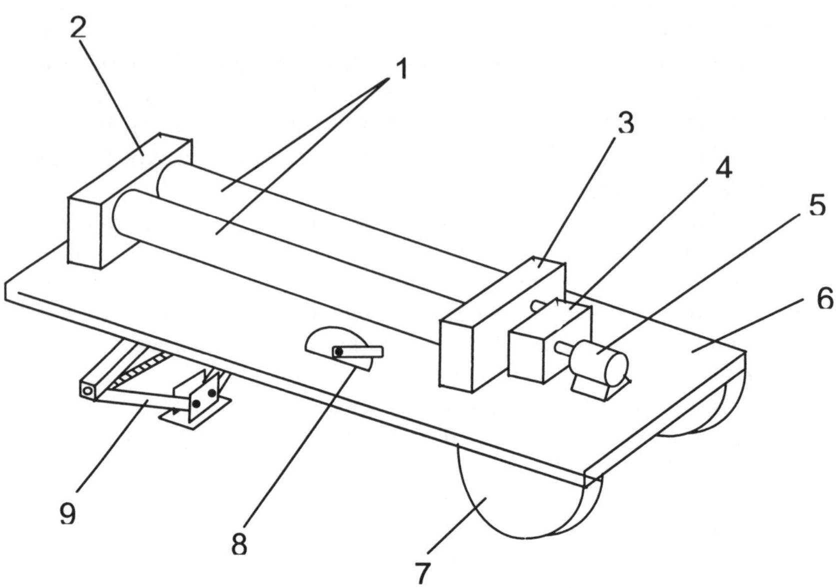

Figure 1. Scheme of a two-roller conveyor for semi-self-flowing transportation of parts

For the first series of experiments two sets of workpieces have been fabricated shaped as headed pins with shank diameter of 11mm. One set had a cylindrical head with diameter varying from 15 to 60 mm in increments of 5mm (see Fig.2, a). The second set had an inverted cone head with the cone angle in the range of 30° to 150° in increments of 10° (see Fig.2, b).

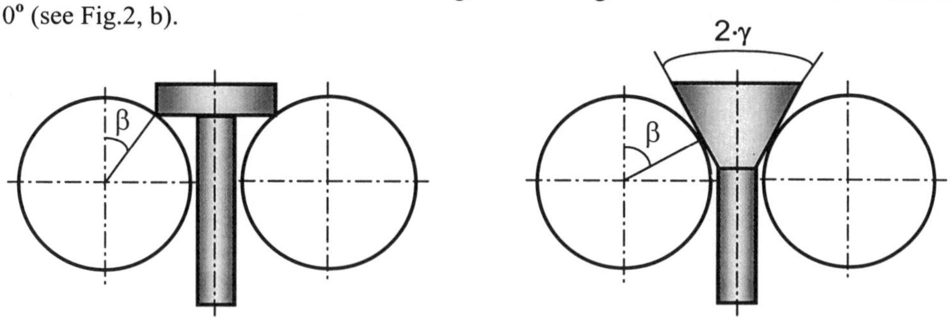

(a) Pins with cylindrical head (b) Pins with inverted cone head

Figure 2. Headed pins on a two-roller conveyor

Preliminary experiments have been conducted to determine the coefficient of friction by means of measuring the angle of repose. For this purpose a workpiece with the largest head was used. The value of the coefficient of kinetic friction was $\mu = 0.18$.

For the first series of experiments the tilt angle was set to $\alpha = 3°$. Workpieces with cylindrical head were used. According to formula (1), for a given tilt angle α and the coefficient of kinetic friction μ, the velocity of conveyance $\mathbf{V_L}$ (in longitudinal direction) is a linear function of peripheral velocity of the rollers $\mathbf{V_p} = \omega \cdot \mathbf{R}$ regardless of the shape of workpieces. However, experimental values of $\mathbf{V_L}$ were different for different workpieces. In particular when the workpiece head diameter was large (55mm), the experimental value of $\mathbf{V_L}$ was very close to that obtained by formula (1), 0.0066 m/sec and 0.0061 m/sec respectively. When the head diameter was small (18.2mm) experimental value of $\mathbf{V_L}$ was 0.0014 /sec. In the meantime formula (1) still suggest the same value of 0.0061 m/sec. This allows us to assume that the head diameter or its shape affects the velocity of conveyance directly or influences the coefficient of friction.

Head diameter **d** (in the case of cylindrical head) or the cone angle $2 \cdot \gamma$ (in the case of inverted cone head) determine the position of the point of contact of head with a roller. Location of this point can be specified by the contact angle β (see Fig.2).

The value of the contact angle can be easily calculated by the following expressions:

- In the case of cylindrical head of diameter **d**

$$\sin\beta = (\mathbf{A} - \mathbf{d}) / (2 \cdot \mathbf{R}) \tag{2}$$

- In the case of inverted cone head with the cone angle $2 \cdot \gamma$

$$\beta = 90° - \gamma \tag{3}$$

The second series of experiments was conducted to determine whether the coefficient of kinetic friction μ is dependent on the contact angle β and to establish relationship between the velocity of conveyance $\mathbf{V_L}$ and peripheral roller velocity $\mathbf{V_p}$ by unitless **K** – factor supposedly dependent on the contact angle.

$$V_L = V_p \cdot K \tag{4}$$

To determine the coefficient of kinetic friction μ for each workpiece and contact angle β the tilt angle was gradually increased, and the conveyor was repeatedly switched on and off to induce sliding of workpieces along the rollers.

To determine the **K**-factor the tilt angle was set to 3° and for each workpiece (and corresponding contact angle β) the velocity of conveyance was determined by counting the time elapsed while the workpiece cowered the distance of 0.5m along the rollers. Results of this experiment are shown in Fig. 3.

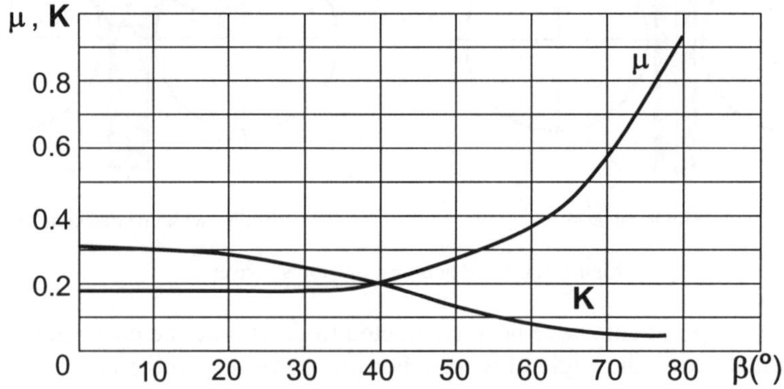

Figure 3. Variation of frictional coefficient μ and **K**-factor with contact angle β

As is seen from the diagram the coefficient of kinetic friction μ is constant within the range of contact angles 0 to 30°, then increases and reaches 0.56 when the contact angle is 70° (three times that for zero contact angle). The **K**-factor has the largest value when the contact angle β is within the range of 0 to 20° (workpieces with large head), then with increasing contact angles declines and approaches zero when β exceeds 70°.

It is impossible to explain this phenomenon on the basis of Amontons-Coulomb hypotheses. Probably a small size head (or head with a small cone angle) behaves like a wedge in a tight slot increasing the coefficient of friction. This allows us to derive an important conclusion that the coefficient of kinetic friction varies with the contact angle. On the basis of experimental data the coefficient of kinetic friction can be expressed as a function of the contact angle β with the following formula

$$\mu = 2.3 \cdot 10^{-10} \cdot (\beta)^5 + 0.18 \tag{5}$$

Here 0.18 represents the coefficient of kinetic friction corresponding to zero contact angle. It is clear that formula (5) will be different for different combination of roller and workpiece materials as well as surface texture.

If using formula (1) recalculate the velocity of conveyance V_L when $\alpha = 3^o$ and $\beta = 60^o$, taking into account variations of the coefficient of friction μ with contact angle β, we will have $V_L = 0.0031$ m/sec, which is much less of the initial value $V_L = 0.0061$ m/sec calculated under the assumption that μ is independent of contact angle β, but still is much greater that the experimental value $V_L = 0.0014$ m/sec.

These two series of experiments demonstrated that it is not sufficient just to take into account the dependence of the coefficient of kinetic friction μ on the contact angle β. It is necessary to derive a new versatile formula that reflects the mechanics of semi-self-flowing transportation of workpieces of different size and shape.

3 Theoretical investigation of semi-self-flowing transportation of workpieces on two-roller conveyors

When rollers are stationary and the tilt angle is below the angle of repose, workpieces do not move. When rollers rotate with peripheral velocity $\mathbf{V_p}$ workpieces move along the rollers with velocity $\mathbf{V_L}$. The point of contact of the workpiece head with the roller leaves on its surface an imaginary track, which is a helix. The helix angle θ defines the relationship between the peripheral roller velocity $\mathbf{V_p}$ and the velocity of conveyance $\mathbf{V_L}$ (see Fig. 4) as follows

$$\mathbf{V_L = V_p \cdot \cot\theta = V_p \cdot K} \tag{6}$$

Since $\mathbf{V_p}$ is given, to determine $\mathbf{V_L}$ it is necessary to derive expression for $\mathbf{\cot\theta}$.

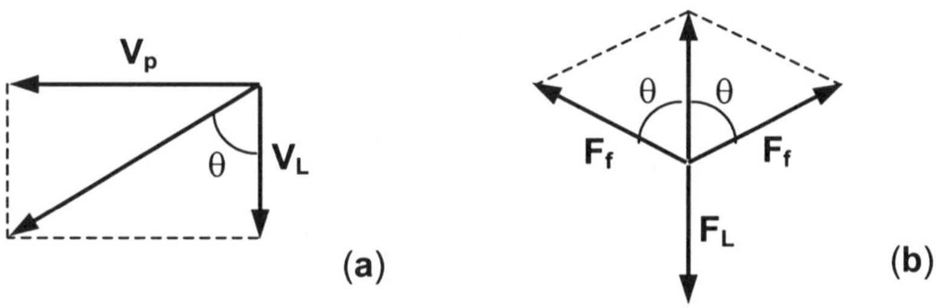

Figure 4. Relationship of velocity and force vectors

Since the workpiece moves with uniform velocity, according to Newton's first low all forces acting on it are in equilibrium. The force $\mathbf{F_L}$ is the component of the gravity force acting along the rollers (see Fig.4, b) which can be expressed as

$$\mathbf{F_L = m \cdot g \cdot \sin\alpha} \tag{7}$$

Here $\mathbf{m \cdot g}$ = weight of a workpiece (N).

Two equal forces $\mathbf{F_f}$ tangential to the helixes on each roller are the frictional forces acting at the points of contact of the workpiece with the rollers. Assuming that the frictional force is proportional to the normal force **N** acting at the point of contact [2], and keeping in mind that the coefficient of kinetic friction μ is dependent on the contact angle β, then

$$\mathbf{F_f = N \cdot \mu} \tag{8}$$

The limited size of this paper does not allow to describe in details the mathematical manipulations. The final formula will be as follows

$$\mathbf{K = \cot\theta = (\tan\alpha \cdot \cos\beta) / \sqrt{[\mu^2 - (\tan\alpha)^2 \cdot (\cos\beta)^2]}} \tag{9}$$

4 Numerical example and discussion

Results of numerical calculations based on the more accurate formula (9) in comparison with experimental results are shown in the table below. As is seen from the table theoretical results are in good agreement with experimental data. This proves that formula (9) is valid for the whole range of contact angles β and tilt angles α. Comparing expressions (1) and (9) we notice that if to put $\beta = 0$ in (9), we will obtain an expression similar to (1) with the only difference being the component $(\mathbf{tan}\alpha)^2$, which is negligible for $\alpha \leq 3^o$.

Table 1 Comparison of experimental and theoretical results

D(mm)	α(Deg)	β(Deg)	μ	$V_{L, theory}$ (m/sec)	$V_{L, exper}$ (m/sec)	% Difference
42	3	20.35	0.181	0.282	0.295	– 4.44
20	3	55.70	0.303	0.098	0.103	– 4.85
42	10	20.35	0.181	2.244	2.37	– 5.32
20	10	55.70	0.303	0.347	0.37	– 6.2
20	25	55.70	0.303	1.742	1.85	– 5.84
16	30	65.93	0.467	0.584	0.63	– 7.30

Thus, we can state that formula (1) is valid when the contact angle β is small (workpieces with large head) and the tilt angle α is under 3°, which has been proved by experimental results as well.

5 Conclusions

On the basis of theoretical and experimental investigation the following conclusions can be drawn:

- Approximate formula for the velocity of conveyance can be used only for small tilt angles $\alpha \leq 3^o$ and small contact angles $\beta \leq 5^o$.
- The coefficient of kinetic friction is dependent on the contact angle β. It can be assumed constant only for contact angles $\beta \leq 40^o$.
- A more accurate formula for the velocity of conveyance was derived which takes into account the tilt angle, the contact angle, peripheral velocity of rollers and variation of the coefficient of friction with the contact angle.
- When the tilt angle is small ($\alpha \leq 7.5^o$), the velocity of conveyance is not sensitive to the contact angle variations.
- When the tilt angle is less than half of the angle of friction (corresponding to the particular contact angle), the velocity of conveyance increases almost linearly with the increase of the tilt angle.
- Two-roller conveyors can be used successfully for the wide range of tilt angles from 1° to 60°.
- Two-roller conveyors provide broad possibilities for control over the velocity of conveyance by varying the tilt angle, the contact angle and the peripheral velocity of rollers.

References

1. **Bobrov V.** and **Chekanov L.**, Conveying and loading devices for transfer lines, *Mashinistroenie, Moscow,* 1980**,** (in Russian).
2. **Riley W. F.** and **Steggers L. D.** Engineering mechanics - Statics, *John Wiley & Sons,* New York, 1993, (p. 378).

DETERMINATION OF THE TOPOLOGY AND STIFFNESS OF A MASS SPRING NETWORK FOR CLOTH DRAPING SIMULATION

C. K. AU
School of Mechanical and Production Engineering
Nanyang Technological University, Singapore
E-mail: mckau@ntu.edu.sg

Z. WU
Department of Mechanical Engineering
Hong Kong University of Science and Technology University

A piece of cloth can be modelled physically by using a mass spring network. The realism and accuracy of the model are largely dependent on the network characteristic, which are defined by the network topology and spring stiffness. This paper presents the derivation of these two quantities based on the strain energy of the cloth. The fabric properties are incorporated into the network characteristic. As a result, distinguishable fabric can be simulated realistically. Draping examples are raised to compare with the real situation.

1 Introduction

Cloth modelling has received considerable attention in computer graphics recently and this deformable modelling technique is mainly for animation and engineering purpose. The detail surveys on the cloth modelling can be found in references [1-3].

Mass-spring network is widely used in modelling deformable objects. Provot [4] used the mass spring network to model a piece of hanging cloth. Howlett [5] extended the data structure of the model by adding the non-active point so that the cloth adapted the shape of the object when draping over it. Hutchinson [6] refined the mass spring network by a hierarchical approach so that it draped over an object with the shape edge.

The realism and the accuracy of cloth model are largely dependent on the network characteristics, which are determined by the network topology and the spring stiffness constants. The network topology refers to the connectivity of the mass points within the network, while the stiffness of the springs characterizes the material properties of the fabric. These two elements in the mass spring network discussed in references [4], [5] and [6] are determined heuristically. Fine tuning and additional constraints are imposed on the network making its appearance looks like the physical phenomenon.

This paper rationalizes the determination of the network topology and the stiffness constants for cloth draping simulation based on the strain energy analysis.

2 Mass Spring Network

A piece of cloth is discretized as a mesh of m×n grid of nodes with the node numbering scheme as shown in figure 1. Applying Newton's second law of motion to the node i in the structure gives:

$$m_i\ddot{\mathbf{x}}_i + \gamma_i\dot{\mathbf{x}} + \sum \mathbf{f}_i = \mathbf{p}_i \tag{1}$$

where m_i is the mass at node i, $\mathbf{x}_i$ is its position, γ_i is the damping coefficient, $\mathbf{f}_i$ is the internal forces exerting on node i, $\mathbf{p}_i$ is the external forces acting on it.

The internal forces $\mathbf{f}_i$ at node i are the in-plane forces and out-of-plane forces, which are represented by the spring forces in the network.

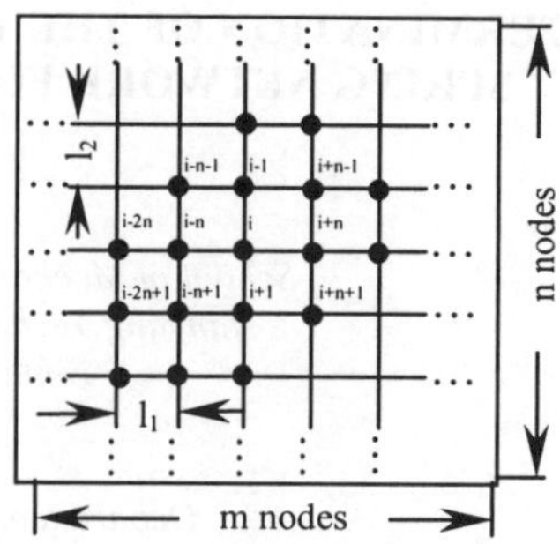

Figure 1: The node numbering scheme

3 Network Topology and Stiffness Constants

Let U_i denotes the strain energy stored in the network at node i during deformation. Assuming that there is no interaction between the in-plane behavior and the out-of-plane behavior, the energy U_i can be decomposed into the in-plane membrane strain energy U_G^i and the out-of-plane bending strain energy U_B^i,

$$U_i = U_G^i + U_B^i \tag{2}$$

Expressing U_G^i and U_B^i at $\mathbf{x}_i$ in terms of the corresponding strain tensors gives

$$U_G^i = \frac{1}{2}A_k \varepsilon_{\alpha\beta}^i C_1^{\alpha\beta\delta\gamma} \varepsilon_{\delta\gamma}^i, \tag{3}$$

$$U_B^i = \frac{1}{2}A_i \kappa_{\alpha\beta}^i C_2^{\alpha\beta\delta\gamma} \kappa_{\delta\gamma}^i \tag{4}$$

where ε and κ are the Green's membrane and bending strain tensors respectively and A_i is the area at node i; $C_1^{\alpha\beta\delta\gamma}$ and $C_2^{\alpha\beta\delta\gamma}$ are the material modulus tensors of the fabric due to in-plane and out-of-plane forces. These elastic modulus tensors are symmetric tensors, hence $C^{\alpha\beta\delta\gamma} = C^{\alpha\beta\gamma\delta} = C^{\beta\alpha\delta\gamma} = C^{\delta\gamma\alpha\beta}$. When the warp and weft directions coincide with the principle coordinate system of the fabric, $C^{1112} = C^{1211} = C^{1222} = C^{2212} = 0$ and only five components of the tensor C^{1111}, C^{1122}, C^{1212}, C^{2211} and C^{2222}, are non-zero[7].

Expressing the Green's membrane and bending strain tensors in terms of metric $G_{\alpha\beta}^i$ and curvature $B_{\alpha\beta}^i$ tensors yields

$$U_G^i = \frac{1}{8}\left(G_{\alpha\beta}^i - G_{\alpha\beta}^{0^i}\right)A_i C_1^{\alpha\beta\delta\gamma}\left(G_{\delta\gamma}^i - G_{\delta\gamma}^{0^i}\right), \tag{5}$$

$$U_B^i = \frac{1}{2}\left(\overline{B}_{\alpha\beta}^i - \overline{B}_{\alpha\beta}^{0^i}\right)A_i C_2^{\alpha\beta\delta\gamma}\left(\overline{B}_{\delta\gamma}^i - \overline{B}_{\delta\gamma}^{0^i}\right) \tag{6}$$

where $\varepsilon_{\alpha\beta}^i = \frac{1}{2}\left(G_{\alpha\beta}^k - G_{\alpha\beta}^{0^k}\right)$; $\kappa_{\alpha\beta}^i = B_{\alpha\beta}^i - B_{\alpha\beta}^{0^i}$ and the superscript 0 denotes the initial state. The discretized metric and curvature tensors are

$$G_{\alpha\beta}^i = \frac{\Delta_\alpha \mathbf{x}_i}{l_\alpha} \cdot \frac{\Delta_\beta \mathbf{x}_i}{l_\beta} \tag{7}$$

$$B^i_{\alpha\beta} = n_i \cdot \frac{\Delta_\alpha \Delta_\beta \mathbf{x}_i}{l_\alpha l_\beta} \tag{8}$$

where $\Delta_\alpha \mathbf{x}_i$ is the change of $\mathbf{x}_i$ with respect to direction α and $\mathbf{n}_i$ is the unit normal vector at $\mathbf{x}_i$. Substituting equation (7) and (8) into equation (5) and (6) generates

$$U^i_G = U^i_G(\mathbf{x}_{i+n}, \mathbf{x}_i, \mathbf{x}_{i-1}) \text{ and} \tag{9}$$

$$U^i_B = U^i_B(\mathbf{x}_{i-n}, \mathbf{x}_{i-1}, \mathbf{x}_i, \mathbf{x}_{i+1}, \mathbf{x}_{i+n-1}, \mathbf{x}_{i+n}) \tag{10}$$

Let $\mathbf{f}^i_G = \sum_j \frac{\partial U^j_G}{\partial \mathbf{x}_i}$ and $\mathbf{f}^i_B = \sum_j \frac{\partial U^j_B}{\partial \mathbf{x}_i}$ be the in-plane and out-of-plane forces acting on the node i respectively ($\mathbf{f}_i = \mathbf{f}^i_G + \mathbf{f}^i_B$), then these forces are written as

$$\mathbf{f}^i_G = \frac{\partial U^i_G}{\partial \mathbf{x}_i} + \frac{\partial U^{i-n}_G}{\partial \mathbf{x}_i} + \frac{\partial U^{i+1}_G}{\partial \mathbf{x}_i} \text{ and} \tag{11}$$

$$\mathbf{f}^i_B = \frac{\partial U^i_B}{\partial \mathbf{x}_i} + \frac{\partial U^{i+n}_B}{\partial \mathbf{x}_i} + \frac{\partial U^{i-1}_B}{\partial \mathbf{x}_i} + \frac{\partial U^{i+1}_B}{\partial \mathbf{x}_i} + \frac{\partial U^{i-n}_B}{\partial \mathbf{x}_i} + \frac{\partial U^{i-n+1}_B}{\partial \mathbf{x}_i} \tag{12}$$

Substituting equation (5) and (6) into equation (11) and (12) gives

$$\begin{aligned}
\mathbf{f}^i_G = \; & k^{11}_G(i)(\mathbf{x}_i - \mathbf{x}_{i+n}) + k^{22}_G(i)(\mathbf{x}_i - \mathbf{x}_{i-1}) + k^{12}_G(i)\big((\mathbf{x}_i - \mathbf{x}_{i+n}) + (\mathbf{x}_i - \mathbf{x}_{i-1})\big) + \\
& k^{11}_G(i-n)(\mathbf{x}_i - \mathbf{x}_{i-n}) + k^{12}_G(i-n)(\mathbf{x}_{i-n-1} - \mathbf{x}_{i-n}) + \\
& k^{22}_G(i+1)(\mathbf{x}_i - \mathbf{x}_{i+1}) + k^{12}_G(i+1)(\mathbf{x}_{i+n+1} - \mathbf{x}_{i+1})
\end{aligned} \tag{13}$$

$$\begin{aligned}
\mathbf{f}^i_B = \; & \big(k^{1111}_B(i) + k^{1122}_B(i) + k^{2211}_B(i) + k^{1212}_B(i) + k^{2222}_B(i)\big)\mathbf{n}_i + \\
& \big(k^{1111}_B(i+n) + k^{2211}_B(i+n)\big)\mathbf{n}_{i+1} + \big(k^{1122}_B(i-1) + k^{2222}_B(i-1)\big)\mathbf{n}_{i-1} + \\
& \big(k^{1122}_B(i+1) + k^{1212}_B(i+1) + k^{2222}_B(i+1)\big)\mathbf{n}_{i+1} + \\
& \big(k^{1111}_B(i-n) + k^{1122}_B(i-n) + k^{1212}_B(i-n)\big)\mathbf{n}_{i-n} + k^{1212}_B(i-n+1)\mathbf{n}_{i-n+1}
\end{aligned} \tag{14}$$

where

$$k^{11}_G(i) = \frac{A_i}{2l_1^2} C_1^{1111}(G^i_{11} - G^{0^i}_{11}) + \frac{A_i}{4l_1 l_2}(C_1^{1122} + C_1^{2211})(G^i_{22} - G^{0^i}_{22});$$

$$k^{22}_G(i) = \frac{A_i}{2l_2^2} C_1^{2222}(G^i_{22} - G^{0^i}_{22}) + \frac{A_i}{4l_1 l_2}(C_1^{1122} + C_1^{2211})(G^i_{11} - G^{0^i}_{11});$$

$$k^{12}_G(i) = \frac{A_i}{4l_1 l_2} C_1^{1212}(G^i_{12} - G^{0^i}_{12});$$

$$k^{1111}_B(i) = -\frac{A_i}{l_1^2} C_2^{1111}(B^i_{11} - B^{0^i}_{11});$$

$$k^{1122}_B(i) = -\frac{A_i}{l_2^2}(C_2^{1122} + C_2^{2211})(B^i_{11} - B^{0^i}_{11});$$

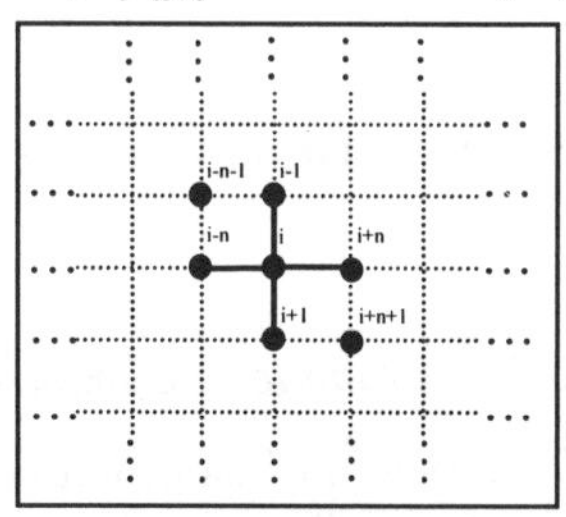

(a) tensile forces

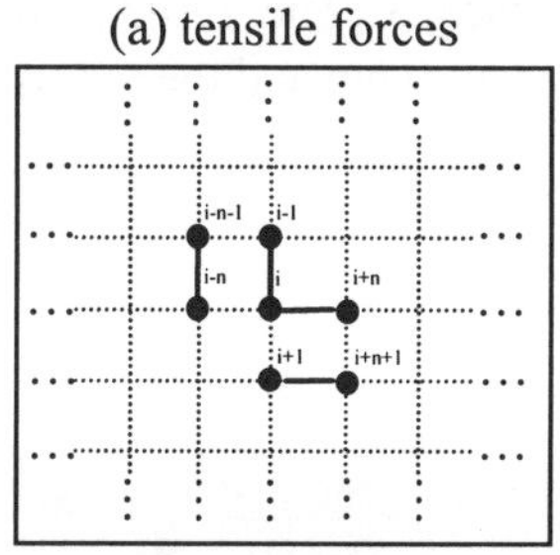

(b) shear forces

$$k_B^{2211}(i) = -\frac{A_i}{l_1^2}(C_2^{1122} + C_2^{2211})(B_{22}^{i} - B_{22}^{0^i});$$

$$k_B^{1212}(i) = \frac{A_i}{l_1 l_2} C_2^{1212}(B_{12}^{i} - B_{12}^{0^i}) \text{ and}$$

$$k_B^{2222}(i) = -\frac{2A_i}{l_2^2} C_2^{2222}(B_{22}^{i} - B_{22}^{0^i})$$

Hence, there are three types of forces acting on each point mass: tensile forces, shear forces and bending forces as depicted in figure 2. Combining the effects together describes how a node is related to its neighbouring nodes.

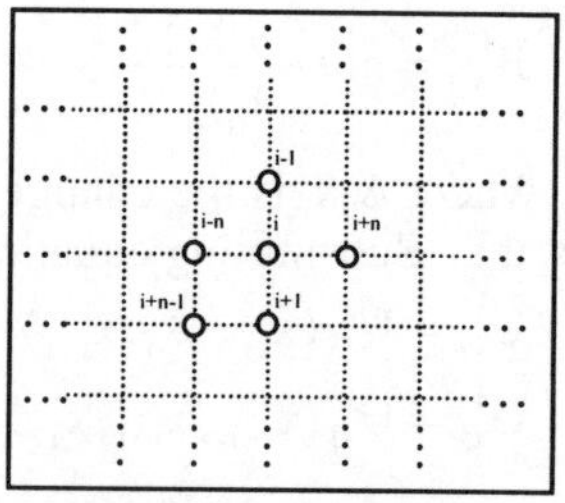

(c) bending forces

Figure 2: Forces on node i

4 Solution

Equation (1) can be solved by using the simple Euler method:

$$\begin{aligned} \ddot{\mathbf{x}}_i(t+\Delta t) &= \frac{1}{m_i}\left(\mathbf{p}_i(t) - \gamma_i \dot{\mathbf{x}}(t) - \sum \mathbf{f}_i(t)\right) \\ \dot{\mathbf{x}}_i(t+\Delta t) &= \dot{\mathbf{x}}_i(t) + \Delta t \cdot \ddot{\mathbf{x}}_i(t) \\ \mathbf{x}_i(t+\Delta t) &= \mathbf{x}_i(t) + \Delta t \cdot \dot{\mathbf{x}}_i(t) \end{aligned} \tag{15}$$

where Δt is a chosen time step.

Collision detection and response are important issue in cloth draping simulation. There are two types of collisions: self-cloth collision and cloth-object collision. Collision detection is performed at each iteration. Once it occurs, the responses are implemented by imposing constraints on the nodal position of the cloth at that instant. For simplicity, the method for collision detection and response proposed by Liu et al[8] is adopted in this implementation.

5 Simulation Examples

5.1 *Example 1: Comparison of the draping simulation with the real object.*

Two pieces of cloth with different fabric properties draping over a sphere is presented in figure 3. The results are compared with the real images. The size of the cloth is 0.6m×0.6m and the radius of the sphere is 0.09m.

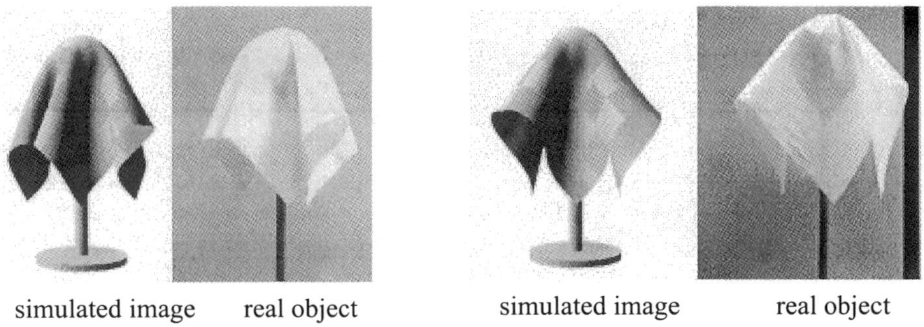

Figure 3: Comparison of the draping behaviour with real image

5.2 Example 2: A piece of cloth falling on a sphere.

The size of the cloth is 1m×1m with meshing size of 50×50 nodes and the radius of the sphere is 0.125m. The animation process is shown in figure 4.

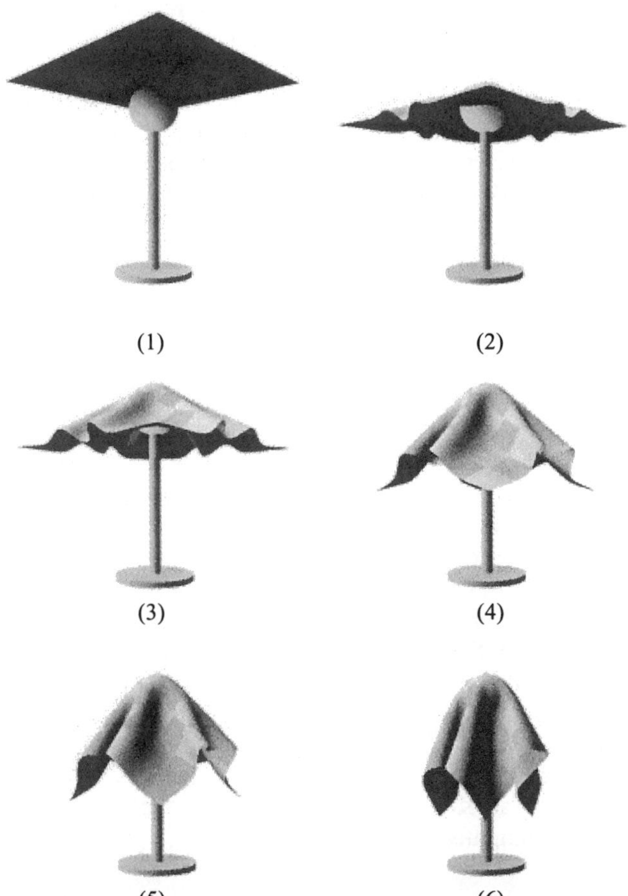

Figure 4: The animation of cloth falling on a sphere

6 Discussion and Conclusion

The network characteristics of the proposed mass spring system is derived based on the strain energy, which is related to the deformation with the fabric material properties. The fabric is an anisotropic material. The material modulus tensors in the stiffness constants can be expressed in terms of the mechanical properties of the fabric such as Young modulus, Poission coefficient and Rigidity modulus. The detail derivation can be found in reference [7]. As a result, cloth with various materials can be modelled efficiently and realistic cloth animation can be created.

Let **M** be the mass matrix, a diagonal (m×n)×(m×n) matrix with the variable mass per unit area ρ_i as diagonal components and **R** be the damping matrix constructed similarly from γ_i. $\mathbf{F}_G$, $\mathbf{F}_B$, **X** and **P** are designated as the m×n vectors corresponding to $\mathbf{f}_G^i$, $\mathbf{f}_B^i$, $\mathbf{x}_i$ and $\rho_i\mathbf{g}$ respectively. Equation (1) can be assembled into the form of matrix

$$\mathbf{M}\ddot{\mathbf{X}} + \mathbf{R}\dot{\mathbf{X}} + K_G\mathbf{X} = K_B\mathbf{N} + \mathbf{P} \tag{16}$$

where K_G and K_B is a m×n in-plane and out-of-plane stiffness matrices; and **N** is a m×n matrix with unit normal vector n_i ($\forall i \in [1, m\times n]$). It can be seen that the fabric has independent tensile stiffness $\mathbf{K}_G$ and bending stiffness $\mathbf{K}_B$. The two stiffness matrices account for the deformation characteristics of the cloth object. The bending deformation is very large compared with the deformation in the membrane direction. This is mainly due to the large values of the stiffness matrix $\mathbf{K}_G$. Relatively, the value of the stiffness matrix $\mathbf{K}_B$ is smaller and both are functions of the nodal point. Hence, the draping behaviour is a non-linear process and is dominated by the bending deformation.

The mass spring network characteristics for cloth modelling is rationalized based on the strain energy due to deformation. The fabric material properties, which affect the draping of the cloth, are incorporated in the characteristics in this approach. This model gives a good and accurate visual effect of cloth draping.

Reference

1. H. N. Ng, R. L. Grimsdale, "Computer Graphics Techniques for modeling cloth" IEEE Computer Graphics and Applications, **16** (5), 28-41, (1996)
2. P. Vilino and Magnenat-Thalmann, "Virtual Clothing", Springer, (2000)
3. D. H. House and D. E. Bree, "Cloth Modeling and animation", A K Peters, (2000)
4. X. Provot, "Deformation constraints in a mass-spring model to describe rigid cloth behavior", Proceedings of Graphics Interface, 147-154 (1995).
5. P. Howelett, "Cloth simulation using mass-spring networks", M.Sc. Dissertation, Department of Computer Science, University of Manchester (1997).
6. D. Hutchinson, M. Preston, T. Hewitt, "Adaptive refinement for mass/spring simulations" 7th Eurographics Workshop on Animation & Simulation, (1997)
7. Z. Wu, "Material-based cloth draping simulation", Ph.D. Dissertation, Hong Kong University of Science and Technology, (1999).
8. Jen-Duo Liu, Ming-Tat Ko, Ruei-Chuan Chang, "Collision avoidance in cloth animation", Visual Computer, **12**, 234-243 (1996).

DETERMINATION OF RAIL CARRIAGE SUSPENSION CHARACTERISTICS BY RESPONSE MEASUREMENTS

R.B. RANDALL, A. LIEW, R.A.J. FORD
School of Mechanical and Manufacturing Engineering
The University of New South Wales, Sydney 2052, Australia
E-mail: b.randall@unsw.edu.au

K. WIWATOWSKI
EDI Rail, 2B Factory Street, Granville,
NSW 2142, Australia
E-mail: K.Wiwatowski@edirail.com.au

The paper describes methods used to determine the rigid body mode shapes and their corresponding natural frequencies and damping values of an inter-urban rail carriage on its own suspension. For bounce, pitch and roll, the method used to induce the motion was to pull specified wheels (in different combinations) up onto chocks of thickness 15 mm, and then let them fall back onto the rail simultaneously. For yaw, the carriage was impacted laterally at one end by a tractor, cushioned by rubber tyres. To analyse the response signals, the effective step responses were treated in two ways. Natural frequencies and (operating) mode shapes were determined approximately by ODS (Operating Deflection Shape) measurement based on measuring relative motions at a specified frequency as close as possible to the natural frequency for that mode. The method used to give the best estimates of mode shapes, frequencies and damping was to treat the effective step responses as though they were impulse responses, since these would still have the same poles. This allowed determination of the global parameters by curve fitting all four response measurements at the same time, and then fitting the mode shapes based on these global parameters. An exponential window had to be used to extract the best part of the response, but the resulting added damping could be subtracted from the final result.

1 Introduction

This paper describes methods used to measure the suspension characteristics of two inter-urban rail vehicles, based on measurements of response only. The main information required was the natural frequencies, and preferably damping, associated with the main rigid body modes of the car, viz, bounce, pitch and roll, and the primary technique was designed to measure these, but another method was used to measure yaw. This method also gave a check on the values for roll.

2 Equipment and Methods

The method devised to measure the first three modes was to run certain wheels up on chocks, of thickness 15 mm, and then by causing them to drop back onto the rail, measure the resulting motion of the whole car vertically in all four corners. This step response would then have to be decomposed into the component modes by analysis. An attempt would be made to isolate the various modes as much as possible by the pattern of wheels raised for each test. For example, for bounce, all wheels would be raised, while for pitch, only the four wheels on the bogie at one end would be raised. This latter would excite both pitch and bounce, but in principle not roll. Raising the wheels on one side would excite roll and bounce, but not pitch. For a so-called “Combined” test, the wheels on one side of one bogie were raised. Figure 1 is a photograph showing the chock under one wheel, this being made by combining three wedges of thickness 5 mm, and progressively shorter length, so as to facilitate pulling the vehicle up onto the chocks. The lower part of Fig. 1 shows a typical group of response time signals on the screen of the analyser used. The method used to excite yaw involved impacting the side of the vehicle at one end using a tractor, cushioned by rubber tyres (see Figure 2). For this measurement, the four accelerometers used to measure the car motion were mounted on the impacted side at each end at two different levels (above and below the windows) and were oriented horizontally.

Measurements were made with a Bruel & Kjaer Pulse system, with eight channels. The four spare channels were used to measure the response of one of the bogies, but these results are not reported here. The accelerometers used were all Bruel & Kjaer type 4396, with built-in preamplifier, powered directly by the Pulse system. All measurements were recorded by throughput-to-disc, with a frequency range of 6.4 kHz. They could then be played back repeatedly for analysis in any desired frequency range. The analysis results can be transferred to the ME-Scope Modal Analysis package (Vibrant Technology, California, USA) giving the possibility for a number of types of further processing.

By measuring frequency response functions (FRFs) between one reference point and all other points it is possible to display the "Operating Deflection Shape" (ODS) at any chosen frequency. Since the total motion can be expressed as a sum of motions in the various natural modes, the closer the frequency is to one of the natural frequencies, the more the ODS will be dominated by that particular mode (in particular in the absence of a continuing forcing function, as in this case). The most likely frequency of a particular mode can be judged as being at the centre of a resonance peak in the autospectrum of a typical response measurement. However, this is not always easy to judge in the case of high damping (broadening the peaks) and closely spaced modes. Moreover, the damping can only be estimated using the "3dB bandwidth", also difficult to determine in the case of closely spaced modes.

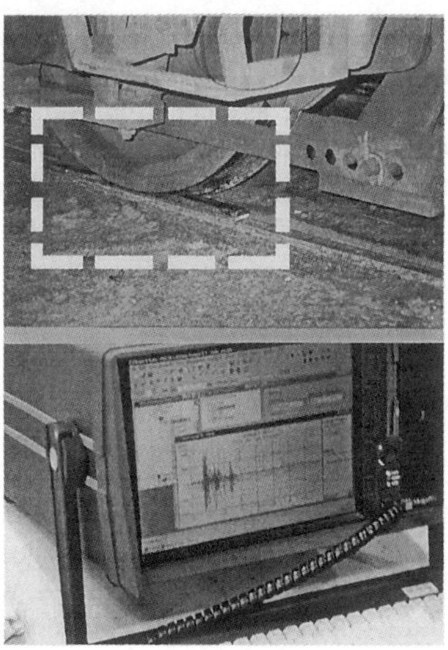

Figure 1 (Upper) Wheel on wedge (Lower) Analyser Screen

Figure 2 Tractor excitation of yaw mode.

Another approach, which was used in this case, was to treat each response signal as a step response, with the same poles (natural frequencies and damping values) as the impulse response. The only difference would be a different weighting on the excitation of modes at different frequencies, but the natural frequency and damping of each mode could be extracted by treating the step response as though it were an impulse response. The complex Fourier spectra of each response were thus analysed by ME-Scope as though they were FRFs based on a measured (impulse) forcing function. This gives the advantage that curve–fitting can be employed to separate closely spaced natural frequencies, and give estimates of the frequency and damping, as global parameters over all four measurements. Note that the yaw measurement was genuinely an impulse response, the impulse being imparted by the tractor over a time period considerably shorter than the response signals analysed. Figure 3 shows the result of a curve fit on data for a drop test measurement. A true impulse response measurement would have time zero exactly at the start of the record. Here, time zero was estimated by trial and error, obtaining a phase spectrum as flat as possible (fig. 4), but the true time zero may have been between two samples. This did not give problems in curve fitting (it would correspond to a small phase shift in the complex mode shape), but does mean that the resonance peaks are not confined to the imaginary part of the complex spectrum representing the FRF. The latter feature can be used by the ME-Scope software to identify significant natural frequencies that occur in all measurements.

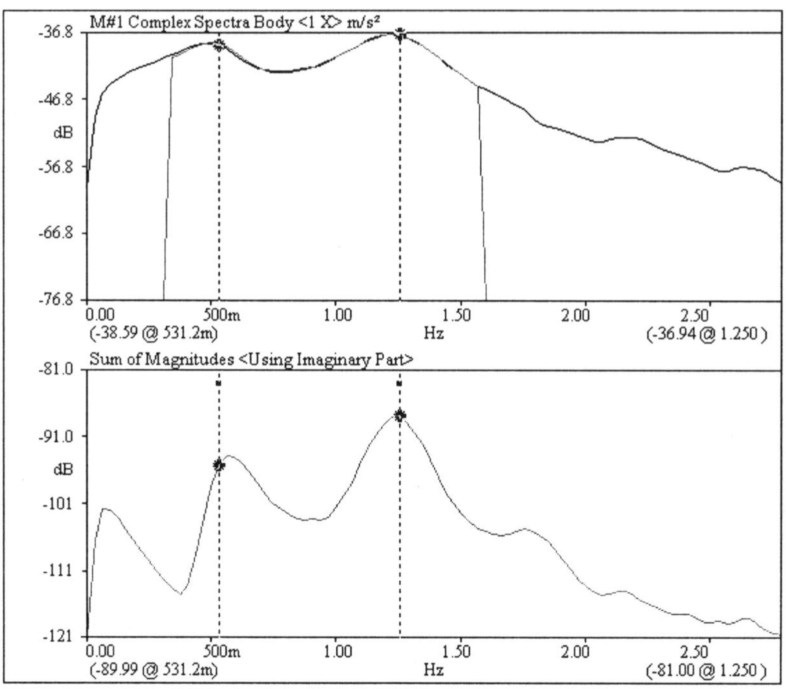

Figure 3 (Upper) Typical response autospectrum and the regenerated section from the curve-fitted parameters. The two peaks represent the resonances detected by the roll measurement, which are well separated. Even so, estimation of the damping from the 3 dB bandwidth would be difficult. (Lower) This curve (involving the imaginary part of the spectra) is provided by the curve fitting program, but can be disregarded as explained in the text.

Most of the response signals were contaminated by various amounts of extraneous disturbances (see screen picture, Fig. 1). Some of these were caused by the fact that not all wheels dropped at the same time (if some wedges slipped a little while the car was pulled up onto them), also by rail roughness on part of the track that was used, and finally by the impact with the loco placed to stop the car after a run of about 4 m. The forward speed of the car on the rails varied depending on the strength of the impulse used to get it in motion, and the time before the final impact was sometimes as short as 4 s. Thus it was necessary to use a "time window" to extract the valid part of each time signal. For the ODS measurements, this typically was a rectangular window, as long as possible, with a half Hanning taper to avoid a sharp discontinuity at each end. With the same window applied to each signal in each group, its effect was cancelled in the FRFs used for ODS. For the step response measurements, an exponential window was used, with a (data dependent) time constant between 1 and 4 s. This corresponds to a precisely known amount of extra damping, which could be subtracted out from the measured value. The exponential decay of the analysed response can be represented by $\exp\{-(\sigma_1 + \sigma_2)t\}$, where σ_1 is the reciprocal of the time constant for the actual damping and σ_2 that for the exponential window. Incidentally, the 3dB bandwidth is given by twice the σ-value, as this determines the frequency spacing on either side of a resonance where the phase change is 45° (ie, the real and imaginary parts of the FRF are equal). The abrupt start of the exponential damping was also used in some cases to remove the first of multiple impact pulses to minimise the "echo" effect in the curve fitting operation.

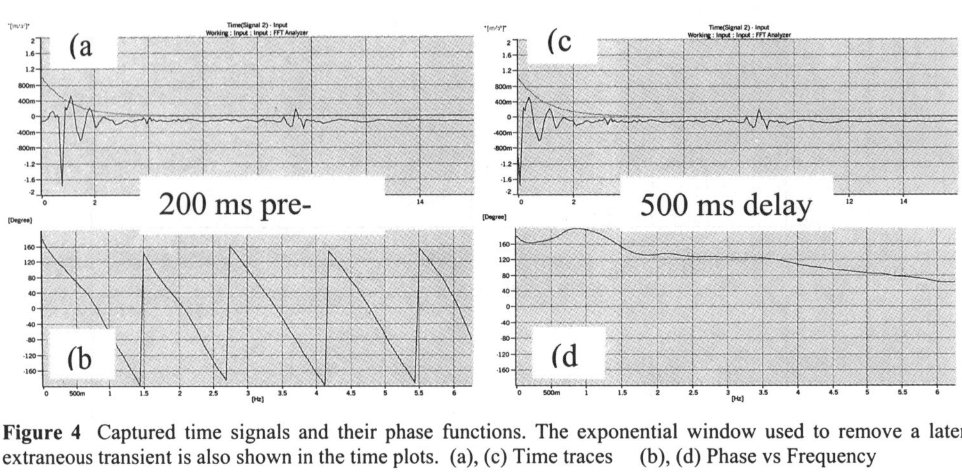

Figure 4 Captured time signals and their phase functions. The exponential window used to remove a later extraneous transient is also shown in the time plots. (a), (c) Time traces (b), (d) Phase vs Frequency

3 Results and Discussion

The results for one of the cars are shown in Table 1. The various rows correspond to the particular test, designed to elicit the nominal response by which they are designated; bounce, pitch, roll etc, and the columns give the frequencies of the various motion types extracted from that measurement. The ODS results at the top of the table are useful mainly for comparison with those extracted by curve fitting, which are thought to be more valid. For the latter, a five star rating system is used to indicate the validity of each result, with ***** being the best quality, and * the poorest.

Table 1 Frequency and Damping Results

EXCITATION MODE	BOUNCE FREQ DAMP (Hz) (%)	PITCH FREQ DAMP (Hz) (%)	ROLL FREQ DAMP (Hz) (%)	YAW FREQ DAMP (Hz) (%)	ANAL. METHOD
Bounce	1.4	2.2	?	N/A	ODS
Pitch	?	1.4?	?	N/A	ODS
Roll	?	?	0.56	N/A	ODS
Combined	?	?	0.59	N/A	ODS
Yaw	N/A	N/A	0.62	1.6	ODS
Bounce	1.26 33.2% **	-	-	-	Curve-Fit
Pitch	-	1.38 26.5% **	-	-	Curve-Fit
Roll	-	-	0.65 -0.2% *	-	Curve-Fit
Combined	1.24 -0.8% **	1.40 18.0% ****	0.69 13.7% ***	-	Curve-Fit
Yaw	-	-	0.63 10.1% ****	1.55 25.9% ***	Curve-Fit

All the mode shapes obtained (by both ODS and curve fitting) were slightly complex (not all points reaching maxima at the same time), but because the damping is localised and not "proportional" this is to be expected. Moreover, because the mass is not distributed symmetrically, the fundamental modes (bounce, pitch etc) are not all uncoupled, and are generally combined, with one or the other typically dominant. The dominant one has been used to designate the mode.

It can be seen that there is reasonable agreement between the results for ODS and curve-fitting, and also for the roll mode estimated by the two methods of step response and impulse response. Because of close coupling, there was a problem in separating the bounce and pitch modes, and therefore some doubt as to the damping of the former. It could reasonably be expected to be similar to that of the pitch mode, as the damping mechanisms in each bogie would be similar, but acting in or out-of phase for the two modes.

4 Conclusion

A method of estimating modal parameters from response measurements using standard curve-fitting software has been found to give good results for frequency and damping estimates.

VIBRATION CONTROL OF A FIVE STOREY BENCHMARK BUILDING EXCITED BY EARTHQUAKE USING LIQUID COLUMN VIBRATION ABSORBERS

B. SAMALI, E. MAYOL AND A. MACK

Centre for Built Infrastructure Research, University of Technology, Sydney
PO Box 123 Broadway NSW 2000, Australia
E-mail: Bijan.Samali@uts.edu.au

K.C.S. KWOK AND P. A. HITCHCOCK

CLP Power Wind and Wave Tunnel Facility, University of Science and Technology
Clear Water Bay, Kowloon, Hong Kong, China
E-mail: kkwok@ust.hk

This paper examines the response reduction capabilities of a passive fluid damper device, known as the Liquid Column Vibration Absorber (LCVA), when attached to a five storey benchmark building structure requiring control. The five storey benchmark building is an experimental model that can be excited at its base level to simulate ground motion. In previous studies the LCVA has been shown to be effective for wind vibration control. In this study the probable structural response reductions that can be achieved due to four benchmark earthquakes is examined. A further investigation is made to the use of multiple LCVA's (MLCVA) for earthquake response suppression. The issue of robustness is also addressed by comparing the response reduction capabilities of the LCVA and MLCVA configuration under earthquake loading. The robustness becomes a primary issue when the LCVA may become largely mistuned in the event of loss of structural stiffness during an earthquake.

1 Introduction

Recent occurrence of fatal earthquakes has resulted in an increased awareness in the need to better control detrimental vibrations to building structures induced by severe earthquakes. As a result, a number of damping devices have been proposed which have been shown to provide suitable means of control. Of such devices, the LCVA has recently been attracting interest due to its inherent simplicity and ease of application.

In prior studies, the LCVA has been shown to effectively reduce wind induced vibrations, and has been traditionally designed for this sole purpose. However, in the event of an earthquake, a structure fitted with an LCVA could be influenced by it, to the extent that the LCVA may possibly cause adverse response. Therefore, consideration to the behaviour of the LCVA and the structure is necessary under the influence of earthquakes.

Much numerical work has been conducted on understanding the behaviour of liquid damping devices under the influence of earthquakes [1], [3], [4]. However, such models have generally over simplified the fluid flow behaviour, and may not depict the more complex response of the fluid. In this study, this is overcome by adopting a detailed analysis using computational fluid dynamics (CFD) to address the fluid flow behaviour. The CFD solution has been validated against experimental results and has been shown to be of sufficient accuracy. The structural solution has been obtained by use of finite elements and has also been validated.

This paper specifically examines the capabilities of a LCVA and MLCVA when installed in the five storey benchmark building and subjected to four benchmark earthquakes. The earthquake records are those adopted by the International Association for Structural Control (IASC), namely El-Centro (1940), Kobe (1995), Hachinoe (1968) and Northridge (1994).

2 Structure

The experimental building structure has been recognised by the International Association for Structural Control (IASC) to be representative for vibration control studies. The structure stands 3.6 m tall and is versatile by allowing use of different mass, floor heights and joint connectivity. Use is therefore made of these features in order to achieve a desired structural stiffness and mass. A prototype of the structure has been modeled using the ANSYS finite element software. Figure 1 illustrates the finite element model used together with dimensional details.

The structure is configured such that it represents a wind sensitive highrise building. Features of such buildings include a low natural frequency, typically below 1 Hz. The nominal frame has a fundamental frequency of 0.76 Hz, making it fall within the realms of wind sensitive structures. A wind sensitive building is chosen for this study as the LCVA or MLCVA is typically augmented on such buildings. The frame considered for investigating mistuning has a frequency of 0.684 Hz representing a 10% reduction in fundamental frequency. A frame with reduced stiffness is considered, in order to simulate loss of structural stiffness in the event of a damaging earthquake.

3 Single and Multiple Liquid Column Vibration Absorber (LCVA, MLCVA)

The LCVA has been developed from the tuned liquid column damper (TLCD). Sakai *et al* [4] initially showed that water within a U tube configuration has potential to stabilise building vibrations. Hitchcock [2] furthered the concept by considering variable area ratios (T_v/T_h), in order to make the LCVA more amenable to frequency tuning and damping. The base configuration of a typical LCVA along with relevant notations is shown in Figure 2. A further adaptation of the LCVA is the use of multiple LCVA's (MLCVA). The MLCVA has also been shown by Hitchcock to give superior response reductions and has also been found to be robust in the event of mis-tuning. Figure 3 illustrates the MLCVA configuration as adopted for this study.

The LCVA adopted has an optimised tuning ratio of 99% and a mass ratio of 1%. The area ratio has been carefully selected in order to have as long as possible horizontal column containing as much mass as practical, for enhanced damping capacity [4].

The MLCVA has a total mass ratio of 1%, to allow for direct comparison to the LCVA. The frequency band of the MLCVA is set at 0.2, with a total of 5 individual units spread over this range. The resulting frequency distribution is therefore spread between 0.9f to 1.1f, with the central LCVA tuned to 99% of the nominal structure frequency. For this total mass ratio and number of units, the critical frequency bandwidth should be about 0.11[1], so as to maximise the damping capacity. A higher bandwidth however has been used in order to allow nesting of the MLCVA configuration. A total of 5 units has generally been shown to be the most effective [1], and hence adopted in this study.

The LCVA and MLCVA have been numerically studied by use of the CFD software CFD-ACE, solving for the Navier Stokes Equation.

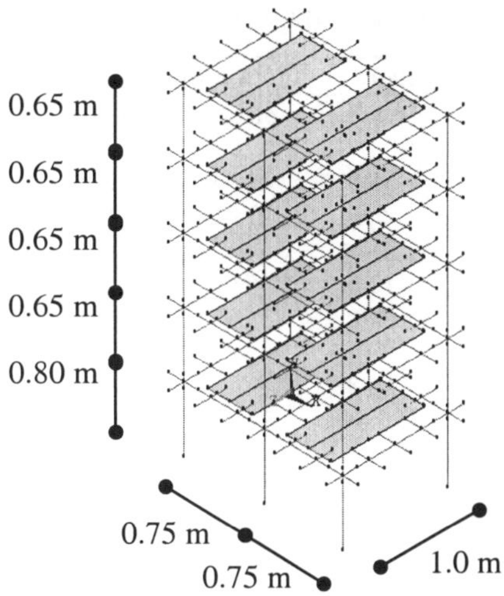

Figure 1. Benchmark building analytical model.

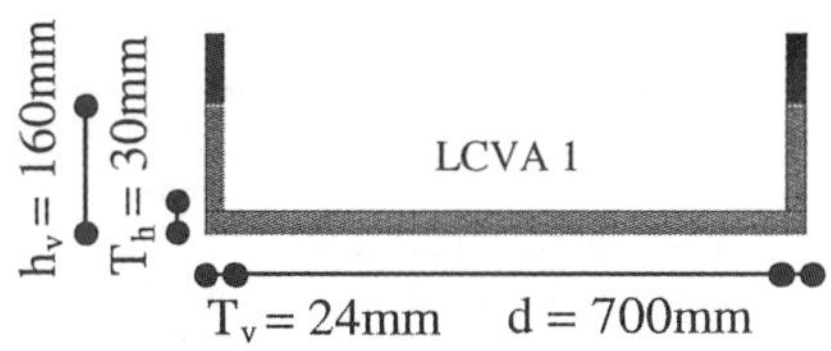

Figure 2. LCVA analytical model and notation.

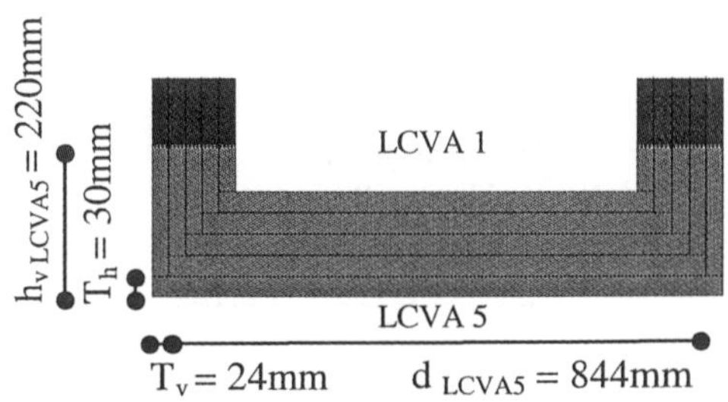

Figure 3. M LCVA analytical model and notation.

4 Earthquake response reductions

For the purpose of earthquake response reduction studies, the intensity of each of the nominated earthquakes was reduced so as to limit the maximum uncontrolled peak displacement of the structure to 15mm at its top level. A further consideration was to limit the time period for rms response calculations to be within the range where the ground motion intensity is within 95% of the peak value. This was done in order to assess the performance of the LCVA over the most intense portion of ground shaking. In order to assess robustness in the event of loss in stiffness, the structural stiffness was altered such that the first mode vibration frequency was 90% of the nominal building frequency.

4.1 Single Liquid Column Vibration Absorber (LCVA)

As part of the study, the structure was fitted with the LCVA specified (Figure 2). The structural response quantities on level 5 for the specified earthquakes are summarised in Table 1.

It can be observed from Table 1, that for Kobe and Northridge the response quantities have been reduced below that of the uncontrolled building. However, for El-Centro and Hachinoe, several of the response quantities have been marginally increased due to use of the LCVA. Both Kobe and Northridge motions are classed as *'near field'* earthquakes, whereas El-Centro and Hachinoe are classed as *'far field'* earthquakes. The LCVA in this case shows superior performance with respect to near field type earthquakes. It can also be noted that the displacement response quantities are better controlled than the acceleration response, and that the rms response is better controlled than the peak response quantities. It was also noted that the rms response reduction capacity would further increase if the full duration of earthquake shaking were considered. This demonstrates the ability of the LCVA to bring the structure quickly to rest after a period of severe shaking.

Table 1. Response of building to earthquake excitations (LCVA)

	uncontrolled		controlled		% reduction	
	peak	**r.m.s.**	**peak**	**r.m.s.**	**peak**	**r.m.s.**
El-Centro earthquake						
acceleration (milli-g)	83.4	34.5	88.9	35.9	*-6.7*	*-3.9*
displacement (mm)	12.9	6.5	13.1	5.6	*-0.9*	*14.0*
Kobe earthquake						
acceleration (milli-g)	52.0	20.8	43.2	12.6	*16.9*	*39.2*
displacement (mm)	15.0	8.2	11.7	3.9	*21.9*	*52.5*
Hachinoe earthquake						
acceleration (milli-g)	63.8	20.8	67.6	16.8	*-6.0*	*19.0*
displacement (mm)	15.0	6.0	9.9	3.4	*34.3*	*44.5*
Northridge earthquake						
acceleration (milli-g)	52.7	19.8	44.2	14.9	*16.1*	*24.8*
displacement (mm)	15.0	7.9	13.8	5.7	*7.9*	*28.4*

4.2 Multiple Liquid Column Vibration Absorber (MLCVA)

It has been found that the MLCVA displays superior performance under conditions of mistuning [2] when excited by a periodic loading. However, further studies have shown that when subjected to earthquake loading there only appears to be a marginal improvement over the LCVA [3]. To examine its potential, the MLCVA is directly compared to the LCVA, for the case of optimally tuned to the structure frequency, and when mistuned (-10%). In all cases the building is excited by the El-Centro earthquake, as it was least controlled with the LCVA.

Table 2 illustrates the response reduction of the building under conditions of mis tuning, for the case of without control, with MLCVA and with LCVA. From Table 2 it can be noted that for –10% mistuning, the building is generally better controlled by the use of a MLCVA. It can also be noted that the response reduction by use of a MLCVA is not greatly increased. A similar finding was made by Sadek *et al* [3] implying that the MLCVA only marginally increased response reduction for small mass ratios. However, it was also noted that if the response were assessed over a longer time period, the rms response reduction greatly improved for the MLCVA. This is shown in Figure 4 for displacement response on level 5. It is also observed that by using the MLCVA several of the response quantities are still excessive, but to a lesser extent.

Table 2. Response of building to earthquake excitations (MLCVA and LCVA)

				MLCVA		LCVA	
frequency change	*(f)*	1.0 f	0.9 f	1.0 f	0.9 f	1.0 f	0.9 f
peak acceleration	(milli-g)	83.4	88.8	85.1	93.9	88.9	92.9
rms acceleration	(milli-g)	34.5	32.9	35.5	33.1	35.9	33.4
peak displacement	(mm)	12.9	17.7	12.5	14.4	13.1	15.1
rms displacement	(mm)	6.5	7.8	5.5	7.3	5.6	7.5

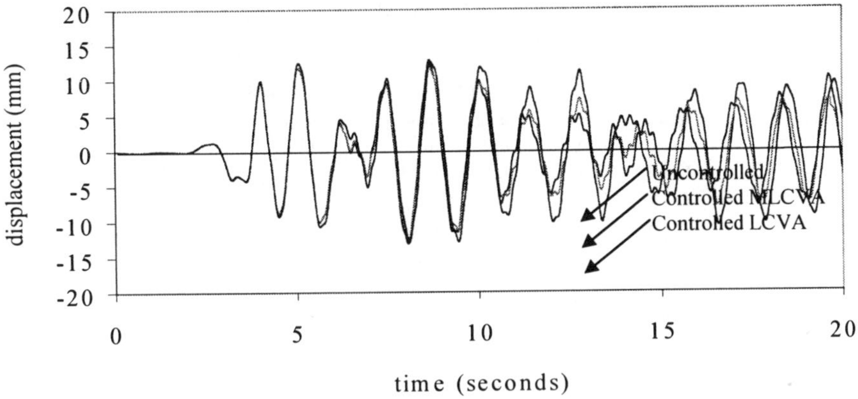

Figure 4. Level 5 displacement under El-Centro earthquake (uncontrolled and controlled – 1.0 f).

Fluid behaviour under excitation

The behaviour of the fluid during the excitation period is of interest. Table 3 shows the fluid force generated in the LCVA and MLCVA individual columns. It can be noted that the maximum force generated for each structural frequency corresponds closely to the individual LCVA that is tuned to it. It can also be noted that the total force generated by the LCVA is greater than the MLCVA, despite producing a lower level of damping.

In addition, the liquid elevation (Table 3) within any particular LCVA is maximised when the structure frequency is closest to its particular frequency. Similar findings were made experimentally by Hitchcock [2]. In his findings it was shown that the motion of several closely tuned LCVA's helps reduce vibration of the structure and prevent re-excitation. This effect would become more prevalent if the structure excitation period were evaluated for a longer duration.

The velocity for each liquid column is shown in Table 4. The liquid velocity is also maximised in the column that has the closest frequency to the structure frequency. The importance of the velocity lies in the inertial damping force generated by the fluid.

Table 3. Force and liquid elevation for LCVA and MLCVA for different tuning ratios

		Fluid force (N)				Liquid elevation (mm)			
		0.9 f		1.0 f		0.9 f		1.0 f	
		peak	rms	peak	rms	peak	rms	peak	rms
MLCVA	– 1	2.8	1.0	3.8	1.4	41.5	18.4	59.2	25.7
	– 2	2.8	1.1	3.7	1.6	47.7	20.2	63.1	31.8
	– 3	3.1	1.2	3.7	1.4	45.9	22.2	56.6	30.2
	– 4	3.7	1.4	3.0	1.1	66.2	26.5	48.3	24.3
	– 5	4.3	1.7	2.9	1.1	83.2	33.5	44.3	21.4
Total		13.3	5.6	12.7	4.8	-	-	-	-
LCVA		16.7	6.5	18.8	6.8	48.7	22.7	55.1	28.4

Table 4. Fluid Velocity for LCVA and MLCVA for different tuning ratios

	MLCVA				LCVA			
	0.9 f		1.0 f		0.9 f		1.0 f	
	peak m/s	rms m/s	peak m/s	rms m/s	peak m/s	rms m/s	peak m/s	rms m/s
– 1	0.17	0.07	0.25	0.11	0.18	0.08	0.22	0.11
– 2	0.18	0.08	0.26	0.13	-	-	-	-
– 3	0.17	0.08	0.23	0.12	-	-	-	-
– 4	0.22	0.09	0.19	0.09	-	-	-	-
– 5	0.28	0.11	0.17	0.08	-	-	-	-

5 Conclusions

This paper has shown that the LCVA strongly influences the behaviour of a structure during an earthquake occurrence. The LCVA, in some instances, may increase response quantities, although not to a great extent. The LCVA's behaviour in this case is strongly dependent upon the type of earthquake, with the near field earthquakes displaying greater response reduction capabilities. It has been found that the MLCVA subjected to earthquake excitation does slightly improve the response reduction capabilities of the LCVA in the case of mistuning (eg. loss of structural stiffness). It was further noted that for a longer duration response, the MLCVA greatly increased rms response reductions, and therefore seems to be superior. This is especially the case in wind applications. By further optimising the MLCVA, it is also anticipated that greater control could be achieved.

6 Acknowledgments

The authors would like to acknowledge the financial support received from the Australian Research Council (ARC) under Large Grant Scheme (Number A89800547), in support of this project.

References

1. Fujino Y. and Sum L.M., Vibration Control by Multiple Tuned Liquid Damper (MTLDs), *Journal of Structural Engineering.* **119,** 12, (1993) pp. 3482-3500
2. Hitchcock P.A., *Vibration control of structures by liquid column vibration absorbers*, Doctoral Thesis, University of Sydney Australia, (1996)
3. Sadek F., Mohraz B. and Lew H.S., Single- and multiple – tuned liquid column dampers for seismic applications, *Earthquake Engineering and Structural Dynamics*, **27**, (1998) pp. 439-463
4. Sakai, F., Takaeda, S. and Tamaki T., Tuned liquid column damper-new type device for suppression of building vibrations, *International Conference on Highrise Buildings*, China, (1989) pp. 926-931

PERFORMANCE OF A FIVE STOREY BENCHMARK MODEL USING AN ACTIVE MASS DRIVER AND A FUZZY CONTROLLER

BIJAN SAMALI, MOHAMMED AL-DAWOD AND JIANCHUN LI

Center for Built Infrastructure Research, Faculty of Engineering, University of Technology Sydney, PO Box 123, B2 L5 RM, NSW 2007, Australia

Email: bijan.samali@uts.edu.au , Maldawod@eng.uts.edu.au & Jianchun.Li@uts.edu.au

This paper presents the second stage of applying an active control scheme to the benchmark model of five stories. The first stage reported the use of an Active Tuned Mass Damper (ATMD) system on the benchmark model, where the control action is achieved by a Fuzzy Logic Controller (FLC). In this stage, the active control scheme is applied to the modified benchmark model equipped with an Active Mass Driver (AMD) system, where the control action is again achieved by using a Fuzzy controller. In addition, the experimental tests on the AMD system is reported. The building model under consideration is the 5-storey, 3.6 meters tall, steel frame designed and manufactured at the University of Technology, Sydney. In order to test, experimentally, the five-storey benchmark model, an Active Mass Driver (AMD) system was designed and manufactured at the University of Technology Sydney and the performance of the AMD hydraulic system was verified analytically. In this stage, the experimental tests are conducted on the AMD system to verify the analytical predictions before using the system on the benchmark model. Performance of the controller is validated through computer simulations. The advantage of the Fuzzy controller is its inherent robustness and ability to handle any non-linear behaviour of the structure. The results of the simulation show good performance of the Fuzzy controller for all four earthquakes used in the simulation.

1 Introduction

A critical aspect of the design of buildings and other civil engineering structures is the reduction of motions, deflections and forces induced by environmental dynamic loading such as winds or earthquakes. Recent technological advances now allow engineers to design intelligent structures capable of counteracting undesirable vibrations by incorporating active control systems in the design of modern structures. Active Mass Driver (AMD) systems have been a popular area of research for the last two decades and significant progress has been made in this area. The most notable installation of AMD's in high rise buildings is the two active mass drivers installed in the 296-meter tall Landmark Tower in Yokohama [1]. The implementation of an AMD is important because it can prevent loss of life and damage to buildings as witnessed by the 1994 Northridge and the 1995 Kobe earthquakes. Active Mass Drivers were studied by Spancer et al [2] for a small-scale three storey, single bay model of 1.58 meters, with a total mass of 227 kg. This study used two earthquake records, the 1940 El Centro and the 1968 Hachinohe earthquakes.

The main objective of this study is to apply a Fuzzy controller to the benchmark model of five stories defined by Samali [3] using an AMD system with the ability to reduce the response of the building under earthquake excitations, specially the severe ones (e.g. the 1994 Northridge and the 1995 Kobe earthquakes). The advantages of Fuzzy Logic are that they can handle nonlinearity, be adaptive, accommodate a high degree of parallel implementation, and tolerate uncertainty in the system. More importantly, human experience and expertise about the system and its performance can be incorporated into the design of a Fuzzy controller. The Fuzzy controller employs, a priori, information on the dynamics of the structure, the empirical knowledge of the structural behaviour, which cannot be formulated mathematically, and the historical data available on major wind and earthquake incidents. A Fuzzy controller is adopted in this study and validated through computer simulation. Fuzzy controller has been investigated for the active control of civil engineering structures [4, 5] and the current study builds on previous work in this area.

In the course of the paper the structural model under consideration will be defined along with the experimental set up of the AMD system. The advantages of employing an intelligent control will also be highlighted. The Fuzzy controller will be described and the results presented and analysed.

2 Structural Model

The model used in this study is the benchmark model of 5-storey, 3.6-meter tall, steel frame designed and manufactured at the University of Technology, Sydney. The total mass of the model is 1,700 kg. The first five natural frequencies of the model are 2.86, 8.89, 15.37, 21.85 and 27.99 Hz, respectively. Damping ratio for all five modes are assumed to be 0.2 % of critical for the proportional damping matrix. A mass driver (MD) with inertial mass of 48.4 kg is installed on the top floor. Experimental tests on the AMD system are conducted in order to determine the parameters of the AMD system through system identification using sweep-sine excitation. The results are as follows: (a) the first natural frequency of the AMD system is 4.6 Hz; (b) the stiffness is 40.4 N/mm and (c) the damping ratio is 14% of critical for the AMD system. The experimental setup involving the AMD system is shown in Figure 1.

Figure 1 – Active Mass Driver (AMD) System

The governing equations of motion using state vectors are:

$$\dot{x} = Ax + Bu + EW$$
$$z = C_z x + D_z u + F_z W$$
$$y = C_y x + D_y u + F_y W + v$$

Where z is the controlled output vector of dimension 18 and includes the displacement, velocity and acceleration of floors 1, 2, 3, 4, 5 and the damper moving mass. The y vector is the measured output vector of dimension 12 and includes the velocities and accelerations defined in vector z. The v vector is the measured noise, x is the state vector, u is the control force, and W is the earthquake excitation. The matrices A, B, E, Cy, Dy, Fy, Cz, Dz and Fz are matrices of appropriate dimensions.

3 Fuzzy Logic Design

Fuzzy logic, introduced by Zadeh, [6], enables the use of linguistic directions as a basis for control. Generally very robust and capable of handling nonlinear systems, Fuzzy logic controllers (FLC) usually require expert knowledge in their construction. In this paper, the preliminary design of the controller will use Larsen's minimum product rule, to combine the membership values for each rule, with the center of gravity (COG) defuzzification scheme, to obtain the output crisp value.

The controller is designed using two input variables (the displacement of the 5th floor and the actuator stroke), each one having seven membership functions, and one output variable (the control force) with nine membership functions. The membership functions chosen for the input and output variables are triangular shaped as illustrated in Figures 2-4, respectively. The Fuzzy variables used to define the Fuzzy space are described in Table 1. The self-organizing Fuzzy logic controller is used to find the final Fuzzy Associative Memory (FAM) as shown in Table 2. The aim of using two input variables for the Fuzzy controller is to show the performance of the Fuzzy approach in the control problem. The small number of feedback variables means the use of fewer sensors; thus a simplification of the control system with advantages in terms of reliability and cost. The Fuzzy controller is implemented into the MATLAB program using an integration time step of 0.005 seconds and the control signal is computed every 0.005 seconds.

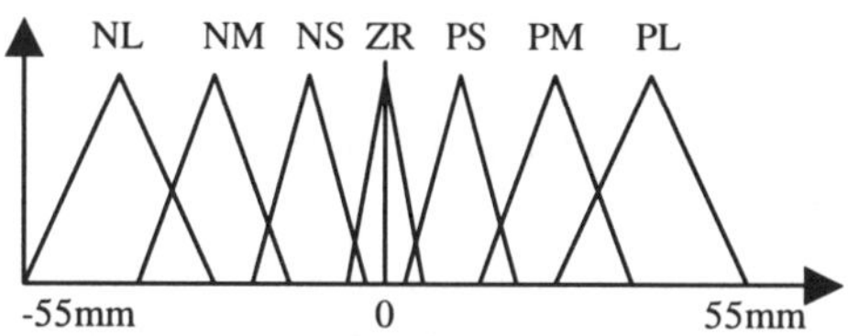

Figure 2 – Membership function for the displacement of 5th floor

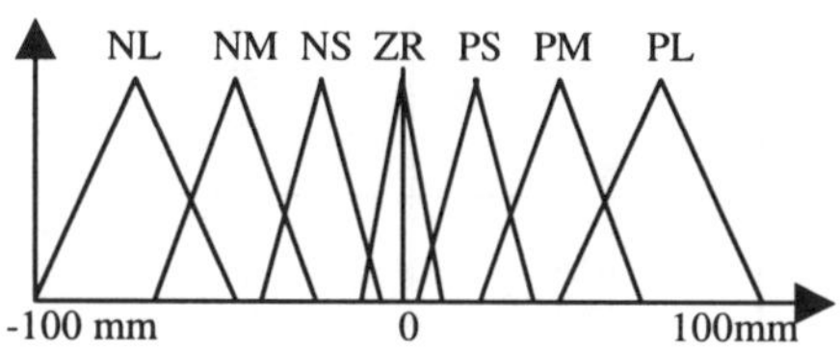

Figure 3 – membership function for the actuator stroke

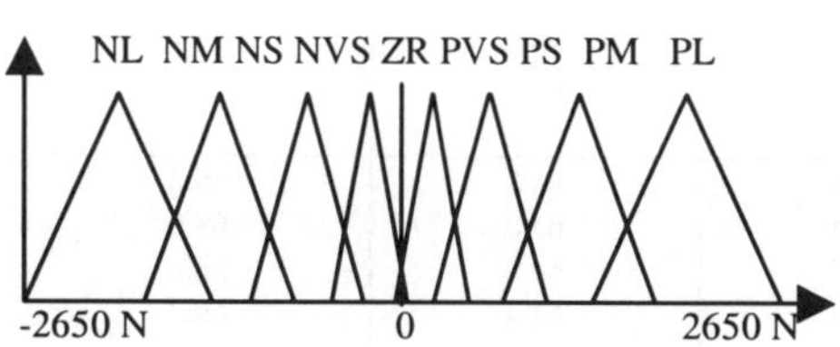

Figure 4 – Membership function for the control force

Table 1- Fuzzy Variables

NL	Negative Large
NM	Negative Medium
NS	Negative Small
NVS	Negative Very Small
ZR	Zero
PVS	Positive Very Small
PS	Positive Small
PM	Positive Medium
PL	Positive Large

Table 2– Fuzzy associative memory (FAM) of the Fuzzy controller

Actuator Stroke (columns); Disp. of 5th floor (rows)

U	NL	NM	NS	ZR	PS	PM	PL
NL	PL	PM	PM	PS	PS	PVS	ZR
NM	PM	PM	PS	PS	PVS	ZR	NVS
NS	PM	PS	PS	PVS	ZR	NVS	NS
ZR	PS	PVS	PVS	ZR	NVS	NVS	NS
PS	PS	PVS	ZR	NVS	NS	NS	NM
PM	PVS	ZR	NVS	NS	NS	NM	NM
PL	ZR	NVS	NS	NS	NM	NM	NL

4 Control Performance

The International Association for Structural Control (IASC) has identified four earthquake records to be used to check the performance of any control system for seismic applications. These earthquake records are El Centro 1940, Hachinohe 1968, Northridge 1994 and Kobe 1995 as shown in Figure 5. In this study, the intensities of these four earthquake records are scaled to 100%, 100%, 30%, and 40% of original intensities, respectively, due to limitations of the actuator velocity of ∓ 1.6 m/sec.

Table 3 shows the variation of the peak and RMS response reductions of all floors under the four earthquake records for displacement and acceleration. Figures 6-9 show the time history responses of the 5th floor for displacement and acceleration due to El Centro, Hachinohe, Northridge and Kobe earthquake records. The results show the ability of the AMD system, using Fuzzy controller, to reduce the response of the building for the four earthquake records but this reduction is different for each earthquake. Figure 10 shows time histories of the control force applied to the building for the four earthquakes.

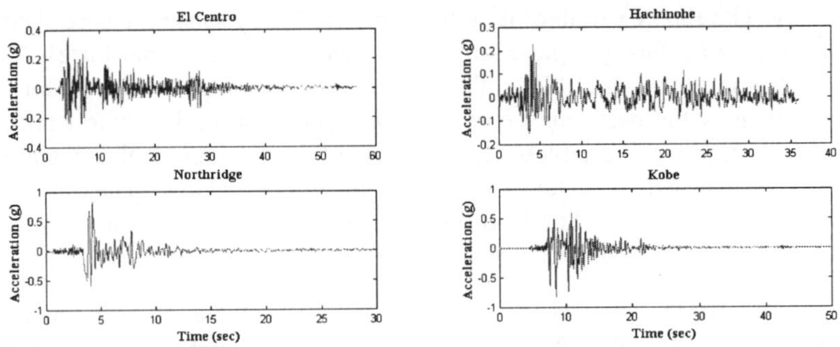

Figure 5 – Time histories of four earthquake records used: El Centro; Hachinohe; Northridge; Kobe

Table 3 – Response reductions of all floors (displacement and acceleration responses)

Earthquake Records	Floor No.	% Reduction			
		Peak Displacement	Peak Acceleration	RMS Displacement	RMS Acceleration
El Centro	1	24.3	0	58.2	50.8
	2	24.8	13.9	58.1	58.3
	3	25.1	8.00	57.9	61.4
	4	24.9	21.4	57.4	62.4
	5	23.8	41.7	56.8	75.7
Hachinohe	1	30.3	16.3	63.8	60.5
	2	30.5	29.3	63.6	65.9
	3	30.4	34.4	63.3	66.5
	4	30.0	39.2	62.9	67.5
	5	29.1	67.3	62.3	80.4
Northridge	1	15.6	0.0	70.3	68.3
	2	18.8	12.3	70.2	71.9
	3	17.7	17.5	70.0	73.1
	4	18.0	27.1	69.7	73.8
	5	17.8	59.3	69.2	84.3
Kobe	1	42.9	28.8	65.2	66.2
	2	43.5	44.1	65.0	68.5
	3	43.7	42.4	64.8	68.6
	4	43.0	46.7	64.4	69.3
	5	41.8	63.4	63.8	82.3

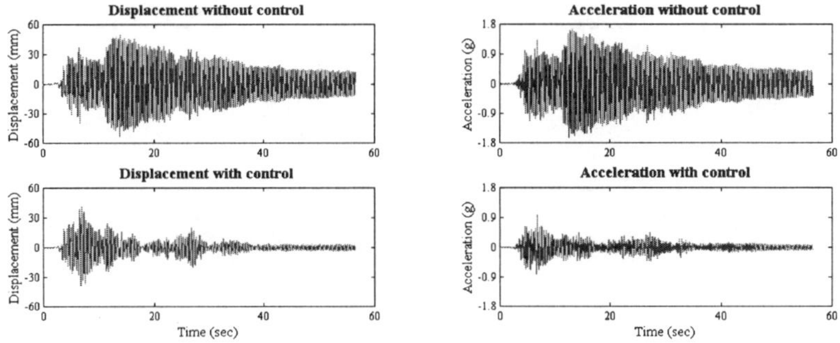

Figure 6 – Time history response of the building under El Centro earthquake

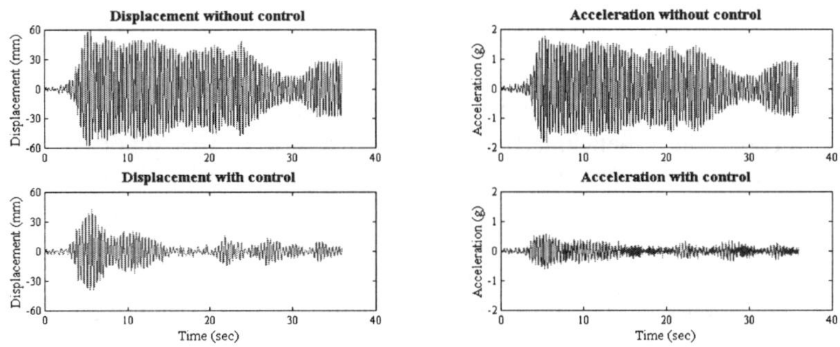

Figure 7 – Time history response of the building under Hachinohe earthquake

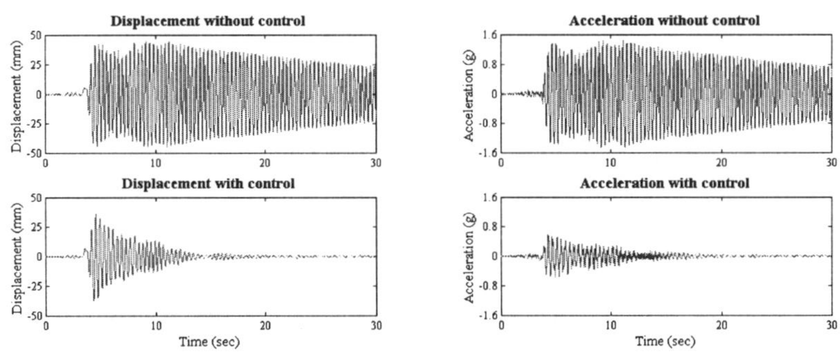

Figure 8 – Time history response of the building under Northridge earthquake

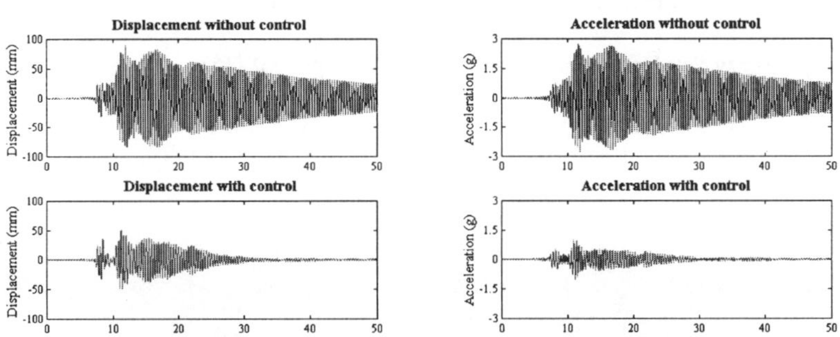

Figure 9 – Time history response of the building under Kobe earthquake

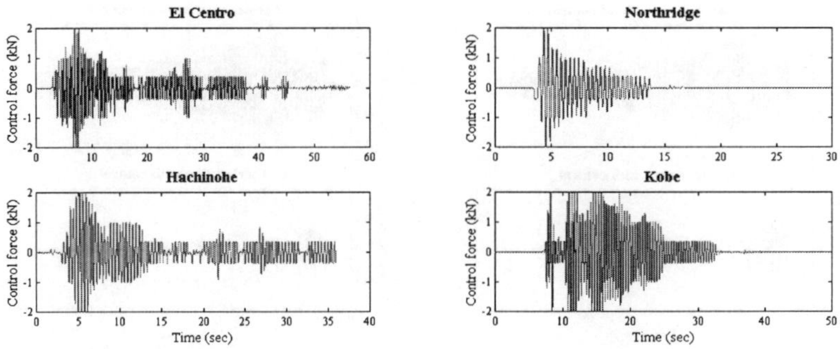

Figure 10 – Time histories of control force for the four earthquake records

5 Conclusions

In this paper, the study on vibration control of seismically excited buildings using actual earthquake records on a benchmark model of five stories was reported. Fuzzy logic controller, for structural control applications, was adopted to drive the AMD and control the system. The results show the ability of the AMD system, using Fuzzy controller, to reduce the response of the building for the four earthquake records. In the next stage of the work, the adopted Fuzzy logic controller will be tested experimentally using a shake table and the benchmark model of 5-stories at the University of Technology, Sydney.

References

1. Yamazaki, S., Tamari, M. and Abiru, H., Full Scale Investigation of Performance of Tuned Mass Dampers Installed in Landmark Tower, Yokohama, Proceedings of 9ICWE, New Delhi, India (1995) pp. 1631-1642.
2. Spencer Jr., B., Dyke, S. and Deoskar, H., Benchmark Problems in Structural Control, Earthquake Engineering and Structural Dynamics, **27** (1998) pp.1127-1247.
3. Samali, B., System Identification of a Five Storey Benchmark Model Using Modal Analysis, Asia-Pacific Vibration Conference '99 Singapore (1999) pp. 701-706.
4. Battaini, M., Casciati, F. and Faravelli L., Fuzzy Control of Structural Vibration. An Active System Driven by a Fuzzy Controller, Earthquake Engineering and Structural Dynamics (1998).
5. Aldawod, M., Samali, B., Naghdy, F. and Kwok K., Application of a Robust Fuzzy Controller to a Wind Excited Structure, Asia-Pacific Vibration Conference '99, Singapore (1999) pp. 812-817.
6. Zadeh, L., Fuzzy Set, Information and Control (1965) pp. 338-353.

VIBRATION CONTROL OF A FIVE-STOREY MODEL WITH SEMI-ACTIVE STIFFNESS DAMPER

SRI LILY DJAJAKESUKMA[1], BIJAN SAMALI[2], HUNG NGUYEN[3]

[1]*PhD Candidate,* [2]*Professor of Structural Engineering,* [3]*Professor of Electrical Engineering*
Faculty of Engineering, University of Technology, Sydney
P.O. Box 123, Broadway NSW 2007, Australia
E-mail: slily@eng.uts.edu.au

This paper presents vibration control of a five-storey model with semi-active stiffness damper subjected to earthquake loads. Semi-active stiffness damper consists of a special type of hydraulic damper connected to the bracing frame and installed in a selected storey unit. Various control laws are applied to demonstrate the system performance. They are resetting control, switching control and linear quadratic regulator (LQR). Three cases of resetting control are developed by selecting control parameters to explore control performance. Moreover, the damper effective stiffness is varied to determine the optimum performance. The results show that semi-active stiffness damper effectively controls the vibration of a five storey model. By selecting an appropriate damper stiffness value, each control law can perform well. Resetting control can achieve maximum displacement reductions of up to 83%, while it is 73% for switching control and 74% for LQR.

1 Introduction

Semi-active stiffness damper (SASD) is one of many semi-active control systems with the capability to reduce structural responses. They will utilize the motion of a structure to develop control actions through the adjustment of damping or stiffness characteristics of the system. Thus, they require only a small amount of energy to operate.

Research and development on semi-active control systems are continuing in order to improve semi active devices or control laws and enhance their performance. A special type of hydraulic damper with a controllable valve has been proposed as an energy dissipation device to reduce structural responses due to earthquake loads. Resetting and switching control were proposed to investigate the system performance. The results show that the performance of each control law depends on the earthquake type and its characteristics [6, 7, 8].

Previous research has applied semi-active stiffness dampers in different types of structures subjected to many types of earthquake motions. Various control laws have been applied to investigate their performance. The results show that most control laws are effective in reducing floor displacements but with varying degree of sensitivity to the earthquake characteristics [1, 4].

This paper focuses on vibration control of a five-storey model with semi-active stiffness damper. Various control laws are proposed to investigate the system performance [2]. Resetting control is derived based on Lyapunov function, whereas switching control is derived based on sliding mode control. They calculate control actions based on local measurements without any type of gain. The LQR utilizes optimal control to minimize a specified mathematical performance criterion. Two benchmark earthquakes, one near-field record and one far-field record, were adopted as input earthquake motions. They are Kobe (1995 NS) scaled to a Peak Ground Acceleration (PGA) of 0.35g and El Centro (1940 NS) scaled to a PGA of 0.20g. Damper effective stiffness is varied to determine the optimum value. The results show a good performance of semi-active stiffness damper in controlling the vibrations of the model.

2 Structural Model

A semi-active stiffness damper is applied to a five-storey model, 3.4 m tall, steel frame designed and manufactured at the University of Technology, Sydney [3]. The height of the first storey is 800 mm with equal heights of 650 mm for other storeys. The short direction of the building with a semi-active stiffness damper was analyzed. The building was modeled as lumped mass system in which the mass of each storey is concentrated at the floor level. The first five natural frequencies of the model are 3.33, 10.56, 18.26, 24.92 and 29.49 Hz, respectively. Damping ratio for all five modes was assumed to be 0.2% of critical for the proportional damping matrix. The first storey stiffness is 4,270 kN/m. The mass of storey units 1 to 5 are m = [0.337 0.33 0.33 0.33 0.37] tonnes, respectively.

For application to structural control, hydraulic dampers are connected to a bracing system and installed in the 1st, 3rd and 5th levels as shown in Figure 1. The control device consists of a cylinder-piston system with a valve in the bypass pipe connecting the two sides of a cylinder filled with hydraulic oil. It can produce a large control force by changing its damping coefficient and stiffness. Control valve will regulate this control action. Damper stiffness is provided by the bulk modulus of the fluid in the cylinder. It is an important physical quantity, which influences the control system performance.

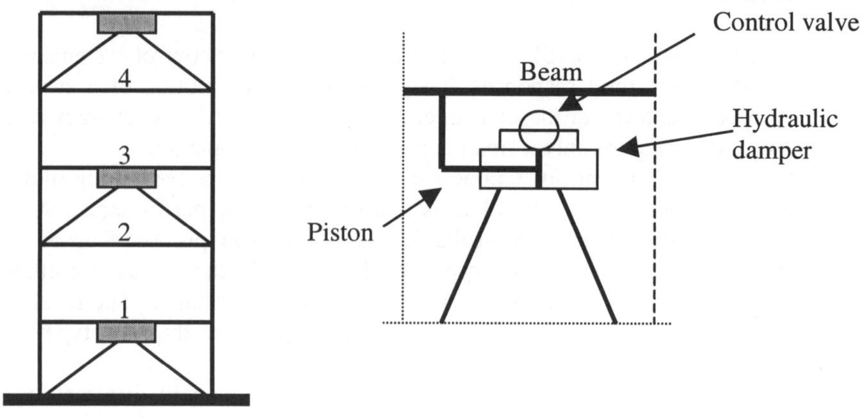

Figure 1. Elevation of the five-storey model with hydraulic dampers installed

A model of the hydraulic damper and bracing system is expressed by the combination of damper stiffness (k_f) and bracing stiffness (k_b), joined in series. The equivalent stiffness of the entire device (damper effective stiffness) is $k_{hi}= k_f\, k_b/(k_f +k_b)$. It is varied between 1,000 – 6,000 kN/m to determine the optimum value.

The system can operate in resetting control mode and switching control mode. Resetting control mode will pulse the valve to open and close quickly at appropriate time instants. When the piston is at resetting position (x_{si}), the energy stored in the damper system is released and converted into heat energy. After this time, there is no force provided into the structure. Hence, by pulsing the valve at appropriate time instants, energy can be taken out of the system to reduce the structural response. Switching control mode will open the valve for certain time intervals and close it during another time interval. The device is used as an on-off device. Therefore, by turning the device off and on in a regulated fashion, the structural response can be reduced.

3 Control Algorithm

Three control laws are derived and applied here to demonstrate the effectiveness and assess the performance of the proposed control system.

3.1 Resetting Control

Resetting control law is derived based on a Lyapunov function V that consists of the potential and kinetic energies [8]. Resetting control is formulated as follows:

1. Reset the SASD in the first storey unit if $\dot{x}_1 + \alpha x_1 + \beta \ddot{x}_0 = 0$
2. Reset the SASD in the ith storey unit if $\dot{x}_i + \alpha x_i = 0$ for $i > 1$

Where α and β are constants, $X(t) = [x_1, x_2, \ldots x_n]^T$ = vector of dimension n of interstorey drifts. $\ddot{x}_0$ is the ground acceleration time history.

Three cases of resetting control are selected to determine the best control performance:

- Resetting1: α=0, β=0 reset the ith SASD device if $\dot{x}_i = 0$ then $x_{si} = x_i$
- Resetting2: α=10, β=0 reset the ith SASD device if $\dot{x}_i + 10x_i = 0$ then $x_{si} = x_i$
- Resetting3: α=0, β=1 reset the ith SASD device :
  - i=1: if $\dot{x}_1 + \ddot{x}_0 = 0$ then $x_{si} = x_i$
  - i>1: if $\dot{x}_i = 0$ then $x_{si} = x_i$

Control Vector U(t) can be calculated from the equivalent stiffness of SASD, interstorey drifts and piston position as follows:

$$U(t) = \sum_{i=1}^{n} K_{di}(X - X_s)$$

Where K_{di} = (n x n) effective stiffness matrix for the SASD damper installed in the ith storey unit. Elements of K_{di} are zero, except K_{di} (i,i) = k_{hi} and K_{di} (i-1, i) = - k_{hi}. And k_{hi} is the effective stiffness of the SASD in the ith storey unit. $X_s(t) = [\, x_{s1}, x_{s2}, \ldots x_{sn}\,]^T$ = vector of dimension n of piston position of the SASD device.

3.2 Switching Control

This control law is derived based on the sliding mode control [6]. The device is operated in switching control mode. When the valve is closed, the effective stiffness k_{hi} is added to the storey unit. On the contrary, it is zero when the valve is open. The control action U(t) is given by:

$$U(t) = \sum_{i=1}^{n} K_{di} v_i(t) X(t) \qquad v_i(t) = \begin{cases} 1 & \text{if} \quad x_i \dot{x}_i \geq 0 \\ 0 & \text{otherwise} \end{cases}$$

Where v_i is the control valve position, $v_i(t) = 1$ if the ith SASD in the ith storey unit is closed, and $v_i(t) = 0$ if it is open.

3.3 LQR Control

A Linear Quadratic Regulator (LQR) is applied to regulate control actions and derived based on optimal control [5]. The equivalent stiffness (k_{hi}) of the hydraulic damper and bracing system is included in the building model. The control action is limited to the maximum capacity of the hydraulic damper and the bracing system (u_{max}). The optimal control force without considering the disturbance is as follows:

$$u(t) = - v_i(t)\, G\, X(t)$$

Where v_i is a condition which regulates the control action at the ith storey unit, G is the LQR feedback gain. The control device can regulate the control force through $v_i(t)$.

$$v_i(t) = \begin{cases} 1 & \text{if} \quad x_i \dot{x}_i \geq 0 \quad \text{then} \quad U = u \quad \text{but} \quad U \leq u_{max} \\ 0 & \text{otherwise} \qquad\qquad \text{then} \quad U = 0 \end{cases}$$

4 Control Performance Results

Vibration control of a five-storey model is investigated by using five control cases under two benchmark earthquakes. The damper effective stiffness is varied to determine the optimum value. Then, the response of each control method, using the optimum damper effective stiffness, is evaluated against the evaluation criteria for peak response quantities and root mean square response.

There are three evaluation criteria for peak response. They are response ratios involving floor displacement ratio (EC1), floor acceleration ratio (EC2) and base shear force ratio (EC3). The other three evaluation criteria are related to root mean square response: floor displacement ratio (EC4), floor acceleration ratio (EC5) and base shear force ratio (EC6). These criteria are calculated as a ratio of the controlled over the uncontrolled responses. They are non-dimensionalized measures, and smaller values indicate a more effective control system.

Figure 2 shows the relative displacement and acceleration of the top floor versus the damper effective stiffness due to Kobe and El Centro earthquakes. In terms of relative displacement, it decreases quickly as the damper effective stiffness is increased from 0 to 2000 kN/m. It then decreases slowly as the stiffness is increased further. But resetting 3 experiences a significant increase for $k_{hi} > 4000$ kN/m. On the other hand, the increase of stiffness generally increases floor acceleration. However, resetting1, resetting2 and LQR can provide a reduction for $k_{hi} = 1000$ kN/m due to El Centro earthquake, while resetting3 increases it rapidly for $k_{hi} > 3000$ kN/m. It is observed that the damper effective stiffness is optimum at about 2000 kN/m and hence used as design stiffness.

The resulting evaluation criteria for five control cases with $k_{hi} = 2000$ kN/m are presented in Table 1. It is observed that resetting3 performs better than other resetting cases and also better than switching and LQR. Resetting3, switching control and LQR algorithms can achieve maximum displacement reductions of up to 83%, 73% and 74%, respectively. In terms of floor acceleration, resetting and switching control could not offer a reduction, especially due to Kobe earthquake. On the other hand, LQR could offer maximum acceleration reduction of up to 20%. All control cases can reduce base shear forces, which is important for safety considerations. It is also noted that the RMS response could achieve a much better level of control than the peak response.

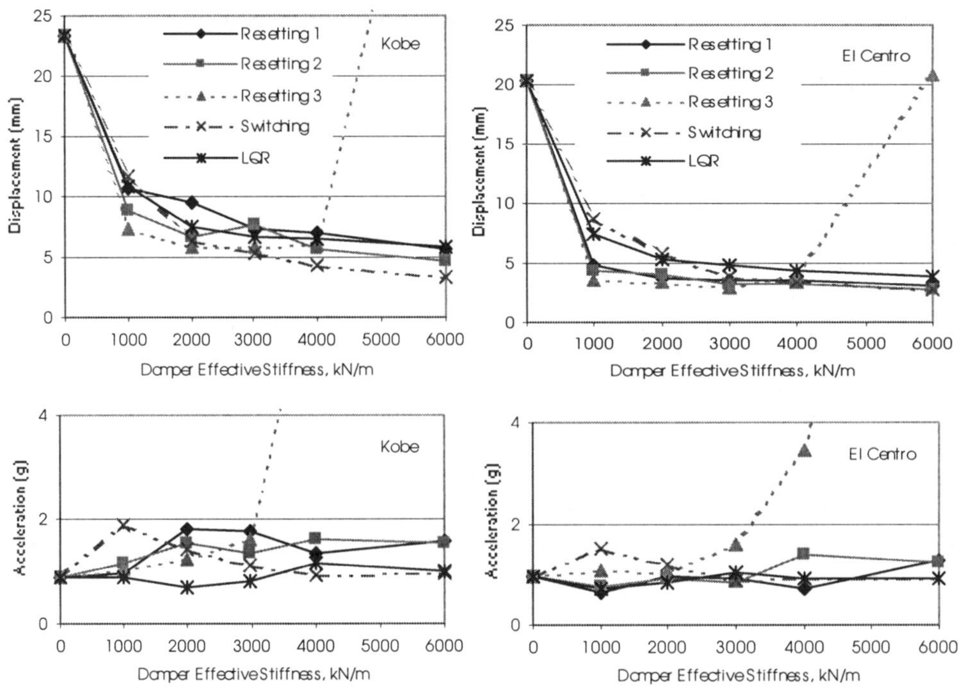

Figure 2. Relative displacement and floor acceleration vs. damper effective stiffness for five control cases

Table 1. Evaluation Criteria for the five-storey model

	Peak Response			RMS Response			Peak Response			RMS Response		
	EC1	EC2	EC3	EC4	EC5	EC6	EC1	EC2	EC3	EC4	EC5	EC6
	Kobe						El Centro					
Resetting1	0.41	2.06	0.55	0.18	0.47	0.15	0.19	1.00	0.38	0.07	0.19	0.07
Resetting2	0.28	1.77	0.39	0.15	0.47	0.13	0.20	0.99	0.42	0.06	0.19	0.08
Resetting3	0.25	1.43	0.34	0.13	0.51	0.13	0.17	1.08	0.39	0.06	0.27	0.08
Switching	0.27	1.61	0.57	0.16	0.31	0.19	0.29	1.27	0.49	0.08	0.20	0.13
LQR	0.32	0.80	0.39	0.21	0.25	0.18	0.26	0.90	0.38	0.09	0.16	0.11

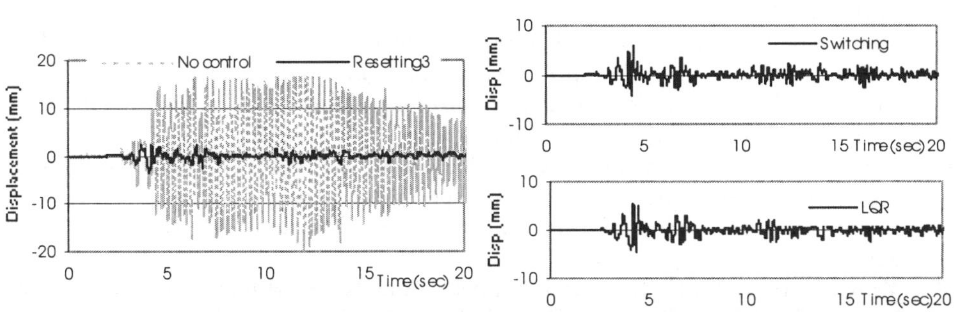

Figure 3 Time histories of top floor displacement due to El Centro earthquake

Time histories of top floor displacement due to El Centro earthquake with k_{hi}=2000 kN/m are displayed in Figure 3. It is clear that all control cases can reduce the relative displacements and add significant damping to the system.

5 Conclusions

Vibration control of a five-storey model with semi-active stiffness damper was investigated by using five control cases and varying damper effective stiffness. Two benchmark earthquakes, Kobe and El Centro, were used as input earthquake motions.

The damper effective stiffness is an important parameter that affects the control system performance. The increase in stiffness will reduce the displacement rapidly until it reaches an optimum value, then the reduction is much slower. On the other hand, the increase in stiffness does not reduce floor acceleration; it can even increase it. Resetting3 is very sensitive to variation in damper effective stiffness. However, resetting3 performs best in reducing floor displacements, while LQR could offer a much better acceleration reduction compared with other control cases.

It has been shown that semi-active stiffness damper effectively controls the vibrations of a five-storey model induced by earthquake loads. The damper effective stiffness must be optimized to achieve best results.

References

1. Djajakesukma, S.L., Samali, B. and Hung Nguyen, Application of a Semi Active Stiffness Damper to a Five Storey Benchmark Model. *The Fifth International Conference on Motion and Vibration Control, Sydney, Australia* **Vol. 2** (2000) pp. 569-574.
2. Nguyen H. T., *Analogue and Digital Control.* Faculty of Engineering, University of Technology, Sydney (1998).
3. Samali B., System Identification of a Five Storey Benchmark Model Using Modal Analysis. *Asia-Pacific Vibration Conference '99 Singapore* (1999) pp.701-706.
4. Samali B., Sri Lily D. and Hung Nguyen, A Comparison of Control Laws for A Semi-Active Stiffness Damper. *The Fifth International Conference on Motion and Vibration Control, Sydney, Australia* **Vol. 2** (2000) pp. 737-742.
5. Soong T. T., *Active Structural Control: Theory and practice*. Longman Scientific & Technical (1990).
6. Yang, Jann N., Kim, Jin-H. and Agrawal, Anil K., Seismic Response Control Using A Semi-Active Stiffness Damper, *Proceeding of International Workshop on Seismic Isolation, Energy Dissipation and Control of Structures*, in Guangzhou, China, published by seismological press, China (1999) pp. 312-319.
7. Yang, Jann N, A. K. Agrawal and W. He, Response Control Of Nonlinear Structures Using A Semi-Active Stiffness Damper. *Proceedings of The International Conference on Advances in Structural Dynamics,* Hongkong, China (2000) pp. 349-356.
8. Yang, Jann N, Jin-H. Kim, and Agrawal, Anil K., Resetting Semiactive Stiffness Damper for Seismic Response Control. *Journal of Structural Engineering*, **Vol. 126**, 12 (2000) pp. 1427-1433.

EARTHQUAKE RESISTANT PERFORMANCE OF LAMINATED AND LEAD CORE RUBBER BEARINGS

YI MIN WU, BIJAN SAMALI AND JIANCHUN LI

Centre for Built Infrastructure Research, Faculty of Engineering, University of Technology, Sydney, Broadway, NSW 2007, Australia E-mail : Yimin.wu@eng.uts.edu.au

Base isolation has been used in many classes of civil engineering structures to reduce the damage from earthquake attacks. The main idea behind this innovative technique is to shift the fundamental natural frequency of the structure to a value much lower than the dominant frequency of the earthquake. In this work, dynamic characteristics of a five-storey benchmark model isolated with laminated rubber bearings (LRB) and Lead core rubber bearings (LCRB) under Kobe and El-Centro ground motions were examined using a shaking table. The earthquake resistant performance of LRB and LCRB isolators was evaluated. The first natural frequency of the model was reduced from 4.7Hz for the unisolated model to 2.9 and 1.7 Hz for the LRB and LCRB isolated models, respectively. It was observed that maximum floor accelerations were significantly reduced with the addition of an LRB or LCRB isolation system regardless of the ground motion input. However, the LRB was identified to be more effective than LCRB in reducing accelerations and model relative displacements, and therefore, provided a better protection of the superstructure and its contents. On the other hand, the LCRB produced smaller base and floor displacements, and hence can ensure structural stability better than the LRB. This property of LCRB stems from high damping and stiffness of the lead core.

1 Introduction

Ground motions, produced by severe earthquakes are often quite damaging to structures and their contents. Conventional earthquake-resistant designs often focus on the strengthening of structures to resist such disturbances and avoid structural collapse, whilst little attention is given to the prevention of damage. Using such design approaches, it is almost impossible to construct completely 'earthquake-proof' structures, which are of reasonable cost and aesthetically acceptable.

Seismic isolation is an efficient design scheme which can successfully resist earthquake loading, and not only provides safety but also prevents damage [1-2]. Seismic isolation is particularly useful for low to medium-rise buildings which happen to have their fundamental frequencies within the "prone-to-resonance" range of dominant earthquake frequencies. Critical facilities, such as emergency response centres, hospitals, fire stations, utilities and communication centres, should remain operational during and after the strongest earthquakes. The disruption to the operation of such essential facilities may be prevented by using seismic isolation to enhance their aseismic capacity by reducing the seismic loads they may experience during a powerful earthquake. This is achieved by mounting the structure on a set of isolators that provide low horizontal stiffness and, consequently, shift the fundamental frequency of the structure to a much lower value. As a result, most deformations occur at the isolation level. The superstructure remains essentially undeformed; moving like a rigid body, which prevents damage to its structural and nonstructural components.

It is necessary to conduct experimental studies on the structural systems with base isolators, because of inelastic behaviour of the base isolation systems. Well- conducted experimentation will provide data for analysis and design of base-isolated structures.

This paper describes a series of shaking table tests for the evaluation of seismic performance of base-isolated structures subjected to various earthquake inputs.

Effectiveness and efficiency of two rubber isolation systems are investigated for further developments of new and more effective isolation systems, and to compare the earthquake resistant performance between the laminated and lead core rubber bearings.

2 Experimental studies

2.1 Five-storey benchmark steel model

The experimental benchmark model designed by Samali [3] offers the flexibility needed to model and test various building configurations. The mass of the model can vary by providing additional mass in the form of plates and/or point masses at pre-designated locations. A three-tonne model is used in this work. The model is designed to have a maximum height of 3 meters without the isolators. In an "equal floor height configuration", the model will have a storey height of 600 mm. The model has a footprint of 1.5m x 1.0m. It consists of two bays in one direction and a single bay in the other.

2.2 Characteristics of the two rubber bearing isolators

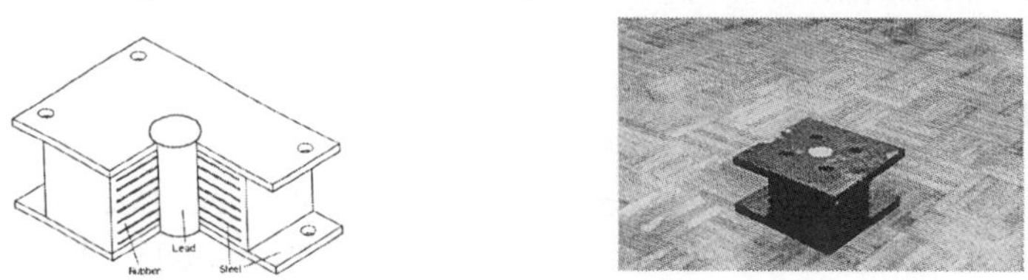

Figure 1. Lead core rubber bearing. Left: schematic, Right: photograph. (The laminated rubber bearing is the same as the lead core rubber bearing but without a lead plug at the center of the bearing.)

The laminated rubber bearing (LRB) used in this study consists of 25 thin rubber sheets with a sheet thickness of 2.2 mm and 25 thin layered steel plates with a thickness of 1.8 mm. The rubber sheets are vulcanized and bonded to the thin steel plates under pressure and heat. The inner thin steel plates prevent lateral bulging of the rubber layers, and produce a large vertical stiffness of about 500 times larger than lateral stiffness of 220 kN/m, and therefore provide a large vertical load carrying capacity. Horizontal flexibility is provided by the shear deformability of the rubber sheets. The overall dimensions of the laminated bearings are 120 x 120 x 100 mm. Two thick mounting steel plates (200 x 200 x 20 mm) were, then, bonded to the bottom and top surfaces of the laminated bearings to allow the isolators to be firmly connected to the shaking table and the superstructure, as shown in Fig. 2.

Configuration and dimensions of lead core rubber bearings (LCRBs) (Fig. 1) are the same as LRB, but a lead plug with a diameter of 30 mm was inserted into a machined hole at the centre of the bearing. In addition to LRB characteristics, a further energy dissipation mechanism can be achieved, via the plastic deformation of the lead plug. A lead rubber bearing also provides initial rigidity under lateral service loads, such as wind loads, due to the presence of the lead plug and its high stiffness prior to yielding. In this case, however, the energy dissipation mechanism is activated only once the lead plug has yielded. The lead core rubber bearing, furthermore provides the greater restoring effect, which is necessary to re-center the isolators to their original locations after the earthquake.

2.3 Shaking table tests

Figure 2. Eexperimental set-up for the LRB-isolated model on the shaking table.

Shaking table tests were carried out using the unidirectional shaking table facility at UTS. The plan dimensions of the table are 3 x 3 m. The shaking table can move in a horizontal direction by a hydraulic actuator with a maximum acceleration of 2.5g (bare table), and a maximum stroke and piston velocity of +-100mm and 550 mm/s, respectively. The instrumentation consists of seven accelerometers and seven LVDTs (displacement transducers) located at each floor, at base level and on the shaking table. A total of 14 channels of data were recorded using two YOKOGAWA Analyzers; two channels for the shaking table responses and 12 channels for the model responses. The shaking table was driven in the longitudinal direction of the five-storey model. As the input excitation to the table, two earthquake records: El Centro (1940) and Kobe (1995) were used. The two earthquake records chosen are among the four records, recommended by the International Association for Structural Control for benchmark seismic studies. Measured maximum accelerations on the shaking table were 0.393g for the El Centro and 0.901g for the Kobe excitation. To maintain dynamic similitude, each record was compressed in time by a factor of 3 to keep the first mode frequency of the model consistent with dominant frequency of the earthquake record. The shaking table tests were conducted on both the unisolated and the base-isolated structures. The base-isolated structure consisted of the superstructure and the base isolation system. Figure 2 shows the experimental set-up for the LRB-isolated five-storey benchmark model on the shaking table.

3 Results and discussion

Natural frequencies of the unisolated model, LRB-isolated and LCRB-isolated models were analyzed using an experimental modal analysis method [3], and the results are summarized in Table 1. The first natural frequency of either LRB or LCRB isolated models is smaller than that of the unisolated model. Because of the smaller lateral stiffness of LRB, the first natural frequency of LRB isolated model is lower than LCBR isolated model and further away from dominant earthquake frequency. Higher mode shapes are mainly influenced by the properties of the superstructure itself. Consequently, the highest frequencies among the three models show little difference (39, 40 and 39 Hz for the unisolated model, LRB-isolated and LCRB-isolated models, respectively).

Table 1. Natural frequencies of the three models tested.

Mode	Natural frequency (Hz)		
	Unisolated	LRB	LCRB
1st	4.7	1.7	2.9
2nd	13.9	8.3	9.6
3rd	20.6	16.3	17.0
4th	33.7	21.3	21.7
5th	38.9	33.7	34.1
6th		40.9	39.0

Table 2. Peak floor accelerations and corresponding RMS values.

Earth-quake	Floor	Peak Acceleration (g) unisolated	LRB	LCRB	RMS Acceleration (g) unisolated	LRB	LCRB
Kobe	isolator		0.67	1.08		0.11	0.08
	1st	1.52	0.66	0.93	0.09	0.12	0.08
	2nd	1.52	0.58	0.90	0.14	0.12	0.08
	3rd	1.74	0.54	0.86	0.19	0.13	0.09
	4th	2.20	0.61	1.04	0.22	0.13	0.10
	5th	2.85	0.63	1.48	0.24	0.12	0.11
El-Centro	isolator		0.22	0.67		0.06	0.05
	1st	0.63	0.21	0.60	0.06	0.06	0.05
	2nd	0.73	0.23	0.64	0.11	0.06	0.05
	3rd	1.21	0.24	0.49	0.14	0.07	0.05
	4th	1.62	0.23	0.71	0.17	0.07	0.05
	5th	1.65	0.25	1.06	0.19	0.06	0.07

Table 3. Floor relative displacements and corresponding RMS values.

Earth-quake	Relative Displacement [mm]						
		Peak value			RMS value		
	Floor	unisolated	LRB	LCRB	unisolated	LRB	LCRB
Kobe	1st	4.37	4.32	5.58	0.63	0.93	0.54
	2nd	8.77	6.59	9.25	1.35	1.44	0.92
	3rd	12.92	9.65	11.89	1.90	1.96	1.18
	4th	16.22	11.42	13.69	2.38	2.41	1.29
	5th	17.62	11.76	15.45	2.55	2.48	1.56
El-Centro	1st	3.42	1.83	2.14	0.52	0.52	0.27
	2nd	7.33	2.65	3.51	1.12	0.84	0.47
	3rd	10.19	3.69	4.42	1.55	1.00	0.58
	4th	13.10	4.60	5.42	1.98	1.29	0.65
	5th	13.77	4.21	6.88	2.11	1.26	0.81

The effectiveness of the base isolation systems was evaluated by comparing the structural responses of the base-isolated and the unisolated models. The maximum floor accelerations for the bareframe and isolated models under strong ground motions are presented in Table 2 and Fig. 3. Model relative displacements, defined as the displacement of the 5th floor relative to the shaking table for the bareframe or relative to the base of column pads for the base isolated model, are summarized in Table 3 and their time histories are presented in Fig. 4. It can be seen that the floor acceleration responses of the unisolated model increase drastically as the floor level increases, compared with the motion of the shaking table. The accelerations associated with LCRB isolated model increase only slightly. On the other hand, the accelerations at all floor levels in the LRB isolated model were found to be very similar to the rubber base acceleration, indicating that the superstructure vibrates approximately as a rigid body. A similar improvement on the model relative displacement was also observed when LRB and LCRB isolators were used. A smaller improvement on the earthquake responses of LCRB-isolated model, compared with those of LRB-isolated model, is attributed to its high initial stiffness and its sudden change at certain base displacements. From Figs. 3 and 4 it is also clear that LRB can filter higher frequency responses, whereas LCRB is not capable of doing so.

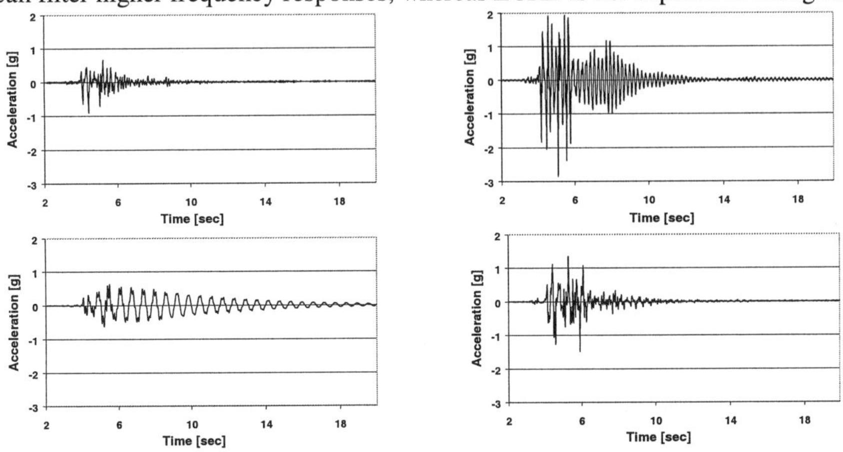

Figure 3. Acceleration time histories of the table and the fifth floor under Kobe earthquake (Top left: table acceleration, Top right: unisolated model, Bottom left: LRB and Bottom right: LCRB).

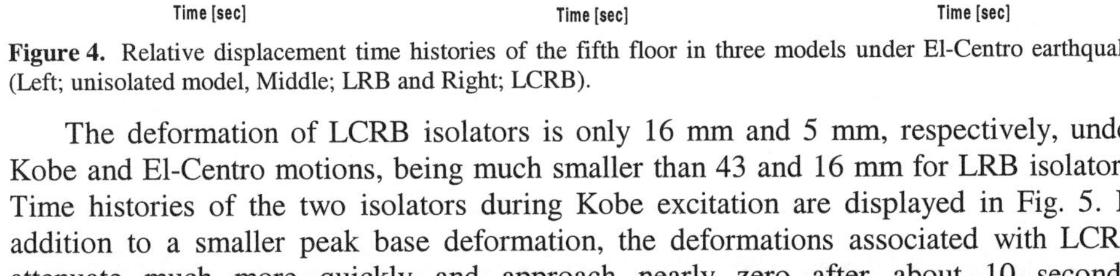

Figure 4. Relative displacement time histories of the fifth floor in three models under El-Centro earthquake (Left; unisolated model, Middle; LRB and Right; LCRB).

The deformation of LCRB isolators is only 16 mm and 5 mm, respectively, under Kobe and El-Centro motions, being much smaller than 43 and 16 mm for LRB isolators. Time histories of the two isolators during Kobe excitation are displayed in Fig. 5. In addition to a smaller peak base deformation, the deformations associated with LCRB attenuate much more quickly and approach nearly zero after about 10 seconds, demonstrating the superior stability of the LCRB-isolated model to that of the LRB-isolated one. A smaller base deformation offers great advantages in the design of base-isolated structures. The damping ratio associated with LCRB isolators is 11.37%, which is much larger than either the unisolated model (0.26%) or the LRB isolator (1.76%). The lead core provides an effective energy absorption mechanism, resulting in a small base deformation but large damping and quick response attenuation with time. Therefore, extra care must be taken if LRB isolators are used to reduce the seismic effect on structures built on soft grounds.

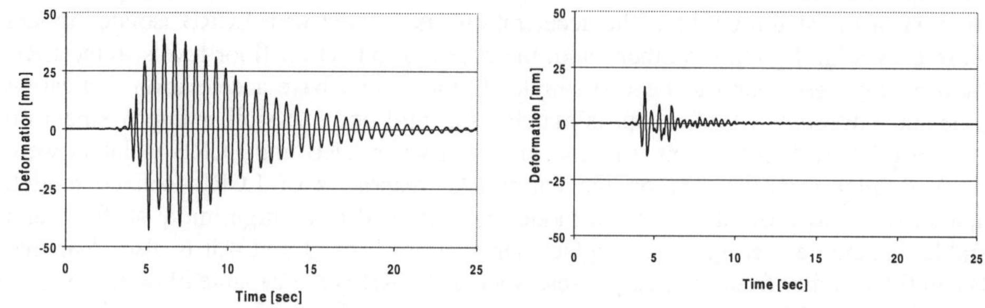

Figure 5. Deformations of laminated rubber bearing (Left) and lead core rubber bearing (Right) under Kobe earthquake.

It is seen from above the discussion that most deformations in a seismically isolated structure occur at the isolation level and are due mostly to the contribution of the fundamental mode in which the superstructure remains essentially rigid. The fundamental mode has the highest participation in the structural response. The deformations of the superstructure are, however, due to higher modes of vibration.

4 Conclusions

In this study, a series of shaking table tests were conducted on the unisolated model, the LRB-isolated and the LCRB-isolated models to evaluate the earthquake resistant performance of LRB and LCRB isolators under strong ground motions. The first natural frequency of the unisolated model, being 4.7 Hz, was reduced to 1.7 and 2.9 Hz, respectively, when LRB and LCRB isolators were used. Both LRB and LCRB can offer great benefits against earthquake attacks. However, some important differences in performance between the two isolators were identified. The LRB was found to be more effective than LCRB in reducing accelerations and model relative displacements, and therefore provided a better protection of the superstructure and its contents. On the other hand, LCRB produced smaller base displacements, and hence a more stable structure.

References

1. P. Komodromos, Seismic Isolation for Earthquake Resistant Structures, *WIT Press* (2000).
2. Yang, J. N., Danielians, A, and Liu, S.C., Aseismic Hybrid Control Systems for Building Structures. *Journal of Engineering Mechanics*, **117**, No.4. (1991), pp.836-853.
3. Samali, B., Li, J., Mayol, E., and Wu, Y. M., System identification of a five- storey benchmark model using modal analysis. *Proceedings of International Conference on Applications of Modal Analysis,* Gold Coast, Queensland, Australia, paper 12. (1999).

VARIABLE STRUCTURE CONTROL OF SEISMICALLY EXCITED STRUCTURE WITH SHAPE MEMORY ALLOY ACTUATORS

JIANCHUN LI AND BIJAN SAMALI
Centre for Built Infrastructure Research
Faculty of Engineering, University of Technology, Sydney, NSW, Australia
email: Bijan.Samali@uts.edu.au

Among the varieties of control algorithms, Variable Structure Control (VSC) or Sliding Mode Control (SMC) is known for its most distinguished feature and ability to result in ultra-robust control systems with respect to parametric uncertainties of structures. In this paper, control methods based on Variable Structure Control (VSC) are presented and applied to a seismically excited civil engineering structure equipped with shape memory alloy actuators. Design of the sliding modes including the eventual sliding mode, was accomplished by using LQR methods. In developing control algorithms, both Lyapunov direct method and reaching law methods were used, which result in four control laws. A comparison of these control laws assisted in further understanding of Sliding Mode Control theory. Simulations were conducted to evaluate the level of vibration reduction using the presented control methods.

1 Introduction

The theory of variable structure control (VSC) or sliding mode control (SMC) was proposed during the 1950s, and now has developed into a general design method for both linear and non-linear automatic control systems. The main advantages of variable structure control are : i) design space (sliding mode space or reaching mode space) has lower order than state space ii) robustness and even invariance can be achieved through proper design. There are three basic types of sliding mode controller design methods including direct switching method, Lyapunov function method and reaching law method [1]. The direct switching method, developed in earlier stage for Single Input/Single Output (SISO), is rarely used now. Lyapunov direct method for controller design is now widely used. Examples of the work include the systematic investigation for the applications of sliding mode control to civil engineering structures by Yang et al. [2]. As a relatively newcomer, the reaching law method exhibits promising potential. It not only establishes the reaching condition but also specifies the dynamic characteristics of the system during the reaching phase. In addition, using reaching law methods, it is simple to solve for sliding mode controller and providing a measure for the reduction of chattering.

On the other hand, using smart materials for structural control has received increasing attention by researchers. Much of the work has focused on development of actuating materials such as piezo-ceramics, etc. Shape Memory Alloy (SMA) has traditionally been used for passive strategies, such as dampers and other types of energy dissipation devices, instead of being used for active control strategies as actuators due to its limited frequency bandwidth. However, most civil engineering related applications deal with low frequency bandwidth, therefore, it is possible to capitalize on the advantages of SMA actuators without much compromise due to their relatively low response time.

This paper presents a technique for designing sliding mode controllers by using both Lyapunov function method and reaching law method. The controllers obtained from two methods are compared and discussed. Numerical simulations were carried out on control of a five-storey building model equipped with SMA actuators.

2 Variable Structure Control

For a general linear time-invariant structure system, the equation of motion can be expressed in state space form as:

$$\dot{z}(t) = Az(t) + Bu(t) + Hf(t) \tag{1}$$

where $z(t)$ is a $2n$ state vector; A is a $(2nx2n)$ system matrix; B is a (nxm) location matrix; and H is a $(2n \times 1)$ excitation matrix.

2.1 Determination of Switching Function

The structure in a VSC system is governed by the sign of a vector-valued function $s(z)$, which is defined to be switching function. A switching function is generally assumed to be m dimensional and linear, ie.,

$$s(z) = Pz \tag{2}$$

where $s(z) = [s_1(z) \;\; s_2(z) \;\; \ldots\ldots \;\; s_m(z)]^T$ and $P = [\, p_1^T \;\; p_2^T \;\; \ldots. \;\; p_m^T \,]^T$

Each scalar switching function $s_i(z)$ describes a linear surface $s_i(z)=0$, which is defined to be a switching surface. Taking derivative of $s(z)$, neglecting the excitations and substituting (1) into it , one has

$$\dot{s}(z) = \frac{\partial s}{\partial z}Az + \frac{\partial s}{\partial z}Bu = 0 \tag{3}$$

equation (3) can be solved for so called equivalent control $u_e = \left[\frac{\partial s}{\partial z}B\right]^{-1}\frac{\partial s}{\partial z}Az$ (4)

If for any initial state z_0 belongs to switching space S, the state of the system $z(t)$ at any given time t also belongs to switching space S, then $z(t)$ is a sliding mode (or sliding motion) of the system. If for every point in S there are trajectories reaching it from both sides of S, then switching surface S is called a sliding surface. The condition under which the state will move toward and reach a sliding surface is called a reaching condition. It is clear that for the definition above, an *nth*-order system with m inputs will have 2^m-1 switching surfaces. However, there is a single switching surface S_E of dimension $(n\text{-}m)$, which is the intersection of all surfaces S_i, $i=1,2,\ldots m$ taken together, called the eventual switching surface to which all trajectories must eventually reach. The surface is given by

$$S_E = \{z \mid s(z) = Pz = 0\} = S_1 \cap S_2 \cap \ldots. \cap S_m \tag{5}$$

On each switching surface, there may be a sliding mode that is described by a differential equation of the same dimension as the switching surface. The sliding mode associated with S_E may be called the eventual sliding mode. There are actually many ways in which a sliding motion can begin. They are called switching scheme. There are two switching schemes of importance for this study, namely i) free-order switching scheme and ii) eventually sliding mode switching scheme.

In free-order switching scheme, the order of sliding modes is not pre-assigned but follows natural trajectory on a first-reach-first-switch scheme. The switching takes place depending on the location of initial state in the state space. The advantages of this switching scheme are i) easy to determine the solution of VSC; ii) reaching mode possessing better dynamic characteristics. In eventual sliding mode switching scheme, the state is driven from any initial state to the eventual sliding surface S_E on which sliding mode control takes place. There may or may not be sliding modes on other switching surfaces. This switching scheme is simple for implementation and is easy to be made smooth. The only drawback is that it does not guarantee good characteristics.

In order to determine switching function *s(z),* it requires determination of *P*. State equation (4) is converted into the regular form (or reduced form):

$$\dot{Y} = \overline{A}Y + \overline{B}U; \qquad s = \overline{P}Y = 0 \tag{6}$$

where

$$\overline{A} = DAD^{-1}; \quad \overline{P} = PD^{-1}; \quad \overline{B} = \begin{bmatrix} 0 \\ \overline{B_2} \end{bmatrix} \tag{7}$$

in which D is a transformation matrix. The design of the switching function s(z) = Pz is obtained by minimising the performance index of the state vector as follows:

$$J = \int_0^\infty z^T(t)Qz(t)dt \tag{8}$$

By solving a matrix Riccati equation for $\overline{P}_1$, the original switching function $s(z) = Pz$ is obtained as

$$P = \overline{P}D = [\overline{P}_1 : I_m]D \tag{9}$$

2.1.1 Design Of Controllers Using Lyapunov Method

Consider a Lyapunov function candidate *V* as follows

$$V(z,t) = \frac{1}{2}s^T s = \frac{1}{2}s^T Pz \tag{10}$$

The global reaching condition, i.e. for the switching function (sliding mode) *s(t)* to be zero as t→ ∞, is given by:

$$\dot{V} = s^T\dot{s} < 0 \qquad \text{when } s \neq 0. \tag{11}$$

taking derivative of s and using the state equation of motion (1),

$$\dot{V} = s^T\frac{\partial s}{\partial z}[Bu - (Az + Hf)] = s^T[PBu - P(Az + Hf)] \tag{12}$$

The finite reaching time is guaranteed by modifying (11)

$\dot{V}(z,t) < -\varepsilon$ when s ≠ 0, where ε is positive.

Clearly, this approach leads to the eventual sliding mode.

Continuous Sliding Mode Control (CSMC)

from (12), let $\varepsilon = s^T\delta s$. A possible continuous controller is given by

$$u = G - (PB)^{-1}\delta s \tag{13}$$

in which $\delta \geq 0$ is referred to as the sliding margin and $G = -(PB)^{-1}P(Az + Hf)$ (14)

Discontinuous Sliding Mode Control (DSMC)

In Addition to the continuous controller represented by (15), a possible discontinuous controller is given:

$$u_i = G_i - (PB)^{-1}\delta H(|s_i| - \varepsilon_0)\mathbf{sgn}(s_i) \tag{15}$$

2.1.2 Design Of Controllers Using Reaching Law Method

Let the dynamics of the switching function be specified by the differential equation called reaching law:

$$\dot{s} = -Q\,\mathbf{sgn}(s) - Kh(s) \tag{16}$$

where gains Q and K are diagonal matrices with positive elements, and

$$\mathbf{sgn}(s) = [\mathbf{sgn}(s_1)\ldots\mathbf{sgn}((s_m)]^T, \qquad h(s) = [h_1(s_1)\ldots h_{m1}(s_m)]^T \tag{17}$$

The scalar function h_i satisfies the condition

$$s_i h_i(s_i) > 0 \quad \text{when} \quad s_i \neq 0,\ i = 1 \text{ to } m \tag{18}$$

Various choices of Q and K specify different rates of *s(z)* and yield different structures in the reaching law. It is clear that the reaching law approach not only establishes the reaching condition but also specifies the dynamic characteristics of the system during the reaching phase. This approach results in the free-order switching scheme. The control law can be obtained directly from reaching law specified by (16):

$$\dot{s} = \frac{\partial s}{\partial z}(Az + Bu + Hf) = -Q\,\mathbf{sgn}(s) - Kh(s) \tag{19}$$

The control law is easily obtained by solving (19):

$$u = \left[\frac{\partial s}{\partial z}B\right]^{-1}\left[\frac{\partial s}{\partial z}(Az + Hf) + Q\,\mathbf{sgn}(s) + Kh(s)\right] \tag{20}$$

Neglecting the external excitation in (20) will reduce it back to the form given by Gao[1]. Two special cases are considered in the following:

The Constant Plus Proportional Rate Reaching Law (CPPRRL)

$$\dot{s} = -Q\,\mathbf{sgn}(s) - Ks \tag{21}$$

then (20) becomes: $$u = [PB]^{-1}[P(Az + Hf) + Q\,\mathbf{sgn}(s) + Ks] \tag{22}$$

The Power Rate Reaching Law (PRRL)

$$\dot{s}_i = -k_i|s_i|^{\alpha}\,\mathbf{sgn}(s_i) \qquad 0 < \alpha < 1 \qquad i = 1 \text{ to } m \tag{23}$$

then (20) becomes: $$u = [PB]^{-1}[PAz + Hf + k_i|s_i|^{\alpha}\,\mathbf{sgn}(s_i)],\quad 0 < \alpha < 1,\quad i = 1 \text{ to } m \tag{24}$$

2.1.3 Comparison of the two controller design methods

By rearrange (22) , the constant plus proportional rate reaching law becomes

$$u = G - [PB]^{-1}Q\,\mathbf{sgn}(s) - [PB]^{-1}Ks \tag{25}$$

Comparing to CSMC (13), it is clear that the CSMC is a type of proportional rate reaching law with $K=\delta$. Comparing to DSMC (15) , it can be seen that DSMC is a type of constant rate reaching law in which $Q = \delta H(s_i - \varepsilon_0)$.

3 Numerical Simulation

3.1 Structural Model And SMA Actuator

A structural model, representing a five-storey building, was designed for this study. The model is 893 mm in height with plan dimensions of 120 mm by 100 mm. The model was designed to possess a first natural frequency of about 2 Hz to ensure the frequency is inside the bandwidth of most earthquakes. The Shape Memory Alloy (SMA) actuators chosen for the study are straight two-way Nitinol wires (diameter = 0.15 mm) with transformation temperature of 36°C. The actuation comes from the unique behaviour of Nitinol alloy when it undergoes a martensitic phase transformation. The actuator wires are installed between two consecutive levels at the inclination angle of 55°. One set of SMA controllers located at the first level, and consisting of two pairs of SMA actuator wires, are arranged in a cross bracing manner connecting adjacent floors. Northridge earthquake of 1994 was used as input excitation in this study. Taking into account the strength of the structural model, PGA of the earthquake was reduced to 0.105g. The same weighting matrix Q = diag[10^6,10^5,10^4,10^3,10^2,1,1,1,1,1] was chosen for switching function design and LQR control.

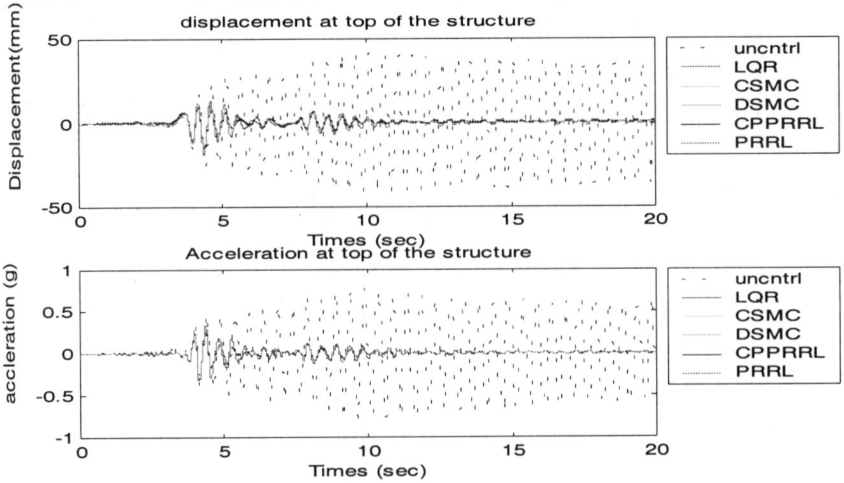

Figure 1. Displacements and accelerations at top floor of the structure model

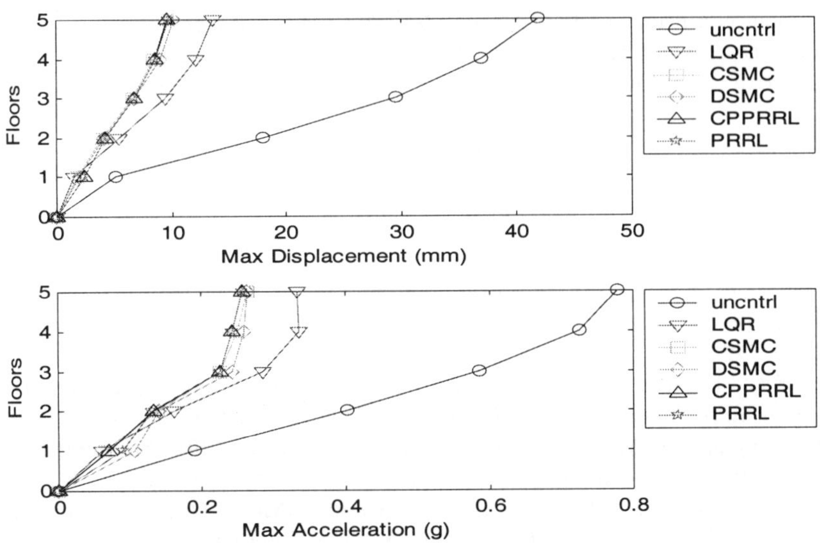

Figure 2. Maximum Responses of each floor (max force = 50N)

The time histories of top floor displacements and accelerations from the simulations are shown Figure 1. Since the structure is lightly damped, the uncontrolled response decays slowly. Meanwhile, all controllers are quite effective, specially the four proposed controllers. This can be seen clearly in Figure 2 and Table 1. For both maximum acceleration and displacement the proposed controllers outperform LQR. They are about 27% better than LQR in terms of maximum displacement reduction and about 21% better in terms of maximum acceleration reduction. Meanwhile, the performance of the four proposed controllers is quite similar, with DSMC being slightly less effective for displacement reduction. The PRRL and CSMC offer a slightly better reduction for acceleration on most floors.

Table 1.Comaprision of maximum responses under different control methods

	max displacement (mm)					**Max acceleration (g)**				
	Floor 1	*Floor 2*	*Floor 3*	*Floor 4*	*Floor 5*	*Floor 1*	*Floor 2*	*Floor 3*	*Floor 4*	*Floor 5*
Uncontrolled	5.1	18.0	29.5	36.9	41.8	0.19	0.40	0.59	0.73	0.78
LQR	1.6	5.5	9.5	12.1	13.7	0.06	0.16	0.29	0.34	0.33
CSMC	1.9	3.9	6.6	8.5	9.6	0.06	0.14	0.23	0.25	0.26
DSMC	2.0	4.0	6.8	8.9	10.2	0.10	0.14	0.24	0.26	0.26
CPPRRL	2.4	4.2	6.8	8.6	9.7	0.07	0.13	0.22	0.24	0.26
PRRL	2.4	4.1	6.7	8.6	9.7	0.09	0.13	0.23	0.24	0.26

4 Conclusions And Some Future Work

The results of the numerical simulation have shown that the proposed controllers derived from Lyapunov direct method and Reaching Law method offer a promising performance in controlling a five storey structural model equipped with SMA. With the same control force, the four proposed controllers clearly outperformed LQR in terms of response reductions under Northridge earthquake of 1994. The additional merit of using VSC is its inherent robustness, which is the subject of further studies. The experimental verification of performance of the proposed controllers is also part of the planned research for future.

References

[1] Gao WB, Hung JC., 'Variable structure control of nonlinear systems: a new approach'. *IEEE Transaction Industrial and Electronics,* Vol. 40, No.1, pp.45-55. 1993.
[2] J. N. Yang, J. C. Wu and A. K. Agrawal, "Sliding Mode Control for Seismically Excited Linear Structures", *Journal of Engineering Mechanics, ASCE*, Vol. 121, No. 12, pp.1386-1390, 1995.
[3] Choi, S.-B and Hwang, J. –H, "Structural Vibration Control Using Shape Memory Actuator", *Journal of Sound and Vibration 230*, pp1168-1174, 1999.
[4] Baz, A. Imam, K., and Mccoy, J., "Active Vibration Control of Flexible Beam Using Shape Memory Alloy", *Journal of Sound and Vibration 140*, p437, 1990.
[5] J., Li and B., Samali, "Control of a Five Story Building Model Under Benchmark Earthquake Using SMA Actuators", Proceedings of the 5th International Conference on Motion and Vibration Control, Dec 4-8 2000, Sydney, Australia, Vol. 1, pp. 173 – 179.

THE APPLICATION OF PERIODIC STRUCTURE THEORY TO VIBRATIONS IN AN INFINITE PERIODICALLY SUPPORTED RAIL

G.P.BROWN AND K.P.BYRNE

School of Mechanical and Manufacturing Engineering, The University of New South Wales, Sydney, Australia.

A model of the free and forced vibrations in a periodically supported infinite rail is developed. Periodic Structure Theory is used to limit the extent of the model to a single rail segment and its support. The model incorporates degrees of freedom internal to the rail segment, allowing the rail support to be represented in detail. The Finite Element method is used to model the stiffness and inertial properties of the rail, enabling the method to be applied to beams of arbitrary cross-section. The dispersion characteristics and driving point receptance for an infinite equal-angle beam are calculated and compare favorably with properties derived using a simple analytical model. The newly developed model is applied to a prototype rail geometry and the results discussed.

1 Introduction

The Bishop Austrans system is being developed in Australia to address the worldwide need for a flexible urban transport system. One of the many novel features of the Bishop Austrans system relates to the wheels and rails, which support nine seat driverless vehicles, each having a mass of about 3000 kg and running at speeds of up to 120 kph. A cross-section of the wheel-rail arrangement is shown in Figure 1. The rail section is approximately a 125 × 50 × 12 mm angle. The upper leg of the resiliently supported steel rail on which the cylindrical steel wheels roll is inclined at 17½° to the horizontal.

The fact that steel wheels roll on steel rails, which are slightly elevated, suggests that so-called rolling noise could be a problem when the routes are in urban areas. The rails and wheels are induced to vibrate and radiate sound as a result of slightly irregular wheels rolling on slightly irregular rails.

Thompson made a major contribution to the study of rolling noise produced by main line rolling stock. One of his papers of particular relevance to the present work [1] related to the vibration of the rail. It is obvious that the first step in predicting the sound radiated by a vibrating rail is knowing how the rail vibrates. Thompson attacked this problem using Periodic Structure Theory (PST) developed by Mead [2]. The model used by Thompson assumed that the rail was continuously supported. Because of this assumption he was able to use an arbitrarily short length of the rail as the periodic element. Using this model he was able to derive the propagation constants and shapes of the waves propagating along the infinite rail and also to find the driving point receptances for the infinite rail. Thompson was able to account for the deformation of the rail cross-section because he could use finite elements to construct a realistic rail cross-section. However, because of the continuous support assumption the fact that rails are periodically supported by the sleepers was not considered.

The rail used in the Bishop Austrans system is supported by a resilient elastomeric mounting every metre and so it was felt that the periodic element that would realistically model this construction would require a one metre length of rail and a single resilient support. Thus the periodic element cannot be made arbitrarily short as was done by Thompson. As a result the finite elements which are used to model the periodic element must have internal coordinates rather than just the external coordinates which connect one periodic element to the next as was done by Thompson. Gry and Gontier [3] have noted that problems arise when internal coordinates are used in periodic elements. It appears that ill-conditioning in the matrices arising from the finite element representation of the periodic element exists and is associated with very short wavelength evanescent waves.

Despite the potential problems, it was decided to pursue the periodic structure theory approach in order to gain a greater understanding of the influence of the elastomeric resilient mountings. It is hoped that by proper selection of the properties of these mountings and even altering the section of the rail that it may be possible to minimise the sound radiated from the vibrating rail.

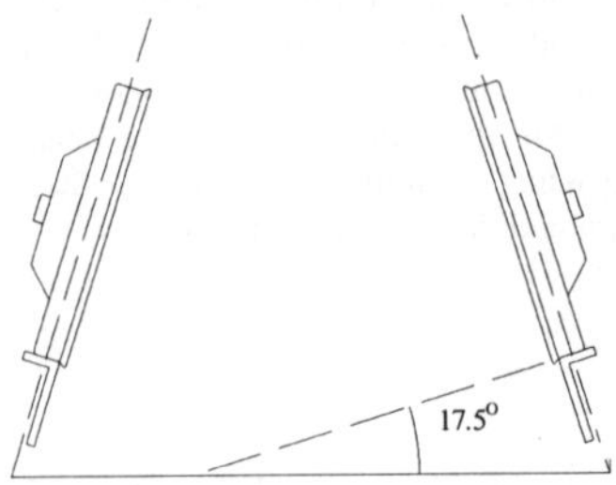

Figure 1. The Bishop Austrans Wheel-Rail Arrangement

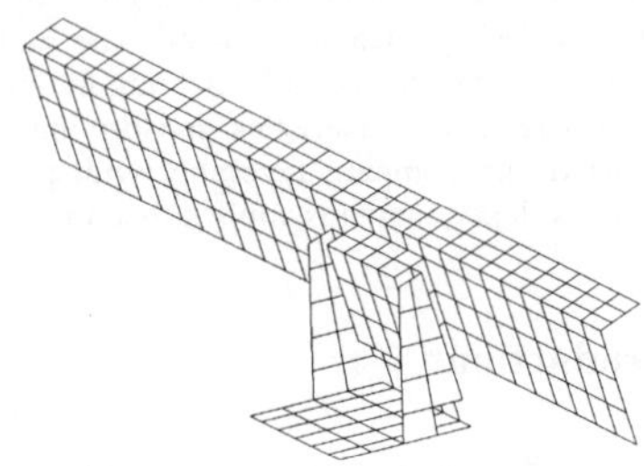

Figure 2. A Periodic Element Representation of the Bishop Austrans Rail

2 Objectives

The primary objective of this work is to develop a means of predicting the driving point and transfer receptances for an infinitely long, straight, resiliently supported rail. These receptances are needed to predict the vibration of the rail in response to the fluctuating contact force arising from the slightly irregular wheel rolling on the slightly irregular rail.

The basic approach to meeting this objective is to first determine the propagation constants and the shapes of the free vibration waves for the infinite rail, which is periodically supported by resilient, elastomeric mountings. The propagation constants for the infinite rail are frequency dependent and define a dispersion relationship for their respective waveshapes. Once this first step has been completed the contributions from as many waveshapes as found necessary can be summed to obtain the receptances. The analogous process with a finite structure is to obtain the natural frequencies and mode shapes and to use modal summation to find the receptances.

3 The Periodic Element

A 350 node model of the periodic element which represents a one metre long section of the Bishop Austrans rail is shown in Figure 2. The rail is formed of plate elements, as is the supporting bracket. The elastomeric, resilient mounting, which is between the rail and the support bracket is formed of brick elements. There are 48 external coordinates at each end of the periodic element and approximately 2000 internal coordinates.

An outline of the theory describing the formulation of equations governing wave propagation along a periodic structure that is infinite in one direction is given in Appendix A. It can be seen from Appendix A that mass and stiffness matrices for the finite element representation of the periodic element must be obtained. The MSC

PATRAN/NASTRAN finite element package along with the NASTRAN language, DMAP, was used to produce the mass and stiffness matrices. Once the mass and stiffness matrices have been formulated, the propagation constants and the free wave shapes can, in principle at least, be found. As shown in Appendix A, the equations which must be solved form a polynomial eigenvalue problem of order two, which is more difficult to solve than the conventional eigenvalue problem. It is known from physical considerations that only 2N meaningful eigenvalues exist for the problem under consideration, where N is the number of coordinates joining two periodic elements.

It is apparent from the dimensions of the matrices and the order of the eigenvalue problem that A-N solutions exist for the present formulation, where A is the total number of coordinates in a single periodic element. This is a considerable number of eigenvalues when large numbers of internal coordinates are present. This makes the calculation of physically meaningful eigenvalues more difficult than the case where no internal coordinates are used.

Fortunately MATLAB has a robust built-in function for solving the polynomial eigenvalue problem ("Polyeig"). All that needs be done is to read the NASTRAN matrices and, for a particular frequency, partition the dynamic matrix, [D], as described in Appendix A, submit the result to the eigenvalue routine and sort the output to find the physically meaningful results. The eigenvalues are related to the propagation constants by a simple expression and the eigenvectors are the free wave shapes.

In order to solve the polynomial eigenvalue problem, MATLAB modifies the matrices into the form of the generalised eigenvalue problem of order one, but of dimension 2(A-N). The QZ algorithm is ultimately used, resulting in a solution time that is proportional to $[2(A-N)]^3$. All of the available eigenvalues are found. In the case of a structure that extends infinitely in one direction whose periodic element has internal coordinates, some of the matrix partitions are null matrices and the polynomial eigenvalue problem can be simplified to a first order problem without increasing the size of the matrices. A significant saving of time results from this simplification, however, the periodic structure problem with internal coordinates is still computationally intensive. In addition to long solution times, there is some evidence of the ill-conditioned matrix problem found by Gry and Gontier [3].

4 Benchmarking

It is prudent, in view of the problem alluded to in the previous section, to attempt to validate of the results obtained using the periodic structure model. The simplest relevant behaviour that could be used as a point of reference is pure flexure about a principal axis of inertia of the beam cross section. The actual Austrans rail angle section is non-symmetrical and therefore its response will likely be dominated by coupled flexural-torsional waveshapes [4]. In order to provide a simple benchmark, the Austrans angle shown in Figure 2 was replaced by an equal angle and a force was applied through the shear centre, perpendicular to the minimum principal bending axis. The benchmarking exercise was carried out with Young's Modulus for the elements that model the resilient elastomeric mounting set to almost zero. The cross-sectional details and properties of the angle used in the validation exercise are given in Figure 3.

The propagation constants derived using the previously described approach can be found and those associated with bending waves about the minimum principal planes, identified. The dispersion curves for both the travelling and evanescent bending waves then can be found.

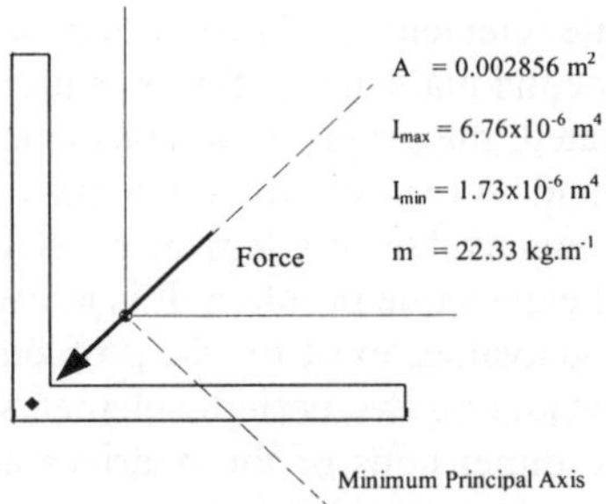

Figure 3. The Equal Angle Rail Cross-Section

The dispersion curve for the bending waves associated with bending about the minimum principal plane of the infinitely long beam also can be found using the well-known Euler-Bernoulli beam model. Ignoring shear deformation, rotary inertia and warping of the section, the model requires only the second moment of area of the beam cross-section about the minimum principal plane, Young's Modulus of the angle's material and the mass per unit length of the beam. This model is of course only valid at low frequencies. The dispersion curves for the two bending waves as found from the periodic structure theory then can be compared with the analytically derived result. A similar approach can be applied to the receptance. The analytically derived driving point receptance for an infinitely long beam can be compared with that derived using periodic structure theory.

The dispersion curve for the waves associated with bending about the minimum principal plane can be found by use of the analytically derived expression for an Euler-Bernoulli beam given in equation (1). The expression relating to the analytically derived driving point receptance is given in equation (2).

$$k_b = \left[\frac{\omega^2 m}{EI}\right]^{1/4} \qquad (1) \qquad\qquad \frac{\eta(0)}{\vec{F}} = -\frac{(1+j)}{4EIk_b^{\,3}} \qquad (2)$$

where 'k_b' is the wavenumber of the bending waves in units of radians/m and 'ω' is the circular frequency in radians/second, 'm' is the mass per unit length in units of kg/m, 'E' is Young's Modulus with units of N/m^2 and 'I' is the second moment of area about a neutral plane with units of m^4.

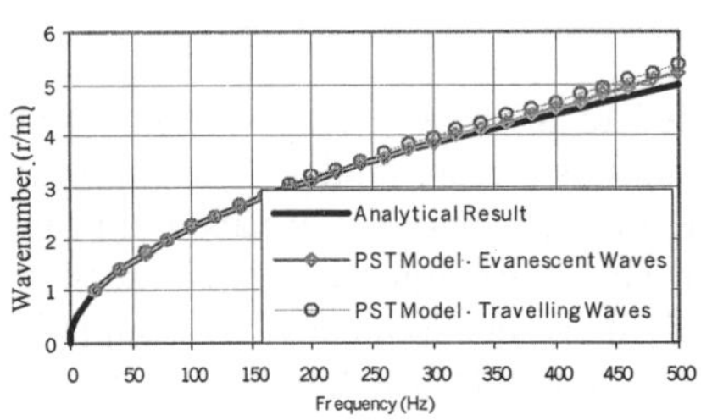

Figure 4. Dispersion Curves for Minimum Principal Axes Bending of the Infinite Equal Angle

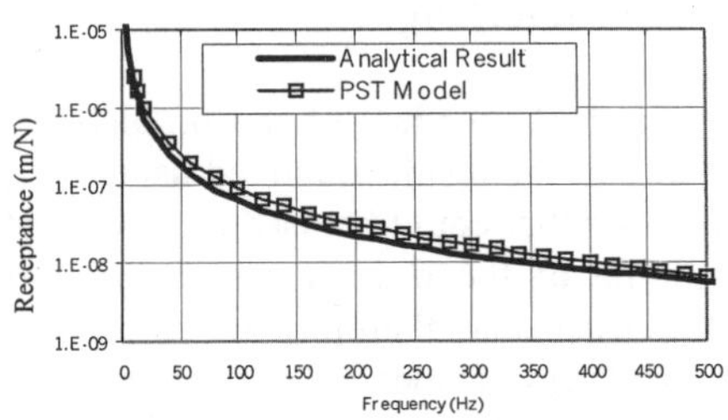

Figure 5. Driving Point Receptance for Minimum Principal Axis Bending of the Infinite Equal Angle

The analytically derived dispersion curve relating to bending waves about the minimum principal plane is plotted in Figure 4 along with curves derived from periodic structure theory. The modulus of the analytically derived driving point receptance associated with the bending waves about the minimum principal plane is shown in Figure 5 along with that derived by the periodic structure theory. Results are presented for 0-

500Hz only as at about 650Hz, waves other than the simple bending wave to which equations (1) and (2) apply, begin to propagate.

The agreement seen in Figures 4 and 5 between the results derived by the two different methods gives some confidence in the use of the periodic structure theory method to find dispersion curves and receptances when the stiffness of the elastomeric, resilient mountings is not zero.

5 Some Results

Bearing in mind the limitations of the validation exercise, it is nevertheless interesting to plot the driving point receptance for the Bishop Austrans rail model as computed using the periodic structure theory. The vertical driving point receptance at a position 3/8 of the distance between supports derived by the previously described theory is plotted in Figure 6. At each frequency, around 16 waveshapes have been included in the receptance summation. A measured receptance function is also given in Figure 6 for comparison. The force was applied through the rail gauge line and a steel spacer has been substituted for the elastomeric rail pad.

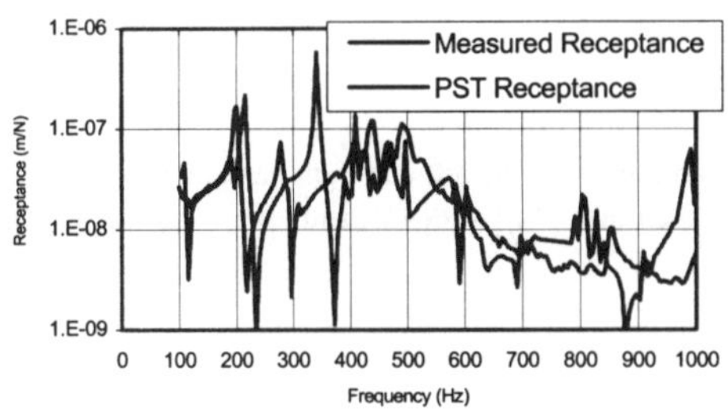

Figure 6. Computed and Measured Driving Point Receptances for the Austrans Rail

The nature of the receptance curve for the periodically supported rail is far more complex than for the benchmark model. There are several peaks in the spectrum that relate to the rail support distance, support properties and cut-on frequencies for various travelling waves.

The PST model receptance function agrees well with the measured result over a significant proportion of the spectrum. However, there remain large differences in the measured and computed values at some frequencies. Refinement of the model is presently being carried out to more accurately represent the support point and its degree of attachment to the rail, together with a review of the tolerances employed at various stages in the PST algorithm.

6 Conclusions

The results of the previously described analysis indicate that it is feasible to use PST with internal coordinates, at least for the particular case of an infinitely long equal angle beam. Results of work being carried out on the Bishop Austrans rail are encouraging, but further work is required to better the agreement between the PST and measured results.

The fact that the cross-section modes are significant at frequencies as low as 500Hz illustrates the need for models of sufficient complexity to represent this behaviour. The need for a model with interior coordinates is also evident, given the sharp peaks in the Austrans rail receptance spectrum between 100 and 500Hz. It would not be possible to predict this spectrum using a model which regards the rail to be continuously supported.

References

1. D.J. THOMPSON 1993 *Journal of Sound and Vibration* **161**, 421-446. Wheel-Rail Noise Generation, Part III: Rail Vibration.
2. D.J. MEAD 1973 *Journal of Sound and Vibration* **27**, 235-260. A General Theory of Harmonic Wave Propagation in Linear Periodic Systems With Multiple Coupling.
3. L. GRY and C. GONTIER 1997 *Journal of Sound and Vibration* **199**, 531-558. Dynamic Modelling of Railway Track: A Periodic Model Based on a Generalised Beam Formulation.
4. P. MULLER 1983 *Journal of Sound and Vibration* **87**, 115-141. Torsional-Flexural Waves in Thin-Walled Open Beams.

Appendix A

The familiar equations of motion for a dynamic system are formulated in terms of mass and stiffness matrices. In the case where ω is unknown, the eigenvalue problem, (A.1) would be generated in terms of ω^2.

$$\left([K]-\omega^2[M]\right)\{q\}=\{F\} \qquad \text{(A.1)}$$

In a periodic system, the displacements and forces at the left and right hand ends of the periodic element are related by a propagation constant μ such that $\{q_r\}=e^{\mu}\{q_l\}$ and $\{F_r\}=-e^{\mu}\{F_l\}$. On substitution, (A.1) becomes:

$$\left([K]-\omega^2[M]\right)\begin{Bmatrix} q_l \\ q_i \\ q_l e^{\mu} \end{Bmatrix}=\begin{Bmatrix} F_l \\ 0 \\ -F_l e^{\mu} \end{Bmatrix} \qquad \text{(A.2)}$$

Putting $[D]=[K]-\omega^2[M]$ and partitioning $[D]$ such that

$$[D]=\begin{bmatrix} D_{ll} & D_{li} & D_{lr} \\ D_{il} & D_{ii} & D_{ir} \\ D_{rl} & D_{ri} & D_{rr} \end{bmatrix}$$

leads to the polynomial eigenvalue problem (A.3) in terms of e^{μ}, given ω, as shown by Mead [2].

$$\left(e^{\mu}\begin{bmatrix} D_{lr} & 0 \\ D_{ir} & 0 \end{bmatrix}+\begin{bmatrix} D_{ll}+D_{rr} & D_{li} \\ D_{il} & D_{ii} \end{bmatrix}+e^{-\mu}\begin{bmatrix} D_{rl} & D_{ri} \\ 0 & 0 \end{bmatrix}\right)\begin{Bmatrix} q_l \\ q_r \end{Bmatrix}=\begin{Bmatrix} 0 \\ 0 \end{Bmatrix} \qquad \text{(A.3)}$$

For a periodic beam element, equation (A.3) can be simplified to a 1st order generalized eigenvalue problem:

$$\left(\begin{bmatrix} D_{ll}+D_{rr} & D_{li} \\ D_{ir} & 0 \end{bmatrix}+e^{\mu}\begin{bmatrix} 0 & D_{ri} \\ D_{il} & D_{ii} \end{bmatrix}\right)\cdot\begin{Bmatrix} q_l \\ q_i \end{Bmatrix}=0 \qquad \text{(A.4)}$$

The propagation constants e^{μ} can be determined and substituted back into the eigenvalue problem to find the eigenvectors or waveshapes $[\Psi]$. The contributions of the waveshapes can be summed to obtain the receptance at the driving point as follows:

$$q_j(0)=\sum_{n=1}^{N} A_n \Psi_{jn} \qquad \text{(A.5)}$$

where A_n is a vector of participation factors.

TRANSVERSE VIBRATIONS OF AN EULER-BERNOULLI BEAM OF CONSTANT DEPTH AND WITH PERIODIC CHANGE IN BREADTH

S. NAGULESWARAN

Department of Mechanical Engineering, University of Canterbury, Christchurch, New Zealand.

A method is developed to obtain the natural frequencies of a beam with any number of changes in cross-section and applied to an n-step beam of constant depth consisting of $(n+1)$ equal portions with the breadth changes at the steps in arithmetic progression. The procedure enabled any number of steps to be tackled without encountering any computational difficulties. Table of the first three frequencies of a clamped-free and a free-clamped stepped beam with 1, 2, 3, 4, 5, 10, 50, 100, 200 steps are tabulated. It is shown that as the number of steps is increased, the frequencies asymptotically approach the frequencies of a 'truncated' beam of constant depth and linearly varying breadth.

Keywords: Vibration, Euler-Bernoulli beam, stepped beam

1 Introduction

Several publications are available on the transverse vibration of beams with one step change in cross-section. Selected publications on this topic are briefly reviewed here. Taleb and Suppiger [1] and Levinson [2] derived the frequency equation of a simply supported stepped beam. Balasubramanian and Subramanian [3, 4] investigated the performance of a four-degree-of-freedom per node element when used in the vibration analysis of a stepped cantilever. The publications by Jang and Bert [5, 6] were the first to list the frequency equations for all combinations of the classical boundary conditions, as 4th order determinants equated to zero. Balasubramanian, Subramanian and Ramani [7] investigated the use of very high order derivatives as nodal degrees-of-freedom. Laura *et al* [8] used the finite element method, Popplewell and Daqing Chang [9] used Rayleigh-Ritz method and Krishnan, Geetha George and Malathi [10] discussed the difficulties of finite difference analysis of a pinned-pinned stepped beam.

The theory developed for Euler-Bernoulli beams with one step change in cross-section may be easily extended to beams with more than one step changes but the writer was not able to locate any publication with more than two step changes in cross-section. One of the reasons is the difficulties with the numerical side of computation. The classical approach to the derivation of the frequency equation of a beam with n step changes in cross-section i.e. $(n+1)$ beam portions, will result in a determinant of order $4(n+1)$ equated to zero. Even for $n = 2$, computational difficulties were encountered with the calculations of the roots. Sato [11] used the finite element method to investigate the vibration of a beam with a narrow groove. Bapat and Bapat [12] suggested the transfer matrices approach for beams with n-steps *but did not present any results*. In the present paper, taking the boundary conditions into account, the mode shape of the first portion of the n-step beam was written in the form $Y_1(X_1)= A\,U_1(X_1)+ B\,V_1(X_1)$. The mode shape of the second portion was expressed as $Y_2(X_2)= A\,U_2(X_2)+ B\,V_2(X_2)$ i.e. with the same constants as in the first portion, taking into account the continuity of deflection, slope, bending moment and shearing force at the first step. This scheme was applied to successive portions of the beam till the last portion was reached. Inclusion of the boundary conditions at the end of the $(n+1)^{th}$ portion resulted in the frequency equation as a 2nd order determinant equated to zero. Computations were successful for any number of steps and any of the classical boundary conditions at the ends.

The constant depth stepped beam considered in this paper had beam portions of equal length and with the breadth changes at the steps in arithmetic progression. The first three frequency parameters are tabulated for beams with 1, 2, 3, 4, 5, 10, 50, 100 and 200 steps. It is shown that with increase in the number of step changes, the frequency parameters asymptotically approached those of a beam of constant depth and linearly varying breadth – the exact frequency of which was presented by Naguleswaran [13].

2 Theoretical Considerations

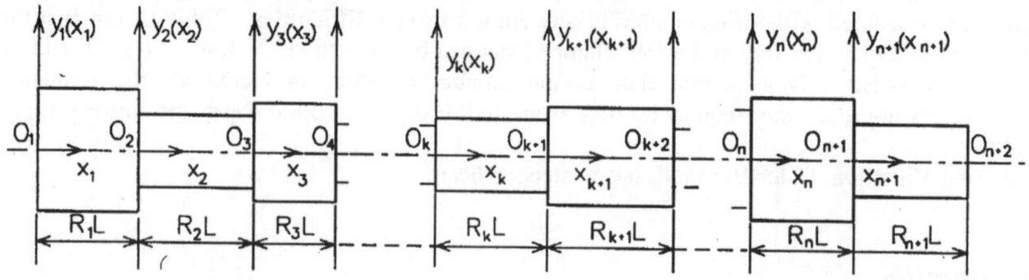

Figure 1. The n-step beam and the co-ordinate systems.

Figure 1 shows a stepped Euler-Bernoulli beam O_1O_{n+2} with n step changes in cross-section at $O_2, O_3,.........O_{n+1}$. For the kth portion of the beam O_kO_{k+1}, the length, flexural rigidity and mass per unit length are R_kL, EI_k, and m_k respectively. Without loss of generality one may choose

$$R_1 + R_2 + ... + R_{n+1} = 1 \tag{1}$$

To write the equation of transverse vibrations, coordinate systems were chosen with origin at $O_1, O_2, \quad O_{n+1}$. For free vibration at frequency ω, if the ordinate $y_k(x_k)$ is the amplitude of vibration at abscissa x_k $(0 \le x_k \le R_kL)$, then the amplitude of bending moment $M_k(x_k)$ and shearing force $Q_k(x_k)$ are:

$$M_k(x_k) = EI_k \frac{d^2 y_k(x_k)}{dx_k^2}, \qquad Q_k(x_k) = -EI_k \frac{d^3 y_k(x_k)}{dx_k^3},$$

$$\frac{dQ_k(x_k)}{dx_k} + m_k\omega^2 y_k(x_k) = 0 \tag{2}$$

The mode shape differential equation of the kth portion of the beam is:

$$EI_k \frac{d^4 y_k(x_k)}{dx_k^4} - m_k\omega^2 y_k(x_k) = 0 \tag{3}$$

To express equations (2), (3) in dimensionless form, one defines the dimensionless abscissa X_k, amplitude $Y_k(X_k)$, the operators D_k, D_k^n, the dimensionless bending moment $M_k(X_k)$, shearing force $Q_k(X_k)$, flexural rigidity ratio $\phi_{k+1,k}$, mass per unit length ratio $\mu_{k+1,k}$ and dimensionless frequency Ω_k as follows:

$$X_k = \frac{x_k}{L}, \quad Y_k(X_k) = \frac{y_k(x_k)}{L}, \quad D_k = \frac{d}{dX_k}, \quad D_k^n = \frac{d^n}{dX_k^n}, \quad M_k(X_k) = \frac{M_k(x_k)L}{EI_k},$$

$$Q_k(X_k) = \frac{Q_k(x_k)L^2}{EI_k}, \quad \phi_{k+1,k} = \frac{EI_{k+1}}{EI_k}, \quad \mu_{k+1,k} = \frac{m_{k+1}}{m_k}. \quad \Omega_1^2 = \frac{m_1\omega^4 L^4}{EI_1} = \alpha_1^4,$$

$$\Omega_k^2 = \alpha_k^4 = \frac{m_k\omega^2 L^4}{EI_k} = \left(\frac{\mu_{k,1}}{\phi_{k,1}}\right)\alpha_1^4 \tag{4}$$

In equations (4), Ω_1 is a dimensionless natural frequency, α_1 is the natural frequency parameter. The nth natural frequency parameter is denoted by $\alpha_{1,n}$.

This beam in the title problem is of constant depth and has n-steps i.e. $(n+1)$ equal portions of length $L/(n+1)$ and the change in breadth at each step is in arithmetic progression. If the breadth at the left end is b_L and the breadth at the right end is b_R, the breadth change between successive steps is $(b_L - b_R)/(n+1)$. For the $(k+1)^{th}$ portion, the breadth b_{k+1}, mass per unit length ratio $\mu_{k+1,k}$, flexural rigidity ratio $\phi_{k+1,k}$ and the frequency parameter α_{k+1} are:

$$b_{k+1} = b_L - (k+1)\frac{b_L - b_R}{n+1}, \quad \phi_{k+1,k} = \frac{b_{k+1}}{b_k}, \quad \mu_{k+1,k} = \frac{b_{k+1}}{b_k}, \quad \alpha_{k+1} = \alpha_1$$

Equations (2) and (3) in dimensionless form are:

$$M_k(X_k) = D_k^2[Y_k(X_k)], \; Q_k(X_k) = -D_k^3[Y_k(X_k)], \; D_k^4[Y_k(X_k)] - \alpha_k^4 Y_k(X_k) = 0 \qquad (5)$$

The solution of the dimensionless mode shape differential equation is:

$$Y_k(X_k) = C_{k,1}\sin\alpha_k X_k + C_{k,2}\cos\alpha_k X_k + C_{k,3}\sinh\alpha_k X_k + C_{k,4}\cosh\alpha_k X_k \qquad (6)$$

where $C_{k,1}$ through to $C_{k,4}$ are the constants of integration.

2.1 The mode shape of first portion

The solution of the mode shape differential equation of the first portion of the beam is:

$$Y_1(X_1) = C_{1,1}\sin\alpha_1 X_1 + C_{1,2}\cos\alpha_1 X_1 + C_{1,3}\sinh\alpha_1 X_1 + C_{1,4}\cosh\alpha_1 X_1$$

The need to satisfy the boundary conditions at O_1 will result in the elimination of two of the constants and the mode shape of the first portion may be expressed as:

$$Y_1(X_1) = A\,U_1(X_1) + B\,V_1(X_1) \qquad (7)$$

where B_{11} and B_{12} are constants and the functions $U_1(X_1)$ and $V_1(X_1)$ are:

if O_1 is clamped: $U_1(X_1) = \sin\alpha_1 X_1 - \sinh\alpha_1 X_1$, $V_1(X_1) = \cos\alpha_1 X_1 - \cosh\alpha_1 X_1$ (8)

or if O_1 is pinned: $U_1(X_1) = \sin\alpha_1 X_1$, $V_1(X_1) = \sinh\alpha_1 X_1$ (9)

or if O_1 is sliding: $U_1(X_1) = \cos\alpha_1 X_1$, $V_1(X_1) = \cosh\alpha_1 X_1$ (10)

or if O_1 is free: $U_1(X_1) = \sin\alpha_1 X_1 + \sinh\alpha_1 X_1$, $V_1(X_1) = \cos\alpha_1 X_1 + \cosh\alpha_1 X_1$ (11)

The derivatives of $U_1(X_1)$ and $V_1(X_1)$ are obtained easily by straight forward differentiation.

2.2 The mode shape of the rest of the beam portions

The mode shape of the second portion of the beam is:

$$Y_2(X_2) = C_{2,1}\sin\alpha_2 X_2 + C_{2,2}\cos\alpha_2 X_2 + C_{2,3}\sinh\alpha_2 X_2 + C_{2,4}\cosh\alpha_2 X_2 \qquad (12)$$

To satisfy the conditions of continuity of deflection and continuity of slope at O_2 one has

$$y_1(R_1 L) = y_2(0), \quad \frac{dy_1(R_1 L)}{dx_1} = \frac{dy_2(0)}{dx_2} \qquad (13)$$

Compatibility of bending moment and of shearing force at the step leads to:

$$EI_1\frac{d^2 y_1(R_1 L)}{dx_1^2} = EI_2\frac{d^2 y_2(0)}{dx_2^2}, \qquad EI_1\frac{d^3 y_1(R_1 L)}{dx_1^3} = EI_2\frac{d^3 y_2(0)}{dx_2^3} \qquad (14)$$

Equations (13) and (14) in dimensionless form are:

$$Y_1(R_1) = Y_2(0), \quad D_1[Y_1(R_1)] = D_2[Y_2(0)],$$

$$D_1^2[Y_1(R_1)] = \phi_{2,1}D_2^2[Y_2(0)], \qquad D_1^3[Y_1(R_1)] = \phi_{2,1}D_2^3[Y_2(0)] \qquad (15)$$

From equations (15), (7) and (12) one gets

$$\begin{aligned}
C_{2,2} + C_{2,4} &= A\,U_1(R_1) + B\,V_1(R_1),\\
\alpha_2(C_{2,1} + C_{2,3}) &= A\,D_1[U_1(R_1)] + B\,D_1[V_1(R_1)],\\
\phi_{2,1}\alpha_2^2(-C_{2,2} + C_{2,4}) &= A\,D_1^2[U_1(R_1)] + B\,D_1^2[V_1(R_1)],\\
\phi_{2,1}\alpha_2^3(-C_{2,1} + C_{2,3}) &= A\,D_1^3[U_1(R_1)] + B\,D_1^3[V_1(R_1)]
\end{aligned} \tag{16}$$

$C_{2,1}$ through to $C_{2,4}$ may now be eliminated. Equation (12) may be now expressed as :

$$Y_2(X_2) = A\,U_2(X_2) + B\,V_2(X_2) \tag{17}$$

where $U_2(X_2)$ and $V_2(X_2)$ are transcendental functions which are derived as follows: :

$$\begin{aligned}
U_2(X_2) &= G_{2,1}\sin\alpha_2X_2 + G_{2,2}\cos\alpha_2X_2 + G_{2,3}\sinh\alpha_2X_2 + G_{2,4}\cosh\alpha_2X_2\\
V_2(X_2) &= H_{2,1}\sin\alpha_2X_2 + H_{2,2}\cos\alpha_2X_2 + H_{2,3}\sinh\alpha_2X_2 + H_{2,4}\cosh\alpha_2X_2
\end{aligned} \tag{18}$$

in which the coefficients $G_{2,1}$ through to $H_{2,4}$ are:

$$\begin{aligned}
G_{2,1} &= \frac{D_1[U_1(R_1)]}{2\alpha_2} - \frac{D_1^3[U_1(R_1)]}{2\alpha_2^3\phi_{2,1}}, & H_{2,1} &= \frac{D_1[V_1(R_1)]}{2\alpha_2} - \frac{D_1^3[V_1(R_1)]}{2\alpha_2^3\phi_{2,1}}\\
G_{2,2} &= \frac{U_1(R_1)}{2} - \frac{D_1^2[U_1(R_1)]}{2\alpha_2^2\phi_{2,1}}, & H_{2,2} &= \frac{V_1(R_1)}{2} - \frac{D_1^2[V_1(R_1)]}{2\alpha_2^2\phi_{2,1}}\\
G_{2,3} &= \frac{D_1[U_1(R_1)]}{2\alpha_2} + \frac{D_1^3[U_1(R_1)]}{2\alpha_2^3\phi_{2,1}}, & H_{2,3} &= \frac{D_1[V_1(R_1)]}{2\alpha_2} + \frac{D_1^3[V_1(R_1)]}{2\alpha_2^3\phi_{2,1}}\\
G_{2,4} &= \frac{U_1(R_1)}{2} + \frac{D_1^2[U_1(R_1)]}{2\alpha_2^2\phi_{2,1}}, & H_{2,4} &= \frac{V_1(R_1)}{2} + \frac{D_1^2[V_1(R_1)]}{2\alpha_2^2\phi_{2,1}}
\end{aligned} \tag{19}$$

By successive application of this procedure, one may express the mode shape of the $(k+1)^{th}$ portion $O_{k+1}O_{k+2}$ as

$$Y_{k+1}(X_{k+1}) = A\,U_{k+1}(X_{k+1}) + B\,V_{k+1}(X_{k+1}) \tag{20}$$

3 The Frequency Equation

The frequency equation is obtained from the boundary conditions at O_{n+2}. For example if the end O_{n+2} is clamped, from equation (20) one has

$$\begin{aligned}
Y_{n+1}(R_{n+1}) &= 0 = A\,U_{n+1}(R_{n+1}) + B\,V_{n+1}(R_{n+1})\\
D_{n+1}[Y_{n+1}(R_{n+1})] &= 0 = A\,D_{n+1}[U_{n+1}(R_{n+1})] + B\,D_{n+1}[V_{n+1}(R_{n+1})]
\end{aligned} \tag{21}$$

The coefficient matrix of equations (21) must be singular. Equations of the same form as equation (21) obtain for other classical boundary conditions and the frequency equations for the four classical cases are:

if beam end O_3 is clamped *(cl)*,

$$U_{n+1}(R_{n+1})\,D_{n+1}[V_{n+1}(R_{n+1})] - D_{n+1}[U_{n+1}(R_{n+1})]\,V_{n+1}(R_{n+1}) = 0 \tag{22}$$

if beam end O_{n+1} is pinned *(pn)*,

$$U_{n+1}(R_{n+1})\,D_{n+1}^2[V_{n+1}(R_{n+1})] - D_{n+1}^2[U_{n+1}(R_{n+1})]\,V_{n+1}(R_{n+1}) = 0 \tag{23}$$

if beam end O_{n+1} is sliding *(sl)*,

$$D_{n+1}[U_{n+1}(R_{n+1})]\,D_{n+1}^3[V_{n+1}(R_{n+1})] - D_{n+1}^3[U_{n+1}(R_{n+1})]\,D_{n+1}[V_{n+1}(R_{n+1})] = 0 \tag{24}$$

if beam end O_{n+1} is free *(fr)*,

$$D_{n+1}^2[U_{n+1}(R_{n+1})]\,D_{n+1}^3[V_{n+1}(R_{n+1})] - D_{n+1}^3[U_{n+1}(R_{n+1})]\,D_{n+1}^2[V_{n+1}(R_{n+1})] = 0 \tag{25}$$

The transcendental equations (22) to (25) will have infinite number of roots.

3.1 The natural frequency calculations

The roots of a frequency equation were determined by a 'search' to bracket an approximate range within which a root is present followed by an iterative procedure based on linear interpolation. The procedure is as follows: For the selected boundary conditions at O_1 , one chose $U_1(X_1)$ and $V_1(X_1)$ from the applicable equations (8) to (11). A trial frequency was assumed ($\Omega_1 = 0.1$ say) and with $X_1 = R_1$ the values of $U_1(R_1)$, $V_1(R_1)$, $D_1[U_1(R_1)]$, $D_1[V_1(R_1)]$, etc. were calculated. One proceeded to calculate the coefficients $G_{2,1}$ through to $H_{2,4}$ from equation (19). The procedure was successively repeated to compute the coefficients $G_{n+1,1}$ through $H_{n+1,4}$. From the boundary conditions at O_{n+2}, one chose and calculated the left hand side of the applicable frequency equations (22) to (25). The value of Ω_1 was increased in steps of 0.1 and the calculations described were repeated till a sign change in the left hand side occurred. The sign change indicated the presence of a root within this range. A 'search' was made within this range but with change of 0.01 in Ω_1 to narrow the range within which the root lies. At this stage an iterative procedure based on linear interpolation was invoked to calculate the root to the pre-set accuracy. The 'search' procedure was continued (from the value of the first root) to locate the second root and so on. Due to space restriction only the first three frequency parameters $\alpha_{1,1}$, $\alpha_{1,2}$ and $\alpha_{1,3}$ for three values of b_R/b_L of clamped-free and of free-clamped beams are tabulated in Table 1 in two sections. The last row of the sections (in bold) are the frequency parameters (where available) of a beam of constant depth and linearly varying breadth from Naguleswaran [13].

Table 1. The first three frequency parameters of clamped-free and free-clamped stepped beam. The number of steps are in the first column. The last row of each section is from reference [13].

	$b_R = 0.25b_L$			$b_R = 0.50b_L$			$b_R = 0.80b_L$		
n	$\alpha_{1,1}$	$\alpha_{1,2}$	$\alpha_{1,3}$	$\alpha_{1,1}$	$\alpha_{1,2}$	$\alpha_{1,3}$	$\alpha_{1,1}$	$\alpha_{1,2}$	$\alpha_{1,3}$
				clamped -free stepped beam					
1	2.4867	4.6552	8.0423	2.1773	4.7271	7.8949	1.9703	4.7181	7.8548
2	2.4301	4.9653	7.7993	2.1477	4.8347	7.8534	1.9609	4.7409	7.8595
3	2.3905	5.0355	7.9475	2.1308	4.8561	7.9066	1.9558	4.7459	7.8717
4	2.3659	5.0513	8.0156	2.1203	4.8620	7.9279	1.9527	4.7473	7.8772
5	2.3495	5.0542	8.0457	2.1133	4.8636	7.9384	1.9506	4.7477	7.8801
10	2.3124	5.0444	8.0772	2.0970	4.8613	7.9516	1.9457	4.7470	7.8840
50	2.2778	5.0204	8.0718	2.0816	4.8529	7.9516	1.9411	4.7445	7.8842
100	2.2732	5.0163	8.0690	2.0795	4.8513	7.9508	1.9405	4.7441	7.8840
200	2.2708	5.0141	8.0675	2.0784	4.8505	7.9503	1.9401	4.7439	7.8838
∞	***2.2684***	***5.0119***	***8.0658***	***2.0773***	***4.8497***	***7.9498***			
				Free-clamped stepped beam					
1	1.3528	4.2381	8.1084	1.5976	4.5314	7.9319	1.7825	4.6565	7.8671
2	1.3941	4.3560	7.7111	1.6233	4.5376	7.8091	1.7916	4.6457	7.8453
3	1.4217	4.3629	7.7159	1.6379	4.5303	7.7982	1.7964	4.6419	7.8375
4	1.4401	4.3619	7.7056	1.6471	4.5280	7.7858	1.7994	4.6409	7.8328
5	1.4529	4.3624	7.6947	1.6533	4.5278	7.7784	1.8014	4.6406	7.8301
10	1.4832	4.3713	7.6756	1.6677	4.5315	7.7679	1.8061	4.6416	7.8266
50	1.5124	4.3911	7.6775	1.6814	4.5398	7.7677	1.8105	4.6441	7.8264
100	1.5163	4.3946	7.6794	1.6832	4.5413	7.7683	1.8111	4.6445	7.8266
200	1.5183	4.3964	7.6805	1.6842	4.5421	7.7687	1.8114	4.6447	7.8267
∞	***1.5203***	***4.3982***	***7.6817***	***1.6851***	***4.5429***	***7.7691***			

4 Conclusions

A scheme is presented to calculate the natural frequencies of a beam with any number of step changes in cross-section. The character of the title problem enabled the frequency equations to be expressed as 2nd order determinants equated to zero. A scheme was presented to progressively calculate the elements of the determinant and a scheme to evaluate the roots of the frequency equation.

The first three frequency parameters are tabulated for clamped-free and a free-clamped beam for 1, 2, 3, 4, 5, 10, 50, 100 and 200 steps. As the number of steps is increased, the frequency parameters approach those of a beam of constant depth and linearly varying breadth.

References

1. Taleb, N. J. and Suppiger, E. W., Vibrations of stepped beams. *Journal of Aerospace Engineering* **28** (1961) pp 295-298.
2. Levinson, M. Vibrations of stepped strings and beams. *Journal of Sound and Vibration* **49**(2) (1976) pp 287-291.
3. Balasubramanian, T. S. and Subramanian, G. On the performance of a four-degree-of-freedom per node element for stepped beam analysis and higher frequency estimation. *Journal of Sound and Vibration* **99**(4) (1985) pp 563-567.
4. Subramanian, G. and Balasubramanian, T. S. Beneficial effects of steps on the free vibration characteristics of beams. *Journal of Sound and Vibration* **118**(3) (1987) pp 555-560.
5. Jang, S. K. and Bert, C. W. Free vibrations of stepped beams: Exact and numerical solutions. *Journal of Sound and Vibration* **130**(2)(1989) pp 342-346. .
6. Jang, S. K. and Bert, C. W. Free vibrations of stepped beams: Higher mode frequencies and effects of steps on frequency. *Journal of Sound and Vibration* **132**(1)(1989) pp 164-168.
7. Balasubramanian, T. S., Subramanian, G. and Ramani, T. S. Significance of very high order derivatives as nodal degrees of freedom in stepped beam vibration analysis. *Journal of Sound and Vibration* **137**(2)(1990) pp 353-356.
8. Laura, P. A. A., Rossi, R. E., Pombo, J. L. and Pasqua, D. Dynamic stiffening of straight beams of rectangular cross-section: a comparison of finite element predictions and experimental results. *Journal of Sound and Vibration* **150**(1)(1991) pp 174-178
9. Popplewell, A. and Daqing Chang Free vibrations of a complex Euler-Bernoulli beam. *Journal of Sound and Vibration* **190**(5)(1996) pp 852-856.
10. Krishnan, A., Geetha George and Malathi, P. Use of finite difference method in the study of stepped beams. *International Journal of Mechanical Engineering Education* **26**(1)(1998) pp 11-24.
11. Sato, H. Free vibration of beams with abrupt changes of cross-section. *Journal of Sound and Vibration* **89**(1)(1983) pp 59-64.
12. Bapat, C. N. and Bapat, C. Natural frequencies of a beam with non-classical boundary conditions and concentrated masses. *Journal of Sound and Vibration* **112**(1)(1987) pp 177-182.
13. Naguleswaran, S. Vibration of an Euler-Bernoulli beam of constant depth and with linearly varying breadth. *Journal of Sound and Vibration* **153**(3)(1992) pp 509-522.

TRANSIENT CHARACTERISTICS OF AN AUTOMATIC TRANSMISSION DURING SHIFT CHANGES

D.K. LIU, N. ZHANG

Dept. of Mechanical & Manufacturing Engineering, Faculty of Engineering, University of Technology, Sydney, P.O.Box 123, Broadway, NSW 2007, Australia
E-mail: dikai.liu@eng.uts.edu.au

J. M. JEYAKUMARAN AND L. VILLANUEVA

Research and Development Department, BTR Automotive, 58 Lisbon Street, Fairfield, NSW 2165, Australia

This study aims to develop a comprehensive mathematical model and a simulation system of an automatic transmission to investigate the transient characteristics during gear changes. In particular, the model is applied to a 4-speed automatic transmission manufactured by BTR Automotive to study the 2-3 shift process. A brief description of the BTR automatic transmission and the 2-3 shift mechanism are presented. A mode description method is used for describing the shift process. The modular structure of the simulation system, which includes engine module, torque converter module, hydraulic system module, geartrain module and modules of clutches and bands, is developed, and the governing equations of motion of the integrated powertrain system are derived. The simulation results are presented and the major parameters affecting the quality of the shift process are discussed.

1 Introduction

In a vehicle fitted with an automatic transmission, frequent stops and starts and rapid acceleration or declaration result in frequent gear changes, which perturb vehicle acceleration and engine speed. These conditions are the main causes of poor shift quality, poor fuel economy and unwanted emissions. In addition, output torque disturbances during gear changes are an important aspect of the customer's perception of overall vehicle quality. With the increasing demand for improved performance of the vehicle such as driveability, passenger comfort and fuel economy, an in-depth knowledge of the dynamic characteristics of automatic transmissions during gear changes becomes very important.

The shifting process in an automatic transmission is complex because many factors affect the shift quality, for example, the magnitude of the gear ratio change, clutch and band configurations, hydraulic control system, operating temperatures, shift scheduling, etc. For smooth gearshift control, the hydraulic pressure of the applying and releasing frictional elements, such as clutches and bands, are precisely controlled with respect to the transient engine torque. A model-based investigation and design of an automatic transmission system, therefore, require complete understanding of the powertrain components including system-level interference forces during gear changes.

Most studies on modelling of an automatic transmission has been carried out by individual transmission manufactures with specific requirements and applications. For example, Kotwicki [1] developed a mathematical model of the vehicle fitted with an automatic transmission based on steady state operations of the torque converter. Pan and Moskwa [2], Kim, *et al.* [3] and Jo *et al.* [4] developed dynamic models of the powertrain system to investigate gear changes in an automatic transmission. Recently, Liu *et al* [5] studied the transient characteristics of the 1-2 shift process of the BTR 4-speed automatic transmission. Simplified dynamic models of the engine, torque converter, and vehicle are often utilized to investigate the transient characteristics of vehicles equipped with automatic transmissions [e.g., 5, 6, 7 and 8].

This study aims to develop a comprehensive mathematical model and a simulation system of the BTR automatic transmission to investigate the transient characteristics during the 2-3 gear change. A brief introduction on the features of the 4-speed automatic transmission and the 2-3 shifting process are presented. Then, the description of the state mode method and the modular structure of the simulation system are presented. Models of each transmission and powertrain module are developed, and the governing equations of motion of the integrated powertrain system are derived. The developed mathematical models are tuned and validated with the data obtained from experiments. The factors affecting the transient performance of the 2-3 shift process are discussed.

2 The BTR 4-speed automatic transmission and the 2-3 shift process

The BTR 4-speed automatic transmission consists of a torque converter with single face lock-up clutch, four multi-plate clutch assemblies, two brake bands, two one-way clutches and a planetary geartrain as shown in Figure 1. The 2nd gear state is achieved by engaging the clutch C2, and applying the clutch C4 to bypass the 3-4 One Way Clutch (3-4 OWC). The brake band B1 holds the Reverse Sun (RS) gear leading to the long pinion 'walking' around the RS gear.

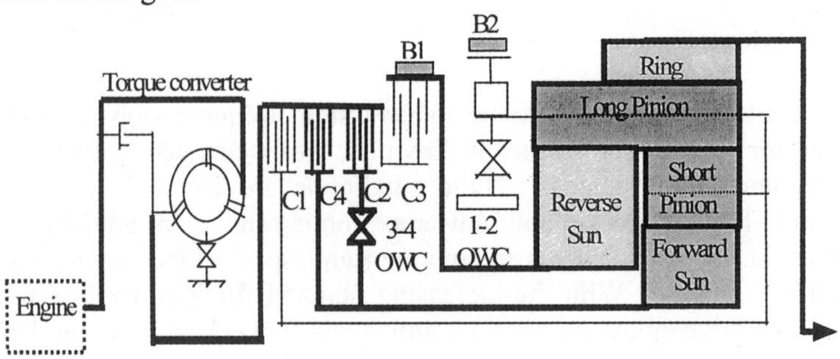

Figure 1. Schematic of the BTR 4-speed automatic transmission

As shown in Figure 1, the torque converter turbine torque, T_t, is transmitted through clutches C2 and C4 to the Forward Sun (FS) gear. The clockwise rotation of the FS gear causes counterclockwise rotation of the short pinions, and because the RS gear is held by band B1, the internal ring gear and output shaft will rotate in a clockwise or forward direction. The torque is transmitted from the FS gear to the output shaft splined to the ring gear. In 2nd gear, the speed ratio from FS gear, $\dot{\theta}_{FS}$, to the ring gear, $\dot{\theta}_R$, is:

$$i_{FS2}=\dot{\theta}_{FS}/\dot{\theta}_R = r_R*(r_{FS}+r_{RS}) / (2*r_{FS}*(r_{RS}+r_{LP})) \quad (1)$$

The speed ratios of other gear elements, such as short pinions (SP), long pinion (LP), reverse sun (RS) gear, and carrier (CR), to the ring gear are listed as follows:

$$i_{SP2}=\dot{\theta}_{SP}/\dot{\theta}_R = r_R*(r_{RS} - r_{SP}) / (2*r_{SP}*(r_{RS}+r_{LP})) \quad (2)$$

$$i_{LP2}=\dot{\theta}_{LP}/\dot{\theta}_R = r_R / (2*r_{LP}) \quad (3)$$

$$i_{CR2}=\dot{\theta}_{CR}/\dot{\theta}_R = r_R/(2*(r_{RS}+r_{LP})) \quad (4)$$

where *i* represents speed ratio, subscripts FS, SP, LP, RS, R and CR represent the gear of forward sun, short pinion, long pinion, reverse sun, ring and carrier, respectively, and subscript 2 represents the 2nd gear state.

During the shift process, the system dynamics changes and the governing equations are determined from the combinations of active elements engaged in each gear state. The 2-3 shift is accomplished by applying the C1 clutch to drive the carrier and releasing the B1 band to release the RS gear. At the end of the 2-3 shift process, the system enters the steady state 3rd gear. The speed ratios in 3rd gear are listed as follows:

$$i_{FS3} = i_{SP3} = i_{LP3} = i_{CR3} = 1.0 \quad (5)$$

where subscript 3 represents the 3rd gear state.

3 State mode description, modular structure and models

The automatic transmission is a parametric system due to discrete speed ratio changes. A particular shift process contains several modes of operation which corresponding to different combinations of active elements in the automatic transmission. The 2-3 up-shift process is divided into four modes according to the state mode description method [5, 9].

$$M=\{M_1, M_2, M_3, M_4\} \quad (6)$$

In Eq. 6, Mode 1, M_1, represents the steady state 2nd gear. The next state, Mode 2, M_2, represents the torque phase where the speed ratios do not change but the output torque decreases with releasing the B1 band and applying the C1 clutch. Mode 3, M_3, is the inertia phase where speeds of elements begins to change and the output torque increases according to the frictional characteristics of B1 band and C1 clutch as well as the inertia properties of the rotating components. The last state, mode 4, M_4, represents the end phase of the shift process in which the carrier rotates at the same speed as the clutches C1, C2, and C4, and FS, and the RS rotates freely. The system enters the steady state 3rd gear.

A modular structure of the vehicle powertrain system is shown in Figure 2. The modular structure consists of engine model, torque converter model, transmission system model, hydraulic system model, control logic model and the models for bands and clutches. Detailed description of these modules can be found in Liu *et al* [5]. The dynamic models of the transmission components are obtained from the free-body diagrams shown in Figure 3. The dynamic equations of motion are as follows:

Engine: $$I_E\, \ddot{\theta}_E = T_E - T_{LU} - T_P \quad (7)$$

Torque converter shaft: $$I_1\, \ddot{\theta}_1 = T_{LU} + T_T - T_{C2} - T_{C4} \quad (8)$$

Forward sun gear: $$I_{FS}\, \ddot{\theta}_{FS} = T_{C2} + T_{C4} - 3F_{SP}\, r_{FS} \quad (9)$$

Short pinion: $$3I_{SP}\, \ddot{\theta}_{SP} = 3\, F_{SP}\, r_{SP} - 3F_{LP}\, r_{SP} \quad (10)$$

Long pinion: $$3I_{LP}\, \ddot{\theta}_{LP} = 3\, F_{LP}\, r_{LP} - 3\, F_R\, r_{LP} - 3\, F_{RS}\, r_{LP} \quad (11)$$

Ring gear: $$I_R\, \ddot{\theta}_R = 3\, F_R\, r_R - T_V \quad (12)$$

Reverse sun gear: $$I_{RS}\, \ddot{\theta}_{RS} = 3\, F_{RS}\, r_{RS} - T_B \quad (13)$$

Carrier: $$A\, \ddot{\theta}_{CR} = 3F_{SP}(r_{FS} + r_{SP}) - 3F_{LP}(r_{SP} + r_{LP}) - 3F_R(r_{RS} + r_{LP}) + 3F_{RS}(r_{RS} + r_{LP}) + T_{C1} \quad (14)$$

where, I is defined as the mass moment of inertia, F represents the force on each gear elements, T is the transmitted torque, and subscripts LU, V, B and OWC represent lock-up clutch, resistance torque, band and one-way-clutch, respectively.

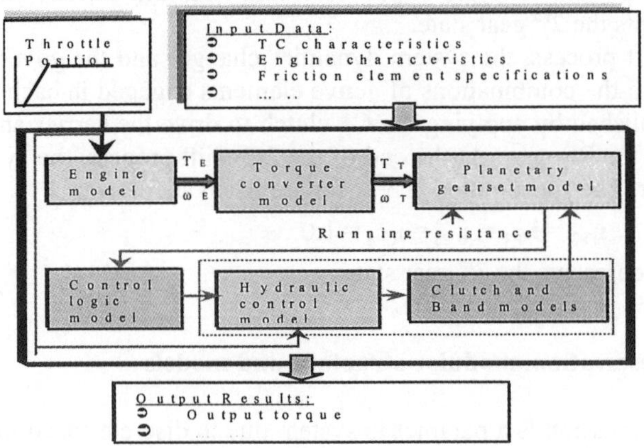

Figure 2. Modular structure of the vehicle powertrain system during shift process

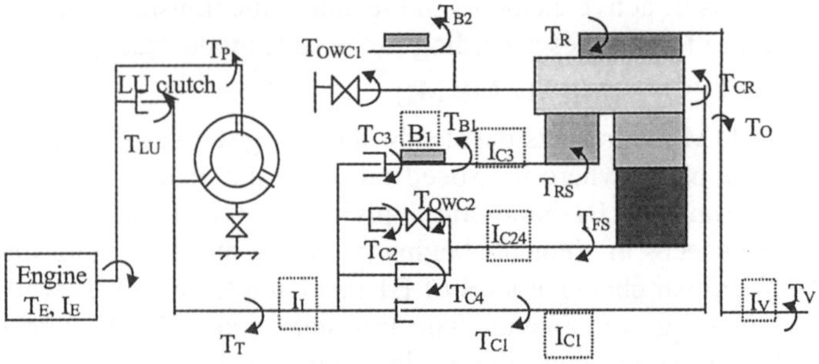

Figure 3. Free-body diagram of the automatic transmission during the 2-3 shift

4 Results and Discussion

Inputs for the simulation system include throttle position, vehicle speed, engine torque characteristics and friction element specifications, and the outputs are the output torque variations, vehicle accelerations/decelerations, reaction forces, and component rotational speeds. Figure 4 shows the typical B1 band force and C1 clutch force applied during the 2-3 shift process at wide open throttle conditions. Figure 5 shows the variation of the dynamic coefficient of friction of the C1 clutch material against the slip speed of the frictional surfaces.

Figures 6 to 9 shows the simulation results of the 2-3 shift performed in wide open throttle conditions at the output shaft speed of 3350 rpm. Figure 6 shows the changes in speed ratios, i_{FS}, i_{CR}, and i_{SP}, together with the distinct features of the torque and inertia phases. As shown in Figure 6, the speed ratios do not change during the torque phase. The speed ratio changes during the inertia phase until the ratio reaches a value 1.0, indicating that the system enters the 3rd gear state. The shift process starts at about 0.45 sec. and ends at about 1.235 sec., with the actual shifting time of about 0.75 sec. rpm.

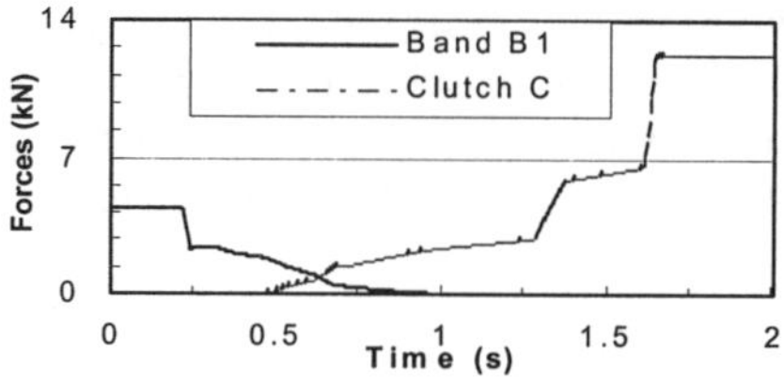

Figure 4. Band B1 and clutch C1 forces

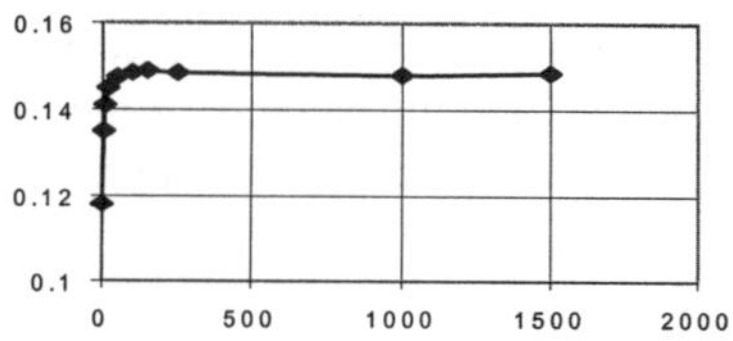

Figure 5.Dynamic coefficient of friction of the clutch C1 material

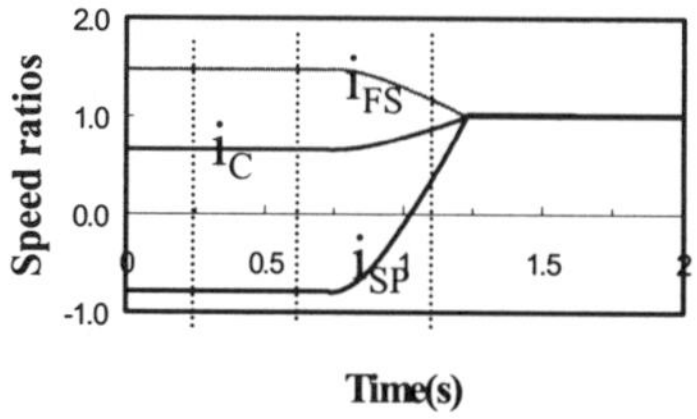

Figure 6. Variation of speed ratios

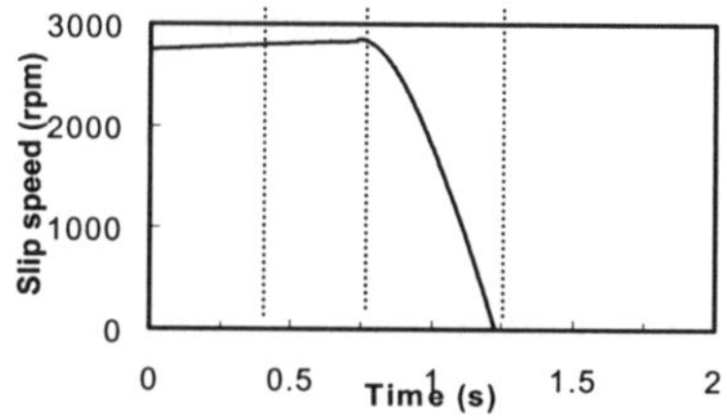

Figure 7. Slip speed of clutch C1

Figure 7 shows the C1 slip speed during the shift process with relatively no change during the torque phase and a sharp drop during the inertia phase. The slip speed becomes zero at the end of inertia phase, indicating that the shift is complete and system enters into the 3rd gear state. The output speed and output torque profiles are also shown in Figures 8 and 9, respectively. The output speed increases steadily with very small variation during the inertial phase. The output torque drops rapidly in the torque phase and then increases sharply in the inertia phase. The features of torque and inertia forces phases reveal the dynamic performance of the upshift process. Theoretically, the longer the torque phase, the less the torque depth, i.e., the torque drop during the torque phase, and the better the shift quality.

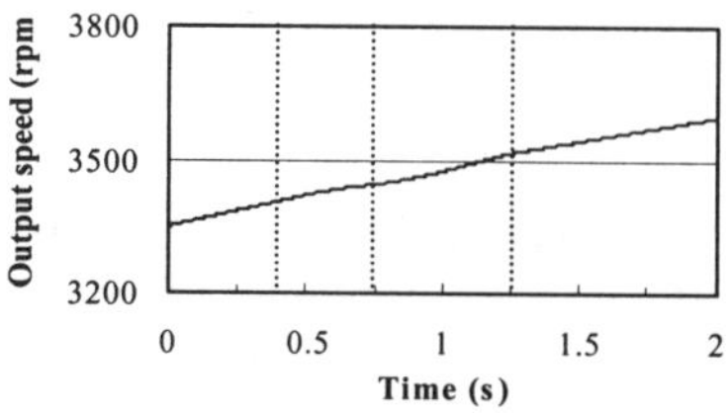

Figure 8. Output shaft speed

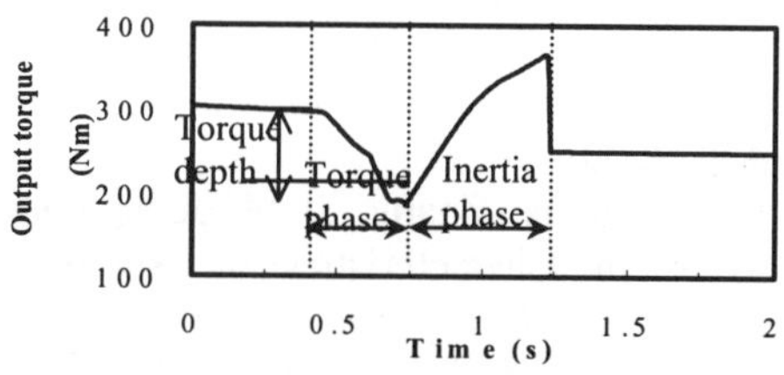

Figure 9. Output shaft torque

Transient characteristics of the 2-3 shift process depend on several design parameters. As an example, the 2-3 simulation is performed with the coefficient of friction of the C1 clutch material is increased by 20%. Results show that the shift process starts at about 0.45 sec. and ends at about 1.1 sec., with the actual shifting time reduced to about 0.65 sec. The output torque profiles show an increase in coefficient of friction of the C1 clutch material will shorten the duration of the inertia phase. Results, however, show that the output torque disturbances increases slightly with the increased coefficient of friction.

The 2-3 simulation is also performed with various operating conditions of the automatic transmission such as different temperatures, shift schedules and shift ramps, and varied vehicle resistance torque. Overall, the output torque profiles predicted by the 2-3 simulation correlates very well with the shift qualities evaluated in vehicle tests.

5 Conclusion

A mathematical model and simulation system has been developed for investigating the transient performance of automatic transmission during the 2-3 shift process. The mode description of the shift process and the modular structure of the powertrain system are presented. The simulation results are consistent with experimental data measured from the vehicle tests. This suggests dynamic model is very useful tool in analyzing the transient phenomena occurring during the shift process and optimizing the design parameters.

Acknowledgement

The financial support for this work is provided jointly by the Australian Research Council and BTR Automotive, Australia.

References

1. Kotwicki, A. J., Dynamic Models for Torque Converter Equipped Vehicles, *SAE Tech. Paper*, No. 820393, 1982.
2. Pan, C.-H., Moskwa, J. J., Dynamic Modeling and Simulation of the Ford Automobile Transmission, *SAE Tech. Paper*, No. 950899, 1995.
3. Kim, Y. H., Yang, J., and Lee, J. M., A Study on the Transient Characteristics of Automatic Transmission with Detailed Dynamic Modelling, *SAE Tech. Paper*, No. 941014, 1994.
4. Jo, H.-S., Park, Y.-I., Lee, J.-M., Jang W.-J., Park, J.-H. and Lim W.-S., A Study on the Improvement of the Shift Characteristics for the Passenger Car Automatic Transmission, *International Journal of Vehicle Design*, Vol.23, Nos.3/4, pp. 307-328, 2000.
5. D.K. Liu, N. Zhang, J. M. Jeyakumaran and L. Villanueva, Dynamic Model of Transient Characteristics of Automatic Transmissions, *Proceedings of the 5th International Conference on Motion and Vibration Control, Dec. 2000*, University of Technology, Sydney, Edited by Bijian Samali, Vol.2, pp. 597-602, 2000.
6. Hwang, S.-J., Chen, J.-S., Liu, L. and Ling, C.-C., Modeling and Simulation of a Powertrain-Vehicle System with Automatic Transmission, *International Journal of Vehicle Design*, Vol.23, Nos.1/2, pp. 145-160, 2000.
7. Megli, T.W., Haghgooie, M. and Colvin, D.S., Shift Characteristics of a 4-Speed Automatic Transmission, *SAE Tech. Paper*, No. 1999-01-1060, 1999.
8. Hojo, Y., Iwatsuki, K., Oba, H. and Ishikawa, K., Toyoto Five-speed Automatic Transmission with Application of Modern Control Theory, *SAE Tech. Paper,* No. 920610, 1992.
9. Munns, S.A., Computer Simulation of Powertrain Components with Methodologies for Generalized System Modeling, M. S. Thesis, University of Wisconsin-Madison, 1996.

USING TRANSIENT ANALYSIS SOFTWARE TO SIMULATE VIBRATION SIGNALS DUE TO ROLLING ELEMENT BEARING DEFECTS

N. S. FENG, E. J. HAHN AND R. B. RANDALL

University of New South Wales, Sydney, 2052, Australia

This paper describes the capability of transient analysis software to simulate rolling element bearing defects in a simple rigid rotor, flexible housing configuration. Numerical examples using a deep groove ball bearing are used to illustrate via rotor displacement and pedestal acceleration amplitude spectra and wave forms the effect of an inner or an outer race defect under gravity and/or unbalance loading. The different effects of inner and outer race defects on the vibration behaviour of the bearing are clearly illustrated.

1 Introduction

Rolling element bearings are widely used in rotating machinery systems where high reliability and low power consumption are primary concerns. Modelling of such bearings taking into account the Hertzian force deformation relationship and possible existence of the radial clearance introduces nonlinearities into the dynamics of the system so that even when the conditions of the bearings are ideal, ie. there are no imperfections, transient analysis is needed to simulate the system responses [1-2]. Contact surface imperfections such as defects on the raceways and/or rolling elements also affect the responses and forces transmitted through bearings, and further complicate analysis of the system responses. There are diagnostic techniques for analyzing the bearing conditions from experimentally measured signals [3]. However, depending on the forces present in the system such as gear forces and unbalance forces, and operating conditions such as running speed and clearances, the same surface defects may provide completely different system response signals. Of benefit would be software which can simulate correctly the response signals emanating from rotating machinery containing gears, bearings etc. which could have a variety of defects, enabling subsequent signal processing to be applied to delineate the actual defects. This investigation uses in-house transient analysis software to model a rolling element bearing defect in a simple rigid rotor, flexible housing configuration; and the capability of this software in simulating the subsequent vibration responses. Though attention is here restricted to deep groove ball bearings with a single defect in either the inner or outer race, further enhancement of the model to cater for multiple defects and/or defects on the rolling element(s) in general roller bearing systems is not expected to prove unduly difficult.

2 Nomenclature

C	radial clearance between rolling elements and raceway
C_b	damping coefficient of bearing
C_d	depth of defect
D_b, D_p	ball and pitch diameters of bearing respectively
$\underline{F}$	vector of excitation and/or nonlinear forces
F_y, F_z	force components transmitted to both the inner and outer races in the y and z directions respectively
K_b	load-deflection factor
K_p, C_p, M_p	stiffness, damping and mass of pedestal respectively
$\mathbf{K}$, $\mathbf{C}$, $\mathbf{M}$	stiffness, damping + gyroscopic, and mass matrices of system respectively

M_r	mass of rotor
N_b	number of balls
t	time
U	unbalance
$\underline{X}$	vector of displacements
y, z	relative displacements of the rotor journal to bearing housing in horizontal and vertical directions respectively
$\Delta\theta_d$	angular extent of defect
θ_d	starting location of defect
θ_{do}	initial starting location of defect
θ_i	angular location of the i'th ball at any instant of time
θ_o	initial location of the first ball
ω	rotor speed
ω_{bi}	ball passing frequency in terms of inner race, $= N_b(\omega-\omega_c)$
ω_{bo}	ball passing frequency in terms of outer race, $= N_b\omega_c$
ω_c	cage speed, $= (1-D_b/D_p)\omega/2$
ω_p	pedestal natural frequency, $= (K_p/M_p)^{1/2}$

3 The Model

The general equations of motion for a rotor bearing system involving nonlinear bearing forces are given as

$$\mathbf{M}\ddot{\underline{X}} + \mathbf{C}\dot{\underline{X}} + \mathbf{K}\underline{X} = \underline{F}(\underline{X}, \dot{\underline{X}}, t)\,. \tag{1}$$

The following assumptions are made regarding the formulation of the force expression for the rolling element bearings:

1. Hertzian contact theory [4] is applicable;
2. Rolling element inertia can be ignored;
3. There is no slip in the bearing, even when contact is lost;
4. There are no contact surface imperfections unless otherwise specified, and only a defect on the inner or the outer race is considered;
5. Linear damping is applicable.

Based on these assumptions and referring to Figure 1, the transmitted force components for deep groove ball bearings are given as:

$$\begin{Bmatrix} F_y \\ F_z \end{Bmatrix} = K_b \sum_{i=1}^{N_b} \gamma_i A^{1.5} \begin{Bmatrix} \cos\theta_i \\ \sin\theta_i \end{Bmatrix} \tag{2}$$

where

$$A = \left(y\cos\theta_i + z\sin\theta_i - C - \beta_i C_d\right) \tag{3}$$

$$\gamma_i = \begin{cases} 1 & \text{if} \quad A > 0 \\ 0 & \text{otherwise} \end{cases} \tag{4}$$

$$\beta_i = \begin{cases} 1 & \text{if} \quad \theta_d < \theta_i < \theta_d + \Delta\theta_d \\ 0 & \text{otherwise} \end{cases} \tag{5}$$

$$\theta_i = \frac{2\pi}{N_b}(i-1) + \omega_c t + \theta_o . \tag{6}$$

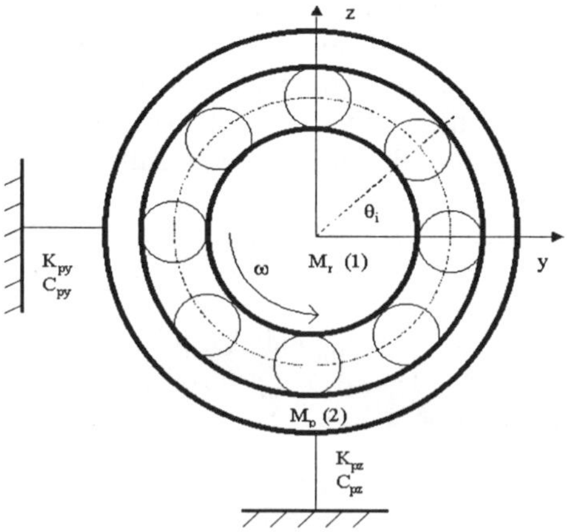

Figure 1. Rotor-rolling element bearing-pedestal schematic

The outer race defect is fixed in location between θ_d to $\theta_d + \Delta\theta_d$, and normally occurs in the unidirectional loading zone; while the inner race defect rotates at the same speed as the rotor ($\theta_d = \theta_{do} + \omega t$) and normally occurs in the unbalance loading zone.

4 The Rotor Bearing System

A simple fictitious horizontal rotor bearing system is chosen for the numerical simulation. A rigid rotor is supported by a single ball bearing which in turn is mounted on a flexible pedestal. The system contains two lumped masses, viz. lumped mass 1 (the rotor mass M_r) and lumped mass 2 (the pedestal mass M_p) as depicted in Figure 1. Each mass is assumed to have two translational degrees of freedom, denoted as y1, z1, y2 and z2 respectively, resulting in a total of four degrees of freedom. Table 1 summarizes the rotor, bearing and pedestal parameters used in this study.

Table 1. Rotor-Bearing-Pedestal Parameters

Rotor	$M_r = 3$kg	$\omega = 1000$rpm	$U = 3\text{x}10^{-4}$ kg.m
Bearing	$D_p = 52$mm	$N_b = 11$	$\omega_c/\omega = 0.38552$
	$D_b = 11.9062$mm	$C = 0$	$\omega_{bo}/\omega = 4.2407$
	$K_b = 13.34\text{x}10^9\text{N/m}^{1.5}$	$C_b = 2940$Ns/m	$\omega_{bi}/\omega = 6.7593$
Pedestal	$M_p = 10$kg	$C_p = 1000$Ns/m	$\omega_p = 40707$rpm
	$K_p = 200$MN/m		

5 Procedures and Results

In-house transient analysis software was enhanced to include the above force expressions and applied to the solution of the system equations of motion (Eq. 1). The software utilizes fixed step 4th order Runge-Kutta integration to determine the accelerations, velocities and displacements of all degrees of freedom of the system at any instant of time. A time step of 4096 points per cycle was chosen for all the results presented here.

The number of points per cycle was occasionally doubled to check the adequacy of the step size and confirm some of the crucial results. The output of the final 32 complete cycles of the data at 256 points per cycle was used to generate the frequency spectra, resulting in 4096 lines of components in which the 33rd is the synchronous. Half of the frequency components are displayed, covering a frequency range up to 1067Hz.

Though there are many parameters which may affect results, this paper is limited to displaying the capability of the software rather than carrying out parametric studies or detailed signal analyses. As a result, only limited results from selected cases are presented here.

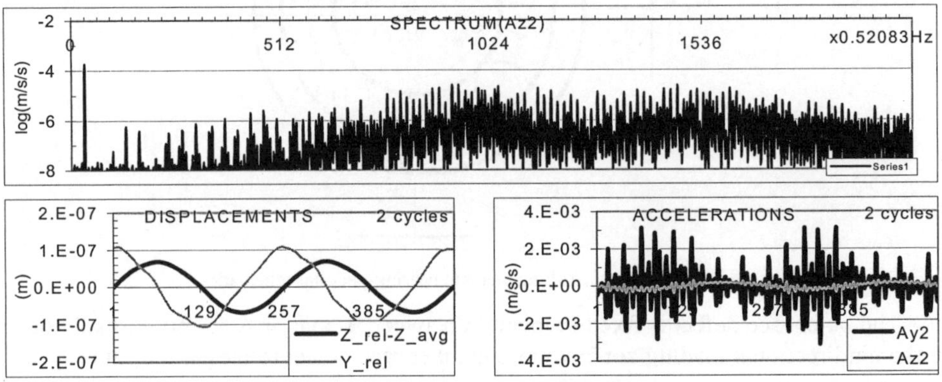

Figure 2. Spectrum and wave forms (no defect)

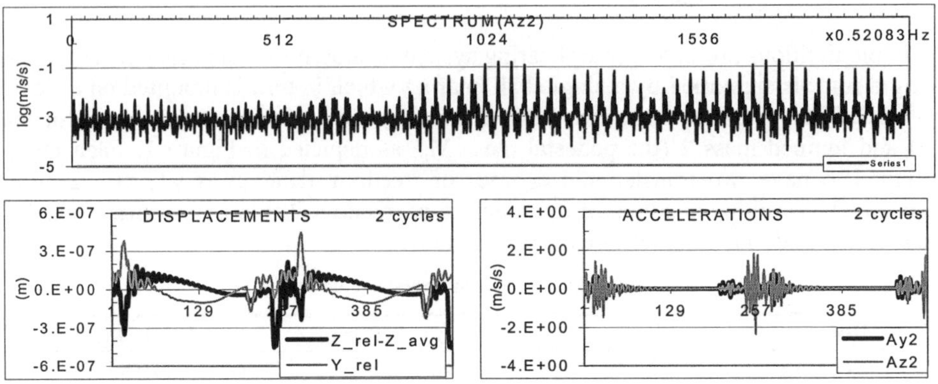

Figure 3. Spectrum and wave forms (inner race defect: $\theta_{do} = -90$ deg., $\Delta\theta_d = 10$ deg.)

For the idealized bearing (no defect) running at the above conditions, Figure 2 shows the selected amplitude spectrum of the pedestal acceleration (Az2), and the wave forms of the relative displacements between the rotor and the pedestal (with the DC component in the vertical direction removed for display purposes) and the pedestal accelerations for two rotor cycles. The corresponding spectra and wave forms for an inner race defect and outer race defect are shown in Figures 3 and 4 respectively. From the wave form signals it can be seen that the inner race defect excites once per revolution but the outer race defect excites at the ball passing frequency ω_{bo}. These frequencies and their harmonics can also be seen in the displacement spectra. Note that because the system is gravity load dominant ($M_r g \approx 10U\omega^2$), the outer race defect can only be detected if it is located in the loading zone near the bottom.

The above cases simulated a typical loading situation where there existed both unidirectional and rotational loads, the latter being one tenth of the magnitude of the former. However, it is possible to encounter loading situations where one or other of the load types is completely dominant. Other parameters also affect the vibration responses. For example, the bearing clearance C and the defect shape display different wave forms and spectra noticeably [5]. Space does not permit the display of the results.

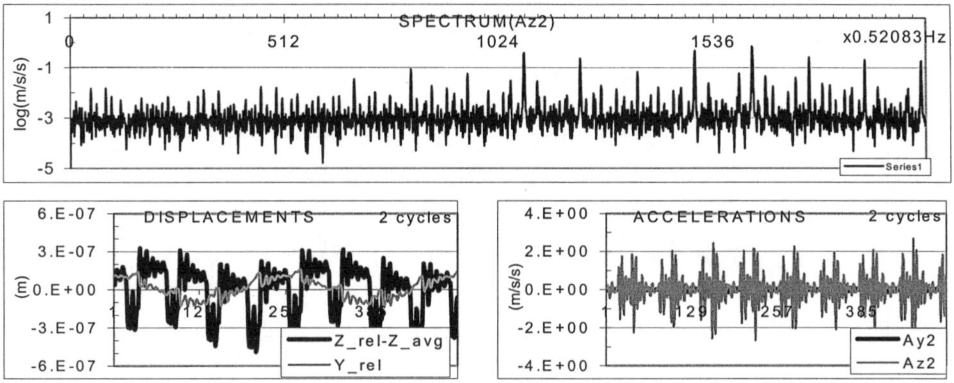

Figure 4. Spectrum and wave forms (outer race defect: θ_d = -90 deg., $\Delta\theta_d$ = 10 deg.)

The above spectra result from the idealized numerical simulation. In practice, however, the measured signals may not be so typical due to some uncertainties such as rolling element slipping and other non-ideal running conditions. As a result, the corresponding raw spectra may not be as clean as expected, in particular in the higher frequency range. In such cases, envelope analysis using Hilbert transform [3] may be applied to extract the signatures of the defects. Thus, upon taking the spectrum segments of 533-800Hz (covering the pedestal resonance of 678Hz) from the spectra Az2 in Figures 2-4 and performing envelope analysis, the corresponding envelope spectra are obtained as shown in Figures 5-7 respectively. It can be seen that the no defect case (Figure 5) produces very low amplitudes for all components; while the inner race defect results in a strong inner race ball passing frequency component (112.7Hz) modulated with the synchronous (16.7Hz) (Figure 6); and the outer race defect gives the outer race ball passing frequency component (70.7Hz) and the associated harmonics (Figure 7).

6 Conclusions

The in-house transient analysis software has been extended to simulate the effects of rolling element bearing race defects. Simple numerical examples using a deep groove ball bearing show that the program is capable of predicting the effects of such defects. Inner or outer race defects affect the vibration behaviour of the bearing differently, depending on the loading and running conditions, and on the defect size and location.

Acknowledgment

This work was supported by the Australian Research Council and the Aeronautical and Maritime Research Laboratory, DSTO.

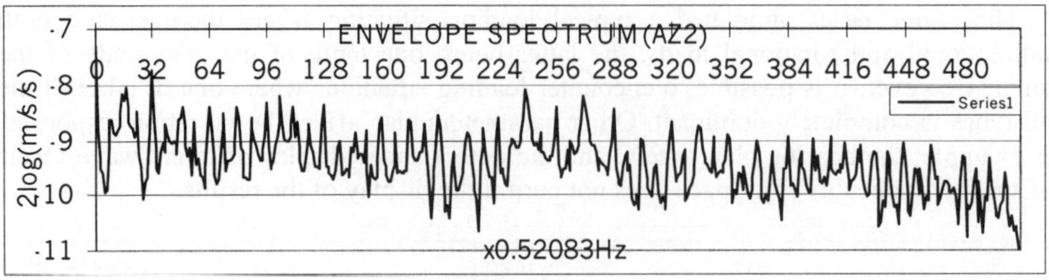

Figure 5. Envelope spectrum (no defect)

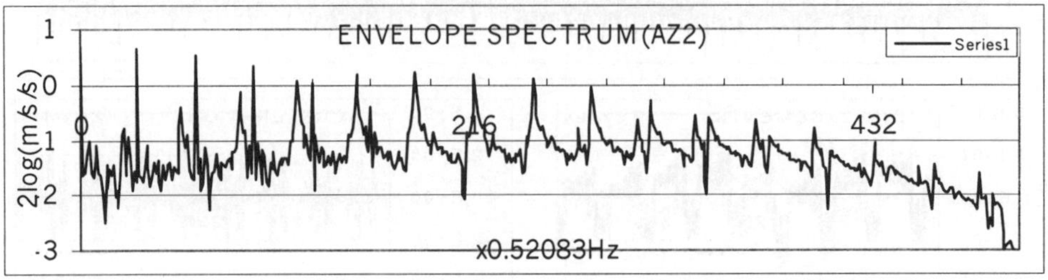

Figure 6. Envelope spectrum (inner race defect: $\theta_d = -90$ deg., $\Delta\theta_d = 10$ deg.)

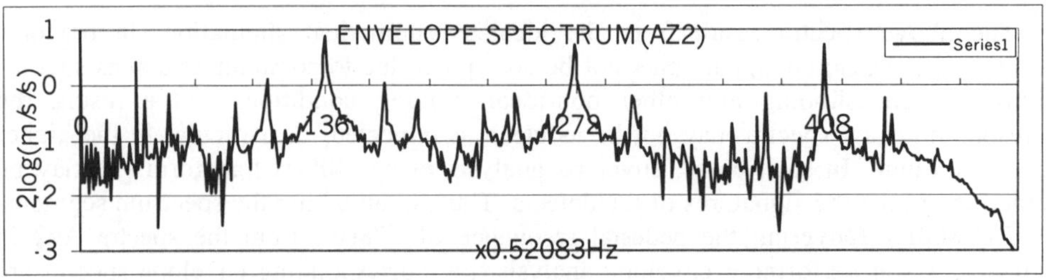

Figure 7. Envelope spectrum (outer race defect: $\theta_d = -90$ deg., $\Delta\theta_d = 10$ deg.)

References

1. Tiwari, M., Gupta, K. and Prakash, Om, "Nonlinear Dynamic Analysis of a Rigid Rotor Supported on Deep Groove Ball Bearing", Proceedings of the 7th International Symposium on Transport Phenomena and Dynamics of Rotating Machinery, Honolulu, Feb. 1998, pp.960-969.
2. Feng, N. S. and Hahn, E. J., "Rolling Element Bearing Non-linearity Effects", ASME paper 2000-GT-391, International Gas Turbine & Aeroengine Congress & Exhibition, Munich, May 2000, 8pp.
3. Randall, R. B., "Frequency Analysis", 3rd ed., K. Larsen & Son, Denmark, 1987, 344pp.
4. Harris, T. A., "Rolling Bearing Analysis", John Wiley & Sons, New York, 1966, 481pp.
5. Feng, N. S., Hahn, E. J. and Randall, R. B., "Simulation of Vibration Signals From a Rolling Element Bearing Defect", International Conference on Health and Usage Monitoring, Melbourne, Feb. 2001, pp.37-52.

DEVELOPMENTS IN REWORK SHAPE OPTIMISATION FOR LIFE EXTENSION OF AGING AIRFRAMES

W. WALDMAN, M. HELLER, M. MCDONALD, G. CHEN

Airframes and Engines Division, DSTO, PO Box 4331, Melbourne VIC 3001, Australia

E-mail: manfred.heller@dsto.defence.gov.au

Shape reworking can be a highly effective repair methodology to extend the life of metallic airframe components suffering fatigue cracking at stress concentrations. In recent years a gradientless shape optimisation technique has been successfully used to determine precise free-form optimal rework shapes, which minimise the peak stresses. In this paper some recent generic and practical applications are first described, such as the optimisation of generic tension fillets, as well as an optimal rework design for the PC9 wing skin. Subsequently the issue of sensitivity of optimal shapes to potential manufacturing errors is discussed, using as an example the case of perturbations to an optimal rework hole in the F-111 wing pivot fitting. The results obtained indicate the suitability of the present approach to yield high quality optimal solutions, noting that typically, optimal shapes are more sensitive to manufacturing error than standard shapes, and hence care must be taken.

1 Introduction

Shape reworking of cracked military airframe components can be a highly effective life-extension practice. The local shape reworking is used to remove cracks, and hence the resultant new uncracked free boundary provides some measure of life extension. Historically, such rework shapes have consisted of circular and straight-line segments, which have been determined by trial-and-error finite element analysis (FEA) methods to evaluate the efficacy of candidate shapes. However, the stress reduction and the associated fatigue life extension obtained is often limited. It is also noted that handbook solutions for optimal shapes are unavailable for typical practical geometric constraints and loading conditions. Hence, due to these limitations, AMRL has undertaken work to develop a capability in determining free-form optimal shapes, to enable the peak stresses to be fully minimised [1–4]. Note that such a free form capability is typically not available in commercial codes. An essential requirement for stress peak optimality is uniformity of stress along that segment of the boundary that is being optimised [1–9].

A further important issue is the robustness of the optimal shapes to variations in conditions for which they were determined. Condition variations that could reduce the optimality of the derived shapes include: (i) variations in the manufactured position with respect to dominant load orientation, and (ii) local variations in the manufactured shape. Hence sensitivity of an optimal shape to such variations will determine the robustness of the design, leading to a compromise between the stress reduction achieved and the range of conditions to which the chosen optimal design is applicable. Hence, in this paper a brief overview of the AMRL gradientless shape optimisation procedure is first presented. Following this, the results of optimisation analyses are presented for the generic design of optimal tension fillets as well as the optimal rework design for the PC9 wing skin. This is followed by an investigation addressing the issue of the sensitivity of optimal shapes to certain potential manufacturing errors.

2 Gradientless shape optimisation method

The gradientless shape optimisation method involves iterative finite element solutions to alter the local geometry to achieve a constant local boundary stress, thereby minimising the peak. Gradientless methods are significantly simpler to implement with standard (i.e. non-optimisation) finite element codes because stress derivatives are not required. The method is based on an analogy with biological growth [1,9], whereby material is added to regions of high stress and removed from regions of low stress. This can achieve a free-form shape that produces a constant stress on the free boundary. Other optimisation algorithms based on biological growth analogies have been utilised experimentally by Heywood [6] and numerically by Mattheck and Burkhardt [9]. In Fig. 1, the boundary movement for any given position i on the boundary is given by

$$d_i = \left(\frac{|\sigma_i| - |\sigma_{th}|}{|\sigma_{th}|} \right) cs$$

where d_i is the movement of the boundary in the direction of the outward normal, σ_i is the tangential stress at position i on the boundary, σ_{th} is a non-zero threshold stress, c is an arbitrary characteristic length and s is an arbitrary step size scaling factor [1–3]. Fig. 2 shows schematically the typical initial non-uniform distribution of tangential stress around a boundary contour. A positive value of d_i indicates material addition. If the magnitude of the local stress σ_i is less than the magnitude of σ_{th}, then material is removed at this position. Otherwise, if it is greater, material is added. The value of σ_{th} is typically chosen to be equal to or less than the peak stress along the region being optimised, and it is updated at each iteration. Convergence of the iterative process can be monitored by looking at the uniformity of stress along the boundary being optimised (e.g. range between the maximum and minimum stresses). Typically, due to geometry constraints, the optimisation renders the stresses uniform only along part of the boundary [1,4,7,8,9].

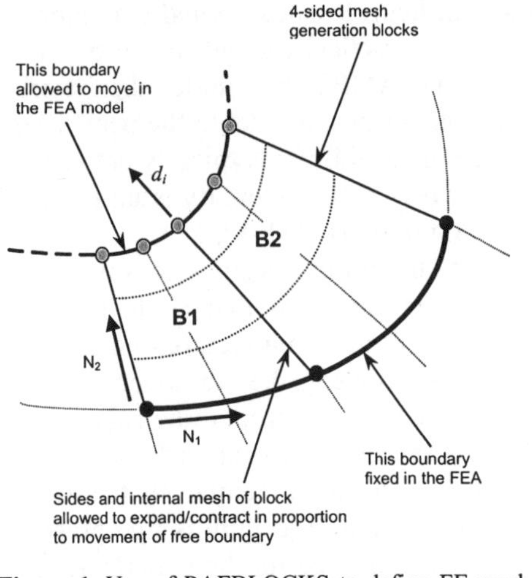

Figure 1. Use of PAFBLOCKS to define FE mesh for shape optimisation of FEA model.

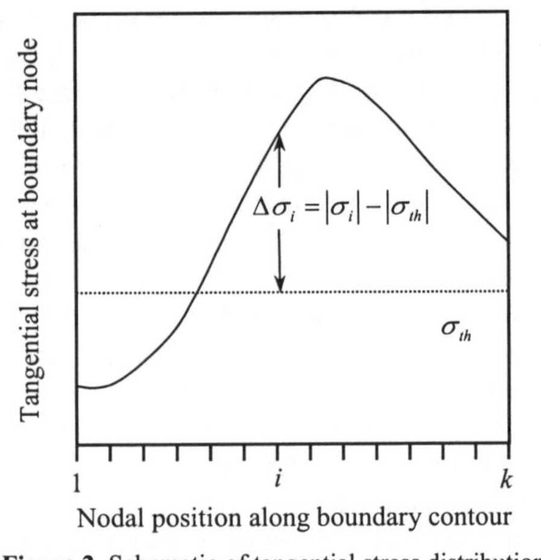

Figure 2. Schematic of tangential stress distribution around a typical boundary contour.

In the analyses, linear elastic FEA has been conducted iteratively using the commercial PAFEC Level 8.5 FEA code. When defining the elements along the moving boundary, the element generation feature called PAFBLOCKs is used. In Fig. 1, these blocks, labelled as B1and B2, are denoted by the thicker lines. The parameters N_1 and N_2 are used to define the number of elements and their spacing within each block. When the nodes used to define the edges of the PAFBLOCKs on the free boundary are moved during optimisation, the internal elements simply shrink or grow in proportion to this nodal movement, and this feature helps avoid local mesh distortion.

3 Optimisation of fillets

3.1 Generic optimal tension fillets

In the literature, considerable interest has been shown in determining optimal shapes of shoulder fillets in flat bars under tension and bending loading. By minimising the peak stress along the fillet contour, large increases in fatigue life for cyclically loaded components can be expected. Recent AMRL work has focussed on determining a family of optimal fillet shapes [3], where these have previously not been available. Fig. 3 shows the generic geometry that was considered. The optimum shapes were determined for a constant prescribed fillet transition length l under tension. The maximum plate width is $2a$ over a length l_a, while the minimum plate width is $2b$ over a length l_b, and the height of the shoulder fillet is h (= a–b). Values of $l_a = l_b = 5l$ were used to keep plate-length effects at a negligible level. For accuracy, highly refined meshes as shown in Fig. 4 were used.

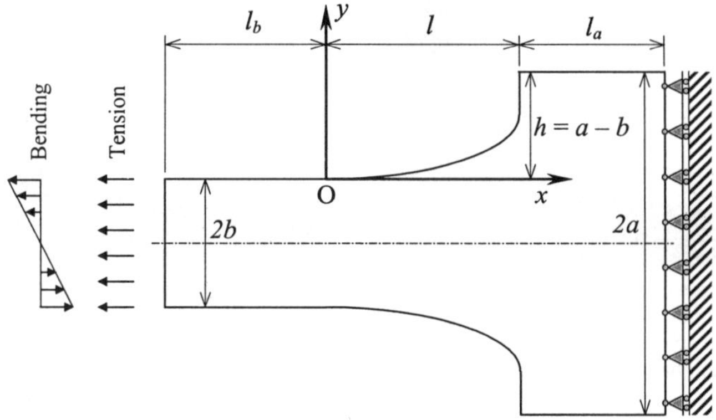

Figure 3. Geometry and notation for plate with fillet subjected to uniaxial tension or bending loading.

Typical solutions had a uniform distribution of tangential stress, with a range that was less than 0.007% of the optimal K_t value. Fig. 4 shows a typical optimal fillet profile for the case l/h = 1.0 and a/b = 1.33, together with stress contours. Fig. 5 shows the variation of K_t obtained for optimal tension fillets as a function of l/h for different a/b ratios. Contrary to common engineering intuition, in optimal shapes the existence of a sharp corner instead of a radiused one does not produce a region of higher stress concentration, as compared to the rest of the optimal fillet, a fact noted in previous work [1,6,9].

Recent enhancements to the gradientless shape optimisation algorithm now allow a constraint on the minimum radius of curvature to be enforced to prevent the formation of sharp corners. Such shapes are desirable since they are more robust against the potential detrimental effects of manufacturing errors and load misalignment. Optimal fillets with minimum radii of curvature corresponding to 30 mm and 60 mm, respectively, are shown in Figs. 6 and 7. Fig. 8 shows the variation in K_t along the fillet boundaries for optimal shapes with different values of minimum radius of curvature. The sharp-corner optimal case (R_{min} = 0 mm) produced a constant K_t = 1.516 along the entire fillet. Fig. 9 shows a plot of $K_t/K_{t(opt)}$ versus R_{min}/l for the case l/h = 1.0 and a/b = 1.33. This demonstrates that there is only a small increase in K_t from its sharp-corner value for values of R_{min} that provide generous fillet radii.

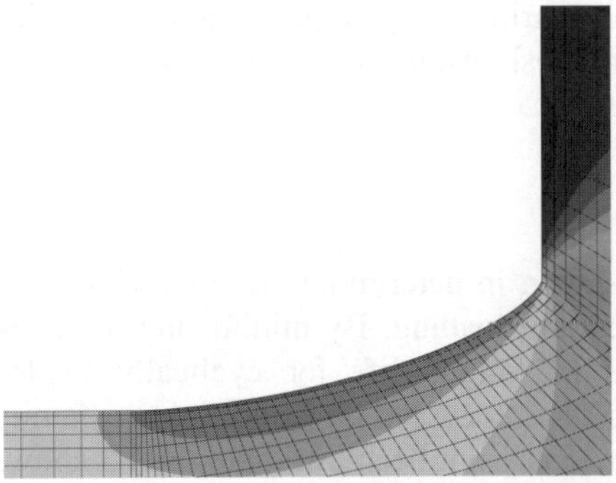

Figure 4. Optimal tension fillet shape and contours of normalised maximum principal stress for case, l/h = 1.0 and a/b = 1.33 (K_t = 1.516).

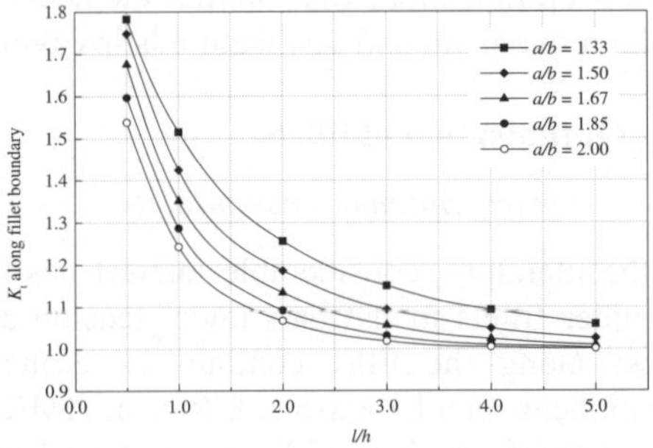

Figure 5. Variation of stress concentration factor K_t for optimal tension fillets.

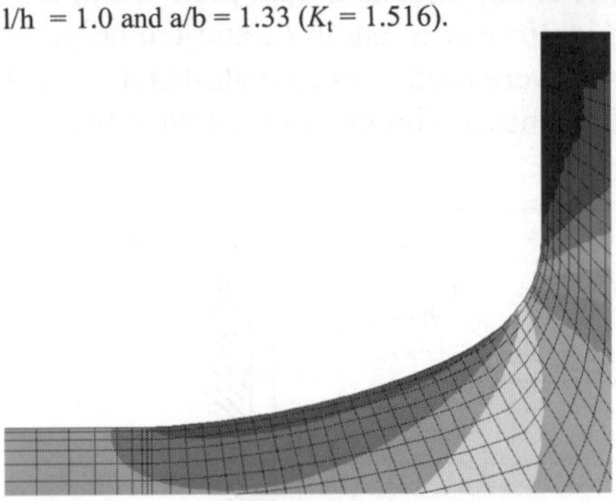

Figure 6. Normalised maximum principal stress contours for optimal tension fillet with R_{min} = 30 mm, for typical case l/h = 1.0 and a/b = 1.33 (K_t = 1.541, 1.6% higher than true sharp-corner optimal).

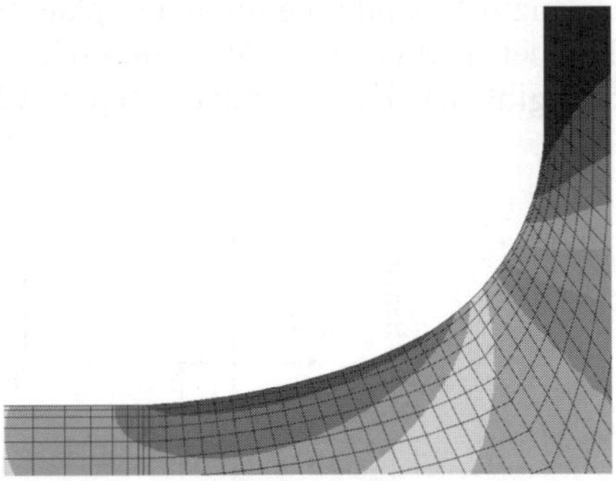

Figure 7. Normalised maximum principal stress contours for optimal tension fillet with R_{min} = 60 mm, for typical case l/h = 1.0 and a/b = 1.33 (K_t = 1.605, 5.9% higher than true sharp-corner optimal).

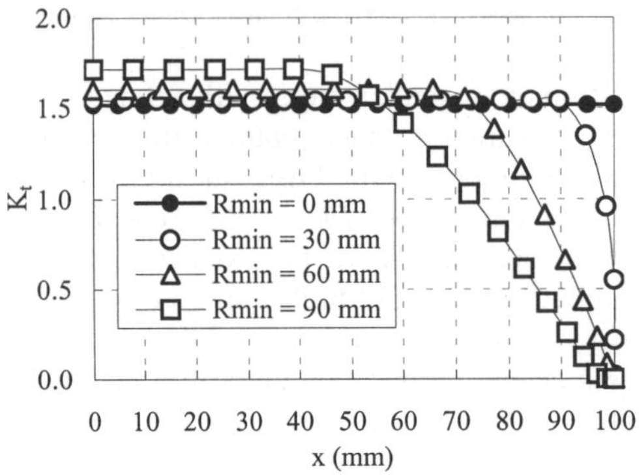

Figure 8. K_t along fillet boundaries for optimal shapes with different values of minimum radius of curvature R_{min} (l = 100 mm, l/h = 1.0, a/b = 1.33).

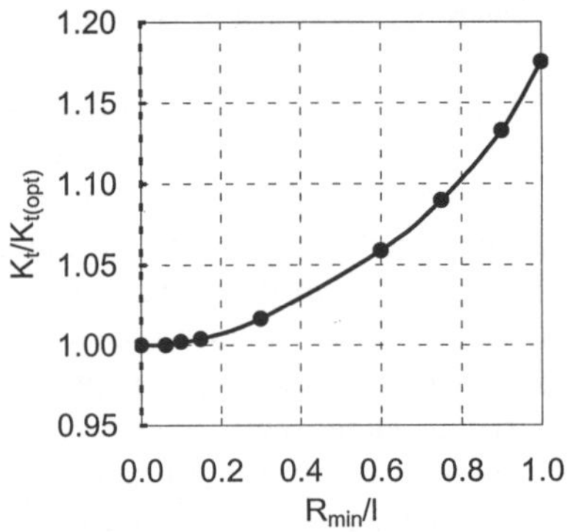

Figure 9. Variation in K_t with different values of minimum radius of curvature R_{min} (l = 100 mm, l/h = 1.0, a/b = 1.33).

3.2 PC9 wing skin rework

The Pilatus PC9/A aircraft used by the RAAF has recently undergone a full-scale fatigue test at AMRL. During testing, cracking at the undercarriage bay door up-lock mechanism mounting holes occurred in the lower wing skin. A standard rework repair was designed and implemented to blend out the region containing the mounting holes by adding two circular fillets with large corner radii. However, cracking again occurred in this region. Shape optimisation was used to develop an improved design of the rework [2]. Initially, a highly-refined FE model of the original lower wing skin was created to analyse the stresses occurring at the two mounting holes for the up-lock mechanism (see Fig. 10), and a peak K_t = 2.93 was calculated. Fig. 11 shows the FE model of the standard rework repair. The peak K_t = 2.44 and occurs at the left-hand side, while on the right-hand side the maximum K_t = 2.34. The reduction in peak SCF produced by this standard repair was 14%, which was less than had originally been envisaged.

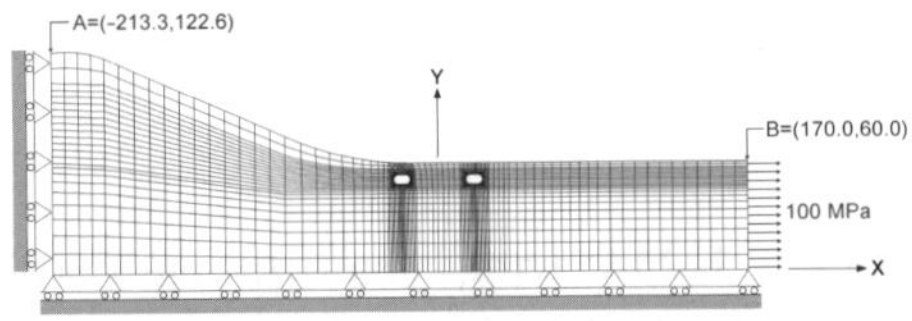

Figure 10. Full finite element mesh of PC9/A lower wing skin and landing gear up-lock mechanism mounting holes showing boundary conditions and the applied uniaxial tension loading.

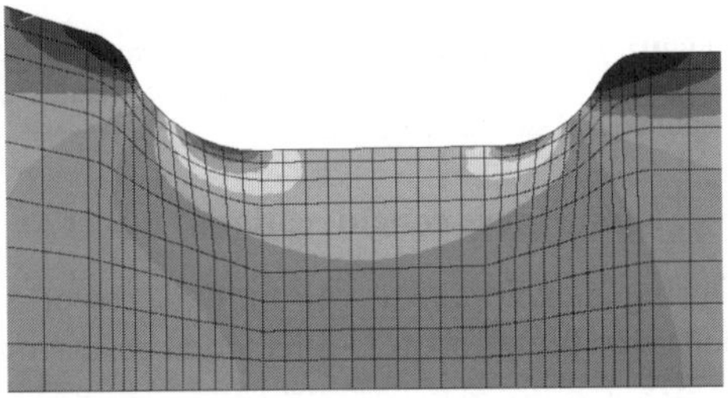

Figure 11. Local stress distribution for original rework repair to PC9/A wing skin.

Optimisation was then performed, and the optimal FE mesh and the contours of normalised von Mises stress are shown in Fig. 11. The optimal rework boundary is a continuous smooth curve, which runs into the vertical faces at each end. The stresses are uniform with K_t = 1.63, with a standard deviation along the optimal curve of only 0.0011. This K_t is 33% less than that in the original rework repair. It is desirable to modify the optimal shape by removing the sharp re-entrant corners to enhance its robustness to slight changes in loading conditions. In Fig. 12, the material located at the corners of the two vertical faces is seen to be very lowly stressed. This redundant material was removed using a smooth curve that was a continuation of the curve at the two corner-notch locations. The resulting rework shape and stress contours are shown in Fig. 13. The peak K_t along the free edge of the rework boundary remained unchanged at K_t = 1.63.

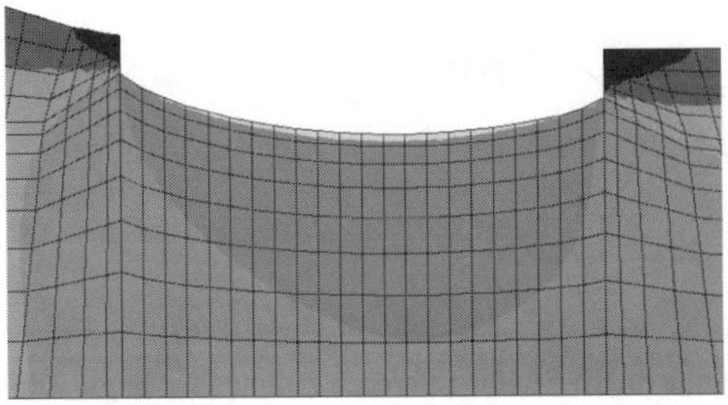

Figure 12. Local stress distribution for the optimal PC9/A wing skin rework shape.

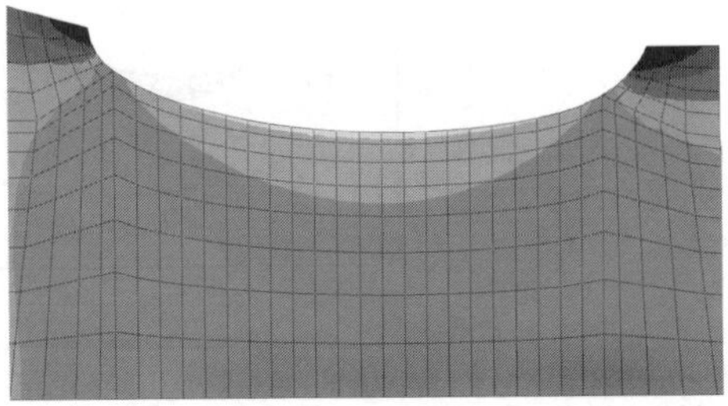

Figure 13. Local stress distribution for the optimal PC9/A wing skin rework shape with smoothed corners after removal of redundant material.

Fig. 14 shows plots of the K_t value around the fillet boundary for the standard rework, the optimal rework and the smoothed optimal rework shape. The uniformity of the stress distribution along the boundaries of the optimised fillets is clearly apparent.

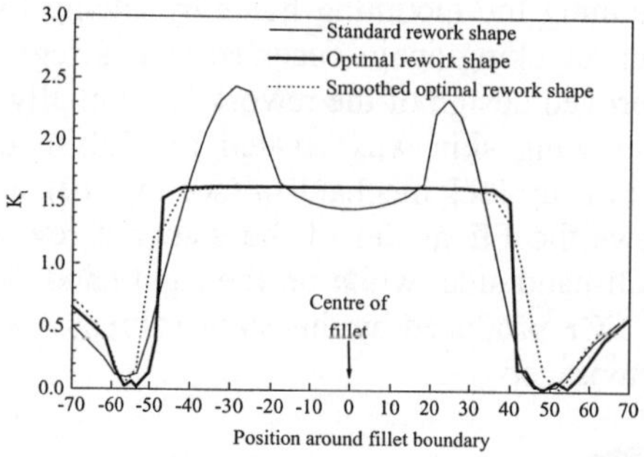

Figure 14. Comparison of K_t variation around fillet boundary for different rework shapes: standard, optimal and smoothed optimal.

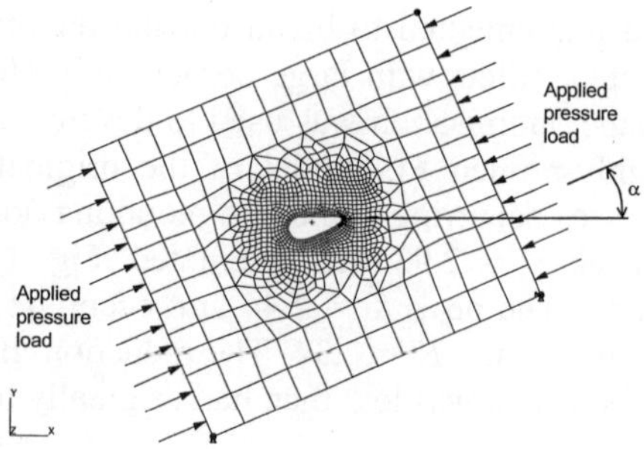

Figure 15. View of 2D MSC.NASTRAN FE model of the FFVH #13 optimal rework shape inclined at α=23 degrees to the remote load direction.

4 Robustness considerations

During development of optimal reworks for the F-111 wing pivot fitting (WPF), considerable effort was directed at assessing/obtaining appropriate robustness levels [10].

4.1 Sensitivity to rotational misalignment of optimal shapes

A simplified 2D FE model was used to study the stress sensitivity to possible rotational variations of an optimal rework for Fuel Flow Vent Hole (FFVH) #13 in the WPF. Fig. 15 shows the FE mesh and the applied compressive loading, where $\alpha = 23°$. Fig. 16 shows the rotated optimal hole corresponding to load misalignments of $\delta\alpha = -6.0°, -3.0°, 0.0°, +1.22°$ and $+5.44°$, noting that the nominal optimal shape has uniform stresses along regions A and B. The sensitivities of the peak compressive stresses in Region A and Region B due to these rotations were determined. The peak stresses in Region A were mainly clustered around an angular position of 240°, while those in Region B were mainly clustered around 0°.

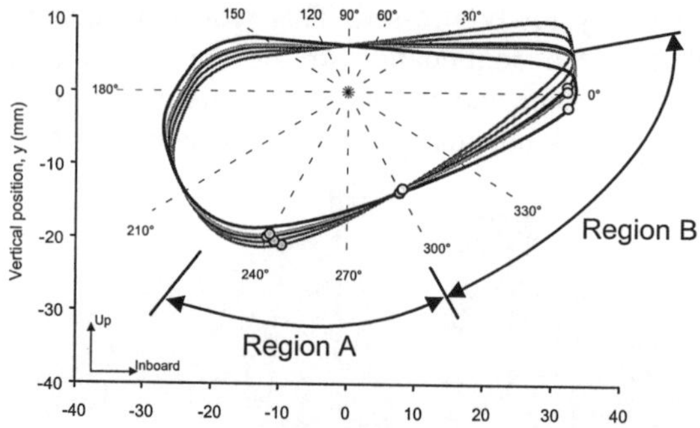

Figure 16. FFVH #13 optimal rework at different orientations relative to the reference remote load angle of α = 23°, showing the positions of the peak stresses within regions A and B at each orientation.

Fig. 17 shows the stress distribution for each region, where it can be seen that for positive δα the stress peak in Region B increases while in Region A it decreases, and vice versa for negative δα. A bilinear behaviour is exhibited, where for negative δα the stress sensitivity is 60 MPa per degree of variation in load angle, while for positive δα it is 126 MPa per degree. The difference in sensitivity in each region is expected, because there is a large difference in the local radius of curvature, *R*, at each end of the nominal constant stress region. Here R = 4.4 mm at the peak stress location in Region B compared to R = 20.8 mm at the peak stress location in Region A. The results indicate that it is desirable to achieve optimal shapes that have similar stress sensitivity to variations at each of the critical regions around an optimal hole. Here control of the local R values at the ends of the constant stress regions is seen to be important.

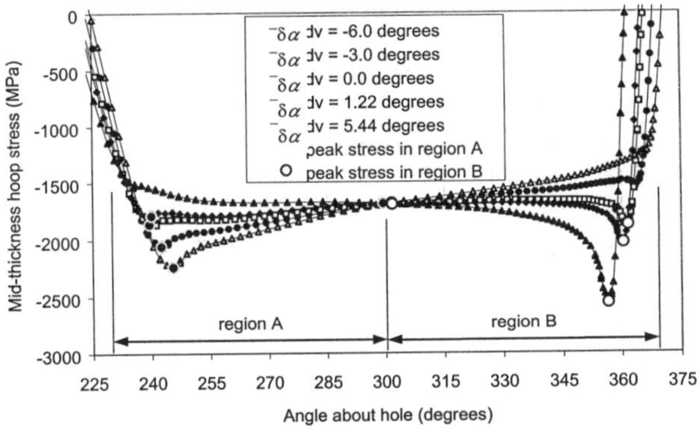

Figure 17. Hoop stress distributions in Regions A and B for FFVH #13 for different perturbation angles.

4.2 *Sensitivity to local perturbation of optimal boundary*

The precise shape of optimal reworks may be perturbed by the possible future requirement to locally remove cracked, corroded or otherwise damaged material from the surface of the rework using a circular grinder. This can result in a significant departure from the optimal shape. The effects of circular grindouts on FFVH #13 were analysed for different radii of curvature R, and carried out to different depths d, as indicated in Fig. 18.

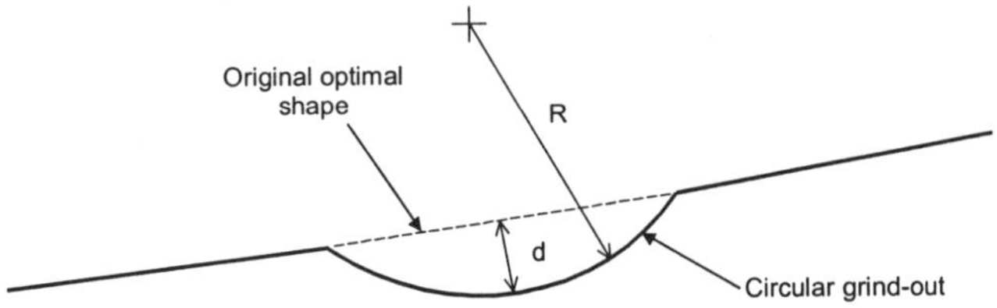

Figure 18. Schematic showing a local circular grindout shape perturbation of an optimal shape.

Referring to Fig. 16, the grindouts were made at an angular position of 291.3° (Point B). Point B is positioned in the middle of the constant-stress optimisation region where the radius of curvature is large (R = 140.1 mm). Three grindout radii were investigated (R = 1.397, 2.794 and 5.588 mm), at each of three grindout depths (d = 25.4, 76.2 and 127.0 μm). The stress distributions around Point B for R = 2.794 mm with different grindout depths are shown in Fig. 19. The baseline stress at Point B is 1710 MPa, and each grindout introduces a significant peak. In order of increasing grindout depth, the corresponding increases in peak stress are 16%, 31%, and 39%, respectively. It is useful to note that a stress solution for the simplified case of a circular notch at the edge of a uniaxially loaded rectangular flat plate is available [11]. Comparison with the current FE results for the above grindouts indicates that the simplified analytical solution gives a good estimate of peak stress in the degraded condition to within 8%. This good agreement is due to the fact that, before the grindout, both the optimal hole shape and the rectangular uniaxially loaded rectangular plate have uniform boundary stress.

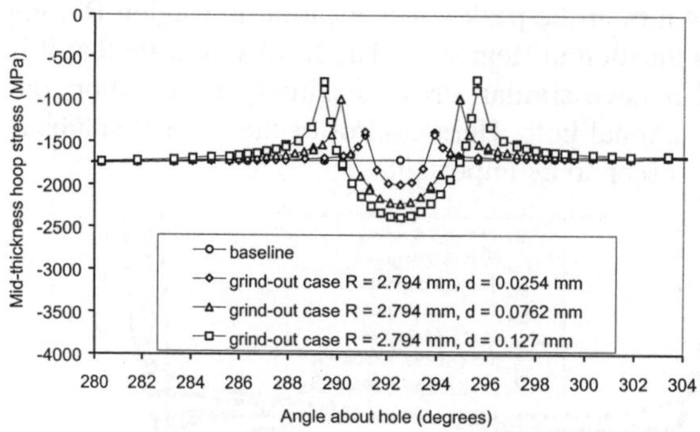

Figure 19. Stress distributions around Point B for FFVH #13 with a grindout radius of R = 2.794 mm for various depths, d.

5 Conclusion

Gradientless shape optimisation methods have been developed and used to design optimal free-form rework shapes that produce minimised peak stresses. These are useful at the design stage, as well as for designing life extension reworks of aging components. The importance of the issue of robustness of optimal shapes has been demonstrated. Work is currently underway to develop methods for the efficient specification of robust optimal shapes. Here it appears desirable to achieve optimal shapes that have similar stress sensitivity to variations at each of the critical regions around the shape.

References

1. Kaye R. and Heller M., *Structural shape optimisation by iterative finite element solution,* (1997) DSTO-RR-0105, AMRL, Melb. Australia.
2. Waldman W. and Heller M., *Shape optimisation of PC9/A lower wing skin rework in vicinity of undercarriage bay door up-lock mechanism mounting holes*. Folio 8, File B2/129 Part 2, (1999) AMRL, Australia.
3. Waldman W. Heller M. and Chen GX., Optimal free-form shapes for shoulder fillets in flat plates under tension and bending. *Int J of Fatigue*, **23** (2001) pp. 509–523.
4. Heller M. McDonald M. Burchill M., Watters KC. *Shape optimisation of critical stiffener runouts in the F-111 wing pivot fitting*. (2001) DSTO-TR-1119, AMRL, Melb. Australia.
5. Baud R. V., Fillet profiles for constant stress. *Prod. Eng.* (April 1934) pp. 133–134.
6. Heywood R. B., *Photoelasticity for Designers* (1969) (Chapter 11: Improvement of Designs), Pergamon Press.
7. Neuber H., Zur optimirung der spannungskonzentration. *Continuum Mechanics and Related Problems of Analysis*, (Nauka Publ. House, Moscow 1972) pp. 375–380.
8. Schnack E., An optimisation procedure for stress concentrations by the finite element technique. *Int J for Numerical Methods in Engineering*, **14** (1979) pp. 115–124.
9. Mattheck C. and Burkhardt S., A new method of structural shape optimisation based on biological growth. *Int J of Fatigue*, **12** No. 3, (1990) pp. 185–190.
10. McDonald M. Heller M., Goldstraw M. and Hew A., *Robustness of the F-111 wing pivot fitting optimal rework shapes*. DSTO-TR-1121, AMRL, Melb. Australia, 2001.
11. Neuber, H., *Theory of notch stresses, principles for exact calculation* (1946), Ann Arbor, Michigan.

THE EFFECTS OF TEMPERATURE VARIATION ON THE CREEP BEHAVIOUR OF PRESSURE VESSELS USING THETA PROJECTION DATA

M LAW, W PAYTEN AND K SNOWDEN

ANSTO (Australian Nuclear Science and Technology Organisation)

In-service piping is often exposed to through-wall temperature gradients. The effects of these on the thermal stresses and creep response are examined using the Theta projection creep algorithm within a finite element code. It was found that the stress and temperature dependence of the creep response may lead to complex stress evolution.

1 Introduction

The effect of thermal gradients on thick walled pressure vessels can be significant. If large thermal gradients exist in steady-state operation, thermally induced stresses [1] will exist in addition to the pressure stresses (fig.1). The creep response of the wall depends on the local stress, temperature, and previous creep history. The combination of the altered initial stress state and the altered creep response may lead to differing final stress states through time and a reduced creep life to that calculated for vessels without temperature variation using a constant secondary creep rate law.

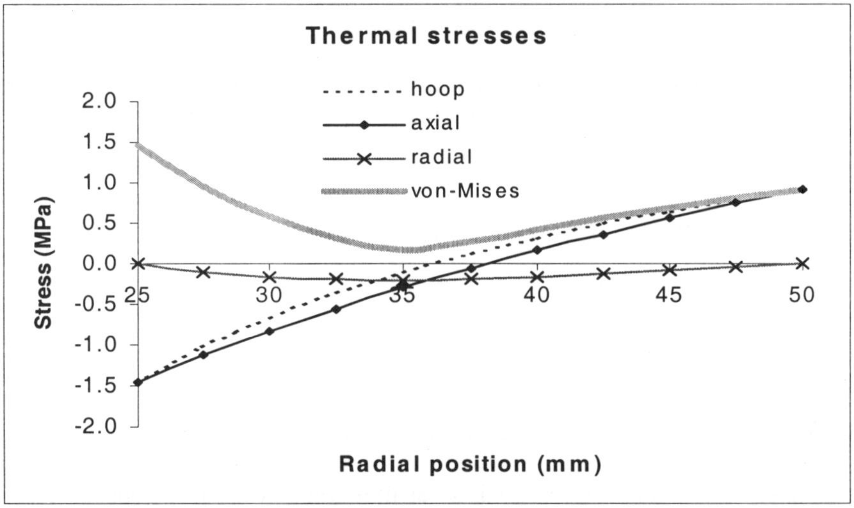

Figure 1 Thermal stresses, inner wall 10°C hotter than outer wall.

The Theta projection method [2] is being more widely used as its advantages and flexibility become appreciated. The Theta projection equation is an expression of creep response over time:

$$\varepsilon_c = \theta_1(1-e^{-\theta_2 t}) + \theta_3(e^{\theta_4 t}-1),$$

where ε_c is the creep strain, t is the time, and the θ terms are experimentally determined constants. This equation is an attempt to both empirically fit the strain-time behaviour of a material during creep, and to provide an insight into the processes occurring during creep [3]. Each Theta term is itself a function of temperature and stress, of the form;

$$\text{Log}\,\theta_i = A_i + B_i\sigma + C_i T + D_i\sigma T,$$

where σ is the stress, T is the temperature and i = 1, 2, 3, 4. The log θ_i values derived from testing vary approximately linearly as functions of stress and temperature[4]. If the θ_i functions can be defined on the basis of short term tests at high temperature, then values for long term creep data can be predicted [5,6].

2 Methods

The θ parameters used in this work were obtained as part of an extensive creep testing program of ex-service $2^1/_4$ Cr – 1 Mo steels. The experimental details have been reported elsewhere [7]. Briefly, the creep specimens were tested at temperature in a vacuum to reduce possible oxidation effects and their extensions were continuously monitored. The creep curve information was fitted to the Theta equation using the method of Evans and Wilshire [2].

2.1 Solution method

To implement the Theta equation within the FEA program (EMRC NISA II), a strain rate procedure was used. The first derivative of the Theta equation is used to calculate the creep rate, and hence the creep strain, for the timestep and element (which in turn depends on the time, the local stress, and the temperature). This is:

$$d\varepsilon/dt = \theta_1\theta_2 \,\mathrm{Exp}\,(-\theta_2\, t) + \theta_3\theta_4 \,\mathrm{Exp}\,(\theta_4\, t)$$

The Theta coefficients given in Table 1 below were derived from 8 creep tests at stresses from 40 to 60 MPa, and at temperatures from 645 ° C to 715° C using regression analysis.

Table 1. Theta Coefficients

	(θ_1)	(θ_2)	(θ_3)	(θ_4)
A	-106.55	18.966	-284.16	-63.664
B (MPa)	2.7948	-2.754	3.6464	8.53E-02
C (K)	1.11E-01	-4.45E-02	2.79E-01	4.88E-02
D (K*Mpa)	-2.98E-03	3.10E-03	-3.66E-03	1.77E-05

3 Modelling of creep under temperature variation

Theta projection and FEA were used to analyse the creep behaviour of cylindrical pressure vessels with the inner wall 10 °C higher than the outer wall. The resulting temperature and stress dependence of the minimum creep rates is shown in fig. 2.

Similar modelling was performed by Loghman and Wahab [8] on a cylindrical pressure vessel which showed a significant variation of the von-Mises stresses with time. The closed form solution employed by them is not useable in complex geometries or multiple material models such as welded joints, whereas the FEA procedure can be used in these situations.

3.1 Results

Thermally induced compressive hoop stresses exist at the point where the temperature is highest; the combination of the thermal stresses, the pressure loading, and differing creep rates across the wall thickness due to temperature variation gives rise to varying stress profiles through time.

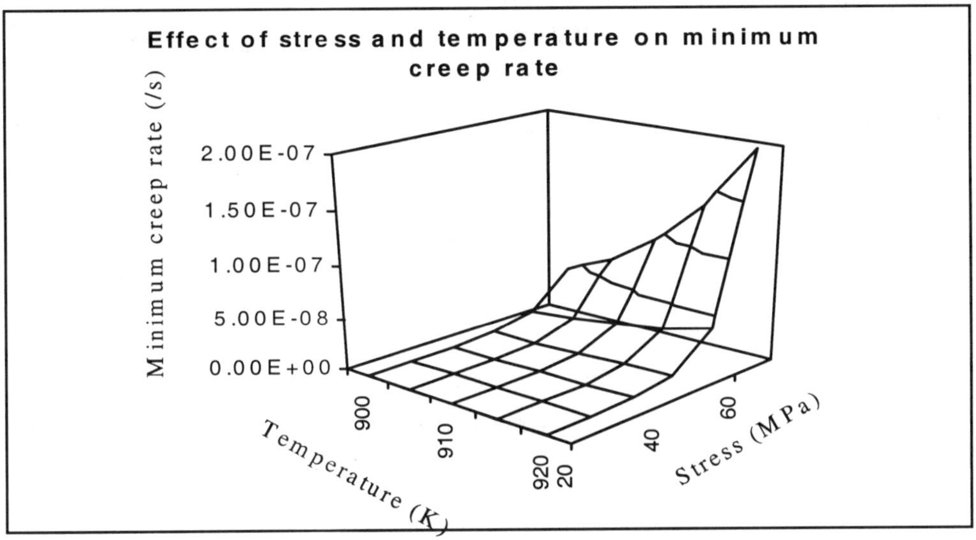

Figure 2 Effect of temperature and stress on minimum creep rate.

There is a thermally induced compressive hoop stress at the inner wall. The combination of the thermal stress, the pressure loading, and differing creep rates across the wall thickness due to temperature variation gives rise to the following stress profiles through time (figs 3, 4).

The von-Mises stresses go through a double reversal (fig.4. also shown in fig. 5 in log time for clarity).

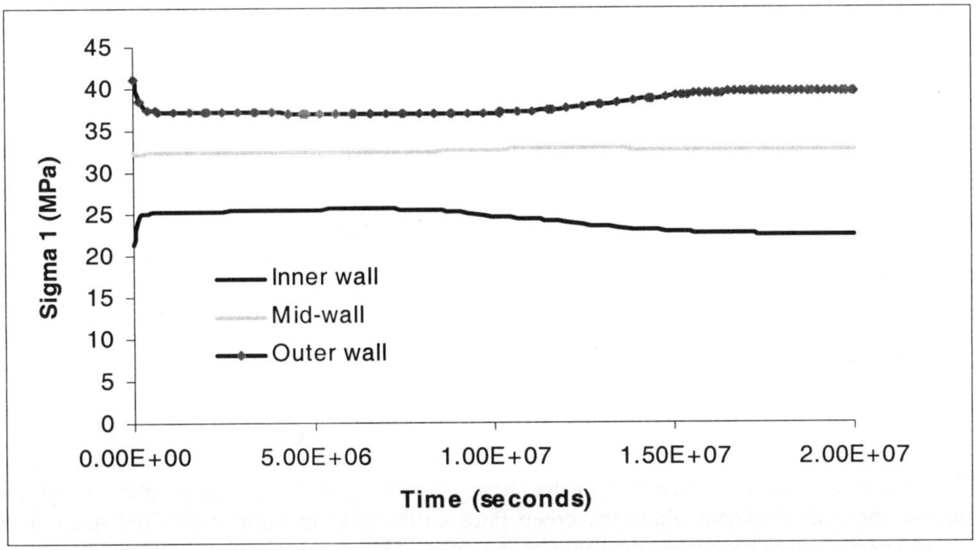

Figure 3 Theta projection modelling of + 10 degree thermal case, first principal stresses.

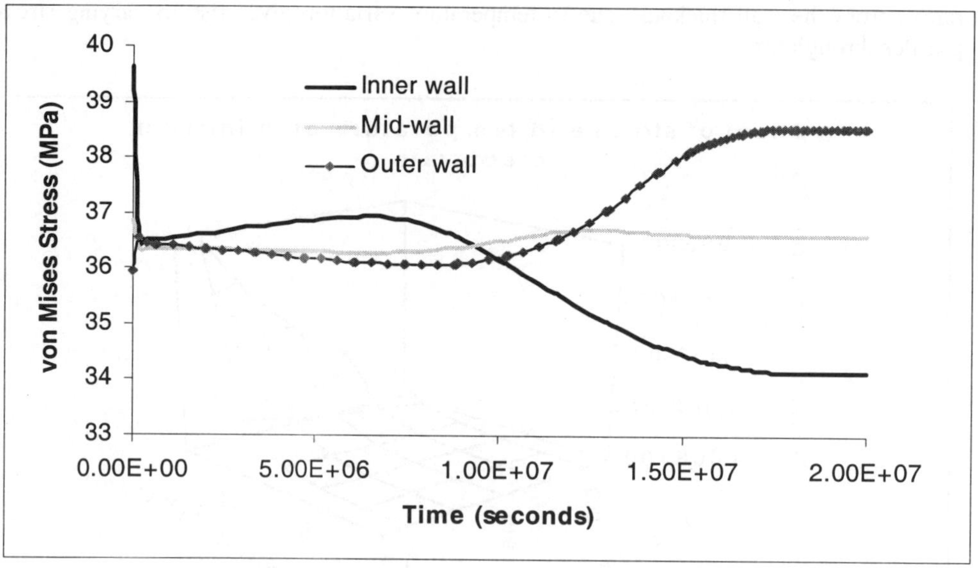

Figure 4 Theta projection modelling of + 10 degree thermal case, von-Mises stresses

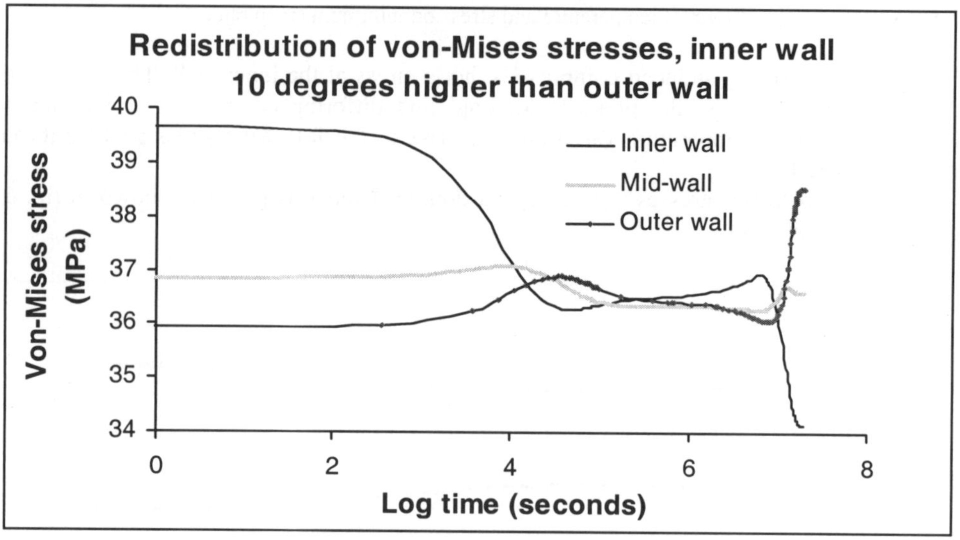

Figure 5 Theta projection modelling of + 10 degree thermal case in log time.

4 Discussion

The stress redistributions show complex behaviour. In addition to the usual creep induced redistribution, coupled with both a thermal and pressure induced initial stress state, there are two linked factors which add to the stress response. Firstly the temperature variation across the wall thickness alters the creep rates so that, at the same stress, the inner wall experiences faster creep rates than that of the outer. Thus the inner wall relaxes offloading stress onto the outer wall. At these higher stresses the outer wall then completes its strain redistribution, this then offloads stress the inner wall. Thus it is some time before the stresses stabilise. The difference in relative creep rates between the inner and outer walls is a function of temperature, local von–Mises stress (fig. 2, 5) and time.

Secondly, the temperature variation means the relative difference in the creep rates may vary through time. For example, the material in one part of the vessel may be entering its tertiary phase while another may still be exhibiting a decreasing or constant creep rate (primary or secondary creep).

5 Conclusion

The redistribution of stresses by creep in a pressure vessel was modelled and compared to results predicted by the Norton equation. The differences found were attributed to the more versatile features of the Theta projection.

The cases modelled included thermal variation across the wall of a pressure vessel. The stress evolution was similar to that found in previously published work. However, more complex interactions were noted as a result of coupled thermal variation of creep rates and thermally induced stresses.

6 Acknowledgments

The authors are grateful to the CRC for Materials Welding and Joining, Pacific Power, and the Australian Institute for Nuclear Science and Engineering (AINSE) for support for this work

References

1. Timoshenko S, Goodier J (1951)"Theory of Elasticity" McGraw Hill, p 413
2. Evans R, Wilshire B (1985) "Creep of Metals and Alloys" Inst of Metals, London 1985
3. Ule B, Rodič T & Šuštar T (1997) " Modification of θ projection creep law by introducing mean stress term" Materials Science & Technology, V13, July 1997, pp 555-559
4. Parker J (1985) "Prediction of Creep deformation and Failure for 1/2Cr-1/4V and 2-1/4Cr-1Mo Steels" ASME J. Pres. Vessel Technology, August 1985, V 107, pp 279-284.
5. Snowden K, Merhtens E (1990) " Analysis of Creep Data for Remaining Life Estimation" Proc. Remaining Life Assessment on High Temp. Plant in Australia" Lucas Heights 1990
6. Evans R, Beden I, Wilshire B (1984) "Creep Life Prediction for $^1/_2$ Cr $^1/_2$ Mo $^1/_4$ V Ferritic Steel" Creep and Fracture of Engineering Materials and Structures, Pineridge Press, Swansea, UK, pp 1277-1290
7. Ripley M, Snowden K (1994) "Remaining Creep Life Assessment of Welds in Steel Pressure Vessels" Proc. Structural Integrity: Experiments, Models and Applications. 10th Biennial Conf. On Fracture, Berlin, Sept 1994, Vol1, pp 761-766
8. Loghman A, Wahab M (1996)"Creep damage simulation of thick walled tubes using the Θ projection concept" Int. J. Pres. Ves & Piping, V 67, pp 105-111

AUTHOR INDEX